A Mathematical
Introduction to General Relativity

Second Edition

Other World Scientific Titles by the Author

A Friendly Approach to Complex Analysis
ISBN: 978-981-4578-98-1
ISBN: 978-981-4578-99-8 (pbk)

Plain Plane Geometry
ISBN: 978-981-4740-43-2
ISBN: 978-981-4740-44-9 (pbk)

A Friendly Approach to Functional Analysis
ISBN: 978-1-78634-333-8
ISBN: 978-1-78634-334-5 (pbk)

A Mathematical Introduction to General Relativity
ISBN: 978-981-12-4377-6
ISBN: 978-981-12-5672-1 (pbk)

A Friendly Approach to Complex Analysis (2nd Edition)
ISBN: 978-981-12-7280-6
ISBN: 978-981-12-7410-7 (pbk)

A Mathematical Introduction to General Relativity

Second Edition

Amol Sasane
London School of Economics, UK

World Scientific

NEW JERSEY • LONDON • SINGAPORE • BEIJING • SHANGHAI • HONG KONG • TAIPEI • CHENNAI

Published by

World Scientific Publishing Co. Pte. Ltd.
5 Toh Tuck Link, Singapore 596224
USA office: 27 Warren Street, Suite 401-402, Hackensack, NJ 07601
UK office: 57 Shelton Street, Covent Garden, London WC2H 9HE

British Library Cataloguing-in-Publication Data
A catalogue record for this book is available from the British Library.

Spacetime Pythagoras Theorem:
With the blue curve representing a worldline and the orange segment depicting a tangent vector, the (Euclidean) area of the rhombus is equal to the difference of the blue square and the red square areas, and is the square of the 'Minkowski-length' of the tangent vector.

A MATHEMATICAL INTRODUCTION TO GENERAL RELATIVITY
Second Edition

ISBN 978-981-98-0219-7 (hardcover)
ISBN 978-981-98-0545-7 (paperback)
ISBN 978-981-98-0220-3 (ebook for institutions)
ISBN 978-981-98-0221-0 (ebook for individuals)

For any available supplementary material, please visit
https://www.worldscientific.com/worldscibooks/10.1142/14082#t=suppl

Desk Editor: Tan Rok Ting

To my family

Preface

The aim of this book is to give a mathematical presentation of the theory of general relativity (i.e., spacetime-geometry-based gravitation theory) to advanced undergraduate or beginning graduate mathematics students. This textbook is intended for self-study, and so the solutions to all the exercises are included at the end of the book.

The prerequisites for reading this book are undergraduate-level mathematical analysis (e.g., having seen the inverse function theorem in a multivariable context, Picard's existence and uniqueness theorem for ordinary differential equations, etc.) undergraduate-level linear algebra, and elementary topology (e.g., in a metric space context). Useful references to look-up these things are e.g. [Apostol (1969)], [Rudin (1983)]. Familiarity with differential geometry is not assumed. A background in physics is not needed, except for a basic university-level general physics course (so that the reader has seen the Newtonian laws of motion and the gravitational law). In case it is needed, we recommend [Feynman (1963)] as reference for this part.

Mathematicians will find spacetime physics presented in the definition-theorem-proof format familiar to them. Precise mathematical definitions of physical notions can be valuable, as they help avoiding pitfalls, especially in dealing with physics describing phenomena that are counter-intuitive to everyday experiences. The definition of physical terms is given using mathematical objects, but I have often tried to supplement these with motivating physical ideas. Thus, for example, first the definition of an 'inertial frame' is given as a special type of chart, and subsequently it is shown that, in this chart, the coordinate representations of geodesics are affinely-parametrised straight lines (i.e., Newton's first law holds: free-fall motion is 'uniform' along a straight line; the 'uniform' part also comes from the geodesy), or an 'observer' is defined as a timelike future-pointing curve, and so on.

We will give a brief overview of the geometry-based gravitation theory here, so as to know roughly what to expect in this book. But before doing that, we quickly recall the classical view of gravitation.

Classically, gravity is an attractive *force* between massive objects. If two particles having masses m, M are separated by a distance d, then they experience forces $\vec{F}$, $-\vec{F}$, directed along the line joining these (point) particles, whose magnitude is directly proportional to the product mM of their masses, and inversely proportional to the square d^2 of the distance between them.

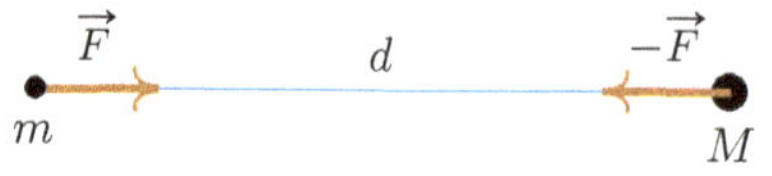

There is a universal constant, G_{N}, called the *gravitational constant*, such that if $|\vec{F}|$ denotes the magnitude of the force, then

$$|\vec{F}| = G_{\mathrm{N}} \frac{mM}{d^2}. \tag{$\blacklozenge$}$$

More generally, if there is mass distribution described by a matter density function[1] $\mathbb{R} \times \mathbb{R}^3 \ni (t, \mathbf{x}) \mapsto \rho(t, \mathbf{x})$, then it creates a gravitational field $\boldsymbol{g}$. The gravitational field $\boldsymbol{g}$ is equal to minus the gradient of a gravitational potential Φ, and Φ is a solution to the Poisson equation

$$\Delta_{\mathbf{x}} \Phi(t, \mathbf{x}) = -4\pi \rho(t, \mathbf{x}), \tag{$\star$}$$

where $\Delta_{\mathbf{x}} = \frac{\partial^2}{\partial x^2} + \frac{\partial^2}{\partial y^2} + \frac{\partial^2}{\partial z^2}$ denotes the Laplacian in the spatial variables (x, y, z) in $\mathbb{R}^3$. (We will derive this in Chapter 13.)

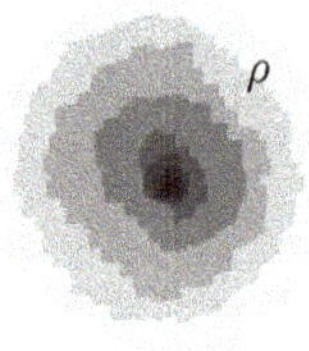

A test particle[2] of mass[3] m at position $\mathbf{x}$ and time t then experiences an attractive force given (with the symbol $\vec{\nabla}$ used to denote the gradient) by

$$\vec{F}(t, \mathbf{x}) = m(-\vec{\nabla} \Phi(t, \mathbf{x})).$$

[1]Here t is 'universal' time, and $\mathbf{x}$ is the position vector in $\mathbb{R}^3$.

[2]That is, matter which responds, but does not influence.

[3]This is to be thought of as 'gravitational' mass, allowing the test particle to 'couple/engage' with the gravitational field.

Knowing the position $\mathbf{x}(t_0)$ and velocity $\frac{d\mathbf{x}}{dt}(t_0)$ of the test particle m at the initial time $t_0 \in \mathbb{R}$ determines the particle trajectory $\mathbf{x}(\cdot)$ via Newton's second law of motion, 'Force is mass[4] times the acceleration', giving

$$\frac{d^2\mathbf{x}}{dt^2}(t) = -\overrightarrow{\nabla}\Phi(t, \mathbf{x}(t)). \qquad (\star\star)$$

At the beginning of the 20th century, the ideas of space and time were brought into question, and an accepted resolution[5] of the conundrums was to agree that nature has a universal speed limit, the speed of light. Thus, no information can travel faster than the speed of light. But then classical gravity is at odds with this idea, since equation ($\blacklozenge$) implies that if one of the masses was a function of time, its variation with time would instantaneously be felt by the other mass. So classical gravity is 'relativistically incorrect'. One of the revisions resulting from the aforementioned revolution was the profound realisation that space and time should be considered as a single whole, a 4-dimensional geometric object called spacetime, rather than two separate entities. Hence began a search for a geometry-based theory of gravitation, met with success[6] in 1915. It was subsequently widely accepted, because

- it subsumed the classical gravitation theory as an approximation,
- could explain observations unexplainable with the classical theory[7], and
- made predictions[8] which were later verified.

It is this geometry-based gravitation theory that we will learn about in this book. The entire book's content can be summarised as follows:

- Spacetime is a 4-dimensional Lorentzian oriented manifold with a time-orientation.
- Spacetime curvature is governed by the matter distribution via the 'field equation'.
- The motion of test matter in spacetime is given by the geodesic equation.

[4]Here, m is the 'inertial' mass. In principle, there is no reason why the inertial mass must equal the gravitational mass. (Just as the coupling with the electrical field is manifested by the electrical charge, which has nothing to do with the inertial mass.) In the classical viewpoint of gravity, the equality of the gravitational mass and the inertial mass is accepted as experimental fact. This equality results in the equation of motion, namely ($\star\star$), to be the same for all test particles, and this is apparent in the purported demonstration by Galileo, where two objects of different masses dropped from the top of the leaning tower of Pisa reached the Earth's surface at the same time.

[5]'Special' theory of relativity (1905).

[6]'General' theory of relativity.

[7]Discrepancy in Mercury's perihelion precession, see Chapter 14.

[8]Deflection of light, see Chapter 8, and the gravitational red-shift, see Chapter 5.

Thus gravity is no longer a 'force' between massive objects, which is the cause of their motion. Rather, massive objects curve spacetime, and all 'freely-falling' test matter[9] moves along geodesics in spacetime. This demystifies why two different objects near the surface of the Earth with identical initial positions and initial velocities have the same trajectories.

In the first part of this book, we develop the differential geometry of Lorentzian manifolds needed to comprehend the above statements. We choose many of our illustrating examples as the Lorentzian manifolds which are later spacetime models. This choice of examples will serve the twofold purpose of making the subsequent physics forthcoming in the second part relatable, and the mathematics learnt in the first part less dry.

Then in the second part, we focus on the physics, covering much of the essential material in the 20th century spacetime-based view of gravity: energy-momentum tensor field of matter, field equation, examples of spacetimes, Newtonian approximation, geodesics, tests of the theory, black holes, and end with a discussion of cosmological models of the universe.

Before we proceed, it is useful to keep the following rough picture in mind. Spacetime is the collection of all 'events'. An event is roughly a point, 'here and now'. Although spacetime $(M, \mathbf{g})$ is a 4-dimensional manifold, we have instead drawn a 2-dimensional surface here:

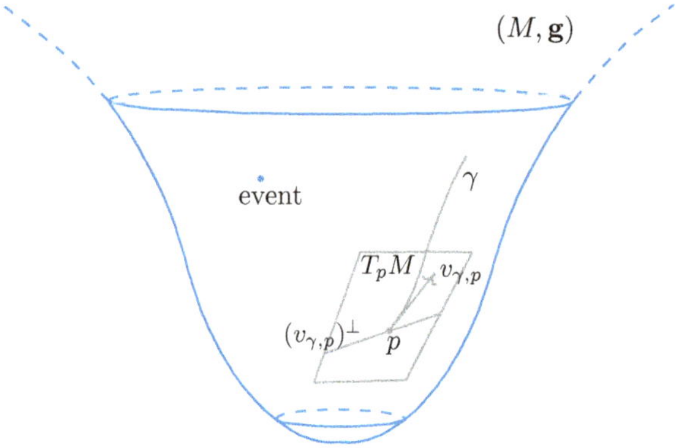

What determines the spacetime $(M, \mathbf{g})$? The answer is the field equation[10]:

$$\text{(geometry)}\ \mathbf{Ric} - \frac{1}{2}\mathbf{S}\mathbf{g} + \Lambda\mathbf{g} = 8\pi\mathbf{T}\ \text{(matter)} \tag{$*$}$$

[9]That is, force-free matter. As gravity is anyway not a force, we assume that there are no other forces either, e.g., even if the matter is charged, there is no electrical field, etc.
[10]Terms appearing here will be explained eventually: ones on the left-hand side (capturing the geometry of M), will be covered in Chapter 9, while the one on the right-hand side (describing the matter content in M), will be specified in Chapter 13.

where the right-hand side $\mathbf{T}$ describes the matter distribution (called the 'energy-momentum tensor field of matter'), and the left-hand side is expressible in terms of the metric $\mathbf{g}$. The metric $\mathbf{g}$ is something that equips the tangent space T_pM at each point of the manifold M with a Lorentzian scalar product, and essentially determines the 'shape' of M. So to obtain M, one needs to find a metric $\mathbf{g}$, which is a solution to this field equation. The field equation $(*)$ is the analogue[11] of the Poisson equation $(\star)$ in the classical viewpoint of gravity. The matter density function ρ on the right-hand side of $(\star)$ is replaced by $\mathbf{T}$, and now instead of finding the unknown gravitational potential Φ, we seek $\mathbf{g}$.

Given the metric $\mathbf{g}$, the motion of test matter (that is, matter whose effect on the $\mathbf{T}$ can be neglected) is given by the geodesic equation[12],

$$\nabla_{V_\gamma} V_\gamma = 0, \qquad (**)$$

where V_γ is the velocity field along γ, and this describes the motion of a particle in the spacetime M by means of a curve $\gamma : [a, b] \to M$, called its 'worldline'. The geodesic equation $(**)$ is the analogue of the equation of motion $(\star\star)$ in classical spacetime.

Does space 'evolve'? Or, what is the reality experienced by us, in terms of our view of the spatial world, with a notion of time passing in our minds? Firstly, we are a 'worldline', namely a curve $\gamma : I \to M$, where $I \subset \mathbb{R}$ is an interval, thought of as the list of events we encounter in our lifetimes—the curve merely describes the order in which we meet these events. The 'time' we experience between two events $p=\gamma(a)$ and $q=\gamma(b)$ along our worldline is the 'proper time', defined by

$$\tau(p, q) := \int_a^b \sqrt{-\mathbf{g}(\gamma(t))(v_{\gamma,\gamma(t)}, v_{\gamma,\gamma(t)})}\, dt.$$

(The notation will be clarified eventually in due course.) We will see that our worldline γ at each event $p=\gamma(t)$ along it, has a 'timelike' tangent vector $v_{\gamma,p}$ in the 4-dimensional tangent space T_pM. This $v_{\gamma,p}$ has a 3-dimensional orthogonal complement $(v_{\gamma,p})^\perp$ (with respect to the scalar product $\mathbf{g}(p)$), and this is roughly the 'space' we perceive at each moment. Thus the time experienced is 'personal' to each observer, and so is the perceived 'space' at each moment. Different observers thus have different ticking clocks (depending on what worldline they have in spacetime, and also different spatial realities.

[11] We will see this in Chapter 13.

[12] The notation which appears here will be clarified in Chapter 7, and we will discuss geodesics in Chapter 8.

How does one use this theory of spacetime in classical (i.e. nonquantum) physics? Given a matter distribution, for example a star, or even the universe, the field equation is solved to obtain the metric $\mathbf{g}$ (analogous to finding the 'gravitational potential' Φ by solving a Poisson equation involving the matter density distribution ρ in classical Newtonian gravity; minus the gradient of the potential then gives the gravitational field $\boldsymbol{g}$). This can then be used to obtain the geodesic equations for test matter (analogous to Newton's second law of motion; knowing the acceleration $\boldsymbol{g}$, and solving the resulting equation describes the particle trajectory).

It is well-known that one road towards the goal of mastering a subject is to teach it, since it makes it necessary to organise what has been learnt, so that the core principles are laid bare, and these notes were originally written with this intention. I hope that readers of this book will find it useful. For readers who want to delve deeper into the subject, some of the sources that have influenced the writing of this book, are recommended. A deeper and more complete treatment is given in the book [Sachs and Wu (1977)], but it assumes prior familiarity with differential geometry, while we develop this language along the way, starting from scratch. A more modern textbook, with an emphasis on Lorentzian geometry, is [O'Neill (1983)]. More elementary than these two books, but also approaching the subject mathematically, are the clear online lectures by Frederic Schuller [Schuller (2015)], and the book [Oloff (2023)]. I have used these and many other sources which are listed in the bibliography, but no claim to originality is made for any of the discussed material, in case there is a missing reference.

I am grateful to José Natário for reading this book, for pointing out various errors and typos, and for suggesting improvements in the text and additional exercises. In particular, an improved proof of Lemma 5.1, and Exercises 5.25, 5.26, 8.12, 9.18, 12.10, 12.15, 13.7, 14.5, 14.6, 14.7, 14.14, 16.4, 16.7, 16.8, 16.9 were suggested by José. I am indebted to Sara Maad Sasane for several useful discussions. This book was used as background for a PhD level course given at the Centre of Mathematical Sciences, Lund University, in Spring 2024. I am grateful to the students attending the lectures for their feedback, in particular, Thomas Munn, Olof Rubin, Alejandro Rodriguez Sponheimer, and Wilhelm Treschow. Thanks are also due to Adam Ostaszewski, Sara Sal Santos and Malik Muhammad Tariq for useful comments. It is a pleasure to thank the World Scientific publishing team: editors Rochelle Kronzek and Rok Ting Tan for their help and enthusiasm, and copy-editors Vanessa Sam and Peng Ah Huay for their meticulous work and comments.

In the second edition I have tried to correct all the mistakes I could find, and also added material covering topics that were omitted in the first edition. The details of the key changes made are listed below.

- Sixty new exercises and their solutions have been added:
  0.1, 0.2, 0.3, 1.3, 1.6, 1.20, 1.22, 1.25, 1.26, 2.9, 2.15, 2.16, 3.4, 3.8, 3.15, 5.2, 5.11, 5.16, 5.21, 5.22, 5.24, 6.9, 6.14, 6.15, 6.16, 6.17, 6.33, 6.34, 8.3, 8.6, 8.9, 8.11, 8.13, 8.14, 8.15, 8.16, 8.17, 9.2, 9.11, 9.12, 9.19, 12.1, 12.2, 12.8, 12.9, 12.10, 12.11, 13.1, 13.3, 13.5, 14.5, 14.8, 14.9, 14.10, 14.12, 15.4, 15.5, 15.6, 16.1, 16.9.
- Remarks 8.2, 8.3, 8.4, 9.1, 12.1, 14.1, 14.2, 14.4, 15.3, 15.4 are new. Remark 7.2 has been revised to include a justifying argument.
- A gap in the proof of Theorem 5.1 has been fixed.
- Finally, this book now includes twenty nine additional figures.

Amol Sasane

Lund, 2024

Notation and terminology

We use $\mathbb{N}$ for the set $\{1, 2, 3, \cdots\}$ of natural numbers, $\mathbb{Z}$ for the set $\{\cdots, -1, 0, 1, \cdots\}$ of integers, and $\mathbb{R}$ for the set of real numbers. The equivalence class of an element $a \in S$ under an equivalence relation on S will be denoted by $[a]$. We denote the identity map on a set S, $S \ni x \mapsto x \in S$, by id_S, and if S is clear from context, simply by id. For a function $f : X \to Y$, and a $y \in Y$, the notation $f \equiv y$ means that for all $x \in X$, $f(x) = y$. RHS means 'right-hand side', and LHS means 'left-hand side'.

Points or vectors in $\mathbb{R}^3$ are often denoted by bold face letters, for example, $\mathbf{x} = (x, y, z) \in \mathbb{R}^3$. The 'vector/cross product' of vectors $\mathbf{x}, \mathbf{y} \in \mathbb{R}^3$ is denoted by $\mathbf{x} \times \mathbf{y} \in \mathbb{R}^3$. The components of a vector $\mathbf{x}$ in $\mathbb{R}^m$ are often denoted by x^i, $1 \leqslant i \leqslant m$, in accordance with the physics literature, and if for example the square of $x^i \in \mathbb{R}$ is meant, we will use parenthesis, writing $(x^i)^2$. The standard basis vectors in $\mathbb{R}^m$ are given by $\mathbf{e}_1 = (1, 0, \cdots, 0), \cdots, \mathbf{e}_m = (0, \cdots, 0, 1)$. We will also use the notation $\langle \mathbf{x}, \mathbf{y} \rangle = x^1 y^1 + \cdots + x^m y^m$ for the Euclidean inner product of $\mathbf{x} = (x^1, \cdots, x^m)$ and $\mathbf{y} = (y^1, \cdots, y^m) \in \mathbb{R}^m$, and $\|\mathbf{x}\| := \sqrt{\langle \mathbf{x}, \mathbf{x} \rangle}$ for the Euclidean 2-norm. Given $\mathbf{x} \in \mathbb{R}^m$ and $r > 0$, the open ball $B(\mathbf{x}, r)$ in the Euclidean topology is $B(\mathbf{x}, r) = \{\mathbf{y} \in \mathbb{R}^m : \|\mathbf{y} - \mathbf{x}\| < r\}$. The closure of a subset S of a topological space will be denoted by $\overline{S}$. Thus the closure of the open ball $B(\mathbf{x}, r)$ in the Euclidean topology is the closed ball $\overline{B(\mathbf{x}, r)} = \{\mathbf{y} \in \mathbb{R}^m : \|\mathbf{y} - \mathbf{x}\| \leqslant r\}$. Throughout this book, unless otherwise stated, repeated 'dummy' indices, mostly one appearing as a superscript and one as a subscript, will be summed over the range of the index in question. Thus here $x^i \mathbf{e}_i$ means $x^1 \mathbf{e}_1 + \cdots + x^m \mathbf{e}_m$. The Kronecker delta symbol is denoted by δ^i_j, or sometimes by δ_{ij}, and means the following:

$$\delta^i_j = \begin{cases} 1 & \text{if } i = j, \\ 0 & \text{if } i \neq j. \end{cases}$$

All vector spaces are assumed to be real. Given a vector space V, its dual vector space consisting of all linear maps $\omega : V \to \mathbb{R}$, will be denoted by V^*, and if $\{e_i,\ i \in I\}$ is a basis for V, the dual basis is denoted by $\{\epsilon^i,\ i \in I\}$, and is defined by $\epsilon^i(e_j) = \delta^i_j$. In case of a finite-dimensional vector space with an ordered basis consisting of vectors $e_1, \cdots, e_m$ (taken in this order), we use the notation $(e_1, \cdots, e_m)$ for the basis instead of the usual set-theoretic notation $\{e_1, \cdots, e_m\}$. If $S \subset V$, then $\operatorname{span} S$ is the subspace consisting of all linear combinations of vectors from S (with an empty sum being defined as the zero vector). Also, $-S := \{-v : v \in S\}$. An invertible linear transformation $T : V \to W$ between vector spaces V, W is referred to as a linear isomorphism.

Let a^i_j, b_{ij}, c^{ij}, $1 \leqslant i \leqslant n$ and $1 \leqslant j \leqslant m$, be collections of numbers or functions. Then writing $A = [a^i_j]$, $B = [b_{ij}]$, $C = [c^{ij}]$ means A, B, C are the $n \times m$ matrices, whose entry in the ith row and j^{th} column is a^i_j, b_{ij}, and c^{ij}, respectively. The identity matrix is denoted by I or sometimes by I_m (to emphasise its size as an $m \times m$ matrix). Thus $I_m = [\delta^i_j]$. The transpose of a matrix M will be denoted by M^{t}. The determinant of a square matrix M is denoted by $\det M$. $\mathbb{R}^{n \times m}$ denotes the set of $n \times m$ matrices with real entries. $\mathrm{GL}_m(\mathbb{R})$ stands for the group of invertible in $\mathbb{R}^{m \times m}$ with the operation of matrix multiplication. The zero entries in a matrix are sometimes left as blanks.

For $n \in \mathbb{N}$ the nth order derivative of an n times differentiable function $f : \mathbb{R} \to \mathbb{R}$ at a point $t \in \mathbb{R}$ will be denoted by $f^{(n)}(t)$. If $U \subset \mathbb{R}^m$ is an open set, then for an $f : U \to \mathbb{R}^n$ having the component functions $f^1, \cdots, f^n$, and differentiable at $p \in \mathbb{R}^m$, the Jacobian matrix of f is given by

$$[f'(p)] = \begin{bmatrix} \frac{\partial f^1}{\partial u^1}(p) & \cdots & \frac{\partial f^1}{\partial u^m}(p) \\ \vdots & & \vdots \\ \frac{\partial f^n}{\partial u^1}(p) & \cdots & \frac{\partial f^n}{\partial u^m}(p) \end{bmatrix}.$$

Continuously differentiable functions on U will be said to be C^1 on U. This is equivalent to the first order partial derivatives of the components of f being continuous on U. If $f^1, \cdots, f^n$ have partial derivatives of all orders on U and they are continuous, then f is said to be C^∞ on U.

We also have the following table of physical constants. We make two remarks: c is not reserved for the speed of light exclusively, as we will also use it to denote suitable real constants in mathematical arguments. Also, we will often use units in which $c = G_{\mathrm{N}} = 1$, where G_{N} is the gravitational constant (see below).

Physical constant	Symbol	Value
Speed of light	c	$3\times10^{8}\,\mathrm{ms}^{-1}$
(Newton) Gravitational constant	G_{N}	$6.67\times10^{-11}\,\mathrm{m}^3\mathrm{kg}^{-1}\mathrm{s}^{-2}$
Planck constant	h	$6.63\times10^{-34}\,\mathrm{m}^2\mathrm{kgs}^{-1}$

Exercise 0.1. Using units where $c = G_{\mathrm{N}} = 1$ allows us to shed the burden of carrying overwhelming dimensions, to facilitate comparisons while performing approximations, and to obtain simpler expressions. Thus we can express some physical quantities, e.g. purely in powers of m (metres). Express in m^n, $n \in \mathbb{Z}$:

(1) Average speed of walking, 5 km/hr.
(2) A day, 24 hours.
(3) Mass of the Earth, $M_{\oplus} = 6 \times 10^{24}$ kg.
(4) Planck's constant, $h = 6.63 \times 10^{-34}$ Joule-second.
(5) Density of water, 10^3 kg m^{-3}.
(6) Acceleration of an apple falling near Earth's surface due to gravity, 9.8 m s^{-2}.

Exercise 0.2. Vice versa, it is important to revert. Convert the following.

(1) Escape velocity from the surface of the Earth, 3.7×10^{-5}, in km s^{-1}.
(2) Estimated age of the universe, 5.4×10^{24}m, into billions of years.
(3) Mass of the Sun, $M_{\oplus} = 1.5$km, into kg.
(4) Power of an LED light bulb, 2.74×10^{-52}, in Watts.
(5) Pressure of one atmosphere, $8.34 \times 10^{-40}\mathrm{m}^{-2}$, into Nm^{-2}.
(6) Acceleration of the Moon with respect to Earth, $3 \times 10^{-20}\mathrm{m}^{-1}$, into ms^{-2}.

Exercise 0.3. Determine the ratio of the magnitude of electrical force to the gravitational force between an electron and a proton. Use the following values:

- Charge on an electron and on a proton is $\mathrm{e} = 1.6 \times 10^{-19}$C.
- Mass of an proton is $m_p = 1.6 \times 10^{-27}$kg.
- Mass of an electron is $m_e = 9.1 \times 10^{-31}$kg.
- The constant in Coulomb's law of electrostatic force is $\frac{1}{4\pi\epsilon_0} = 9 \times 10^9 \frac{\mathrm{N\,m}^2}{\mathrm{C}^2}$.

Analogous to the above large dimensionless number of order 10^{40}, one can build yet another one by taking the age of the universe (about 13.7 billion years), the speed of light, and the classical radius of the electron ($\frac{\mathrm{e}^2}{4\pi\epsilon_0 m_e c^2} = 3.7 \times 10^{-16}$m). Determine this value, and show that it is in the order of 10^{42}.
Dirac interpreted such large number coincidences to have deeper cosmological significance, for example that perhaps the gravitational constant G_{N} varies with time t as $G_{\mathrm{N}} \propto \frac{1}{t}$, see, e.g., [Dirac (1974)].

Contents

Chapter 1

Smooth manifolds

As mentioned in the preface, spacetime is, on the set-theoretic level, a collection of points, called events. We want to do calculus on spacetime, because firstly, the field equation determining the geometry of spacetime is a partial differential equation on spacetime, and secondly, the geodesic equation describing motion of matter in spacetime is an ordinary differential equation on spacetime. It was discovered in the early 20th century that the appropriate structure to be used to describe spacetime is that of a Lorentzian smooth manifold. In this chapter, we will first learn about the notion of a smooth manifold.

Roughly speaking, an m-dimensional smooth manifold is a set M which can be covered by patches such that in each patch one can introduce coordinates and use them to do calculus. As coordinates are ad-hoc, we need to make sure that only certain 'admissible' coordinates are allowed, so that the definition of smoothness is independent of the choice of coordinates. It turns out that this also endows the manifold with a topology, so that smooth manifolds are special types of topological spaces.

1.1 Charts and atlases

We have prior experience with using coordinates, for example, for the surface of a sphere, one could use polar and azimuthal angles. But then we realise that in order to obtain an injective mapping from the sphere to the set of parameter ranges of the coordinates, we must work with patches on the sphere, instead of the whole sphere in one go. So we anticipate that also in the case of a manifold, it will need to be covered by patches, and coordinates need to be set up in each patch.

Before we introduce the notion of a manifold, we explain what we mean by putting coordinates in a patch. The aim of this section is to introduce the notions of a 'chart' and an 'atlas'. Roughly speaking, a chart describes the local coordinates set up in a patch of the manifold, and an atlas is a collection of such charts so that the union of the patches of the charts covers the whole manifold. In order that there is no conflict later on when defining smooth objects on the manifold, we will demand that the charts that make up an atlas are 'compatible'.

First, let us recall the notion of a topological space.

Definition 1.1. (Topological space.)
A *topological space* $(M, \mathcal{O})$ is a set M together with a collection $\mathcal{O}$ of subsets of M, such that the following hold:

(T1) $\varnothing, M \in \mathcal{O}$.

(T2) Whenever U_i, $i \in I$, belong to $\mathcal{O}$, also $\bigcup_{i \in I} U_i \in \mathcal{O}$.

(T3) Whenever $U, V \in \mathcal{O}$, also $U \cap V \in \mathcal{O}$.

The elements of $\mathcal{O}$ are called *open sets*, and $\mathcal{O}$ itself is referred to as a *topology on* M.

Recall that $\mathbb{R}^m$ can be equipped with its usual Euclidean topology.

Example 1.1. (Euclidean space $\mathbb{R}^m$.)
Let $\mathbb{R}^m = \{\mathbf{x} = (x^1, \cdots, x^m) : x^\ell \in \mathbb{R},\ 1 \leqslant \ell \leqslant m\}$ be the real vector space with componentwise operations of vector addition and multiplication by real scalars. For vectors $\mathbf{x}, \mathbf{y} \in \mathbb{R}^m$, their Euclidean inner product is given by $\langle \mathbf{x}, \mathbf{y} \rangle = x^1 y^1 + \cdots + x^m y^m$. For $\mathbf{x} \in \mathbb{R}^m$, we define the ball $B(\mathbf{x}, r)$ with center $\mathbf{x}$ and radius $r > 0$ by $B(\mathbf{x}, r) = \{\mathbf{y} \in \mathbb{R}^m : \langle \mathbf{x} - \mathbf{y}, \mathbf{x} - \mathbf{y} \rangle < r^2\}$. We define $\mathcal{O}$ to be the collection of subsets U of $\mathbb{R}^m$ with the property that whenever $\mathbf{x} \in U$, there exists an $r > 0$ such that $B(\mathbf{x}, r) \subset U$. Then $\mathcal{O}$ is a topology on $\mathbb{R}^m$. ◇

Definition 1.2. (Chart.)
Let M be a set. An *m-chart* on M is a pair (U, φ), where $U \subset M$, the map $\varphi : U \to \mathbb{R}^m$ is injective, and $\varphi(U)$ is an open subset of $\mathbb{R}^m$.

Henceforth, we will often drop the specification 'm' in 'm-chart', and simply refer to 'charts' for M, with the understanding that for a given M, the m is fixed. A chart allows us to talk about the *coordinates* of a point $p \in U \subset M$, with respect to the chart (U, φ), as the m-tuple of numbers $\varphi(p) \in \mathbb{R}^m$.

Example 1.2. (A chart for $\mathbb{R}^m$.) Let $U = \mathbb{R}^m$, and $\varphi : \mathbb{R}^m \to \mathbb{R}^m$ be the identity map id. Then $(\mathbb{R}^m, \mathrm{id})$ is a chart on $\mathbb{R}^m$. ◇

We will see later on that locally a manifold looks like $\mathbb{R}^m$ (its tangent space).

In Example 1.1, there is a distinguished point, namely the origin, but as nature does not provide natural coordinate systems, we also introduce $\mathbb{R}^m$ which has forgotten its origin, namely an affine space.

Example 1.3. (Affine space.)
An *affine space of dimension* m consists of

- a set M (of 'points'),
- an m-dimensional vector space V (whose vectors 'translate' points of M),
- a map $M \times V \ni (p, \mathbf{v}) \mapsto p + \mathbf{v} \in M$,

such that the following hold:

(A1) for all $p \in M$, $\mathbf{u}, \mathbf{v} \in V$, $p + (\mathbf{u} + \mathbf{v}) = (p + \mathbf{u}) + \mathbf{v}$,

(A2) for all $p \in M$, $p + \mathbf{0} = p$,

(A3) for all $p, q \in M$, there exists a unique $\mathbf{v}_{pq} \in V$ such that $q = p + \mathbf{v}_{pq}$.

Let $p \in M$ be fixed and let $\{\mathbf{e}_1, \cdots, \mathbf{e}_m\}$ be a basis for V. Given any $q \in M$, there exists a unique vector $\mathbf{v}_{pq} \in V$ such that $q = p + \mathbf{v}_{pq}$. This vector $\mathbf{v}_{pq}$ can be expressed in terms of the basis vectors, giving unique coordinates $\varphi(q) := (x^1, \cdots, x^m) \in \mathbb{R}^m$. Thus, $q = p + \mathbf{v}_{pq} = p + x^i \mathbf{e}_i$. Clearly, φ is injective, and with $U := M$, $\varphi(U) = \mathbb{R}^m$. So (M, φ) is a chart on M. $\diamond$

Exercise 1.1. Show that if $p, q, r \in M$, then $\mathbf{v}_{pr} = \mathbf{v}_{pq} + \mathbf{v}_{qr}$.

Example 1.4. (Charts on the sphere S^2.) Let

$$S^2 := \{(x, y, z) \in \mathbb{R}^3 : x^2 + y^2 + z^2 = 1\},$$

and $\mathbf{n} = (0, 0, 1)$, $\mathbf{s} = (0, 0, -1)$ denote the north and south poles in S^2. Set $U_\mathbf{n} = S^2 \backslash \{\mathbf{n}\}$, $U_\mathbf{s} = S^2 \backslash \{\mathbf{s}\}$, and define the 'stereographic' projections

$$S^2 \backslash \{\mathbf{n}\} = U_\mathbf{n} \ni (x, y, z) \longmapsto \varphi_\mathbf{n}(x, y, z) = \frac{1}{1 - z}(x, y) \in \mathbb{R}^2,$$

$$S^2 \backslash \{\mathbf{s}\} = U_\mathbf{s} \ni (x, y, z) \longmapsto \varphi_\mathbf{s}(x, y, z) = \frac{1}{1 + z}(x, y) \in \mathbb{R}^2.$$

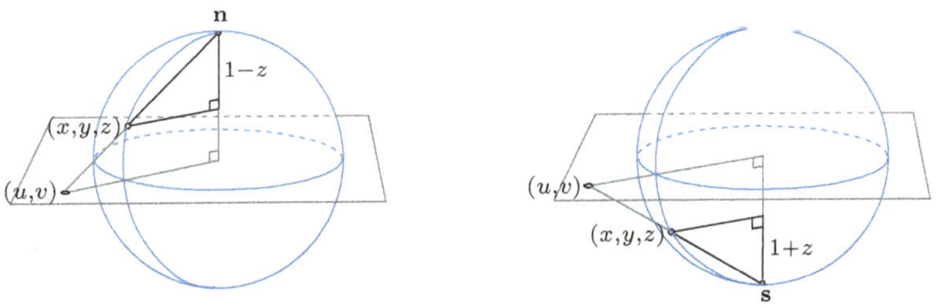

Charts $(U_\mathbf{n}, \varphi_\mathbf{n})$ and $(U_\mathbf{s}, \varphi_\mathbf{s})$ on S^2. Looking at the two similar triangles in the left picture, we have $(1 - z) : 1 = x : u$ and $(1 - z) : 1 = y : v$. Analogously, from the two similar triangles in the right picture, $(1 + z) : 1 = x : u$ and $(1 + z) : 1 = y : v$. This holds irrespective of the sign of z.

Then $(U_\mathbf{n}, \varphi_\mathbf{n})$ and $(U_\mathbf{s}, \varphi_\mathbf{s})$ are charts on S^2. $\diamond$

Exercise 1.2. Show that the inverse $\varphi_{\mathbf{n}}^{-1}:\mathbb{R}^2 \to S^2\backslash\{\mathbf{n}\}$ of the map $\varphi_{\mathbf{n}}$ is given by $(u,v) \overset{\varphi_{\mathbf{n}}^{-1}}{\longmapsto} \left(\frac{2u}{u^2+v^2+1}, \frac{2v}{u^2+v^2+1}, \frac{u^2+v^2-1}{u^2+v^2+1}\right)$. *Hint:* $u^2+v^2 = \frac{x^2+y^2}{(1-z)^2} = \frac{1-z^2}{(1-z)^2}$.

Exercise 1.3. Let $H^2 = \{(x,y,t) \in \mathbb{R}^3 : x^2+y^2-t^2 = -1, t>0\}$. The line joining the south pole $\mathbf{s} = (0,0,-1)$ to a point $\mathbf{p} \in H^2$ meets the xy-plane at $\mathbf{x}(\mathbf{p}) \in \mathbb{R}^2$. Show that $\varphi_{\mathbf{s}}(\mathbf{p}) = \frac{1}{1+t}(x,y)$, $\mathbf{p} = (x,y,t) \in H^2$, and $(H^2, \varphi_{\mathbf{s}})$ is a chart on H^2.

Exercise 1.4. Show that $(\mathbb{R}, x \mapsto x^3)$ is a chart on $\mathbb{R}$.

Exercise 1.5. (Cylinder.)
Consider the cylinder C in $\mathbb{R}^3$ given by $C = S^1 \times \mathbb{R} = \{(x,y,z) \in \mathbb{R}^3 : x^2+y^2 = 1\}$. Let $U_{x+} := \{(x,y,z) \in C : x > 0\}$, $\varphi_{x+}(x,y,z) = (y,z)$. Show that (U_{x+}, φ_{x+}) is a chart on C. Similar charts (U_{x-}, φ_{x-}), (U_{y+}, φ_{y+}) and (U_{y-}, φ_{y-}) can be defined analogously, so that the union of $U_{x+}, U_{x-}, U_{y+}, U_{y-}$ contains C.

A collection of charts for M (with the same m) will form an atlas provided they cover the set M and satisfy a compatibility condition.

Definition 1.3. (Atlas.)
Let M be a set. A collection of m-charts $\{(U_i, \varphi_i) : i \in I\}$ on M is called an *m-atlas* if it has the following properties:

(A1) $\bigcup_{i \in I} U_i = M$.

(A2) For all $i, j \in I$, $\varphi_i(U_i \cap U_j)$ is open in $\mathbb{R}^m$.

(A3) For all $i, j \in I$, $\varphi_j \circ \varphi_i^{-1} : \varphi_i(U_i \cap U_j) \to \varphi_j(U_i \cap U_j)$ is C^∞.
(The maps $\varphi_j \circ \varphi_i^{-1}$ are called *chart transition maps*.)

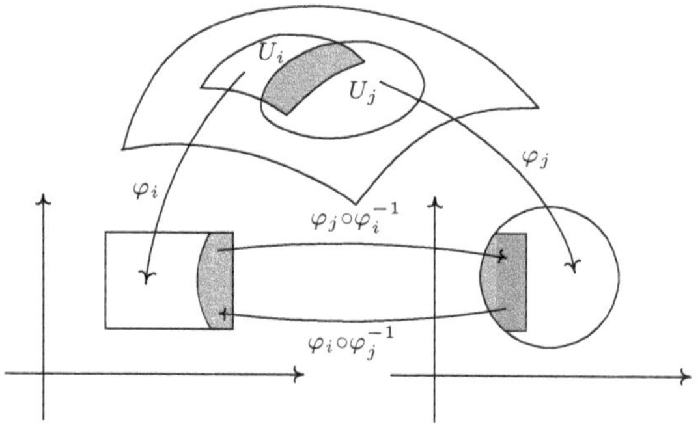

Compatible charts (U_i, φ_i) and (U_j, φ_j).

Just as with charts, we will often drop the m, and speak simply of an atlas, instead of an m-atlas. Note that $\varphi_j \circ \varphi_i^{-1} : \varphi_i(U_i \cap U_j) \to \varphi_j(U_i \cap U_j)$ is a bijection. Recall that a function $f : U \to \mathbb{R}^n$ is C^∞ on an open subset U of $\mathbb{R}^m$ if the components f^i of f, $1 \leqslant i \leqslant n$, have at each point of U, all partial derivatives of all orders with respect to the variables x^j, $1 \leqslant j \leqslant m$, which are also continuous on U.

The single charts in Examples 1.2, 1.3 and Exercise 1.4 are all atlases in a trivial manner.

Example 1.5. (S^2 revisited.)
The charts $(U_\mathbf{n}, \varphi_\mathbf{n})$ and $(U_\mathbf{s}, \varphi_\mathbf{s})$ from Example 1.4 form an atlas for S^2. Firstly, $U_\mathbf{n} \cup U_\mathbf{s} = S^2$. Secondly, $\varphi_\mathbf{n}(U_\mathbf{n} \cap U_\mathbf{s}) = \mathbb{R}^2 \backslash \{(0,0)\} = \varphi_\mathbf{s}(U_\mathbf{n} \cap U_\mathbf{s})$. Finally, $\varphi_\mathbf{s} \circ \varphi_\mathbf{n}^{-1}$, $\varphi_\mathbf{n} \circ \varphi_\mathbf{s}^{-1} : \mathbb{R}^2 \backslash \{(0,0)\} \to \mathbb{R}^2 \backslash \{(0,0)\}$, the chart transition maps, are both given by $(u,v) \mapsto \frac{(u,v)}{u^2+v^2}$, which is C^∞. ◇

Exercise 1.6. We revisit Exercise 1.3. Defining the projection map $p : H^2 \to \mathbb{R}^2$ by $p(x,y,t) = (x,y)$ for all $(x,y,t) \in H^2$, it is easy to see that p is injective and $p(H^2) = \mathbb{R}^2$. Thus (H^2, p) is a chart for H^2. Show that the charts $(H^2, \varphi_\mathbf{s})$ and (H^2, p), form an atlas for H^2.

Exercise 1.7. Show that the four charts in Exercise 1.5 form an atlas for the cylinder C in $\mathbb{R}^3$.

Example 1.6. (A non-atlas.) With $m = 1$ in Example 1.2, we get the chart $(\mathbb{R}, \varphi_1)$, with the chart map $\varphi_1(x) := x$ for $x \in \mathbb{R}$. On the other hand, in Exercise 1.4, we had seen that $(\mathbb{R}, \varphi_2 := (x \mapsto x^3))$ is yet another chart on $\mathbb{R}$. While they individually form atlases $\mathcal{A}_1 := \{(\mathbb{R}, \varphi_1)\}$ and $\mathcal{A}_2 := \{(\mathbb{R}, \varphi_2)\}$, their union $\mathcal{A} := \mathcal{A}_1 \cup \mathcal{A}_2 = \{(\mathbb{R}, \varphi_1), (\mathbb{R}, \varphi_2)\}$ does not form an atlas. Indeed, not all chart transition maps are smooth. Although $\varphi_2 \circ \varphi_1^{-1} = (x \mapsto x^3)$ is smooth on $\mathbb{R}$, $\varphi_1 \circ \varphi_2^{-1} = (x \mapsto x^{1/3})$ is not C^∞ everywhere on $\mathbb{R}$ since it is not differentiable at $x = 0$. ◇

The previous example motivates the following definition.

Definition 1.4. (Compatible atlases.)
Let M be a set. Two m-atlases $\mathcal{A}_1, \mathcal{A}_2$ are *compatible* if $\mathcal{A}_1 \cup \mathcal{A}_2$ is also an m-atlas on M.

Example 1.7. (Affine space revisited.) It is clear from the chart map definition given in Example 1.3, that different choices of points p, and of bases $\{\mathbf{e}_1, \cdots, \mathbf{e}_m\}$, will lead to different coordinates. Consider points $p, p' \in M$, and bases $B = \{\mathbf{e}_1, \cdots, \mathbf{e}_m\}$, $B' = \{\mathbf{e}'_1, \cdots, \mathbf{e}'_m\}$ for V, giving the chart maps φ, φ'. Are the atlases $\{(M, \varphi)\}$ and $\{(M, \varphi')\}$ compatible? To investigate this, we compute the chart transition map $\varphi' \circ \varphi^{-1} : \mathbb{R}^m \to \mathbb{R}^m$.

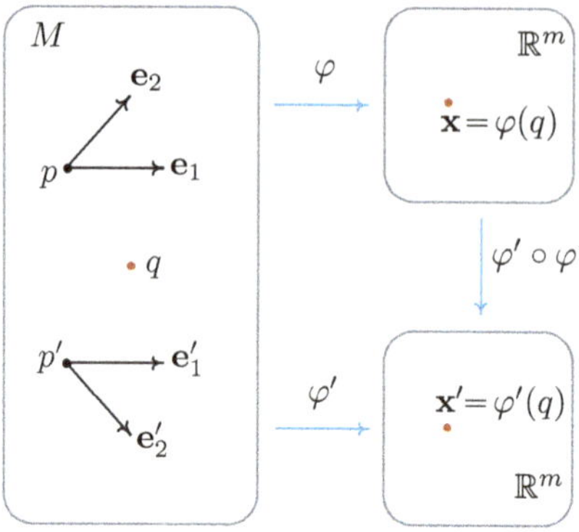

Change of coordinates.

Given $\mathbf{x} = (x^1, \cdots, x^m) \in \mathbb{R}^m$, $q := \varphi^{-1}(\mathbf{x}) \in M$, and we wish to find the coordinates $\mathbf{x}'$ of this q using the point p' and the basis B'. So we need to write $q = p' + \mathbf{v}_{p'q}$, and find $\mathbf{x}'$ by expanding $\mathbf{v}_{p'q}$ using the basis B'. We have $\mathbf{v}_{p'q} = \mathbf{v}_{p'p} + \mathbf{v}_{pq}$, and as $q = \varphi^{-1}(\mathbf{x})$, also $\mathbf{v}_{pq} = x^i\mathbf{e}_i$. Introduce the change of basis matrix $A = [A_i^j] \in \mathrm{GL}_m(\mathbb{R})$, where A_i^j denotes the entry in the jth row and i^{th} column of A, defined by $\mathbf{e}_i = A_i^j \mathbf{e}'_j$. Also, let $\mathbf{b} = (b^1, \cdots, b^m) \in \mathbb{R}^m$ be defined by $\mathbf{v}_{p'p} = b^j\mathbf{e}'_j$. Then

$$\mathbf{v}_{p'q} = \mathbf{v}_{p'p} + \mathbf{v}_{pq} = b^j\mathbf{e}'_j + x^i\mathbf{e}_i = b^j\mathbf{e}'_j + x^i(A_i^j\mathbf{e}'_j) = (b^j + A_i^j x^i)\mathbf{e}'_j,$$

and so the chart transition map $\varphi' \circ \varphi^{-1}$ is the affine linear map given by $\varphi' \circ \varphi^{-1}(\mathbf{x}) = \mathbf{b} + A\mathbf{x}$, for all $\mathbf{x} \in \mathbb{R}^m$, which is C^∞. Its inverse $\varphi \circ (\varphi')^{-1}$ is given by $\varphi \circ (\varphi')^{-1}(\mathbf{x}) = -A^{-1}\mathbf{b} + A^{-1}\mathbf{x}$, $\mathbf{x} \in \mathbb{R}^m$. So the atlases $\{(M, \varphi)\}$ and $\{(M, \varphi')\}$ are compatible. ◇

Exercise 1.8. Show that the set $\mathbb{R}^m \times \mathrm{GL}_m(\mathbb{R})$ is a group with the composition $(\mathbf{b}_2, A_2) \cdot (\mathbf{b}_1, A_1) = (\mathbf{b}_2 + A_2\mathbf{b}_1, A_2A_1)$ for $(\mathbf{b}_2, A_2), (\mathbf{b}_1, A_1)$ in $\mathbb{R}^m \times \mathrm{GL}_m(\mathbb{R})$.

Exercise 1.9. Prove that compatibility is an equivalence relation on the collection of all atlases on a set M.

Thus to specify a manifold[1], we should work with compatible atlases (so that in hindsight, there will be no conflict when we define smooth objects on the manifold), and given that compatibility is an equivalence relation, we just need to commit to one particular atlas for the set M at hand. We make this a definition.

---

[1]The name 'manifold' comes from the German word 'mannigfaltigkeit' used by Riemann in his doctoral thesis, which contained, among other things, a discussion of multi-valued complex functions and their (now called) Riemann-surfaces.

Definition 1.5. (Smooth manifold, dimension of a manifold.)
A *smooth manifold* $(M, [\mathcal{A}])$ is a set M together with an equivalence class $[\mathcal{A}]$ of compatible m-atlases on M. We call m the *dimension* of the smooth manifold M. A chart from an atlas in $[\mathcal{A}]$ is said to be *admissible* for the smooth manifold. We refer to $[\mathcal{A}]$ as a *smooth structure on* M.

Thus the atlases given in Examples 1.2, 1.3, 1.4 and Exercise 1.5 can be used to make the respective set M into a smooth manifold. As we will use it frequently, we will call $\mathbb{R}^m$ with the atlas comprising the single chart $(\mathbb{R}^m, \text{id})$ as the *smooth manifold* $\mathbb{R}^m$ *with the standard smooth structure.*

In Example 1.6, the two atlases $\mathcal{A}_1$ and $\mathcal{A}_2$ are not compatible. Hence, $(\mathbb{R}, [\mathcal{A}_1])$ and $(\mathbb{R}, [\mathcal{A}_2])$ are two different smooth manifolds.

Example 1.8. (Sphere.) Consider the sphere as a smooth manifold with the smooth structure given by the atlas in Example 1.5. In this example, we give a different compatible atlas, using one of the charts as the familiar one with spherical polar coordinates. It can be shown that the map

$$(0, \pi) \times (0, 2\pi) \ni (\theta, \phi) \mapsto \big((\sin\theta)\cos\phi, (\sin\theta)\sin\phi, \cos\theta\big) \in S^2 \subset \mathbb{R}^3$$

is injective, and hence a bijection onto its image

$$U = S^2 \backslash \{(x, y, z) \in \mathbb{R}^3 : y = 0 \text{ and } x \geqslant 0\}.$$

For a point $p \in U$, the angle $\theta(p)$ is called the *polar angle of* p, and the angle $\phi(p)$ is called the *azimuthal angle of* p. For a point $p \in U$, we define the map φ on U by $\varphi(p) := (\theta(p), \phi(p)) \in (0, \pi) \times (0, 2\pi)$, where if $p = (x, y, z)$, then $\theta(p) = \cos^{-1} z$ and

$$\phi(p) := \measuredangle(x, y) := \begin{cases} \cos^{-1} \dfrac{x}{\sqrt{x^2+y^2}} & \text{if } y>0, \\ \pi - \sin^{-1} \dfrac{y}{\sqrt{x^2+y^2}} & \text{if } x<0, \\ 2\pi - \cos^{-1} \dfrac{x}{\sqrt{x^2+y^2}} & \text{if } y<0. \end{cases}$$

Here $\cos^{-1} : (-1, 1) \to (0, \pi)$ and $\sin^{-1} : (-1, 1) \to (-\frac{\pi}{2}, \frac{\pi}{2})$ are the inverse trigonometric functions. It can be checked that ϕ is well-defined and that the map $\mathbb{R}^2 \backslash \{(x, y) \in \mathbb{R}^2 : x \geqslant 0\} \ni (x, y) \mapsto \measuredangle(x, y)$ is C^∞. Using this, it can be checked that (U, φ) is an admissible chart: e.g., if $v > 0$, then

$$\varphi_{\mathbf{n}}(U_{\mathbf{n}} \cap U) \ni (u, v) \overset{\varphi \circ \varphi_{\mathbf{n}}^{-1}}{\longmapsto} \Big(\cos^{-1} \frac{u^2+v^2-1}{u^2+v^2+1}, \cos^{-1} \frac{u}{\sqrt{u^2+v^2}}\Big)$$

is C^∞, and for all $(\theta, \phi) \in (0, \pi) \times (0, 2\pi)$, we have

$$\varphi(U \cap U_{\mathbf{n}}) \ni (\theta, \phi) \overset{\varphi_{\mathbf{n}} \circ \varphi^{-1}}{\longmapsto} \Big(\frac{(\sin\theta)\cos\phi}{1 - \cos\theta}, \frac{(\sin\theta)\sin\phi}{1 - \cos\theta}\Big)$$

is C^∞.

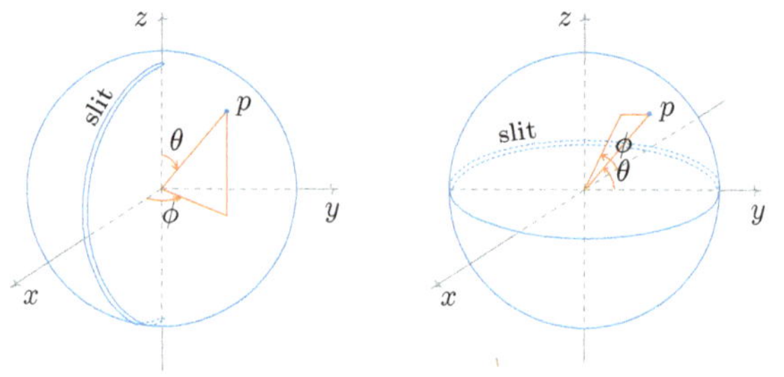

The chart U covers S^2 except for a 'slit', namely the intersection of S^2 with the half plane $\{(x, y, z) : y = 0,\ x \geqslant 0\}$. In order to cover S^2, we can take another chart (V, ψ), defined in a similar manner, by taking a differently placed slit, in a plane perpendicular to the one containing the original slit. Then V together with U, covers S^2. More explicitly, V covers S^2 except for the intersection of S^2 with the half plane $\{(x, y, z) : z = 0,\ x \leqslant 0\}$. V is the image of the map

$$(0, \pi) \times (0, 2\pi) \ni (\theta, \phi) \mapsto \big(-(\sin\theta)\cos\phi, \cos\theta, (\sin\theta)\sin\phi\big) \in S^2 \subset \mathbb{R}^3,$$

and this map is obtained by taking the polar angle with the positive y-axis, and the azimuthal angle with the negative x-axis counterclockwise in the $y = 0$ plane; see the picture above. ◇

Exercise 1.10. Consider the square $S := \{(x, y) \in \mathbb{R}^2 : |x| + |y| = 1\}$. Show that $\mathcal{A} := \{(U_+, \varphi_+), (U_-, \varphi_-), (V_+, \psi_+), (V_-, \psi_-)\}$ is an atlas for S, where

$$\begin{aligned} U_+ &:= \{(x, y) \in S : x > 0\} & \varphi_+(x, y) &= y, \\ U_- &:= \{(x, y) \in S : x < 0\} & \varphi_-(x, y) &= y, \\ V_+ &:= \{(x, y) \in S : y > 0\} & \psi_+(x, y) &= x, \\ V_- &:= \{(x, y) \in S : y < 0\} & \psi_-(x, y) &= x. \end{aligned}$$

Thus $(S, [\mathcal{A}])$ is a smooth manifold. So a smooth manifold may not necessarily 'appear' smooth.

We now discuss four 'spacetimes' by just looking at the underlying smooth structure. Later on, our examples below will be made 'Lorentzian' manifolds, that is, smooth manifolds with some added structure. While we are not yet ready to specify the added structure, we nevertheless introduce these by only describing their smooth structures. Thus these are 'pre-' spacetimes for now, ripe for becoming legitimate spacetimes later.

Example 1.9. (Minkowski spacetime.) Let V be a 4-dimensional real vector space. Suppose that M is an affine space over V. Take any $p \in M$, and a basis $\{\mathbf{e}_1, \mathbf{e}_2, \mathbf{e}_3, \mathbf{e}_4\}$ for V, and let $\varphi : M \to \mathbb{R}^4$ be the corresponding chart as in Example 1.3. Let $\mathcal{A} := \{(M, \varphi)\}$. Then $(M, [\mathcal{A}])$ is a smooth manifold, referred to as the *Minkowski spacetime*. ◇

Example 1.10. (Cylindrical spacetime.) Let $M = \mathbb{R} \times S^1$ with the atlas $\mathcal{A}$ comprising the charts defined in Exercise 1.5. Then $(M, [\mathcal{A}])$ is a smooth manifold, referred to as the *cylindrical spacetime.* ◇

Example 1.11. (FLRW spacetime.) Let $I := (0, \infty)$. Then $M := I \times \mathbb{R}^3$, with the atlas $\{(I \times \mathbb{R}^3, \mathrm{id}_{I\times\mathbb{R}^3})\}$ is a 4-dimensional smooth manifold, called the FLRW *spacetime* (after Friedman, Lemaitre, Robertson, Walker). ◇

Exercise 1.11. (Product of smooth manifolds.) Let M be an m-dimensional smooth manifold, with an atlas $\mathcal{A}_M = \{(U_i, \varphi_i), i \in I\}$. Let N be an n-dimensional smooth manifold, with an atlas $\mathcal{A}_N = \{(V_j, \psi_j), j \in J\}$. Define for $i \in I$, $j \in J$, the maps $\varphi_i \times \psi_j : U_i \times V_j \to \mathbb{R}^{m+n}$ by $(\varphi_i \times \psi_j)(p, q) = (\varphi_i(p), \psi_j(q))$ for all $p \in U_i$, $q \in V_j$. Show that $\{(U_i \times V_j, \varphi_i \times \psi_j), i \in I, j \in J\}$ is an atlas for $M \times N$, making it an $(m + n)$-dimensional smooth manifold.

Example 1.12. (Schwarzschild[2] spacetime.) Let $m > 0$, and $I := (2m, \infty)$. Let $M = \mathbb{R} \times I \times S^2$, where S^2 is the unit sphere in $\mathbb{R}^3$. Taking the atlases $\{(\mathbb{R}, \mathrm{id}_\mathbb{R})\}$, $\{(I, \mathrm{id}_I)\}$, and $\{(U, \varphi), (V, \psi)\}$ (Example 1.8), for $\mathbb{R}, I, S^2$, respectively, we see that M is a smooth manifold using the construction based on Exercise 1.11. We call this 4-dimensional smooth manifold the *Schwarzschild spacetime.* ◇

1.2 Topology on a smooth manifold

We will want to talk about continuous maps between smooth manifolds, for example a 'worldline' in a spacetime (Definition 1.2). The way we equip a smooth manifold with a topology is by insisting that the chart maps are homeomorphisms (Theorem 1.2). This is the motivation for the following definition.

Definition 1.6. (Open set in a smooth manifold.)
Let $(M,[\mathcal{A}])$ be an m-dimensional smooth manifold and $\{(U_i, \varphi_i), i \in I\} \in [\mathcal{A}]$. A set $U \subset M$ is *open* if for all $i \in I$, $\varphi_i(U \cap U_i)$ is open in $\mathbb{R}^m$, where $\mathbb{R}^m$ is given its standard Euclidean topology, described by the Euclidean metric

$$d(\mathbf{x}, \mathbf{y}) := \sqrt{\sum_{i=1}^{m} (x^i - y^i)^2} \quad \mathbf{x} = (x^1, \cdots, x^m),\ \mathbf{y} = (y^1, \cdots, y^m) \in \mathbb{R}^m.$$

Proposition 1.1. *Definition 1.6 of an open set is well-defined, that is, it does not depend on the choice of the atlas in* $[\mathcal{A}]$.

Proof. Let $\mathcal{A}_1 = \{(U_i, \varphi_i), i \in I\}$ and $\mathcal{A}_2 = \{(V_j, \psi_j), j \in J\}$ be atlases in $[\mathcal{A}]$. Let $U \subset M$, and suppose for each $i \in I$, $A_i := \varphi_i(U \cap U_i)$ is open in

[2]After Karl Schwarzschild (1873–1916), a German physicist and astronomer.

$\mathbb{R}^m$. Let $j \in J$. We must show that $\psi_j(U \cap V_j)$ is open in $\mathbb{R}^m$. We have

$$\psi_j(U \cap V_j) = \psi_j\big((U \cap M) \cap V_j\big) = \psi_j\big(U \cap \big(\bigcup_i U_i\big) \cap V_j\big) = \psi_j\big(\bigcup_i (U \cap U_i \cap V_j)\big)$$
$$= \bigcup_i \psi_j(U \cap U_i \cap V_j). \qquad (\star)$$

Set $B_i := \varphi_i(U_i \cap V_j)$. Then B_i is open, since the charts (U_i, φ_i) and (V_j, ψ_j) belong to the atlas $\mathcal{A}_1 \cup \mathcal{A}_2$. The intersection of this open B_i with the open set $A_i = \varphi_i(U \cap U_i)$, is open. Now $A_i \cap B_i = \varphi_i(U \cap U_i \cap V_j)$. (Indeed, $\supset$ is trivially true, and $\subset$ follows from the injectivity of φ_i on U_i.) Consider the C^∞ (and in particular, continuous) map $\varphi_i \circ \psi_j^{-1} : \psi_j(U_i \cap V_j) \to \varphi_i(U_i \cap V_j)$. As the open set $A_i \cap B_i = \varphi_i(U \cap U_i \cap V_j)$ is contained in the open set $\varphi_i(U_i \cap V_j) \subset \mathbb{R}^m$, it follows that $(\varphi_i \circ \psi_j^{-1})^{-1}(A_i \cap B_i)$ is an open subset of the open set $\psi_j(U_i \cap V_j) \subset \mathbb{R}^m$, that is,

$$(\varphi_i \circ \psi_j^{-1})^{-1}(A_i \cap B_i) = \psi_j(\varphi_i^{-1}(\varphi_i(U \cap U_i \cap V_j))) = \psi_j(U \cap U_i \cap V_j)$$

is open in $\mathbb{R}^m$. So $\psi_j(U \cap V_j) \overset{(\star)}{=} \bigcup_i \psi_j(U \cap U_i \cap V_j)$ is open in $\mathbb{R}^m$. □

We show that calling such sets 'open' is justified, as they form a topology on the manifold.

Theorem 1.1. *Let $(M, [\mathcal{A}])$ be an m-dimensional smooth manifold. Then the collection $\mathcal{O} := \{U \subset M : U$ is open in $M\}$ is a topology on M.*

Proof. Let $\{(U_i, \varphi_i), i \in I\} \in [\mathcal{A}]$. Then $\varnothing = \varphi_i(\varnothing \cap U_i)$ is open in $\mathbb{R}^m$ for all $i \in I$, and so $\varnothing \in \mathcal{O}$. Also, for all $i \in I$, $\varphi_i(U_i) = \varphi_i(M \cap U_i)$ is open in $\mathbb{R}^m$ since (U_i, φ_i) is a chart, and so $M \in \mathcal{O}$.

Let $U, V \in \mathcal{O}$. Then for all $i \in I$, $\varphi_i((U \cap V) \cap U_i) = \varphi_i(U \cap U_i) \cap \varphi_i(V \cap U_i)$ ($\subset$ is always true for any map, and $\supset$ holds by the injectivity of φ_i). Being the intersection of open sets, $\varphi_i((U \cap V) \cap U_i)$ is open in $\mathbb{R}^m$ for all $i \in I$, and consequently, $U \cap V \in \mathcal{O}$.

Let $V_j \in \mathcal{O}$ for all $j \in J$. Then we have that for all $i \in I$,

$$\varphi_i\big(\big(\bigcup_j V_j\big) \cap U_i\big) = \bigcup_j \varphi_i(V_j \cap U_i),$$

is open in $\mathbb{R}^m$, as it is the union of open sets $\varphi_i(V_j \cap U_i)$ in $\mathbb{R}^m$. Hence, $\bigcup_j V_j \in \mathcal{O}$. □

Definition 1.7. (Topology induced by a smooth structure.)
Let $(M, [\mathcal{A}])$ be an m-dimensional smooth manifold. Then the collection $\mathcal{O} := \{U \subset M : U$ is open (Definition 1.6) in $M\}$ is called the *topology induced on M by the smooth structure* $[\mathcal{A}]$.

Remark 1.1. Often in the literature, a smooth manifold is defined by first introducing the concept of a 'topological manifold', where one starts with a topological space which can be covered by charts which are homeomorphisms to open subsets of $\mathbb{R}^m$. We have not adopted this route, since such an approach forces one to begin with a topology. But we now reconcile our definition with this prevalent one in the following result. $\ast$

Theorem 1.2.
Let $(M, [\mathcal{A}])$ be an m-dimensional smooth manifold, and let

$$\mathcal{O} := \{U \subset M : U \text{ is open in } M\}$$

be the topology induced on M by the smooth structure $[\mathcal{A}]$. Suppose that $\{(U_i, \varphi_i), i \in I\} \in [\mathcal{A}]$. Then for each $i \in I$, $\varphi_i : U_i \to \varphi_i(U_i)$ is a homeomorphism.

Proof. Let $i \in I$. As (U_i, φ_i) is a chart, we know that $\varphi_i(U_i)$ is open in $\mathbb{R}^m$ and that $\varphi_i : U_i \to \varphi_i(U_i)$ is a bijection. We only need to show the continuity of φ_i and φ_i^{-1}. Let $V \subset \varphi_i(U_i)$ be open. Then $\varphi_i^{-1}V \subset U_i$. We must show that this is an open set in M. For any $j \in I$, we have $\varphi_j((\varphi_i^{-1}V) \cap U_j) = (\varphi_j \circ \varphi_i^{-1})(V \cap \varphi_i(U_i \cap U_j))$. As V and $\varphi_i(U_i \cap U_j)$ are open in $\mathbb{R}^m$, so is their intersection. Thus $(\varphi_j \circ \varphi_i^{-1})(V \cap \varphi_i(U_i \cap U_j))$, being the inverse image under the (C^∞ and hence) continuous map $(\varphi_j \circ \varphi_i^{-1})^{-1}$ of the open set $V \cap \varphi_i(U_i \cap U_j)$ $(\subset \varphi_i(U_i \cap U_j))$, is open. Hence $\varphi_j((\varphi_i^{-1}V) \cap U_j)$ is open for all $j \in I$, that is, $\varphi_i^{-1}V$ is open in M. So $\varphi_i : U_i \to \varphi_i(U_i)$ is continuous.

Let $U \subset U_i$ be open. We want to show that $\varphi_i(U) = (\varphi_i^{-1})^{-1}U$ is open in $\mathbb{R}^m$ (and hence open in $\varphi_i(U_i)$). The fact that U is open means in particular that $\varphi_i(U \cap U_i)$ is open in $\mathbb{R}^m$. But $\varphi_i(U \cap U_i) = \varphi_i(U)$, since $U \subset U_i$. Thus the inverse map $\varphi_i^{-1} : \varphi_i(U_i) \to U_i$ is also continuous. $\square$

Exercise 1.12. Let $\mathbb{R}^m$ be equipped with the standard smooth structure. Show that the topology induced by this smooth structure coincides with the standard Euclidean topology.

Exercise 1.13. Consider the double cone $C = \{(x, y, z) \in \mathbb{R}^3 : x^2 + y^2 = z^2\} \subset \mathbb{R}^3$. Show that C cannot carry a smooth structure $[\mathcal{A}]$ making it a 2-dimensional smooth manifold such that the topology induced by $[\mathcal{A}]$ on C coincides with the subspace topology on C (as a subset of $\mathbb{R}^3$ with its standard Euclidean topology). If we delete the point $\mathbf{0} = (0, 0, 0)$ from C, i.e., we consider $C_* := C \backslash \{\mathbf{0}\}$, then we do get a smooth manifold, for example by taking an atlas comprising two charts, namely $(\{(x, y, z) \in C_* : z > 0\}, \pi)$ and $(\{(x, y, z) \in C_* : z < 0\}, \pi)$, where the chart map π in each case is just the restriction to these chart domains of the projection map onto the xy-plane: $\mathbb{R}^2 \ni (x, y, z) \mapsto (x, y) \in \mathbb{R}^2$.

Remark 1.2. (Hausdorff and second countable assumptions on $\mathcal{O}$.)
In order to do analysis, it is desirable to have two additional properties enjoyed by the topology $\mathcal{O}$:

(H) A topology $\mathcal{O}$ on a set M is *Hausdorff* if for every $p, q \in M$, there exist $U, V \in \mathcal{O}$ such that $p \in U$, $q \in V$, and $U \cap V = \varnothing$. Thus distinct points possess disjoint neighbourhoods, a type of 'separation axiom'. Such a property is quite basic, since otherwise limits of sequences are not guaranteed to be unique.

(S) A *basis for* $\mathcal{O}$ is a collection $\mathcal{B} = \{B_i : i \in I\}$ of open sets such that every open set in $\mathcal{O}$ is a union of elements from $\mathcal{B}$. A topology $\mathcal{O}$ on a set M is *second countable* if there exists a countable basis for $\mathcal{O}$. When wanting to do 'integration' on manifolds, this property will be needed in order to construct a so-called 'partition of unity', which will essentially mean that we can use m-charts to set up Riemann integrals of functions defined on the manifold, and patch these contributions to obtain an integral of the function defined on the whole manifold.

Unfortunately, for a smooth manifold, neither of these properties are guaranteed to hold for the topology $\mathcal{O}$ from Theorem 1.1. So, in order to proceed without pitfalls, we will make a standing assumption that whenever we talk of a smooth manifold in this book, we will mean in addition that the associated topology $\mathcal{O}$ is Hausdorff and second countable. The standard topology of the Euclidean space $\mathbb{R}^m$ generated by the 2-norm $\|\cdot\|$ satisfies the second countability assumption since the open balls with centers all of whose components are rational numbers, and whose radius is also a rational number, form a countable basis. Now, if the manifold can be covered by an atlas in the smooth structure containing countably many charts, then it follows that (since the chart maps are homeomorphisms) the images of members of the countable basis for $\mathbb{R}^m$ under the inverse of the chart maps will form a countable basis for the topology of the manifold. All the examples of smooth manifolds considered in this book will be of this type. $\ast$

Exercise 1.14. Let U be an open subset of a smooth manifold M given by an atlas $\mathcal{A}$. Let $\mathcal{A}_U := \{(U \cap V, \psi|_{U \cap V}) : (V, \psi) \in \mathcal{A}\}$. Show that $\mathcal{A}_U$ is an atlas for U. Prove that if (W, σ) is admissible for M, then $(U \cap W, \sigma|_{U \cap W})$ is admissible for $(U, [\mathcal{A}_U])$. U is then said to be given the *smooth structure induced by* $(M, [\mathcal{A}])$. In particular, if (U, φ) is an admissible chart for M, then $[\mathcal{A}_U] = [\{(U, \varphi)\}]$.

As a spacetime M is the collection of all events, the life of a particle can be modelled by a curve in M by stringing together all the events encountered by the particle in its lifetime. Let $I \subset \mathbb{R}$ be an interval and M be a smooth manifold. A continuous map $\gamma : I \to M$ is called a *curve* or a *worldline*.

1.3 Smooth maps

The point of the definition of a smooth manifold is to enable the consideration of smooth objects on it, for example, a real-valued smooth function (think of temperature), a 'vector field', etc., in an unambiguous way.

Definition 1.8. (Smooth map.)
Let M, N be smooth manifolds, with dimensions m, n, respectively. A map $f : M \to N$ is said to be *smooth* if for all $p \in M$,

- there exists an admissible chart (U, φ) for M such that $p \in U$,
- there exists an admissible chart (V, ψ) for N such that $f(U) \subset V$ (in particular $f(p) \in V$),
- $\psi \circ f \circ \varphi^{-1} : \varphi(U) \to \mathbb{R}^n$ is C^∞ on $\varphi(U) \subset \mathbb{R}^m$.

If M is a smooth manifold, and $\mathbb{R}$ has the standard smooth structure, then we use the notation $C^\infty(M)$ to denote the set of all smooth maps $f : M \to \mathbb{R}$.

For a smooth manifold M, the identity map $\mathrm{id}_M : M \to M$ is smooth.

Example 1.13. (Chart maps are smooth.)
Let (U, φ) be a chart from an atlas defining the smooth manifold M. We now consider U itself to be a smooth manifold, described by the trivial atlas $\{(U, \varphi)\}$. Then $\varphi(U) \subset \mathbb{R}^m$ is an open subset of $\mathbb{R}^m$. We consider $\varphi(U)$ as a smooth manifold described by the atlas comprising the single chart $(\varphi(U), \mathrm{id}_{\varphi(U)})$. We claim that the chart map $\varphi : U \to \varphi(U)$ is smooth. For each $p \in U$, we take the admissible chart (U, φ) for U containing p, and the admissible chart $(V := \varphi(U), \mathrm{id}_{\varphi(U)})$ for the smooth manifold $\varphi(U)$. Then $\varphi(U) = V$. Moreover, $\mathrm{id}_{\varphi(U)} \circ \varphi \circ \varphi^{-1} = \mathrm{id}_{\varphi(U)} : \varphi(U) \to \varphi(U) \subset \mathbb{R}^m$, which is clearly C^∞. As $p \in U$ was arbitrary, $\varphi : U \to \varphi(U)$ is smooth. ◇

Exercise 1.15. Let M, N be smooth manifolds and $f : M \to N$ be a smooth map. Show that f is continuous.

Exercise 1.16. Let M_1, M_2, M_3 be smooth manifolds, and let $f_{12} : M_1 \to M_2$, $f_{23} : M_2 \to M_3$ be smooth maps. Prove that $f_{23} \circ f_{12} : M_1 \to M_3$ is smooth.

Exercise 1.17. Let M, N be smooth manifolds, and $M \times N$ be the smooth manifold described in Exercise 1.11. Let the projection map $\pi_M : M \times N \to M$ be given by $M \times N \ni (p, q) \mapsto p \in M$. Given a $q \in N$, let the injection map $i_q : M \to M \times N$ be given by $M \ni p \mapsto (p, q) \in M \times N$.

- Show that π_M is smooth. (Similarly, $M \times N \ni (p, q) \mapsto q \in N$ is smooth.)
- Show that i_q is smooth. (Also, for $p \in M$, $N \ni q \mapsto (p, q) \in M \times N$ is smooth.)

In particular, Exercise 1.17 has the following consequences. Firstly, given any $g \in C^\infty(M)$, the map $M \times N \ni (p,q) \mapsto g(p) \in \mathbb{R}$, is an element of $C^\infty(M \times N)$, as it is the composition of the smooth maps g and π_M. Secondly, given an $f \in C^\infty(M \times N)$ and a $q \in N$, the 'slice map' f_q, given by $M \ni p \mapsto f(p,q) \in \mathbb{R}$ is smooth too, since $f_q = f \circ i_q$. We will use these observations later on to show that the 'tangent space of $M \times N$ at (p,q)' can be identified with $T_pM \times T_qN$ in Exercise 2.8.

Exercise 1.18. (Smoothness is a local property.) Let M, N be smooth manifolds. Show that $f : M \to N$ is smooth if and only if for every U open in M, $f|_U : U \to N$ is smooth. Here U has the induced smooth structure from that of M.

The operations $+, \cdot : C^\infty(M) \times C^\infty(M) \to C^\infty(M)$ are defined pointwise:

$$\left.\begin{aligned}(f+g)(p) &= f(p)+g(p)\\ (f\cdot g)(p) &= f(p)\cdot g(p)\end{aligned}\right\} \text{ for all } p \in M.$$

It can be checked that $f+g$, $f \cdot g \in C^\infty(M)$, and that with these operations, $(C^\infty(M), +, \cdot)$ is a ring, with the additive identity being the zero function $\mathbf{0} \in C^\infty(M)$ (given by $M \ni p \mapsto \mathbf{0}(p) := 0 \in \mathbb{R}$), and the multiplicative identity $\mathbf{1} \in C^\infty(M)$ (given by $M \ni p \mapsto \mathbf{1}(p) := 1 \in \mathbb{R}$). However, $C^\infty(M)$ is not a field, since not every[3] $f \in C^\infty(M)\backslash\{\mathbf{0}\}$ will have a multiplicative inverse. We will see later that the set of 'smooth vector fields' on a manifold has the natural structure of a module over the ring $C^\infty(M)$.

We will meet geodesics later on, which will be the 'straightest' possible curves in the Lorentzian manifold, describing paths of 'freely falling' particles. The straight lines in Euclidean space and great circles on the sphere S^2 are geodesics. In any case, they are 'smooth' curves.

Definition 1.9. (Smooth curve.)
A smooth map $\gamma : I \to M$, where I is an open interval in $\mathbb{R}$, is called a *smooth curve*. If $I \subset \mathbb{R}$ is any interval, not necessarily open, then a curve $\gamma : I \to M$ is a *smooth curve* if there exists an open interval $\widetilde{I} \supset I$, and a smooth curve $\widetilde{\gamma} : \widetilde{I} \to M$ such that $\widetilde{\gamma}|_I = \gamma$.

Just like in linear algebra, where one aim is to classify vector spaces up to isomorphisms, in differential geometry, the notion analogous to an isomorphism is that of a diffeomorphism.

Definition 1.10. (Diffeomorphism.)
Let M, N be smooth manifolds. A bijection $f : M \to N$ such that f and $f^{-1} : N \to M$ are both smooth, is called a *diffeomorphism*, and M and N are then said to be *diffeomorphic*.

[3]Consider an $f \in C^\infty(M)\backslash\{\mathbf{0}\}$ that has a zero at some point. In fact, in Chapter 2, we will construct nonzero functions that vanish outside a neighbourhood of a point.

Example 1.14. (Chart maps are diffeomorphisms.) Let (U, φ) be a chart from an atlas defining the smooth manifold M, and consider U as a smooth manifold with the atlas $\{(U, \varphi)\}$. Recall from Example 1.13 that the chart map $\varphi : U \to \varphi(U)$ is smooth. Also, it is a bijection onto the open set $\varphi(U)$. We show that its inverse $\varphi^{-1} : \varphi(U) \to U$ is smooth too. For all $\varphi(p) \in \varphi(U)$, with $p \in U$, we take the admissible chart $(\varphi(U), \mathrm{id}_{\varphi(U)})$ for $\varphi(U)$ containing $\varphi(p)$, and take the admissible chart (U, φ) for U. Then $\varphi^{-1}(\varphi(U)) = U$. Moreover, $\varphi \circ \varphi^{-1} \circ (\mathrm{id}_{\varphi(U)})^{-1} = \mathrm{id}|_{\varphi(U)} : \varphi(U) \to \varphi(U)$, which is clearly C^∞. As this happens with each point in $\varphi(U)$, we conclude that $\varphi^{-1} : \varphi(U) \to U$ is smooth. ◇

Exercise 1.19. Let M be an affine space over V, considered as a smooth manifold in the usual way. For a $\mathbf{v} \in V$, define $\gamma_{\mathbf{v}} : \mathbb{R} \to M$ by $\gamma_{\mathbf{v}}(t) = p + t\mathbf{v}$, $t \in \mathbb{R}$. Show that $\gamma_{\mathbf{v}}$ is a smooth curve.

Exercise 1.20. Let U, V be open subsets of $\mathbb{R}^m, \mathbb{R}^n$, respectively. We consider U, V as smooth manifolds with the smooth structures $[\{(U, \mathrm{id}_U)\}], [\{(V, \mathrm{id}_V)\}]$, respectively. Show that $f : U \to V$ is smooth if and only if f is C^∞.

Exercise 1.21. Let $\mathbb{R}$ be equipped with the two incompatible atlases $\mathcal{A}_1$ and $\mathcal{A}_2$ given in Example 1.6. Prove that $(\mathbb{R}, [\mathcal{A}_1])$ is diffeomorphic to $(\mathbb{R}, [\mathcal{A}_2])$. (From our earlier considerations, the incompatibility of $\mathcal{A}_1$ with $\mathcal{A}_2$ can be expressed by saying that the *identity map* fails to be a diffeomorphism between the smooth manifolds $(\mathbb{R}, [\mathcal{A}_1])$ and $(\mathbb{R}, [\mathcal{A}_2])$. However, this exercise shows that there may nevertheless be other maps which serve as a diffeomorphism.)

Exercise 1.22. Let M, N be smooth manifolds, and $f : M \to N$ be a diffeomorphism. If (U, φ) is an admissible chart for M, then it is easy to see that $(f(U), \varphi \circ f^{-1})$ is a chart for N. Show that $(f(U), \varphi \circ f^{-1})$ is an admissible chart for N.

Exercise 1.23. Let M be a smooth manifold. Show that the set

$$\mathrm{Diff}(M) := \{f : M \to M \mid f \text{ is a diffeomorphism}\},$$

together with the operation $\circ$ of composition of maps, forms a group.

Exercise 1.24. (Lie group and left-translation diffeomorphisms.) A *Lie group* is a group $(G, \cdot)$ equipped with a smooth structure, such that the multiplication map $G \times G \ni (p, q) \mapsto p \cdot q \in G$, and the inverse map $G \ni q \mapsto q^{-1} \in G$, are smooth. Given $p \in G$, the *left-translation by p* is the map $L_p : G \to G$ defined by $G \ni q \mapsto p \cdot q$. Show that L_p is a diffeomorphism for each $p \in G$.

Exercise 1.25. (Submanifolds.) Suppose that M is an m-dimensional smooth manifold. A subset $N \subset M$ is said to be a *submanifold* of dimension $n \leqslant m$ if for each $p \in N$, there exists an admissible chart (U, φ) of M such that $p \in U$, and $\varphi(U \cap N) = \widetilde{U} \times \{\mathbf{0}\} \subset \mathbb{R}^n \times \mathbb{R}^{m-n} = \mathbb{R}^m$, where $\widetilde{U}$ is an open subset of $\mathbb{R}^n$. Then (U, ϕ) is called an *allowed chart for N*. Let $\pi : \mathbb{R}^m \to \mathbb{R}^n$ be the projection map onto the first n components.

Prove that

$$\mathcal{A}_N := \{(U \cap N, \pi \circ \varphi|_{U \cap N}) : (U, \varphi) \text{ is an allowed chart for } N\}$$

is an atlas for N. So N is a smooth manifold with the smooth structure $[\mathcal{A}_N]$. Show that the inclusion map $i : N \hookrightarrow M$ is smooth.

Exercise 1.26. Let M_1, M_2 be smooth manifolds and N_1, N_2 be submanifolds of M_1, M_2, respectively. If $f : M_1 \to M_2$ is a smooth map such that $f(N_1) \subset N_2$, then show that $f|_{N_1} : N_1 \to N_2$ is also smooth.

Before beginning with the second chapter, we make a remark on some notation which will be used from now on. For a smooth manifold M, we will often take for granted that its dimension is denoted by m. Charts will often be denoted by (U, φ), but also by $(U, \mathbf{x})$, where the understanding is that the component functions of the map $\mathbf{x} : U \to \mathbb{R}^m$ are denoted by $x^i : U \to \mathbb{R}$, $1 \leqslant i \leqslant m$. Moreover, given a function $f : M \to \mathbb{R}$, a point $p \in M$, and an admissible chart (U, φ), we will denote the partial derivative of $f \circ \varphi^{-1} : \varphi(U) \to \mathbb{R}$ with respect to the ith variable at the point $\varphi(p)$ by

$$\frac{\partial (f \circ \varphi^{-1})}{\partial u^i}(\varphi(p)).$$

Chapter 2

Tangent and cotangent spaces

Intuitively, a tangent space at a point $\mathbf{p}$ on a surface M in $\mathbb{R}^3$ is the plane at $\mathbf{p}$ tangential to the surface, consisting of all tangent vectors. Tangent vectors are the 'velocities' of curves passing through $\mathbf{p}$. Thus we imagine a picture like this:

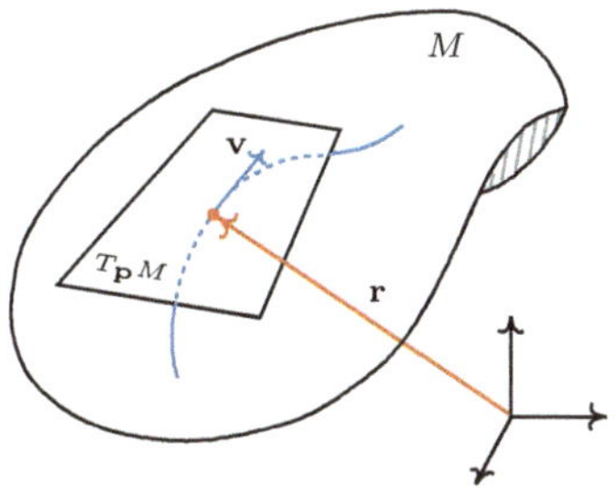

Tangent plane $T_{\mathbf{p}}M$ and tangent vector $\mathbf{v}$ at a point $\mathbf{p}$ on the surface M.

However, when we have only an abstract manifold M at hand, this is no longer an adequate definition, since there is no 'natural'[1] embedding in $\mathbb{R}^d$. We ought to keep the definitions 'intrinsic', not relying on any embedding, and using just the manifold structure. The way around the obstacle is that we begin by revisiting the familiar notion above, and express it in a way which lends itself to an appropriate generalisation.

It will turn out that the 'right' way to think about tangent vectors at a point $p \in M$ is as linear maps $v : C^\infty(M) \to \mathbb{R}$ that obey a Leibniz rule, given by $v(f \cdot g) = f(p)v(g) + g(p)v(f)$ for all $f, g \in C^\infty(M)$.

This may appear strange, but it is not too far-fetched. Indeed, imagine what a child running in a playground experiences as its own velocity vector in spacetime: The child perceives the changes in (the smooth functions on spacetime such as) temperature, pressure and so on. The faster these

[1]There is a result of Whitney, saying that an m-dimensional smooth manifold can be embedded in $\mathbb{R}^{2m}$, but being a purely mathematical construct, it is devoid of any immediate physical relevance.

change, the larger the perceived velocity. Thus if $\gamma : (-\epsilon, \epsilon) \to M$ describes the child's worldline, with $\gamma(0) = p$, then it makes sense to think of its velocity at a point $p \in M$ as a map

$$f \mapsto \frac{d(f \circ \gamma)}{dt}(0), \quad f \in C^\infty(M),$$

since the above expression registers how fast f (temperature, pressure, etc.) is changing at p, as the child goes about playing!

2.1 Tangent vectors to a surface as derivations

A (parametrised) *surface*[2] M in $\mathbb{R}^3$ is a map $(u, v) \mapsto \mathbf{r}(u, v) : D \to \mathbb{R}^3$ such that the '*normal vector*'

$$\mathbf{n} := \frac{\partial \mathbf{r}}{\partial u} \times \frac{\partial \mathbf{r}}{\partial v} \neq \mathbf{0} \text{ everywhere in } D,$$

where the parameter domain D is an open set in $\mathbb{R}^2$. Here $\times$ is the usual cross product of vectors in $\mathbb{R}^3$. Let $\mathbf{r}(u_0, v_0) = \mathbf{p} \in M$. Our multivariable-calculus understanding of a tangent vector $\mathbf{v}$ at $\mathbf{p}$ is that there is curve $(-\epsilon, \epsilon) \ni t \overset{\gamma}{\longmapsto} \mathbf{r}(u(t), v(t))$ passing through $\mathbf{p} = \gamma(0)$, such that we have $\dot{\gamma}(0) := \frac{d\gamma}{dt}(0) = \mathbf{v}$. We note that

$$\dot{\gamma}(0) = \dot{u}(0)\frac{\partial \mathbf{r}}{\partial u}(u_0, v_0) + \dot{v}(0)\frac{\partial \mathbf{r}}{\partial v}(u_0, v_0). \qquad (\star)$$

As several different curves may have the same tangent vector, it is not appropriate to associate a tangent vector with a special curve. The tangent space $T_{\mathbf{p}}M$ is the set of all tangent vectors, built by considering all curves passing through $\mathbf{p}$, and taking the corresponding tangent vectors to these curves at $\mathbf{p}$. But we see from the above expression, that by changing the maps $t \mapsto u(t), v(t)$, we can get all possible coefficients $\dot{u}(0)$ and $\dot{v}(0)$ in the linear combination on the right-hand side of $(\star)$ above, and so

$$T_{\mathbf{p}}M = \text{span}\Big\{\frac{\partial \mathbf{r}}{\partial u}(u_0, v_0), \frac{\partial \mathbf{r}}{\partial v}(u_0, v_0)\Big\}.$$

We visualise $T_{\mathbf{p}}M$ as a plane placed at $\mathbf{p}$, tangential to the surface, and having the normal

$$\mathbf{n}_{\mathbf{p}} := \frac{\partial \mathbf{r}}{\partial u}(u_0, v_0) \times \frac{\partial \mathbf{r}}{\partial v}(u_0, v_0).$$

As mentioned in the introductory remarks to this chapter, when working in an abstract manifold, using such a definition of a tangent vector and tangent space is not feasible. We will now see[3] that the correct notion is that of a 'derivation'.

[2]See e.g. [Apostol (1969), Chapter 12].

[3]As this section is just meant to provide some motivation for our definition to follow in the next section, we will allow some sloppiness in our arguments.

A *derivation* v *at* $\mathbf{p} \in M$ is a linear map $v : C^\infty(M) \to \mathbb{R}$ which satisfies the Leibniz rule $v(f \cdot g) = f(\mathbf{p})v(g) + g(\mathbf{p})v(f)$ for all $f, g \in C^\infty(M)$. In our setting of surfaces, let us first show that every tangent vector gives rise to a derivation. So suppose that we have a tangent vector arising from a curve $\boldsymbol{\gamma} : (-\epsilon, \epsilon) \to M$ such that $\boldsymbol{\gamma}(0) = \mathbf{p}$. We define

$$v_{\boldsymbol{\gamma},\mathbf{p}} f := \frac{d(f \circ \gamma)}{dt}(0), \quad f \in C^\infty(M).$$

Then for $c \in \mathbb{R}$ and $f, g \in C^\infty(M)$, we have

$$\begin{aligned} v_{\boldsymbol{\gamma},\mathbf{p}}(f+g) &= \frac{d((f+g)\circ\gamma)}{dt}(0) = \frac{d(f\circ\gamma)}{dt}(0) + \frac{d(g\circ\gamma)}{dt}(0) = v_{\boldsymbol{\gamma},\mathbf{p}}(f) + v_{\boldsymbol{\gamma},\mathbf{p}}(g), \\ v_{\boldsymbol{\gamma},\mathbf{p}}(c \cdot f) &= \frac{d((c\cdot f)\circ\gamma)}{dt}(0) = c\frac{d(f\circ\gamma)}{dt}(0) = c\, v_{\boldsymbol{\gamma},\mathbf{p}}(f), \end{aligned}$$

showing that $v_{\boldsymbol{\gamma},\mathbf{p}}$ is linear. It also satisfies the Leibniz rule since

$$\begin{aligned} v_{\boldsymbol{\gamma},\mathbf{p}}(f \cdot g) &= \frac{d((f\cdot g)\circ\gamma)}{dt}(0) = \frac{d((f\circ\gamma)\cdot(g\circ\gamma))}{dt}(0) \\ &= (f\circ\gamma)(0)\,\frac{d(g\circ\gamma)}{dt}(0) + (g\circ\gamma)(0)\,\frac{d(f\circ\gamma)}{dt}(0) \\ &= f(\mathbf{p})\, v_{\boldsymbol{\gamma},\mathbf{p}}(g) + g(\mathbf{p})\, v_{\boldsymbol{\gamma},\mathbf{p}}(f). \end{aligned}$$

Now let us show that every derivation arises from a tangent vector to some curve. To do this, we will first prove the following 'division' lemma.

Lemma 2.1. *Let B be an open ball in $\mathbb{R}^m$ centered at $\mathbf{0}$. Suppose that $f \in C^\infty(B)$ and $f(\mathbf{0}) = 0$. Then there exist $g_1, \cdots, g_m \in C^\infty(B)$ such that $f = x^1 g_1 + \cdots + x^m g_m$ in B.*

Proof. For $\mathbf{x} \in B$, $1 \leqslant k \leqslant m$, with $g_k(\mathbf{x}) := \int_0^1 \frac{\partial f}{\partial x^k}(t\mathbf{x})\,dt$, we have

$$x^k g_k = \int_0^1 \frac{\partial f}{\partial x^k}(t\mathbf{x})\,x^k\,dt - \int_0^1 \frac{d}{dt}(f(t\mathbf{x}))\,dt = f(\mathbf{x}) - f(\mathbf{0}) = f(\mathbf{x}). \qquad \square$$

Either from the formula for g_k in the proof, or upon differentiating both sides of $f = x^1 g_1 + \cdots + x^m g_m$ with respect to x_k, we see that

$$g_k(\mathbf{0}) = \frac{\partial f}{\partial x^k}(\mathbf{0}), \quad 1 \leqslant k \leqslant m.$$

We will also need the following two lemmas saying that derivations annihilate constant functions and that they are 'local'.

Lemma 2.2. *Let M be a surface, v be a derivation at $\mathbf{p} \in M$, and let $f \in C^\infty(M)$ be constant. Then $vf = 0$.*

Proof. Define $\mathbf{1} \in C^\infty(M)$ by $\mathbf{1}(p) = 1$ for all $p \in M$. If f is the constant function c, then $f = c\mathbf{1}$. We have $v(\mathbf{1}) = v(\mathbf{1}\cdot\mathbf{1}) = 1\,v(\mathbf{1}) + 1\,v(\mathbf{1}) = 2v(\mathbf{1})$, giving $v(\mathbf{1}) = 0$. Consequently, $v(f) = v(c\mathbf{1}) = cv(\mathbf{1}) = c0 = 0$. $\square$

Lemma 2.3. *Let M be a surface, and v be a derivation at $\mathbf{p} \in M$. If $f, g \in C^\infty(M)$ agree in a neighbourhood of $\mathbf{p}$, then $vf = vg$.*

Proof. Suppose that f, g coincide in U. Let χ be a 'bump function', namely a function which is identically equal to 1 in a neighbourhood of $\mathbf{p}$, and is identically zero outside a neighbourhood V, such that the closure $\overline{V}$ of V is contained in U. The existence of such a function is plausible, but we postpone the proof until Section 2.3. Thus $(f-g)\cdot\chi \equiv 0$ on M, giving $0 = v((f-g)\cdot\chi) = (vf - vg)\chi(\mathbf{p}) + (f(\mathbf{p}) - g(\mathbf{p}))v\chi = (vf - vg)1 + 0$. □

Let[4] v be a derivation at $\mathbf{p} \in M$. We justify below that there exists an $\epsilon > 0$ and a smooth curve $\boldsymbol{\gamma} : (-\epsilon, \epsilon) \to M$ such that $\boldsymbol{\gamma}(0) = \mathbf{p}$ and $v_{\boldsymbol{\gamma},\mathbf{p}} = \mathrm{v}$.

The map $M \ni (x, y, z) = \mathbf{q} \overset{x}{\longmapsto} x \in \mathbb{R}$ is smooth, and so by acting with v, we get a number $\mathrm{v}x$. Similarly, we define $\mathrm{v}y$ and $\mathrm{v}z$. This gives a vector $(\mathrm{v}x, \mathrm{v}y, \mathrm{v}z) \in \mathbb{R}^3$. We first show that this vector belongs to the span of the vectors $\frac{\partial \mathbf{r}}{\partial u}(u_0, v_0)$ and $\frac{\partial \mathbf{r}}{\partial v}(u_0, v_0)$.

Suppose that $\mathbf{p} = \mathbf{r}(u_0, v_0)$. For a point $(M \ni)\, \mathbf{q} = \mathbf{r}(u, v)$, we write $(u, v) = (\varphi^1(\mathbf{q}), \varphi^2(\mathbf{q}))$. Moreover, let $\mathbf{r}(u, v) =: (X(u, v), Y(u, v), Z(u, v))$. By Lemma 2.1, $X(u, v) - X(u_0, v_0) = (u - u_0)\xi(u, v) + (v - v_0)\eta(u, v)$, in a small ball around (u_0, v_0) for some smooth functions ξ and η that satisfy

$$\xi(\varphi^1(\mathbf{p}), \varphi^2(\mathbf{p})) = \xi(u_0, v_0) = \frac{\partial X}{\partial u}(u_0, v_0) \quad \text{and}$$

$$\eta(\varphi^1(\mathbf{p}), \varphi^2(\mathbf{p})) = \eta(u_0, v_0) = \frac{\partial X}{\partial v}(u_0, v_0).$$

We have

$$\begin{aligned} x(\mathbf{q}) - x(\mathbf{p}) &= x(\mathbf{r}(u, v)) - x(\mathbf{r}(u_0, v_0)) = X(u, v) - X(u_0, v_0) \\ &= (u - u_0)\xi(u, v) + (v - v_0)\eta(u, v) \\ &= (\varphi^1(\mathbf{q}) - \varphi^1(\mathbf{p}))\xi(\varphi^1(\mathbf{q}), \varphi^2(\mathbf{q})) \\ &\quad + (\varphi^2(\mathbf{q}) - \varphi^2(\mathbf{p}))\eta(\varphi^1(\mathbf{q}), \varphi^2(\mathbf{q})). \end{aligned}$$

Operating by v on both sides, we obtain, using the Leibniz rule, that

$$\begin{aligned} \mathrm{v}x - 0 &= \mathrm{v}(\mathbf{q} \mapsto \varphi^1(\mathbf{q}))\xi(\varphi^1(\mathbf{p}), \varphi^2(\mathbf{p})) \\ &\quad + (\varphi^1(\mathbf{p}) - \varphi^1(\mathbf{p}))\mathrm{v}(\mathbf{q} \mapsto \xi(\varphi^1(\mathbf{q}), \varphi^2(\mathbf{q}))) \\ &\quad + \mathrm{v}(\mathbf{q} \mapsto \varphi^2(\mathbf{q}))\eta(\varphi^1(\mathbf{p}), \varphi^2(\mathbf{p})) \\ &\quad + (\varphi^2(\mathbf{p}) - \varphi^2(\mathbf{p}))\mathrm{v}(\mathbf{q} \mapsto \eta(\varphi^1(\mathbf{q}), \varphi^2(\mathbf{q}))) \\ &= \mathrm{v}(\mathbf{q} \mapsto \varphi^1(\mathbf{q}))\frac{\partial X}{\partial u}(u_0, v_0) + 0 + \mathrm{v}(\mathbf{q} \mapsto \varphi^2(\mathbf{q}))\frac{\partial X}{\partial v}(u_0, v_0) + 0. \end{aligned}$$

[4]We use the notation v for the derivation here, instead of v, in order to avoid confusion with the parameter v from $(u, v) \in D$ in the parametrisation of the surface M.

With similar computations carried out for vy and vz, we get

$$\begin{bmatrix} \mathrm{v}x \\ \mathrm{v}y \\ \mathrm{v}z \end{bmatrix} = \begin{bmatrix} \frac{\partial X}{\partial u}(u_0,v_0) & \frac{\partial X}{\partial v}(u_0,v_0) \\ \frac{\partial Y}{\partial u}(u_0,v_0) & \frac{\partial Y}{\partial v}(u_0,v_0) \\ \frac{\partial Z}{\partial u}(u_0,v_0) & \frac{\partial Z}{\partial v}(u_0,v_0) \end{bmatrix} \begin{bmatrix} U \\ V \end{bmatrix} = \begin{bmatrix} \frac{\partial \mathbf{r}}{\partial u}(u_0,v_0) & \frac{\partial \mathbf{r}}{\partial v}(u_0,v_0) \end{bmatrix} \begin{bmatrix} U \\ V \end{bmatrix},$$

where $U := \mathrm{v}(\mathbf{q} \mapsto \varphi^1(\mathbf{q}))$ and $V := \mathrm{v}(\mathbf{q} \mapsto \varphi^2(\mathbf{q}))$. Since the normal vector $\mathbf{n_p} \neq \mathbf{0}$, the above U, V are unique. Recall that $(u_0, v_0) \in D$ is such that $\mathbf{r}(u_0, v_0) = \mathbf{p}$. Define $u(t) = u_0 + Ut$, $v(t) = v_0 + Vt$ for $t \in (-\epsilon, \epsilon)$, where the $\epsilon > 0$ is taken small enough so that $(u(t), v(t))$ lies in a ball centered at (u_0, v_0) contained in D. Define $\gamma(t) = \mathbf{r}(u(t), v(t))$, $t \in (-\epsilon, \epsilon)$. Then

$$\dot{\gamma}(0) = \frac{d\mathbf{r}(u_0+Ut, v_0+Vt)}{dt}(0) = \frac{\partial \mathbf{r}}{\partial u}(u_0,v_0)U + \frac{\partial \mathbf{r}}{\partial v}(u_0,v_0)V = \begin{bmatrix} \mathrm{v}x \\ \mathrm{v}y \\ \mathrm{v}z \end{bmatrix}.$$

Let $D_0 \subset D$ be an open ball containing (u_0, v_0) and $\delta > 0$ be such that the 'cylinder' $\Omega = \{\mathbf{q} + t\,\mathbf{n_p} : \mathbf{q} \in \mathbf{r}(D_0),\ |t| < \delta\}$ is an open set in $\mathbb{R}^3$, that is, the surface patch obtained from D_0 can be fattened by displacing the surface along $\mathbf{n_p}$. If $f \in C^\infty(M)$, then we may extend f as a constant in the direction of normal $\mathbf{n_p}$ by setting $f(\mathbf{q} + t\,\mathbf{n_p}) := f(\mathbf{q})$, $\mathbf{q} \in \mathbf{r}(D_0)$. Then $f \in C^\infty(\Omega)$. The translated set $\{\mathbf{x} : \mathbf{x} + \mathbf{p} \in \Omega\}$ is open and contains the origin $\mathbf{0}$, and we can find a small ball B contained within it. Then for $\mathbf{x} \in B$, define φ by $\varphi(\mathbf{x}) = f(\mathbf{x}+\mathbf{p}) - f(\mathbf{p})$. Then $\varphi(\mathbf{0}) = 0$. By Lemma 2.1, there are g_k, $k = 1, 2, 3$, such that $\varphi = x^k g_k$ on B and

$$\frac{\partial f}{\partial x^\ell}(\mathbf{p}) = \frac{\partial \varphi}{\partial x^\ell}(\mathbf{0}) = g_\ell(\mathbf{0}).$$

For $\mathbf{q} \in \mathbf{r}(D_0)$,

$$f(\mathbf{q}) = f(\mathbf{q} - \mathbf{p} + \mathbf{p}) = \varphi(\mathbf{q} - \mathbf{p}) + f(\mathbf{p}) = (q^k - p^k)\, g_k(\mathbf{q} - \mathbf{p}) + f(\mathbf{p}),$$

and so

$$\begin{aligned} \mathrm{v}f &= \mathrm{v}(f(\mathbf{p}) + (q^k - p^k)\, g_k(\mathbf{q} - \mathbf{p})) \\ &= \mathrm{v}(f(\mathbf{p})) + (p^k - p^k)\,\mathrm{v}(g_k(\cdot - \mathbf{p})) + g_k(\mathbf{p} - \mathbf{p})\,(\mathrm{v}(q^k) - 0) \\ &= 0 + 0 + \mathrm{v}x\,\frac{\partial f}{\partial x}(\mathbf{p}) + \mathrm{v}y\,\frac{\partial f}{\partial y}(\mathbf{p}) + \mathrm{v}z\,\frac{\partial f}{\partial z}(\mathbf{p}). \end{aligned} \qquad (\star\star)$$

On the other hand,

$$\begin{aligned} v_{\gamma,\mathbf{p}} f &= \frac{d(f \circ \gamma)}{dt}(0) = \begin{bmatrix} \frac{\partial f}{\partial x}(\mathbf{p}) & \frac{\partial f}{\partial y}(\mathbf{p}) & \frac{\partial f}{\partial z}(\mathbf{p}) \end{bmatrix} \dot{\gamma}(0) \\ &= \mathrm{v}x\,\frac{\partial f}{\partial x}(\mathbf{p}) + \mathrm{v}y\,\frac{\partial f}{\partial y}(\mathbf{p}) + \mathrm{v}z\,\frac{\partial f}{\partial z}(\mathbf{p}). \end{aligned}$$

Consequently, $\mathrm{v}f = v_{\gamma,\mathbf{p}} f$ for all $f \in C^\infty(M)$, that is, $\mathrm{v} = v_{\gamma,\mathbf{p}}$.

Thus in the context of surfaces, we have realised that the classical view of tangent vectors at a point $\mathbf{p} \in M$ as being the velocity vector of some curve γ at $\mathbf{p}$ can be replaced by the view that tangent vectors are simply 'derivations', namely linear maps $\mathrm{v} : C^\infty(M) \to \mathbb{R}$ that obey the Leibniz rule. This completes our discussion of the motivation for defining tangent vectors as derivations in the abstract setting of smooth manifolds.

2.2 Tangent vector definition

Definition 2.1. (Tangent vector, tangent space.)
Let M be a smooth manifold and $p \in M$. A *tangent vector* v *at* p is a map $v : C^\infty(M) \to \mathbb{R}$ such that for all $f, g \in C^\infty(M)$ and $c \in \mathbb{R}$,

- (linear) $v(f + cg) = v(f) + cv(g)$, and
- (Leibniz rule) $v(f \cdot g) = f(p)v(g) + g(p)v(f)$.

The set of all tangent vectors at p is the *tangent space* T_pM at p.

Exercise 2.1. Show that T_pM is a vector space with addition and scalar multiplication defined pointwise. (In Theorem 2.1, we will show that the dimension of the vector space T_pM is equal to the dimension of the manifold M.)

Exercise 2.2. Let $v \in T_pM$. Prove that if $f \in C^\infty(M)$ is constant, then $vf = 0$.

Example 2.1. (Tangent vectors to curves as vectors.)
Let $(0 \in) I \subset \mathbb{R}$ be an open interval and $\gamma : I \to M$ be a smooth curve such that $\gamma(0) = p \in M$. Define $v_{\gamma,p} : C^\infty(M) \to \mathbb{R}$ by

$$v_{\gamma,p} f = \frac{d(f \circ \gamma)}{dt}(0) \text{ for all } f \in C^\infty(M). \tag{2.1}$$

We claim that $v_{\gamma,p} \in T_pM$. Indeed, for $f, g \in C^\infty(M)$ and $c \in \mathbb{R}$, we have

$$\begin{aligned}
v_{\gamma,p}(f+cg) &= \frac{d((f+cg)\circ\gamma)}{dt}(0) = \frac{d(f\circ\gamma + c(g\circ\gamma))}{dt}(0) \\
&= \frac{d(f\circ\gamma)}{dt}(0) + c\frac{d(g\circ\gamma)}{dt}(0) = v_{\gamma,p}(f) + cv_{\gamma,p}(g), \quad \text{and} \\
v_{\gamma,p}(f\cdot g) &= \frac{d((f\cdot g)\circ\gamma)}{dt}(0) = \frac{d((f\circ\gamma)\cdot(g\circ\gamma))}{dt}(0) \\
&= (f\circ\gamma)(0)\frac{d(g\circ\gamma)}{dt}(0) + (g\circ\gamma)(0)\frac{d(f\circ\gamma)}{dt}(0) \\
&= f(p)v_{\gamma,p}(g) + g(p)v_{\gamma,p}(f).
\end{aligned}$$

So $v_{\gamma,p} \in T_pM$. ◇

Exercise 2.3. (Different curves with a common tangent vector.)
Consider $\mathbb{R}$ as a smooth manifold with the standard smooth structure. For $t \in \mathbb{R}$, define $\gamma_1(t) = t - 1$, and $\gamma_2(t) = t + 3t^2 - 1$. Show that $v_{\gamma_1,-1} = v_{\gamma_2,-1}$.

Exercise 2.4. Let M be a smooth manifold. Let I, J be open subsets of $\mathbb{R}$, and $J \ni u \mapsto h(u) \in I$ be a C^∞ function. Let $\gamma : I \to M$ be a smooth curve. Show that the tangent vectors of the smooth curve $\gamma \circ h : J \to M$ satisfy

$$v_{\gamma \circ h, (\gamma \circ h)(s)} = \dot{h}(s)\, v_{\gamma, \gamma(h(s))} \in T_{\gamma(h(s))}M, \text{ for all } s \in J,$$

where $\dot{h}(s) := \frac{dh}{du}(s)$.

2.3 Bump functions

A tool which will prove to be handy in the sequel is 'bump-function technology'. In a nutshell, it is a result on the existence of smooth functions which are identically 1 in the neighbourhood of a point and vanish outside a bigger neighbourhood. This will be very useful when we want to extend smooth objects beyond a chart in a smooth manner, or while patching stuff together on a manifold having obtained contributions on charts.

The following exercise will be used in the construction of a bump function χ around a point p in a smooth manifold M (i.e., a $C^\infty(M)$ function χ which is identically 1 in a neighbourhood U of p and is zero outside a somewhat bigger neighbourhood $V \supset U$).

Exercise 2.5. Let $f(t) = \left\{ \begin{array}{ll} e^{-1/t} & \text{for } t > 0 \\ 0 & \text{for } t \leqslant 0 \end{array} \right\}$. We will show that $f \in C^\infty(\mathbb{R})$.

(1) Let $g : \mathbb{R} \to \mathbb{R}$ be continuous on $\mathbb{R}$, continuously differentiable on $\mathbb{R}_* := \mathbb{R} \backslash \{0\}$, and such that $\lim_{t \to 0} g'(t)$ exists. Show g is continuously differentiable on $\mathbb{R}$.

(2) Suppose $n \in \mathbb{N}$. Let $g : \mathbb{R} \to \mathbb{R}$ be continuously differentiable $n-1$ times on $\mathbb{R}$, and n times on $\mathbb{R}_*$, and such $\lim_{t \to 0} g^{(n)}(t)$ exists. Prove that g is n times continuously differentiable on $\mathbb{R}$.

(3) Show that f is infinitely many times differentiable.

Hint: Using induction on n, show that for $t > 0$, $f^{(n)} = R_n f$, where R_n is a rational function. Conclude that $\lim_{t \searrow 0} t^{-n} f(t) = 0$.

Lemma 2.4. (Existence of a bump function.)
Let $U \subset \mathbb{R}^m$ be an open set, and let $\mathbf{0} \in U$. Then given any $R > 0$ such that $\overline{B(\mathbf{0}, R)} \subset U$, there exists a pointwise-nonnegative $\chi \in C^\infty(U)$ such that χ vanishes outside $B(\mathbf{0}, R)$ and is identically 1 on a ball $B(\mathbf{0}, r)$ centered at $\mathbf{0}$ with radius $r < R$.

Proof. By composing the function $f \in C^\infty(\mathbb{R})$ constructed in Exercise 2.5 with $t \mapsto 1 - t^2$, we get a function $\varphi \in C^\infty(\mathbb{R})$ that vanishes outside $[-1, 1]$. See the left picture in the first figure below. With $\psi(x) := \int_{-\infty}^{x} \varphi(t)dt$, we have $\psi \in C^\infty(\mathbb{R})$, ψ is 0 on $(-\infty, -1)$, and is a constant on $(1, \infty)$.

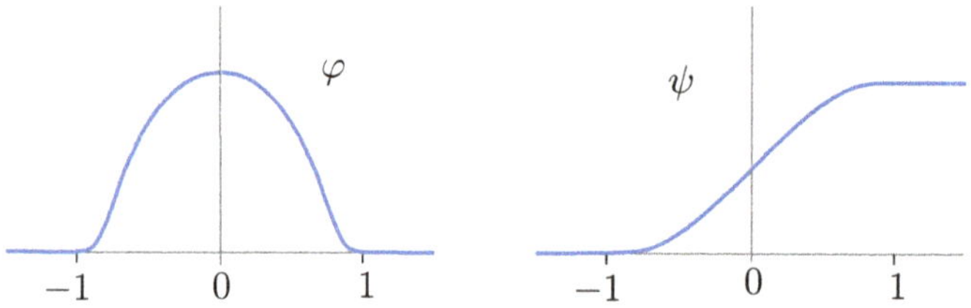

By scaling and shifting, we get a function depicted in the top left-hand side of the following figure, and by reflecting, the one in the bottom left-hand side. Their pointwise product yields the bump function σ in one variable, shown on the right in the following picture.

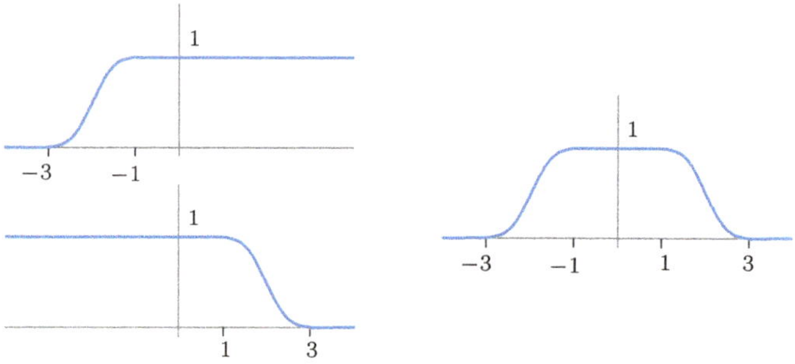

Defining $F(\mathbf{x}) = \sigma(x^1) \cdots \sigma(x^m)$, we see that $F \equiv 1$ on $B(\mathbf{0}, 1)$, and $F \equiv 0$ outside $B(\mathbf{0}, 3\sqrt{m})$ (because if $(x^1)^2 + \cdots + (x^m)^2 \geqslant 3^2 m$, then it cannot be the case that each x^i satisfies $|x^i| < 3$, and so for some i, $|x^i| \geqslant 3$, giving $\sigma(x^i) = 0$). Using suitable 'dilation', namely $\chi = F(\lambda \cdot)$, with $\lambda := 3\sqrt{m}/R$, we have χ is identically 1 on a ball $B(\mathbf{0}, r)$ with radius $r := R/(3\sqrt{m}) < R$, and is zero in U outside $B(\mathbf{0}, R)$. $\square$

Corollary 2.1. (Existence of a bump function.) *Let $U \subset M$ be an open set, and let $p \in U$. Then there exists a pointwise-nonnegative $f \in C^\infty(M)$ such that f is identically equal to 1 in a neighbourhood of p, and is zero outside a neighbourhood V such that $\overline{V} \subset U$.*

Proof. Let (W, φ) be an admissible chart map such that $p \in W \subset U$, and also assume, without loss of generality, that $\varphi(p) = \mathbf{0}$. Let $R > 0$ be such that $\overline{B(\mathbf{0}, R)} \subset \varphi(W)$. By the previous Lemma 2.4, there exists a $\chi \in C^\infty(\mathbb{R}^m)$ such that $\chi \equiv 1$ in a ball $B(\mathbf{0}, r) \subset \varphi(W)$, where $r < R$, and $\chi \equiv 0$ outside the bigger ball $B(\mathbf{0}, R)$. Define $V = \varphi^{-1} B(\mathbf{0}, R)$. As the chart map φ is a homeomorphism, $\overline{V} = \varphi^{-1} \overline{B(\mathbf{0}, R)} \subset W \subset U$. Define $f = \chi \circ \varphi$ on W, and identically zero outside it. $\square$

Exercise 2.6. Consider the smooth manifold $\mathbb{R}^2$ with the standard smooth structure. Construct a smooth curve $\gamma : \mathbb{R} \to \mathbb{R}^2$ such that $\gamma(0) = (0,0)$, the range of γ is as shown below, and is traversed in the directions indicated by the arrows.

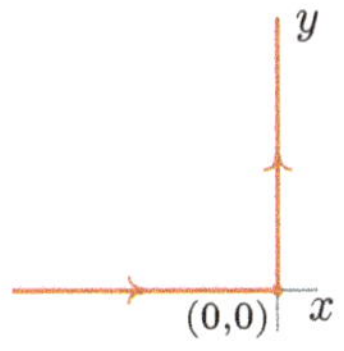

2.4 Chart-induced basis for the tangent space

We will now show the following crucial result, which will allow tangent vectors to act on smooth functions defined 'locally'.

Lemma 2.5. *Let M be a smooth manifold, $p \in M$, and $v \in T_pM$. If $f, g \in C^\infty(M)$ agree in a neighbourhood of p, then $vf = vg$.*

Proof. Let U be an open subset of M containing p, such that f, g coincide on U. By Corollary 2.1, there exists a bump function χ which is identically 1 in a neighbourhood of p, and zero outside a neighbourhood V of p such that $\overline{V} \subset U$. Then $(f-g)\cdot\chi \equiv 0$ on M, and so

$$\begin{aligned} 0 &= v((f-g)\cdot\chi) = (v(f) - v(g))1 + (f(p) - g(p))v(\chi) \\ &= v(f) - v(g) + 0v(\chi) = v(f) - v(g). \end{aligned}$$

□

Remark 2.1. (Extension of the definition of tangent vectors.)
The above lemma can be used to extend the action of tangent vectors $v \in T_pM$ to functions $f \in C^\infty(U)$, where U is any open set in M containing the point p. Indeed, if $\chi \in C^\infty(M)$ is a bump function which is zero outside a neighbourhood V of p with $\overline{V} \subset U$ and is identically 1 in a neighbourhood of p, then we define F_χ to be $\chi\cdot f$ in U and 0 outside U. Then $F_\chi \in C^\infty(M)$, and we set $v(f) := v(F_\chi)$. The definition makes sense, for if σ is a different bump function with the same properties, then since $\sigma\cdot f$ matches with $\chi\cdot f$ in a neighbourhood of p (both being equal to f), it follows from Lemma 2.5 that $v(F_\chi) = v(F_\sigma)$. *

Definition 2.2. (Chart-induced tangent vectors.)
Let M be a smooth manifold of dimension m, and $(U, \mathbf{x})$ be a chart in a defining atlas for M. Let $p \in U$, and let x^i be the coordinate functions, that is $\mathbf{x} = (x^1, \cdots, x^m)$. Define $\partial_{x^i,p} \in T_pM$ by

$$\partial_{x^i,p} f := \frac{\partial (f|_U \circ \mathbf{x}^{-1})}{\partial u^i}(\mathbf{x}(p)), \quad f \in C^\infty(M).$$

Then $\partial_{x^1,p}, \cdots, \partial_{x^m,p}$ are the *chart-induced basis tangent vectors* by $(U, \mathbf{x})$.

It is clear that $\partial_{x^i,p} \in T_pM$: It is linear, since for $f, g \in C^\infty(M)$ and $c \in \mathbb{R}$,

$$\begin{aligned}\partial_{x^i,p}(f+cg) &= \frac{\partial((f+cg)\circ \mathbf{x}^{-1})}{\partial u^i}(\mathbf{x}(p))\\ &= \frac{\partial(f\circ \mathbf{x}^{-1})}{\partial u^i}(\mathbf{x}(p)) + c\frac{\partial(g\circ \mathbf{x}^{-1})}{\partial u^i}(\mathbf{x}(p)) = (\partial_{x^i,p}f) + c(\partial_{x^i,p}g).\end{aligned}$$

Moreover, the Leibniz rule holds:

$$\begin{aligned}\partial_{x^i,p}(f\cdot g) &= \frac{\partial((f\cdot g)\circ \mathbf{x}^{-1})}{\partial u^i}(\mathbf{x}(p)) = \frac{\partial((f\circ \mathbf{x}^{-1})\cdot(g\circ \mathbf{x}^{-1}))}{\partial u^i}(\mathbf{x}(p))\\ &= (f\circ \mathbf{x}^{-1})(\mathbf{x}(p))\frac{\partial(g\circ \mathbf{x}^{-1})}{\partial u^i}(\mathbf{x}(p)) + (g\circ \mathbf{x}^{-1})(\mathbf{x}(p))\frac{\partial(f\circ \mathbf{x}^{-1})}{\partial u^i}(\mathbf{x}(p))\\ &= f(p)\,\partial_{x^i,p}(g) + g(p)\,\partial_{x^i,p}(f).\end{aligned}$$

Theorem 2.1. *$\partial_{x^1,p}, \cdots, \partial_{x^m,p}$ form a basis for T_pM.*

Thus, the vector-space-dimension of T_pM matches the smooth-manifold-dimension of M. This makes sense, since in the case of surfaces in $\mathbb{R}^3$, the tangent plane at a point on the surface locally looks like the surface around that point. In the case of spacetime M, we will later see that at an event $p \in M$, the instantaneous 3-dimensional space we see around, is infinitesimally the orthogonal complement (with respect to the 'Lorentzian scalar product $\mathbf{g}(p)$' in the 4-dimensional T_pM), to the tangent vector $v_{\gamma,p}$ to our worldline γ at p.

Proof. We show that $\{\partial_{x^1,p}, \cdots, \partial_{x^m,p}\}$ is independent and spans T_pM.

(Independence.) Let χ be a bump function around p. Set $f = x^i\chi \in C^\infty(M)$. Then

$$\begin{aligned}\partial_{x^j,p}f &= \frac{\partial((x^i\chi)\circ \mathbf{x}^{-1})}{\partial u^j}(\mathbf{x}(p)) = \chi(p)\frac{\partial(x^i\circ \mathbf{x}^{-1})}{\partial u^j}(\mathbf{x}(p)) + x^i(p)\frac{\partial(\chi\circ \mathbf{x}^{-1})}{\partial u^j}(\mathbf{x}(p))\\ &= 1\cdot\frac{\partial u^i}{\partial u^j}(\mathbf{x}(p)) + x^i(p)\cdot 0 = \delta^i_j.\end{aligned}$$

If $c^j\partial_{x^j,p} = 0$, then acting on $f = x^i\chi$ yields $c^i = 0$, showing independence.

(Spanning.) Let $f \in C^\infty(M)$. Choose an $r > 0$ so that $B(\mathbf{x}(p), r) \subset \mathbf{x}(U)$. Define $h \in C^\infty(B(\mathbf{0}, r))$ by $h(\mathbf{u}) = (f\circ\mathbf{x}^{-1})(\mathbf{u}+\mathbf{x}(p)) - f(p)$. Then $h(\mathbf{0}) = 0$. By Lemma 2.1, $h = u^i g_i$, for some $g_i \in C^\infty(B(\mathbf{0}, r))$ satisfying

$$g_i(\mathbf{0}) = \frac{\partial h}{\partial u^i}(\mathbf{0}) = \frac{\partial(f\circ \mathbf{x}^{-1})}{\partial u^i}(\mathbf{x}(p)) = \partial_{x^i,p}f.$$

In the neighbourhood $\mathbf{x}^{-1}(B(\mathbf{x}(p), r))$ of p, $f(\cdot) = h(\mathbf{x}(\cdot) - \mathbf{x}(p)) + f(p)$. So

$$\begin{aligned}
vf &= v(h(\mathbf{x}(\cdot) - \mathbf{x}(p)) + f(p)) = v(h(\mathbf{x}(\cdot) - \mathbf{x}(p))) + v(f(p)) \\
&= v(h(\mathbf{x}(\cdot) - \mathbf{x}(p))) + 0 = v((x^i(\cdot) - x^i(p))g_i(\mathbf{x}(\cdot) - \mathbf{x}(p))) \\
&= (x^i(p) - x^i(p))v(g_i(\mathbf{x}(\cdot) - \mathbf{x}(p))) + g_i(\mathbf{x}(p) - \mathbf{x}(p))v(x^i(\cdot) - x^i(p)) \\
&= 0 \cdot v(g_i(\mathbf{x}(\cdot) - \mathbf{x}(p))) + g_i(\mathbf{0}) \cdot v(x^i(\cdot) - x^i(p)) \\
&= (\partial_{x^i,p} f)(v(x^i) - v(x^i(p))) = (\partial_{x^i,p} f)(v(x^i) - 0) = (v(x^i)\partial_{x^i,p})f.
\end{aligned}$$

So $v = v(x^i)\partial_{x^i,p}$. Consequently, $\partial_{x^1,p}, \cdots, \partial_{x^m,p}$ span T_pM. □

Example 2.2. ($T_pM \simeq V$ for an affine space M over V.)
Let M be an affine space over V. For $p \in M$ and $\mathbf{v} \in V$, define the smooth curve $\gamma_{\mathbf{v}}(t) = p + t\mathbf{v}$, $t \in \mathbb{R}$ (Exercise 1.19). Then $v_{\gamma_{\mathbf{v}},p} \in T_pM$. Let $\{\mathbf{e}_1, \cdots, \mathbf{e}_m\}$ be a basis for V, and $\mathbf{v} = v^i\mathbf{e}_i$. The chart defined in Example 1.3 induces a basis $\{\partial_{x^1,p}, \cdots, \partial_{x^m,p}\}$ for T_pM. Let us find the coordinates of $v_{\gamma_{\mathbf{v}},p}$ with respect to this basis. From the end of the proof of Theorem 2.1, $v_{\gamma_{\mathbf{v}},p} = v_{\gamma_{\mathbf{v}},p}(x^i)\,\partial_{x^i,p}$. Thus the ith component is

$$v_{\gamma_{\mathbf{v}},p}(x^i) = \frac{d(x^i \circ \gamma_{\mathbf{v}})}{dt}(0) = \frac{d(tv^i)}{dt}(0) = v^i,$$

which coincides with the components of $\mathbf{v}$ with respect to the basis $\{\mathbf{e}_1, \cdots, \mathbf{e}_m\}$ for V. Consider the isomorphism $\mathrm{I} : V \to T_pM$ that sends the basis vector $\mathbf{e}_i$ to the basis vector $\partial_{x^i,p}$ for all $1 \leqslant i \leqslant m$. Then $v^i\mathbf{e}_i = \mathbf{v}$ is sent to $v^i\partial_{x^i,p} = v_{\gamma_{\mathbf{v}},p}$. Thus our isomorphism I is the map $\mathbf{v} \mapsto v_{\gamma_{\mathbf{v}},p}$. ◇

Example 2.3. (Physicist's definition of a vector.)
In physics literature sometimes, a vector, based at a point p in an m-dimensional manifold M, is defined as an 'abstract object' with the following behaviour specified for its components: When one commits to an admissible chart/coordinate system $(U, \mathbf{x})$ on M, the abstract object produces an m-tuple $(v^1, \cdots, v^m)$, and if one has two admissible coordinate systems containing p, say $(U, \mathbf{x})$ and $(U', \mathbf{x}')$, then the resulting m-tuples are related by

$$v'^i = \frac{\partial x'^i}{\partial x^j}(\boldsymbol{\xi})\, v^j,$$

where $\dfrac{\partial x'^i}{\partial x^j}(\boldsymbol{\xi})$ is an abbreviation for the partial derivative

$$\frac{\partial(x'^i \circ \mathbf{x}^{-1})}{\partial u^j}(\mathbf{x}(p))$$

of the ith component of chart transition map (sometimes called 'change of coordinates' in the literature) $\mathbf{x}' \circ \mathbf{x}^{-1} : \mathbf{x}(U \cap U') \to \mathbf{x}'(U \cap U')$ with respect to the jth variable, at the point $\boldsymbol{\xi} := \mathbf{x}(p)$. We show below that this way of thinking coincides with our notion of a vector. Thus, starting with a tangent vector $v \in T_pM$, and using its representations obtained via the bases $\{\partial_{x^1,p}, \cdots, \partial_{x^m,p}\}$ and $\{\partial_{x'^1,p}, \cdots, \partial_{x'^m,p}\}$, we will establish the above transformation rule for the components of v. Indeed, we have

$$v'^i = v(x'^i) = v^j \partial_{x^j,p}(x'^i) = v^j \frac{\partial(x'^i \circ \mathbf{x}^{-1})}{\partial u^j}(\mathbf{x}(p)),$$

as required.

Having chosen a basis, if we represent the vector as a column vector of its components with respect to the basis, then the above transformation expression can be rewritten as matrix multiplication as follows:

$$\begin{bmatrix} v'^1 \\ \vdots \\ v'^m \end{bmatrix} = \begin{bmatrix} \frac{\partial(x'^1 \circ \mathbf{x}^{-1})}{\partial u^1}(\mathbf{x}(p)) & \cdots & \frac{\partial(x'^1 \circ \mathbf{x}^{-1})}{\partial u^m}(\mathbf{x}(p)) \\ \vdots & \ddots & \vdots \\ \frac{\partial(x'^m \circ \mathbf{x}^{-1})}{\partial u^1}(\mathbf{x}(p)) & \cdots & \frac{\partial(x'^m \circ \mathbf{x}^{-1})}{\partial u^m}(\mathbf{x}(p)) \end{bmatrix} \begin{bmatrix} v^1 \\ \vdots \\ v^m \end{bmatrix}.$$

In particular, since $\partial_{x^i,p} = \delta_i^k \partial_{x^k,p}$, it follows that

$$\partial_{x^i,p} = \left(\delta_i^k \frac{\partial x'^j}{\partial x^k}(\boldsymbol{\xi})\right) \partial_{x'^j,p} = \frac{\partial x'^j}{\partial x^i}(\boldsymbol{\xi})\, \partial_{x'^j,p}.$$

Swapping the roles of $\mathbf{x}$ and $\mathbf{x}'$, $\partial_{x'^i,p} = \frac{\partial x^j}{\partial x'^i}(\boldsymbol{\xi}')\, \partial_{x^j,p}$, where $\boldsymbol{\xi}' = \mathbf{x}'(p)$. ◇

Lemma 2.6. (Vectors as the tangent vectors of curves.)
Let $v \in T_pM$. Then there exists an $\epsilon > 0$ and a smooth curve $\gamma : (-\epsilon, \epsilon) \to M$ such that $\gamma(0) = p$ and $v_{\gamma,p} = v$.

Proof. Let $(U, \mathbf{x})$ be an admissible chart containing p. If $\partial_{x^i,p}$, $1 \leqslant i \leqslant m$, denote the chart-induced tangent vectors, then we can write $v = v^i \partial_{x^i,p}$ for some numbers v^i. Let $\epsilon > 0$ be small enough so that $\mathbf{x}(p) + tv^i\mathbf{e}_i \in \mathbf{x}(U)$ for $|t| < \epsilon$, where $\mathbf{e}_1, \cdots, \mathbf{e}_m$ are the standard basis vectors in $\mathbb{R}^m$. Set $\gamma(t) = \mathbf{x}^{-1}(\mathbf{x}(p) + tv^i\mathbf{e}_i)$, $t \in (-\epsilon, \epsilon)$. Then $\gamma(0) = \mathbf{x}^{-1}(\mathbf{x}(p)) = p$, and for all $f \in C^\infty(M)$,

$$\begin{aligned} v_{\gamma,p} f &= \frac{d(f \circ \gamma)}{dt}(0) = \frac{d(f(\mathbf{x}^{-1}(\mathbf{x}(p) + tv^i\mathbf{e}_i)))}{dt}(0) \\ &= \frac{\partial(f \circ \mathbf{x}^{-1})}{\partial u^i}(\mathbf{x}(p)) v^i = v^i \partial_{x^i,p} f = vf. \end{aligned}$$

Consequently, $v_{\gamma,p} = v$. □

In particular, from the above, $\partial_{x^i,p}$ is the velocity vector $v_{\gamma^i,p}$ of a locally defined curve γ^i passing through $p = \gamma^i(0)$: $\gamma^i(t) = \mathbf{x}^{-1}(\mathbf{x}(p) + t\mathbf{e}_i)$, for all $|t| < \epsilon$, with a small enough $\epsilon > 0$.

Example 2.4. (Curve tangent vectors in terms of chart-induced basis.) Let $\gamma : I \to M$ be a smooth curve, where I is an open interval, and suppose that $\gamma(I) \subset U$, where $(U, \mathbf{x})$ is an admissible chart for M. We claim that

$$v_{\gamma,\gamma(t)} = \frac{d(\mathbf{x} \circ \gamma)^i}{dt}(t)\, \partial_{x^i,\gamma(t)}, \quad t \in I.$$

Indeed, for all $f \in C^\infty(M)$, we have

$$\begin{aligned} v_{\gamma,\gamma(t)} f &= \frac{d(f \circ \gamma)}{dt}(t) = \frac{d(f \circ \mathbf{x}^{-1} \circ \mathbf{x} \circ \gamma)}{dt}(t) \\ &= \frac{\partial(f \circ \mathbf{x}^{-1})}{\partial u^i}(\mathbf{x}(\gamma(t)))\frac{d(\mathbf{x} \circ \gamma)^i}{dt}(t) = \frac{d(\mathbf{x} \circ \gamma)^i}{dt}(t)\, \partial_{x^i,\gamma(t)} f, \end{aligned}$$

and so we see that the components of $v_{\gamma,\gamma(t)}$ with respect to the chart-induced tangent vectors are just the derivatives of components of the chart representation of the curve. $\diamond$

Exercise 2.7. Consider the smooth manifold $\mathbb{R}^2$ with the standard smooth structure. Consider the admissible chart (V, ψ), where $V = \{(x, y) : y > 0\}$, $\psi = (r, \theta)$, $r(x, y) = \sqrt{x^2 + y^2}$ and $\theta(x, y) = \cos^{-1}(x/\sqrt{x^2 + y^2})$. Let $p = (x, y) \in V$. Express $\partial_{r,p}$ and $\partial_{\theta,p}$ in terms of $\partial_{x,p}$ and $\partial_{y,p}$.

Exercise 2.8. Let $M \times N$ be the product of the smooth manifolds M, N, and let $(p, q) \in M \times N$. For $v \in T_pM$ and $w \in T_qN$, define $v \oplus w : C^\infty(M \times N) \to \mathbb{R}$ by $(v \oplus w)f = v(f(\cdot, q)) + w(f(p, \cdot))$ for all $f \in C^\infty(M \times N)$. Show that $v \oplus w$ belongs to $T_{(p,q)}(M \times N)$. Prove that the map $T_pM \times T_qN \ni (v, w) \mapsto v \oplus w \in T_{(p,q)}(M \times N)$ is linear, injective, and surjective. Thus $T_{(p,q)}(M \times N)$ is isomorphic to the direct sum of T_pM and T_qN, written as $T_{(p,q)}(M \times N) \simeq T_pM \times T_qN$.

Exercise 2.9. Suppose N is a submanifold (Exercise 1.25, p.15) of the smooth manifold M. Show that for each $p \in N$, T_pN is isomorphic to a subspace of T_pM.

2.5 Derivatives of smooth maps

If $\gamma : \mathbb{R} \to M$ is a smooth curve, and $f : M \to N$ is a smooth map, then $f \circ \gamma : \mathbb{R} \to N$ is a smooth curve. For $p = \gamma(0) \in M$ and $g \in C^\infty(N)$,

$$v_{f\circ\gamma,f(p)} g = \frac{d(g \circ (f \circ \gamma))}{dt}(0) = \frac{d((g \circ f) \circ \gamma))}{dt}(0) = v_{\gamma,p}(g \circ f).$$

Hence $f : M \to N$ induces a mapping sending tangent vectors at $p \in M$ to tangent vectors at $f(p) \in N$. This motivates the following definition.

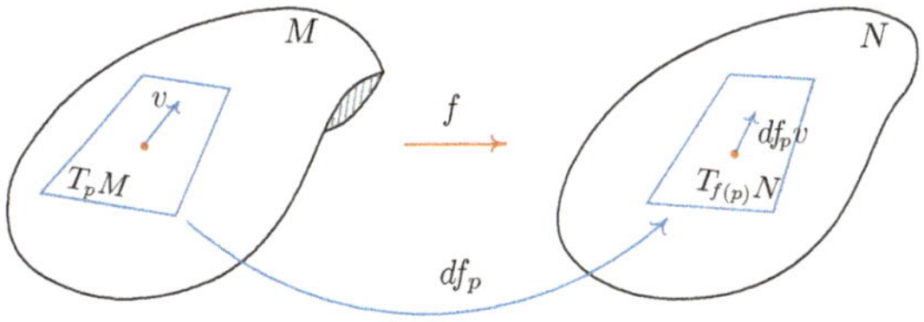

Definition 2.3. (Derivative of a smooth map.)
Let M, N be smooth manifolds, and $f : M \to N$ be a smooth map.
If $p \in M$, then define $df_p : T_pM \to T_{f(p)}N$ by $(df_p(v))(g) = v(g \circ f)$ for all $g \in C^\infty(N)$ and $v \in T_pM$. The map df_p is called the *derivative of f at p*.

For $v \in T_pM$, $df_p(v)$ is a tangent vector at $f(p) \in N$: For all $g, h \in C^\infty(N)$ and $c \in \mathbb{R}$,

$$\begin{aligned}(df_p(v))(g+ch) &= v((g+ch) \circ f) = v(g \circ f + c(h \circ f)) = v(g \circ f) + cv(h \circ f)\\ &= (df_p(v))(g) + c(df_p(v))(h), \text{ and}\\ (df_p(v))(g \cdot h) &= v((g \cdot h) \circ f) = v((g \circ f) \cdot (h \circ f))\\ &= (g \circ f)(p)v(h \circ f) + (h \circ f)(p)v(g \circ f)\\ &= g(f(p))(df_p(v))(h) + h(f(p))(df_p(v))(g).\end{aligned}$$

Exercise 2.10. Let M, N be smooth manifolds, $p \in M$, and $f : M \to N$ be a smooth map. Prove that $df_p : T_pM \to T_{f(p)}N$ is a linear map.

Exercise 2.11. Let $\gamma : \mathbb{R} \to M$ be a smooth curve passing through the point $p = \gamma(0)$. Let $\partial_{t,0}$ denote the tangent vector at $t = 0 \in \mathbb{R}$ induced by the chart $(\mathbb{R}, \text{id})$. Let $v_{\gamma,p}$ be defined by (2.1) (p.22). Show that $v_{\gamma,p} = d\gamma_0(\partial_{t,0})$.

Theorem 2.2. *Let M, N be smooth manifolds, $f : M \to N$ be a smooth map, and $p \in M$. Let $(U, \mathbf{x})$ be an admissible chart for M containing p, and let $(V, \mathbf{y})$ be an admissible chart for N containing $f(p)$. Then the matrix for the linear transformation $df_p : T_pM \to T_{f(p)}N$ with respect to the bases $\{\partial_{x^i,p}, 1 \leqslant i \leqslant m\}$ and $\{\partial_{y^j,f(p)}, 1 \leqslant j \leqslant n\}$ is given by the Jacobian matrix of $\mathbf{y} \circ f \circ \mathbf{x}^{-1}$ at the point $\mathbf{x}(p)$.*

Proof. For $g \in C^\infty(N)$, we have

$$\begin{aligned}(df_p(\partial_{x^i,p}))g &= \partial_{x^i,p}(g \circ f) = \frac{\partial(g \circ f \circ \mathbf{x}^{-1})}{\partial u^i}(\mathbf{x}(p))\\ &= \frac{\partial(g \circ \mathbf{y}^{-1} \circ \mathbf{y} \circ f \circ \mathbf{x}^{-1})}{\partial u^i}(\mathbf{x}(p))\\ &= \frac{\partial(g \circ \mathbf{y}^{-1})}{\partial v^j}(\mathbf{y}(f(p)))\,\frac{\partial(\mathbf{y} \circ f \circ \mathbf{x}^{-1})^j}{\partial u^i}(\mathbf{x}(p))\\ &= \frac{\partial(\mathbf{y} \circ f \circ \mathbf{x}^{-1})^j}{\partial u^i}(\mathbf{x}(p))\,\partial_{y^j,f(p)}g.\end{aligned}$$

Thus $df_p(\partial_{x^i,p}) = \dfrac{\partial(\mathbf{y} \circ f \circ \mathbf{x}^{-1})^j}{\partial u^i}(\mathbf{x}(p))\,\partial_{y^j,f(p)}$. □

Exercise 2.12. Consider $\mathbb{R}^2$ as a smooth manifold with the standard smooth structure. For $t \in \mathbb{R}$, let $\psi_t : \mathbb{R}^2 \to \mathbb{R}^2$ be given by $(x, y) \mapsto (x - yt, y)$. Show that $(d\psi_t)_p(\partial_{y,p}) = -t\,\partial_{x,\psi_t(p)} + \partial_{y,\psi_t(p)}$.

Exercise 2.13. Let M_1, M_2, M_3 be smooth manifolds. Suppose that $f: M_1 \to M_2$ and $g: M_2 \to M_3$ are smooth maps. Let $p \in M_1$. Show $d(g \circ f)_p = dg_{f(p)} \circ df_p$.

Exercise 2.14. Let $f : M \to N$ be a diffeomorphism between the smooth manifolds M and N. Let $p \in M$. Show that $df_p : T_pM \to T_{f(p)}N$ is an isomorphism. Conclude that M and N have the same dimension.

We end this section with a useful result, namely, the inverse function theorem, in the smooth manifold setting. First we give a C^∞ version in the Euclidean setting, as a consequence of the 'usual' continuously differentiable version; see for example [Rudin (1983), Theorem 9.24, p. 221].

Proposition 2.1. (Euclidean, C^∞-inverse function theorem.)
Let $U, V \subset \mathbb{R}^m$ be open, $f : U \to V$ be C^∞, and $f'(p)$ be invertible at $p \in U$. Then there exist neighbourhoods $U_0 \subset U$ of p and $V_0 \subset V$ of $f(p)$ such that $f|_{U_0} : U_0 \to V_0$ is a diffeomorphism.

Proof. There are the claimed neighbourhoods such that $f^{-1} : V_0 \to U_0$ is C^1 on V_0 (see, e.g., [Rudin (1983), Theorem 9.24]). As $f \circ f^{-1} = \mathrm{id}_{V_0}$, we obtain by the chain rule that $f'(f^{-1}(x)) \cdot (f^{-1})'(x) = \mathrm{id}_{\mathbb{R}^m}$ for all $x \in V_0$, and so $f'(f^{-1}(x))$ is an invertible matrix, and $(f^{-1})'(x) = (f'(f^{-1}(x)))^{-1}$. By Cramer's rule [Artin (1991), Theorem 5.7], the (i, j) entry of the matrix for $(f^{-1})'(x)$ is given by a polynomial combination of the partials of f, divided by the nonzero determinant of the Jacobian matrix of f, and as these are C^∞, f^{-1} is C^∞. In light of Exercise 1.20 (p.15), we are done. □

Theorem 2.3. (Smooth manifold-inverse function theorem.)
Let M, N be smooth manifolds, $f : M \to N$ be smooth, and df_p be invertible at a point $p \in M$. Then there exist neighbourhoods U_0 of p, and V_0 of $f(p)$, such that $f|_{U_0} : U_0 \to V_0$ is a diffeomorphism.

Proof. As df_p is bijective, M and N have the same dimension, say m. Since f is smooth, there exist admissible charts (U, φ) for M and (V, ψ) for N, such that the point $p \in U$, $f(U) \subset V$, and $\widetilde{f} = \psi \circ f \circ \varphi^{-1} : \widetilde{U} \to \widetilde{V}$ is C^∞, where $\widetilde{U} = \varphi(U)$ and $\widetilde{V} = \psi(V)$ are the open sets in $\mathbb{R}^m$. As φ, ψ are diffeomorphisms, $d\widetilde{f}_{\varphi(p)} = d\psi_{f(p)} \circ df_p \circ d(\varphi^{-1})_{\varphi(p)}$ is invertible. By Prop. 2.1, there exists a neighbourhood $\widetilde{U}_0 \subset \widetilde{U}$ containing $\varphi(p)$ and a neighbourhood $\widetilde{V}_0 \subset \widetilde{V}$ containing $\psi(f(p))$ such that $\widetilde{f}|_{\widetilde{U}_0} : \widetilde{U}_0 \to \widetilde{V}_0$ is a diffeomorphism. Then $U_0 := \varphi^{-1}\widetilde{U}_0 \subset U$ is a neighbourhood of p, and $V_0 := \psi^{-1}\widetilde{V}_0 \subset V$ is a neighbourhood of $f(p)$. Moreover, $f|_{U_0} = \psi^{-1} \circ \widetilde{f}|_{\widetilde{U}_0} \circ \varphi$, that is, $f|_{U_0} : U_0 \to V_0$ is the composition of the diffeomorphism $\widetilde{f}|_{\widetilde{U}_0}$ with the diffeomorphisms ψ^{-1} and ϕ. So $f|_{U_0} : U_0 \to V_0$ is a diffeomorphism. □

Remark 2.2. We now give the justification of a technical fact, as an application of the Euclidean C^∞ inverse function theorem. This fact will be used later in Remark 7.1.

Fact: Let M be a smooth manifold, I be an open interval containing 0, and $\gamma : I \to M$ be a smooth curve such that $v_{\gamma,\gamma(0)} \neq 0$.

Then there is an open interval $J \subset I$ containing 0, and admissible chart $(U, \mathbf{x})$ containing $\gamma(J)$, such that $(\mathbf{x} \circ \gamma)(t) = (t, 0, \cdots, 0)$ for all $t \in J$.

Proof: Let $(V, \mathbf{y})$ be an admissible chart containing $p := \gamma(0)$. Then

$$0 \neq v_{\gamma,p} = \frac{d(\mathbf{y} \circ \gamma)^i}{dt}(0)\, \partial_{y^i,p}.$$

So there exists an index i such that

$$\frac{d(\mathbf{y} \circ \gamma)^i}{dt}(0) \neq 0.$$

Without loss of generality, let $i = 1$. By the inverse function theorem applied to the smooth map $\gamma^1 = (\mathbf{y} \circ \gamma)^1 : I \to \mathbb{R}$, there exists an open interval $\widetilde{J} \subset I$ containing 0 and an open interval $K \subset \mathbb{R}$ containing $y^1(p)$, such that the smooth map $(\mathbf{y} \circ \gamma)^1 : \widetilde{J} \to K$ is a bijection with a smooth inverse $\eta : K \to \widetilde{J}$. Let $\widetilde{U}$ be the open set $\widetilde{U} = V \cap (y^1)^{-1}K$, which is nonempty as $p \in \widetilde{U}$. As $\mathbf{y} : V \to \mathbf{y}(V)$ is a diffeomorphism, the nonempty set $\mathbf{y}(\widetilde{U})$ is open. Now for $\boldsymbol{\beta} = (\beta^1, \cdots, \beta^m) \in \mathbf{y}(\widetilde{U})$, we have $\beta^1 \in K$. Define the map $g : \mathbf{y}(\widetilde{U}) \to \mathbb{R}^m$ by $g(\boldsymbol{\beta}) = (\eta(\beta^1), \beta^2 - \gamma^2(\eta(\beta^1)), \cdots, \beta^m - \gamma^m(\eta(\beta^1)))$ for all $\boldsymbol{\beta} = (\beta^1, \cdots, \beta^m) \in \mathbf{y}(\widetilde{U})$. We have

$$g'(\boldsymbol{\beta}) = \begin{bmatrix} \eta'(\beta^1) & \mathbf{0} \\ * & I_{m-1} \end{bmatrix},$$

which is invertible in $\mathbf{y}(\widetilde{U})$, and in particular, at $\mathbf{y}(p)$. By the inverse function theorem, there exists a neighbourhood $W \subset \mathbf{y}(\widetilde{U})$ of $\mathbf{y}(p)$, such that $g : W \to g(W)$ is a diffeomorphism. Set $U = \mathbf{y}^{-1}W \subset \widetilde{U} \subset V$, and $\mathbf{x} = g \circ \mathbf{y} : U \to \mathbb{R}^m$. This is a diffeomorphism onto its image. As $0 \in \widetilde{J}$, and $\gamma(0) = p \in U$, it follows by the continuity of γ that we can choose an open interval $J \subset \widetilde{J}$ containing 0, such that $\gamma(t) \in U$ for all $t \in J$. For $t \in J \subset \widetilde{J}$, we have

$$\begin{aligned}(\mathbf{x} \circ \gamma)(t) &= (g \circ \mathbf{y} \circ \gamma)(t) \\ &= (\eta(\gamma^1(t)), \gamma^2(t) - \gamma^2(\eta(\gamma^1(t))), \cdots, \gamma^m(t) - \gamma^m(\eta(\gamma^1(t)))) \\ &= (t, \gamma^2(t) - \gamma^2(t), \cdots, \gamma^m(t) - \gamma^m(t)) = (t, 0, \cdots, 0).\end{aligned}$$

This completes the proof of the claimed fact. $*$

Exercise 2.15. Let $m > n$, $f : \mathbb{R}^m \to \mathbb{R}^n$ be a C^∞ map, and $q \in f(\mathbb{R}^m)$ be such that for all $p \in M := f^{-1}\{q\}$, $f'(p)$ is surjective. Show that M is a submanifold of $\mathbb{R}^m$ (where the latter is given the standard smooth structure) of dimension $m - n$.

Exercise 2.16. Show that $\mathrm{SL}_n(\mathbb{R}) = \{A \in \mathbb{R}^{n\times n} : \det A = 1\}$ is a smooth submanifold of $\mathbb{R}^{n\times n} \simeq \mathbb{R}^{n^2}$ (with the standard smooth structure) of dimension $n^2 - 1$. *Hint:* $\det' I = \text{trace}$. For $A \in \mathrm{SL}_n(\mathbb{R})$, $(\det' A)(h) = \text{trace}(A^{-1}h)$, $h \in \mathbb{R}^{n\times n}$.

2.6 Cotangent space

In this section, we will consider the dual space $(T_pM)^*$ of T_pM consisting of linear maps $\omega : T_pM \to \mathbb{R}$. Elements of $(T_pM)^*$ will be called cotangent vectors or 1-forms[5]. Later on, we will also learn that the instantaneous momentum is a 1-form. This can be intuitively understood as follows. We know that force is the rate of change of momentum, and so as a geometric object, the nature of the force is the same as that of momentum. We know that work, which is a scalar, is obtained by taking the 'line integral' of the force. But when doing so, we take the dot product of the force with a little vectorial displacement along the curve, namely a tangent vector. So the force/momentum acts on tangent vectors and produces scalars.

We will see later on that tangent spaces together with their duals (called cotangent spaces) at a point $p \in M$ can be used to build the space of *tensors* at the point p, and when we consider a varying p, one gets a map $p \mapsto$ (tensor at p), which is referred to as a tensor *field.* The field equations of spacetime will involve such tensor fields. The modest aim in this section, though, is to introduce the cotangent space as the dual of the tangent space.

Definition 2.4. (1-forms, cotangent space.)
The dual space $(T_pM)^*$ of T_pM is called the *cotangent space* to M at $p \in M$. Each element of $(T_pM)^*$ is called a 1-*form* or a *cotangent vector.*

Example 2.5. (Gradient of a function.) Let $f \in C^\infty(M)$ and $p \in M$. Define the 1-form $df_p \in (T_pM)^*$ by $df_p(v) = v(f)$ for all $v \in T_pM$. For $v, w \in T_pM$ and $c \in \mathbb{R}$, $df_p(v + cw) = (v + cw)(f) = v(f) + cw(f) = df_p(v) + c\,df_p(w)$. The 1-form df_p is called the *gradient of* f at p. ◇

Example 2.6. (Gradient revisited.) Let $\mathbb{R}$ be given the standard smooth structure $[\{(\mathbb{R}, u \mapsto u)\}]$. If $q \in \mathbb{R}$, then $T_q\mathbb{R}$ can be identified with $\mathbb{R}$, by mapping $w = w(u)\,\partial_{u,q}$ to $w(u)$. Given an $f \in C^\infty(M)$, we have the derivative $df_p : T_pM \to T_{f(p)}\mathbb{R}$ of the smooth map f (Definition 2.3). But by identification of $T_{f(p)}\mathbb{R}$ with $\mathbb{R}$, we may consider the derivative df_pv acting on $v \in T_pM$, as a real number, rather than a vector in $T_{f(p)}\mathbb{R}$. In fact, this number is the action $df_p(v)$ of the 1-form df_p on v, because if the vector $df_p(v) = c\,\partial_{u,f(p)}$, then $c = (df_p(v))(u \mapsto u) = v(\mathrm{id} \circ f) = v(f)$.

[5] This terminology will become clear after we study more general k-forms in Chapter 10.

So it makes sense to use the same notation $df_p(v)$, both for the action of derivative df_p on v (which is a vector in $T_{f(p)}\mathbb{R}$), and the action of the 1-form df_p on v (which is a real number). ◇

Definition 2.5. (Chart-induced 1-forms.)
Let M be a smooth manifold of dimension m, and $(U, \mathbf{x})$ be a chart in a defining atlas for M. Let $p \in U$, and let x^i be the coordinate functions, that is, $\mathbf{x} = (x^1, \cdots, x^m)$. Define the 1-forms $(dx^1)_p, \cdots (dx^m)_p \in (T_pM)^*$ by $(dx^i)_p(v) = v(x^i)$ for all $v \in T_pM$. Then $(dx^1)_p, \cdots, (dx^m)_p$ are called the *chart-induced 1-forms* by the chart $(U, \mathbf{x})$.

Theorem 2.4. *$(dx^1)_p, \cdots, (dx^m)_p$ form a basis for $(T_pM)^*$.*

Proof. In fact, $\{(dx^1)_p, \cdots, (dx^m)_p\}$ is the dual basis to the chart-induced basis $\{\partial_{x^1,p}, \cdots, \partial_{x^m,p}\}$:

$$(dx^i)_p(\partial_{x^j,p}) = \partial_{x^j,p}x^i = \frac{\partial(x^i \circ \mathbf{x}^{-1})}{\partial u^j}(\mathbf{x}(p)) = \frac{\partial u^i}{\partial u^j}(\mathbf{x}(p)) = \delta^i_j.$$

If $c_j(dx^j)_p = 0$, then by acting on $\partial_{x^i,p}$, we obtain $c_i = \delta^i_j c_j = 0$. Thus, $(dx^1)_p, \cdots, (dx^m)_p$ are independent. Also, for $\omega \in (T_pM)^*$ and $v \in T_pM$, we can write $v = v^j \partial_{x^j,p}$, and so

$$\begin{aligned}\big(\omega - \omega(\partial_{x^i,p})(dx^i)_p\big)(v) &= \big(\omega - \omega(\partial_{x^i,p})(dx^i)_p\big)(v^j\partial_{x^j,p}) \\ &= v^j\omega(\partial_{x^j,p}) - \omega(\partial_{x^i,p})v^j\delta^i_j = 0,\end{aligned}$$

showing that $\omega = \omega(\partial_{x^i,p})(dx^i)_p$. Hence the 1-forms $(dx^1)_p, \cdots, (dx^m)_p$ span $(T_pM)^*$. □

In particular, if $f \in C^\infty(M)$, and $(U, \mathbf{x})$ is an admissible chart for M containing p, then the proof of Theorem 2.4 shows that the components of the 1-form df_p, with respect to the basis $\{(dx^1)_p, \cdots, (dx^m)_p\}$ for $(T_pM)^*$, are given by the m-tuple $(\partial_{x^1,p}f, \cdots, \partial_{x^m,p}f)$.

Exercise 2.17. (Physicist's definition of a 'covector'.)
In physics literature sometimes, a covector, based at a point p in an m-dimensional manifold M, is defined as an 'abstract object' with the following behaviour specified for its components: When one commits to an admissible chart/coordinate system $(U, \mathbf{x})$ on M, the abstract object produces an m-tuple $(\omega_1, \cdots, \omega_m)$, and if one has two admissible coordinate systems, say $(U, \mathbf{x})$ and $(U', \mathbf{x}')$, containing the point p, then the resulting m-tuples are related by

$$\omega'_i = \frac{\partial x^j}{\partial x'^i}(\boldsymbol{\xi}')\omega_j, \text{ where } \frac{\partial x^j}{\partial x'^i}(\boldsymbol{\xi}') := \frac{\partial(x^j \circ \mathbf{x}'^{-1})}{\partial u^i}(\mathbf{x}'(p)) \text{ and } \boldsymbol{\xi}' := \mathbf{x}'(p).$$

The aim of this exercise is to show that this way of thinking coincides with our notion of a 1-form. Thus, starting with a 1-form $\omega \in (T_pM)^*$, and using its representations in the bases $\{(dx^1)_p, \cdots, (dx^m)_p\}$ and $\{(dx'^1)_p, \cdots, (dx'^m)_p\}$, establish the above transformation rule for the components of ω. We remark that, having chosen a basis, if we represent the 1-form as row vector of its components with respect to the basis, then the above transformation expression can be rewritten as matrix multiplication as follows:

$$\begin{bmatrix} \omega_1' \cdots \omega_m' \end{bmatrix} = \begin{bmatrix} \omega_1 \cdots \omega_m \end{bmatrix} \begin{bmatrix} \frac{\partial(x^1 \circ \mathbf{x}'^{-1})}{\partial u^1}(\mathbf{x}'(p)) & \cdots & \frac{\partial(x^1 \circ \mathbf{x}'^{-1})}{\partial u^m}(\mathbf{x}'(p)) \\ \vdots & \ddots & \vdots \\ \frac{\partial(x^m \circ \mathbf{x}'^{-1})}{\partial u^1}(\mathbf{x}'(p)) & \cdots & \frac{\partial(x^m \circ \mathbf{x}'^{-1})}{\partial u^m}(\mathbf{x}'(p)) \end{bmatrix}.$$

In particular, as $(dx^i)_p = \delta^i_k (dx^k)_p$, we obtain

$$(dx^i)_p = \Big(\delta^i_k \frac{\partial x^k}{\partial x'^j}(\boldsymbol{\xi}')\Big)(dx'^j)_p = \frac{\partial x^i}{\partial x'^j}(\boldsymbol{\xi}')(dx'^j)_p.$$

Swapping the roles of $\mathbf{x}$ and $\mathbf{x}'$, $(dx'^i)_p = \dfrac{\partial x'^i}{\partial x^j}(\boldsymbol{\xi})(dx^j)_p$, where $\boldsymbol{\xi} := \mathbf{x}(p)$.

Exercise 2.18. (1-forms on $\mathbb{R}^3$: Cartesian to spherical coordinates.) Consider the smooth manifold $\mathbb{R}^3$ with the standard smooth structure, and the admissible chart $(U, \mathbf{x})$, where $U := \mathbb{R}^3 \backslash \{(x, y, z) \in \mathbb{R}^3 : x \geqslant 0,\ y = 0\}$, $\mathbf{x} = (r, \theta, \phi)$, $\mathbf{x}(U) = (0, \infty) \times (0, \pi) \times (0, 2\pi)$, and $\mathbf{x}^{-1} : \mathbf{x}(U) \to U$ is given by

$$\mathbf{x}^{-1}(r, \theta, \phi) = (r(\cos\phi)\sin\theta,\ r(\sin\phi)\sin\theta,\ r\cos\theta) \text{ for } (r, \theta, \phi) \in \mathbf{x}(U).$$

Let $p \in U$ be such that $\mathbf{x}(p) = (r, \theta, \phi)$. Prove that

$$\begin{aligned} (dx)_p &= (\cos\phi)(\sin\theta)(dr)_p + r(\cos\phi)(\cos\theta)(d\theta)_p - r(\sin\phi)(\sin\theta)(d\phi)_p \\ (dy)_p &= (\sin\phi)(\sin\theta)(dr)_p + r(\sin\phi)(\cos\theta)(d\theta)_p + r(\cos\phi)(\sin\theta)(d\phi)_p \\ (dz)_p &= (\cos\theta)(dr)_p - r(\sin\theta)(d\theta)_p. \end{aligned}$$

Exercise 2.19. Let $(U, \mathbf{x}), (U', \mathbf{x}')$ be admissible charts for a smooth manifold M. Show that the matrices describing the change of components of tangent vectors and 1-forms, with respect to the chart-induced bases, are inverses of each other: If

$$A = [A^i_j] := \begin{bmatrix} \frac{\partial(x'^1 \circ \mathbf{x}^{-1})}{\partial u^1}(\mathbf{x}(p)) & \cdots & \frac{\partial(x'^1 \circ \mathbf{x}^{-1})}{\partial u^m}(\mathbf{x}(p)) \\ \vdots & \ddots & \vdots \\ \frac{\partial(x'^m \circ \mathbf{x}^{-1})}{\partial u^1}(\mathbf{x}(p)) & \cdots & \frac{\partial(x'^m \circ \mathbf{x}^{-1})}{\partial u^m}(\mathbf{x}(p)) \end{bmatrix} \text{ and}$$

$$B = [B^i_j] := \begin{bmatrix} \frac{\partial(x^1 \circ \mathbf{x}'^{-1})}{\partial u^1}(\mathbf{x}'(p)) & \cdots & \frac{\partial(x^1 \circ \mathbf{x}'^{-1})}{\partial u^m}(\mathbf{x}'(p)) \\ \vdots & \ddots & \vdots \\ \frac{\partial(x^m \circ \mathbf{x}'^{-1})}{\partial u^1}(\mathbf{x}'(p)) & \cdots & \frac{\partial(x^m \circ \mathbf{x}'^{-1})}{\partial u^m}(\mathbf{x}'(p)) \end{bmatrix},$$

then $AB = I_m = BA$. So $A^i_j B^j_k = \delta^i_k$, and $B^i_j A^j_k = \delta^i_j$.

Exercise 2.20. Let M be a smooth manifold. Let $f, g \in C^\infty(M)$ and $p \in M$. Prove the Leibniz rule $d(f \cdot g)_p = f(p)\, dg_p + g(p)\, df_p$.

2.7 Pull-back of 1-forms

We have seen that the derivative of a smooth map $f : M \to N$ between smooth manifolds M, N can be used to 'push forward' tangent vectors at a point $p \in M$ to tangent vectors at $f(p) \in N$. We now study the 'dual' of this map, $(df_p)^* : (T_{f(p)}N)^* \to (T_pM)^*$, which can be used to 'pull back' 1-forms.

Definition 2.6. (Pull-back of a 1-form.)
Let M and N be smooth manifolds, $f : M \to N$ be a smooth map, and $p \in M$. Define $(df_p)^* : (T_{f(p)}N)^* \to (T_pM)^*$ by $((df_p)^*\omega)(v) = \omega(df_p(v))$ for all $\omega \in (T_{f(p)}N)^*$ and all $v \in T_pM$. The 1-form $(df_p)^*\omega \in (T_pM)^*$ on M is called the *pull-back of* ω *under* f.

First, we note that for $v, \widetilde{v} \in T_pM$ and $c \in \mathbb{R}$,

$$\begin{aligned}((df_p)^*\omega)(v + c\widetilde{v}) &= \omega(df_p(v + c\widetilde{v})) = \omega(df_p(v) + c\,df_p(\widetilde{v})) \\ &= \omega(df_p(v)) + c\omega(df_p(\widetilde{v})) \\ &= ((df_p)^*\omega)(v) + c((df_p)^*\omega)(\widetilde{v}),\end{aligned}$$

showing that $(df_p)^*\omega \in (T_pM)^*$. Moreover, the map $(df_p)^*$ is linear, since if $\omega, \theta \in (T_{f(p)}N)^*$ and $c \in \mathbb{R}$, then for all $v \in T_pM$, we have

$$\begin{aligned}((df_p)^*(\omega + c\theta))(v) &= (\omega + c\theta)(df_p(v)) = \omega(df_p(v)) + c\theta(df_p(v)) \\ &= ((df_p)^*\omega)(v) + c((df_p)^*\theta)(v) \\ &= ((df_p)^*(\omega) + c(df_p)^*(\theta))(v),\end{aligned}$$

and so

$$(df_p)^*(\omega + c\theta) = (df_p)^*(\omega) + c(df_p)^*(\theta).$$

Thus, $(df_p)^* : (T_{f(p)}N)^* \to (T_pM)^*$ is linear.

Summarising, given a smooth map $f : M \to N$,

- tangent vectors can be pushed forward, and
- 1-forms can be pulled back.

Later, we will see how these can be used to define the push-forward of vector fields or pull-back of 1-form fields, from one manifold to another using a map linking the two manifolds.

Exercise 2.21. Let $\mathbb{R}, \mathbb{R}^2$ be equipped with their standard smooth structures, and let $f : \mathbb{R} \to \mathbb{R}^2$ be the smooth map given by $f(t) = (\cos t, \sin t)$, $t \in \mathbb{R}$. Consider the chart $(\mathbb{R}^2, (x, y) \mapsto (x, y))$ on $\mathbb{R}^2$. Let $t \in \mathbb{R}$, $q = f(t) \in \mathbb{R}^2$, and $\omega \in (T_q\mathbb{R}^2)^*$ be the 1-form $\omega = -y(q)(dx)_q + x(q)(dy)_q$. Determine $(df_t)^*\omega \in (T_t\mathbb{R})^*$.

Chapter 3

Vector fields and 1-form fields

In the previous chapter, we defined the notion of a tangent vector. We now want to talk about '(smooth) vector fields'. Intuitively, a vector field on a manifold is thought of as a distribution of tangent vectors on the manifold so that they change 'smoothly' from point to point. The wind velocity on the surface of the Earth is an example of a vector field. But we wish to define vector fields 'intrinsically' (i.e., built using the abstract manifold), and moreover, the vector field should be 'smooth' (to do calculus).

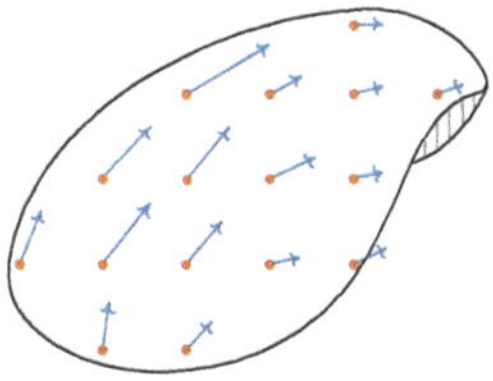

The first hurdle, which is to intrinsically define a vector field on a smooth manifold M, does not seem to be problematic. We want to associate to each point $p \in M$ a tangent vector $V_p \in T_pM$. So the vector field must be a map V with domain M. But what should the target space be? At the point p, we need to pick a tangent vector in T_pM, and not from some other tangent space T_qM with $q \neq p$. Thus, we consider the 'disjoint union' of all the tangent spaces,

$$TM = \bigsqcup_{p \in M} T_pM = \{(p, v) : p \in M \text{ and } v \in T_pM\} = \textstyle\bigcup_{p \in M} (\{p\} \times T_pM).$$

We think of this as a bundle of tangent vectors, aptly called 'tangent bundle'. An obvious surjection is the map $\pi : TM \to M$ given by $\pi((p, v)) = p$, for all $(p, v) \in TM$. Thus every 'vector' in the tangent bundle 'remembers its base point' $p \in M$. Then a vector field is firstly a map $V : M \to TM$ such that $\pi \circ V = \mathrm{id}_M$. This is referred to as a 'section' of the tangent bundle. So the following diagram (on the left) commutes.

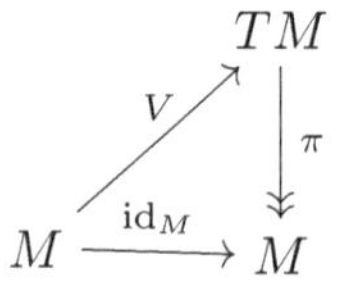

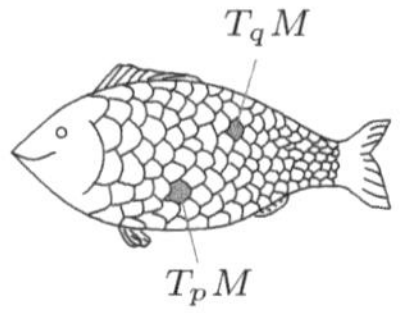

In order to talk about smoothness, we would like make TM a smooth manifold, and then a vector field would be a section from M to TM, which is also a smooth map between these smooth manifolds. Hence, we want to give an atlas for TM, that is, specify a collection of compatible charts on TM which cover TM. Intuitively, given any point $(p, v) \in TM$, we want to assign coordinates to it. Now the point p, being in M, has its natural coordinates $\varphi(p)$, arising from some chart (U, φ) for M that contains p. But this also gives natural coordinates to v, since we can write $v = v(x^i)\,\partial_{x^i,p}$ using the chart-induced basis vectors $\partial_{x^1,p}, \cdots, \partial_{x^m,p}$. So we can give (p, v) the $2m$ coordinates $(\varphi(p), v(x^1), \cdots, v(x^m))$. It turns out that such charts do give an atlas for TM. We will make these intuitive ideas precise in the next section.

In this chapter, we will also define 1-form fields, that is, smooth maps $M \ni p \mapsto \Omega_p \in (T_pM)^*$, by constructing, analogous to the tangent bundle, a cotangent bundle TM^*. These two constructions (of tangent and cotangent bundles) together form the building blocks for tensor fields. Indeed, at the end of this chapter, with TM and TM^* in hand, we will be able to talk about tensor fields in the next chapter.

3.1 Tangent bundle

Definition 3.1. (Tangent bundle.)
Let $(M, [\mathcal{A}])$ be a smooth manifold. Define the set

$$TM = \bigsqcup_{p \in M} T_pM = \{(p, v) : p \in M \text{ and } v \in T_pM\} = \bigcup_{p \in M} (\{p\} \times T_pM).$$

For each chart (U, φ) in an atlas $\mathcal{A}$ belonging to $[\mathcal{A}]$, define $\mathbf{U}$ by

$$\mathbf{U} = \{(p, v) : p \in U \text{ and } v \in T_pM\} = \bigcup_{p \in U} (\{p\} \times T_pM),$$

and $\boldsymbol{\varphi} : \mathbf{U} \to \varphi(U) \times \mathbb{R}^m$ by $\boldsymbol{\varphi}(p, v) = (\varphi(p), v(x^1), \cdots, v(x^m)) \in \varphi(U) \times \mathbb{R}^m$, where the component functions of φ are denoted by $x^1, \cdots, x^m$.

We call $(\mathbf{U}, \boldsymbol{\varphi})$ the *lift* of the chart (U, φ). Let $\boldsymbol{\mathcal{A}}$ be the collection $\{(\mathbf{U}, \boldsymbol{\varphi}) : (U, \varphi) \in \mathcal{A}\}$. Define the *tangent bundle* to be the smooth manifold $(TM, [\boldsymbol{\mathcal{A}}])$. The map $\pi : TM \to M$ sending $(p, v) \in TM$ to $p \in M$, is called the *canonical projection*.

As the sets U from $\mathcal{A}$ cover M, the corresponding sets $\mathbf{U}$ from $\boldsymbol{\mathcal{A}}$ cover TM. Also, as $\varphi : U \to \varphi(U)$ is surjective, and as every m-tuple $(v^1, \cdots, v^m) \in \mathbb{R}^m$ results in a vector $v = v^i \, \partial_{x^i,p} \in T_pM$, $\boldsymbol{\varphi}$ is a surjective map. It is also injective: If $\boldsymbol{\varphi}(p,v) = \boldsymbol{\varphi}(q,w)$, then $\varphi(p) = \varphi(q)$, so that $p = q$, thanks to the injectivity of φ, and moreover, $v = v(x^i)\, \partial_{x^i,p} = w(x^i)\, \partial_{x^i,q} = w$. Thus $\boldsymbol{\varphi} : \mathbf{U} \to \varphi(U) \times \mathbb{R}^m$ is a bijective map onto the open subset $\varphi(U) \times \mathbb{R}^m$ of $\mathbb{R}^{2m}$, and so $(\mathbf{U}, \boldsymbol{\varphi})$ is a chart for TM.

Moreover, if $(\mathbf{U}, \boldsymbol{\varphi}), (\mathbf{V}, \boldsymbol{\psi})$ are lifts of the charts $(U, \varphi), (V, \psi)$ for M, then we have $\boldsymbol{\varphi}(\mathbf{U} \cap \mathbf{V}) = \varphi(U \cap V) \times \mathbb{R}^m$, which is open in $\mathbb{R}^{2m}$.

Let $(\mathbf{U}, \boldsymbol{\varphi})$, $(\mathbf{V}, \boldsymbol{\psi})$ be lifts of admissible charts (U, φ), (V, ψ) for M. We will show that the chart transition maps corresponding to $(\mathbf{U}, \boldsymbol{\varphi})$, $(\mathbf{V}, \boldsymbol{\psi})$ are smooth. Besides completing the verification that $\boldsymbol{\mathcal{A}}$ is an atlas, this will also show that the construction of the smooth structure on TM is independent of the choice of the defining atlas $\mathcal{A}$ from the smooth structure $[\mathcal{A}]$ on M. Let $(\boldsymbol{\alpha}, \boldsymbol{\beta}) \in \boldsymbol{\varphi}(\mathbf{U} \cap \mathbf{V})$. With $\boldsymbol{\beta} = (\beta^1, \cdots, \beta^m)$, $\varphi = (x^1, \cdots, x^m)$ and $\psi = (y^1, \cdots, y^m)$, we have

$$\begin{aligned}(\boldsymbol{\psi} \circ \boldsymbol{\varphi}^{-1})(\boldsymbol{\alpha}, \boldsymbol{\beta}) &= \boldsymbol{\psi}(\varphi^{-1}\boldsymbol{\alpha}, \beta^i \, \partial_{x^i, \varphi^{-1}\boldsymbol{\alpha}}) \\ &= ((\psi \circ \varphi^{-1})(\boldsymbol{\alpha}), \beta^i \, \partial_{x^i,\varphi^{-1}\boldsymbol{\alpha}} y^1, \cdots, \beta^i \, \partial_{x^i,\varphi^{-1}\boldsymbol{\alpha}} y^m).\end{aligned}$$

The map $\varphi(U \cap V) \ni \boldsymbol{\alpha} \mapsto (\psi \circ \varphi^{-1})(\boldsymbol{\alpha})$ is C^∞. Also, the map

$$\boldsymbol{\alpha} \mapsto \partial_{x^i,\varphi^{-1}\boldsymbol{\alpha}} y^j = \frac{\partial (y^j \circ \varphi^{-1})}{\partial u^i}(\varphi(\varphi^{-1}\boldsymbol{\alpha})) = \frac{\partial (\psi \circ \varphi^{-1})^j}{\partial u^i}(\boldsymbol{\alpha})$$

is a C^∞ function (since the chart transition map $\psi \circ \varphi^{-1}$ is C^∞). It follows that the map $(\boldsymbol{\alpha}, \boldsymbol{\beta}) \mapsto ((\psi \circ \varphi^{-1})(\boldsymbol{\alpha}), \beta^i \, \partial_{x^i,\varphi^{-1}(\boldsymbol{\alpha})} y^1, \cdots, \beta^i \, \partial_{x^i,\varphi^{-1}(\boldsymbol{\alpha})} y^m)$ is C^∞, as wanted.

Consequently, TM is a smooth manifold of dimension $2m$, where m is the dimension of the smooth manifold M (and also the vector space dimension of each tangent space T_pM).

Exercise 3.1. Let M be a smooth manifold. Show that the canonical projection $\pi : TM \to M$ is a smooth map.

Exercise 3.2. Let M, N be smooth manifolds and let $f : M \to N$ be a smooth map. Define $\mathbf{df} : TM \to TN$ by $\mathbf{df}(p, v) = (f(p), df_p(v))$ for all $(p, v) \in TM$. Show that $\mathbf{df}$ is smooth.

3.2 Vector fields

Now that we have a tangent bundle with the structure of a smooth manifold, we can talk about its sections which are smooth, and these are vector fields.

Definition 3.2. (Vector field, and its action on a smooth function.) Let M be a smooth manifold. A smooth map $V : M \to TM$ satisfying $\pi \circ V = \mathrm{id}_M$ is called a *vector field.* For $p \in M$, we write $V(p) = (p, V_p)$, where $V_p \in T_pM$. The set of all vector fields on M is denoted[1] by T_0^1M.

If V is a vector field and $f \in C^\infty(M)$, then the function $Vf : M \to \mathbb{R}$ is defined by $(Vf)(p) = V_p(f)$ for all $p \in M$.

The following gives useful criteria for checking if a section is smooth.

Theorem 3.1. *Let M be a smooth manifold, and $V : M \to TM$ satisfy $\pi \circ V = \mathrm{id}_M$. Then the following are equivalent:*
(1) *V is smooth. (So V is a vector field.)*
(2) *For every $f \in C^\infty(M)$, $Vf \in C^\infty(M)$.*
(3) *For all charts $(U, \mathbf{x})$ from an atlas defining the smooth manifold M, the maps given by $U \ni p \mapsto V_p x^i \in \mathbb{R}$, $1 \leqslant i \leqslant m$, belong to $C^\infty(U)$.*

Proof.
(1)⇒(2): Let (U, φ) be an admissible chart for M. For $\boldsymbol{\alpha} \in \varphi(U)$, we have

$$((Vf) \circ \varphi^{-1})(\boldsymbol{\alpha}) = V_{\varphi^{-1}\boldsymbol{\alpha}} f = V_{\varphi^{-1}\boldsymbol{\alpha}}(x^i)\, \partial_{x^i, \varphi^{-1}\boldsymbol{\alpha}} f.$$

But by virtue of the fact that f is smooth, the following map is C^∞:

$$\varphi(U) \ni \boldsymbol{\alpha} \mapsto \partial_{x^i, \varphi^{-1}\boldsymbol{\alpha}} f = \frac{\partial (f \circ \varphi^{-1})}{\partial u^i}(\boldsymbol{\alpha}).$$

Let $(\mathbf{U}, \boldsymbol{\varphi})$ be the lift of (U, φ). As V is smooth, it follows that the map $\boldsymbol{\varphi} \circ V \circ \varphi^{-1} : \varphi(U) \mapsto \mathbb{R}^{2m}$ is C^∞, that is, the map

$$\varphi(U) \ni \boldsymbol{\alpha} \mapsto (\boldsymbol{\alpha}, V_{\varphi^{-1}\boldsymbol{\alpha}}(x^1), \cdots, V_{\varphi^{-1}\boldsymbol{\alpha}}(x^m))$$

is C^∞, and in particular, the projection to any of the last m components, $\boldsymbol{\alpha} \mapsto V_{\varphi^{-1}\boldsymbol{\alpha}}(x^i)$, is C^∞. So $(Vf) \circ \varphi^{-1}$ is C^∞ on $\varphi(U)$. But $\varphi : U \to \varphi(U)$ is a diffeomorphism. Thus $(Vf)|_U$ is smooth for each admissible chart (U, φ) for M. By Exercise 1.18 (p.14), Vf is smooth on M.

(2)⇒(3): Let $p \in U$. Let $\chi \in C^\infty(M)$ be a bump function which is identically equal to 1 in a neighbourhood $U_0 \subset U$ of p, and is 0 outside as slightly bigger neighbourhood U_1, with $\overline{U_1} \subset U$. Then $\chi x^i \in C^\infty(M)$, and so $V(\chi x^i)$ is smooth. In particular, the map given by

$$\varphi(U_0) \ni \boldsymbol{\alpha} \mapsto (V(\chi x^i))(\varphi^{-1}\boldsymbol{\alpha}) = V_{\varphi^{-1}\boldsymbol{\alpha}}(\chi x^i) = V_{\varphi^{-1}\boldsymbol{\alpha}}(x^i)$$

belongs to $C^\infty(\varphi(U_0))$. So $U \ni p \mapsto V_p x^i \in \mathbb{R}$, $1 \leqslant i \leqslant m$, belong to $C^\infty(U)$.

(3)⇒(1): For $\boldsymbol{\alpha} \in \varphi(U)$, $(\boldsymbol{\varphi} \circ V \circ \varphi^{-1})(\boldsymbol{\alpha}) = (\boldsymbol{\alpha}, V_{\varphi^{-1}\boldsymbol{\alpha}}(x^1), \cdots, V_{\varphi^{-1}\boldsymbol{\alpha}}(x^m))$ For $1 \leqslant i \leqslant m$, since $U \ni p \mapsto V_p x^i$ is smooth, $\varphi(U) \ni \boldsymbol{\alpha} \mapsto V_{\varphi^{-1}\boldsymbol{\alpha}}(x^i)$ is C^∞. Hence $\varphi(U) \ni \boldsymbol{\alpha} \mapsto (\boldsymbol{\varphi} \circ V \circ \varphi^{-1})(\boldsymbol{\alpha})$ is C^∞. Consequently, V is smooth. □

[1]The somewhat odd notation will become clear when we discuss tensor fields (Chap. 4).

We will henceforth identify $V(p) = (p, V_p)$ with V_p itself, and so the vector field V sends $p \in M$ to $V_p \in T_pM$.

Clearly, the result of pointwise adding a pair of vector fields is a new vector field, and so is the result of scaling a vector field pointwise. This makes the set T_0^1M of all vector fields into a real vector space.

However, given a vector field V, rather than just producing a uniform scaling everywhere by a factor $c \in \mathbb{R}$, that is, $M \ni p \mapsto c\,V_p \in T_pM$, we can consider a less restrictive possibility of having the c depend on p in a smooth manner. Thus, we want to consider not just multiplication by real scalars, $\cdot : \mathbb{R} \times T_0^1M \to T_0^1M$, but rather, replace $\mathbb{R}$ by $C^\infty(M)$, and have a 'scalar' multiplication with smooth functions, $\cdot : C^\infty(M) \times T_0^1M \to T_0^1M$. The set $C^\infty(M)$ with its pointwise operations is not a field, but only a ring. We will see that scalar multiplication by elements of $C^\infty(M)$ makes T_0^1M a $C^\infty(M)$-module. First, we give the definition of multiplication of arbitrary vector fields by elements of the ring $C^\infty(M)$.

Definition 3.3. (Multiplication of vector fields by smooth functions.)
Let $V \in T_0^1M$ and $f \in C^\infty(M)$. Define $f \cdot V \in T_0^1M$ by $(f \cdot V)(p) = f(p)\,V_p$, for all $p \in M$.

T_0^1M is a module over the ring $C^\infty(M)$. Thus the usual vector space axioms with respect to vector addition hold, with vector addition on T_0^1M defined pointwise: For $V, W \in T_0^1M$, $V + W$ is given by $(V + W)_p = V_p + W_p$, for all $p \in M$. Let $\mathbf{1} \in C^\infty(M)$ denote the constant function taking value 1 everywhere on M. For all $f, g \in C^\infty(M)$, and all $V, W \in T_0^1M$,

$$\mathbf{1}\cdot V = V,\ \ (fg)\cdot V = f\cdot(g\cdot V),\ \ (f+g)\cdot V = f\cdot V + g\cdot V,\ \ f\cdot(V+W) = f\cdot V + f\cdot W.$$

Exercise 3.3. (Vector fields are derivations[2].) Let M be a smooth manifold. Show that $V(fg) = f\cdot(Vg) + g\cdot(Vf)$ for all $f, g \in C^\infty(M)$ and $V \in T_0^1M$.

Example 3.1. (Chart-induced vector fields.)
Let $(U, \mathbf{x})$ be an admissible chart for a smooth manifold with the coordinate functions denoted by x^i, $1 \leqslant i \leqslant m$. Then $(U, [\mathcal{A}])$, where $\mathcal{A} = \{(U, \mathbf{x})\}$, is a smooth manifold. By Theorem 3.1, it follows that for all $1 \leqslant i \leqslant m$, the section $\partial_{x^i} : U \to TU$ given by $p \mapsto \partial_{x^i,p}$, is a vector field $\partial_{x^i} \in T_0^1U$, since $U \ni p \mapsto \partial_{x^i,p}x^j = \delta_i^j$, $1 \leqslant j \leqslant m$, are constant functions (and so smooth). $\diamond$

Lemma 3.1. (Extension of a vector to a vector field.)
Let $v \in T_pM$. Then there exists a vector field $V \in T_0^1M$ such that $V_p = v$.

[2] A *derivation* δ on an algebra A over $\mathbb{R}$ is a linear map $\delta : A \to A$ satisfying the Leibniz rule $\delta(ab) = a\,\delta b + b\,\delta a$ for all $a, b \in A$. An *algebra* A over $\mathbb{R}$ is a real vector space with a bilinear product $\cdot : A \times A \to A$.

Proof. Let $(U, \mathbf{x})$ be an admissible chart containing p. Let $\chi \in C^\infty(M)$ be a bump function that is 1 in a neighbourhood of p, and identically 0 outside a neighbourhood W with $\overline{W} \subset U$. Write $v = v^i \partial_{x^i,p}$, where $\partial_{x^i,p}$ denote the chart-induced tangent vectors, and $v^i := v(x^i)$.

Define the section V by $V_q = \begin{cases} v^i \chi(q) \partial_{x^i,q} & \text{if } q \in U, \\ \mathbf{0} \in T_q M & \text{if } q \in M \backslash U. \end{cases}$

Then $V_p = v^i \, 1 \, \partial_{x^i,p} = v$. By Theorem 3.1, V is smooth. □

Exercise 3.4. Let M be a connected smooth manifold, and $f \in C^\infty(M)$ be such that $Vf = 0$ for all $V \in T_0^1 M$. Show that f is constant.

Exercise 3.5. Let $\mathbb{R}^2, \mathbb{R}$ be given standard smooth structures. We use the global charts $(\mathbb{R}^2, (x,y) \mapsto (x,y))$ and $(\mathbb{R}, u \mapsto u)$. Suppose that $V = xy\,\partial_x + x^2 \partial_y \in T_0^1 \mathbb{R}^2$, $f = x^2 y \in C^\infty(\mathbb{R}^2)$ and $p = (1,1) \in \mathbb{R}^2$. Determine $(fV)_p$, $(Vf)(p)$, $df_p(V_p)$.

3.3 Push forward of vector fields under diffeomorphisms

We know that if $f : M \to N$ is a smooth map, then a vector $v \in T_p M$ can be 'pushed forward' to a vector $df_p(v) \in T_{fp} N$. One might hope then that a whole vector *field* V can also be pushed forward using this mechanism for pushing *vectors* forward. However, we see that there are two problems:

- The map $f : M \to N$ may not be surjective, and so not every point $q \in N$ might be $f(p)$ for some $p \in N$. As the action of df_p only produces vectors at $T_{f(p)} N$, we cannot define a vector field on N, since a vector field on N has to be defined everywhere on N.
- The map $f : M \to N$ may not be injective. This gives rise to the problem that if $f(m_1) = f(m_2)$, then we have a conflict: should the evaluation of the vector field at this point be the push forward of V_{m_1} or of V_{m_2}? They can be different, as shown in the following example.

Example 3.2. (Non-identical push forwards of vectors under a non-injective map.) Let $\mathbb{R}$ be considered as a smooth manifold with the standard smooth structure. Let $f : \mathbb{R} \to \mathbb{R}$ be a non-injective, but surjective function, for example, the cubic $f(x) = (x-1)x(x+1)$. Note that f is surjective, but not injective (e.g. $f(0) = f(1) = 0$). Let $V \in T_0^1 \mathbb{R}$ be the vector field $V = u\,\partial_u$. Then note that $V_0 = \mathbf{0} \in T_0 \mathbb{R}$, and $V_1 = \partial_{u,1} \in T_1 \mathbb{R}$. But $df_0(V_0) \neq df_1(V_1)$ since if we take $g = [u \mapsto u] \in C^\infty(\mathbb{R})$, then we have $(df_0(V_0))g = 0$, while

$$(df_1(V_1))g = V_1(g \circ f) = \frac{d((u-1)u(u+1))}{du}(1) = (3u^2 - 1)|_{u=1} = 2. \qquad \diamond$$

The problems mentioned above disappear if we have a diffeomorphism f, and although a bijective smooth map would suffice, we will use push forwards of vector fields in the context of the 'flow' of a vector field, where we *will* have diffeomorphisms.

Definition 3.4. (Push forward of a vector field under a diffeomorphism.) Let $f : M \to N$ be a diffeomorphism, and $V \in T_0^1 M$. We define the *push forward* $f_* V$ of V under f by $(f_* V)_{f(p)} = df_p(V_p)$ for all $p \in M$.

If $V, W \in T_0^1 M$, and $c \in \mathbb{R}$, then $f_*(V + cW) = f_* V + c f_* W$.

Exercise 3.6. Let $f : M \to N$ be a diffeomorphism, and $V \in T_0^1 M$ be a vector field. Show that $f_* V \in T_0^1 N$.

Exercise 3.7. (Left-invariant vector field on a Lie group, Lie algebra of a Lie group.) Let G be a Lie group with identity element $e \in G$. For $p \in G$, recall (see Exercise 1.24, p.15) the left translation map, $G \ni q \mapsto L_p q = p \cdot q \in G$. A vector field $V \in T_0^1 G$ is called *left-invariant* if $(dL_p)_q V_q = V_{pq}$ for all $p, q \in G$. Show that $V \in T_0^1 G$ is left-invariant if and only if $(L_p)_* V = V$ for all $p \in G$. The set of all left-invariant vector fields is denoted by $\mathfrak{g}$. With pointwise operations, $\mathfrak{g}$ forms a vector space, called the *Lie algebra*[3] *of* G. The aim of this exercise is to show that $T_e G \simeq \mathfrak{g}$ as vector spaces. For a vector $v \in T_e G$, define $\mathbf{I}(v) = V$, where $V : G \to TG$ is defined by $V(p) = (p, (dL_p)_e v)$ for all $p \in G$. It can be shown that the section V is smooth, so that $V \in T_0^1 G$. Show that V is left-invariant. Prove that the linear map $\mathbf{I} : T_e G \to \mathfrak{g}$, sending $v \in T_e G$ to $\mathbf{I} v \in T_0^1 G$, is bijective.

Exercise 3.8. Recall the smooth manifold $\mathrm{SL}_n(\mathbb{R})$ from Exercise 2.16 (p.33). With matrix multiplication, $\mathrm{SL}_n(\mathbb{R})$ forms a group. Using Exercise 1.26 (p.16), it can be shown that $\mathrm{SL}_n(\mathbb{R})$ is a Lie group. Define the vector space

$$\mathfrak{sl}_n(\mathbb{R}) := \{\alpha \in \mathbb{R}^{n \times n} : \operatorname{trace} \alpha = 0\}$$

with the usual matrix operations. The aim of this exercise is to identify $\mathfrak{sl}_n(\mathbb{R})$ as the Lie algebra of $\mathrm{SL}_n(\mathbb{R})$.

(1) If $\alpha \in \mathfrak{sl}_n(\mathbb{R})$, then show that $e^\alpha \in \mathrm{SL}_n(\mathbb{R})$. (For $\alpha \in \mathfrak{sl}_n(\mathbb{R})$, define the curve $\gamma : \mathbb{R} \to \mathrm{SL}_n(\mathbb{R})$ by $\gamma(t) = e^{t\alpha}$, $t \in \mathbb{R}$. Then γ is smooth by Exercise 1.26.)

(2) Show that if $\gamma : (-\epsilon, \epsilon) \to \mathbb{R}^{n \times n}$ ($\epsilon > 0$) is a smooth curve such that $\gamma(0) = I_n$ and $\gamma(t) \in \mathrm{SL}_n(\mathbb{R})$ for all $t \in (-\epsilon, \epsilon)$, then $\dot\gamma(0) \in \mathfrak{sl}_n(\mathbb{R})$.

(3) Let $\mathbb{R}^{n \times n} \simeq \mathbb{R}^{n^2}$ be endowed with the standard smooth structure. Then we can identify the tangent vector $v \in T_{I_n}(\mathbb{R}^{n \times n})$ with a matrix, denoted by $[v]$, using the chart-induced components from the global chart $(\mathbb{R}^{n \times n}, \mathrm{id})$.
In particular, if $\gamma : (-\epsilon, \epsilon) \to \mathbb{R}^{n^2}$ is a smooth curve such that $\gamma(0) = I_n$ and $\det(\gamma(t)) = 1$ for all $t \in (-\epsilon, \epsilon)$, then then $[v_{\gamma, I_n}] = \dot\gamma(0) \in \mathfrak{sl}_n(\mathbb{R})$.
From Exercise 1.25 (p.15), the inclusion map $\iota : \mathrm{SL}_n(\mathbb{R}) \to \mathbb{R}^{n \times n}$ is smooth. Show that $T_{I_n}(\mathrm{SL}_n(\mathbb{R})) \ni v \mapsto [(d\iota)_{I_n}(v)] \in \mathfrak{sl}_n(\mathbb{R})$ is an isomorphism.

Exercise 3.9. Let M, N be smooth manifolds, and $f : M \to N$ be a diffeomorphism. Let $V \in T_0^1 M$ and $\varphi \in C^\infty(N)$. Show that $((f_* V)\varphi) \circ f = V(\varphi \circ f)$.

Exercise 3.10. Consider $\mathbb{R}^2$ with the standard smooth structure, and the global chart $(\mathbb{R}^2, (x, y) \mapsto (x, y))$. For $t \in \mathbb{R}$, let $\psi_{-t} : \mathbb{R}^2 \to \mathbb{R}^2$ be the diffeomorphism given by $(x, y) \mapsto (x \cos t - y \sin t, x \sin t + y \cos t)$. Let $V = \partial_y$. Determine $(\psi_{-t})_* V$.

---

[3]The term 'algebra' (rather than 'vector space') is used because it turns out that the Lie bracket defines a multiplication on $\mathfrak{g}$; see Exercise 3.12, p.45.

3.4 Lie algebra of vector fields

In this section, we will equip T_0^1M with a (anti-commutative) 'multiplication' operation, called the 'Lie bracket', which assigns to two vector fields V, W, a new vector field $[V, W]$. With $[\cdot, \cdot]$, the real vector space T_0^1M forms an algebra over $\mathbb{R}$. We will see later on that there is a geometric motivation[4] for the Lie bracket. The Lie bracket will play a role later while discussing symmetries of spacetime.

If $V, W \in T_0^1M$ and $f \in C^\infty(M)$, then $Wf \in C^\infty(M)$, and so for each $p \in M$, V_p can act on it, yielding a real number $V_p(Wf)$. It is clear that $f \mapsto V_p(Wf)$ is linear, but it turns out the Leibniz rule fails in general, preventing this map from being a vector in T_pM. However, if we subtract a corresponding product in the other order, namely $W_p(Vf)$, then the problematic terms cancel each other, and give rise to a vector in T_pM. If we now allow the point p to vary in M, then we obtain a vector field on M.

Definition 3.5. (Lie bracket.)
Let M be a smooth manifold. Define the *Lie bracket of* $V, W \in T_0^1M$ to be the vector field $[V, W] \in T_0^1M$ given by $[V, W]_p(f) = V_p(Wf) - W_p(Vf)$ for all $f \in C^\infty(M)$ and all $p \in M$.

We now show that the Lie bracket of vector fields is indeed a vector field.

Lemma 3.2. *If* $V, W \in T_0^1M$, *then* $[V, W] \in T_0^1M$.

Proof. First, we show that at each $p \in M$, $[V, W]_p \in T_pM$, so that $[V, W]$ is at least a section, and then show that it is also smooth using Theorem 3.1 (p.40). The map $[V, W]_p : C^\infty(M) \to \mathbb{R}$ is linear: for $f, g \in C^\infty(M)$, $c \in \mathbb{R}$,

$$\begin{aligned} [V, W]_p(f+cg) &= V_p(W(f+cg)) - W_p(V(f+cg)) \\ &= V_p(Wf+cWg) - W_p(Vf+cVg) \\ &= V_p(Wf) - W_p(Vf) + c(V_p(Wg) - W_p(Vg)) \\ &= [V, W]_pf + c[V, W]_pg. \end{aligned}$$

To show the Leibniz rule, we have, for $f, g \in C^\infty(M)$,

$$\begin{aligned} [V, W]_p(fg) &= V_p(W(fg)) - W_p(V(fg)) \\ &= V_p(g(Wf) + f(Wg)) - W_p(g(Vf) + f(Vg)) \\ &= g(p)V_p(Wf) + \cancel{(Wf)(p)V_pg} + f(p)V_p(Wg) + \cancel{(Wg)(p)V_pf} \\ &\quad - g(p)W_p(Vf) - \cancel{(Vf)(p)W_pg} - f(p)W_p(Vg) - \cancel{(Vg)(p)W_pf} \\ &= g(p)[V, W]_pf + f(p)[V, W]_pg. \end{aligned}$$

[4]Propositions 3.1, 3.2, 6.6, Theorem 6.3.

Finally, we show smoothness of $p \mapsto [V,W]_p$. If $f \in C^\infty(M)$, then as V, W are vector fields, Theorem 3.1 implies $Wf, Vf \in C^\infty(M)$, and moreover $V(Wf), W(Vf) \in C^\infty(M)$ too. So $[V,W]f = W(Vf) - V(Wf) \in C^\infty(M)$ for all $f \in C^\infty(M)$, and again by Theorem 3.1, it follows that $[V,W]$ is a vector field. □

Now we show that the Lie bracket of chart-induced vector fields vanishes.

Lemma 3.3. *Let $(U, \mathbf{x})$ be an admissible chart for a smooth manifold, and let $\partial_{x^i} \in T_0^1 U$, $1 \leqslant i \leqslant m$, be the coordinate induced vector fields on U. Then $[\partial_{x^i}, \partial_{x^j}] = 0$ for all $1 \leqslant i, j \leqslant m$.*

Proof. For $f \in C^\infty(U)$, we have

$$\begin{aligned}
\partial_{x^i}(\partial_{x^j} f) &= \partial_{x^i}\big(p \mapsto \frac{\partial (f \circ \mathbf{x}^{-1})}{\partial u^j}(\mathbf{x}(p))\big) \\
&= \big(p \mapsto \frac{\partial}{\partial u^i}\big(\frac{\partial (f \circ \mathbf{x}^{-1})}{\partial u^j} \circ \mathbf{x} \circ \mathbf{x}^{-1}\big)(\mathbf{x}(p))\big) \\
&= \big(p \mapsto \frac{\partial^2 (f \circ \mathbf{x}^{-1})}{\partial u^i \partial u^j}(\mathbf{x}(p))\big) \\
&= \big(\frac{\partial^2 (f \circ \mathbf{x}^{-1})}{\partial u^i \partial u^j}\big) \circ \mathbf{x}.
\end{aligned}$$

$\partial_{x^j}(\partial_{x^i} f)$ is obtained by the swap $i \leftrightarrow j$. As $f \circ \mathbf{x}^{-1}$ is C^∞, thus also C^2, its second order partials do not depend on the order of taking partial derivatives (see e.g. [Apostol (1969), §8.23]). So

$$\frac{\partial^2 (f \circ \mathbf{x}^{-1})}{\partial u^i \partial u^j} = \frac{\partial^2 (f \circ \mathbf{x}^{-1})}{\partial u^j \partial u^i}.$$

Consequently, $[\partial_{x^i}, \partial_{x^j}] = 0$. □

Exercise 3.11. Let M, N be smooth manifolds, $f : M \to N$ be a diffeomorphism, and $X, Y \in T_0^1 M$. Show that $f_*[X,Y] = [f_* X, f_* Y]$.

Exercise 3.12. Let G be a Lie group, and $\mathfrak{g}$ its Lie algebra.
Show that if $V, W \in \mathfrak{g}$, then $[V,W] \in \mathfrak{g}$.

Exercise 3.13. Let M be a smooth manifold, $X, Y \in T_0^1 M$ and $f \in C^\infty(M)$.
Show that $[fX, Y] = f[X,Y] - (Yf)X$.

Consider the smooth manifold $\mathbb{R}^2$ with the standard smooth structure.
Let ∂_x, ∂_y be the vector fields induced by the global chart $(\mathbb{R}^2, (x,y) \mapsto (x,y))$.
Show that $[(1+y)\partial_x, \partial_y] = -\partial_x$.

Exercise 3.14. Suppose that M is a smooth manifold, $(U, \mathbf{x})$ is an admissible chart for M, and $\partial_{x^1}, \cdots, \partial_{x^m} \in T_0^1 U$ are the chart-induced vector fields on U. Let $V = V^i \partial_{x^i}$ and $W = W^j \partial_{x^j}$, where $V^i, W^j \in C^\infty(U)$. Show that

$$[V,W] = (V^i \partial_{x^i} W^j - W^i \partial_{x^i} V^j)\,\partial_{x^j}.$$

Exercise 3.15. The general linear group

$$\mathrm{GL}_n(\mathbb{R}) = \{\alpha \in \mathbb{R}^{n\times n} : \det \alpha \neq 0\} = \mathbb{R}^{n\times n}\backslash(\det^{-1}\{0\})$$

is an open subset of $\mathbb{R}^{n\times n} \simeq \mathbb{R}^{n^2}$, with the smooth structure induced from the standard smooth structure of $\mathbb{R}^{n^2}$. As matrix multiplication and inversion are smooth operations, it forms a Lie group. Using the chart $(\mathrm{GL}_n(\mathbb{R}), \mathrm{id})$, $T_{I_n}(\mathrm{GL}_n(\mathbb{R}))$ can be identified with $\mathbb{R}^{n\times n}$: For $1 \leqslant i, j \leqslant n$, let $x^{ij} := ([\alpha^{ij}] \mapsto \alpha^{ij})$ denote the component functions of id. Then $v = v^{ij}\partial_{x^{ij}} \in T_{I_n}(\mathrm{GL}_n(\mathbb{R}))$ corresponds to the matrix $[v] := [v^{ij}] \in \mathbb{R}^{n\times n}$. For $v \in T_{I_n}(\mathrm{GL}_n(\mathbb{R}))$, define the left-invariant vector field V extending v by $V_p = (dL_p)_{I_n}(v)$ for all $p \in \mathrm{GL}_n(\mathbb{R})$ (see Exercise 3.7, p.43). Show that if V, W are left-invariant vector fields extending $v, w \in T_{I_n}(\mathrm{GL}_n(\mathbb{R}))$, then $[V, W]_{I_n} = \alpha^{ij}\partial_{x^{ij}}$, where $[\alpha^{ij}]$ is the matrix commutator $[v, w] := [v][w] - [w][v]$.

Theorem 3.2. *Let M be a smooth manifold, $X, Y, Z \in T_0^1 M$, and $c \in \mathbb{R}$. Then the following hold*:
(1) $[X + cY, Z] = [X, Z] + c[Y, Z]$.
(2) $[X, Y] = -[Y, X]$.
(3) (Jacobi identity) $[X, [Y, Z]] + [Y, [Z, X]] + [Z, [X, Y]] = 0$.

Proof. (1), (2) follow immediately from the definition. We verify (3) below. Each term on the left-hand side gives 4 terms, and altogether we get twelve terms, which will cancel in pairs, thanks to the cyclicity of the expression on the left-hand side: For $f \in C^\infty(M)$,

$$\begin{aligned}[X,[Y,Z]]f &= X([Y,Z]f) - [Y,Z](Xf)\\ &= X(Y(Zf) - Z(Yf)) - Y(Z(Xf)) + Z(Y(Xf))\\ &= X(Y(Zf)) - X(Z(Yf)) - Y(Z(Xf)) + Z(Y(Xf)).\end{aligned}$$

By cyclically permuting the vector fields, we can read off that

$$\begin{aligned}[Y,[Z,X]]f &= Y(Z(Xf)) - Y(X(Zf)) - Z(X(Yf)) + X(Z(Yf)),\\ [Z,[X,Y]]f &= Z(X(Yf)) - Z(Y(Xf)) - X(Y(Zf)) + Y(X(Zf)).\end{aligned}$$

Adding these, we see that the result is the zero function. □

3.5 Cotangent bundle

Definition 3.6. (Cotangent bundle.)
Let $(M, [\mathcal{A}])$ be a smooth manifold. Define the set

$$TM^* = \bigsqcup_{p\in M} (T_pM)^* = \{(p,\omega) : p \in M, \text{ and } \omega \in (T_pM)^*\} = \bigcup_{p\in M} (\{p\} \times (T_pM)^*).$$

For each chart (U, φ) in an atlas $\mathcal{A}$ belonging to $[\mathcal{A}]$, define $\mathbf{U}$ by

$$\mathbf{U} = \{(p,\omega) : p \in U,\ \omega \in (T_pM)^*\} = \bigcup_{p\in U} (\{p\} \times (T_pM)^*),$$

and $\boldsymbol{\varphi} : \mathbf{U} \to \varphi(U) \times \mathbb{R}^m$ by $\boldsymbol{\varphi}(p,\omega) = (\varphi(p), \omega(\partial_{x^1,p}), \cdots, \omega(\partial_{x^m,p})) \in \varphi(U) \times \mathbb{R}^m$, where the component functions of φ are denoted by $x^1, \cdots, x^m$.

We call $(\mathbf{U}, \boldsymbol{\varphi})$ the *lift* of the chart (U, φ). Let $\boldsymbol{\mathcal{A}} := \{(\mathbf{U}, \boldsymbol{\varphi}) : (U, \varphi) \in \mathcal{A}\}$. Define the *cotangent bundle* to be the smooth manifold $(TM^*, [\boldsymbol{\mathcal{A}}])$.

The map $\pi : TM^* \to M$, $TM^* \ni (p, \omega) \mapsto p \in M$, is called the *canonical projection.*

Since the sets U from $\mathcal{A}$ cover M, the corresponding $\mathbf{U}$ from $\boldsymbol{\mathcal{A}}$ cover TM^*. Also, as $\varphi : U \to \varphi(U)$ is surjective, and because every m-tuple $(\omega_1, \cdots, \omega_m) \in \mathbb{R}^m$ results in a 1-form $\omega = \omega_i (dx^i)_p \in (T_pM)^*$, $\boldsymbol{\varphi}$ is a surjective map. The map $\boldsymbol{\varphi}$ is also injective, since if $\boldsymbol{\varphi}(p, \omega) = \boldsymbol{\varphi}(q, \nu)$, then we have $\varphi(p) = \varphi(q)$, so that $p = q$, thanks to the injectivity of φ, and moreover, we have $\omega = \omega(\partial_{x^i,p})(dx^i)_p = \nu(\partial_{x^i,q})(dx^i)_q = \nu$. Hence, $\boldsymbol{\varphi} : \mathbf{U} \to \varphi(U) \times \mathbb{R}^m$ is a bijective map onto the open subset $\varphi(U) \times \mathbb{R}^m$ of $\mathbb{R}^{2m}$, and so $(\mathbf{U}, \boldsymbol{\varphi})$ is a chart for TM.

Moreover, if $(\mathbf{U}, \boldsymbol{\varphi}), (\mathbf{V}, \boldsymbol{\psi})$ are the lifts of the charts $(U, \varphi), (V, \psi)$ for M, then we have $\boldsymbol{\varphi}(\mathbf{U} \cap \mathbf{V}) = \varphi(U \cap V) \times \mathbb{R}^m$, which is open in $\mathbb{R}^{2m}$.

Let $(\mathbf{U}, \boldsymbol{\varphi})$ and $(\mathbf{V}, \boldsymbol{\psi})$ be the lifts of admissible charts (U, φ) and (V, ψ) for M. We will show that the chart transition maps corresponding to $(\mathbf{U}, \boldsymbol{\varphi}), (\mathbf{V}, \boldsymbol{\psi})$ are smooth. Besides completing the verification that $\boldsymbol{\mathcal{A}}$ is an atlas, this will also show that the construction of the smooth structure on TM^* is independent of the choice of the defining atlas $\mathcal{A}$ from the smooth structure $[\mathcal{A}]$ for M. Let the point $(\boldsymbol{\alpha}, \boldsymbol{\beta}) \in \boldsymbol{\varphi}(\mathbf{U} \cap \mathbf{V})$. With $\boldsymbol{\beta} = (\beta_1, \cdots, \beta_m)$, $\varphi = (x^1, \cdots, x^m)$ and $\psi = (y^1, \cdots, y^m)$, we have

$$
\begin{aligned}
&(\boldsymbol{\psi} \circ \boldsymbol{\varphi}^{-1})(\boldsymbol{\alpha}, \boldsymbol{\beta}) = \boldsymbol{\psi}(\varphi^{-1}\boldsymbol{\alpha}, \beta_i (dx^i)_{\varphi^{-1}\boldsymbol{\alpha}}) \\
&= ((\psi \circ \varphi^{-1})(\boldsymbol{\alpha}), \beta_i (dx^i)_{\varphi^{-1}\boldsymbol{\alpha}}(\partial_{y^1,\varphi^{-1}\boldsymbol{\alpha}}), \cdots, \beta_i (dx^i)_{\varphi^{-1}\boldsymbol{\alpha}}(\partial_{y^m,\varphi^{-1}\boldsymbol{\alpha}})).
\end{aligned}
$$

The map $U \cap V \ni \boldsymbol{\alpha} \mapsto (\varphi \circ \psi^{-1})(\boldsymbol{\alpha})$ is C^∞. Also,

$$
\begin{aligned}
(dx^i)_{\varphi^{-1}\boldsymbol{\alpha}}(\partial_{y^j,\varphi^{-1}\boldsymbol{\alpha}}) = \partial_{y^j,\varphi^{-1}\boldsymbol{\alpha}}(x^i) &= \frac{\partial(x^i \circ \psi^{-1})}{\partial u^j}(\psi(\varphi^{-1}\boldsymbol{\alpha})) \\
&= \frac{\partial(\varphi \circ \psi^{-1})^i}{\partial u^j}((\psi \circ \varphi^{-1})(\boldsymbol{\alpha})) \\
&= \Big(\Big(\frac{\partial(\varphi \circ \psi^{-1})^i}{\partial u^j}\Big) \circ (\psi \circ \varphi^{-1})\Big)(\boldsymbol{\alpha}).
\end{aligned}
$$

Hence the map sending $\boldsymbol{\alpha}$ to $(dx^i)_{\varphi^{-1}\boldsymbol{\alpha}}(\partial_{y^j,\varphi^{-1}\boldsymbol{\alpha}})$ is a C^∞ function. Thus

$$
(\boldsymbol{\alpha}, \boldsymbol{\beta}) \mapsto ((\psi \circ \varphi^{-1})(\boldsymbol{\alpha}), \beta_i (dx^i)_{\varphi^{-1}\boldsymbol{\alpha}}(\partial_{y^1,\varphi^{-1}\boldsymbol{\alpha}}), \cdots, \beta_i (dx^i)_{\varphi^{-1}\boldsymbol{\alpha}}(\partial_{y^m,\varphi^{-1}\boldsymbol{\alpha}}))
$$

is C^∞, as wanted.

Consequently, TM^* is a smooth manifold of dimension $2m$, where m is the dimension of the smooth manifold (and also the dimension of each cotangent space $(T_pM)^*$).

Exercise 3.16. Let M be a smooth manifold. Show that the canonical projection $\pi : TM^* \to M$ is a smooth map.

3.6 1-form fields

In this section, we will introduce the 'dual' object to a vector field, namely a '1-form field', which is, roughly speaking, a smooth distribution of 1-forms on the manifold. We will see that, just like the $C^\infty(M)$-module $T_0^1 M$ of vector fields, the collection $T_1^0 M$ of all 1-form fields will also have the structure of a $C^\infty(M)$-module. Together, $T_0^1 M$ and $T_1^0 M$ form the building blocks for (r, s)-tensor fields (which will be '$C^\infty(M)$-multilinear maps' from r copies of $T_1^0 M$ and s copies of $T_0^1 M$ to $C^\infty(M)$, as we will learn in the next chapter). These tensor fields form the vocabulary to discuss geometric notions in Lorentzian geometry, such as the metric $\mathbf{g}$ (a $(0,2)$-tensor field), the curvature tensor field $\mathbf{R}$ (a $(1,3)$-tensor field), etc. The field equation for spacetime is an equality between two $(0,2)$-tensor fields.

Recall that in the previous section, using the smooth manifold structure on M, we made TM^* a smooth manifold by gluing together the cotangent spaces smoothly. Hence we can talk about smooth 'TM^*-sections', which are called 1-form fields.

Definition 3.7. (1-form field.)
Let M be a smooth manifold, and TM^* be its cotangent bundle.
A 1-*form field* is a smooth map $\Omega : M \to (TM)^*$ such that $\pi \circ \Omega = \mathrm{id}_M$.
For $p \in M$, we write $\Omega(p) = (p, \Omega_p)$, where $\Omega_p \in (T_pM)^*$.
The set of all 1-form fields on M is denoted by $T_1^0 M$.

Definition 3.8. (1-form field action on a vector field.)
Let $V \in T_0^1 M$ be a vector field and $\Omega \in T_1^0 M$ a 1-form field. Then the function $\Omega V : M \to \mathbb{R}$ is defined by $(\Omega V)(p) = \Omega_p(V_p)$ for all $p \in M$.

Exercise 3.17. Let $\Omega \in T_1^0 M$. Show that the map $T_0^1 M \ni V \mapsto \Omega V \in C^\infty(M)$ is a $C^\infty(M)$-linear map: for all vector fields $V, W \in T_0^1 M$ and all functions $f \in C^\infty(M)$, we have $\Omega(V + fW) = \Omega V + f \Omega W$.

The following result gives useful criteria for checking if a TM^*-section is smooth.

Theorem 3.3. *Let M be a smooth manifold, and $\Omega : M \to TM^*$ satisfy $\pi \circ \Omega = \mathrm{id}_M$. Then the following are equivalent:*

(1) *Ω is smooth. (So $\Omega \in T_1^0 M$.)*
(2) *For every $V \in T_0^1 M$, $\Omega V \in C^\infty(M)$.*
(3) *For all charts $(U, \mathbf{x})$ from an atlas defining the smooth manifold M, the maps given by $U \ni p \mapsto \Omega_p(\partial_{x^i,p}) \in \mathbb{R}$, $1 \leqslant i \leqslant m$, belong to $C^\infty(U)$.*

Proof.
(1)⇒(2): Let (U, φ) be an admissible chart for M. For $\boldsymbol{\alpha} \in \varphi(U)$, we have

$$\begin{aligned}((\Omega V) \circ \varphi^{-1})\boldsymbol{\alpha} &= \Omega_{\varphi^{-1}\boldsymbol{\alpha}} V_{\varphi^{-1}\boldsymbol{\alpha}} = \Omega_{\varphi^{-1}\boldsymbol{\alpha}}(V_{\varphi^{-1}\boldsymbol{\alpha}}(x^i)\partial_{x^i,\varphi^{-1}\boldsymbol{\alpha}}) \\ &= V_{\varphi^{-1}\boldsymbol{\alpha}}(x^i)\,\Omega_{\varphi^{-1}\boldsymbol{\alpha}}\partial_{x^i,\varphi^{-1}\boldsymbol{\alpha}}.\end{aligned}$$

As V is smooth, $\boldsymbol{\alpha} \mapsto V_{\varphi^{-1}\boldsymbol{\alpha}}(x^i)$ is C^∞ on $\varphi(U)$. Also, since Ω is smooth, $\boldsymbol{\varphi} \circ \Omega \circ \varphi^{-1}$ is C^∞ on $\varphi(U)$, that is,

$$\varphi(U) \ni \boldsymbol{\alpha} \mapsto (\boldsymbol{\alpha}, \Omega_{\varphi^{-1}\boldsymbol{\alpha}}(\partial_{x^1,\varphi^{-1}\boldsymbol{\alpha}}), \cdots, \Omega_{\varphi^{-1}\boldsymbol{\alpha}}(\partial_{x^m,\varphi^{-1}\boldsymbol{\alpha}}))$$

is C^∞. So the map $\varphi(U) \ni \boldsymbol{\alpha} \mapsto \Omega_{\varphi^{-1}\boldsymbol{\alpha}}\partial_{x^i,\varphi^{-1}\boldsymbol{\alpha}}$ is C^∞ for all i. Thus the map $\varphi(U) \ni \boldsymbol{\alpha} \mapsto ((\Omega V) \circ \varphi^{-1})\boldsymbol{\alpha}$ is C^∞. Hence $\Omega V \in C^\infty(M)$.

(2)⇒(3): Let $p \in U$. Let $\chi \in C^\infty(M)$ be a bump function which is identically 1 in a neighbourhood $U_0 \subset U$ of p, and is 0 outside a slightly bigger neighbourhood U_1 with $\overline{U_1} \subset U$. Then we have that $\chi\, \partial_{x^i} \in T_0^1 M$, and so $\Omega(\chi\,\partial_{x^i})$ is smooth. Hence the map

$$\begin{aligned}\varphi(U_0) \ni \boldsymbol{\alpha} \mapsto (\Omega(\chi\,\partial_{x^i}))(\varphi^{-1}\boldsymbol{\alpha}) = \Omega_{\varphi^{-1}\boldsymbol{\alpha}}(\chi\,\partial_{x^i})_{\varphi^{-1}\boldsymbol{\alpha}} &= \Omega_{\varphi^{-1}\boldsymbol{\alpha}}(1\,\partial_{x^i,\varphi^{-1}\boldsymbol{\alpha}}) \\ &= \Omega_{\varphi^{-1}\boldsymbol{\alpha}}\partial_{x^i,\varphi^{-1}\boldsymbol{\alpha}}\end{aligned}$$

belongs to $C^\infty(\varphi(U_0))$. So $U \ni p \mapsto \Omega_p(\partial_{x^i,p})$, $1 \leqslant i \leqslant m$, belong to $C^\infty(U)$.

(3)⇒(1): For $\boldsymbol{\alpha} \in \varphi(U)$,

$$(\boldsymbol{\varphi} \circ \Omega \circ \varphi^{-1})\boldsymbol{\alpha} = (\boldsymbol{\alpha}, \Omega_{\varphi^{-1}\boldsymbol{\alpha}}(\partial_{x^1,\varphi^{-1}\boldsymbol{\alpha}}), \cdots, \Omega_{\varphi^{-1}\boldsymbol{\alpha}}(\partial_{x^m,\varphi^{-1}\boldsymbol{\alpha}})).$$

As $U \ni p \mapsto \Omega_p(\partial_{x^i,p})$ are smooth, the maps $\varphi(U) \ni \boldsymbol{\alpha} \mapsto \Omega_{\varphi^{-1}\boldsymbol{\alpha}}(\partial_{x^i,\varphi^{-1}\boldsymbol{\alpha}})$, $1 \leqslant i \leqslant m$, are all C^∞. Hence $\varphi(U) \ni \boldsymbol{\alpha} \mapsto (\boldsymbol{\varphi} \circ \Omega \circ \varphi^{-1})(\boldsymbol{\alpha})$ is C^∞. So Ω is smooth. □

Exercise 3.18. (Gradient field.) Let M be a smooth manifold, and $f \in C^\infty(M)$. Define $df : M \to TM^*$ by $(df)(p) = (p, df_p)$, for $p \in M$, where $df_p \in (T_pM)^*$ is the 1-form defined in Example 2.5. Prove that $df \in T_1^0 M$.

We identify $\Omega(p) = (p, \Omega_p)$ with $\Omega_p \in (T_pM)^*$, and so the 1-form field Ω sends points $p \in M$ to corresponding 1-forms $\Omega_p \in (T_pM)^*$.

Clearly, the result of pointwise adding a pair of 1-form fields is a new 1-form field, and so is the result of scaling a 1-form field pointwise. This makes the set $T_1^0 M$ of all 1-form fields into a vector space. However, given a 1-form field Ω, rather than just uniformly producing a scaling with a factor $c \in \mathbb{R}$, that is, $M \ni p \mapsto c\,\Omega_p \in (T_pM)^*$, we can consider a less restrictive possibility of having the c depend on p in a smooth manner. We will see that this makes $T_1^0 M$ into a $C^\infty(M)$-module. First, we give the definition of multiplication of arbitrary 1-form fields by elements of the ring $C^\infty(M)$.

Definition 3.9. (Multiplication of 1-form fields by smooth functions.)
Let M be a smooth manifold. Let $\Omega \in T_1^0 M$ and $f \in C^\infty(M)$.
Define $f \cdot \Omega \in T_1^0 M$ by $(f \cdot \Omega)(p) = f(p)\Omega_p$ for all $p \in M$.

$T_1^0 M$ is a module over the ring $C^\infty(M)$. In particular, besides the usual vector space axioms with respect to vector addition in $T_1^0 M$, we have that for all $f, g \in C^\infty(M)$, and all $\Omega, \Theta \in T_1^0 M$, the following hold:
$\mathbf{1} \cdot \Omega = \Omega$, $(fg) \cdot \Omega = f \cdot (g \cdot \Omega)$, $(f+g) \cdot \Omega = f \cdot \Omega + g \cdot \Omega$, $f \cdot (\Omega + \Theta) = f \cdot \Omega + f \cdot \Theta$.
Here $\mathbf{1} \in C^\infty(M)$ denotes the constant function equal to 1 everywhere.

Example 3.3. (Chart-induced 1-form fields.)
Let $(U, \mathbf{x})$ be an admissible chart for a smooth manifold with the coordinate functions denoted by x^i, $1 \leqslant i \leqslant m$. Then $(U, [\mathcal{A}])$, where $\mathcal{A} = \{(U, \mathbf{x})\}$, is a smooth manifold. Since $(dx^i)_p \partial_{x^j,p} = \delta_j^i$, Theorem 3.3 implies that $U \ni p \mapsto (dx^i)_p$, $1 \leqslant i \leqslant m$, are all elements of $T_1^0 U$. ◇

Lemma 3.4. (Extension of a 1-form to a 1-form field.)
Let M be a smooth manifold. Let $\omega \in (T_p M)^*$. Then there exists a 1-form field $\Omega \in T_0^1 M$ such that $\Omega_p = \omega$.

Proof. Let $(U, \mathbf{x})$ be an admissible chart containing p. Let $\chi \in C^\infty(M)$ be a bump function that is identically 1 in a neighbourhood of p, and identically 0 outside a neighbourhood V with $\overline{V} \subset U$. Write $\omega = \omega_i (dx^i)_p$. Define the TM^*-section Ω by

$$\Omega_q = \begin{cases} \omega_i \chi(q)(dx^i)_q & \text{if } q \in U, \\ 0 \in (T_q M)^* & \text{if } q \in M \backslash U. \end{cases}$$

Then $\Omega_p = \omega_i 1 (dx^i)_p = \omega$. By Theorem 3.3, Ω is smooth. □

Exercise 3.19. (A 1-form field which is not a gradient field.)
Consider $\mathbb{R}^2$ with the standard smooth structure. Show that not every 1-form field Ω on $\mathbb{R}^2$ is df for some $f \in C^\infty(\mathbb{R}^2)$ as follows. If $\Omega = df$ for some $f \in C^\infty(\mathbb{R}^2)$, then in the chart $(\mathbb{R}^2, (x, y) \mapsto (x, y))$, writing $\Omega = (\partial_x f)dx + (\partial_y f)dy$, we get $\Omega \partial_x = \partial_x f$ and $\Omega \partial_y = \partial_y f$, implying $\partial_y(\Omega \partial_x) = \partial_x(\Omega \partial_y)$. Construct a 1-form field Ω, which is not a gradient field, by choosing $\Omega \partial_x, \Omega \partial_y$ appropriately.

3.7 Pull-back of a 1-form field

We had seen that given a diffeomorphism $f : M \to N$ between two smooth manifolds M and N, any vector field $V \in T_0^1 M$ can be pushed forward to a vector field $f_* V \in T_0^1 N$. We had also seen that a 1-form ω at $f(p)$ can be pulled back to a 1-form $(df_p)^* \omega$ at $p \in M$. Now we will see that we can in fact pull back a whole 1-form field in this manner, and what we need from f is that it only be smooth (so f need not be a diffeomorphism).

Definition 3.10. (Pull-back of a 1-form field.)
Let M, N be a smooth manifolds, and $f : M \to N$ be a smooth map. The *pull-back* $f^*\Omega \in T_1^0 M$ *of a* 1*-form field* $\Omega \in T_1^0 N$ *under* f is defined by $(f^*\Omega)_p = (df_p)^*\Omega_{f(p)}$ for all $p \in M$.

To check that this does define a 1-form field on M, it suffices to prove that for every $V \in T_0^1 M$, $(f^*\Omega)V \in C^\infty(M)$. We check this locally around a point p by making use of a admissible charts $(U, \mathbf{x})$, $(W, \mathbf{y})$, for M, N, respectively, containing p, $f(p)$, respectively, and such that $f(U) \subset W$. For all $q \in U$, writing $V_q = V_q(x^i)\,\partial_{x^i,q}$ and $\Omega_{f(q)} = \Omega_{f(q)}(\partial_{y^j,f(q)})\,(dy^j)_{f(q)}$, we have

$$\begin{aligned}((f^*\Omega)V)(q) &= ((df_q)^*\Omega_{f(q)})V_q = \Omega_{f(q)}(df_q(V_q)) \\ &= \Omega_{f(q)}(\partial_{y^j,f(q)})(dy^j)_{f(q)}(df_q(V_q(x^i)\,\partial_{x^i,q})) \\ &= \Omega_{f(q)}(\partial_{y^j,f(q)}) \cdot (dy^j)_{f(q)}(V_q(x^i) \cdot df_q(\partial_{x^i,q})) \\ &= \Omega_{f(q)}(\partial_{y^j,f(q)}) \cdot V_q(x^i) \cdot (dy^j)_{f(q)}(((df_q)(\partial_{x^i,q}))(y^k)\,\partial_{y^k,f(q)}) \\ &= \Omega_{f(q)}(\partial_{y^j,f(q)}) \cdot V_q(x^i) \cdot (dy^j)_{f(q)}(\partial_{x^i,q}(y^k \circ f) \cdot \partial_{y^k,f(q)}) \\ &= \Omega_{f(q)}(\partial_{y^j,f(q)}) \cdot V_q(x^i) \cdot \partial_{x^i,q}(y^j \circ f).\end{aligned}$$

Since the maps $U \ni q \mapsto \Omega_{f(q)}\partial_{y^j,f(q)}$, $V_q(x^i)$, $\partial_{x^i,q}(y^j \circ f)$ are smooth, so is their pointwise product.

Exercise 3.20. Let M, N be smooth manifolds, $f : M \to N$ be a smooth map, $g \in C^\infty(N)$, and $\Omega \in T_1^0 N$. Prove that $f^*(g \cdot \Omega) = (g \circ f) \cdot (f^*\Omega)$.

Exercise 3.21. Let M, N be smooth manifolds, $f : M \to N$ be a smooth map, and $g \in C^\infty(N)$. Prove that $f^*(dg) = d(g \circ f)$.

The previous two exercises give a recipe for the computation of the pull-back in any chart. Let $p \in M$, and $(U, \mathbf{x})$, $(V, \mathbf{y})$ be admissible charts for M, N, respectively, containing p, $f(p)$, and such that $f(U) \subset V$. If $\Omega \in T_1^0 N$ and we define $\Omega_j = \Omega(\partial_{y^j}) \in C^\infty(V)$, then in U,

$$\begin{aligned}f^*\Omega = f^*(\Omega_j\,dy^j) &= (\Omega_j \circ f)\cdot f^*(dy^j) = (\Omega_j \circ f)\cdot d(y^j \circ f) \\ &= (\Omega_j \circ f)\cdot((d(y^j \circ f))(\partial_{x^i}))\cdot dx^i = (\Omega_j \circ f)\cdot(\partial_{x^i}(y^j \circ f))\,dx^i.\end{aligned}$$

Example 3.4. Let $M = \mathbb{R}^3$ and $N = \mathbb{R}^2$ be equipped with the standard smooth structures. Consider the admissible charts:

$$\begin{aligned}&(U = \mathbb{R}^3, (x, y, z) \mapsto (x, y, z)) \text{ for } \mathbb{R}^3, \text{ and} \\ &(V = \mathbb{R}^2, (u, v) \mapsto (u, v)) \text{ for } \mathbb{R}^2.\end{aligned}$$

Let $f : \mathbb{R}^3 \to \mathbb{R}^2$ be the smooth map $f(x, y, z) = (x + y, yz)$ for $(x, y, z) \in \mathbb{R}^3$. Let $\Omega = v\,du + u\,dv \in T_1^0 N$.

We have

$$\begin{aligned}f^*\Omega &= (v\circ f)\cdot(\partial_x(u\circ f)\,dx + \partial_y(u\circ f)\,dy + \partial_z(u\circ f)\,dz)\\ &\quad +(u\circ f)\cdot(\partial_x(v\circ f)\,dx + \partial_y(v\circ f)\,dy + \partial_z(v\circ f)\,dz)\\ &= (yz)\cdot(\partial_x(x+y)\,dx + \partial_y(x+y)\,dy + \partial_z(x+y)\,dz)\\ &\quad +(x+y)\cdot(\partial_x(yz)\,dx + \partial_y(yz)\,dy + \partial_z(yz)\,dz)\\ &= (yz)\cdot(dx+dy) + (x+y)(z\,dy + y\,dz)\\ &= yz\,dx + (x+2y)z\,dy + (x+y)y\,dz.\end{aligned}$$

◇

3.8 Integral curves and the flow of vector fields

A vector field describes a 'flow' on a manifold. For example, consider the wind velocity vector field on the surface of the Earth. Then the path traced by a dust particle describes an 'integral curve' γ, namely, at a point p along γ, the velocity of the speck of dust (which is the tangent vector to the curve), is the wind velocity at that point.

Definition 3.11. (Integral curve.)
Let $V\in T_0^1M$. A smooth curve $\gamma: I\to M$, where $I\subset\mathbb{R}$ is an open interval, is called an *integral curve of* V if $V_{\gamma(t)} = v_{\gamma,\gamma(t)}$ for all $t\in I$.

Using a chart-induced basis representation (see Example 2.4, p.29), the integral curve condition $V_{\gamma(t)} = v_{\gamma,\gamma(t)}$ $(t\in I)$ can be locally expressed as a system of first order ordinary differential equations: if $V = v^i\partial_{x^i}$ in a chart $(U,\mathbf{x})$, where $v^i = V(x^i)\in C^\infty(U)$, then

$$v^i(\gamma(t))\,\partial_{x^i,\gamma(t)} = \frac{d(\mathbf{x}\circ\gamma)^i}{dt}(t)\,\partial_{x^i,\gamma(t)},\quad t\in I.$$

Thus component-wise, we have:

$$\frac{d(\mathbf{x}\circ\gamma)^i}{dt}(t) = v^i(\gamma(t)),\quad 1\leqslant i\leqslant m,\ t\in I. \qquad (\star)$$

Define the m functions $\gamma^i: I\to\mathbb{R}$ by $\gamma^i(t) = x^i(\gamma(t)) = (\mathbf{x}\circ\gamma)^i(t)$, for $t\in I$, and $1\leqslant i\leqslant m$. Suppose that $0\in I$, and $\gamma(0)=p$. Set $\xi^i = x^i(p)$. Writing $v^i(\gamma(t)) = v^i(\mathbf{x}^{-1}\circ(\mathbf{x}\circ\gamma)(t))$ on the right-hand side of $(\star)$, we obtain the initial value problem:

$$\left.\begin{array}{ll}\dfrac{d\gamma^1}{dt}(t) = v^1(\mathbf{x}^{-1}(\gamma^1(t),\cdots,\gamma^m(t))) & \gamma^1(0)=\xi^1\\ \vdots & \vdots\\ \dfrac{d\gamma^m}{dt}(t) = v^m(\mathbf{x}^{-1}(\gamma^1(t),\cdots,\gamma^m(t))) & \gamma^m(0)=\xi^m\end{array}\right\},\ t\in I. \qquad (\star\star)$$

Example 3.5. Consider $\mathbb{R}^2$ with the standard smooth structure given by the global chart $(\mathbb{R}^2, (x,y) \mapsto (x,y))$. Let $V = x\,\partial_y - y\,\partial_x$. The differential equations describing an integral curve $\gamma(t) = (x(t), y(t))$ of V are given by

$$\begin{cases} \dot{x}(t) = -y(t) \\ \dot{y}(t) = x(t) \end{cases}$$

where $\dot{} = \frac{d}{dt}$. If the curve starts at $t = 0$ from (x_0, y_0), then the unique solution is

$$\begin{bmatrix} x(t) \\ y(t) \end{bmatrix} = \begin{bmatrix} \cos t & -\sin t \\ \sin t & \cos t \end{bmatrix} \begin{bmatrix} x_0 \\ y_0 \end{bmatrix}, \quad t \in \mathbb{R}.$$

The resulting curve γ in $\mathbb{R}^2$ (thought of as a plane) describes the path of a counterclockwise moving particle in a circular path with centre $(0,0)$, starting at (x_0, y_0). ◇

Exercise 3.22. In contrast to the above example, where the integral curve existed for all $t \in \mathbb{R}$, the curve may exist only on a finite interval. As an example, consider $V = u^2\,\partial_u \in T_0^1\mathbb{R}$ on the smooth manifold $\mathbb{R}$ with the standard smooth structure given by the global chart $(\mathbb{R}, u \mapsto u)$. Show that if the integral curve starts at $t = 0$ at $p = \gamma(0) > 0$, then $\gamma(t) = p/(1 - pt)$ for all $t < 1/p$, which 'escapes to ∞' in the finite time $1/p < \infty$.

From the theory of differential equations, it follows that (since the v^i are all smooth) there exists a neighbourhood $(-\epsilon, \epsilon)$ of 0 and a neighbourhood N of $\mathbf{x}(p)$ in $\mathbf{x}(U)$, such that for each $\boldsymbol{\xi} = (\xi^1, \cdots, \xi^m) \in N$, the initial value problem for the system of ordinary differential equations $(\star\star)$ on p.52 has a unique smooth solution $(-\epsilon, \epsilon) \ni t \mapsto (\gamma^1(t; \boldsymbol{\xi}), \cdots, \gamma^m(t; \boldsymbol{\xi}))$. Also, the solution to $(\star\star)$ depends smoothly on the initial condition $\boldsymbol{\xi}$. We will not prove this here, but refer the interested reader to [Hartman (2002), Thm. 4.1, p.100]. Thus equation $(\star)$ on p.52 has a unique smooth solution $\gamma(t; \mathbf{x}(q))$ for each initial condition q in the neighbourhood $\mathbf{x}^{-1}N$ of p, living on the time interval $(-\epsilon, \epsilon)$, and the map $\mathbf{x}^{-1}N \ni q \mapsto \gamma(t; \mathbf{x}(q))$ is smooth.

Let us suppose from now on that for our vector field on the smooth manifold M, the integral curves exist for all $t \in \mathbb{R}$, and the pathological cases as in Exercise 3.22 do not arise. We will call such vector fields as *complete* vector fields.

Definition 3.12. (Complete vector field.)
Let M be a smooth manifold. A vector field $V \in T_0^1 M$ is *complete* if for all $p \in M$, the integral curve γ_p passing through p is defined for all $t \in \mathbb{R}$.

Suppose that a smooth manifold admits a complete vector field V. Then we obtain for each frozen $t \in \mathbb{R}$, a map ψ_t, given by $p \overset{\psi_t}{\mapsto} \gamma_p(t) : M \to M$, sending the initial condition $\gamma_p(0) = p$ to the evaluation-at-time t of the integral curve γ_p of V. In other words, we imagine a whole bunch of particles starting at different places being transported along the manifold by the vector field V, and then looking at the result (taking a snapshot) at time t. Then $p \mapsto \gamma_p(t) : M \to M$ is smooth, thanks to the smooth dependence of the initial value problem solution on the initial condition. The collection of maps $\{\psi_t, t \in \mathbb{R}\}$ ('snapshots' of the ensemble of particles at various times $t \in \mathbb{R}$) describe the 'flow' of the vector field V.

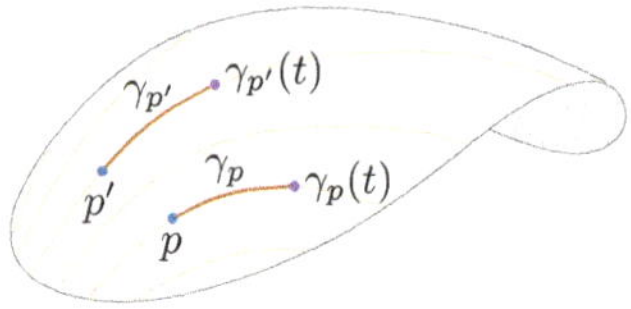

Definition 3.13. (Flow of a vector field.)
Let $V \in T_0^1 M$ be a complete vector field on the smooth manifold M.
Let $\gamma_p : \mathbb{R} \to M$ be the integral curve of V satisfying $\gamma_p(0) = p \in M$.
For $t \in \mathbb{R}$, the map $\psi_t : M \to M$ given by $\psi_t(p) = \gamma_p(t)$ for all $p \in M$, is called a *flow map of* V. The collection $\mathcal{F}_V = \{\psi_t : M \to M,\ t \in \mathbb{R}\}$ of all flow maps, is called the *flow of* V.

Since for each $p \in M$, $\gamma_p(0) = p$, we have $\psi_0 = \mathrm{id}_M$. The following result makes composition a group operation on $\mathcal{F}_V$.

Lemma 3.5. *Let $V \in T_0^1 M$ be a complete vector field on the smooth manifold M with the flow $\{\psi_t : M \to M,\ t \in \mathbb{R}\}$. Then $\psi_{t+s} = \psi_t \circ \psi_s = \psi_s \circ \psi_t$ for all $t, s \in \mathbb{R}$.*

Proof. For $p \in M$, let γ_p be the integral curve of V such that $\gamma_p(0) = p$. Let $s \in \mathbb{R}$. Set $q = \gamma_p(s) = \psi_s(p)$.

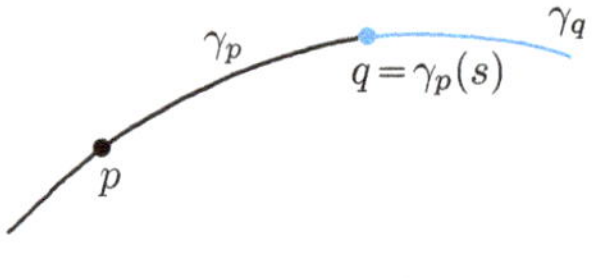

Consider the curve $\gamma : \mathbb{R} \to M$, given by $\gamma(t) = \gamma_p(s+t)$ for all $t \in \mathbb{R}$. Then $\gamma(0) = \gamma_p(s) = q$, and $V_{\gamma(t)} = V_{\gamma_p(s+t)} = v_{\gamma_p,\gamma_p(s+t)}$. But for $f \in C^\infty(M)$,

$$\begin{aligned} v_{\gamma_p,\gamma_p(s+t)} f &= \frac{d(f \circ \gamma_p)}{dt}(s+t) = \lim_{h \to 0} \frac{f(\gamma_p(s+t+h)) - f(\gamma_p(s+t))}{h} \\ &= \lim_{h \to 0} \frac{f(\gamma(t+h)) - f(\gamma(t))}{h} = \frac{d(f \circ \gamma)}{dt}(t) = v_{\gamma,\gamma(t)} f. \end{aligned}$$

Hence $V_\gamma(t) = v_{\gamma_p,\gamma_p(s+t)} = v_{\gamma,\gamma(t)}$, that is, γ is an integral curve of V such that $\gamma(0) = q$. By the uniqueness of solutions, $\gamma = \gamma_q$. So

$$\psi_{t+s}p = \gamma_p(s+t) = \gamma(t) = \gamma_q(t) = \psi_t q = \psi_t(\psi_s p)$$

for all $p \in M$. Thus $\psi_{t+s} = \psi_t \circ \psi_s$. Also, $\psi_{t+s} = \psi_{s+t} = \psi_s \circ \psi_t$. □

Composition of maps is associative. Also, $\mathrm{id}_M = \psi_0 \in \mathcal{F}_V$ serves as the identity element with composition. By Lemma 3.5, every $\psi_t \in \mathcal{F}_V$ has an inverse with respect to composition: $\psi_{-t} \circ \psi_t = \psi_t \circ \psi_{-t} = \psi_{t-t} = \psi_0 = \mathrm{id}_M$. As each $\psi_t \in \mathcal{F}_V$ is smooth, and $(\psi_t)^{-1} = \psi_{-t}$, we conclude that every flow map is a diffeomorphism.

Exercise 3.23. Consider $\mathbb{R}^2$ with the standard smooth structure given by the global chart $(\mathbb{R}^2, (x,y) \mapsto (x,y))$. Determine the flow $\{\psi_t : \mathbb{R}^2 \to \mathbb{R}^2,\ t \in \mathbb{R}\}$ of the vector field $V = (1+y)\,\partial_x$. If $p = (x_0, y_0) \in \mathbb{R}^2$, then find $(d\psi_{-t})_p \partial_{y,p}$.

Exercise 3.24. Consider $\mathbb{R}^3$ with the standard smooth structure given by the global chart $(\mathbb{R}^3, (x,y,z) \mapsto (x,y,z))$. Let $L_x, L_y, L_z \in T_0^1\mathbb{R}^3$ be the vector fields given by $L_x = y\,\partial_z - z\,\partial_y$, $L_y = z\,\partial_x - x\,\partial_z$, $L_z = x\,\partial_y - y\,\partial_x$. Compute $[L_a, L_b]$ for $a, b = x, y, z$. Find the flow maps for L_x, L_y, L_z.

Finally, we give the geometric meaning of the Lie bracket. More generally, we define a Lie derivative below, which measures the rate of change of the quantity at hand (function, vector field, and later on even tensor fields) along the flow of a vector field.

Definition 3.14. (Lie derivative.)
Let M be a smooth manifold. Let $V \in T_0^1 M$. The *Lie derivative* $\mathcal{L}_V f$ of a function $f \in C^\infty(M)$ with respect to V is the function $Vf \in C^\infty(M)$. The *Lie derivative* $\mathcal{L}_V W$ of a vector field $W \in T_0^1 M$ with respect to V is the vector field $[V, W] \in T_0^1 M$.

Proposition 3.1. *Let M be a smooth manifold, and $V \in T_0^1 M$ be a complete vector field with the flow $\{\psi_t : M \to M,\ t \in \mathbb{R}\}$. Then for all $p \in M$ and all $f \in C^\infty(M)$, we have* $(\mathcal{L}_V f)(p) = \dfrac{d((f \circ \psi_t)(p))}{dt}(0)$.

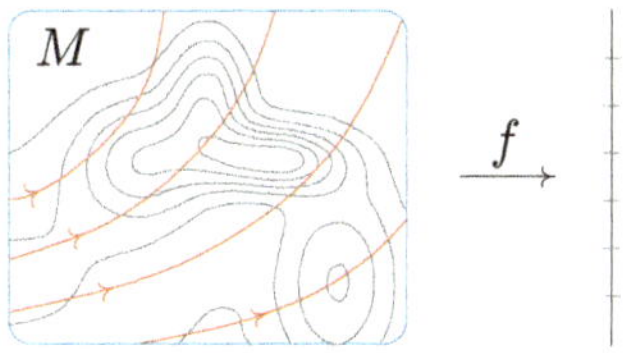

Proof. Let $\gamma_p(t) = \psi_t(p)$, $t \in \mathbb{R}$, be the integral curve of V such that $\gamma_p(0) = p$. Then $(f \circ \psi_t)(p) = f(\psi_t p) = f(\gamma_p(t)) = (f \circ \gamma_p)(t)$. Thus

$$\begin{aligned}\frac{d((f \circ \psi_t)(p))}{dt}(0) &= \frac{d(f \circ \gamma_p)}{dt}(0) = v_{\gamma_p, \gamma_p(0)} f = v_{\gamma_p, p} f \\ &= V_p f = (Vf)(p) = (\mathcal{L}_V f)(p).\end{aligned}$$

□

Proposition 3.2. *Let M be a smooth manifold, and $V \in T_0^1 M$ be a complete vector field with the flow $\{\psi_t : M \to M, t \in \mathbb{R}\}$. Then for all vector fields $W \in T_0^1 M$,*

$$\mathcal{L}_V W = [V, W] = \frac{d((d\psi_{-t})_{\psi_t \cdot} W_{\psi_t \cdot})}{dt}(0).$$

Here $(d\psi_{-t})_{\psi_t \cdot} W_{\psi_t \cdot}$ is the vector field $M \ni q \mapsto (d\psi_{-t})_{\psi_t q} W_{\psi_t q} \in T_q M$. The smoothness follows by noting that this vector field is $(\psi_{-t})_* W$: For $q \in M$,

$$((\psi_{-t})_* W)(q) = ((\psi_{-t})_* W)(\mathrm{id}_M q) = ((\psi_{-t})_* W)(\psi_{-t}(\psi_t q)) = (d\psi_{-t})_{\psi_t q} W_{\psi_t q}.$$

Thus for each $p \in M$,

$$\frac{d((d\psi_{-t})_{\psi_t p} W_{\psi_t p})}{dt}(0) = \lim_{t \to 0} \frac{(d\psi_{-t})_{\psi_t p} W_{\psi_t p} - W_p}{t}$$

makes sense as an element[5] of $T_p M$ because both W_p and $(d\psi_{-t})_{\psi_t p} W_{\psi_t p}$ belong to $T_p M$ (and provided we accept that the derivative exists).

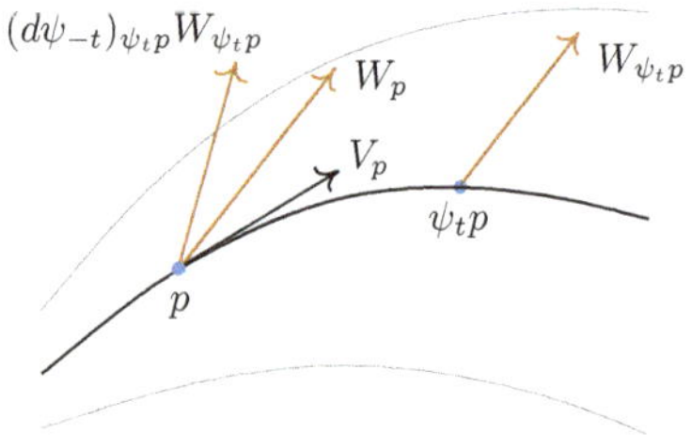

Proof. Let $\gamma_q(t) = \psi_t(q)$ for $t \in \mathbb{R}$ and $q \in M$. Let $p \in M$. Define H by $H(t, s) = W_{\psi_t p}(f \circ \psi_{-s})$. Let $f \in C^\infty(M)$. Then

$$\begin{aligned}(V(Wf))(p) &= V_p(Wf) = v_{\gamma_p, \gamma_p(0)}(Wf) = \frac{d((Wf) \circ \gamma_p)}{dt}(0) \\ &= \frac{d((Wf)(\psi_t p))}{dt}(0) = \frac{d(W_{\psi_t p} f)}{dt}(0) = \frac{\partial H}{\partial t}(0, 0).\end{aligned}$$

[5]Here we may consider the question of the existence of the limit defining the derivative by equipping the m-dimensional vector space $T_p M$ with *any* norm, since all norms on finite-dimensional vector spaces are equivalent.

With $h(t) = -t$ for all $t \in \mathbb{R}$, we have $(\gamma_q \circ h)(t) = \psi_{-t}(q)$, and

$$\begin{aligned}(Vf)(q) &= V_q f = v_{\gamma_q,q} f = \frac{d(f \circ \gamma_q)}{dt}(0) = -(-1)\frac{d(f \circ \gamma_q)}{dt}(0) \\ &= -\dot{h}(0)\frac{d(f \circ \gamma_q)}{dt}(0) = -\frac{d(f \circ \gamma_q \circ h)}{dt}(0) = -\frac{d((f \circ \psi_{-t})(q))}{dt}(0).\end{aligned}$$

We have[6]

$$\begin{aligned}(W(Vf))(p) &= W_p(Vf) = -W_p\Big(M \ni q \mapsto \frac{d((f \circ \psi_{-t})(q))}{dt}(0)\Big) \\ &\overset{(*)}{=} -\frac{d(W_p(f \circ \psi_{-t}))}{dt}(0) = -\frac{\partial H}{\partial s}(0,0). \end{aligned} \tag{3.1}$$

Thus

$$\begin{aligned}[V,W]_p f &= ([V,W]f)(p) = (V(Wf))(p) - (W(Vf))(p) \\ &= \frac{\partial H}{\partial t}(0,0) + \frac{\partial H}{\partial s}(0,0) = \frac{d(t \mapsto H(t,t))}{dt}(0) \\ &= \frac{d(W_{\psi_t p}(f \circ \psi_{-t}))}{dt}(0).\end{aligned}$$

Consequently, $[V,W] = \dfrac{d((d\psi_{-t})_{\psi_t \cdot} W_{\psi_t \cdot})}{dt}(0)$. □

[6]The equality $(*)$ in (3.1) needs justification. Formally, it is the process of taking W_p inside the limit:

$$W_p \frac{d((f \circ \psi_{-t})(\cdot))}{dt}(0) = W_p \lim_{t \to 0} \frac{(f \circ \psi_{-t})(\cdot) - f}{t} = \lim_{t \to 0} \frac{W_p((f \circ \psi_{-t})(\cdot)) - W_p f}{t} = \frac{d(W_p(f \circ \psi_{-t}))}{dt}(0).$$

We refer the reader to [Lee (2013), Theorem 9.38] for a detailed proof Proposition 3.2.

Chapter 4

Tensor fields

In this chapter, on smooth manifolds M, using the building blocks $T^1_0 M$ and $T^0_1 M$, namely the $C^\infty(M)$-modules of all vector fields and of all 1-form fields, we will introduce (r, s)-tensor fields, which will be $C^\infty(M)$-multilinear maps defined on the Cartesian products of these modules to $C^\infty(M)$. As we had mentioned earlier, tensor fields are fundamental in Lorentzian geometry and in spacetime physics. The Lorentzian geometry will be specified by the metric $\mathbf{g}$ on a smooth manifold, where the metric will be a certain type of a $(0, 2)$-tensor field. This metric will induce a curvature tensor field $\mathbf{R}$ (via something called the Levi-Civita connection), and $\mathbf{R}$ will be a $(1, 3)$-tensor field. Vector field and 1-form fields are themselves simple examples of tensor fields: vector fields are $(1, 0)$-tensor fields, and 1-form fields are $(0, 1)$-tensor fields.

4.1 (r, s)-tensor fields

We are familiar with multilinear maps in the context of vector spaces (for example, the inner product on $\mathbb{R}^m$ is a bilinear map, and the determinant on $\mathbb{R}^m$ is an m-linear form with some extra properties). An (r, s)-tensor field is just a special type of a multilinear map, except that the field $\mathbb{R}$ of scalars is replaced by the ring $C^\infty(M)$ consisting of all smooth functions on a smooth manifold M.

Definition 4.1. ((r, s)-tensor field.)
Let M be a smooth manifold. Let $r, s \geqslant 0$ be integers. An *(r, s)-tensor field* is a map

$$T : \underbrace{T^0_1 M \times \cdots \times T^0_1 M}_{r \text{ copies}} \times \underbrace{T^1_0 M \times \cdots \times T^1_0 M}_{s \text{ copies}} \to C^\infty(M),$$

which is $C^\infty(M)$-multilinear, that is, it is $C^\infty(M)$-linear in each slot.

More explicitly,

$$\left.\begin{array}{l}\text{for all } 1\leqslant i\leqslant r, \text{ for all } \Omega,\Theta \in T_1^0M,\ f\in C^\infty(M), \text{ and}\\ \text{for all } \Omega^1,\cdots,\Omega^{i-1},\Omega^{i+1},\cdots,\Omega^r\in T_1^0M,\ V_1,\cdots,V_s\in T_0^1M,\\ \quad T(\Omega^1,\cdots,\Omega^{i-1},\Omega+f\Theta,\Omega^{i+1},\cdots,\Omega^r,V_1,\cdots,V_s)\\ \quad= \quad T(\Omega^1,\cdots,\Omega^{i-1},\Omega,\Omega^{i+1},\cdots,\Omega^r,V_1,\cdots,V_s)\\ \qquad + fT(\Omega^1,\cdots,\Omega^{i-1},\Theta,\Omega^{i+1},\cdots,\Omega^r,V_1,\cdots,V_s),\end{array}\right\} \tag{4.1}$$

and

$$\left.\begin{array}{l}\text{for all } 1\leqslant j\leqslant s, \text{ for all } V,W \in T_0^1M,\ f\in C^\infty(M), \text{ and}\\ \text{for all } \Omega^1,\cdots,\Omega^r\in T_1^0M,\ V_1,\cdots,V_{j-1},V_{j+1},\cdots,V_s\in T_0^1M,\\ \quad T(\Omega^1,\cdots,\Omega^r,V_1,\cdots,V_{j-1},V+fW,V_{j+1},\cdots,V_s)\\ \quad= \quad T(\Omega^1,\cdots,\Omega^r,V_1,\cdots,V_{j-1},V,V_{j+1},\cdots,V_s)\\ \qquad + fT(\Omega^1,\cdots,\Omega^r,V_1,\cdots,V_{j-1},W,V_{j+1},\cdots,V_s).\end{array}\right\} \tag{4.2}$$

The set of all (r,s)-tensor fields is denoted by T_s^rM.

If $r>0$ and $s=0$, then only (4.1) is required to hold (with (4.2) satisfied vacuously). If $r=0$ and $s>0$, then only (4.2) is required to hold. If $r=0=s$, then $T_0^0M:=C^\infty(M)$.

Example 4.1. Let M be a smooth manifold. Then $T: T_1^0M\times T_0^1M\to M$, given by $T(\Omega,V)=\Omega V$ for all $\Omega\in T_1^0M$ and all $V\in T_0^1M$, is $C^\infty(M)$-linear in each slot. So T is a $(1,1)$-tensor field. ◇

As 1-form fields act $C^\infty(M)$-linearly on vector fields (see Exercise 3.17, p.48), every 1-form field is a $(0,1)$-tensor field. Note that the notation for the set of $(0,1)$-tensor fields is T_1^0M, which coincides with the notation for 1-form fields we had used in Chapter 3. Likewise, if $V\in T_0^1M$ is a vector field, we can define a $(1,0)$-tensor field T_V by setting $T_V(\Omega)=\Omega(V)$ for all $\Omega\in T_1^0M$. Thus every vector field can be considered as a $(1,0)$-tensor field.

Exercise 4.1. Suppose that $V\in T_0^1M$. Let T_V be the $(1,0)$-tensor field given by $T_1^0M\ni\Omega\mapsto\Omega(V)\in C^\infty(M)$. Suppose $T_V=0$ (the map $T_1^0M\ni\Omega\mapsto\mathbf{0}\in C^\infty(M)$, where $\mathbf{0}(p)=0$ for all $p\in M$). Prove that $V=0$ (the zero vector field), that is, for all $p\in M$, $V_p=0$, the zero vector in T_pM.

Example 4.2. (A non-tensor field.) Let $W\in T_0^1M$ be a fixed vector field on M, and $g\in C^\infty(M)$ be a fixed function. Define $T: T_0^1(M)\to C^\infty(M)$ by $T(V)=W(Vg)$ for all $V\in T_0^1M$. Then T is additive, since for $V_1,V_2\in T_0^1M$,

$$T((V_1+V_2)g) = W(V_1g+V_2g) = W(V_1g)+W(V_2g) = T(V_1)+T(V_2).$$

However, T is not $C^\infty(M)$-linear in general, since for $V \in T_0^1M$, and for $f \in C^\infty(M)$,

$$T(fV) = W(fVg) = fW(Vg)+(Vg)Wf = fT(V)+(Vg)Wf \neq fT(V),$$

whenever $(Vg)Wf \neq 0$. ◇

Exercise 4.2. Let $W \in T_0^1M$ and $g \in C^\infty(M)$. Define $T : T_0^1M \to C^\infty(M)$ by $T(V) = V(Wg)$ for all $V \in T_0^1M$. Show that T is a $(0,1)$-tensor field on M.

T_s^rM is a $C^\infty(M)$-module with pointwise addition, and scalar multiplication by $C^\infty(M)$ elements defined as follows: For $T \in T_s^rM$ and $f \in C^\infty(M)$,

$$(f\cdot T)(\Omega^1, \cdots, \Omega^r, V_1, \cdots, V_s) = f\cdot(T(\Omega^1, \cdots, \Omega^r, V_1, \cdots, V_s)),$$

for all 1-form fields $\Omega^1, \cdots, \Omega^r \in T_1^0M$, and vector fields $V_1, \cdots V_s \in T_0^1M$. The multiplication $\cdot$ on the right-hand side is the multiplication in $C^\infty(M)$.

Exercise 4.3. ($\otimes$ and $\wedge$.) Let $\Omega, \Theta \in T_1^0M$. Define the 'tensor product' of the fields Ω and Θ, $\Omega\otimes\Theta : T_0^1M \times T_0^1M \to C^\infty(M)$, by $(\Omega\otimes\Theta)(V,W) = (\Omega V)(\Theta W)$ for all $V, W \in T_0^1M$. Show that $\Omega\otimes\Theta$ is a $(0,2)$-tensor field. Now define the $(0,2)$-tensor field $\Omega\wedge\Theta$ ('wedge product') by $\Omega\wedge\Theta = \Omega\otimes\Theta - \Theta\otimes\Omega$. Show that $\Omega\wedge\Theta = -\Theta\wedge\Omega$ and $(\Omega\wedge\Theta)(V,V) = 0$ for all $V \in T_0^1M$.

Exercise 4.4. (Bilinearity of $\otimes$.) With the same notation as in the previous exercise, show that for any $C^\infty(M)$ function f, $f(\Omega\otimes\Theta) = (f\Omega)\otimes\Theta = \Omega\otimes(f\Theta)$. Also, prove that for $\Omega_1, \Omega_2, \Theta \in T_1^0M$, $(\Omega_1+\Omega_2)\otimes\Theta = \Omega_1\otimes\Theta + \Omega_2\otimes\Theta$ and $\Theta\otimes(\Omega_1+\Omega_2) = \Theta\otimes\Omega_1 + \Theta\otimes\Omega_2$.

Let $T \in T_s^rM$ with $r \geqslant 1$. Let $\Omega \in T_1^0M$. Then evaluation, say, in the first slot, with Ω,

$$\begin{array}{rcl}(T_1^0M)^{r-1}\times(T_0^1M)^s & \to & C^\infty(M)\\ (\Omega^1, \cdots, \Omega^{r-1}, V_1, \cdots, V_s) & \mapsto & T(\Omega, \Omega^1, \cdots, \Omega^{r-1}, V_1, \cdots, V_s)\end{array}$$

is an $(r-1, s)$-tensor field. Similarly, evaluating T on any $(1\leqslant)$ i $(\leqslant r)$ fixed 1-form fields, and any $(1\leqslant)$ j $(\leqslant s)$ fixed vector fields, gives an $(r-i, s-j)$-tensor field.

4.2 Point evaluation of tensor fields

We know that a 1-form field Ω evaluated at a point $p \in M$ gives a 1-form $\Omega_p \in (T_pM)^*$. Similarly, a vector field V evaluated at a point p gives an element of the vector space T_pM. Now we will show that an (r,s)-tensor field at a point p gives rise to an '(r,s)-tensor'. Thus, an (r,s)-tensor field can be viewed as a mapping which assigns to each point p of the manifold, an (r,s)-tensor. We first define the linear-algebraic notion of a tensor[1].

[1]The stress in continuum mechanics is an example, and the word 'tensor' comes from the Latin word *'tendere'*, meaning 'stretch'.

Definition 4.2. ((r,s)-tensor.)
Let V be an m-dimensional real vector space and V^* denote its dual. An (r,s)-*tensor on* V is an $\mathbb{R}$-multilinear map

$$\tau : \underbrace{V^* \times \cdots \times V^*}_{r \text{ copies}} \times \underbrace{V \times \cdots \times V}_{s \text{ copies}} \to \mathbb{R}$$

that is, τ is $\mathbb{R}$-linear in each slot. More explicitly,

$$\left.\begin{array}{l} \text{for all } 1 \leqslant i \leqslant r, \text{ for all } \omega, \theta \in V^*, \ c \in \mathbb{R}, \text{ and} \\ \text{for all } \omega^1, \cdots, \omega^{i-1}, \omega^{i+1}, \cdots, \omega^r \in V^*, \ v_1, \cdots, v_s \in V, \text{ we have} \\ \quad \tau(\omega^1, \cdots, \omega^{i-1}, \omega + c\theta, \omega^{i+1}, \cdots, \omega^r, v_1, \cdots, v_s) \\ \quad = \quad \tau(\omega^1, \cdots, \omega^{i-1}, \omega, \omega^{i+1}, \cdots, \omega^r, v_1, \cdots, v_s) \\ \qquad + c\tau(\omega^1, \cdots, \omega^{i-1}, \theta, \omega^{i+1}, \cdots, \omega^r, v_1, \cdots, v_s), \end{array}\right\} \quad (4.3)$$

and

$$\left.\begin{array}{l} \text{for all } 1 \leqslant j \leqslant s, \text{ for all } v, w \in V, \ c \in \mathbb{R}, \text{ and} \\ \text{for all } \omega^1, \cdots, \omega^r \in V^*, \ v_1, \cdots, v_{j-1}, v_{j+1}, \cdots, v_s \in V, \text{ we have} \\ \quad \tau(\omega^1, \cdots, \omega^r, v_1, \cdots, v_{j-1}, v + cw, v_{j+1}, \cdots, v_s) \\ \quad = \quad \tau(\omega^1, \cdots, \omega^r, v_1, \cdots, v_{j-1}, v, v_{j+1}, \cdots, v_s) \\ \qquad + c\tau(\omega^1, \cdots, \omega^r, v_1, \cdots, v_{j-1}, w, v_{j+1}, \cdots, v_s). \end{array}\right\} \quad (4.4)$$

The set of all (r,s)-tensors on V is denoted by $T^r_s V$. We set $T^1_0 V = (V^*)^*$, $T^0_1 V = V^*$ and $T^0_0 V = \mathbb{R}$. If $V = T_pM$, then we denote the set of all (r,s)-tensors by $T^r_s M(p)$.

Since $(V^*)^* \simeq V$, we identify $v \in V$ as an element of $T^1_0 V$ with the action $v(\omega) = \omega(v)$ for all $\omega \in V^*$. With pointwise addition and scalar multiplication, $T^r_s V$ is a real vector space.

Exercise 4.5. Let $V = \mathbb{R}^3$. Define $\tau : V^* \times V \times V \to \mathbb{R}$ by $\tau(\boldsymbol{\omega}, \mathbf{v}, \mathbf{w}) = \boldsymbol{\omega}(\mathbf{v} \times \mathbf{w})$, for $\boldsymbol{\omega} \in (\mathbb{R}^3)^*$, $\mathbf{v}, \mathbf{w} \in \mathbb{R}^3$, and where $\mathbf{v} \times \mathbf{w}$ denotes the cross product in $\mathbb{R}^3$ of $\mathbf{v}, \mathbf{w}$. Show that τ is a $(1,2)$-tensor on $\mathbb{R}^3$.

Exercise 4.6. (Tensor product of tensors.) Let V be a real vector space. Let τ be an (r,s)-tensor on V and τ' be an (r',s')-tensor on V. Define their *tensor product* $\tau \otimes \tau' : (V^*)^{r+r'} \times V^{s+s'} \to \mathbb{R}$ by

$$\begin{aligned} &(\tau \otimes \tau')(\omega^1, \cdots, \omega^{r+r'}, v_1, \cdots, v_{s+s'}) \\ &\qquad = \tau(\omega^1, \cdots, \omega^r, v_1, \cdots, v_s)\, \tau'(\omega^{r+1}, \cdots, \omega^{r+r'}, v_{s+1}, \cdots, v_{s+s'}). \end{aligned}$$

Then $\tau \otimes \tau'$ is an $(r+r', s+s')$-tensor on V.
Prove that the tensor product $\otimes$ is associative.

Exercise 4.7. Let V be an m-dimensional real vector space, with a basis $\{e_1, \cdots, e_m\}$, and let $\{\epsilon^1, \cdots, \epsilon^m\}$ be the corresponding dual basis for the dual vector space V^*. Show that the set

$$B = \{e_{i_1} \otimes \cdots \otimes e_{i_r} \otimes \epsilon^{j_1} \otimes \cdots \otimes \epsilon^{j_s} : 1 \leqslant i_1, \cdots, i_r, j_1, \cdots, j_s \leqslant m\}$$

is a basis for $T_s^r V$. Conclude that $\dim T_s^r V = m^{r+s}$.

We will now show that every (r, s)-tensor field T can be viewed as a map

$$\begin{aligned} M \ni p \mapsto (p, T(p)) \in \bigsqcup_{p \in M} T_s^r M(p) &= \{(p, \tau) : p \in M \text{ and } \tau \in T_s^r M(p)\} \\ &= \bigcup_{p \in M} (\{p\} \times T_s^r M(p)), \end{aligned}$$

where $T(p) \in T_s^r M(p)$ is an (r, s)-tensor on $T_p M$. Before giving the definition of the (r, s)-tensor $T(p)$ arising from a tensor field T, we will prove the following result.

Theorem 4.1. *Let M be a smooth manifold, $T \in T_s^r M$, and $p \in M$. Suppose that:*

- *$\Omega^1, \cdots, \Omega^r \in T_1^0 M$ and $\Theta^1, \cdots, \Theta^r \in T_1^0 M$ are such that $(\Omega^i)_p = (\Theta^i)_p$, $i = 1, \cdots, r$*
- *$V_1, \cdots, V_s \in T_0^1 M$ and $W_1, \cdots W_s \in T_0^1 M$ are such that $(V_j)_p = (W_j)_p$, $j = 1, \cdots, s$.*

Then $(T(\Omega^1, \cdots, \Omega^r, V_1, \cdots, V_s))(p) = (T(\Theta^1, \cdots, \Theta^r, W_1, \cdots, W_s))(p)$.

Proof. Let $(U, \mathbf{x})$ be an admissible chart containing p. Let χ be a bump function that is identically 1 in a neighbourhood of p, and is identically zero outside an open set U_0, such that $\overline{U_0} \subset U$. Then[2] $\chi dx^i \in T_1^0 M$ for $1 \leqslant i \leqslant m$, and $\chi \partial_{x^j} \in T_0^1 M$ for $1 \leqslant j \leqslant m$.

For ease of exposition, we will prove this just in the case when[3] we have $r = s = 1$. First, $\chi^2 \Omega = \chi^2 \Omega(\partial_{x^k}) dx^k = \Omega(\chi \partial_{x^k})(\chi dx^k) = f_k(\chi dx^k)$, with $f_k := \Omega(\chi \partial_{x^k}) \in C^\infty(M)$. Similarly, we have that $\chi^2 \Theta = \widetilde{f}_k(\chi dx^k)$, where $\widetilde{f}_k := \Theta(\chi \partial_{x^k}) \in C^\infty(M)$. We note that

$$\begin{aligned} f_k(p) &= (\Omega(\chi \partial_{x^k}))(p) \\ &= \Omega_p(\chi \partial_{x^k})_p \\ &= \Theta_p(\chi \partial_{x^k})_p \\ &= (\Theta(\chi \partial_{x^k}))(p) \\ &= \widetilde{f}_k(p). \end{aligned}$$

[2]Here by χdx^i, we mean the 1-form Ω defined by $\Omega_q = \chi(q)(dx^i)_q$ if $q \in U$, and $\Omega_q = 0$ if $q \notin U$. The notation $\chi \partial_{x^i}$ has a similar connotation.

[3]The proof in the general case is the same, except notationally messier, and the factor $\chi^{2(1+1)} = \chi^4$, used below in (4.5), should be replaced by $\chi^{2(r+s)}$.

Moreover, we have

$$\chi^2 V = \chi^2 V(x^\ell)\partial_{x^\ell} = V(\chi x^\ell)(\chi\partial_{x^\ell}) = g^\ell(\chi\partial_{x^\ell}),$$

where $g^\ell := V(\chi x^\ell) \in C^\infty(M)$. Similarly, $\chi^2 W = W(\chi x^\ell)(\chi\partial_{x^\ell}) = \widetilde{g}^\ell(\chi\partial_{x^\ell})$, where $\widetilde{g}^\ell := W(\chi x^\ell) \in C^\infty(M)$. Note that

$$\begin{aligned} g^\ell(p) &= (V(\chi x^\ell))(p) = V_p(\chi x^\ell) = 1\,V_p x^\ell + 0 \\ &= 1\,W_p x^\ell + 0 = W_p(\chi x^\ell) = (W(\chi x^\ell))(p) = \widetilde{g}^\ell(p). \end{aligned}$$

So

$$\begin{aligned} \chi^{2(1+1)}T(\Omega, V) &= T(\chi^2\Omega, \chi^2 V) \\ &= T(f_k(\chi dx^k), g^\ell(\chi\partial_{x^\ell})) \\ &= f_k g^\ell T(\chi dx^k, \chi\partial_{x^\ell}). \end{aligned} \tag{4.5}$$

Similarly, $\chi^{2(1+1)}T(\Theta, W) = \widetilde{f}_k\,\widetilde{g}^\ell\,T(\chi\,dx^k, \chi\,\partial_{x^\ell})$. Consequently,

$$\begin{aligned} (T(\Omega, V))(p) &= 1\,(T(\Omega, V))(p) = (\chi(p))^{2(1+1)}(T(\Omega, V))(p) \\ &= (\chi^{2(1+1)}T(\Omega, V))(p) \\ &= (f_k g^\ell T(\chi dx^k, \chi\partial_{x^\ell}))(p) \\ &= f_k(p)\,g^\ell(p)(T(\chi dx^k, \chi\partial_{x^\ell}))(p) \\ &= \widetilde{f}_k(p)\,\widetilde{g}^\ell(p)(T(\chi dx^k, \chi\partial_{x^\ell}))(p) \\ &= \cdots (\text{retrace the steps}) \cdots \\ &= (T(\Theta, W))(p). \end{aligned}$$

This completes the proof. □

In light of this, we can now introduce a 'point evaluation' of a tensor field.

Definition 4.3. (Evaluation of a tensor field at a point.)
Let M be a smooth manifold, $T \in T^r_s M$, and $p \in M$. We define the tensor $T(p) \in T^r_s M(p)$ as follows: For all $\omega^1, \cdots, \omega^r \in (T_pM)^*$ and all $v_1, \cdots, v_s \in T_pM$, we set

$$T(p)(\omega^1, \cdots, \omega^r, v_1, \cdots, v_s) := (T(\Omega^1, \cdots, \Omega^r, V_1, \cdots, V_s))(p),$$

where $\Omega^1, \cdots, \Omega^r \in T^0_1 M$ are any 1-form fields such that $(\Omega^i)_p = \omega^i$, for all $1 \leqslant i \leqslant r$, and $V_1, \cdots, V_s \in T^1_0 M$ are any vector fields such that $(V_j)_p = v_j$, for all $1 \leqslant j \leqslant s$.

By Theorem 4.1, the notion above is well-defined, since it depends neither on the choice of the 1-form field extensions of the 1-forms $\omega^1, \cdots, \omega^r$, nor on the vector field extensions of the vectors $v_1, \cdots, v_s$. It is also clear that $T(p)$ is an (r, s)-tensor: additivity is immediate, and we note that constants can be considered as constant $C^\infty(M)$ functions.

Remark 4.1. (Tensor field as a smooth section of a tensor bundle.) One can use the language of bundles in order to define an (r, s)-tensor bundle (analogous to the tangent/cotangent bundle), where to each point $p \in M$, one assigns in a smooth manner an (r, s)-tensor. Then a tensor field can be defined as 'section' of this (r, s)-tensor bundle, and this notion coincides with our notion of tensor fields from Definition 4.1. $\ast$

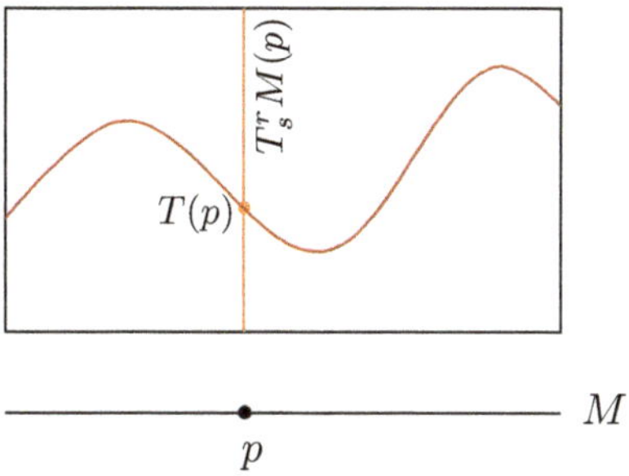

Proposition 4.1. *Suppose for each p in the smooth manifold M, there exists a tensor $T(p) \in T^r_s M(p)$ such that for all $\Omega^1, \cdots, \Omega^r \in T^0_1 M$ and all $V_1, \cdots, V_s \in T^1_0 M$, the map*

$$M \ni p \mapsto (T(\Omega^1, \cdots, \Omega^r, V_1, \cdots, V_s))(p) := T(p)((\Omega^1)_p, \cdots, (\Omega^r)_p, (V_1)_p, \cdots, (V_s)_p)$$

is an element of $C^\infty(M)$. Then $T \in T^r_s M$.

Proof. For ease of exposition, we just consider the case when $r = s = 1$. We only need to show the $C^\infty(M)$-multilinearity of $T : T^0_1 M \times T^1_0 M \to C^\infty(M)$. For $\Omega^1, \Omega^2 \subset T^0_1 M$, $V \in T^1_0 M$, and $f \in C^\infty(M)$, we have for all $p \in M$ that

$$\begin{aligned}
(T(\Omega^1 + f\Omega^2, V))(p) &= T(p)((\Omega^1 + f\Omega^2)_p, V_p) \\
&= T(p)((\Omega^1)_p + f(p)(\Omega^2)_p, V_p) \\
&= T(p)((\Omega^1)_p, V_p) + f(p)T(p)((\Omega^2)_p, V_p) \\
&= (T(\Omega^1, V))(p) + (f \cdot T(\Omega^2, V))(p) \\
&= (T(\Omega^1, V) + f \cdot T(\Omega^2, V))(p).
\end{aligned}$$

Thus

$$T(\Omega^1 + f\Omega^2, V) = T(\Omega^1, V) + f(T(\Omega^2, V)).$$

Similarly, for $\Omega \in T^0_1 M$, $V_1, V_2 \in T^1_0 M$, and $f \in C^\infty(M)$,

$$T(\Omega, V_1 + f V_2) = T(\Omega, V_1) + f(T(\Omega, V_2)).$$

This completes the proof. $\square$

4.3 Tensor field components, tensor product and contraction

Just like a vector field and a 1-form field have chart-induced components that represent the field at hand, we can represent a tensor field using its chart-induced components.

Definition 4.4. (Chart-induced tensor field components at a point.)
Let $(U, \mathbf{x})$ be an admissible chart for an m-dimensional smooth manifold M, $p \in M$, and $T \in T^r_s M$. We define the *chart-induced tensor components of T at p* by $T(p)^{i_1\cdots i_r}_{j_1\cdots j_s} = T(p)((dx^{i_1})_p, \cdots, (dx^{i_r})_p, \partial_{x^{j_1},p}, \cdots, \partial_{x^{j_s},p})$, where $1 \leqslant i_1, \cdots, i_r, j_1, \cdots, j_s \leqslant m$.

Exercise 4.8. Let $(U, \mathbf{x})$ be an admissible chart for a smooth manifold M, and $T \in T^r_s M$. The m^{r+s} chart-induced tensor field components $T^{i_1\cdots i_r}_{j_1\cdots j_s}$ are the maps $U \ni p \mapsto T(p)^{i_1\cdots i_r}_{j_1\cdots j_s}$. Show that $T^{i_1\cdots i_r}_{j_1\cdots j_s} \in C^\infty(U)$. (It is then customary in the physics literature to denote the chart induced components for another chart $(U', \mathbf{x}')$ by $T'^{i_1\cdots i_r}_{j_1\cdots j_s}$.)

Exercise 4.9. Let $T \in T^1_1 M$ be the $(1,1)$-tensor field on a smooth manifold M, given by $T(\Omega, V) = \Omega V$ for all $\Omega \in T^0_1 M$ and all $V \in T^1_0 M$. Suppose that $(U, \mathbf{x})$ is an admissible chart for M. Determine the chart-induced components of T.

Exercise 4.10. (Transformation rule for components.)
Let M be a smooth manifold and $T \in T^r_s M$. Prove that if $(U, \mathbf{x})$ and $(U', \mathbf{x}')$ are two overlapping admissible charts, then the chart-induced tensor field components $T'^{i_1\cdots i_r}_{j_1\cdots j_s}$, $T^{i_1\cdots i_r}_{j_1\cdots j_s}$ with respect to $(U', \mathbf{x}')$, $(U, \mathbf{x})$, respectively, transform as follows. For $p \in U \cap U'$,

$$T'^{i_1\cdots i_r}_{j_1\cdots j_s}(p) = \frac{\partial x'^{i_1}}{\partial x^{i'_1}}(\mathbf{x}(p)) \cdots \frac{\partial x'^{i_r}}{\partial x^{i'_r}}(\mathbf{x}(p)) \frac{\partial x^{j'_1}}{\partial x'^{j_1}}(\mathbf{x}'(p)) \cdots \frac{\partial x^{j'_s}}{\partial x'^{j_s}}(\mathbf{x}'(p)) T^{i'_1\cdots i'_r}_{j'_1\cdots j'_s}(p)$$

where $\dfrac{\partial x^j}{\partial x'^i}(\mathbf{x}'(p)) := \dfrac{\partial(x^j \circ \mathbf{x}'^{-1})}{\partial u^i}(\mathbf{x}'(p))$ and $\dfrac{\partial x'^j}{\partial x^i}(\mathbf{x}(p)) := \dfrac{\partial(x'^j \circ \mathbf{x}^{-1})}{\partial u^i}(\mathbf{x}(p))$.

Exercise 4.11. (Physicist's definition of a tensor field.)
Let M be a smooth manifold described by an atlas $\mathcal{A}$. Suppose that for each chart $(U, \mathbf{x})$ in $\mathcal{A}$, there exist functions $T^{i_1\cdots i_r;U}_{j_1\cdots j_s} \in C^\infty(\mathbf{x}(U))$, such that the following transformation rule holds for any two overlapping charts $(U, \mathbf{x})$ and $(U', \mathbf{x}')$:

$$\begin{aligned} &T^{i_1\cdots i_r;U'}_{j_1\cdots j_s}(\mathbf{x}'(p)) \\ &= \frac{\partial x'^{i_1}}{\partial x^{i'_1}}(\mathbf{x}(p)) \cdots \frac{\partial x'^{i_r}}{\partial x^{i'_r}}(\mathbf{x}(p)) \frac{\partial x^{j'_1}}{\partial x'^{j_1}}(\mathbf{x}'(p)) \cdots \frac{\partial x^{j'_s}}{\partial x'^{j_s}}(\mathbf{x}'(p)) T^{i'_1\cdots i'_r;U}_{j'_1\cdots j'_s}(\mathbf{x}(p)). \end{aligned}$$

Let T be given as follows: For $p \in (U, \mathbf{x})$, $\Omega^1, \cdots, \Omega^r \in T^0_1 M$, $V_1, \cdots, V_s \in T^1_0 M$,

$$\begin{aligned} &(T(\Omega^1, \cdots, \Omega^r, V_1, \cdots, V_s))(p) \\ &:= T^{i_1\cdots i_r;U}_{j_1\cdots j_s}(\mathbf{x}(p)) \cdot (\Omega^1)_p(\partial_{x^{i_1},p}) \cdots (\Omega^r)_p(\partial_{x^{i_r},p}) \cdot (dx^{j_1})_p((V_1)_p) \cdots (dx^{j_s})_p((V_s)_p). \end{aligned}$$

Then prove that T is well-defined, that is, it does not depend on the chosen chart containing the point $p \in M$, and that $T \in T^r_s M$.

An admissible chart has the nice feature that there are available chart-induced 1-form fields dx^i and vector fields ∂_{x^j}. This suggests that we could use the tensor field component functions of T to analyse the 'restriction $T|_U$' to the chart U by defining, for example in the $r=s=1$ case,

$$T|_U(\Omega_i dx^i, V^j\partial_{x^j}) = T^i_j \Omega_i V^j,$$

where $\Omega = \Omega_i dx^i \in T^0_1 U$, $V = V^j\partial_{x^j} \in T^1_0 M$. But if U is the intersection of two admissible charts, then on U there are two available coordinate chart maps, say $\mathbf{x}$ and $\mathbf{x}'$. Let us show that the above is well-defined, that is, it does not depend on the choice of the chart map: We have

$$\begin{aligned} T'^i_j\,\Omega'_i\,V'^j &= \frac{\partial x'^i}{\partial x^{i'}}(\mathbf{x}(\cdot))\,\frac{\partial x^{j'}}{\partial x'^j}(\mathbf{x}'(\cdot))\,T^{i'}_{j'}\,\frac{\partial x^k}{\partial x'^i}(\mathbf{x}'(\cdot))\,\Omega_k\,\frac{\partial x'^j}{\partial x^\ell}(\mathbf{x}(\cdot))\,V^\ell \\ &= \delta^k_{i'}\,\Omega_k\,\delta^{j'}_\ell\,V^\ell\,T^{i'}_{j'} = \Omega_{i'}V^{j'}T^{i'}_{j'}. \end{aligned}$$

So we give the following definition.

Definition 4.5. (Restriction of a tensor field to an admissible chart.) Let M be a smooth manifold. Let $T \in T^r_s M$. Let $(U,\mathbf{x})$ be an admissible chart for M. We define the *restriction* $T|_U \in T^r_s U$ *of* T *to* U by

$$T|_U(dx^{k_1},\cdots,dx^{k_r},\partial_{x^{\ell_1}},\cdots,\partial_{x^{\ell_s}}) := T^{k_1\cdots k_r}_{\ell_1\cdots\ell_s},$$

and by extending $C^\infty(U)$-multilinearly, that is, given $\Omega^i = \Omega^i_k\,dx^k \in T^0_1 U$ $(1 \leqslant i \leqslant r)$ and $V_j = V^\ell_j\,\partial_{x^\ell} \in T^1_0 U$ $(1 \leqslant j \leqslant s)$,

$$T|_U(\Omega^1,\cdots,\Omega^r,V_1,\cdots,V_s) := \Omega^1_{k_1}\cdots\Omega^r_{k_r}V^{\ell_1}_1\cdots V^{\ell_s}_s\,T^{k_1\cdots k_r}_{\ell_1\cdots\ell_s}.$$

The following result is key to the evaluation of tensor fields in practice, for example when the chart-induced components of the various objects are known.

Proposition 4.2. *Let M be a smooth manifold, $T \in T^r_s M$, $(U,\mathbf{x})$ be an admissible chart, $p \in U$, $\omega^1,\cdots,\omega^r \in (T_pM)^*$ and $v_1,\cdots,v_s \in T_pM$. Set*

$$\begin{aligned} \omega^k_i &= \omega^k(\partial_{x^i,p}) \quad (1 \leqslant k \leqslant r,\ 1 \leqslant i \leqslant m), \text{ and} \\ v^j_\ell &= (dx^j)_p v_\ell \quad (1 \leqslant \ell \leqslant s,\ 1 \leqslant j \leqslant m). \end{aligned}$$

If $T(p)^{i_1,\cdots,i_r}_{j_1,\cdots,j_s}$ are the chart-induced components of T at p, then

$$\begin{aligned} T(p)(\omega^1,\cdots,\omega^r,v_1,\cdots,v_s) &= T(p)^{i_1,\cdots,i_r}_{j_1,\cdots,j_s}\,\omega^1_{i_1}\cdots\omega^r_{i_r}\,v^{j_1}_1\cdots v^{j_s}_s \\ &= T|_U(p)(\omega^1,\cdots,\omega^r,v_1,\cdots,v_s). \end{aligned}$$

Proof. For ease of exposition, we show this just for a $(1,1)$-tensor field. Let χ be a bump function around p, that is, a $C^\infty(M)$ element that is identically 1 in a neighbourhood of p, and identically zero outside U_0, with $\overline{U_0} \subset U$. Then (viewing ω_i, v^j as constant maps $M \ni p \mapsto \omega_i, v^j$), we have

$$\begin{aligned} T(p)(\omega, v) &= (T(\chi \omega_i \, dx^i, \chi v^j \, \partial_{x^j}))(p) = \omega_i \, v^j (T(\chi \, dx^i, \chi \, \partial_{x^j}))(p) \\ &= \omega_i \, v^j T(p)((dx^i)_p, \partial_{x^j,p}) = \omega_i \, v^j T(p)^i_j. \end{aligned}$$

Also, if we define $\Omega = \omega_i \, dx^i \in T^0_1 U$ and $V = v^j \, \partial_{x^j} \in T^1_0 U$, then we have

$$\begin{aligned} T|_U(p)(\omega, v) &= (T|_U(\Omega, V))(p) = (T|_U(\omega^i \, dx^i, v^j \, \partial_{x^j}))(p) \\ &= \omega_i \, v^j (T|_U(dx^i, \partial_{x^j}))(p) = \omega_i \, v^j T^i_j(p) = \omega_i \, v^j T(p)^i_j. \end{aligned}$$

This completes the proof. □

In order to get the representation of the tensor field using its chart-induced components, we will produce a ($C^\infty(U)$-module) spanning set for $T^r_s U$ using the building blocks $\{dx^1, \cdots, dx^m\}$ and $\{\partial_{x^1}, \cdots, \partial_{x^m}\}$. To this end, we first introduce the notion of the tensor product of two tensor fields.

Definition 4.6. (Tensor product of tensor fields.)
Let M be a smooth manifold, $T \in T^r_s M$, $T' \in T^{r'}_{s'} M$. The *tensor product* $T \otimes T' \in T^{r+r'}_{s+s'} M$ *of* T *and* T' is defined as follows: for $\Omega^1, \cdots, \Omega^{r+r'} \in T^0_1 M$ and for $V_1, \cdots, V_{s+s'} \in T^1_0 M$,

$$\begin{aligned} &(T \otimes T')(\Omega^1, \cdots, \Omega^{r+r'}, V_1, \cdots, V_{s+s'}) \\ &:= T(\Omega^1, \cdots, \Omega^r, V_1, \cdots, V_s) \cdot T'(\Omega^{r+1}, \cdots, \Omega^{r+r'}, V_{s+1}, \cdots, V_{s+s'}), \end{aligned}$$

where the $\cdot$ on the right-hand side is multiplication in $C^\infty(M)$.

Exercise 4.12. Let M be a smooth manifold, $T \in T^r_s M$, and $T' \in T^{r'}_{s'} M$. Recall Exercise 4.6 (p.62), where the tensor product of tensors was defined. Show that for all $p \in M$, $(T \otimes T')(p) = T(p) \otimes T'(p)$.

The tensor product is 'bilinear': For $T, \widetilde{T} \in T^r_s M$, $S, \widetilde{S} \in T^{r'}_{s'} M$, $f \in C^\infty(M)$,

$$\begin{aligned} (T + f\widetilde{T}) \otimes S &= T \otimes S + f(\widetilde{T} \otimes S) \text{ and} \\ T \otimes (S + f\widetilde{S}) &= T \otimes S + f(T \otimes \widetilde{S}). \end{aligned}$$

In particular, $f(T \otimes S) = (f\,T) \otimes S = T \otimes (f\,S)$. It can be seen that the tensor product is associative.

By the associativity of $\otimes$, if $(U, \mathbf{x})$ is an admissible chart, then, even without the use of parentheses, $\partial_{x^{i_1}} \otimes \cdots \otimes \partial_{x^{i_r}} \otimes dx^{j_1} \otimes \cdots \otimes dx^{j_s}$ makes sense as an element of $T^r_s U$. We show that these can be used to decompose any tensor field locally in a chart using its chart-induced components.

Let $\Omega^1, \cdots, \Omega^r \in T_1^0 U$ and $V_1, \cdots, V_s \in T_0^1 U$. Then we have $\Omega^i = \Omega^i_k dx^k$ $(1 \leqslant i \leqslant r)$ and $V_j = V_j^\ell \partial_{x^\ell} \in T_0^1 U$ $(1 \leqslant j \leqslant s)$, where $\Omega^i_k := \Omega^i \partial_{x^k} = \partial_{x^k} \Omega_i$ and $V_j^\ell := V_j(x^\ell) = (dx^\ell) V_j$. Thus

$$\begin{aligned}
& T|_U(\Omega^1, \cdots, \Omega^r, V_1, \cdots, V_s) \\
&= T^{i_1 \cdots i_r}_{j_1 \cdots j_s} \Omega^1_{i_1} \cdots \Omega^r_{i_r} V_1^{j_1} \cdots V_s^{j_s} \\
&= T^{i_1 \cdots i_r}_{j_1 \cdots j_s} (\partial_{x^{i_1}} \Omega^1) \cdots (\partial_{x^{i_r}} \Omega^r)((dx^{j_1}) V_1) \cdots ((dx^{j_s}) V_s) \\
&= T^{i_1 \cdots i_r}_{j_1 \cdots j_s} (\partial_{x^{i_1}} \otimes \cdots \otimes \partial_{x^{i_r}} \otimes dx^{j_1} \otimes \cdots \otimes dx^{j_s})(\Omega^1, \cdots, \Omega^r, V_1, \cdots, V_s) \\
&= (T^{i_1 \cdots i_r}_{j_1 \cdots j_s} \partial_{x^{i_1}} \otimes \cdots \otimes \partial_{x^{i_r}} \otimes dx^{j_1} \otimes \cdots \otimes dx^{j_s})(\Omega^1, \cdots, \Omega^r, V_1, \cdots, V_s).
\end{aligned}$$

Exercise 4.13. (Noncommutativity of $\otimes$.)
Consider $\mathbb{R}^2$ with the standard smooth structure given by the global chart $(\mathbb{R}^2, (x,y) \mapsto (x,y))$. Calculate $(dx \otimes dy)(\partial_x, \partial_y)$ and $(dy \otimes dx)(\partial_x, \partial_y)$.

Exercise 4.14. Let M be a smooth manifold, $T \in T^r_s M$, $S \in T^{r'}_{s'} M$, and $(U, \mathbf{x})$ be an admissible chart for M. Show that the chart-induced components of $T \otimes S$ are given by $(T \otimes S)^{i_1 \cdots i_{r+r'}}_{j_1 \cdots j_{s+s'}} = T^{i_1 \cdots i_r}_{j_1 \cdots j_s} S^{i_{r+1} \cdots i_{r+r'}}_{j_{s+1} \cdots j_{s+s'}}$.

The tensor product produces new tensor fields from existing ones. We end this section by learning about yet another operation, called 'contraction'. If T is an (r,s)-tensor field, where $r \geqslant 1$ and $s \geqslant 1$, then a contraction $\mathbf{C}^i_j T$ of T will be an $(r-1, s-1)$ tensor field. This will be useful later to understand complicated tensor fields like the curvature tensor field $\mathbf{R}$, and with appropriate contractions, we will obtain simpler curvatures such as the Ricci curvature tensor field **Ric**. We begin with the following crucial result, which gives an intrinsic function associated with a $(1,1)$-tensor field, akin to the trace of a linear transformation $T : V \to V$ on a finite-dimensional vector space V.

Proposition 4.3. *Suppose that M is a smooth manifold, and $T \in T^1_1 M$. Define $\mathbf{C}T : M \to \mathbb{R}$ by $(\mathbf{C}T)(p) = T(p)((dx^i)_p, \partial_{x^i,p})$, where $(U, \mathbf{x})$ is any admissible chart containing $p \in M$. Then $\mathbf{C}T$ is well-defined, that is, it does not depend on the choice of the admissible chart. Moreover, $\mathbf{C}T \in C^\infty(M)$.*

Proof. With a different admissible chart $(U', \mathbf{x}')$ containing p, we have

$$\begin{aligned}
T(p)((dx'^i)_p, \partial_{x'^i,p}) &= T(p)\Big(\frac{\partial x'^i}{\partial x^j}(\mathbf{x}(p))(dx^j)_p, \frac{\partial x^k}{\partial x'^i}(\mathbf{x}'(p))\, \partial_{x^k,p}\Big) \\
&= \frac{\partial x'^i}{\partial x^j}(\mathbf{x}(p)) \frac{\partial x^k}{\partial x'^i}(\mathbf{x}'(p))\, T(p)((dx^j)_p, \partial_{x^k,p}) \\
&= \delta^k_j\, T(p)((dx^j)_p, \partial_{x^k,p}) \\
&= T(p)((dx^j)_p, \delta^k_j\, \partial_{x^k,p}) = T(p)((dx^j)_p, \partial_{x^j,p}).
\end{aligned}$$

The smoothness of $\mathbf{C}T$ follows from Exercises 4.8 (p.66) and 1.18 (p.14). □

Now we are ready to define the operation of contraction. Given an (r,s)-tensor field T with r and s both at least 1, we note that if we fill in all the 1-form field slots except for one, and all the vector field slots except for one, then we are left with a $(1,1)$-tensor field. We can then apply the contraction map $\mathbf{C}$ to this. The result is a C^∞ function. But clearly this depends $C^\infty(M)$-multilinearly on the initially chosen 1-form fields and the vector fields. Thus we have a $(r-1,s-1)$-tensor field. We give the precise formulation below.

Definition 4.7. (Contraction.)
Let M be a smooth manifold. Let $1 \leqslant i \leqslant r$, $1 \leqslant j \leqslant s$. Let $T \in T^r_s M$.
Define the *contraction* $\mathbf{C}^i_j T \in T^{r-1}_{s-1} M$ *with respect to the* i *and* j *indices* by

$$\begin{aligned}&(\mathbf{C}^i_j T)(\Omega^1, \cdots, \Omega^{r-1}, V_1, \cdots, V_{s-1})\\ &= \mathbf{C}(T(\Omega^1, \cdots, \Omega^{i-1}, \cdot, \Omega^i, \cdots \Omega^{r-1}, V_1, \cdots, V_{j-1}, \cdot, V_j, \cdots, V_{s-1})),\end{aligned}$$

for all $\Omega^1, \cdots, \Omega^{r-1} \in T^0_1 M$ and all $V_1, \cdots, V_{s-1} \in T^1_0 M$.

Exercise 4.15. Let M be a smooth manifold. Let $T \in T^1_3 M$, and $(U, \mathbf{x})$ be an admissible chart for M. Show that $\mathbf{C}^1_2 T$ has the chart-induced components given by $(\mathbf{C}^1_2 T)_{jk} = T^i_{jik}$.

Exercise 4.16. Consider $M = \mathbb{R}^m$ with the standard smooth structure given by the global chart $(\mathbb{R}^m, \mathrm{id})$. Let $\tau = [\tau^i_j]$ be an $m \times m$ matrix. Then matrix multiplication of column vectors in $\mathbb{R}^m$ by τ induces a (smooth) map $\tau : \mathbb{R}^m \to \mathbb{R}^m$. Let the tensor field $T \in T^1_1 M$ be defined by

$$(T(\Omega, V))(p) = \Omega_{\tau(p)}((d\tau)_p V_p) \text{ for } p \in M,\ \Omega \in T^0_1 M \text{ and } V \in T^1_0 M.$$

Show that $\mathbf{C}T \equiv \mathrm{trace}(\tau)$.

The contraction operator can be applied repeatedly. For example, if we have $T \in T^2_2 M$, then $\mathbf{C}(\mathbf{C}^1_2 T)$ is an element of $C^\infty(M)$, and in an admissible chart $(U, \mathbf{x})$, $\mathbf{C}(\mathbf{C}^1_2 T)|_U = T^{ij}_{ji}$.

Exercise 4.17. Let M be a smooth manifold. How many different contractions does tensor field $T \in T^2_3 M$ have in general?

Exercise 4.18. Let M be a smooth manifold. Suppose that $r, s \geqslant 1$, $1 \leqslant i \leqslant r$, and $1 \leqslant j \leqslant s$. Show that $\mathbf{C}^i_j : T^r_s M \to T^{r-1}_{s-1}$ is $C^\infty(M)$-linear, that is, for all $T, S \in T^r_s M$, and all $f \in C^\infty(M)$, $\mathbf{C}^i_j(T + fS) = (\mathbf{C}^i_j T) + f\,\mathbf{C}^i_j S$.

4.4 Pull-back of $(0, s)$-tensor fields

Recall that given a smooth map $f : M \to N$ between two smooth manifolds M and N, any 1-form field $\Omega \in T^0_1 N$ can be pulled back under f to a 1-form field $f^*\Omega \in T_1^0 M$. Now we will see that this operation of pulling back can be extended to $(0,s)$-tensor fields[4] on N.

Definition 4.8. (Pull-back of $(0,s)$-tensor fields.)
Let M, N be smooth manifolds, and let $f : M \to N$ be a smooth map. For $T \in T_s^0 N$, the *pull-back* $f^*T \in T_s^0 M$ of T under f is defined by

$$((f^*T)(V_1, \cdots, V_s))(p) = T(f(p))(df_p((V_1)_p), \cdots, df_p((V_s)_p)),$$

for all $p \in M$, and all $V_1, \cdots, V_s \in T_0^1 M$.

Let us check that f^*T is well-defined. First, for each $p \in M$, the map sending $v_1, \cdots, v_s \in T_p M$ to $T(f(p))(df_p(v_1), \cdots, df_p(v_s))$ is $\mathbb{R}$-multilinear, and thus it defines a $(0,s)$-tensor in $T_s^0 M(p)$. Next, we will use Proposition 4.1 (p.65) to show that f^*T is a tensor field. We need to show that for all vector fields $V_1, \cdots, V_s \in T_0^1 M$, the map $p \mapsto T(f(p))(df_p((V_1)_p), \cdots, df_p((V_s)_p))$ is smooth. We check this locally around a point p by making use of a admissible charts $(U, \mathbf{x})$, $(W, \mathbf{y})$, for M, N, respectively, containing the points p and $f(p)$, respectively, and such that $f(U) \subset W$. We decompose $V_j = V_j(x^i)\,\partial_{x^i}$ $(1 \leqslant j \leqslant s)$, and $T = T_{j_1 \cdots j_s} dy^{j_1} \otimes \cdots \otimes dy^{j_s}$. Then for $q \in U$, we have

$$\begin{aligned}
&((f^*T)(V_1, \cdots, V_s))(q)\\
&= T(f(q))(df_q((V_1)_q), \cdots, df_q((V_s)_q))\\
&= T_{j_1 \cdots j_s}(f(q))\cdot\\
&\quad (dy^{j_1} \otimes \cdots \otimes dy^{j_s})(f(q))\big(df_q((V_1)_q(x^{i_1})\,\partial_{x^{i_1},q}), \cdots, df_q((V_s)_q(x^{i_s})\,\partial_{x^{i_s},q})\big)\\
&= T_{j_1 \cdots j_s}(f(q)) \cdot (V_1)_q(x^{i_1}) \cdots (V_s)_q(x^{i_s})\cdot\\
&\quad (dy^{j_1} \otimes \cdots \otimes dy^{j_s})(f(q))\big(\partial_{x^{i_1},q}(y^{\ell_1} \circ f)\,\partial_{y^{\ell_1},f(q)}, \cdots, \partial_{x^{i_s},q}(y^{\ell_s} \circ f)\,\partial_{y^{\ell_s},f(q)}\big)\\
&= T_{j_1 \cdots j_s}(f(q)) \cdot (V_1)_q(x^{i_1}) \cdots (V_s)_q(x^{i_s}) \cdot \partial_{x^{i_1},q}(y^{j_1} \circ f) \cdots \partial_{x^{i_s},q}(y^{j_s} \circ f),
\end{aligned}$$

and so $U \ni q \mapsto ((f^*T)(V_1, \cdots, V_s))(q)$, being the product of smooth functions, is smooth in U, as wanted.

Exercise 4.19. Let $\mathbb{R}^2$ be equipped with the standard smooth structure, and let the open subsets

$$\begin{aligned}
M &= \{(r,\theta) : r > 0,\ -\tfrac{\pi}{2} < \theta < \tfrac{\pi}{2}\},\\
N &= \{(x,y) : x > 0\}
\end{aligned}$$

be equipped with the induced smooth structure. Let $f : M \to N$ be the smooth map $f(r,\theta) = (r\cos\theta, r\sin\theta)$ for all $(r,\theta) \in M$. Let $T \in T_2^0 N$ be the $(0,2)$-tensor field given by $T = \frac{1}{x^2} dy \otimes dy$ in the chart $(N, (x,y) \mapsto (x,y))$. Determine f^*T.

---

[4]In particular, for example, in the context of Lorentzian manifolds M, N, with their respective metrics $\mathbf{g}, \mathbf{h}$ (which will be certain $(0,2)$-tensor fields), we will be able to talk about the pull-back $f^*\mathbf{h}$ of $\mathbf{h}$ (and if f is a diffeomorphism satisfying $f^*\mathbf{h} = \mathbf{g}$, then we will call f an 'isometry', see Exercise 5.13 on p.83).

Chapter 5

Lorentzian manifolds

So far, we have introduced notions using only the smooth structure on the manifold. But the smooth structure is still rather flexible. Within this picture, two spheres in $\mathbb{R}^3$ of different radii would be considered the same, and even the surface of a potato would be the same smooth manifold as a sphere. In this sense, the smooth manifold has not yet acquired a 'shape'. We now introduce the notion of a metric, which is roughly speaking, a $(0, 2)$-tensor field $\mathbf{g}$ on a manifold M, whose evaluation at any point $p \in M$ equips the tangent space T_pM with a scalar product $\mathbf{g}(p)$ (analogous to the dot product of vectors in the plane). This structure allows us to assign lengths to tangent vectors and angles between them. It will also enable us to assign lengths to curves.

In the mathematical subject of Riemannian geometry, one studies smooth manifolds equipped with a metric, which in each tangent space T_pM gives an 'inner product' $\mathbf{g}(p)$, namely a 'positive definite' scalar product. On the other hand, in spacetime physics, the metric $\mathbf{g}$ on spacetime is such that the scalar product $\mathbf{g}(p)$ on T_pM at each $p \in M$ is 'indefinite', and in fact 'Lorentzian', which means that the scalar product in each tangent space T_pM has the 'index 1' (roughly speaking, there exists a vector $v \in T_pM$ so that the scalar product $\mathbf{g}(p)(v, v) < 0$, and the restriction of $\mathbf{g}(p)$ to the 'orthogonal complement' $v^\perp \subset T_pM$ of v is positive definite). We will see what this means precisely, in the course of this chapter. This structure allows us to talk about observers, the observer's instantaneous perception of space, and of time. We can then introduce the notion of geodesics, which are, loosely speaking, the straightest curves. We will also see in later chapters that the matter content in the spacetime essentially 'curves' spacetime, and determines a tensor field derived from metric $\mathbf{g}$. Test matter moves along geodesics in this curved spacetime.

5.1 Scalar product

We begin by introducing the notion of a scalar product on a real vector space V. Later on, the vector space V will be replaced by particular tangent spaces T_pM, for points p belonging to a smooth manifold M. So we will always consider only finite-dimensional, real vector spaces V.

Definition 5.1. (Scalar product, inner product.)
Let V be a real vector space.
A map $g : V \times V \to \mathbb{R}$ is a *scalar product on* V if it satisfies:
- (Symmetric) For all $v, w \in V$, $g(v,w) = g(w,v)$.
- (Bilinear) For all $u, v, w \in V$ and all $c \in \mathbb{R}$, $g(u+cv, w) = g(u,w) + cg(v,w)$.
- (Nondegenerate) If $v \in V$ is such that for all $w \in V$, $g(v,w) = 0$, then $v = \mathbf{0}$.

Let U be a subspace of V. A symmetric, bilinear map $g : V \times V \to \mathbb{R}$ is
- *positive definite on* U if for all $u \in U \setminus \{\mathbf{0}\}$, $g(u,u) > 0$.
- *negative definite on* U if for all $u \in U \setminus \{\mathbf{0}\}$, $g(u,u) < 0$.
- an *inner product on* V if g is positive definite on V.

If g is positive (or negative) definite on U, then g is nondegenerate on U.

Example 5.1. (Euclidean inner product.)
Let $V = \mathbb{R}^2$, and define $g : \mathbb{R}^2 \times \mathbb{R}^2 \to \mathbb{R}$ by

$$g(\mathbf{v}_1, \mathbf{v}_2) = x_1 x_2 + y_1 y_2 \text{ for all } \mathbf{v}_1 = (x_1, y_1),\ \mathbf{v}_2 = (x_2, y_2) \in \mathbb{R}^2.$$

Then g is an inner product on $\mathbb{R}^2$. Let q denote the 'quadratic form' $\mathbb{R}^2 \ni \mathbf{v} \mapsto g(\mathbf{v}, \mathbf{v})$. The only vector $\mathbf{v} \in \mathbb{R}^2$ such that $q(\mathbf{v}) = 0$ is $\mathbf{v} = \mathbf{0}$. The level sets $\{(x,y) \in \mathbb{R}^2 : x^2 + y^2 = c\}$ of the quadratic form q are circles, as shown on the left in the following picture. ◇

Example 5.2. (Minkowski scalar product.)
Let $V = \mathbb{R}^2$, and define $g : \mathbb{R}^2 \times \mathbb{R}^2 \to \mathbb{R}$ by

$$g(\mathbf{v}_1, \mathbf{v}_2) = x_1 x_2 - t_1 t_2 \text{ for all } \mathbf{v}_1 = (x_1, t_1),\ \mathbf{v}_2 = (x_2, t_2) \in \mathbb{R}^2.$$

We will check that g is a scalar product. But g cannot be an inner product, since, for example, $g(\mathbf{e}_2, \mathbf{e}_2) = -1 < 0$, where $\mathbf{e}_2 = (0,1)$. The bilinearity and symmetry of g are clear. Moreover, if $\mathbf{v} = (x,t) \in \mathbb{R}^2$ is such that $g(\mathbf{v}, \mathbf{w}) = 0$ for all $\mathbf{w} \in \mathbb{R}^2$, then taking in particular $\mathbf{w} = \mathbf{e}_1$ and $\mathbf{w} = \mathbf{e}_2$, we obtain $x = 0$ and $t = 0$, so that $\mathbf{v} = (x,t) = (0,0) = \mathbf{0}$. This shows that g is nondegenerate. The level sets $\{(x,t) \in \mathbb{R}^2 : x^2 - t^2 = c\}$ of the quadratic form $\mathbb{R}^2 \ni \mathbf{v} \mapsto g(\mathbf{v}, \mathbf{v})$, are hyperbolas, as shown on the right in the following picture, with the degenerate case $c = 0$ giving the two lines at an angle of 45° with the coordinate axes. ◇

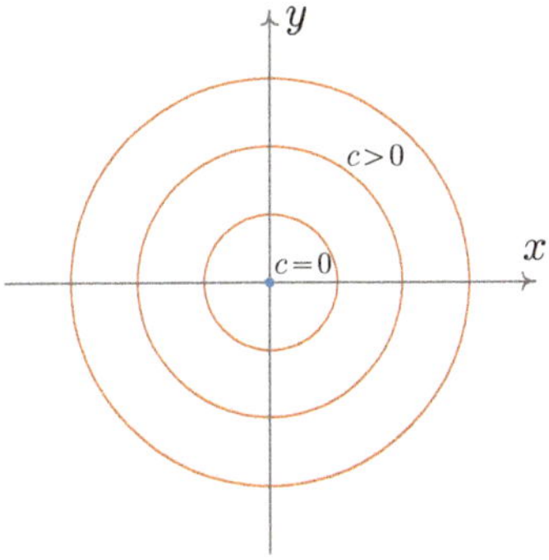

Euclidean plane

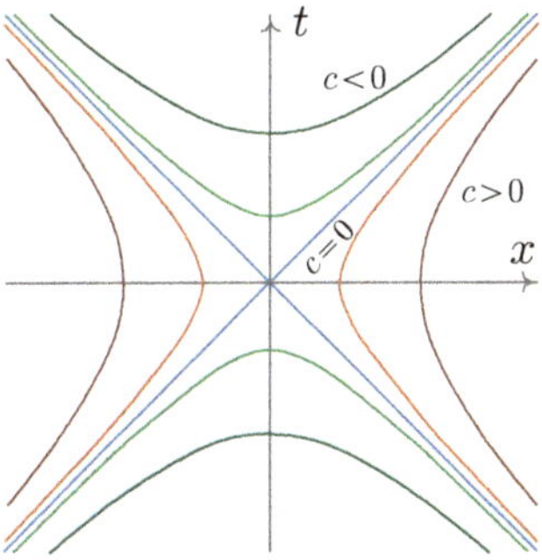

Minkowski plane

Exercise 5.1. Let V be a finite-dimensional vector space with a scalar product g. Let V^* be the dual space of V. Show that the map $V \ni v \mapsto v^\flat := g(v, \cdot) \in V^*$ is a linear isomorphism.

Exercise 5.2. Consider Example 5.2 again. Let $\mathbf{v} = (a, b) \in \mathbb{R}^2$ be such that $b > a > 0$. Depict the set $\mathbf{v}^\perp := \{\mathbf{w} \in \mathbb{R}^2 : g(\mathbf{w}, \mathbf{v}) = 0\}$ in the Minkowski plane.

Definition 5.2. (Index of a scalar product.)
Let V be a finite-dimensional vector space with the scalar product g. Define the set $\mathcal{N}_g = \{U : U$ is a subspace of V, and g is negative definite on $U\}$. The *index of the scalar product* g is defined by $\mathfrak{i}_g = \max\limits_{U \in \mathcal{N}_g} \dim U$.

If $\mathcal{N}_g = \{\mathbf{0}\}$, where $\mathbf{0}$ denotes the subspace $\{0\}$, and $0 \in V$ is the zero vector, then $\mathfrak{i}_g = 0$.

Definition 5.3. (Orthogonal vectors, orthogonal complement.)
Let V be a finite-dimensional vector space with the scalar product g.
Two vectors $v, w \in V$ are said to be *orthogonal* if $g(v, w) = 0$.
Let U be a subset of V. The *orthogonal complement* $U^\perp$ *of* U is the subspace $U^\perp := \{v \in V : g(u, v) = 0$ for all $u \in U\}$. In the case that $U = \operatorname{span}\{v\}$ for a vector $v \in V$, we denote $U^\perp$ simply by $v^\perp$.

Exercise 5.3. Let V be a finite-dimensional vector space with a scalar product g. Let U be a subspace of V. Show that $U \subset U^{\perp\perp}$.

Theorem 5.1.
Let V be a finite-dimensional vector space, and g a scalar product on V. Let U be a subspace of V. Then the following are equivalent:

(1) $V = U + U^\perp$ *and* $U \cap U^\perp = \{0\}$. (*We write* $V = U \oplus U^\perp$.)

(2) $g : U \times U \to \mathbb{R}$ *is nondegenerate,*
i.e., the restriction $g|_{U \times U}$ *is nondegenerate.*

Proof. (1) ⇒ (2): If $u \in U$ is such that for all $u' \in U$, $g(u,u') = 0$, then we have $(U \ni) u \in U^\perp$. So $u \in U \cap U^\perp = \{0\}$.

(2) ⇒ (1): If $u \in U \cap U^\perp$, then for all $u' \in U$, $g(u,u') = 0$. As $g : U \times U \to \mathbb{R}$ is nondegenerate, we conclude that $u = 0$. So $U \cap U^\perp = \{0\}$. Let $k := \dim U$, and $B = \{e_1, \cdots, e_k\}$ be a basis for U. Let $m := \dim V$, and extend B to a basis $B' = \{e_1, \cdots, e_k, f_1, \cdots, f_{m-k}\}$ for V. Let $G := [g(e_i, e_j)] \in \mathbb{R}^{k\times k}$. We claim that G is invertible. Indeed, if the column vector $(c^1, \cdots, c^m)$ is in the kernel of G, then the vector $u := c^i e_i \in U$ is such that $g(e_j, u) = 0$ for each $j \in \{1, \cdots, m\}$, and so $g(u', u) = 0$ for all $u' \in U$. As $g|_{U\times U}$ is nondegenerate, $u = 0$. Thus $c^i = 0$, by the independence of B.

Let $F = [g(e_i, f_j)] \in \mathbb{R}^{k\times(m-k)}$ be the matrix with the entry in the ith row and j^{th} column equal to $g(e_i, f_j)$. Define $A = G^{-1}F \in \mathbb{R}^{k\times(m-k)}$, and denote its entry in the ith row and j^{th} column by α^i_j. Set $f'_j = f_j - \alpha^i_j e_i$, $1 \leqslant j \leqslant m-k$. As $GG^{-1} = I_k$, $[\, g(e_\ell, e_1) \cdots g(e_\ell, e_k)\,]G^{-1} = [\,\delta^1_\ell \cdots \delta^k_\ell\,]$, $1 \leqslant \ell \leqslant k$. Each $f'_j \in U^\perp$ because for all $\ell \in \{1, \cdots, k\}$, we have

$$\begin{aligned} g(f'_j, e_\ell) &= g(f_j - \alpha^i_j e_i, e_\ell) = g(f_j, e_\ell) - g(e_i, e_\ell)\alpha^i_j = g(f_j, e_\ell) - g(e_\ell, e_i)\alpha^i_j \\ &= g(f_j, e_\ell) - \big[\, g(e_\ell, e_1) \cdots g(e_\ell, e_k)\,\big]G^{-1}\begin{bmatrix} g(e_1, f_j) \\ \vdots \\ g(e_k, f_j) \end{bmatrix} \\ &= g(f_j, e_\ell) - \delta^i_\ell g(e_i, f_j) = 0. \end{aligned}$$

Let $v \in V$. As B' is a basis for V, there exist $c^1, \cdots, c^k, d^1, \cdots, d^{m-k} \in \mathbb{R}$ such that $v = c^i e_i + d^j f_j = c^i e_i + d^j(f'_j + \alpha^i_j e_i) = (c^i + \alpha^i_j d^j)e_i + d^j f'_j = u + w$, where $u = (c^i + \alpha^i_j d^j)e_i \in \operatorname{span} B = U$, and $w = d^j f'_j \in \operatorname{span}\{f'_1, \cdots, f'_{m-k}\} \subset U^\perp$. So $U + U^\perp = V$. □

In particular, if $U \in \mathcal{N}_g$, then $V = U \oplus U^\perp$.

Theorem 5.2. *Let V be a finite-dimensional vector space, and g a scalar product on V. If $U \in \mathcal{N}_g$ and $\dim U = \mathfrak{i}_g$, then $g : U^\perp \times U^\perp \to \mathbb{R}$ is positive definite.*

Proof. Suppose that $v \in U^\perp \setminus \{0\}$ is such that $g(v,v) \leqslant 0$.

Let us show we can assume with no loss of generality that $g(v,v) < 0$. Suppose that $g(v,v) = 0$. Since $v \neq 0$, and as g is nondegenerate on V, there exists a $w \in V$ such that $g(v,w) \neq 0$, and we can choose[1] $w \in U^\perp$. For all $\lambda \in \mathbb{R}$, we have $v_\lambda := \lambda v + w \in U^\perp$, and

$$g(v_\lambda, v_\lambda) = \lambda^2 g(v,v) + g(w,w) + 2\lambda g(v,w) = 0 + g(w,w) + 2\lambda g(v,w).$$

[1] As $U \in \mathcal{N}_g$, we have $V = U \oplus U^\perp$. So by decomposing $w = w' + w''$, $w' \in U$, $w'' \in U^\perp$, we have that $0 \neq g(v,w) = g(v,w') + g(v,w'') = 0 + g(v,w'')$.

As $\lambda \mapsto g(v_\lambda, v_\lambda)$ a linear function of λ with a nonzero slope, it is negative for some $\lambda \in \mathbb{R}$.

We now show $g(v,v)<0$ is impossible. Set $U' = \{u + \alpha v : \alpha \in \mathbb{R},\, u \in U\}$. For $u' = u + \alpha v \in U' \setminus \{0\}$,

$$g(u', u') = g(u,u) + \alpha^2 g(v,v) + 2\alpha g(u,v) = g(u,u) + \alpha^2 g(v,v) < 0,$$

as both summands are nonpositive, and $u' \neq 0$ (which implies $u \neq 0$ or $\alpha \neq 0$). But then $U' \in \mathcal{N}_g$, contradicting the definition of $\mathfrak{i}_g$, since we have that $\dim U' = \dim U + 1 > \dim U = \mathfrak{i}_g$. □

We recall the Gram-Schmidt orthonormalisation procedure for producing an orthonormal set out of an independent set, while preserving the span.

Proposition 5.1. (Gram-Schmidt orthonormalisation.)
Let V be a vector space, and let g be a scalar product on V. Let U be a subspace of V such that g is positive definite on U. Let $\{u_1, \cdots, u_k\}$ be an independent set of vectors in U. With $\|u\| := \sqrt{g(u,u)}$ for $u \in U$, set

$$e_1 = \frac{u_1}{\|u_1\|}, \quad e_\ell = \frac{u_\ell - g(u_\ell, e_1)e_1 - \cdots - g(u_\ell, e_{\ell-1})e_{\ell-1}}{\|u_\ell - g(u_\ell, e_1)e_1 - \cdots - g(u_\ell, e_{\ell-1})e_{\ell-1}\|} \text{ for } 2 \leqslant \ell \leqslant k.$$

Then $\{e_1, \cdots, e_k\}$ is an orthonormal set in U, that is, $g(e_i, e_j) = \delta_{ij}$ for $1 \leqslant i, j \leqslant k$, and for all $1 \leqslant \ell \leqslant k$, $\operatorname{span}\{u_1, \cdots, u_\ell\} = \operatorname{span}\{e_1, \cdots, e_\ell\}$.

Proof. We proceed inductively. If $k = 1$, then, as $\{u_1\}$ is independent, $u_1 \neq 0$. Thus the vector $e_1 = u_1/\|u_1\|$ is well-defined, $g(e_1, e_1) = 1$, and $\operatorname{span}\{u_1\} = \operatorname{span}\{e_1\}$. Suppose the claim holds for $k = \ell$ for some ℓ. Let $\{u_1, \cdots, u_{\ell+1}\}$ be an independent set in U. Then also $\{u_1, \cdots, u_\ell\}$ is independent. By the induction hypothesis, there exists an orthonormal set $\{e_1, \cdots, e_\ell\}$ such that $\operatorname{span}\{u_1, \cdots, u_\ell\} = \operatorname{span}\{e_1, \cdots, e_\ell\}$. As $\{u_1, \cdots, u_{\ell+1}\}$ is independent, $u_{\ell+1} \notin \operatorname{span}\{u_1, \cdots, u_\ell\} = \operatorname{span}\{e_1, \cdots, e_\ell\}$. It follows that the vector $v := u_{\ell+1} - (g(u_{\ell+1}, e_1)e_1 + \cdots + g(u_{\ell+1}, e_\ell)e_\ell) \neq 0$. So $e_{\ell+1} = v/\|v\| \in U$ is well-defined. Trivially $g(e_{\ell+1}, e_{\ell+1}) = 1$. For all $i \leqslant \ell$, $g(v, e_i) = g(u_{\ell+1}, e_i) - g(u_{\ell+1}, e_i)g(e_i, e_i) = 0$, and so $g(e_{\ell+1}, e_i) = 0$. Hence $\{e_1, \cdots, e_\ell, e_{\ell+1}\}$ is an orthonormal set. Finally, as

$$e_{\ell+1} = \frac{1}{\|v\|}\Big(u_{\ell+1} - \underbrace{\boxed{\quad\cdots\quad}}_{\substack{\in \operatorname{span}\{e_1, \cdots, e_\ell\} \\ = \operatorname{span}\{u_1, \cdots, u_\ell\}}}\Big),$$

we have

$$\operatorname{span}\{e_1, \cdots, e_\ell, e_{\ell+1}\} = \operatorname{span}\{u_1, \cdots, u_\ell, e_{\ell+1}\} = \operatorname{span}\{u_1, \cdots, u_\ell, u_{\ell+1}\}.$$

This completes the proof. □

Theorem 5.3. *Each scalar product g on an m-dimensional vector space V admits a basis $\{e_1, \cdots, e_m\}$, satisfying*

$$g(e_i, e_j) = \begin{cases} -\delta_{ij} & \text{if } i \leqslant \mathfrak{i}_g, \\ \delta_{ij} & \text{if } i > \mathfrak{i}_g. \end{cases}$$

Such a basis is called an orthonormal basis.

Proof. If $U \in \mathcal{N}_g$ and $\dim U = \mathfrak{i}_g$, then we can choose an orthonormal basis $\{e_1, \cdots, e_{\mathfrak{i}_g}\}$ with respect to the positive definite scalar product $-g : U \times U \to \mathbb{R}$ on U. Then we have $g(e_i, e_j) = -\delta_{ij}$ for all $1 \leqslant i, j \leqslant \mathfrak{i}_g$. Moreover, we can construct an orthonormal basis $\{e_{\mathfrak{i}_g+1}, \cdots, e_m\}$ with respect to the positive definite scalar product $g : U^\perp \times U^\perp \to \mathbb{R}$ on $U^\perp$, i.e., $g(e_i, e_j) = \delta_{ij}$, for all $\mathfrak{i}_g + 1 \leqslant i, j \leqslant m$. As $V = U \oplus U^\perp$, $\{e_1, \cdots, e_m\}$ is a basis for V. □

The following result will be useful for determining the index of a scalar product.

Theorem 5.4. *Let V be an m-dimensional vector space, and g a scalar product on V. Let $\{e_1, \cdots, e_m\}$ be a basis for V, such that there exists an $i_* \in \{0, 1, \cdots, m\}$ such that*

$$g(e_i, e_j) = \begin{cases} -\delta_{ij} & \text{if } i \leqslant i_*, \\ \delta_{ij} & \text{if } i > i_*. \end{cases}$$

Then $i_* = \mathfrak{i}_g$.

Proof. Let the subspace $U \in \mathcal{N}_g$ be such that $\dim U = \mathfrak{i}_g$. Define the map $L : U \to \operatorname{span}\{e_1, \cdots, e_{i_*}\}$ as follows: For $u = u^i e_i$, $Lu := u^1 e_1 + \cdots + u^{i_*} e_{i_*}$. Then L is easily seen to be linear. We now show it is injective. If $Lu = 0$, that is, $u^i = 0$ for $i = 1, \cdots, i_*$, then we have

$$g(u, u) = g(u^i e_i, u^j e_j) = \sum_{i=i_*+1}^{m} (u^i)^2 \geqslant 0.$$

But since $-g$ is positive definite on $U \in \mathcal{N}_g$, we can now conclude that $u = 0$. Hence L is injective. It follows by the rank-nullity theorem that

$$\mathfrak{i}_g = \dim U = \text{rank } L + \dim\ker L = \text{rank } L + 0 \leqslant \dim \operatorname{span}\{e_1, \cdots, e_{i_*}\} = i_*.$$

So $\mathfrak{i}_g \leqslant i_*$.

For the reverse inequality, we'll show that $U' := \operatorname{span}\{e_1, \cdots, e_{i_*}\} \in \mathcal{N}_g$. Then $\dim U' = i_* \leqslant \mathfrak{i}_g$ by the definition of $\mathfrak{i}_g$. If $u' = c^1 e_1 + \cdots + c^{i_*} e_{i_*}$ for some $c^1, \cdots, c^{i_*} \in \mathbb{R}$, then we have

$$g(u', u') = -((c^1)^2 + \cdots + (c^{i_*})^2) \leqslant 0,$$

and $g(u', u') = 0$ only if $c^1 = \cdots = c^{i_*} = 0$, that is, $u' = 0$. Thus $\mathfrak{i}_g = i_*$. □

Definition 5.4. (Minkowski scalar product.)
Let V be a finite-dimensional real vector space. A scalar product g on V, with index $\mathfrak{i}_g = 1$ is called a *Minkowski scalar product.*

Besides the Minkowski plane considered earlier in Example 5.2, the following higher-dimensional generalisation is an example of a vector space with a Minkowski scalar product.

Example 5.3. Let $\langle \cdot, \cdot \rangle$ be the Euclidean inner product on $\mathbb{R}^{m-1}$. On the vector space $V = \mathbb{R}^m = \mathbb{R} \times \mathbb{R}^{m-1}$, define the scalar product η by

$$\eta((u_0, \mathbf{u}), (v_0, \mathbf{v})) = -u_0 v_0 + \langle \mathbf{u}, \mathbf{v} \rangle.$$

Then η is a Minkowski scalar product on $\mathbb{R}^m$. The standard basis vectors form an orthonormal basis for $\mathbb{R}^m$ with respect to η:

$$e_0 = \begin{bmatrix} 1 \\ 0 \\ 0 \\ \vdots \\ 0 \end{bmatrix}, \quad e_1 = \begin{bmatrix} 0 \\ 1 \\ 0 \\ \vdots \\ 0 \end{bmatrix}, \quad \cdots, \quad e_{m-1} = \begin{bmatrix} 0 \\ 0 \\ \vdots \\ 0 \\ 1 \end{bmatrix}.$$

In Example 5.5, we give many more orthonormal bases. ◇

The general case of a Minkowski scalar product can be reduced to the above example by choosing an orthonormal basis $\{e_0, e_1, \cdots, e_{m-1}\}$ such that $g(e_0, e_0) = -1$ and $g(e_i, e_i) = 1$ for $1 \leqslant i \leqslant m-1$. Then for

$$u = u^0 e_0 + \sum_{i=1}^{m-1} u^i e_i, \qquad v = v^0 e_0 + \sum_{i=1}^{m-1} v^i e_i, \tag{5.1}$$

with $\mathbf{u} := (u^1, \cdots, u^{m-1}) \in \mathbb{R}^{m-1}$, $\mathbf{v} := (v^1, \cdots, v^{m-1}) \in \mathbb{R}^{m-1}$, we have that $g(u, v) = -u^0 v^0 + \langle \mathbf{u}, \mathbf{v} \rangle$.

Definition 5.5. (Timelike/spacelike/lightlike vector, causal vector.)
Let V be a real finite-dimensional vector space, and g a Minkowski scalar product on V. A vector $v \in V$ is called

- *timelike* if $g(v, v) < 0$
- *spacelike* if $g(v, v) > 0$
- *lightlike* if $v \neq 0$ and $g(v, v) = 0$.

A vector $v \in V$ is called *causal* if it is timelike or lightlike.

So $V = \{\text{spacelike vectors}\} \cup \{\text{timelike vectors}\} \cup \{\text{lightlike vectors}\} \cup \{0\}$.

Example 5.4. Let us revisit Example 5.3. We consider $\mathbb{R}^m$ as a topological space with its usual Euclidean topology (this is the induced topology when

$\mathbb{R}^m$ is considered as a smooth manifold with its standard smooth structure). A vector $(v^0, \mathbf{v}) \in \mathbb{R} \times \mathbb{R}^{m-1}$ is timelike if $-(v^0)^2 + \|\mathbf{v}\|^2 < 0$. Thus $v^0 \neq 0$. For a fixed $v^0 \neq 0$, the collection of all vectors $\mathbf{v} \in \mathbb{R}^{m-1}$ such that $(v^0, \mathbf{v})$ is timelike, is an open ball $B(\mathbf{0}, |v^0|) = \{\mathbf{v} \in \mathbb{R}^{m-1} : \|\mathbf{v}\| < |v^0|\}$. The set of all timelike vectors has two connected components C and $-C$, where

$$C = \{(v^0, \mathbf{v}) \in \mathbb{R}^m : v^0 > 0, \ \|\mathbf{v}\|^2 < (v^0)^2\} \text{ (a solid open spherical cone).}$$

We note that $\text{span}((-C) \cup C) = V$ and also that $(-C) \cap C = \varnothing$. The set of lightlike vectors is the set $\{(v^0, \mathbf{v}) \in \mathbb{R}^m : \|\mathbf{v}\| = |v^0| \neq 0\}$. ◇

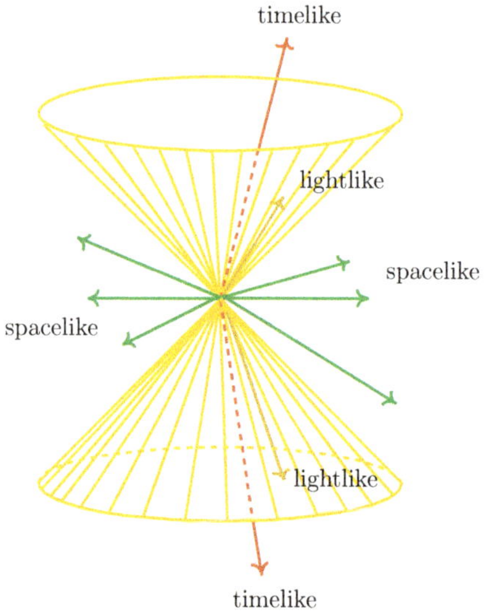

Exercise 5.4. Let V be a finite-dimensional vector space and g a Minkowski scalar product on V. Let $u, v \in V$ be two lightlike vectors. Prove that u, v are orthogonal if and only if there exists a $t \in \mathbb{R} \setminus \{0\}$ such that $u = tv$. *Hint:* Use the decomposition (5.1), using an orthonormal basis for V. For the 'only if' part, use the equality condition in the Cauchy-Schwarz inequality for the Euclidean inner product.

Lemma 5.1. *Let V be a finite-dimensional vector space, and g a Minkowski scalar product on V. Let $v \in V \setminus \{0\}$ be orthogonal to a timelike vector u. Then v is spacelike.*

Proof. This follows from Theorem 5.2 (p.76), by taking $U = \text{span}\{u\}$. Then as the vector u is timelike, $U \in \mathcal{N}_g$, and moreover, $\dim U = 1 = \mathfrak{i}_g$. So $g : U^\perp \times U^\perp \to \mathbb{R}$ is positive definite. As $v \in U^\perp \setminus \{0\}$, it follows that $g(v, v) > 0$, that is, v is spacelike. □

Exercise 5.5. Let g be a Minkowski scalar product on a finite-dimensional vector space V. On the set T of timelike vectors in V, define the relation $\sim$ as follows: For $v, w \in T$, $v \sim w$ if $g(v,w) < 0$. Prove that $\sim$ is an equivalence relation[2].
Hint: For transitivity, proceed as follows. If $v \sim w$ and $w \sim x$, then $g(v,w) < 0$ and $g(x,w) < 0$. So there is a $k > 0$ such that $g(v,w) = kg(x,w)$, i.e., $g(w, v - kx) = 0$. By Lemma 5.1, conclude $v - kx$ is either 0 or spacelike, and in each case $v \sim x$.

Proposition 5.2. (Reversed Cauchy-Schwarz.)
Let g be a Minkowski scalar product on a finite-dimensional vector space V. If $v, w \in V$ are timelike, then $(g(v,w))^2 \geqslant g(v,v)\,g(w,w)$.
Equality holds if and only if v, w are dependent.

Proof. Let us define $u := av - bw$, $a := g(v,w)$, and $b := g(v,v) < 0$. Then we have $g(u,v) = ag(v,v) - bg(w,v) = 0$. So u, v are orthogonal, and as v is timelike, by Lemma 5.1, either $u = 0$ or u is spacelike. In either case, we have $g(u,u) \geqslant 0$, i.e., $a^2 g(v,v) + b^2 g(w,w) \geqslant 2ab\, g(v,w) = 2(g(v,w))^2 g(v,v)$. Dividing by $g(v,v) < 0$, we obtain $(g(v,w))^2 + g(v,v)\,g(w,w) \leqslant 2(g(v,w))^2$, giving the desired inequality.

If equality holds, then $a^2 = (g(v,w))^2 = g(v,v)\,g(w,w)$, and so

$$\begin{aligned} g(u,u) &= a^2 g(v,v) + b^2 g(w,w) - 2ab\, g(v,w) = a^2 b + b^2 g(w,w) - 2aba \\ &= b^2 g(w,w) - a^2 b = (g(v,v))^2 g(w,w) - g(v,v)\, g(w,w)\, g(v,v) = 0. \end{aligned}$$

But, as observed above, u is spacelike or $u = 0$. But u is not spacelike since $g(u,u) = 0$. So $u = 0$, that is, $av - bw = 0$. Since $b < 0$, v, w are dependent.

If v, w are dependent, then since each is nonzero, we have $v = tw$ for some $t \in \mathbb{R}\setminus\{0\}$, and then

$$(g(v,w))^2 = (g(tw,w))^2 = t^2 (g(w,w))^2 = g(tw,tw)\, g(w,w) = g(v,v)\, g(w,w).$$

This completes the proof. □

Exercise 5.6. (Reversed triangle inequality.)
Let g be a Minkowski scalar product on a finite-dimensional vector space V. For timelike $w \in V$, define $\tau(w) = \sqrt{-g(w,w)}$. Let $u, v \in V$ be timelike vectors such that $g(u,v) < 0$. Show that $u + v$ is timelike and that $\tau(u+v) \geqslant \tau(u) + \tau(v)$. (In Theorem 5.5, we will see that this inequality is related to the 'twin paradox'.)

Definition 5.6. (Lorentz transformation.)
Let η be the Minkowski scalar product on $\mathbb{R}^m$ given in Example 5.3 (p.79). A matrix $\Lambda \in \mathbb{R}^{m \times m}$ is a *Lorentz transformation* if for all $u, v \in \mathbb{R}^m$,

$$\eta(\Lambda u, \Lambda v) = \eta(u, v).$$

The set of all Lorentz transformations is denoted by $O(1, m-1)$.

[2]We remark that in particular, $\sim$ is transitive. After we discuss time-orientation on a Lorentzian manifold, this will imply that if we have two future-pointing timelike vectors at a point, then their scalar product will be negative.

Exercise 5.7. Let $[\eta] \in \mathbb{R}^{m\times m}$ be the diagonal matrix with the entries $-1, 1, \cdots, 1$ along the diagonal. The entry in the ith row and j^{th} column of $[\eta]$ (respectively Λ) is denoted by η_{ij} (respectively Λ^i_j), the indices being counted from 0, i.e., $0 \leqslant i, j \leqslant m-1$. Show that the following are equivalent:

(1) $\Lambda \in O(1, m-1)$.

(2) $\Lambda^{\mathrm{t}}[\eta]\Lambda = [\eta]$.

(3) $\eta_{ij} = \Lambda^k_i \Lambda^\ell_j \eta_{k\ell}$.

Exercise 5.8. Show the inclusion $O(1, m-1) \subset \mathrm{GL}_m(\mathbb{R})$, that is, every Lorentz transformation is invertible.

Exercise 5.9. Show that $O(1, m-1)$ is a subgroup of the group $\mathrm{GL}_m(\mathbb{R})$.

Exercise 5.10. Show that in $\mathbb{R}^4$, the following are Lorentz transformations. These are examples of a 'boost', a spatial rotation, 'time-reversal', and a spatial reflection, respectively. Here $\phi, \theta \in \mathbb{R}$.

$$\begin{bmatrix} \cosh\phi & \sinh\phi & \\ \sinh\phi & \cosh\phi & \\ & & I_2 \end{bmatrix}, \quad \begin{bmatrix} 1 & & & \\ & \cos\theta & -\sin\theta & \\ & \sin\theta & \cos\theta & \\ & & & 1 \end{bmatrix}, \quad \begin{bmatrix} -1 & \\ & I_3 \end{bmatrix}, \quad \begin{bmatrix} 1 & & \\ & -1 & \\ & & I_2 \end{bmatrix}.$$

Exercise 5.11. Show that $O(1, m-1)$ is a submanifold of dimension $m(m-1)/2$ of the real vector space $\mathbb{R}^{m\times m} \simeq \mathbb{R}^{m^2}$ (with the standard smooth structure).

Proposition 5.3. *Let V be an m-dimensional vector space, with a Minkowski scalar product g. Let $B = \{e_0, \cdots, e_{m-1}\}$ be an orthonormal basis for V with respect to g such that $g(e_0, e_0) = -1$, $g(e_i, e_i) = 1$ for $1 \leqslant i \leqslant m-1$. Let $B' = \{f_0, \cdots, f_{m-1}\}$ be another basis for V, and Λ be the transformation matrix taking B to B', i.e., $f_j = \Lambda^i_j e_i$. Then the following are equivalent:*

(1) B' is an orthonormal basis for V, $g(f_0, f_0) = -1, g(f_i, f_i) = 1\,(1 \leqslant i \leqslant m-1)$.

(2) $\Lambda \in O(1, m-1)$.

Proof. Let $[\eta_{ij}]$ be the diagonal matrix with the diagonal entries $-1, 1, \cdots, 1$. Then

$$g(f_i, f_j) = g(\Lambda^k_i e_k, \Lambda^\ell_j e_\ell) = \Lambda^k_i \Lambda^\ell_j g(e_k, e_\ell) = \Lambda^k_i \Lambda^\ell_j \eta_{k\ell}. \qquad (\star)$$

(1)⇒(2): $\eta_{ij} \overset{(1)}{=} g(f_i, f_j) \overset{(\star)}{=} \Lambda^k_i \Lambda^\ell_j \eta_{k\ell}$, and by Exercise 5.7, $\Lambda \in O(1, m-1)$.

(2)⇒(1): $g(f_i, f_j) \overset{(\star)}{=} \Lambda^k_i \Lambda^\ell_j \eta_{k\ell} = \eta_{ij}$. The last equality follows from the fact that $\Lambda \in O(1, m-1)$, and using Exercise 5.7. □

Example 5.5. In Example 5.3, we had seen that $\{e_0, e_1, \cdots, e_{m-1}\}$ forms an orthonormal basis with respect to the Minkowski scalar product η on $\mathbb{R}^m$. For any matrix $\Lambda \in O(1, m-1)$, define the vectors $f_j = \Lambda^i_j e_i$ for all $j = 0, 1, \cdots, m-1$. Then $\{f_0, f_1, \cdots, f_{m-1}\}$ is an orthonormal basis. We note that $f_0, f_1, \cdots, f_{m-1}$ are the m columns of the matrix Λ. ◇

5.2 Semi-Riemannian, Riemannian, Lorentzian manifolds

Definition 5.7. (Metric, Semi-Riemannian/Lorentzian manifold.)
Let M be an m-dimensional smooth manifold. A *metric* $\mathbf{g}$ *on* M is a $(0,2)$-tensor field $\mathbf{g} \in T_2^0 M$, such that there exists an $\iota(\mathbf{g}) \in \{0, 1, \cdots, m\}$ so that for all points $p \in M$, $\mathbf{g}(p) : T_pM \times T_pM \to \mathbb{R}$ is a scalar product with the index $\mathfrak{i}_{\mathbf{g}(p)} = \iota(\mathbf{g})$. We call $\iota(\mathbf{g})$ the *metric index*.

The pair $(M, \mathbf{g})$, where M is smooth manifold, and $\mathbf{g}$ a metric on M, is called a *semi-Riemannian manifold*. A semi-Riemannian manifold $(M, \mathbf{g})$ is said to be a

- *Lorentzian manifold* if $\iota(\mathbf{g}) = 1$ and $m \geqslant 2$.
- *Riemannian manifold* if $\iota(\mathbf{g}) = 0$.

If $\mathbf{g}$ is a metric, then so is $-\mathbf{g}$, and $\iota(-\mathbf{g}) = m - \iota(\mathbf{g})$. So if $\iota(\mathbf{g}) = m-1 \geqslant 1$, then up to a sign on the metric, we have essentially a Lorentzian manifold. Similarly if $\iota(\mathbf{g}) = m$, then we have Riemannian manifold. Throughout this book, we will arrange $\iota(\mathbf{g}) = 1$ for our Lorentzian manifolds, rather than $\iota(\mathbf{g}) = m - 1$ (although some literature uses this).

Exercise 5.12. Let $(M, \mathbf{g})$ be a semi-Riemannian manifold and let $V, \widetilde{V} \in T_0^1 M$. Suppose for all $W \in T_0^1 M$, $\mathbf{g}(V, W) = \mathbf{g}(\widetilde{V}, W)$. Prove that $V = \widetilde{V}$.
Hint: Show pointwise equality using the nondegeneracy of $\mathbf{g}(p)$, and the extendibility of each $w \in T_pM$ to a $W \in T_0^1 M$ (Lemma 3.1, p.41). (This will be used when discussing the 'Koszul formula' defining the Levi-Civita connection.)

Exercise 5.13. Suppose that $(M, \mathbf{g})$ and $(N, \mathbf{h})$ are semi-Riemannian manifolds. We call a map $f : M \to N$ an *isometry* if f is a diffeomorphism, and $f^*\mathbf{h} = \mathbf{g}$. Show that $\text{Iso}(M) = \{f : M \to M \,|\, f \text{ is an isometry}\}$ is a subgroup of $(\text{Diff}(M), \circ)$ (the group of diffeomorphisms; see Exercise 1.23, p.15).

Exercise 5.14. Consider $\mathbb{R}^3$ as a smooth manifold with the standard smooth structure. Equip $\mathbb{R}^3$ with the metric $\mathbf{g}$, given in the chart $(\mathbb{R}^3, (x, y, z) \mapsto (x, y, z))$ by $\mathbf{g} = dx \otimes dx + dy \otimes dy + dz \otimes dz$. Determine $\mathbf{g}$ in the 'spherical coordinates' chart $(U, \mathbf{x})$ described in Exercise 2.18 (p.35). For $t \in \mathbb{R}$, let $\psi_t : \mathbb{R}^3 \to \mathbb{R}^3$ be the map $\psi_t(x, y, z) = (x, (\cos t)y - (\sin t)z, (\sin t)y + (\cos t)z)$ for $(x, y, z) \in \mathbb{R}^3$. Show that ψ_t is an isometry.

Example 5.6. (Minkowski spacetime.) Let M be an affine space over a 4-dimensional vector space V, with the smooth structure as in Example 1.9 (p.8). Suppose that V has been equipped with a Minkowski scalar product g. Let $\{\mathbf{f}_0, \mathbf{f}_1, \mathbf{f}_2, \mathbf{f}_3\}$ be any orthonormal basis for V with respect to g, with $g(\mathbf{f}_0, \mathbf{f}_0) = -1$, $g(\mathbf{f}_i, \mathbf{f}_i) = 1$ $(i = 1, 2, 3)$. Fix any $p \in M$. This induces an admissible chart covering all of M, and is called an *inertial frame*. The chart map (x^0, x^1, x^2, x^3) is such that for any $q \in M$, $q = p + x^i(q)\mathbf{f}_i$ (summation over $i = 0, 1, 2, 3$). Define the metric $\mathbf{g}$ by specifying its chart-induced components:

$$\mathbf{g}_{ij} = \mathbf{g}(\partial_{x^i}, \partial_{x^j}) \equiv \eta_{ij},$$

where $\eta_{00} = -1$, $\eta_{11} = \eta_{22} = \eta_{33} = 1$, and whenever $i \neq j$, $\eta_{ij} = 0$. Hence $\mathbf{g}(q) : T_qM \times T_qM \to \mathbb{R}$ is given by

$$\mathbf{g}(q)(v^i \partial_{x^i,q}, w^j \partial_{x^j,q}) = \eta_{ij} v^i w^j = -v^0 w^0 + v^1 w^1 + v^2 w^2 + v^3 w^3.$$

By Proposition 4.1 (p.65), it follows that $\mathbf{g} \in T_2^0 M$. Hence $\mathbf{g}$ is a metric with $\iota(\mathbf{g}) = 1$.

Furthermore, the metric is independent of the original choice of the orthonormal basis and of the 'base' point p, and this can be seen as follows. Let $\widetilde{p} \in M$, and let $\widetilde{B} = \{\widetilde{\mathbf{f}}_0, \widetilde{\mathbf{f}}_1, \widetilde{\mathbf{f}}_2, \widetilde{\mathbf{f}}_3\}$ be another orthonormal basis, with $g(\widetilde{\mathbf{f}}_0, \widetilde{\mathbf{f}}_0) = -1$, $g(\widetilde{\mathbf{f}}_i, \widetilde{\mathbf{f}}_i) = 1$ $(i = 1, 2, 3)$. We denote the corresponding induced chart map by $\widetilde{\mathbf{x}} : M \to \mathbb{R}^4$. Then by Proposition 5.3 (p.82), there exists a Lorentz transformation Λ such that

$$\mathbf{f}_j = \Lambda^i_j \widetilde{\mathbf{f}}_i.$$

By Examples 1.7 (p.5) and 2.3 (p.27), if $T_qM \ni v = v^i \partial_{x^i,q} = \widetilde{v}^i \partial_{\widetilde{x}^i,q}$, then

$$\widetilde{v}^i = \Lambda^i_j v^j.$$

Thus if $\widetilde{\mathbf{g}}$ is the metric obtained by the global chart induced by $\widetilde{p}$ and $\widetilde{B}$, then for $v, w \in T_qM$,

$$\widetilde{\mathbf{g}}(q)(v, w) = \eta_{ij} \widetilde{v}^i \widetilde{w}^j = \eta_{ij} \Lambda^i_k v^k \Lambda^j_\ell w^\ell = \eta_{k\ell} v^k w^\ell = \mathbf{g}(q)(v, w).$$

The Lorentzian manifold $(M, \mathbf{g})$ thus obtained will be called *Minkowski spacetime.* $\diamond$

For a surface M in $\mathbb{R}^3$, any tangent vector $v \in T_pM$ can act on the three smooth functions $x, y, z : M \to \mathbb{R}$, and produce a vector (vx, vy, vz) in $\mathbb{R}^3$. If $\mathbb{R}^3$ is equipped with a scalar product g (for example the Euclidean inner product), then for $p \in M$, we define $\mathbf{g}(p) : T_pM \times T_pM \to \mathbb{R}$ by

$$\mathbf{g}(p)(v, w) = g((vx, vy, vz), (wx, wy, wz)) \text{ for } v, w \in T_pM.$$

Then $\mathbf{g}(p) : T_pM \times T_pM \to \mathbb{R}$ is bilinear, and symmetric. In particular, $\mathbf{g}(p)$ is a $(0, 2)$-tensor on T_pM. Using Proposition 4.1 (p.65), it can be shown that $\mathbf{g} \in T_2^0 M$. Suppose that $\mathbf{g}(p)$ is nondegenerate on T_pM for each $p \in M$, and that the index of $\mathbf{g}(p)$ is constant. Then $\mathbf{g}$ is a metric on M, called the *induced metric on* M *from* $(\mathbb{R}^3, g)$, and $(M, \mathbf{g})$ is a semi-Riemannian manifold. We note that if we use the Euclidean inner product on $\mathbb{R}^3$, then $\mathbf{g}(p)$ will be positive definite on T_pM for each $p \in M$, and so $(M, \mathbf{g})$ is a Riemannian manifold in this case. We will now consider an example.

Example 5.7. (Sphere.)
Consider the sphere as a smooth manifold with the smooth structure given by the atlas in Example 1.5 (p.5). We use the compatible atlas of spherical polar coordinates described in Example 1.8 (p.7). The image of the injective map $(0,\pi)\times(0,2\pi) \ni (\theta,\phi) \mapsto ((\sin\theta)\cos\phi,\, (\sin\theta)\sin\phi,\, \cos\theta) \in S^2 \subset \mathbb{R}^3$ is denoted by U, and the inverse of this map is denoted by φ. Then (U,φ) is an admissible chart. We equip $\mathbb{R}^3$ with the Euclidean inner product. Then the chart-induced component $\mathbf{g}(p)_{\theta\theta}$ of $\mathbf{g}$ at $p \in S^2$, where $\varphi(p) = (\theta_0, \phi_0)$ is

$$\begin{aligned}\mathbf{g}(p)_{\theta\theta} &= \mathbf{g}(p)(\partial_{\theta,p}, \partial_{\theta,p}) = (\partial_{\theta,p}x)^2 + (\partial_{\theta,p}y)^2 + (\partial_{\theta,p}z)^2 \\ &= \Big(\frac{\partial((\sin\theta)\cos\phi)}{\partial\theta}(\theta_0,\phi_0)\Big)^2 + \Big(\frac{\partial((\sin\theta)\sin\phi)}{\partial\theta}(\theta_0,\phi_0)\Big)^2 + \Big(\frac{\partial\cos\theta}{\partial\theta}(\theta_0,\phi_0)\Big)^2 \\ &= (\cos\theta_0)^2(\cos\phi_0)^2 + (\cos\theta_0)^2(\sin\phi_0)^2 + (-\sin\theta_0)^2 = 1.\end{aligned}$$

Similarly,

$$\begin{aligned}\mathbf{g}(p)_{\theta\phi} &= \mathbf{g}(p)_{\phi\theta} = \mathbf{g}(p)(\partial_{\theta,p}, \partial_{\phi,p}) \\ &= (\partial_{\theta,p}x)(\partial_{\phi,p}x) + (\partial_{\theta,p}y)(\partial_{\phi,p}y) + (\partial_{\theta,p}z)(\partial_{\phi,p}z) \\ &= (\cos\theta_0)(\cos\phi_0)(\sin\theta_0)(-\sin\phi_0) \\ &\quad + (\cos\theta_0)(\sin\phi_0)(\sin\theta_0)(\cos\phi_0) + (-\sin\theta_0)0 \\ &= 0, \\ \mathbf{g}(p)_{\phi\phi} &= \mathbf{g}(p)(\partial_{\phi,p}, \partial_{\phi,p}) = (\partial_{\phi,p}x)^2 + (\partial_{\phi,p}y)^2 + (\partial_{\phi,p}z)^2 \\ &= (\sin\theta_0)^2(-\sin\phi_0)^2 + (\sin\theta_0)^2(\cos\phi_0)^2 + 0^2 = (\sin\theta_0)^2.\end{aligned}$$

Summarising, we arrange the components in a matrix, describing the metric $\mathbf{g}$ in the chart (U,φ): If $p \in S^2$ is such that $\varphi(p) = (\theta_0, \phi_0)$, then

$$\begin{bmatrix} \mathbf{g}(p)_{\theta\theta} & \mathbf{g}(p)_{\theta\phi} \\ \mathbf{g}(p)_{\phi\theta} & \mathbf{g}(p)_{\phi\phi} \end{bmatrix} = \begin{bmatrix} 1 & 0 \\ 0 & (\sin\theta_0)^2 \end{bmatrix}.$$

We see that this is a positive definite invertible matrix, as expected. The chart U covers S^2 except for the intersection of S^2 with the half plane $\{(x,y,z) : y = 0,\ x \geqslant 0\}$. We can take another chart (V,ψ), defined in a similar manner, which, together with U, covers S^2: V covers S^2 except for the intersection of S^2 with the half plane $\{(x,y,z) : z = 0,\ x \leqslant 0\}$, and is the image of the map

$$(0,\pi)\times(0,2\pi) \ni (\theta,\phi) \mapsto (-(\sin\theta)(\cos\phi), \cos\theta, (\sin\theta)(\sin\phi)) \in S^2 \subset \mathbb{R}^3.$$

An expression for the component matrix of $\mathbf{g}(p)$ can be obtained in the chart V in a similar manner as above, and in fact it so happens that we get the same component matrix. We note that, as expected, the component matrices in both charts are positive definite. We have that $(S^2, \mathbf{g})$ is a Riemannian manifold. ◇

Exercise 5.15. Consider the surface M in $\mathbb{R}^3$, where $M = \{(x, y, z) \in \mathbb{R}^3 : z = xy\}$, as a smooth manifold with the smooth structure $[\mathcal{A}]$, where $\mathcal{A}$ consists of the single chart M with the chart map $M \ni p = (u, v, uv) \mapsto (u, v) \in \mathbb{R}^2$. Let $\mathbb{R}^3$ be equipped with the Euclidean inner product $\langle \cdot, \cdot \rangle$, and M be equipped with the induced metric $\mathbf{g}$ from $(\mathbb{R}^3, \langle \cdot, \cdot \rangle)$. Determine the component matrix for $\mathbf{g}$ using the chart-induced basis. Show that $\mathbf{g}$ is a Riemannian metric on M.

Example 5.8. (Cylindrical spacetime.)
Consider the cylinder as a smooth manifold with the smooth structure given by the atlas in Exercise 1.7 (p.5). It can be shown that the map

$$(0, 2\pi) \times \mathbb{R} \ni (\theta, t) \mapsto (\cos\theta, \sin\theta, t) \in S^1 \times \mathbb{R} =: M \subset \mathbb{R}^3$$

is injective, and hence a bijection onto its image, denoted, say, by U. Denoting the inverse of this map by φ, it can be checked that (U, φ) is an admissible chart. Equip $\mathbb{R}^3$ with the Minkowski scalar product η, given by $\eta((x_1, y_1, t_1), (x_2, y_2, t_2)) = x_1x_2 + y_1y_2 - t_1t_2$, for $(x_1, y_1, t_1), (x_2, y_2, t_2) \in \mathbb{R}^3$. The chart-induced component $\mathbf{g}(p)_{\theta\theta}$ at $p \in M$, where $\varphi(p) = (\theta_0, t_0)$, is

$$\begin{aligned} \mathbf{g}(p)_{\theta\theta} &= \mathbf{g}(p)(\partial_{\theta,p}, \partial_{\theta,p}) = (\partial_{\theta,p}x)^2 + (\partial_{\theta,p}y)^2 - (\partial_{\theta,p}z)^2 \\ &= (-\sin\theta_0)^2 + (\cos\theta_0)^2 - 0^2 = 1. \end{aligned}$$

Similarly,

$$\begin{aligned} \mathbf{g}(p)_{\theta t} &= \mathbf{g}(p)_{t\theta} = \mathbf{g}(p)(\partial_{\theta,p}, \partial_{t,p}) \\ &= (\partial_{\theta,p}x)(\partial_{t,p}x) + (\partial_{\theta,p}y)(\partial_{t,p}y) - (\partial_{\theta,p}z)(\partial_{t,p}z) \\ &= (-\sin\theta_0)0 + (\cos\theta_0)0 - (0)1 = 0, \quad \text{and} \\ \mathbf{g}(p)_{tt} &= \mathbf{g}(p)(\partial_{t,p}, \partial_{t,p}) = (\partial_{t,p}x)^2 + (\partial_{t,p}y)^2 - (\partial_{t,p}z)^2 \\ &= 0^2 + 0^2 - 1^2 = -1. \end{aligned}$$

Summarising, the (U, φ)-induced metric component matrix is given by

$$\begin{bmatrix} \mathbf{g}(p)_{\theta\theta} & \mathbf{g}(p)_{\theta t} \\ \mathbf{g}(p)_{t\theta} & \mathbf{g}(p)_{tt} \end{bmatrix} = \begin{bmatrix} 1 & 0 \\ 0 & -1 \end{bmatrix},$$

where $p \in U \subset M$ is such that $\varphi(p) = (\theta_0, t_0)$. We see that this is an invertible matrix. Hence $\mathbf{g}(p)$ is nondegenerate for all $p \in U$. Also the index is 1. In particular, $(U, \mathbf{g})$ is a Lorentzian manifold. Taking V to be the image of $(-\pi, \pi) \times \mathbb{R} \ni (\theta, t) \mapsto (\cos\theta, \sin\theta, t) \in S^1 \times \mathbb{R} = M \subset \mathbb{R}^3$, we obtain another admissible chart (V, ψ), which together with U, covers M. In a manner similar to the above, we can compute component matrix for the induced metric on V from $(\mathbb{R}^3, \eta)$, and in fact it turns out to be identical to the one above for the chart (U, φ). From these component matrices, we see that the metric has index $\iota(\mathbf{g}) = 1$ everywhere on C, and so $(M, \mathbf{g})$ is a Lorentzian manifold. ◇

Exercise 5.16. Consider Exercise 1.3 (p.4) again, where H^2 is given the smooth structure $[\{(H^2, \varphi_{\mathbf{s}})\}]$. Equip H^2 with the metric $\mathbf{g}$ induced from $(\mathbb{R}^3, \eta)$, where

η is Minkowski scalar product as in the previous example. Find the component matrix of $\mathbf{g}$ in the chart $(H^2, \varphi_{\mathbf{s}})$, and show that $\mathbf{g}$ is Riemannian on H^2.

Example 5.9. (FLRW spacetime.)
Consider the FLRW spacetime $M = I \times \mathbb{R}^3$, where $I = (0, \infty)$, as a smooth manifold with the smooth structure given by the atlas comprising the single chart $(M, \mathrm{id}_{I \times \mathbb{R}^3})$; see Example 1.11 (p.9). Let $a : I \to I$ be a C^∞ function. For any $p \in M$, we have the basis $B = \{\partial_{t,p}, \partial_{x,p}, \partial_{y,p}, \partial_{z,p}\}$ for T_pM. Let $\mathbf{g}$ be the metric on M which has the metric component matrix for $\mathbf{g}(p)$ with respect to this B given by

$$\begin{bmatrix} \mathbf{g}(p)_{tt} & \mathbf{g}(p)_{tx} & \mathbf{g}(p)_{ty} & \mathbf{g}(p)_{tz} \\ \mathbf{g}(p)_{xt} & \mathbf{g}(p)_{xx} & \mathbf{g}(p)_{xy} & \mathbf{g}(p)_{xz} \\ \mathbf{g}(p)_{yt} & \mathbf{g}(p)_{yx} & \mathbf{g}(p)_{yy} & \mathbf{g}(p)_{yz} \\ \mathbf{g}(p)_{zt} & \mathbf{g}(p)_{zx} & \mathbf{g}(p)_{zy} & \mathbf{g}(p)_{zz} \end{bmatrix} = \begin{bmatrix} -1 & 0 & 0 & 0 \\ 0 & (a(t))^2 & 0 & 0 \\ 0 & 0 & (a(t))^2 & 0 \\ 0 & 0 & 0 & (a(t))^2 \end{bmatrix},$$

where $p = (t, x, y, z) \in M$. Then $(M, \mathbf{g})$ is a Lorentzian manifold. ◇

Exercise 5.17. Let $(M_1, \mathbf{g}_1)$ be a Lorentzian manifold and $(M_2, \mathbf{g}_2)$ be a Riemannian manifold. Show that the product $M_1 \times M_2$ of these smooth manifolds equipped with the metric $\mathbf{g}$ described below, makes $(M_1 \times M_2, \mathbf{g})$ a Lorentzian manifold. First recall from Exercise 2.8 (p.29), that for all $(p_1, p_2) \in M_1 \times M_2$, we have $T_{(p_1,p_2)}(M_1 \times M_2) \simeq T_{p_1}M_1 \times T_{p_2}M_2$. Define the metric $\mathbf{g}$ by

$$\mathbf{g}((p_1, p_2))(v_1 \oplus w_1, v_2 \oplus w_2) = \mathbf{g}_1(p_1)(v_1, w_1) + \mathbf{g}_2(p_2)(v_2, w_2)$$

for $v_1 \oplus w_1, v_2 \oplus w_2 \in T_{(p_1,p_2)}(M_1 \times M_2)$. Here we use the notation $v \oplus w$ from Exercise 2.8 (p.29).

Example 5.10. (Schwarzschild spacetime.) Recall Schwarzschild spacetime $M = \mathbb{R} \times (2m, \infty) \times S^2$ as a smooth manifold from Example 1.12 (p.9). Let (U, φ) be the chart for S^2 from Example 1.8 (p.7). For any $p = (t_0, r_0, \mathbf{p}_0) \in M$, with $t_0 \in \mathbb{R}$, $r_0 \in (2m, \infty)$ and $\mathbf{p}_0 \in U \subset S^2$, a basis for T_pM is given by the tangent vectors[3] $v_t, v_r, v_\theta, v_\phi$, given as follows. For $f \in C^\infty(M)$, using the notation from Exercise 2.8 (p.29),

$$\begin{aligned} v_t &= \partial_{t,t_0} \oplus 0 \oplus 0, & v_t f &= \partial_{t,t_0}(f(\cdot, r_0, \mathbf{p}_0)) \\ v_r &= 0 \oplus \partial_{r,r_0} \oplus 0, & v_r f &= \partial_{r,r_0}(f(t_0, \cdot, \mathbf{p}_0)) \\ v_\theta &= 0 \oplus 0 \oplus \partial_{\theta,p_0}, & v_\theta f &= \partial_{\theta,\mathbf{p}_0}(f(t_0, r_0, \cdot)) \\ v_\phi &= 0 \oplus 0 \oplus \partial_{\phi,p_0}, & v_\phi f &= \partial_{\phi,\mathbf{p}_0}(f(t_0, r_0, \cdot)). \end{aligned}$$

Let $\mathbf{g}_{S^2}$ be the Riemannian metric from Example 5.7 (p.85). Define

$$\begin{aligned} &\mathbf{g}(p)(\alpha v_t \oplus \beta v_r \oplus v, \alpha' v_t \oplus \beta' v_r \oplus v') \\ &= -\Big(1 - \frac{2m}{r_0}\Big)\alpha\alpha' + \Big(1 - \frac{2m}{r_0}\Big)^{-1}\beta\beta' + r_0^2\, \mathbf{g}_{S^2}(\mathbf{p}_0)(v, v'), \end{aligned}$$

where $\alpha, \alpha', \beta, \beta' \in \mathbb{R}$, $v, v' \in T_{\mathbf{p}_0}S^2$. Then $\mathbf{g}$ is a metric on M.

[3] We remark that in the sequel, we use the more natural notation $\partial_{t,p}$, $\partial_{r,p}$, $\partial_{\theta,p}$, $\partial_{\phi,p}$ for $v_t, v_r, v_\theta, v_\phi$, respectively.

The metric component matrix for $\mathbf{g}(p)$ in the chart $\mathbb{R}\times(2m,\infty)\times U$, with respect to the basis vectors $v_t, v_r, v_\theta, v_\phi$, in this order, is given by:

$$\begin{bmatrix} -(1-\frac{2m}{r_0}) & 0 & 0 & 0 \\ 0 & (1-\frac{2m}{r_0})^{-1} & 0 & 0 \\ 0 & 0 & r_0^2 & 0 \\ 0 & 0 & 0 & r_0^2(\sin\theta_0)^2 \end{bmatrix}.$$

This has determinant $-r_0^4(\sin\theta_0)^2 \neq 0$, and so $\mathbf{g}(p)$ is nondegenerate, and moreover, the index is 1, since for $r_0 > 2m$, the first diagonal entry is negative, while the others are positive. We obtain the same expression for the metric component matrix for $\mathbf{g}(p)$ in the chart $\mathbb{R}\times(2m,\infty)\times V$, where V is as in Example 1.8 (p.7). ◇

Exercise 5.18. Let $(M,\mathbf{g})$ be a Lorentzian manifold, and $(U,\mathbf{x})$ be an admissible chart for M. Let $G=[\mathbf{g}_{ij}]$ be the $m\times m$ matrix of chart-induced component functions $\mathbf{g}_{ij}\in C^\infty(U)$, $1\leqslant i,j\leqslant m$. Prove that $\det G(p)<0$ for all $p\in U$.
Hint: Choose an orthonormal basis $B=\{e_i : 1\leqslant i\leqslant m\}$ for T_pM with respect to $\mathbf{g}(p)$. Write $\partial_{x^i,p}$ using B, and find an expression for $\mathbf{g}_{ij}(p)$.

Exercise 5.19. Let $(M,\mathbf{g})$ be a semi-Riemannian m-dimensional manifold. Let $(U,\mathbf{x})$ be an admissible chart for M. Let $G=[\mathbf{g}_{ij}]\in\mathbb{R}^{m\times m}$ be the matrix of chart-induced component functions $\mathbf{g}_{ij}$, $1\leqslant i,j\leqslant m$. Show that for each $p\in U$, $G(p)$ is symmetric, invertible. Also show that $p\mapsto \mathbf{g}^{ij}=[(G(p))^{-1}]_{ij}$ are elements of $C^\infty(U)$, where $[(G(p))^{-1}]_{ij}$ denotes the entry in the ith row and j^{th} column of the matrix $(G(p))^{-1}$. We call $\mathbf{g}^{ij}$ the chart-induced *inverse metric* components.

Exercise 5.20. Let $(M,\mathbf{g})$ be a semi-Riemannian m-dimensional manifold, and $(U,\mathbf{x}),(U',\mathbf{x}')$ be admissible charts for M, with chart-induced inverse metric components denoted by $\mathbf{g}^{ij}$, $\mathbf{g}'^{ij}$, respectively. Show that

$$\mathbf{g}'^{ij} = \frac{\partial x'^i}{\partial x^k}\mathbf{g}^{k\ell}\frac{\partial x'^j}{\partial x^\ell}, \text{ where } \frac{\partial x'^i}{\partial x^j} := \frac{\partial(x'^i\circ\mathbf{x}^{-1})}{\partial u^j}(\mathbf{x}(\cdot)).$$

Exercise 5.21. Recall the Lie group structure on $\mathrm{GL}_n(\mathbb{R})$ from Exercise 3.15 (p.46). Define the *trace metric* $\mathbf{g}$ by $\mathbf{g}(p)(v,w)=\mathrm{trace}(p^{-1}vp^{-1}w)$ for $p\in\mathrm{GL}_n(\mathbb{R})$ and $v,w\in T_p(\mathrm{GL}_n(\mathbb{R}))\simeq\mathbb{R}^{n\times n}$. Check that at each $p\in\mathrm{GL}_n(\mathbb{R})$, $\mathbf{g}(p)$ is a scalar product on $\mathbb{R}^{n\times n}$, of index $n(n-1)/2$. (*Hint*: Every matrix is the sum of a symmetric and an skew-symmetric matrix.) So $\mathbf{g}$ is a metric on $\mathrm{GL}_n(\mathbb{R})$ with index $n(n-1)/2$. If $n=2$, then $\mathrm{GL}_2(\mathbb{R})$ is a 4-dimensional Lorentzian manifold.

5.3 Time-orientation, observers, and proper time

In each tangent space T_pM of a Lorentzian manifold $(M,\mathbf{g})$, where $p\in M$, the metric $\mathbf{g}(p)$ at p gives a scalar product with index 1, resulting in a 'causal structure', partitioning nonzero vectors of T_pM into spacelike, lightlike and timelike vectors. Thus we may imagine the Lorentzian manifold as a distribution of light cones (depicted in the following picture on a surface M, with $m-2$ dimensions suppressed).

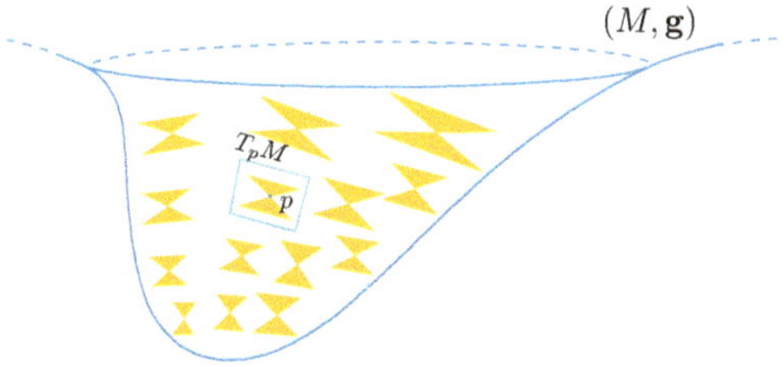

We wish to be able to speak of 'future' light cones, where e.g. for an observer, time 'only moves forwards'[4]. This amounts to a smoothly varying choice of one of the halves of the light cones at points of the Lorentzian manifold M. This smoothly varying choice is encoded in the existence of an everywhere timelike vector field V, where we declare the timelike value V_p of the vector field at a point $p \in M$, to be future-pointing (thus making a choice of the cone; note that $-V_p$ would then be deemed 'past-pointing'.)

Definition 5.8. (Time-orientation, future-pointing vectors).
A Lorentzian manifold $(M, \mathbf{g})$ is *time-oriented* if there exists a vector field $V \in T_0^1 M$ such that at each $p \in M$, V_p is timelike.
The vector field V is then referred to as a *time-orientation for* $(M, \mathbf{g})$.
Let $p \in M$. A timelike or lightlike tangent vector $v \in T_pM$ is *future-pointing* if $\mathbf{g}(p)(v, V_p) < 0$, and $-v$ is then *past-pointing.*

Example 5.11. (Minkowski spacetime.)
Consider the Minkowski spacetime $(M, \mathbf{g})$ from Example 5.6 (p.83). For each $p \in M$, recall from Example 2.2 (p.27), the identification of T_pM with V via the isomorphism $\mathrm{I} : V \to T_pM$ given by $V \ni \mathbf{v} \mapsto v_{\gamma_{\mathbf{v}},p}$, where $\gamma_{\mathbf{v}}(t) = p + t\mathbf{v}$ for all $t \in \mathbb{R}$. Suppose that $\mathbf{v} \in V$ is any timelike vector. Give $(M, \mathbf{g})$ the time-orientation[5] $\widetilde{V} \in T_0^1 M$, where $\widetilde{V}_p := \mathrm{I}(\mathbf{v})$, $p \in M$. Then $(M, \mathbf{g})$ is said to be given the *time-orientation induced by* $\mathbf{v}$. ◇

[4]This is for example based on our everyday experience of thermodynamical processes, where we can distinguish between the past and the future by observing the change in entropy: the second law of thermodynamics implies that the entropy increases.

[5]To see that $\widetilde{V}$ is smooth, we show that its component functions in a global chart are smooth. Consider the chart $\varphi = (x^0, x^1, x^2, x^3)$ (Example 1.3, p.3), induced by a $p_* \in M$ and a basis $\{\mathbf{e}_0, \mathbf{e}_1, \mathbf{e}_2, \mathbf{e}_3\}$ for V. Let $\mathbf{v} = v^i \mathbf{e}_i$. Then
$$\widetilde{V}_p(x^i) = v_{\gamma_{\mathbf{v}},p} x^i = \tfrac{d(x^i \circ \gamma_{\mathbf{v}})}{dt}(0) = \tfrac{d(b^i + tv^i)}{dt}(0) = v^i,$$
where $\gamma_{\mathbf{v}}(t) = p + tv^j \mathbf{e}_j = p_* + \mathbf{v}_{p_* p} + tv^j \mathbf{e}_j = p_* + b^j \mathbf{e}_j + tv^j \mathbf{e}_j$ and $\mathbf{v}_{p_* p} = b^j \mathbf{e}_j$. So $\widetilde{V} = v^i \partial_{x^i}$. As the constant functions $M \ni p \mapsto v^i$ are smooth, $\widetilde{V}$ is smooth.

Example 5.12. (Cylindrical spacetime.)
The cylindrical spacetime $(M, \mathbf{g})$ from Example 5.8 (p.86) can be given a time-orientation $W \in T_0^1 M$, where W is defined as follows: In the chart (U, φ), $W_p := \partial_{t,p}$ $(p \in U)$, and in the chart (V, ψ), $W_p := \partial_{t,p}$ $(p \in V)$. Then W is a well-defined element of $T_0^1 M$, and is timelike everywhere on M. ◇

Example 5.13. (FLRW spacetime.)
The FLRW spacetime $(M, \mathbf{g})$ from Example 5.9 (p.87) can be given the time-orientation $V \in T_0^1 M$, where the vector field V is defined as follows: In the global chart $(M, \mathrm{id}_{I \times \mathbb{R}^3})$, $V_p = \partial_{t,p}$, $p \in M$. ◇

Example 5.14. (Schwarzschild spacetime.)
The Schwarzschild spacetime $(M, \mathbf{g})$ from Example 5.10 (p.87) can be given the time-orientation $W \in T_0^1 M$, where W is defined as follows: For p in the chart domain $\mathbb{R} \times (2m, \infty) \times U$, $W_p = \partial_{t,p}$, and for p belonging to the chart domain $\mathbb{R} \times (2m, \infty) \times V$, $W_p = \partial_{t,p}$. Then W is well-defined and timelike everywhere. ◇

Example 5.15. (A non-time-orientable spacetime.)
We remark that a Lorentzian manifold does not have to be time-orientable. A simple visual example is the following. Consider the smooth manifold C, which is the same as the smooth manifold in our cylindrical spacetime example, but now we consider a different metric $\widetilde{\mathbf{g}}$. In our cylindrical spacetime example, the metric $\mathbf{g}$ was such that all the light cones had their axis along the z-axis in $\mathbb{R}^3$. But now suppose that the light cone tilts as we go along a circle γ round the cylinder, such that it makes a $180°$ rotation as it traverses this circle γ. See the following picture.

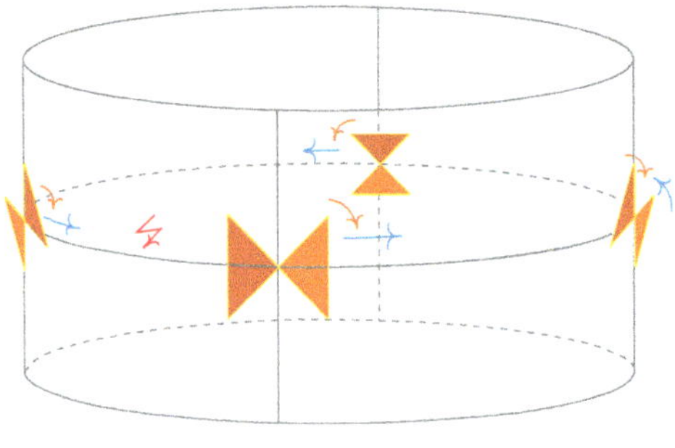

Intuitively, we expect that a smoothly varying choice of one half of the light cone is not possible. The technical details of this are relegated to an appendix to this chapter. ◇

Exercise 5.22. Let $(M, \mathbf{g})$ be a 4-dimensional Lorentzian manifold with a time-orientation $V \in T_0^1 M$, and let $p \in T_p M$. Determine if the following statements are true or false.

(1) The sum of two spacelike vectors in $T_p M$ is spacelike.
(2) The sum of two lightlike vectors in $T_p M$ is lightlike.
(3) Every timelike vector in $T_p M$ is the sum of two lightlike vectors.
(4) The sum of two timelike vectors in $T_p M$ is timelike.
(5) The sum of two timelike and future-pointing vectors in $T_p M$ is timelike and future-pointing.

Definition 5.9. (Observer, proper time.)
Let $(M, \mathbf{g})$ be a Lorentzian manifold with a time-orientation $V \in T_0^1 M$.

An *observer* is a smooth curve $\gamma : I \to M$, where I is an interval in $\mathbb{R}$, such that for all $t \in I$, $v_{\gamma,\gamma(t)}$ is timelike and future-pointing.

The *proper time* $\tau_\gamma(p, q)$ *experienced by the observer between the events* $p = \gamma(a)$ *and* $q = \gamma(b)$, where $a, b \in I$ and $b > a$, is defined by

$$\tau_\gamma(p, q) = \int_a^b \sqrt{-\mathbf{g}(\gamma(t))(v_{\gamma,\gamma(t)}, v_{\gamma,\gamma(t)})}\, dt.$$

The following result shows that the proper time is independent of the parametrisation of the curve. This is expected, since time experienced by a worldline between two events is physical, and should not depend on the parametrisation, which is somewhat arbitrary, as long as it strings together the events in the order in which they were encountered by the worldline.

Proposition 5.4. *Let*

- $\gamma : [a, b] \to M$ *be smooth and* $v_{\gamma,\gamma(t)}$ *be timelike for all* $t \in [a, b]$,
- $\varphi : [c, d] \to [a, b]$ *be smooth,* $\varphi(c) = a$, $\varphi(d) = b$, $\varphi'(t) > 0$ *for all* $t \in [c, d]$,
- $\widetilde{\gamma} : [c, d] \to M$ *be given by* $\widetilde{\gamma} = \gamma \circ \varphi$.

Then $\tau(\widetilde{\gamma}(c), \widetilde{\gamma}(d)) = \tau(\gamma(a), \gamma(b))$.

Proof. First, we note that for any $f \in C^\infty(M)$, and for $c \leqslant t \leqslant d$,

$$\begin{aligned} v_{\widetilde{\gamma},\widetilde{\gamma}(t)} f &= \frac{d(f \circ \widetilde{\gamma})}{dt}(t) = \frac{d(f \circ \gamma \circ \varphi)}{dt}(t) = \frac{d(f \circ \gamma)}{dt}(\varphi(t))\,\varphi'(t) \\ &= (\varphi'(t)\, v_{\gamma,\gamma(\varphi(t))}) f. \end{aligned}$$

Thus $v_{\widetilde{\gamma},\widetilde{\gamma}(t)} = \varphi'(t)\, v_{\gamma,\gamma(\varphi(t))}$. So $v_{\widetilde{\gamma},\widetilde{\gamma}(t)}$ is timelike for all $t \in [c, d]$, thanks to the facts that $v_{\gamma,\gamma(t)}$ is timelike for all $t \in [a, b]$, $\varphi'(t) > 0$ for all $t \in [a, b]$, and the relation

$$\mathbf{g}(\widetilde{\gamma}(t))(v_{\widetilde{\gamma},\widetilde{\gamma}(t)}, v_{\widetilde{\gamma},\widetilde{\gamma}(t)}) = (\varphi'(t))^2 \mathbf{g}(\gamma(\varphi(t)))(v_{\gamma,\gamma(\varphi(t))}, v_{\gamma,\gamma(\varphi(t))}).$$

Let $p=\gamma(a)=\widetilde{\gamma}(c)$ and $q=\gamma(b)=\widetilde{\gamma}(d)$. Then we have

$$\begin{aligned}
\tau_{\widetilde{\gamma}}(p,q) &= \int_c^d \sqrt{-\mathbf{g}(\widetilde{\gamma}(t))(v_{\widetilde{\gamma},\widetilde{\gamma}(t)}, v_{\widetilde{\gamma},\widetilde{\gamma}(t)})}\, dt \\
&= \int_c^d \sqrt{-\mathbf{g}(\gamma(\varphi(t)))(\varphi'(t)\, v_{\gamma,\gamma(\varphi(t))}, \varphi'(t)\, v_{\gamma,\gamma(\varphi(t))})}\, dt \\
&= \int_c^d \sqrt{-\mathbf{g}(\gamma(\varphi(t)))(v_{\gamma,\gamma(\varphi(t))}, v_{\gamma,\gamma(\varphi(t))})}\, \varphi'(t)\, dt \quad (\text{as } \varphi'(t)>0) \\
&\overset{s=\varphi(t)}{=} \int_a^b \sqrt{-\mathbf{g}(\gamma(s))(v_{\gamma,\gamma(s)}, v_{\gamma,\gamma(s)})}\, ds = \tau_\gamma(p,q).
\end{aligned}$$

This completes the proof. □

So physically, the parametrisation of the worldline of an observer just tells us about the order of occurrence of the events along the worldline, and does not have any real physical meaning otherwise. We can choose the parametrisation arbitrarily/conveniently as long as the order of events is preserved (guaranteed by the change of variables function φ satisfying $\varphi'>0$ in Proposition 5.4). On the other hand, the proper time is something which is physically real (the time experienced/recorded by a ticking clock carried by the observer). When discussing proper times, there is no loss of generality in assuming that the worldlines of observers are parametrised on a convenient interval, such as $[0,1]$.

Exercise 5.23. (Life of μ: Muon decay and time dilation.)
Consider the Minkowski spacetime $(M,\mathbf{g})$ from Example 5.6 (p.83). Let $(U,\mathbf{x})$ be the inertial frame induced by $p\in M$, and an orthonormal basis $\{\mathbf{e}_0,\mathbf{e}_1,\mathbf{e}_2,\mathbf{e}_3\}$ for V with respect to for g, such that $g(\mathbf{e}_0,\mathbf{e}_0)=-1$. Define the time-orientation on $(M,\mathbf{g})$ as the one induced by $\mathbf{e}_0$ (Example 5.11, p.89). If $d>0$ and $u\in(0,1)$, consider the curve γ, $\gamma(t)=p+(1-t)d\mathbf{e}_1+t\frac{d}{u}\mathbf{e}_0$, for all $t\in\mathbb{R}$. Set $P=\gamma(0)=p+d\mathbf{e}_1$ and $Q=\gamma(1)=p+\frac{d}{u}\mathbf{e}_0$.

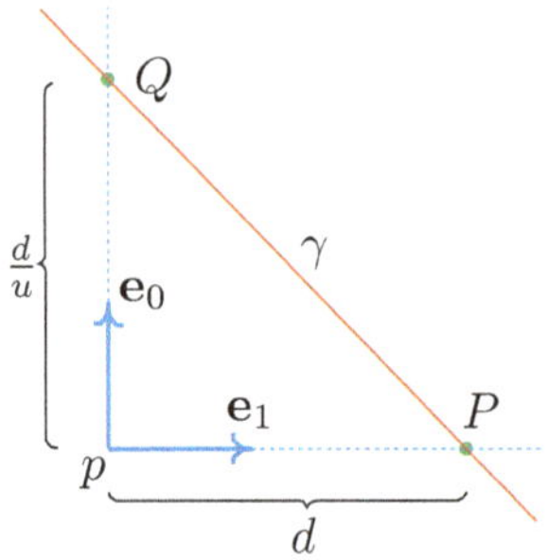

Let $T>0$. Show that if $1>u>\dfrac{1}{\sqrt{1+\frac{T^2}{d^2}}}$, then γ is an observer and $\tau_\gamma(P,Q)<T$.

Remark 5.1. A muon is a particle, present in cosmic radiation, which decays in its proper time $T = 2.2 \times 10^{-6}$ sec. Thus, $T = 660$m, in units in which the speed c of light is 1. Let $d = 10$km, the thickness of the Earth's atmosphere, and $u = 0.998$ (99.8% of the speed of light). Then the inequality above is satisfied. Hence, if a muon just outside the Earth's atmosphere is travelling towards the Earth along γ ('having speed u in the inertial frame $(U, \mathbf{x})$'), then it can traverse, without decaying, the thickness of the Earth's atmosphere. Note that in our inertial frame coordinate time, a particle can travel at most 660m in the coordinate time 2.2×10^{-6} sec. So how does the muon manage to travel a considerably larger distance (bigger than the thickness d of the Earth's atmosphere)? The answer is that for the muon decay, only its own proper time matters, not the coordinate time in some (quite arbitrarily chosen) inertial frame. Cosmic radiation muons can indeed be detected on the surface of the Earth. *

Exercise 5.24. The same proof, mutatis mutandis, as that of Proposition 5.4 shows that the length of a curve in a Riemannian manifold $(M, \mathbf{g})$ is independent of the parametrisation of the curve. Here the *length* of a smooth curve $\gamma : [a, b] \to M$ is

$$L(\gamma) := \int_a^b \sqrt{\mathbf{g}(\gamma(t))(v_{\gamma,\gamma(t)}, v_{\gamma,\gamma(t)})}\, dt.$$

Now let $M = \{(x, y) \in \mathbb{R}^2 : y > 0\}$, with the smooth structure $[\{(M, \mathrm{id})\}]$, be endowed with the Riemannian metric $\mathbf{g}$ given in the global chart (M, id) by

$$\mathbf{g}(\mathbf{p})(v^x \partial_{x,\mathbf{p}} + v^y \partial_{y,\mathbf{p}}, w^x \partial_{x,\mathbf{p}} + w^y \partial_{y,\mathbf{p}}) = \frac{v^x w^x + v^y w^y}{y^2} \text{ for all } \mathbf{p} = (x, y) \in M.$$

(1) Let $x_0 \in \mathbb{R}$. Define the vertical line $\lambda : (0, \infty) \to M$ by $\lambda(y) = (x_0, y)$, $y > 0$. Let $0 < y < y_0$. Determine the length of λ between y and y_0. What happens when $y \to 0$?

(2) Let $x_0 \in \mathbb{R}$, and $r > 0$. Define the semicircular arc $\sigma : (0, \pi) \to M$ by $\sigma(\theta) = (x_0 + r\cos\theta, r\sin\theta)$, $\theta \in (0, \pi)$. Let $0 < \theta < \theta_0 < \pi$. Determine the length of σ between θ and θ_0. What happens when $\theta \to 0$?

Theorem 5.5. (Twin paradox in Minkowski spacetime.)
In Minkowski spacetime $(M, \mathbf{g})$, let $p \in M$, and $\mathbf{v} \in V$ be any timelike vector. Let $(M, \mathbf{g})$ be given the time-orientation $\widetilde{V} \in T_0^1 M$ induced by $\mathbf{v}$. Consider two observers, $\gamma_A, \gamma_B : [0, 1] \to M$, where

- $\gamma_A(t) = p + t\mathbf{v}$, $0 \leqslant t \leqslant 1$,
- $\gamma_B(0) = p = \gamma_A(0)$, *and* $\gamma_B(1) = p + \mathbf{v} = \gamma_A(1)$.

Then with $q := p + \mathbf{v}$, we have $\tau_{\gamma_A}(p, q) \geqslant \tau_{\gamma_B}(p, q)$.

Thus if γ_A, γ_B are imagined to be the worldlines of two twins, then twin A ages faster. We will see later on that γ_A is a 'geodesic', implying that the motion of twin A is 'unaccelerated', and in this sense twin A can be thought of as 'resting', while the other twin undergoes 'accelerated motion'. The inequality in their experienced proper times can then be summarised by saying: Resting is rusting!

Intuitively, the twin paradox inequality follows from a repeated application of the reversed triangle inequality by a 'discretisation' process. A picture in the Minkowski plane with a coarse discretisation is shown below.

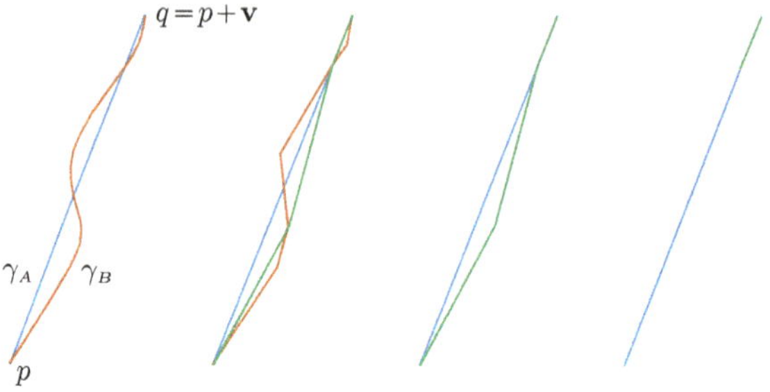

Proof. We will use a basis induced by a global chart. From the proof of Theorem 5.3, we can construct a basis $\mathcal{B}=\{\mathbf{e}_0,\mathbf{e}_1,\mathbf{e}_2,\mathbf{e}_3\}$ for V, orthonormal with respect to the Minkowski scalar product g on V, such that

$$\mathbf{e}_0 = \frac{\mathbf{v}}{\sqrt{-g(\mathbf{v},\mathbf{v})}}.$$

We use the chart (M,φ) induced by[6] p and $\mathcal{B}$. For all $t \in [0,1]$, we have $\gamma_B(t)=p+\gamma_B^i(t)\mathbf{e}_i$, where $\gamma_B^i:=(\varphi\circ\gamma_B)^i$ are real-valued smooth functions of $t\in[0,1]$. Then $v_{\gamma_B,\gamma_B(t)}=\dot{\gamma}_B^i\partial_{x^i,\gamma_B(t)}$, where $\dot{} = \frac{d}{dt}$. For a point $x\in M$, recall[7] the isomorphism $\mathrm{I}: V \to T_xM$, which sends $\mathbf{w}\mapsto v_{\gamma_\mathbf{w},x}$, where $\gamma_\mathbf{w}(t) := x+t\mathbf{w}$, $t\in\mathbb{R}$. The time-orientation $\widetilde{V}\in T_0^1M$ on M induced by $\mathbf{v}$ is given by $\widetilde{V}_x = \mathrm{I}\mathbf{v}$, $x\in M$. Since γ_B is an observer, $v_{\gamma_B,\gamma_B(t)}$ is future-pointing:

$$\begin{aligned}0 > \mathbf{g}(\gamma_B(t))(v_{\gamma_B,\gamma_B(t)},\mathrm{I}\mathbf{v}) &= \mathbf{g}(\gamma_B(t))(\dot{\gamma}_B^i(t)\partial_{x^i,\gamma_B(t)},\sqrt{-g(\mathbf{v},\mathbf{v})}\,\partial_{x^0,\gamma_B(t)})\\ &= -\sqrt{-g(\mathbf{v},\mathbf{v})}\,\dot{\gamma}_B^0(t),\end{aligned}$$

and so $\dot{\gamma}_B^0(t)>0$ for all $t\in[0,1]$. We have

$$\begin{aligned}-\mathbf{g}(\gamma_B(t))(v_{\gamma_B,\gamma_B(t)},v_{\gamma_B,\gamma_B(t)}) &= (\dot{\gamma}_B^0(t))^2-(\dot{\gamma}_B^1(t))^2-(\dot{\gamma}_B^2(t))^2-(\dot{\gamma}_B^3(t))^2\\ &\leqslant (\dot{\gamma}_B^0(t))^2,\end{aligned}$$

and so we obtain

$$\begin{aligned}\tau_{\gamma_B}(p,p+\mathbf{v}) &= \int_0^1\sqrt{-\mathbf{g}(\gamma_B(t))(v_{\gamma_B,\gamma_B(t)},v_{\gamma_B,\gamma_B(t)})}\,dt\\ &\leqslant \int_0^1\sqrt{(\dot{\gamma}_B^0(t))^2}\,dt\\ &= \int_0^1\dot{\gamma}_B^0(t)\,dt = \gamma_B^0(1)-\gamma_B^0(0) = -g(\mathbf{v},\mathbf{e}_0)-0=\sqrt{-g(\mathbf{v},\mathbf{v})}.\end{aligned}$$

[6]See Example 1.3 (p.3).
[7]See Example 2.2 (p.27).

But $v_{\gamma_A,\gamma_A(t)} = v^i \partial_{x^i,\gamma_A(t)}$, and so

$$-\mathbf{g}(\gamma_A(t))(v_{\gamma_A,\gamma_A(t)}, v_{\gamma_A,\gamma_A(t)}) = (v^0)^2 - (v^1)^2 - (v^2)^2 - (v^2)^2 = -g(\mathbf{v}, \mathbf{v}).$$

Thus,

$$\tau_{\gamma_A}(p, p+\mathbf{v}) = \int_0^1 \sqrt{-\mathbf{g}(\gamma_A(t))(v_{\gamma_A,\gamma_A(t)}, v_{\gamma_A,\gamma_A(t)})}\, dt = \sqrt{-g(\mathbf{v}, \mathbf{v})}.$$

Consequently, $\tau_{\gamma_B}(p, p+\mathbf{v}) \leqslant \sqrt{-g(\mathbf{v}, \mathbf{v})} = \tau_{\gamma_A}(p, p+\mathbf{v})$. □

One might mistakenly attribute the twin paradox, that is, the inequality above, to be an artefact of the accelerated motion of twin B. This is somewhat misleading, since the acceleration is only needed to make the two twins meet again. The *source* of the inequality is the reversed triangle inequality. To emphasise this, we show that the inequality can also be brought to bear by taking two *geodesics* albeit in a modified spacetime, namely the Minkowski plane made into a cylinder. The reason we need to modify the Minkowski plane to the cylindrical spacetime is as follows. We will see later on that geodesics in the Minkowski plane are straight lines, and straight lines originating at a point won't ever meet again. To make them meet again, we can 'roll up' a Minkowski plane to make a cylinder.

Theorem 5.6. (Twin paradox in cylindrical spacetime.)
Let $(M, \mathbf{g})$ *be the cylindrical spacetime from* Example 5.8 (p.86), *with the time-orientation* W *as given in* Example 5.12 (p.90).
Consider the two observers $\gamma_A, \gamma_B : [0, 2\pi] \to M$ *given by*

- $\gamma_A(t) = (1, 0, 2t)$, $0 \leqslant t \leqslant 2\pi$,
- $\gamma_B(t) = (\cos t, \sin t, 2t)$, $0 \leqslant t \leqslant 2\pi$.

Then with $p = (1, 0, 0)$ *and* $q = (1, 0, 4\pi)$, *we have* $\tau_{\gamma_A}(p, q) \geqslant \tau_{\gamma_B}(p, q)$.

Proof. Using the chart (V, ψ) from Example 5.8, it can be seen that the proper time experienced by observer γ_A is given by

$$\tau_{\gamma_A}(p, q) = \int_0^{2\pi} \sqrt{-\mathbf{g}(\gamma_A(t))(2\,\partial_{t,\gamma_A(t)}, 2\,\partial_{t,\gamma_A(t)})}\, dt = \int_0^{2\pi} \sqrt{4}\, dt = 2\cdot(2\pi).$$

For the observer γ_B, we will use the chart (U, φ), which covers $\gamma_B|_{(0,2\pi)}$, and only the endpoints p, q do not belong to U. We have

$$\begin{aligned}\tau_{\gamma_B}(p, q) &= \int_0^{2\pi} \sqrt{-\mathbf{g}(\gamma_B(t))(\partial_{\theta,\gamma_B(t)} + 2\,\partial_{t,\gamma_B(t)}, \partial_{\theta,\gamma_B(t)} + 2\,\partial_{t,\gamma_B(t)})}\, dt \\ &= \int_0^{2\pi} \sqrt{-(1-4)}\, dt = \sqrt{3}\cdot(2\pi).\end{aligned}$$

Thus $\tau_{\gamma_A}(p, q) = 2\cdot(2\pi) \geqslant \sqrt{3}\cdot(2\pi) = \tau_{\gamma_B}(p, q)$. □

We reiterate that in Minkowski space, the symmetry among the two twins was broken because one of them was a geodesic while the other one wasn't.

So the non-resting twin 'feels' accelerated. In the case of the cylindrical spacetime, the symmetry is broken by the fact that one of them winds around the cylinder, while the other doesn't. However, neither twin 'feels' accelerated! It is the global topology which distinguishes the two geodesics and the resulting difference in the experienced proper times. See also Exercise 5.25 (p.99), where in Schwarzschild spacetime, we construct the worldlines of two twins, where the accelerated twin ages more!

The Schwarzschild spacetime models the spacetime outside a spherically symmetric body of mass m, in an otherwise empty universe. It turns out that realistically, this is a good approximation near the body (where the effect of the body is strongly felt, and the rest of the universe doesn't matter). In particular, it can be used for modelling the spacetime geometry near the Earth. We end this section with the discussion of a result in Schwarzschild spacetime, which says roughly speaking that objects near the surface of the Earth age slower, and that the light emanating from the surface of the Earth, observed high above, is red-shifted. But first we give the following definition.

Definition 5.10.
(Energy/frequency of a light signal measured by an observer at an event.)
Let $(M, \mathbf{g})$ be a Lorentzian manifold with a time-orientation.
A *light signal* is a smooth curve[8] $\lambda : I \to M$, where $I \subset \mathbb{R}$ is an interval, such that for each $t \in I$, $v_{\lambda,\lambda(t)}$ is lightlike.
Let γ be an observer, and suppose that $p \in M$ belongs to the intersection of the images of γ and λ. Then the *energy/frequency of λ measured by γ at p* is defined to be

$$-\frac{\mathbf{g}(p)(v_{\gamma,p}, v_{\lambda,p})}{\sqrt{-\mathbf{g}(p)(v_{\gamma,p}, v_{\gamma,p})}}.$$

Example 5.16. (Gravitational red-shift.)
Recall the Schwarzschild spacetime $(M, \mathbf{g})$ from Example 5.10 (p.87), and the time-orientation given in Example 5.14 (p.90). Let $r_2 > r_1 > 2m$, and let $\mathbf{p} \in S^2$. Consider the two observers $\gamma_1(t) = (t, r_1, \mathbf{p})$ and $\gamma_2(t) = (t, r_2, \mathbf{p})$, $t \in \mathbb{R}$. Suppose that at the event $P_1 = (t_1, r_1, \mathbf{p})$ on the worldline γ_1, a 'radial light signal' $\lambda : [a, b] \to M$ is sent towards the other observer, reaching it at the event $P_2 = (t_2, r_2, \mathbf{p})$. We write $\lambda(\tau) = (t(\tau), r(\tau), \mathbf{p})$, $\tau \in [a, b]$, where $\lambda(a) = P_1$ and $\lambda(b) = P_2$. Since λ is the worldline of light, $v_{\lambda,\lambda(t)}$ is lightlike for all t.

[8]The curve should be a 'geodesic', which will be defined in Chapter 8. Right now we just assume that we have a special type of smooth curve given, and carry on.

Thus if $\cdot' = \frac{d}{d\tau}$, then

$$\begin{aligned} 0 &= \mathbf{g}(\lambda(\tau))(v_{\lambda,\lambda(\tau)}, v_{\lambda,\lambda(\tau)}) \\ &= \mathbf{g}(\lambda(\tau))(t'(\tau)\,\partial_{t,\lambda(\tau)} + r'(\tau)\,\partial_{r,\lambda(\tau)}, t'(\tau)\,\partial_{t,\lambda(\tau)} + r'(\tau)\,\partial_{r,\lambda(\tau)}) \\ &= -\Big(1-\frac{2m}{r(\tau)}\Big)\,(t'(\tau))^2 + \frac{1}{1-\frac{2m}{r(t)}}\,(r'(\tau))^2. \end{aligned}$$

As the light signal travels radially outwards, we suppose that $r'(\tau)>0$ and $t'(\tau)>0$. Let $\tau := t^{-1} : [t_1, t_2] \to [a,b]$ be the inverse of $t : [a,b] \to [t_1, t_2]$. With $h := r \circ \tau : [t_1, t_2] \to (2m, \infty)$, we get the differential equation

$$\frac{dh}{dt}(t) = \frac{d(r\circ\tau)}{dt}(t) = \frac{r'(\tau(t))}{t'(\tau(t))} = 1-\frac{2m}{r(\tau(t))} = 1-\frac{2m}{h(t)}.$$

We have $h(t_1) = r(\tau(t_1)) = r(a) = r_1$, and similarly, $h(t_2) = r_2$. Integrating,

$$t_2 - t_1 = \int_{t_1}^{t_2} 1\,dt = \int_{t_1}^{t_2} \frac{1}{1-\frac{2m}{h(t)}}\frac{dh}{dt}(t)\,dt \overset{r=h(t)}{=} \int_{r_1}^{r_2} \frac{r}{r-2m}\,dr.$$

This expression for $t_2 - t_1$ depends only on r_1, r_2 and is independent of t_1.

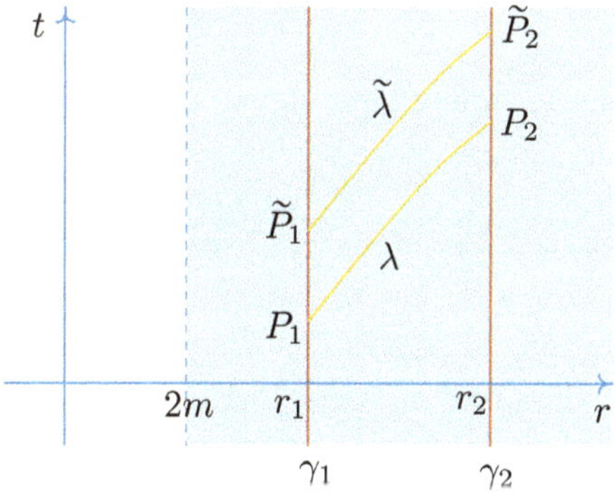

Suppose another radial light signal $\widetilde{\lambda}$ is sent by the first observer at the event $\widetilde{P}_1 = (\widetilde{t}_1, r_1, \mathbf{p})$, with $\widetilde{t}_1 > t_1$, towards the second observer, reaching it at the event $\widetilde{P}_2 = (\widetilde{t}_2, r_2, \mathbf{p})$. Then it follows from the above that $\widetilde{t}_2 - \widetilde{t}_1 = t_2 - t_1$, and so also $\widetilde{t}_2 - t_2 = \widetilde{t}_1 - t_1$. We have that the proper time experienced by the first observer between the events P_1 and $\widetilde{P}_1$ is

$$\begin{aligned} \tau_{\gamma_1}(P_1, \widetilde{P}_1) &= \int_{t_1}^{\widetilde{t}_1} \sqrt{-\mathbf{g}(\gamma_1(t))(v_{\gamma_1,\gamma_1(t)}, v_{\gamma_1,\gamma_1(t)})}\,dt \\ &= \int_{t_1}^{\widetilde{t}_1} \sqrt{-\mathbf{g}(\gamma_1(t))(\partial_{t,\gamma_1(t)}, \partial_{t,\gamma_1(t)})}\,dt = \sqrt{1-\frac{2m}{r_1}}\,(\widetilde{t}_1 - t_1). \end{aligned}$$

Similarly, the proper time experienced by the second observer between P_2 and $\widetilde{P}_2$ is

$$\tau_{\gamma_2}(P_2, \widetilde{P}_2) = \sqrt{1-\frac{2m}{r_2}}\,(\widetilde{t}_2 - t_2).$$

Thanks to $\widetilde{t}_2 - t_2 = \widetilde{t}_1 - t_1 =: \Delta t$, and the inequality $r_2 > r_1$, we get

$$\tau_{\gamma_2}(P_2, \widetilde{P}_2) = \sqrt{1 - \frac{2m}{r_2}}\,\Delta t > \sqrt{1 - \frac{2m}{r_1}}\,\Delta t = \tau_{\gamma_1}(P_1, \widetilde{P}_1).$$

If $P_1, \widetilde{P}_1$ were the subsequent ticks of a clock carried by the first observer, then the information about these ticks is 'seen' and recorded by the second observer along the light signals $\lambda, \widetilde{\lambda}$ at the events $P_2, \widetilde{P}_2$. By the above inequality, the second observer reckons that the first observer's clock runs slow (as compared to the clock carried by him/her). We say that 'a clock in a stronger gravitational field is seen as running slow by an observer in a weaker gravitational field'. So near the surface of the Earth, we age slower as compared to someone higher up.

Red-shift: The frequency of the outgoing light signal λ measured by the first observer at P_1 is

$$\begin{aligned}\nu_1 &= \frac{-\mathbf{g}(P_1)(v_{\gamma_1,P_1}, v_{\lambda,P_1})}{\sqrt{-\mathbf{g}(P_1)(v_{\gamma_1,P_1}, v_{\gamma_1,P_1})}} = \frac{-\mathbf{g}(P_1)(\partial_{t,P_1}, t'(a)\,\partial_{t,P_1} + r'(a)\,\partial_{r,P_1})}{\sqrt{-\mathbf{g}(P_1)(\partial_{t,P_1}, \partial_{t,P_1})}} \\ &= t'(a)\sqrt{1 - \frac{2m}{r_1}}.\end{aligned}$$

The frequency of the incoming light signal λ measured by the second observer at P_2 is

$$\begin{aligned}\nu_2 &= \frac{-\mathbf{g}(P_2)(v_{\gamma_2,P_2}, v_{\lambda,P_2})}{\sqrt{-\mathbf{g}(P_2)(v_{\gamma_2,P_2}, v_{\gamma_2,P_2})}} = \frac{-\mathbf{g}(P_2)(\partial_{t,P_2}, t'(b)\,\partial_{t,P_2} + r'(b)\,\partial_{r,P_2})}{\sqrt{-\mathbf{g}(P_2)(\partial_{t,P_2}, \partial_{t,P_2})}} \\ &= t'(b)\sqrt{1 - \frac{2m}{r_2}}.\end{aligned}$$

Later (see (15.6) in the proof of Prop. 15.1, p.336) we will see that as λ is a lightlike geodesic, there exists a constant E such that for all $\tau \in [a, b]$,

$$t'(\tau)\Big(1 - \frac{2m}{r(\tau)}\Big) = E.$$

So

$$\frac{\nu_2}{\nu_1} = \frac{t'(b)\sqrt{1 - \frac{2m}{r_2}}}{t'(a)\sqrt{1 - \frac{2m}{r_1}}} = \frac{\sqrt{1 - \frac{2m}{r_1}}}{\sqrt{1 - \frac{2m}{r_2}}}.$$

As $r_2 > r_1$, we have $\nu_2 < \nu_1$, that is, the light observed by the second observer is 'redder' than that observed by the first observer. The *red-shift* z is defined by

$$\mathrm{z} := \frac{\nu_1 - \nu_2}{\nu_2} = \frac{\sqrt{1 - \frac{2m}{r_2}}}{\sqrt{1 - \frac{2m}{r_1}}} - 1.$$

We say that 'light climbing out of a gravitational field is red-shifted'. ◇

Exercise 5.25. (Twin paradox in Schwarzschild spacetime.)
The spacetime around Earth is well-described by the Schwarzschild spacetime $(M, \mathbf{g})$ where $m = M_\oplus$, the mass of Earth. Let (U, φ) be the spherical coordinate chart given in Example 1.8 (p.7). Let the curve $\gamma : \mathbb{R} \to M$ be a timelike 'geodesic' in the equatorial plane ($\theta = \frac{\pi}{2}$) having a constant r-coordinate r_0: $\gamma(s) = (\mathrm{t}(s), r_0, \varphi^{-1}(\frac{\pi}{2}, \phi(s)))$, $s \in \mathbb{R}$. It can be shown[9] that since γ is a geodesic,

$$(\phi')^2 = \frac{m}{r_0^3}(\mathrm{t}')^2,$$

where $\cdot' := \frac{d}{ds}$. Let $\phi(\gamma(0)) = \phi_0$, and let $s_0 > 0$ be the smallest parameter value such that $\phi(\gamma(s_0)) = \phi_0$. Let T_{orbit} be the proper time experienced by γ between the events $p := \gamma(0)$ and $q := \gamma(s_0)$. Consider also the stationary timelike observer $\widetilde{\gamma}(s) = (s, r_0, \varphi^{-1}(\frac{\pi}{2}, \phi_0))$, $s \in \mathbb{R}$. Let $T_{\text{stationary}}$ be the proper time experienced by $\widetilde{\gamma}$ between the events p, q. Show that $T_{\text{orbit}} < T_{\text{stationary}}$. The stationary observer $\widetilde{\gamma}$ is not a geodesic (Exercise 8.5, p.154), so that his/her motion in spacetime is 'accelerated'. In contrast to Theorem 5.5 (p.93), now the accelerated twin ages more! However the proper time maximising curve is always a timelike geodesic (Theorem 8.5, p.175), which in this case happens not to be the circular orbit, but a radial up-and-down motion. This is explored in the next exercise.

Exercise 5.26. (Radial timelike geodesic in Schwarzschild spacetime.)
In continuation of the previous exercise, now consider a radial timelike geodesic γ_*, 'parametrised[10] by proper time', as described in Exercise 14.11 (p.321). Thus $\gamma_* : I \to M$, $I \ni \tau \mapsto (t(\tau), r(\tau), \mathbf{p})$, where $\mathbf{p} \in S^2$ is fixed. In Exercise 14.11, the following equations are derived (which we accept for now):

$$-\big(1 - \tfrac{2m}{r}\big)(t')^2 + \big(1 - \tfrac{2m}{r}\big)^{-1}(r')^2 = -1, \quad t'' = -\frac{2m}{r^2}\big(1 - \tfrac{2m}{r}\big)^{-1} t' r', \quad r'' = -\frac{m}{r^2}.$$

Here $\cdot' := \frac{d}{d\tau}$. Using these, prove that there exists a constant $E > 0$ such that

$$\big(1 - \tfrac{2m}{r}\big)t' = E, \qquad (r')^2 = E^2 - 1 + \frac{2m}{r}.$$

Show that a free-falling particle that is thrown upwards from $r = r_0$, reaches a maximum height $r = r_1$, and falls back to $r = r_0$, experiences a proper time

$$\Delta\tau = 2\int_{r_0}^{r_1} \frac{1}{\sqrt{\frac{2m}{r} - \frac{2m}{r_1}}}\,dr,$$

corresponding to a 'coordinate time' lapse $\Delta t = 2\displaystyle\int_{r_0}^{r_1} \frac{\sqrt{1 - \frac{2m}{r_1}}}{\big(1 - \frac{2m}{r}\big)\sqrt{\frac{2m}{r} - \frac{2m}{r_1}}}\,dr.$

Use this to show $\Delta\tau > \Delta t\sqrt{1 - \dfrac{2m}{r_0}}$.

(The particle experiences more proper time than the stationary observer at $r = r_0$.)
Hint: Compare the derivatives of $\Delta\tau$ and $\Delta t\sqrt{1 - \frac{2m}{r_0}}$ with respect to r_0 using

$$\Delta t \geqslant 2\int_{r_0}^{r_1} \frac{r_0}{\sqrt{1 - \frac{2m}{r_1}}\sqrt{2m}\sqrt{r_1 - r}}\,dr \geqslant \frac{2r_0^2}{m\sqrt{1 - \frac{2m}{r_1}}}\sqrt{\frac{2m}{r_0} - \frac{2m}{r_1}}.$$

[9]Follows from the r-component of the geodesic equation, using $r' \equiv 0$, $\theta' \equiv 0$; see the solution (p.450) to Exercise 8.12 (p.162).
[10]See Lemma 12.1, p.267.

Exercise 5.27. (Unruh effect.)
Let $(M, \mathbf{g})$ be the 2-dimensional Minkowski spacetime over $V = \mathbb{R}^2$. Let $(U, \mathbf{x})$ be the inertial frame induced by a $p \in M$, and an orthonormal basis $\{\mathbf{e}_0, \mathbf{e}_1\}$ for V with respect to $g := \eta$, where η is as in Example 5.3 (p.79), and let $\eta(\mathbf{e}_0, \mathbf{e}_0) = -1$. The time-orientation on $(M, \mathbf{g})$ is the one induced by $\mathbf{e}_0$. Given an $a > 0$, consider the observer $\gamma : \mathbb{R} \to M$ given by

$$\gamma(t) = p + \frac{1}{a}(\sinh(at)\mathbf{e}_0 + \cosh(at)\mathbf{e}_1), \quad t \in \mathbb{R}.$$

The boundary $\partial H = \{p + t(\mathbf{e}_0 + \mathbf{e}_1) : t \in \mathbb{R}\}$ of the half-plane

$$H = \{p + v^0\mathbf{e}_0 + v^1\mathbf{e}_1 : v^0 \geqslant v_1\}$$

forms a 'horizon'[11] for the observer, since an event cannot send information to the observer from within H. Thus any light signal from the observer entering the horizon is lost to him irretrievably. Let $E > 0$, and λ_+, λ_- be a 'photon pair' created at an event $\gamma(t_0)$ on the observer's worldline with 'energies' $E, -E$:

$$\left.\begin{aligned} \lambda_+(t) &= \gamma(t_0) + E\,(t - t_0)\,(\mathbf{e}_0 + \mathbf{e}_1) \\ \lambda_-(t) &= \gamma(t_0) + E\,(t - t_0)\,(\mathbf{e}_0 - \mathbf{e}_1) \end{aligned}\right\}, \quad t \geqslant t_0.$$

Suppose that the photon λ_- with energy $-E$ crosses the horizon ∂H. Determine the coordinate time Δt in the chart $(U, \mathbf{x})$ it takes for this photon to reach the horizon.

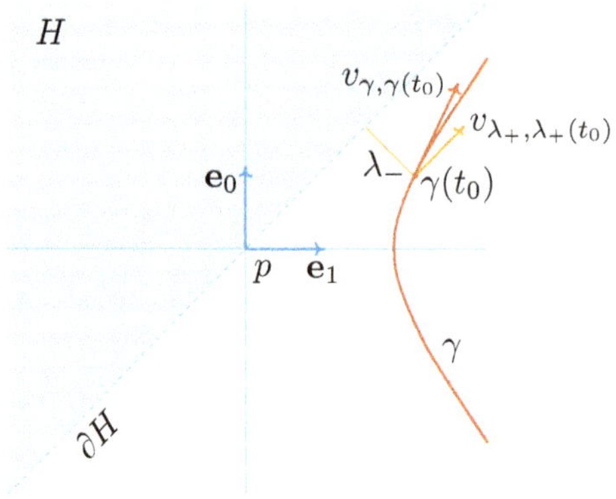

Assuming the uncertainly relation $E\Delta t = \frac{\hbar}{2}$, find E, and also show that the energy E_{observed} of the photon λ_+ measured by the observer is given by

$$E_{\text{observed}} = \hbar a.$$

(This can be related to an effective temperature of the horizon perceived by the observer,

$$T = \frac{E_{\text{observed}}}{2\pi k_{\text{B}}} = \frac{\hbar a}{2\pi k_{\text{B}}},$$

where k_B is the Boltzmann constant.)

[11]Roughly speaking, a *horizon* is the boundary of a region $\mathcal{R}$ of spacetime such that the events from within $\mathcal{R}$ cannot send a light signal to observers (that is, paths whose range lies) outside $\mathcal{R}$.

5.4 Raising and lowering indices or musicality

In the setting of a semi-Riemannian manifold, a 1-form field can be converted into a vector field and vice versa, using the so-called musical isomorphisms, as we shall see in this section. We will learn about two maps,

$$\cdot^\flat : T_0^1 M \to T_1^0 M \text{ ('flat') and } \cdot^\sharp : T_1^0 M \to T_0^1 M \text{ ('sharp')},$$

which will be inverses of each other. If one looks at chart-induced components, the flat map 'lowers' the index (which, here, we think of as lowering the pitch, hence 'flat'), while the sharp map 'raises' the index.

Definition 5.11. (Flat map.)
Let $(M, \mathbf{g})$ be a semi-Riemannian manifold. Define $\cdot^\flat : T_0^1 M \to T_1^0 M$ by

$$(V^\flat)W = \mathbf{g}(V, W) \text{ for all } V, W \in T_0^1 M.$$

The $C^\infty(M)$-linearity of $V^\flat$ follows from the $C^\infty(M)$-linearity of $\mathbf{g}$ in its second slot. We now show that $\cdot^\flat$ is a $C^\infty(M)$-module isomorphism.

Theorem 5.7. *The map* $\cdot^\flat : T_0^1 M \to T_1^0 M$ *is* $C^\infty(M)$*-linear and bijective.*

Proof. The $C^\infty(M)$ linearity of $\cdot^\flat$ follows from the $C^\infty(M)$-linearity of $\mathbf{g}$ in its first slot.
Injectivity: Suppose that $V \in T_0^1 M$ and $V^\flat = 0$, that is, for all vector fields $W \in T_0^1 M$, $0 = V^\flat W = \mathbf{g}(V, W)$. Let $p \in M$ and $w \in T_p M$. By Lemma 3.1 (p.41), there exists a vector field $W \in T_0^1 M$ such that $W_p = w$. So we have $0 = (\mathbf{g}(V, W))(p) = \mathbf{g}(p)(V_p, W_p) = \mathbf{g}(p)(V_p, w)$. As $w \in T_p M$ was arbitrary, and since $\mathbf{g}(p)$ is non-degenerate, it follows that $V_p = 0$. Since $p \in M$ was arbitrary, $V = 0$. Thanks to the linearity of $\cdot^\flat$, we can conclude that $\cdot^\flat$ is injective.
Surjectivity: Suppose that $\Omega \in T_1^0 M$. Let $(U, \mathbf{x})$ be any admissible chart. Then we have $\Omega|_U = \Omega_i dx^i$, where $\Omega_i = \Omega\, \partial_{x^i}$. Define a vector field $V|_U$ by $V|_U = \mathbf{g}^{ij} \Omega_i \partial_{x^j}$, where we use the notation $\mathbf{g}^{ij}$ from Exercise 5.19 (p.88) for the chart-induced inverse metric components. As the charts cover M, this procedure allows us to define a vector in $T_p M$ at each $p \in M$. To check well-definition, we note that if $(U', \mathbf{x}')$ were an overlapping chart, then using the transformation properties of the 1-form components, of the basis elements of $T_p M$, and of the inverse metric components (Exercise 5.20, p.88), we obtain in $U \cap U'$ that

$$\begin{aligned} V|_{U'} &= \mathbf{g}'^{ij} \Omega_i' \partial_{x'^j} = \frac{\partial x'^i}{\partial x^k} \mathbf{g}^{k\ell} \frac{\partial x'^j}{\partial x^\ell} \Omega_r \frac{\partial x^r}{\partial x'^i} \frac{\partial x^s}{\partial x'^j} \partial_{x^s} = \frac{\partial x'^i}{\partial x^k} \frac{\partial x^r}{\partial x'^i} \frac{\partial x'^j}{\partial x^\ell} \frac{\partial x^s}{\partial x'^j} \mathbf{g}^{k\ell} \Omega_r \partial_{x^s} \\ &= \delta_k^r \delta_\ell^s \mathbf{g}^{k\ell} \Omega_r \partial_{x^s} = \mathbf{g}^{rs} \Omega_r \partial_{x^s} = V|_U. \end{aligned}$$

Here we also used the result from Exercise 2.19 (p.35). The smoothness of V follows from Theorem 3.1 (p.40), since we know that the chart-induced components are smooth. To complete the proof of the surjectivity, we will now show that for this V, we have that $V^\flat = \Omega$. Indeed, for any $W \in T_0^1 M$, we have in any admissible chart $(U, \mathbf{x})$ that

$$V^\flat W = \mathbf{g}(V, W) = \mathbf{g}_{ij} V^i W^j = \mathbf{g}_{ij} \mathbf{g}^{ki} \Omega_k W^j = \delta_j^k \Omega_k W^j = \Omega_j W^j = \Omega W.$$

As the charts cover M, $V^\flat W = \Omega W$ on M. Since $W \in T_0^1 M$ was arbitrary, $V^\flat = \Omega$. □

The map $\cdot^\sharp : T_1^0 M \to T_0^1 M$, defined as the inverse of the bijective map $\cdot^\flat : T_0^1 M \to T_1^0 M$, is also seen to be a $C^\infty(M)$-linear map as follows: Indeed, let $\Omega, \Theta \in T_1^0 M$ and $f \in C^\infty(M)$. By the bijectivity of $\cdot^\flat$, there exist unique $V, W \in T_0^1 M$ such that $V^\flat = \Omega$ and $W^\flat = \Theta$. The $C^\infty(M)$-linearity of $\cdot^\flat$ implies that $(fV + W)^\flat = fV^\flat + W^\flat = f\Omega + \Theta$. Now the injectivity of $\cdot^\flat$ gives, in light of the definition of $\cdot^\sharp$, that $(f\Omega + \Theta)^\sharp = fV + W = f\Omega^\sharp + \Theta^\sharp$.

Definition 5.12. (Sharp isomorphism.)
Let $(M, \mathbf{g})$ be a semi-Riemannian manifold. The inverse of the $C^\infty(M)$-linear bijective map $\cdot^\flat : T_0^1 M \to T_1^0 M$ is the $C^\infty(M)$-linear bijective map $\cdot^\sharp : T_1^0 M \to T_0^1 M$, and is called the *sharp isomorphism*.

The surjectivity part of the proof of Theorem 5.7 shows that if $\Omega \in T_0^1 M$, then in any admissible chart $(U, \mathbf{x})$, $\Omega^\sharp|_U = \mathbf{g}^{ij} \Omega_i \partial_{x^j}$, where $\Omega_i = \Omega\, \partial_{x^i}$.

Exercise 5.28. (Gradient field as a vector field.)
Let M be a smooth manifold and $f \in C^\infty(M)$. We had defined the 1-form field $df \in T_1^0 M$ in Exercise 3.18 (p.49). If M is equipped with a metric $\mathbf{g}$, then using the musical isomorphism $\cdot^\sharp$, we can convert the 1-form field df into a vector field grad $f \in T_0^1 M$, defined by grad $f = (df)^\sharp$. Prove that for any $V \in T_0^1 M$, $\mathbf{g}(\text{grad}\, f, V) = Vf$.

Now let $M = \mathbb{R}^m$ with the standard smooth structure, and let the metric $\mathbf{g}$ be the usual Euclidean inner product $\langle \cdot, \cdot \rangle$. Consider the global chart $(\mathbb{R}^m, \text{id})$. Show that the chart induced components $(\text{grad}\, f)^i$ are given by $\partial_{x^i} f$.

Exercise 5.29. (Musicality in a vector space with a scalar product.)
Let V be an m-dimensional vector space with a scalar product g. Recall that in Exercise 5.1 (p.75), we had seen that $\cdot^\flat : V \to V^*$, where $v^\flat := g(v, \cdot)$ $(v \in V)$, is an $\mathbb{R}$-linear isomorphism. Denote the inverse of $\cdot^\flat$ by $\cdot^\sharp : V^* \to V$. Let $\{v_1, \cdots, v_m\}$ be a basis for V and let $\{\omega^1, \cdots, \omega^m\}$ be the corresponding dual basis for V^*, that is, $\omega^i v_j = \delta_j^i$ for all $1 \leqslant i, j \leqslant m$. Show that $(v_i)^\flat = g(v_i, v_j)\omega^j$. We conclude that if $\{v_1, \cdots, v_m\}$ is an orthonormal basis, then

$$(v_i)^\flat = \begin{cases} \omega^i & \text{if } \iota(v_i) := g(v_i, v_i) = +1, \\ -\omega^i & \text{if } \iota(v_i) := g(v_i, v_i) = -1. \end{cases}$$

In this case (i.e., when $\{v_1, \cdots, v_m\}$ is an orthonormal), show that

$$\omega^\sharp = \sum_{i=1}^{m} \omega(v_i) g(v_i, v_i) v_i.$$

Exercise 5.30. (Trace.)
Suppose that $(M, \mathbf{g})$ is a Lorentzian manifold.
A tensor field $T \in T_2^0 M$ is *symmetric* if $T(V, W) = T(W, V)$ for all $V, W \in T_0^1 M$.
For a symmetric $T \in T_2^0 M$, define $T^\sharp \in T_1^1 M$ by

$$T^\sharp(\Omega, V) = T(\Omega^\sharp, V) \text{ for all } \Omega \in T_1^0 M \text{ and } V \in T_0^1 M,$$

and the *trace of* T by

$$\operatorname{trace} T := \mathbf{C}(T^\sharp) \in C^\infty(M).$$

Show that if M is 4-dimensional, then $\operatorname{trace} \mathbf{g} = 4$.

Appendix: Example 5.15

Here we give an argument justifying the impossibility of time-orientability. First, consider the admissible charts $(U, \mathbf{x})$ and $(V, \mathbf{y})$ covering C given by the usual cylindrical coordinates:

- $U = \{p = (\cos\theta, \sin\theta, z) : 0 < \theta < 2\pi, \ z \in \mathbb{R}\}$, with $U \ni p = (\cos\theta, \sin\theta, z) \overset{\mathbf{x}}{\mapsto} (\theta, z)$, and
- $V = \{q = (\cos\varphi, \sin\varphi, z) : -\pi < \varphi < \pi, \ z \in \mathbb{R}\}$, with $V \ni q = (\cos\varphi, \sin\varphi, z) \overset{\mathbf{y}}{\mapsto} (\varphi, z)$.

Then the chart transition maps are given by $(\mathbf{y} \circ \mathbf{x}^{-1})(\theta, z) = (a(\theta), z)$, and $(\mathbf{x} \circ \mathbf{y}^{-1})(\varphi, z) = (b(\varphi), z)$, where

$$a(\theta) := \left\{ \begin{array}{ll} \theta & \text{if } 0 < \theta < \pi \\ \theta - 2\pi & \text{if } \pi < \theta < 2\pi \end{array} \right\},$$

$$b(\varphi) := \left\{ \begin{array}{ll} \varphi & \text{if } \ 0 < \varphi < \pi \\ \varphi + 2\pi & \text{if } -\pi < \varphi < 0 \end{array} \right\}.$$

We define $\widetilde{\mathbf{g}} \in T_2^0 M$ on C by giving its the chart-induced component matrices G^U, G^V with respect to the ordered bases $(\partial_t, \partial_\theta)$, $(\partial_t, \partial_\varphi)$, and using Exercise 4.11 (p.66) to show well-definition. Set

$$\left. \begin{array}{l} G^U(\theta, z) = G(\theta) \ \text{ for } \ (\theta, z) \in \mathbf{x}(U) \\ G^V(\varphi, z) = G(\varphi) \ \text{ for } \ (\varphi, z) \in \mathbf{y}(V) \end{array} \right\}, \ \text{where } G(\alpha) = \begin{bmatrix} \cos\alpha & \sin\alpha \\ \sin\alpha & -\cos\alpha \end{bmatrix}.$$

Since the Jacobian matrices of the chart transition maps are equal to the identity matrix, and also as

$$G^U(\varphi + 2\pi, z) = G^U(\varphi, z) = G^V(\varphi, z),$$
$$G^V(\theta - 2\pi, z) = G^V(\theta, z) = G^U(\theta, z),$$

it follows that G^U, G^V define a $(0, 2)$-tensor field $\widetilde{\mathbf{g}}$ on C.

$\widetilde{\mathbf{g}}$ is a metric: Firstly, the matrices G^U, G^V, are both pointwise symmetric and invertible, giving the symmetry and nondegeneracy of $\widetilde{\mathbf{g}}(p)$ for each $p \in C$. To see that the metric index of $\widetilde{\mathbf{g}}$ is 1, we note that with

$$\mathbf{v} = \left(\cos\frac{\alpha}{2}\right)\mathbf{e}_1 + \left(\sin\frac{\alpha}{2}\right)\mathbf{e}_2, \qquad \mathbf{w} = -\left(\sin\frac{\alpha}{2}\right)\mathbf{e}_1 + \left(\cos\frac{\alpha}{2}\right)\mathbf{e}_2,$$

we have $\mathbf{v}^{\mathrm{t}} G(\alpha)\mathbf{v} = 1$, $\mathbf{w}^{\mathrm{t}} G(\alpha)\mathbf{w} = -1$, $\mathbf{v}^{\mathrm{t}} G(\alpha)\mathbf{w} = 0$. Here $\mathbf{e}_1 = (1,0)$ and $\mathbf{e}_2 = (0,1)$ in $\mathbb{R}^2$.

Consider the smooth curve $\gamma : (0, 2\pi) \to C$ given by

$$\gamma(\theta) = (\cos\theta, \sin\theta, 0) \text{ for all } \theta \in (0, 2\pi).$$

Let $W \in T_0^1 U$ be the vector field

$$W_p = -\left(\sin\frac{\theta}{2}\right)\partial_{t,p} + \left(\cos\frac{\theta}{2}\right)\partial_{\theta,p}$$

for all points $p = (\cos\theta, \sin\theta, z) \in U$, $0 < \theta < 2\pi$, $z \in \mathbb{R}$. Then W_p is timelike for all $p \in U$ since $g(p)(W_p, W_p) = -1 < 0$. Now suppose $X \in T_0^1 C$ is a time-orientation for $(C, \widetilde{\mathbf{g}})$, and write $X = X^t\partial_t + X^\varphi\partial_\varphi$ in V, for some $X^t, X^\varphi \in C^\infty(V)$. Then for all $p \in U \cap V$, we have

$$\partial_{\varphi,p} = \frac{\partial b}{\partial\varphi}(\mathbf{y}(p))\,\partial_{\theta,p} + \frac{\partial z}{\partial\varphi}(\mathbf{y}(p))\,\partial_{z,p} = 1\cdot\partial_{\theta,p} + 0\cdot\partial_{z,p} = \partial_{\theta,p}.$$

Thus in $U \cap V$, we have $X = X^t\partial_t + X^\varphi\partial_\varphi = X^t\partial_t + X^\varphi\partial_\theta$. Define smooth function $f : (0, 2\pi) \to \mathbb{R}$ by

$$f(\theta) = \widetilde{\mathbf{g}}(\gamma(\theta))(W_{\gamma(\theta)}, X_{\gamma(\theta)}) \text{ for all } \theta \in (0, 2\pi).$$

Then by the continuity of f, and the fact that W is timelike everywhere, we must have f either everywhere positive, or everywhere negative. (Indeed, otherwise f will be zero at some θ_*, giving a point $\gamma(\theta_*) \in C$, where the nonzero timelike vectors $X_{\gamma(\theta_*)}, W_{\gamma(\theta_*)}$ are orthogonal, a contradiction to Lemma 5.1, p.80.) Set $p = (1, 0, 0) \in C$. Then as X_p is timelike, we obtain $(X^t(p))^2 - (X^\varphi(p))^2 < 0$. This implies that $X^\varphi(p) \neq 0$. But we have

$$\lim_{\theta \searrow 0} f(\theta) = \begin{bmatrix} -\sin\frac{\theta}{2} & \cos\frac{\theta}{2} \end{bmatrix} \begin{bmatrix} \cos\theta & \sin\theta \\ \sin\theta & -\cos\theta \end{bmatrix} \begin{bmatrix} X^t(\mathbf{y}^{-1}(\theta, 0)) \\ X^\varphi(\mathbf{y}^{-1}(\theta, 0)) \end{bmatrix} = -X^\varphi(p),$$

while

$$\lim_{\theta \nearrow 2\pi} f(\theta) = \begin{bmatrix} -\sin\frac{\theta}{2} & \cos\frac{\theta}{2} \end{bmatrix} \begin{bmatrix} \cos\theta & \sin\theta \\ \sin\theta & -\cos\theta \end{bmatrix} \begin{bmatrix} X^t(\mathbf{y}^{-1}(\theta - 2\pi, 0)) \\ X^\varphi(\mathbf{y}^{-1}(\theta - 2\pi, 0)) \end{bmatrix} = X^\varphi(p),$$

a contradiction to the fact that f is either everywhere positive or everywhere negative. So $(C, \widetilde{\mathbf{g}})$ is not time-orientable.

Chapter 6

Levi-Civita connection

In this chapter, we will introduce additional structure on a smooth manifold, called a connection. This will allow us to calculate the directional derivative of a vector field in a given direction. While there are infinitely many connections definable on a smooth manifold, in the context of a semi-Riemannian manifold, there is a certain natural one, called the Levi-Civita connection induced by the metric $\mathbf{g}$. For spacetimes $(M, \mathbf{g})$, we will always use the Levi-Civita connection induced by $\mathbf{g}$.

We know that if $v \in T_pM$, and $f \in C^\infty(M)$, then the directional derivative of f in the 'direction' provided by the tangent vector v is simply $v(f)$. But now suppose that we replace f by a vector field $W \in T_0^1 M$. Can we still 'differentiate W in the direction of v'?

Before we address this question, we remark that wanting a notion of a directional derivative $\nabla_v W$ of a vector field W in a direction given by v is quite natural. For example if we are moving along a curve $\gamma : \mathbb{R} \to M$, we might want to know the rate of change $\nabla_{v_{\gamma,\gamma(t)}} W$ of a vector field W at $\gamma(t)$ in the direction of the tangent vector $v_{\gamma,\gamma(t)}$. In particular, we even wish to replace the vector field W by simply the map $t \mapsto v_{\gamma,\gamma(t)}$ so that we have a notion of acceleration $\mathfrak{a} := \nabla_{v_{\gamma,\gamma(t)}} v_{\gamma,\gamma(\cdot)}$. We will then be able to discuss the 'straight' curves as ones for which the acceleration is identically zero along the points of the curve, and these will be called geodesics, which is the subject of a later chapter.

Returning back to the question about the possibility of differentiating W in the direction of v, there is a fundamental obstacle. Ideally, we want to consider a curve $\gamma : (-\epsilon, \epsilon) \to M$ so that $p = \gamma(0)$, $v = v_{\gamma,\gamma(0)}$, and take the limit as $t \to 0$ of the difference quotient

$$\frac{W_{\gamma(t)} - W_{\gamma(0)}}{t}.$$

The difficulty is that $W_{\gamma(t)} \in T_{\gamma(t)}M$, while $W_{\gamma(0)} \in T_{\gamma(0)}M$, and since the spaces $T_{\gamma(t)}M$, $T_{\gamma(0)}M$ are different, we cannot make sense of $W_{\gamma(t)} - W_{\gamma(0)}$.

It would be great if we could somehow 'connect' the tangent spaces $T_{\gamma(t)}M$, $T_{\gamma(0)}M$, as in the happy situation of $\mathbb{R}^m$, where (as we shall see below), there is a standard way of doing so (using the vector space structure of $\mathbb{R}^m$), making the above idea work. In a general manifold, however, there is no such 'connection' available from the smooth structure alone.

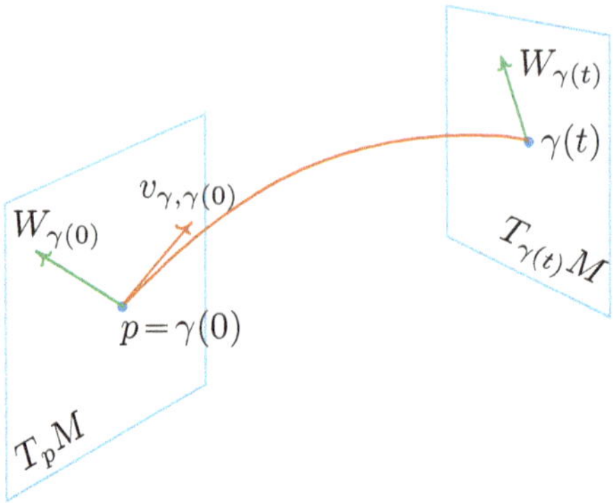

We will have to put such an additional structure 'by hand', and this is exactly the notion of a 'connection', which we will soon introduce. There are infinitely many connections available on a given smooth manifold, and so one has to make a choice in order to be able to do the aforementioned differentiation. However, for a semi-Riemannian manifold, there is a certain natural choice, induced from the metric, called the Levi-Civita connection. The Levi-Civita connection allows the definition of a geometric object called the curvature tensor field $\mathbf{R}$. The spacetime field equation involves tensor fields derived from $\mathbf{R}$.

6.1 Motivation for connections: the case of $\mathbb{R}^m$

The smooth manifold $M = \mathbb{R}^m$, with its standard smooth structure, has a global chart provided by the identity map, and so we can identify each tangent space T_pM with $\mathbb{R}^m$, using the map

$$T_pM \ni v = v^i \partial_{x^i,p} \mapsto (v^1, \cdots, v^m) \in \mathbb{R}^m.$$

With this identification, given a $v \in T_pM$ and a $W \in T_0^1 M$, we can define the derivative of W in the direction of v as follows. Let $\gamma : \mathbb{R} \to \mathbb{R}^m$ be any smooth curve such that $\gamma(0)=p$ and $v_{\gamma,p}=v$, e.g., $\gamma(t)=p+tv$. Then

$$\overline{\nabla}_v W := \lim_{t\to 0} \frac{W_{\gamma(t)} - W_{\gamma(0)}}{t}.$$

We now show that $\overline{\nabla}_v W = v(W^i)\partial_{x^i}$, that is, the derivative of W in the direction of v is simply obtained by taking the directional derivatives of the chart-induced component functions of W in the direction of v. This proves in particular that $\overline{\nabla}_v W$ does not depend on the choice of γ (so long as $\gamma(0) = p$ and $v_{\gamma,p} = v$). If $W = W^i\partial_{x^i}$, then keeping the above identification in mind,

$$\begin{aligned}\overline{\nabla}_v W &= \lim_{t\to 0} \frac{W_{\gamma(t)} - W_{\gamma(0)}}{t} = \lim_{t\to 0} \frac{W^i(\gamma(t)) - W^i(\gamma(0))}{t}\,\partial_{x^i,p} \\ &= \frac{d(W^i\circ\gamma)}{dt}\Big|_{t=0}\partial_{x^i,p} = (v_{\gamma,\gamma(0)}W^i)\,\partial_{x^i,p} = (vW^i)\,\partial_{x^i,p}.\end{aligned}$$

We have $\overline{\nabla}_v W \in T_pM$. If instead of v, we have a vector field at hand, and we set $(\overline{\nabla}_V W)(p) = \overline{\nabla}_{V_p} W$, then we obtain a map

$$\begin{aligned}\overline{\nabla} : T_0^1M \times T_0^1M &\longrightarrow T_0^1M \\ (V, W) &\longmapsto \overline{\nabla}_V W = V(W^i)\partial_{x^i}.\end{aligned}$$

From the formula $\overline{\nabla}_V W = V(W^i)\partial_{x^i}$, we obtain for $c \in \mathbb{R}$, $f \in C^\infty(M)$, $V, W, X \in T_0^1M$:

(C1) ($\mathbb{R}$-linearity of $\overline{\nabla}_V$) $\quad \overline{\nabla}_V(W + cX) = \overline{\nabla}_V W + c\overline{\nabla}_V X$

(C2) (Leibniz) $\quad \overline{\nabla}_V(fW) = (Vf)W + f\overline{\nabla}_V W$

(C3) ($C^\infty(M)$-linearity of $\overline{\nabla}_{\cdot}W$) $\quad \overline{\nabla}_{fV+X} W = f\overline{\nabla}_V W + \overline{\nabla}_X W$.

A *connection on* $M = \mathbb{R}^m$ is a map $\nabla : T_0^1M \times T_0^1M \to T_0^1M$ that satisfies (C1)-(C3). Thus $\overline{\nabla}$ is a connection on $\mathbb{R}^m$, and we call it the *flat connection on* $\mathbb{R}^m$. However, there are infinitely many other connections on $\mathbb{R}^m$; see Exercise 6.3 (p.108).

What makes $\overline{\nabla}$ special among all connections is that besides the mandatory requirements (C1), (C2), (C3), $\overline{\nabla}$ also satisfies (C4) and (C5) given below. In (C5), we consider $\mathbb{R}^m$ as being endowed with the metric $\langle\cdot,\cdot\rangle$ defined in each $T_pM \simeq \mathbb{R}^m$ to be the usual Euclidean inner product on $\mathbb{R}^m$. Thus the formulation of (C5) needs extra structure on $\mathbb{R}^m$ (besides its standard smooth structure), namely a metric.

(C4) (Torsion-free) $\quad \overline{\nabla}_V W - \overline{\nabla}_W V - [V, W] = 0$.

(C5) (Metric-compatible) $\quad V\langle W, X\rangle = \langle\overline{\nabla}_V W, X\rangle + \langle W, \overline{\nabla}_V X\rangle$.

The reason behind calling (C4) the 'torsion-free' property is that the map

$$T_0^1M \times T_0^1M \ni (V, W) \mapsto B(V, W) := \overline{\nabla}_V W - \overline{\nabla}_W V - [V, W] \in T_0^1M$$

is $C^\infty(M)$-bilinear (see Exercise 6.4, p.109), and so the map

$$\begin{aligned}&T : T_1^0M \times T_0^1M \times T_0^1M \to \mathbb{R} \\ &\qquad (\Omega, V, W) \mapsto \Omega(B(V, W)) = \Omega(\overline{\nabla}_V W - \overline{\nabla}_W V - [V, W])\end{aligned}$$

is a $(1, 2)$-tensor field, which is called the *torsion tensor-field*.

Exercise 6.1. Check (C1)–(C5) for the flat connection $\overline{\nabla}$ on $(\mathbb{R}^m, \mathbf{g})$.

Exercise 6.2. Consider $\mathbb{R}^3$ with the standard smooth structure and the flat connection. Let $V = yz\partial_x + zx\partial_y + xy\partial_z$, and $W = (xy^2+z)\partial_x + (y^2-x)\partial_y + (x+z^3)\partial_z$. Determine $\overline{\nabla}_V W$, $\overline{\nabla}_{V_p} W$, and $(\overline{\nabla}_V W)(p)$, where $p = (1,1,1)$.

Exercise 6.3. Consider the smooth manifold $\mathbb{R}^m$ with the standard smooth structure. Suppose that $\Gamma_{ij}^k \in C^\infty(\mathbb{R}^m)$ are m^3 arbitrary smooth functions on $\mathbb{R}^m$. Let $\nabla : T_0^1 M \times T_0^1 M \to T_0^1 M$ be given by $(V, W) \mapsto \nabla_V W = (V^j W^i \Gamma_{ij}^k + VW^k)\partial_{x^k}$, for all $V = V^j \partial_{x^j}$, $W = W^i \partial_{x^i} \in T_0^1 \mathbb{R}^m$ (x^i are the component functions of $\mathrm{id}_{\mathbb{R}^m}$). Show that ∇ satisfies (C1), (C2), (C3), and so is a connection on $\mathbb{R}^m$. (The flat connection $\overline{\nabla}$ is then a special case where all $\Gamma_{ij}^k \equiv 0$ on $\mathbb{R}^m$.)

Example 6.1. Consider $\mathbb{R}^2$ with its standard smooth structure, and define the connection ∇ as in Exercise 6.3, where

$$\Gamma_{xx}^x = \Gamma_{yy}^x = \Gamma_{xx}^y = \Gamma_{yy}^y = 0, \quad \Gamma_{xy}^x = \Gamma_{yx}^x = \frac{y}{1+x^2+y^2}, \quad \Gamma_{xy}^y = \Gamma_{yx}^y = \frac{x}{1+x^2+y^2}.$$

Then ∇ defines a connection on $\mathbb{R}^2$. In Example 6.2 (p.110), we will see that this connection gives a 'non-flat saddle shape' to $\mathbb{R}^2$. ◇

6.2 Definition of a connection

Based on the considerations above, we give the following definition.

Definition 6.1. (Connection on a smooth manifold, covariant derivative.) Let M be a smooth manifold. A map

$$\begin{aligned} \nabla : T_0^1 M \times T_0^1 M &\longrightarrow T_0^1 M \\ (V, W) &\longmapsto \nabla_V W \end{aligned}$$

is a *connection* if it satisfies for all $c \in \mathbb{R}$, $f \in C^\infty(M)$, and $V, W, X \in T_0^1 M$:

(C1) ($\mathbb{R}$-linearity of ∇_V) $\quad \nabla_V (W + cX) = \nabla_V W + c\nabla_V X$

(C2) (Leibniz) $\quad \nabla_V (fW) = (Vf)W + f\nabla_V W$

(C3) ($C^\infty(M)$-linearity of $\nabla_\cdot W$) $\quad \nabla_{fV+X} W = f\nabla_V W + \nabla_X W$.

The vector field $\nabla_V W$ is sometimes referred to as the *covariant*[1] *derivative of W with respect to the vector field V.*

For a fixed $W \in T_0^1 M$, $T_1^0 M \times T_0^1 M \ni (\Omega, V) \mapsto \Omega(\nabla_V W) \in C^\infty(M)$ is a $(1,1)$-tensor field. (C3) provides the linearity in the vector field slot.

[1]For an (r, s)-tensor field, the top indices on its components are historically referred to as *contravariant*, and the ones below as *covariant*. Fixing W, if we associate $\nabla_\cdot W$ with the $(1,1)$-tensor field sending $(\Omega, V) \in T_1^0 M \times T_0^1 M$ to $\Omega(\nabla_V W) \in C^\infty(M)$, then starting with a $(1,0)$-tensor field W, and we have obtained a $(1,1)$-tensor field, with an increase in the covariance index from 0 to 1. Thus $W \mapsto \nabla_\cdot W$ is referred to as 'taking the covariant derivative'.

Exercise 6.4. (Torsion tensor field.) Let M be a smooth manifold with a connection ∇. Define $B : T_0^1M \times T_0^1M \to T_0^1M$ by $B(V,W) = \nabla_V W - \nabla_W V - [V,W]$ for $V,W \in T_0^1M$. Show that B is $C^\infty(M)$-bilinear, that is, it is $C^\infty(M)$-linear in each slot by verifying that

- $[fV,W] = f[V,W] - (Wf)V$
- $B(fV + X, W) = f \cdot B(V,W) + B(X,W)$
- $B(V,W) = -B(W,V)$.

Define $T(\Omega, V, W) := \Omega(B(V,W)) = \Omega(\nabla_V W - \nabla_W V - [V,W])$ for all $\Omega \in T_1^0M$, $V,W \in T_0^1M$. Show that T is a $(1,2)$-tensor field, called the *torsion tensor field*[2].

Exercise 6.5. Suppose that ∇, $\widetilde{\nabla}$ are connections on a smooth manifold M, and that $\rho \in C^\infty(M)$. Prove that $(1-\rho)\nabla + \rho\widetilde{\nabla}$ is also a connection on M.

Exercise 6.6. Let $\mathbb{R}^2$ be given the standard smooth structure, and suppose it is also equipped with the flat connection $\overline{\nabla}$. Show that $2\overline{\nabla}$ is not a connection on $\mathbb{R}^2$. *Hint:* The Leibniz rule (C2) can be shown to fail, e.g. by taking $f = x$ and $V = \partial_x = W$ in the admissible chart $(\mathbb{R}^2, (x,y) \mapsto (x,y))$.

Exercise 6.7. Suppose that $\nabla, \widetilde{\nabla}$ are two connections on a smooth manifold M. Prove that their difference defines a $(1,2)$-tensor field T on M, where T is given by $T(\Omega, V, W) := \Omega(\nabla_V W - \widetilde{\nabla}_V W)$ for $V,W \in T_0^1M$ and $\Omega \in T_1^0M$.

Exercise 6.8. (Lie derivative revisited.) We'd seen in Definition 3.14 (p.55) that on a smooth manifold M, the Lie derivative $\mathcal{L} : T_0^1M \times T_0^1M \to T_0^1M$ can be defined using simply the smooth structure of M, by $\mathcal{L}_V W = [V,W]$ for all $V,W \in T_0^1M$. Prove that $\mathcal{L}$ satisfies (C1) and (C2). However (C3) fails in general, and show this concretely, taking $M = \mathbb{R}^2$ with the standard smooth structure, $f = x$, and the vector fields $W = \partial_x = V$, $X = 0$, in the chart $(\mathbb{R}^2, (x,y) \mapsto (x,y))$.

Remark 6.1. So the Lie derivative $\mathcal{L}_V W$ is not the appropriate notion of the 'directional derivative of W in the direction of V' we seek. Indeed, the sought-after directional derivative at a point should depend only on the value V_p of V at p, and not on what V does in the vicinity of p. But the Lie derivative *does* take into account how V changes. To see this, consider a V of the form $V = fX$. Then we have $\mathcal{L}_{fX} W = f\mathcal{L}_X W - (Wf)X$. The presence of the extra term $(Wf)X$ shows that, if we use two different functions f that match at p, so that the V_p is the same, the different values of f in the neighbourhood of p will have an effect, even at the single point p being considered (since the W acts on f as a differential operator; so although the two fs match at p, if their derivatives differ, then this will be detected by a suitable differential operator W). Also, one of the applications of the connection we want to have, is to be able to define the acceleration of a curve by calculating the derivative $\nabla_{v_{\gamma,\gamma(\cdot)}} v_{\gamma,\gamma(\cdot)}$ of the velocity vector to the curve in the direction of the velocity vector of the curve, and again the Lie derivative is not useful since $\mathcal{L}_V V = 0$ for any vector field V which is an extension of the $v_{\gamma,\gamma(\cdot)}$. *

[2]The name originates from the fact that this tensor field describes how a tangent space twists along a curve when it is 'parallelly transported' along the curve. The notion of parallel transport will be discussed in Chapter 7.

Example 6.2. Consider the graph of $z = f(x, y) = xy$, which has a 'saddle' shape, as a parametrised surface $M = \{\mathbf{r}(u, v) := (u, v, uv) \in \mathbb{R}^3 : u, v \in \mathbb{R}\}$.

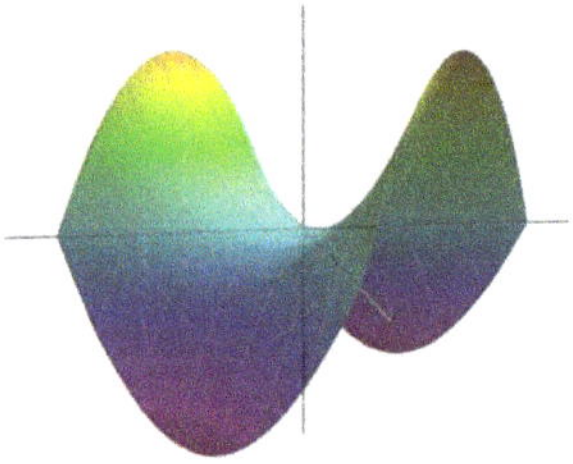

We will view a tangent vector $v \in T_pM$ at any point $p \in M$ as a tangent vector $\widetilde{v} \in T_p\mathbb{R}^3$, and by identifying[3] $T_p\mathbb{R}^3$ with $\mathbb{R}^3$, also as a vector $(\widetilde{v}x, \widetilde{v}y, \widetilde{v}z) \in \mathbb{R}^3$. (Indeed, given a vector $v \in T_pM$, there exists a smooth curve $\gamma : (-\epsilon, \epsilon) \to M$ such that $\gamma(0) = p$ and $v_{\gamma,p} = v$. We simply consider γ to be a curve in $\mathbb{R}^3$, and use its tangent vector $\widetilde{v} \in T_p\mathbb{R}^3$. Thus $\widetilde{v} = v_{\gamma,p}(x)\partial_{x,p} + v_{\gamma,p}(y)\partial_{y,p} + v_{\gamma,p}(z)\partial_{z,p} = (vx)\partial_{x,p} + (vy)\partial_{y,p} + (vz)\partial_{z,p}$.)

Let $\mathbb{R}^3$ be equipped with its usual Euclidean inner product $\langle \cdot, \cdot \rangle$ with the induced norm denoted by $\| \cdot \|$. Given a vector field $X \in T_0^1M$, we can extend it to a vector field $X \in T_0^1\mathbb{R}^3$. For example, it can be extended by translating the surface along the z-axis: $X_{(x,y,z+a)} = X_{p=(x,y,z)}$ for all $p \in M$ and all $a \in \mathbb{R}$. So if we are given $V, W \in T_0^1M$, we can extend them to vector fields $V, W \in T_0^1\mathbb{R}^3$. Keeping in mind that the calculation of $\overline{\nabla}_V W$ at a point p only requires the values of W along a curve γ in M such that $\gamma(0) = p$ and $v_{\gamma,p} = V_p$, the manner of extension does not matter, and so we can determine $\overline{\nabla}_V W$ at all points $p \in M$. At each such p, $(\overline{\nabla}_V W)(p) \in T_0^1\mathbb{R}^3$, but may not necessarily belong to T_pM. To remedy this, we subtract its component in $(T_pM)^\perp$.

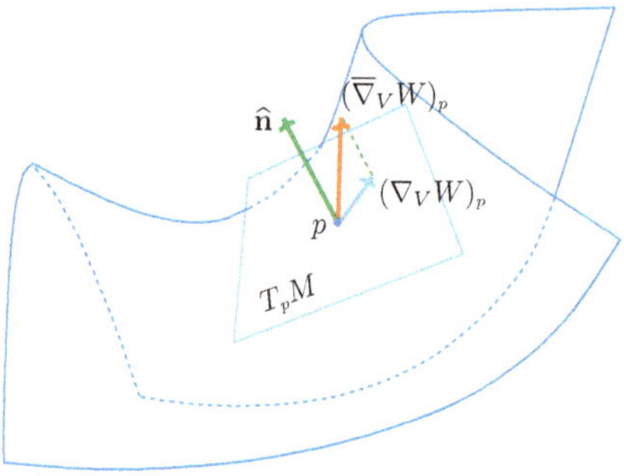

[3]Via the map sending a tangent vector $\widetilde{v} \in T_p\mathbb{R}^3$ to the triple $(\widetilde{v}x, \widetilde{v}y, \widetilde{v}z)$ of its chart-induced components in the global chart $(\mathbb{R}^3, \mathrm{id})$.

Thus we define $\nabla : T_0^1 M \times T_0^1 M \to T_0^1 M$ by $\nabla_V W = \overline{\nabla}_V W - \langle \overline{\nabla}_V W, \widehat{\mathbf{n}} \rangle \widehat{\mathbf{n}}$, where $\widehat{\mathbf{n}}$ is the 'unit normal field' to M, given by

$$\widehat{\mathbf{n}} = \frac{\frac{\partial \mathbf{r}}{\partial u} \times \frac{\partial \mathbf{r}}{\partial v}}{\left\| \frac{\partial \mathbf{r}}{\partial u} \times \frac{\partial \mathbf{r}}{\partial v} \right\|} = \frac{(-v, -u, 1)}{\|(-v, -u, 1)\|} = \frac{1}{\sqrt{1+u^2+v^2}}(-v, -u, 1).$$

Using the properties of $\overline{\nabla}$, it can be shown that ∇ is a connection on M, but we will not do this here. Rather, we will first compute it explicitly, and show that we get the same expression for the connection, in terms of the chart-induced coordinates, as described in Example 6.1 (p.108) for $\mathbb{R}^2$. Thus we may intuitively think of the connection as something which gives 'shape' to a smooth manifold: at the level of the smooth structure on $\mathbb{R}^2$, $\mathbb{R}^2$ should not be thought of as a plane, since it does not have any shape. Only if we endow the smooth manifold $\mathbb{R}^2$ with the flat connection does it become a plane, and only if we endow it with the connection from Example 6.1 does it become the saddle, etc.

Identifying tangent vectors to M as corresponding vectors in $\mathbb{R}^3$, we get (with $x, y, z \in C^\infty(M)$ defined by $x(p) = u$, $y(p) = v$, $z(p) = uv$, for all $p = (u, v, uv) \in M$) that: $\partial_u = (\partial_u x)\partial_x + (\partial_u y)\partial_y + (\partial_u z)\partial_z = 1\,\partial_x + 0\,\partial_y + y\,\partial_z$, corresponding to $(1, 0, y) \in \mathbb{R}^3$. Similarly, $\partial_v = \partial_y + x\,\partial_z \equiv (0, 1, x) \in \mathbb{R}^3$. So

$\overline{\nabla}_{\partial_u} \partial_u = \overline{\nabla}_{\partial_x + y\partial_z}(\partial_x + y\,\partial_z) = \overline{\nabla}_{\partial_x} \partial_x + \overline{\nabla}_{\partial_x}(y\,\partial_z) + y\overline{\nabla}_{\partial_z} \partial_x + y\overline{\nabla}_{\partial_z}(y\,\partial_z) = 0.$

Similarly, $\overline{\nabla}_{\partial_v} \partial_v = \overline{\nabla}_{\partial_y + x\,\partial_z}(\partial_y + x\,\partial_z) = 0$. The nonzero contributions are

$$\overline{\nabla}_{\partial_v} \partial_u = \overline{\nabla}_{\partial_y} \partial_x + \overline{\nabla}_{\partial_y}(y\,\partial_z) + x\overline{\nabla}_{\partial_z} \partial_x + x\overline{\nabla}_{\partial_z}(y\,\partial_z) = 0 + \partial_z + 0 + 0 = \partial_z,$$
$$\overline{\nabla}_{\partial_u} \partial_v = \overline{\nabla}_{\partial_x} \partial_y + \overline{\nabla}_{\partial_x}(x\,\partial_z) + y\overline{\nabla}_{\partial_z} \partial_y + y\overline{\nabla}_{\partial_z}(x\,\partial_z) = 0 + \partial_z + 0 + 0 = \partial_z.$$

Thus for $V = V^u \partial_u + V^v \partial_v$ and $W = W^u \partial_u + W^v \partial_v$, we obtain

$$\overline{\nabla}_V W = V(W^u)\partial_u + V(W^v)\partial_v + W^u V^v \partial_z + W^v V^u \partial_z.$$

As at each $p \in M$, $\widehat{\mathbf{n}}(p)$ is orthogonal in $\mathbb{R}^3$ to $\frac{\partial \mathbf{r}}{\partial u}(p)$ and also to $\frac{\partial \mathbf{r}}{\partial v}(p)$,

$$\begin{aligned}
\nabla_V W &= \overline{\nabla}_V W - (W^u V^v + W^v V^u)\langle (0,0,1), \widehat{\mathbf{n}} \rangle \widehat{\mathbf{n}} \\
&= V(W^u)\partial_u + V(W^v)\partial_v + (W^u V^v + W^v V^u)\partial_z \\
&\quad - (W^u V^v + W^v V^u)\langle (0,0,1), \widehat{\mathbf{n}} \rangle \widehat{\mathbf{n}} \\
&= V(W^u)\partial_u + V(W^v)\partial_v + (W^u V^v + W^v V^u)(\partial_z - \langle (0,0,1), \widehat{\mathbf{n}} \rangle \widehat{\mathbf{n}}).
\end{aligned}$$

We have

$$\begin{aligned}
\partial_z - \langle (0,0,1), \widehat{\mathbf{n}} \rangle \widehat{\mathbf{n}} &= (0,0,1) - \frac{1}{1+u^2+v^2}(-v, -u, 1) = \frac{1}{1+u^2+v^2}(v, u, u^2+v^2) \\
&= \frac{v}{1+u^2+v^2}(1, 0, v) + \frac{u}{1+u^2+v^2}(0, 1, u) = \Gamma^u_{uv} \partial_u + \Gamma^v_{uv} \partial_v.
\end{aligned}$$

Here, Γ^u_{uv} and Γ^v_{uv} are the expressions from Example 6.1: Recall that

$$\Gamma^u_{uu} = \Gamma^u_{vv} = \Gamma^v_{uu} = \Gamma^v_{vv} = 0,\ \Gamma^u_{uv} = \Gamma^u_{vu} = \frac{v}{1+u^2+v^2},\ \Gamma^v_{uv} = \Gamma^v_{vu} = \frac{u}{1+u^2+v^2}.$$

We have

$$\nabla_V W = (V^u W^u \Gamma^u_{uu} + V^v W^u \Gamma^u_{uv} + V^u W^v \Gamma^u_{vu} + V^v W^v \Gamma^u_{vv} + V(W^u))\partial_u$$
$$+ (V^u W^u \Gamma^v_{uu} + V^v W^u \Gamma^v_{uv} + V^u W^v \Gamma^v_{vu} + V^v W^v \Gamma^v_{vv} + V(W^v))\partial_v.$$

By Exercise 6.3, ∇ satisfies (C1)–(C3). So ∇ is a connection on M. ◇

Exercise 6.9. Repeat the calculation done in Example 6.2 for the parametrised surface $M = \{\mathbf{r}(u,v) = (u,v,0) \in \mathbb{R}^3 : u,v \in \mathbb{R}\}$, and show that the resulting connection in chart-induced coordinates is just the flat connection $\overline{\nabla}$ on $\mathbb{R}^2$.

6.3 Locality and connection coefficients/Christoffel symbols

In this section, we will learn that the covariant derivative can be computed locally, and the connection information is encoded in any admissible chart by m^3 functions Γ^k_{ij} called the 'connection coefficients' or 'Christoffel symbols'. We begin with the following crucial fact.

Lemma 6.1. (Locality of the connection.)
Let M be a smooth manifold with a connection ∇ on M, $p \in M$, and $U \subset M$ be an open set containing p. Let $V, \widetilde{V}, W, \widetilde{W} \in T^1_0 M$ be such that in U, $V = \widetilde{V}$ and $W = \widetilde{W}$. Then $(\nabla_V W)_p = (\nabla_{\widetilde{V}} \widetilde{W})_p$.

Proof. We prove this in two steps. First, keeping V fixed, and then replacing also V. Let $\chi \in C^\infty(M)$ be a bump function such that $\chi \equiv 1$ in a neighbourhood U_0 of p, and $\chi \equiv 0$ outside a neighbourhood U_1 where $\overline{U_1} \subset U$. Then we have $\chi(W - \widetilde{W}) = 0 \in T^1_0 M$. With $c = 0 \in \mathbb{R}$, we have $\nabla_V(\chi(W - \widetilde{W})) = \nabla_V 0 = \nabla_V(c\,0) = c\,\nabla_V 0 = 0\nabla_V 0 = 0$. But the left-hand side $\nabla_V(\chi(W - \widetilde{W}))$ can be expanded using the Leibniz rule, and so

$$(V\chi)(W - \widetilde{W}) + \chi\nabla_V(W - \widetilde{W}) = 0.$$

Evaluation at p yields $(V\chi)(p)\cdot 0 + 1\cdot((\nabla_V W)_p - (\nabla_V \widetilde{W})_p) = 0$, as wanted.

Now we show $(\nabla_V \widetilde{W})_p = (\nabla_{\widetilde{V}} \widetilde{W})_p$. With χ as above, $\chi(V - \widetilde{V}) = 0$. So

$$\chi(\nabla_V \widetilde{W} - \nabla_{\widetilde{V}} \widetilde{W}) \overset{\text{(C3)}}{=} \nabla_{\chi(V - \widetilde{V})} \widetilde{W} = \nabla_0 \widetilde{W} = \nabla_{0\cdot 0} \widetilde{W} = 0\nabla_0 \widetilde{W} = 0.$$

Evaluating at p, $1((\nabla_V \widetilde{W})_p - (\nabla_{\widetilde{V}} \widetilde{W})_p) = 0$, completing the proof. □

The result above allows us to 'restrict' a connection to open subsets of M.

Proposition 6.1. (Restriction of a connection to an open subset.)
Let M be a smooth manifold with a connection ∇. Let U be an open subset of M, with the smooth structure induced from M. Then there exists a unique connection ∇^U on U such that for all vector fields $V, W \in T^1_0 M$, we have $\nabla^U_{V|_U}(W|_U) = (\nabla_V W)|_U$.

Proof. *Uniqueness*: Suppose that $\nabla^U, \widetilde{\nabla}^U$ are two such connections on U. Let $V, W \in T^1_0 U$. Let $p \in U$. Take a bump function χ such that $\chi \equiv 1$ in a neighbourhood U_0 of p, and $\chi \equiv 0$ outside a neighbourhood U_1 where $\overline{U_1} \subset U$. Then $\chi V, \chi W \in T^1_0 M$, and they agree with V, W in the neighbourhood $U_0 \subset U$ of p. Thus by Lemma 6.1,

$$\begin{aligned}(\nabla^U_V W)_p &= (\nabla^U_{(\chi V)|_U}(\chi W)|_U)_p = ((\nabla_{(\chi V)}(\chi W))|_U)_p \\ &= (\widetilde{\nabla}^U_{(\chi V)|_U}(\chi W)|_U)_p = (\widetilde{\nabla}^U_V W)_p.\end{aligned}$$

As the point $p \in U$ was arbitrary, and since the vector fields $V, W \in T^1_0 U$ were arbitrary, it follows that ∇^U coincides with $\widetilde{\nabla}^U$.

Existence: Let $V, W \in T^1_0 U$, and $p \in U$. To define $(\nabla^U_V W)_p$, we first take any bump function χ as above, and define $(\nabla^U_V W)_p = (\nabla_{(\chi V)}(\chi W))_p$. This is well-defined by Lemma 6.1. The smoothness of $p \mapsto (\nabla^U_V W)_p \in T_p U$ follows from the fact that for all $p \in U_0$, we can use the same χ, so that we have $(\nabla^U_V W)_p = (\nabla_{(\chi V)}(\chi W))_p$ for all $p \in U_0$, and the right-hand side depends smoothly on p in U_0. We now check the connection axioms.

Let $V, W, X \in T^1_0 U$, $c \in \mathbb{R}$, and $f \in C^\infty(U)$. Then for all $p \in U$:

$$\begin{aligned}(\nabla^U_V (W + cX))_p &= (\nabla_{\chi V}(\chi(W + cX)))_p \\ &= (\nabla_{\chi V}(\chi W))_p + c(\nabla_{\chi V}(\chi X))_p \\ &= (\nabla^U_V W + c\nabla^U_V X)_p,\end{aligned}$$

showing (C1).
Also,

$$\begin{aligned}(\nabla^U_V (fW))_p &= (\nabla_{\chi^2 V}(\chi^2 f W))_p \\ &= ((\chi^2 V)(\chi f))(p)\cdot(\chi W)_p + (\chi f)(p)\cdot(\nabla_{\chi^2 V}(\chi W))_p \\ &= (Vf)(p)\cdot(\chi W)_p + (\chi f)(p)\cdot\chi(p)\cdot(\nabla_{\chi V}(\chi W))_p \\ &= ((Vf)W + f\cdot\nabla^U_V W)_p,\end{aligned}$$

proving (C2). Finally,

$$\begin{aligned}(\nabla^U_{fV+X} W)_p &= (\nabla_{\chi^2(fV+X)}(\chi^2 W))_p \\ &= (\chi f)(p)\cdot(\nabla_{\chi V}(\chi^2 W))_p + \chi(p)(\nabla_{\chi X}(\chi^2 W))_p \\ &= (\chi f)(p)\cdot((\chi V)\chi)(p)\cdot(\chi W)_p + (\chi f)(p)\cdot\chi(p)\cdot(\nabla_{\chi V}(\chi W))_p \\ &\quad +1\cdot((\chi X)\chi)(p)\cdot(\chi W)_p + 1\cdot\chi(p)\cdot(\nabla_{\chi X}(\chi W))_p \\ &= 1\cdot f(p)\cdot 0\cdot W_p + f(p)\cdot 1\cdot(\nabla^U_V W)_p + 1\cdot 0\cdot W_p + 1\cdot 1\cdot(\nabla^U_X W)_p \\ &= (f\,\nabla^U_V W + \nabla^U_X W)_p,\end{aligned}$$

showing (C3). This completes the proof. □

Exercise 6.10. The aim of this exercise is to show the following stronger version of Lemma 6.1, which, roughly speaking, says that $\nabla_V W$ is determined at a point p by just knowing the value V_p of V at p and the values of W in any arbitrarily small neighbourhood of p.
Let M be a smooth manifold with a connection ∇ on M. Suppose that $p \in M$ and $U \subset M$ be an open set containing p. Let the vector fields $V, \widetilde{V}, W, \widetilde{W} \in T_0^1 M$ be such that $V_p = \widetilde{V}_p$, and $W = \widetilde{W}$ in U. Then $(\nabla_V W)_p = (\nabla_{\widetilde{V}} \widetilde{W})_p$.
Hint: Proceed as follows. Keeping $\widetilde{V}$ fixed, the claim about replacing $\widetilde{W}$ by W was proved in Lemma 6.1. It remains to show $(\nabla_{V-\widetilde{V}} W)_p = 0$. Set $X = V - \widetilde{V}$. Choose a chart $(U, \mathbf{x})$ containing p, write $X = X^i \partial_{x^i}$, and note that $X^i(p) = 0$. Using Proposition 6.1, we may use the restriction of the connection to U. Note that $(\nabla_X W)_p = (\nabla^U_{X^i \partial_{x^i}} W|_U)_p$.

Henceforth, when considering the restriction to U, we will simply use the notation ∇ instead of ∇^U. In light of the above, we can now give the following definition.

Definition 6.2. (Connection coefficients/Christoffel symbols.)
Let M be an m-dimensional smooth manifold with a connection ∇, and $(U, \mathbf{x})$ be an admissible chart. The *connection coefficients/Christoffel symbols of* ∇ *with respect to* $(U, \mathbf{x})$ are the m^3 functions Γ^k_{ij}, $1 \leqslant i, j, k \leqslant m$, defined by $\nabla_{\partial_{x^j}} \partial_{x^i} = \Gamma^k_{ij} \partial_{x^k}$. More explicitly, $\Gamma^k_{ij} = dx^k(\nabla_{\partial_{x^j}} \partial_{x^i})$.

We can calculate the covariant derivatives using chart-induced components: if $V = V^j \partial_{x^j}$ and $W = W^i \partial_{x^i}$ in a chart $(U, \mathbf{x})$, then

$$\nabla_V W = \nabla_{V^j \partial_{x^j}} (W^i \partial_{x^i}) = V^j \nabla_{\partial_{x^j}} (W^i \partial_{x^i}) = V^j (\partial_{x^j} W^i) \partial_{x^i} + V^j W^i \nabla_{\partial_{x^j}} \partial_{x^i}.$$

Relabelling the dummy summation index i in the first summand by k, and substituting $\nabla_{\partial_{x^j}} \partial_{x^i} = \Gamma^k_{ij} \partial_{x^k}$ in the second summand, yields

$$\nabla_V W = (V^j \partial_{x^j}(W^k) + V^j W^i \Gamma^k_{ij}) \partial_{x^k} = (V(W^k) + V^j W^i \Gamma^k_{ij}) \partial_{x^k}. \quad (6.1)$$

Remark 6.2. (Physicist's notation for covariant differentiation.)
In a chart $(U, \mathbf{x})$, given an $f \in C^\infty(U)$, one sometimes meets the notation $f_{,i}$ for $\partial_{x^i} f$. If $V \in T_0^1 M$, then in U, $\nabla_{\partial_{x^i}} V = (\partial_{x^i} V^k + \Gamma^k_{ji} V^j) \partial_{x^k}$. So the kth component of $\nabla_{\partial_{x^i}} V$ is $(\nabla_{\partial_{x^i}} V)^k = V^k{}_{,i} + \Gamma^k_{ji} V^j$, which is written as

$$V^k{}_{;i} = V^k{}_{,i} + \Gamma^k_{ji} V^j.$$

The notation $V^k{}_{;i}$ for $(\nabla_{\partial_{x^i}} V)^k$ may cause confusion, as the latter means we take the kth component of the vector field $\nabla_{\partial_{x^i}} V$, while the former suggests that we are taking a new sort of derivative of *only* the kth component (and this is clearly wrong, since $(\nabla_{\partial_{x^i}} V)^k$ does involve all the other components of V too, as is evident by the right-hand side of the above display). We will not use the notation $V^k{}_{;i}$. $\ast$

Proposition 6.2. *Let M be a smooth manifold with a connection ∇. Let $(V, \mathbf{y})$, $(U, \mathbf{x})$ be admissible charts for M, with connection coefficients denoted by $\Gamma_{ij}^{k;V}$, $\Gamma_{ij}^{k;U}$, respectively. Then in $U \cap V$, we have*

$$\Gamma_{ij}^{k;V} = (\partial_{x^\ell} y^k)(\partial_{y^j} x^r)(\partial_{y^i} x^s)\Gamma_{sr}^{\ell;U} + (\partial_{x^\ell} y^k)(\partial_{y^j}(\partial_{y^i}(x^\ell)).$$

Proof. Let us first show that for any function $f \in C^\infty(M)$, we have $(\partial_{y^j} x^r)\partial_{x^r} f = \partial_{y^j} f$. Indeed, we have that the right-hand side is

$$\begin{aligned}\partial_{y^j} f &= \frac{\partial(f \circ \mathbf{y}^{-1})}{\partial v^j}(\mathbf{y}(\cdot)) = \frac{\partial(f \circ \mathbf{x}^{-1} \circ \mathbf{x} \circ \mathbf{y}^{-1})}{\partial v^j}(\mathbf{y}(\cdot)) \\ &= \frac{\partial(f \circ \mathbf{x}^{-1})}{\partial u^r}(\mathbf{x}(\cdot))\,\frac{\partial(\mathbf{x} \circ \mathbf{y}^{-1})^r}{\partial v^j}(\mathbf{y}(\cdot)) = (\partial_{x^r} f)(\partial_{y^j} x^r),\end{aligned}$$

which matches with the left-hand side of the equality we wanted to prove. In order to show the claimed transformation law for the connection coefficients, we use the definition of the connection coefficients in the chart $(V, \mathbf{y})$, and convert all the coordinate expressions relying on the $(V, \mathbf{y})$ chart using the basis elements induced by the $(U, \mathbf{x})$ chart:

$$\begin{aligned}\Gamma_{ij}^{k;V} &= dy^k(\nabla_{\partial_{y^j}} \partial_{y^i}) = (dy^k(\partial_{x^\ell})\,dx^\ell)(\nabla_{(\partial_{y^j} x^r)\partial_{x^r}}((\partial_{y^i} x^s)\partial_{x^s})) \\ &= (\partial_{x^\ell} y^k)\,dx^\ell((\partial_{y^j} x^r)\nabla_{\partial_{x^r}}((\partial_{y^i} x^s)\partial_{x^s})) \\ &= (\partial_{x^\ell} y^k)(\partial_{y^j} x^r)\,dx^\ell(\nabla_{\partial_{x^r}}((\partial_{y^i} x^s)\partial_{x^s})) \\ &= (\partial_{x^\ell} y^k)(\partial_{y^j} x^r)\,dx^\ell(\partial_{x^r}(\partial_{y^i} x^s)\,\partial_{x^s} + (\partial_{y^i} x^s)\Gamma_{sr}^{t;U}\,\partial_{x^t}) \\ &= (\partial_{x^\ell} y^k)(\partial_{y^j} x^r)\,\partial_{x^r}(\partial_{y^i} x^s)\,\delta_s^\ell + (\partial_{x^\ell} y^k)(\partial_{y^j} x^r)(\partial_{y^i} x^s)\Gamma_{sr}^{t;U}\,\delta_t^\ell \\ &= (\partial_{x^\ell} y^k)(\partial_{y^j} x^r)\,\partial_{x^r}(\partial_{y^i} x^\ell) + (\partial_{x^\ell} y^k)(\partial_{y^j} x^r)(\partial_{y^i} x^s)\Gamma_{sr}^{\ell;U}.\end{aligned}$$

Using our observation above with $f = \partial_{y^i} x^\ell$ yields

$$\Gamma_{ij}^{k;V} = (\partial_{x^\ell} y^k)(\partial_{y^j} x^r)(\partial_{y^i} x^s)\Gamma_{sr}^{\ell;U} + (\partial_{x^\ell} y^k)\,\partial_{y^j}(\partial_{y^i} x^\ell).$$

This completes the proof. □

Thus Γ_{ij}^k cannot be the components of some $(1,2)$-tensor field. Indeed, the presence of the second summand in the above transformation law prevents them from being tensor field components. For a tensor field, if all the tensor field components vanish at a point p in a chart $(U, \mathbf{x})$, then thanks to the multilinearity (the transformation law from Exercise 4.10, p.66), all the tensor field components with respect to any other chart $(V, \mathbf{y})$ covering p would also vanish at p. But even if all the connection coefficients with respect to $(U, \mathbf{x})$ vanished at p, then the presence of the second summand above, can prevent the connection coefficients with respect to a different overlapping chart $(V, \mathbf{y})$, from vanishing at p. The following example illustrates this.

Example 6.3. Consider $\mathbb{R}^2$ with its standard smooth structure and the flat connection $\overline{\nabla}$. Let us determine the connection coefficients with respect to the chart $(U=\mathbb{R}^2, \mathrm{id}_{\mathbb{R}^2})$. We have $\overline{\nabla}_{\partial_{x^j}} \partial_{x^i} = 0$, for all $i, j \in \{1,2\}$, and so it follows that $\Gamma_{ij}^{k;U} = 0$ for all $i, j, k \in \{1,2\}$.

However, the connection coefficients do not necessarily vanish in other charts, e.g. with polar coordinates, as we shall see explicitly now. Suppose that (for ease of handling the angular coordinate) $V = \{(x,y) \in \mathbb{R}^2 : y > 0\}$, and $\mathbf{y}(x,y) = (r,\theta)$, where

$$r := \sqrt{x^2+y^2} \quad \text{and} \quad \theta := \cos^{-1}\frac{x}{\sqrt{x^2+y^2}},$$

and $\cos^{-1} : (-1,1) \to (0,\pi)$ is the inverse of the cosine function. Then, using the above transformation rule, we have for example (where we use the more suggestive θ, r to label the Γ-symbols rather than number indices)

$$\begin{aligned}\Gamma^r_{\theta\theta} &= 0 + (\partial_x r)\,\partial_\theta(\partial_\theta x) + (\partial_y r)\,\partial_\theta(\partial_\theta y)\\ &= \frac{x}{\sqrt{x^2+y^2}}(-x) + \frac{y}{\sqrt{x^2+y^2}}(-y) = -\sqrt{x^2+y^2} = -r.\end{aligned}$$

Similarly, $\Gamma^r_{rr} = \Gamma^r_{r\theta} = \Gamma^r_{\theta r} = 0$, $\Gamma^\theta_{r\theta} = \Gamma^\theta_{\theta r} = \dfrac{1}{\sqrt{x^2+y^2}} = \dfrac{1}{r}$, and $\Gamma^\theta_{rr} = \Gamma^\theta_{\theta\theta} = 0$. ◇

Exercise 6.11. (Divergence of a vector field, given a connection.)
Let ∇ be a connection on a smooth manifold M. Let $V \in T^1_0 M$. Then we know

$$T^0_1 M \times T^1_0 M \ni (\Omega, X) \mapsto \Omega(\nabla_X V) \in C^\infty(M)$$

is a $(1,1)$-tensor field, say T. Contracting T then yields a $C^\infty(M)$ function

$$\mathbf{C}T =: \operatorname{div} V,$$

called the *divergence of* V. Show that in a chart $(U, \mathbf{x})$, $\operatorname{div} V = \partial_{x^i} V^i + V^j \Gamma^i_{ji}$. (In particular, if $M = \mathbb{R}^3$, equipped with the standard smooth structure and the flat connection, then the connection coefficients all vanish in the global chart $(\mathbb{R}^3, \mathrm{id})$, and so for a vector field $V = V^x \partial_x + V^y \partial_y + V^z \partial_z \in T^1_0 \mathbb{R}^3$, we obtain $\operatorname{div} V = \partial_x V^x + \partial_y V^y + \partial_z V^z$.)

Exercise 6.12. Recall from Exercise 6.4 the definition of the torsion tensor field corresponding to a connection ∇ on a smooth manifold M. Prove that if ∇ satisfies the torsion freeness condition

(C4) $\nabla_V W - \nabla_W V - [V, W] = 0$ for all $V, W \in T^1_0 M$,

then the Christoffel symbols with respect to any admissible chart satisfy $\Gamma^k_{ij} = \Gamma^k_{ji}$.

Exercise 6.13. (Connection 1-forms.)
Let ∇ be a connection on a smooth manifold M, and let $(U, \mathbf{x})$ be an admissible chart. Show that there exists a unique matrix $[\Omega^i_j]$ of 1-form fields $\Omega^i_j \in T^0_1 U$ such that $\nabla_V \partial_{x^i} = (\Omega^j_i V)\,\partial_{x^j}$ for all $V \in T^1_0 M$. *Hint:* Act by dx^k to get $\Omega^k_i = \Gamma^k_{ij}\, dx^j$. Show that the connection coefficients can be recovered from the connection 1-form field by the formula $\Gamma^k_{ij} = \Omega^k_i\, \partial_{x^j}$.

6.4 The Levi-Civita connection

In this section, we will learn about the 'miracle' of semi-Riemannian geometry, namely that there is a unique torsion-free and metric-compatible connection on a given semi-Riemannian manifold $(M, \mathbf{g})$.

Theorem 6.1. (Fundamental theorem of semi-Riemannian geometry.) *Let $(M, \mathbf{g})$ be a semi-Riemannian manifold. Then there exists a unique connection ∇ on M that satisfies for all $V, W, X \in T_0^1 M$*

(C4) (Torsion-free) $\quad \nabla_V W - \nabla_W V - [V, W] = 0$

(C5) (Metric-compatible) $\quad V(\mathbf{g}(W, X)) = \mathbf{g}(\nabla_V W, X) + \mathbf{g}(W, \nabla_V X)$.

Proof. *Uniqueness*: Suppose that a connection ∇ satisfying (C1) – (C5) exists. Using the metric-compatibility and the torsion-freeness, we have for any three vector fields X, Y, Z

$$\begin{aligned} X\mathbf{g}(Y, Z) &= \mathbf{g}(\nabla_X Y, Z) + \mathbf{g}(Y, \nabla_X Z) = \mathbf{g}(\nabla_X Y, Z) + \mathbf{g}(Y, \nabla_Z X + [X, Z]) \\ &= \mathbf{g}(\nabla_X Y, Z) + \mathbf{g}(\nabla_Z X, Y) + \mathbf{g}([X, Z], Y). \end{aligned}$$

Cyclically permuting the X, Y, Z, we have:

$$\begin{aligned} X\mathbf{g}(Y, Z) &= \mathbf{g}(\nabla_X Y, Z) + \mathbf{g}(\nabla_Z X, Y) + \mathbf{g}([X, Z], Y) \\ Y\mathbf{g}(Z, X) &= \mathbf{g}(\nabla_Y Z, X) + \mathbf{g}(\nabla_X Y, Z) + \mathbf{g}([Y, X], Z) \\ Z\mathbf{g}(X, Y) &= \mathbf{g}(\nabla_Z X, Y) + \mathbf{g}(\nabla_Y Z, X) + \mathbf{g}([Z, Y], X). \end{aligned}$$

Subtracting the third from the sum of the first two gives

$$\begin{aligned} &X\mathbf{g}(Y, Z) + Y\mathbf{g}(Z, X) - Z\mathbf{g}(X, Y) \\ &= 2\mathbf{g}(\nabla_X Y, Z) + \mathbf{g}([X, Z], Y) + \mathbf{g}([Y, X], Z) - \mathbf{g}([Z, Y], X). \end{aligned}$$

Consequently, $\mathbf{g}(\nabla_X Y, Z)$ is given by the following *Koszul formula*[4]:

$$\begin{aligned} \mathbf{g}(\nabla_X Y, Z) = \tfrac{1}{2}\big(&X\mathbf{g}(Y, Z) + Y\mathbf{g}(Z, X) - Z\mathbf{g}(X, Y) \\ &- \mathbf{g}([X, Z], Y) - \mathbf{g}([Y, X], Z) + \mathbf{g}([Z, Y], X)\big). \end{aligned}$$

As $Z \in T_0^1 M$ was arbitrary, this uniquely determines $\nabla_X Y$ (Exercise 5.12).

Existence: For fixed $X, Y \in T_0^1 M$, the map sending Z to the right-hand side of Koszul's formula is $C^\infty(M)$-linear:

$$\begin{aligned} Z \mapsto \Omega(X, Y)(Z) := \tfrac{1}{2}\big(&X\mathbf{g}(Y, Z) + Y\mathbf{g}(Z, X) - Z\mathbf{g}(X, Y) \\ &- \mathbf{g}([X, Z], Y) - \mathbf{g}([Y, X], Z) + \mathbf{g}([Z, Y], X)\big). \end{aligned}$$

Indeed, the terms $-Z\mathbf{g}(X, Y)$ and $-\mathbf{g}([Y, X], Z)$ are $C^\infty(M)$-linear in Z. The combination of $X\mathbf{g}(Y, Z)$ with $-\mathbf{g}([X, Z], Y)$ is $C^\infty(M)$-linear too: the additivity is obvious, and we have homogeneity in scaling by $f \in C^\infty(M)$

[4] After Jean-Louis Koszul (1921 – 2018), a French mathematician.

because

$$
\begin{aligned}
&X\mathbf{g}(Y, fZ) - \mathbf{g}([X, fZ], Y) \\
&= (Xf)\mathbf{g}(Y, Z) + fX\mathbf{g}(Y, Z) - \mathbf{g}(f[X, Z] + (Xf)Z, Y) \\
&= \cancel{(Xf)\mathbf{g}(Y, Z)} + fX\mathbf{g}(Y, Z) - f\,\mathbf{g}([X, Z], Y) - \cancel{(Xf)\mathbf{g}(Z, Y)}. \qquad (\star)
\end{aligned}
$$

Similarly, $Y\mathbf{g}(Z, X) + \mathbf{g}([Z, Y], X) = Y\mathbf{g}(X, Z) - \mathbf{g}([Y, Z], X)$ is $C^\infty(M)$-linear in Z. So $\Omega(X, Y) \in T_1^0 M$. By Theorem 5.7, there exists a vector field $\nabla_X Y \in T_0^1 M$ such that $(\nabla_X Y)^\flat = \Omega(X, Y)$, i.e., $\nabla_X Y = (\Omega(X, Y))^\sharp$. We need to check that this ∇ satisfies (C1)–(C5). This is straightforward, albeit a bit tedious. (C1) is immediate. To check (C2), we have firstly

$$
\begin{aligned}
\Omega(X, fY)(Z) &= \tfrac{1}{2}(X\mathbf{g}(fY, Z) + fY\mathbf{g}(Z, X) - Z\mathbf{g}(X, fY) \\
&\qquad - \mathbf{g}([X, Z], fY) - \mathbf{g}([fY, X], Z) + \mathbf{g}([Z, fY], X)) \\
&= \tfrac{1}{2}((Xf)\mathbf{g}(Y, Z) - (Zf)\mathbf{g}(X, Y) + (Xf)\mathbf{g}(Y, Z) + (Zf)\mathbf{g}(Y, X)) \\
&\qquad + f\Omega(X, Y)(Z) \\
&= f\,\Omega(X, Y)(Z) + (Xf)\mathbf{g}(Y, Z) = f\,\Omega(X, Y)(Z) + (Xf)Y^\flat(Z).
\end{aligned}
$$

Hence $\Omega(X, fY) = f\ \Omega(X, Y) + (Xf)Y^\flat$. Applying $\cdot^\sharp$ on both sides, we obtain $\nabla_X(fY) = f\nabla_X Y + (Xf)Y$.

For (C3), we first check that $X \mapsto \Omega(X, Y)(Z)$ is $C^\infty(M)$-linear. Additivity is clear. Homogeneity in scaling by $f \in C^\infty(M)$ can be seen as follows. The terms $X\mathbf{g}(Y, Z)$ and $\mathbf{g}([Z, Y], X)$ clearly respect the scaling of X. The combination of $Y\mathbf{g}(Z, X)$ with $-\mathbf{g}([Y, X], Z)$ works as in $(\star)$ above (and also similarly the combination $-Z\mathbf{g}(X, Y) - \mathbf{g}([X, Z], Y)$, seen by exchanging Y, Z):

$$
\begin{aligned}
&Y\mathbf{g}(Z, fX) - \mathbf{g}([Y, fX], Z) \\
&= (Yf)\mathbf{g}(Z, X) + fY\mathbf{g}(Z, X) - \mathbf{g}(f[Y, X] + (Yf)X, Z) \\
&= (Yf)\mathbf{g}(Z, X) + fY\mathbf{g}(Z, X) - f\mathbf{g}([Y, X], Z) - (Yf)\mathbf{g}(X, Z).
\end{aligned}
$$

So $\Omega(fX + \widetilde{X}, Y)(Z) = f\,\Omega(X, Y)(Z) + \Omega(\widetilde{X}, Y)(Z)$ for all $Z \in T_0^1 M$. Hence we conclude that $\Omega(fX + \widetilde{X}, Y) = f\,\Omega(X, Y) + \Omega(\widetilde{X}, Y)$. As $\cdot^\sharp$ is $C^\infty(M)$-linear, (C3) holds.

Next, we show (C4): We have

$$
\begin{aligned}
\Omega(X, Y)(Z) - \Omega(Y, X)(Z) &= \tfrac{1}{2}\big(\underline{X\mathbf{g}(Y, Z)} + Y\mathbf{g}(Z, X) - Y\mathbf{g}(X, Z) - \underline{X\mathbf{g}(Z, Y)} \\
&\qquad - \mathbf{g}([X, Z], Y) - \mathbf{g}([Y, X], Z) + \mathbf{g}([Z, Y], X) \\
&\qquad + \mathbf{g}([Y, Z], X) + \mathbf{g}([X, Y], Z) - \mathbf{g}([Z, X], Y)\big) \\
&= \mathbf{g}([X, Y], Z) = [X, Y]^\flat Z.
\end{aligned}
$$

Thus we have $\Omega(X, Y) - \Omega(Y, X) = [X, Y]^\flat$, and applying $\cdot^\sharp$, we obtain $\nabla_X Y - \nabla_Y X = [X, Y]$, that is, (C4) holds.

Finally, we verify (C5): Firstly,

$$\mathbf{g}(\nabla_X Y, Z) = (\nabla_X Y)^\flat(Z) = (\Omega(X,Y)^\sharp)^\flat(Z) = \Omega(X,Y)(Z),$$

which is precisely the right-hand side of the Koszul formula. Hence we have

$$\begin{aligned}\mathbf{g}(\nabla_X Y, Z)+\mathbf{g}(Y,\nabla_X Z) &= \tfrac{1}{2}(X\mathbf{g}(Y,Z) + \underline{Y\mathbf{g}(Z,X)} - \underline{Z\mathbf{g}(X,Y)} \\ &\quad - \underline{\mathbf{g}([X,Z],Y)} - \underline{\mathbf{g}([Y,X],Z)} + \underline{\underline{\mathbf{g}([Z,Y],X)}} \\ &\quad + X\mathbf{g}(Z,Y) + \underline{Z\mathbf{g}(Y,X)} - \underline{Y\mathbf{g}(X,Z)} \\ &\quad - \underline{\mathbf{g}([X,Y],Z)} - \underline{\mathbf{g}([Z,X],Y)} + \underline{\underline{\mathbf{g}([Y,Z],X)}}) \\ &= X\mathbf{g}(Y,Z),\end{aligned}$$

proving (C5). This completes the proof. □

Exercise 6.14. Suppose that $(M, \mathbf{g})$ is a semi-Riemannian manifold with the metric $\mathbf{g}$ and the Levi-Civita connection ∇. Let $f \in C^\infty(M)$ be such that $\mathbf{g}((df)^\sharp, (df)^\sharp)$ is constant on M. Prove that $\nabla_{(df)^\sharp}((df)^\sharp) = 0$.

Exercise 6.15. Let $(M, \mathbf{g}), (N, \mathbf{h})$ be semi-Riemannian manifolds with corresponding Levi-Civita connections $\nabla^{\mathbf{g}}, \nabla^{\mathbf{h}}$. Let $f : M \to N$ be an isometry. Show that $f_*(\nabla^{\mathbf{g}}_X Y) = \nabla^{\mathbf{h}}_{f_* X}(f_* Y)$ for all $X, Y \in T_0^1 M$. *Hint:* Use the Koszul formula, and Exercise 3.11 (p.45).

Exercise 6.16. Let $(G, \mathbf{g})$ be a Lie group with a metric $\mathbf{g}$ which is *left-invariant*, that is, for all $p \in G$, the left-translation L_p is an isometry. Show that for all $X, Y, Z \in \mathfrak{g}$, $2\mathbf{g}(\nabla_X Y, Z) = \mathbf{g}([X,Y],Z) + \mathbf{g}([Z,X],Y) + \mathbf{g}(X,[Z,Y])$. Suppose additionally that for all $X, Y, Z \in \mathfrak{g}$, $\mathbf{g}([Z,X],Y) = -\mathbf{g}(X,[Z,Y])$. Show that $2\nabla_X Y = \mathcal{L}_X Y$ for all $X, Y \in \mathfrak{g}$.

Exercise 6.17. Let $\mathbf{g}, \widetilde{\mathbf{g}}$ be two semi-Riemannian metrics on a smooth manifold M such that $\widetilde{\mathbf{g}} = f\mathbf{g}$ for some pointwise positive function $f \in C^\infty(M)$. Let $\nabla, \widetilde{\nabla}$ denote Levi-Civita connections corresponding to $\mathbf{g}, \widetilde{\mathbf{g}}$, respectively. Show that for all $X, Y, Z \in T_0^1 M$, $\widetilde{\nabla}_X Y = \nabla_X Y + \frac{1}{2f}((Xf)Y + (Yf)X - \mathbf{g}(X,Y)\mathrm{grad}\, f)$, where $\mathrm{grad}\, f \in T_0^1 M$ is defined by $\mathbf{g}(\mathrm{grad}\, f, Z) = Zf$ for all $Z \in T_0^1 M$ (see Exercise 5.28, p.102).

Proposition 6.3. *Let $(M, \mathbf{g})$ be a semi-Riemannian manifold, and ∇ denote the Levi-Civita connection. Then the connection coefficients with respect to an admissible chart $(U, \mathbf{x})$ are given by*

$$\Gamma^k_{ij} = \Gamma^k_{ji} = \frac{\mathbf{g}^{\ell k}}{2}\left(\partial_{x^i}\mathbf{g}_{j\ell} + \partial_{x^j}\mathbf{g}_{\ell i} - \partial_{x^\ell}\mathbf{g}_{ij}\right).$$

Proof. Recall that $[\partial_{x^i}, \partial_{x^j}] = 0$ for all i, j (Lemma 3.3, p.45). Taking $X = \partial_{x^i}$, $Y = \partial_{x^j}$, $Z = \partial_{x^\ell}$ in the Koszul formula, we thus obtain,

$$\begin{aligned}\mathbf{g}(\nabla_{\partial_{x^i}}\partial_{x^j}, \partial_{x^\ell}) &= \frac{1}{2}(\partial_{x^i}\mathbf{g}(\partial_{x^j}, \partial_{x^\ell}) + \partial_{x^j}\mathbf{g}(\partial_{x^\ell}, \partial_{x^i}) - \partial_{x^\ell}\mathbf{g}(\partial_{x^i}, \partial_{x^j})) \\ &= \frac{1}{2}(\partial_{x^i}\mathbf{g}_{j\ell} + \partial_{x^j}\mathbf{g}_{\ell i} - \partial_{x^\ell}\mathbf{g}_{ij}).\end{aligned}$$

Substituting $\nabla_{\partial_{x^i}}\partial_{x^j} = \Gamma^r_{ji}\partial_{x^r}$ and multiplying throughout by the inverse metric component $\mathbf{g}^{\ell k}$ (and summing over the dummy index ℓ) yields

$$\begin{aligned}\Gamma^k_{ij} = \Gamma^k_{ji} &= \Gamma^r_{ji}\delta^k_r = \Gamma^r_{ji}\mathbf{g}_{r\ell}\mathbf{g}^{\ell k} = \mathbf{g}(\Gamma^r_{ji}\partial_{x^r}, \partial_{x^\ell})\mathbf{g}^{\ell k} \\ &= \mathbf{g}(\nabla_{\partial_{x^i}}\partial_{x^j}, \partial_{x^\ell})\mathbf{g}^{\ell k} = \frac{\mathbf{g}^{\ell k}}{2}(\partial_{x^i}\mathbf{g}_{j\ell} + \partial_{x^j}\mathbf{g}_{\ell i} - \partial_{x^\ell}\mathbf{g}_{ij}). \qquad \square\end{aligned}$$

Exercise 6.18. Let $(M, \mathbf{g})$ be an m-dimensional Lorentzian manifold with the Levi-Civita connection ∇. Let $(U, \mathbf{x})$ be an admissible chart for M, and define $G = [\mathbf{g}_{ij}]$. Show that $\partial_{x^k} \det G = (-1)^{i+j}\det[G^{ij}]\,\partial_{x^k}\mathbf{g}_{ij}$. Here G^{ij} denotes the (i, j)-minor of G, that is, the matrix obtained by deleting the ith row and j^{th} column of G; see for instance [Artin (1991), §1.5].

Hint: Use the product rule to write the derivative as a sum of determinants of matrices in which the ith row of G has been replaced by the row comprising the derivatives of the entries of that row. Then in each determinant summand, expand the determinant by going along the row.

Using the symmetry and invertibility of $G(p)$ ($p \in U$), show that

$$\partial_{x^k} \det G = (\det G)\,\mathbf{g}^{ij}\partial_{x^k}\mathbf{g}_{ij},$$

where $[\mathbf{g}^{ij}]$ is the inverse of G. *Hint:* Use Cramer's formula [Artin (1991), §1.5].

Prove that $\partial_{x^k}(\log\sqrt{-\det G}) = \Gamma^j_{jk}$.

Example 6.4. (Sphere.) Recall Example 5.7 (p.85), where we considered the sphere S^2 as a Riemannian manifold using the metric $\mathbf{g}$ induced on the tangent spaces from the usual Euclidean inner product on $\mathbb{R}^3$. We know that in the chart (U, φ) described there,

$$G = \begin{bmatrix} \mathbf{g}_{\theta\theta} & \mathbf{g}_{\theta\phi} \\ \mathbf{g}_{\phi\theta} & \mathbf{g}_{\phi\phi}\end{bmatrix} = \begin{bmatrix} 1 & 0 \\ 0 & (\sin\theta)^2\end{bmatrix}, \quad G^{-1} = \begin{bmatrix} \mathbf{g}^{\theta\theta} & \mathbf{g}^{\theta\phi} \\ \mathbf{g}^{\phi\theta} & \mathbf{g}^{\phi\phi}\end{bmatrix} = \begin{bmatrix} 1 & 0 \\ 0 & (\sin\theta)^{-2}\end{bmatrix}.$$

We use the letters θ, ϕ of the variables to label the connection coefficients of the Levi-Civita connection. As $\mathbf{g}^{\theta\phi} = 0$, only one summand survives in the expression for $\Gamma^\theta_{\phi\phi}$, and

$$\Gamma^\theta_{\phi\phi} = 0 + \frac{\mathbf{g}^{\theta\theta}}{2}(\partial_\phi\mathbf{g}_{\theta\phi} + \partial_\phi\mathbf{g}_{\theta\phi} - \partial_\theta\mathbf{g}_{\phi\phi}) = \frac{1}{2}(0 + 0 - \partial_\theta(\sin\theta)^2) = -(\sin\theta)\cos\theta.$$

Similarly, $\Gamma^\theta_{\theta\theta} = 0$, $\Gamma^\theta_{\theta\phi} = \Gamma^\theta_{\phi\theta} = 0$, $\Gamma^\phi_{\theta\phi} = \Gamma^\phi_{\phi\theta} = \frac{\cos\theta}{\sin\theta} = \cot\theta$, $\Gamma^\phi_{\theta\theta} = \Gamma^\phi_{\phi\phi} = 0$. ◇

Example 6.5. (Minkowski spacetime.) Recall the Minkowski spacetime $(M, \mathbf{g})$ described in Example 5.6 (p.83). Consider an inertial frame induced by the choice of a $p \in M$ and an orthonormal basis $\{\mathbf{f}_0, \mathbf{f}_1, \mathbf{f}_2, \mathbf{f}_3\}$ for V with respect to g. The chart-induced component matrix $[\mathbf{g}_{ij}]$ of the metric is the diagonal matrix with diagonal entries $-1, 1, 1, 1$. So all the connection coefficients Γ^k_{ij} of the Levi-Civita connection in this chart are zeroes. ◇

Example 6.6. (Cylindrical spacetime.) Recall the cylindrical spacetime from Example 5.8 (p.86) and the chart (U, φ) described there. In this chart, the component matrix $G = [\mathbf{g}_{ij}]$ of the metric is the constant diagonal matrix with diagonal entries $1, -1$. So all the connection coefficients Γ^k_{ij} of the Levi-Civita connection in this chart are zeroes. ◇

Example 6.7. (FLRW spacetime.) Recall the FLRW spacetime from Example 5.9 (p.87), namely $M = I \times \mathbb{R}^3$, where $I := (0, \infty)$. For the admissible global chart $(M, \mathrm{id}_{I\times\mathbb{R}^3})$, the chart-induced component matrix $G = [\mathbf{g}_{ij}]$ of the metric is the diagonal matrix with diagonal entries $-1, a^2, a^2, a^2$. Thus G^{-1} is also a diagonal matrix, with the diagonal entries $-1, \frac{1}{a^2}, \frac{1}{a^2}, \frac{1}{a^2}$. With the indexing $0, 1, 2, 3$ used for the variables t, x, y, z, we have, as an example, the following connection coefficient of the Levi-Civita connection:

$$\Gamma^0_{11} = \frac{\mathbf{g}^{00}}{2}(\partial_{x^1}\mathbf{g}_{10} + \partial_{x^1}\mathbf{g}_{10} - \partial_{x^0}\mathbf{g}_{11}) + 0 + 0 + 0 = \frac{-1}{2}(0 + 0 - \partial_0 a^2) = a\dot{a},$$

where $\dot{a} := \frac{da}{dt}$. Similarly, for $1 \leqslant i, j, k \leqslant 3$, $\Gamma^0_{00} = 0$, $\Gamma^0_{0i} = \Gamma^0_{i0} = 0$, $\Gamma^0_{ij} = \delta_{ij} a\dot{a}$, $\Gamma^i_{00} = 0$, $\Gamma^i_{0j} = \Gamma^i_{j0} = \delta_{ij}\frac{\dot{a}}{a}$, and $\Gamma^i_{jk} = 0$. ◇

Example 6.8. (Schwarzschild spacetime.) Recall the Schwarzschild spacetime from Example 5.10 (p.87). In the chart $\mathbb{R} \times (2m, \infty) \times U$ described there, the metric component matrix $G = [\mathbf{g}_{ij}]$ is diagonal with the diagonal entries $-(1 - \frac{2m}{r})$, $(1 - \frac{2m}{r})^{-1}$, r^2, $r^2(\sin\theta)^2$. Thus for example, we have the following connection coefficient of the Levi-Civita connection:

$$\Gamma^r_{tt} = \frac{\mathbf{g}^{rr}}{2}(\partial_t \mathbf{g}_{rt} + \partial_t \mathbf{g}_{rt} - \partial_r \mathbf{g}_{tt}) = -\frac{1}{2}\Big(1 - \frac{2m}{r}\Big)\partial_r\Big(-\Big(1 - \frac{2m}{r}\Big)\Big) = \frac{m}{r^2}\Big(1 - \frac{2m}{r}\Big).$$

Similarly,

$$\begin{aligned}
\Gamma^t_{tr} &= \Gamma^t_{rt} = \frac{m}{r^2(1 - \frac{2m}{r})},\\
\Gamma^r_{rr} &= -\frac{m}{r^2(1 - \frac{2m}{r})}, \quad \Gamma^r_{\theta\theta} = -r\Big(1 - \frac{2m}{r}\Big), \quad \Gamma^r_{\phi\phi} = -r\Big(1 - \frac{2m}{r}\Big)(\sin\theta)^2,\\
\Gamma^\theta_{r\theta} &= \Gamma^\theta_{\theta r} = \frac{1}{r}, \quad \Gamma^\theta_{\phi\phi} = -(\sin\theta)(\cos\theta),\\
\Gamma^\phi_{\theta\phi} &= \Gamma^\phi_{\phi\theta} = \frac{\cos\theta}{\sin\theta}, \quad \Gamma^\phi_{r\phi} = \Gamma^\phi_{\phi r} = \frac{1}{r},
\end{aligned}$$

and all other Γ-symbols are zeroes. ◇

Exercise 6.19. For the saddle surface M considered in Exercise 5.15 (p.86), with the induced metric $\mathbf{g}$ on M from the Euclidean inner product on $\mathbb{R}^3$, determine the Christoffel symbols for the Levi-Civita connection with respect to the global chart given there. Note that these coincide with the connection coefficients determined in Example 6.2 (p.110), showing that the connection considered in Example 6.2 coincides with the Levi-Civita connection corresponding to $\mathbf{g}$.

Exercise 6.20. (Curvature tensor field.)
Let M be a smooth manifold with a connection ∇.
For $X, Y, Z \in T_0^1 M$, define $R(X,Y)Z \in T_0^1 M$ by

$$R(X,Y)Z = \nabla_X \nabla_Y Z - \nabla_Y \nabla_X Z - \nabla_{[X,Y]} Z.$$

(1) Show that
- $T_0^1 M \ni X \mapsto R(X,Y)Z \in T_0^1 M$ (for fixed Y, Z)
- $T_0^1 M \ni Y \mapsto R(X,Y)Z \in T_0^1 M$ (for fixed Z, X)
- $T_0^1 M \ni Z \mapsto R(X,Y)Z \in T_0^1 M$ (for fixed X, Y)

are all $C^\infty(M)$-linear.

(2) For any $p \in M$, and $x, y, z \in T_pM$, set $R(p)(x,y)z = (R(X,Y)Z)_p$, where $X, Y, Z \in T_0^1 M$ are any vector fields such that $X_p = x$, $Y_p = y$ and $Z_p = z$. Use the $C^\infty(M)$-linearity of R to show that $R(p) : (T_pM)^3 \to T_pM$ is well-defined.

(3) Via R, we obtain a $(1,3)$-tensor field $\mathbf{R}$, called the *curvature tensor field*, defined as follows: $\mathbf{R}(\Omega, X, Y, Z) := \Omega(R(X,Y)Z)$ for $X, Y, Z \in T_0^1 M$ and for $\Omega \in T_1^0 M$. An admissible chart $(U, \mathbf{x})$ is called an *affine chart* if all the connection coefficients with respect to $(U, \mathbf{x})$ are identically zero. We call M *locally flat* if for every point $p \in M$, there exists an affine chart $(U, \mathbf{x})$ such that $p \in U$. Show that $\mathbf{R} = \mathbf{0}$ for a locally flat manifold. (In Chapter 9, we will learn that the converse also holds.)

(4) Examples of locally flat semi-Riemannian manifolds are the Minkowski spacetime given in Example 5.6 (p.83), and the cylindrical spacetime given in Example 5.8 (p.86). However, for the sphere S^2 considered as a Riemannian manifold as in Example 5.7 (p.85), show that in the chart (U, φ) given there, $R(\partial_\theta, \partial_\phi)\partial_\theta = \partial_\phi \neq 0$, so that $\mathbf{R}(d\phi, \partial_\theta, \partial_\phi, \partial_\theta) = -1 \neq 0$.

6.5 Covariant derivative of tensor fields

We started our discussion accepting that we know how to differentiate functions $f \in C^\infty(M)$ in a direction $v \in T_pM$, resulting in $vf \in \mathbb{R}$. If we have a vector field V, providing a 'distribution' of such directions, then we obtain $Vf \in C^\infty(M)$. We wanted to do a similar thing replacing f by a vector field W. We realised that to do so we need additional structure, namely a connection ∇ on M. How about wanting to differentiate a 1-form field Ω in the direction provided by V? And more generally even an (r,s)-tensor field? It turns out that once we have a connection ∇, we are able to differentiate any (r,s)-tensor field. The key to this is that we want a sensible Leibniz rule, and this will motivate Definitions 6.3 and 6.4.

For a $(0,0)$-tensor field, namely a $C^\infty(M)$ element, and a $V \in T_0^1 M$, we set $\nabla_V f = Vf \in C^\infty(M)$. Note that this definition uses only the smooth structure on M. We wish to define $\nabla_V \Omega$, where $\Omega \in T_1^0 M$ is a 1-form field.

Just as

for a $(0,0)$-tensor field f, $\nabla_V f$ is a $(0,0)$-tensor field,
for a $(1,0)$-tensor field W, $\nabla_V W$ is a $(1,0)$-tensor field,

we want $\nabla_V \Omega$ to be a 1-form field. We first note that the action of Ω on a vector field W is a smooth function, and so there is no problem defining $\nabla_V(\Omega(W))$: it is $V(\Omega(W)) \in C^\infty(M)$. But now suppose that, however we define $\nabla_V \Omega \in T_1^0 M$ to be, we want the Leibniz rule:

$$(V(\Omega(W))=)\;\; \nabla_V(\Omega(W)) = (\nabla_V \Omega)(W) + \Omega(\nabla_V W).$$

Then we can 'solve for' $\nabla_V \Omega$, and obtain

$$(\nabla_V \Omega)(W) = V(\Omega(W)) - \Omega(\nabla_V W).$$

So this seems to give the action of our desired 1-form field $\nabla_V \Omega$ on any vector field W, and the right hand side is an element of $C^\infty(M)$. In other words, it defines the 1-form field $\nabla_V \Omega$! In order to use this as the definition of $\nabla : T_1^0 M \to T_1^0 M$, we need to make sure that

$$T_0^1 M \ni W \mapsto V(\Omega(W)) - \Omega(\nabla_V W) \in C^\infty(M)$$

is a 1-form field. We do so below: For $f \in C^\infty(M)$, $W, X \in T_0^1 M$, we have

$$\begin{aligned}
&(\nabla_V \Omega)(fW+X) \\
&= V(\Omega(fW+X)) - \Omega(\nabla_V(fW+X)) \\
&= V(f\Omega W + \Omega X) - \Omega((Vf)W + f\nabla_V W + \nabla_V X) \\
&= (Vf)\Omega W + fV(\Omega W) + V(\Omega X) - (Vf)\Omega W - f\Omega\nabla_V W - \Omega\nabla_V X \\
&= f(V(\Omega W) - \Omega(\nabla_V W)) + V(\Omega X) - \Omega(\nabla_V X) \\
&= f(\nabla_V \Omega)(W) + (\nabla_V \Omega)(X).
\end{aligned}$$

Definition 6.3. (Covariant derivative of functions and of 1-form fields.)
Let M be a smooth manifold with a connection ∇, and $V \in T_0^1 M$.
If $f \in C^\infty(M)$, then its *covariant derivative* $\nabla_V f \in C^\infty(M)$ is $\nabla_V f = Vf$.
If $\Omega \in T_1^0 M$, then its *covariant derivative* $\nabla_V \Omega \in T_1^0 M$ is defined by

$$(\nabla_V \Omega)(W) = V(\Omega W) - \Omega\nabla_V W \text{ for all } W \in T_0^1 M.$$

Exercise 6.21. Let M be a smooth manifold with a connection ∇.
Let $f \in C^\infty(M)$, $\Omega, \Theta \in T_1^0 M$ and $V, W \in T_0^1 M$.
Show that $\nabla_V(f\Omega + \Theta) = (Vf)\Omega + f\nabla_V \Omega + \nabla_V \Theta$, and $\nabla_{fV+W}\Omega = f\nabla_V \Omega + \nabla_W \Omega$.

Exercise 6.22. ($\cdot^\flat$ commutes with ∇.) Let $(M, \mathbf{g})$ be a semi-Riemannian manifold with the metric $\mathbf{g}$ and the Levi-Civita connection ∇.
For all vector fields $V, W \in T_0^1 M$, show that $\nabla_V(W^\flat) = (\nabla_V W)^\flat$.

A similar strategy works with any (r, s)-tensor field T.

By the desired Leibniz rule, for 1-form fields $\Omega^1, \cdots, \Omega^r \in T_1^0 M$ and vector fields $V, W_1, \cdots, W_s \in T_0^1 M$, we have

$$\begin{aligned}
&V(T(\Omega^1, \cdots, \Omega^r, W_1, \cdots, W_s)) \\
&= \nabla_V(T(\Omega^1, \cdots, \Omega^r, W_1, \cdots, W_s)) \\
&= (\nabla_V T)(\Omega^1, \cdots, \Omega^r, W_1, \cdots, W_s) \\
&\quad + \sum_{i=1}^{r} T(\Omega^1, \cdots, \Omega^{i-1}, \nabla_V \Omega^i, \Omega^{i+1}, \cdots, \Omega^r, W_1, \cdots, W_s) \\
&\quad + \sum_{j=1}^{s} T(\Omega^1, \cdots, \Omega^r, W_1, \cdots, W_{j-1}, \nabla_V W_j, W_{j+1}, \cdots, W_s),
\end{aligned}$$

and as before we solve for $\nabla_V T$, yielding

$$\begin{aligned}
&(\nabla_V T)(\Omega^1, \cdots, \Omega^r, W_1, \cdots, W_s) \\
&= V(T(\Omega^1, \cdots, \Omega^r, W_1, \cdots, W_s)) \\
&\quad - \sum_{i=1}^{r} T(\Omega^1, \cdots, \Omega^{i-1}, \nabla_V \Omega^i, \Omega^{i+1}, \cdots, \Omega^r, W_1, \cdots, W_s) \\
&\quad - \sum_{j=1}^{s} T(\Omega^1, \cdots, \Omega^r, W_1, \cdots, W_{j-1}, \nabla_V W_j, W_{j+1}, \cdots, W_s).
\end{aligned}$$

Again, we must check

$$(\Omega^1, \cdots, \Omega^r, W_1, \cdots, W_s) \mapsto [\text{the RHS of the above}],$$

is $C^\infty(M)$-multilinear. We check this only in the case of a $(1, 1)$-tensor field T, with the proof in the general case being analogous. We have, for $f \in C^\infty(M)$, $\Theta, \Omega \in T_1^0 M$, $V, W \in T_0^1 M$, that

$$\begin{aligned}
&(\nabla_V T)(f\Omega + \Theta, W) \\
&= V(T(f\Omega + \Theta, W)) - T(\nabla_V(f\Omega + \Theta), W) - T(f\Omega + \Theta, \nabla_V W) \\
&= V(fT(\Omega, W) + T(\Theta, W)) - T((Vf)\Omega + f\nabla_V\Omega + \nabla_V\Theta, W) \\
&\quad - fT(\Omega, \nabla_V W) - T(\Theta, \nabla_V W) \\
&= (Vf)T(\Omega, W) + fV(T(\Omega, W)) + V(T(\Theta, W)) \\
&\quad -(Vf)T(\Omega, W) \\
&\quad - f(T(\nabla_V\Omega, W)) - T(\nabla_V\Theta, W) - fT(\Omega, \nabla_V W) - T(\Theta, \nabla_V W) \\
&= f(V(T(\Omega, W)) - T(\nabla_V\Omega, W) - T(\Omega, \nabla_V W)) \\
&\quad + V(T(\Theta, W)) - T(\nabla_V\Theta, W) - T(\Theta, \nabla_V W) \\
&= f(\nabla_V T)(\Omega, W) + (\nabla_V T)(\Theta, W).
\end{aligned}$$

$C^\infty(M)$-linearity in the second slot follows similarly. The above motivates the following definition.

Definition 6.4. (Covariant derivative of (r,s)-tensor fields.)
Let M be a smooth manifold with a connection ∇.
Let $r \geqslant 2$ or $s \geqslant 1$, and $V \in T_0^1 M$.
If $T \in T_s^r M$, then the *covariant derivative* $\nabla_V T \in T_s^r M$ is defined by

$$\begin{aligned}
&(\nabla_V T)(\Omega^1, \cdots, \Omega^r, W_1, \cdots, W_s) \\
&= V(T(\Omega^1, \cdots, \Omega^r, W_1, \cdots, W_s)) \\
&\quad - \sum_{i=1}^{r} T(\Omega^1, \cdots, \Omega^{i-1}, \nabla_V \Omega^i, \Omega^{i+1}, \cdots, \Omega^r, W_1, \cdots, W_s) \\
&\quad - \sum_{j=1}^{s} T(\Omega^1, \cdots, \Omega^r, W_1, \cdots, W_{j-1}, \nabla_V W_j, W_{j+1}, \cdots, W_s),
\end{aligned}$$

for $\Omega^1, \cdots, \Omega^r \in T_1^0 M$ and $W_1, \cdots, W_s \in T_0^1 M$ (and where an empty sum is taken as the zero function).

From the definition we see that $\nabla_V(T+S) = \nabla_V T + \nabla_V S$ for (r,s)-tensor fields T and S.

Exercise 6.23. Let M be a smooth manifold with a connection ∇. Let $\mathbf{R}$ be a $(1,3)$-tensor-field such that it has the symmetry $\mathbf{R}(\Omega, X, Y, Z) = -\mathbf{R}(\Omega, Y, X, Z)$ for all $\Omega \in T_1^0 M$ and all $X, Y, Z \in T_0^1 M$. Show that the covariant derivative inherits this symmetry, that is, for $\Omega \in T_1^0 M$ and all $V, X, Y, Z \in T_0^1 M$, $(\nabla_V \mathbf{R})(\Omega, X, Y, Z) = -(\nabla_V \mathbf{R})(\Omega, Y, X, Z)$.

Exercise 6.24. Let M be a smooth manifold with a connection ∇.
Let $f \in C^\infty(M)$, $T \in T_s^r M$ and $V \in T_0^1 M$. Prove that $\nabla_V(fT) = (Vf)T + f(\nabla_V T)$.

Exercise 6.25. Let M be a smooth manifold with a connection ∇.
Let $T \in T_s^r M$, $V, W \in T_0^1 M$, $f \in C^\infty(M)$. Show that $\nabla_{fV+W}(T) = f(\nabla_V T) + \nabla_W T$.

A manifestation of the metric-compatibility property (C5) of the Levi-Civita connection on a smooth manifold with a metric, is the following result. In the next chapter on 'parallel transport', we will see yet another description of (C5), namely that the scalar product remains constant under parallel transport induced by the Levi-Civita connection.

Theorem 6.2. *Let $(M, \mathbf{g})$ be a semi-Riemannian manifold, ∇ be the Levi-Civita connection on M. Then*

$$\nabla_{\cdot} \mathbf{g} = 0,$$

that is, for all $V \in T_0^1 M$, $\nabla_V \mathbf{g} = 0$.

Proof. For $V, W, X \in T_0^1 M$,

$$(\nabla_V \mathbf{g})(W, X) = V(\mathbf{g}(W, X)) - \mathbf{g}(\nabla_V W, X) - \mathbf{g}(W, \nabla_V X) \overset{\text{(C5)}}{=} 0.$$

As W, X were arbitrary, we conclude that $\nabla_V \mathbf{g} = 0$. □

If M is a smooth manifold with a connection ∇, and $V, W \in T_0^1 M$, then we had seen that $\nabla_V W$ can be found in an admissible chart $(U, \mathbf{x})$ using the components of V, W and the Christoffel symbols:

$$\nabla_V W = (V^j \partial_{x^j} W^k + V^j W^i \Gamma_{ij}^k)\partial_{x^k} = (V(W^k) + V^j W^i \Gamma_{ij}^k)\partial_{x^k}.$$

A similar calculation can be done for the covariant derivative of 1-form fields, and of more general tensor fields.

Example 6.9. (Covariant derivative of 1-form fields using chart-induced components.) Let M be a smooth manifold with a connection ∇, and let $(U, \mathbf{x})$ be an admissible chart. Let $V \in T_0^1 M$ and $\Omega \in T_1^0 M$. Then $\nabla_V \Omega \in T_1^0 M$, and so we can write $\nabla_V \Omega = (\nabla_V \Omega)_i dx^i$ in U. To determine the components $(\nabla_V \Omega)_i$, we compute

$$\begin{aligned}(\nabla_V \Omega)_i &= (\nabla_V \Omega)(\partial_{x^i}) = V((\Omega_k dx^k)\partial_{x^i}) - \Omega_k dx^k(\nabla_{V^j \partial_{x^j}} \partial_{x^i}) \\ &= V(\Omega_k \delta_i^k) - \Omega_k dx^k(V^j \nabla_{\partial_{x^j}} \partial_{x^i}) = V\Omega_i - \Omega_k V^j dx^k(\Gamma_{ij}^\ell \partial_{x^\ell}) \\ &= V\Omega_i - \Omega_k V^j \Gamma_{ij}^\ell \delta_\ell^k = V\Omega_i - \Omega_k V^j \Gamma_{ij}^k = V^j \partial_{x^j}\Omega_i - \Omega_k V^j \Gamma_{ij}^k. \quad \diamond\end{aligned}$$

Exercise 6.26. Suppose that M is a smooth manifold with a connection ∇, $(U, \mathbf{x})$ is an admissible chart, $V \in T_0^1 M$, and $T \in T_2^1 M$. Show that

$$(\nabla_V T)_{jk}^i = V(T_{jk}^i) + \Gamma_{rs}^i T_{jk}^r V^s - \Gamma_{js}^r T_{rk}^i V^s - \Gamma_{ks}^r T_{jr}^i V^s.$$

We now show that in a smooth manifold M with a connection ∇, the operations of taking the covariant derivative and that of contraction commute.

Proposition 6.4. *Let M be an m-dimensional smooth manifold with a connection ∇. Let $1 \leqslant r, s \leqslant m$ and let $T \in T_s^r M$. Then for all $V \in T_0^1 M$, we have $\mathbf{C}_j^i(\nabla_V T) = \nabla_V(\mathbf{C}_j^i T)$.*

Proof. We will prove this only in the case when $r = s = 1$. The proof in the general case is analogous. In an admissible chart $(U, \mathbf{x})$, we have $\nabla_V(\mathbf{C}T) = V(\mathbf{C}T) = V(T(dx^i, \partial_{x^i})) = V(T_i^i)$. On the other hand,

$$\begin{aligned}\mathbf{C}(\nabla_V T) &= (\nabla_V T)(dx^i, \partial_{x^i}) \\ &= V(T(dx^i, \partial_{x^i})) - T(\nabla_V dx^i, \partial_{x^i}) - T(dx^i, \nabla_V \partial_{x^i}) \\ &= V(T_i^i) - T((V\delta_j^i - \delta_\ell^i V^k \Gamma_{jk}^\ell) dx^j, \partial_{x^i}) - T(dx^i, V^k \Gamma_{ik}^j \partial_{x^j}) \\ &= V(T_i^i) + V^k \Gamma_{jk}^i T_i^j - V^k \Gamma_{ik}^j T_j^i = V(T_i^i).\end{aligned}$$

Consequently, $\mathbf{C}(\nabla_V T) = \nabla_V(\mathbf{C}T)$. □

∇ satisfies a Leibniz product rule over the tensor product $\otimes$ of tensor fields.

Proposition 6.5. *Let M be a smooth manifold with a connection ∇. If $S \in T_s^r M$ and $T \in T_v^u M$, then for all $V \in T_0^1 M$, we have*

$$\nabla_V(S \otimes T) = (\nabla_V S) \otimes T + S \otimes \nabla_V T.$$

Proof. We show this for $S = \Omega \in T_1^0 M$ and $T \in T_1^1 M$ (and we will also see this when $S \in T_0^1 M$ in Exercise 6.27 below). The general case follows analogously by an inductive argument (and by working in an admissible chart $(U, \mathbf{x})$, decomposing S as $S_{j_1 \cdots j_s}^{i_1 \cdots i_r} \partial_{x^{i_1}} \otimes \cdots \otimes \partial_{x^{i_r}} \otimes dx^{j_1} \otimes \cdots \otimes dx^{j_s}$ and also using Exercise 6.24, p.125).

For $\Theta \in T_1^0 M$, we have and $W, X \in T_0^1 M$

$$\begin{aligned}
&(\nabla_V(\Omega \otimes T))(\Theta, W, X) \\
&= V(\Omega(W)T(\Theta, X)) - (\Omega \otimes T)(\nabla_V \Theta, W, X) \\
&\quad - (\Omega \otimes T)(\Theta, \nabla_V W, X) - (\Omega \otimes T)(\Theta, W, \nabla_V X) \\
&= \underline{V(\Omega W)}T(\Theta, X) + \Omega(W)V(T(\Theta, X)) - \Omega(W)T(\nabla_V \Theta, X) \\
&\quad \underline{- \Omega(\nabla_V W)}T(\Theta, X) - \Omega(W)T(\Theta, \nabla_V X) \\
&= \underline{(V(\Omega W) - \Omega(\nabla_V W))}T(\Theta, X) \\
&\quad + \Omega(W)(V(T(\Theta, X)) - T(\nabla_V \Theta, X) - T(\Theta, \nabla_V X)) \\
&= (\nabla_V \Omega)(W)T(\Theta, X) + \Omega(W)(\nabla_V T)(\Theta, X) \\
&= ((\nabla_V \Omega) \otimes T)(\Theta, W, X) + (\Omega \otimes \nabla_V T)(\Theta, W, X) \\
&= ((\nabla_V \Omega) \otimes T + \Omega \otimes \nabla_V T)(\Theta, W, X).
\end{aligned}$$

Thus $\nabla_V(\Omega \otimes T) = (\nabla_V \Omega) \otimes T + \Omega \otimes \nabla_V T$. □

Exercise 6.27. Let M be a smooth manifold with a connection ∇. Let $W \in T_0^1 M$ and $T \in T_1^1 M$. For all $V \in T_0^1 M$, show that

$$\nabla_V(W \otimes T) = (\nabla_V W) \otimes T + W \otimes \nabla_V T.$$

Exercise 6.28. Let $(M, \mathbf{g})$ be a semi-Riemannian manifold with the Levi-Civita connection ∇. Let $T \in T_s^r M$ and $V \in T_0^1 M$. Show that

$$\begin{aligned}
\nabla_V(\mathbf{g} \otimes T) &= \mathbf{g} \otimes \nabla_V T, \text{ and} \\
\nabla_V(T \otimes \mathbf{g}) &= (\nabla_V T) \otimes \mathbf{g}.
\end{aligned}$$

Exercise 6.29. (∇ commutes with $\cdot^\flat$ and $\cdot^\sharp$).
The results of this exercise will be useful later (in Proposition 9.6, p.191). Suppose that $(M, \mathbf{g})$ is a semi-Riemannian manifold with the metric $\mathbf{g}$ and the Levi-Civita connection ∇. For $T \in T_2^0 M$, define $T^\sharp \in T_1^1 M$ by $T^\sharp(\Omega, V) = T(\Omega^\sharp, V)$ for all $\Omega \in T_1^0 M$, $V \in T_0^1 M$. Furthermore, for $T \in T_1^1 M$, define $T^\flat \in T_2^0 M$ by $T^\flat(V, W) = T(V^\flat, W)$ for $V, W \in T_0^1 M$.

(1) Show that for all $T \in T_1^1 M$, $(T^\flat)^\sharp = T$.
(2) Show that for all $T \in T_2^0 M$, $(T^\sharp)^\flat = T$.
(3) Show that for all $T \in T_1^1 M$, $T^\flat = \mathbf{C}_2^1(\mathbf{g} \otimes T)$.
(4) Show that for all $T \in T_1^1 M$, and all $V \in T_0^1 M$, $\nabla_V(T^\flat) = (\nabla_V T)^\flat$.
(5) Show that for all $T \in T_2^0 M$, and all $V \in T_0^1 M$, $\nabla_V(T^\sharp) = (\nabla_V T)^\sharp$.
Hint: Use (1), (2), (4).

6.6 Lie derivative of tensor fields

Let M be a smooth manifold. Recall that if $V \in T_0^1 M$, then the Lie derivative $\mathcal{L}_V f := Vf$ for $f \in C^\infty(M)$, and $\mathcal{L}_V W := [V, W]$ if $W \in T_0^1 M$ (Definition 3.14, p.55). We extend this now, and define the Lie derivative of arbitrary tensor fields by demanding a Leibniz rule, akin to what was done in the previous section for the covariant derivative. We begin by extending the definition of the Lie derivative to 1-form fields.

Definition 6.5. (Lie derivative of functions and of 1-form fields.)
Let M be a smooth manifold, and let $V \in T_0^1 M$. If $f \in C^\infty(M)$, then the *Lie derivative* $\mathcal{L}_V f \in C^\infty(M)$ of f is defined by $\mathcal{L}_V f = Vf$. If $\Omega \in T_1^0 M$, then the *Lie derivative* $\mathcal{L}_V \Omega \in T_1^0 M$ of Ω is defined by

$$(\mathcal{L}_V \Omega)(W) = V(\Omega W) - \Omega[V, W] \text{ for all } W \in T_0^1 M.$$

Let us check $\mathcal{L}_V \Omega$ is a 1-form field. Clearly, $T_0^1 M \ni W \mapsto V(\Omega W) - \Omega[V, W]$ is $\mathbb{R}$-linear. The map is also $C^\infty(M)$-linear: For $f \in C^\infty(M)$, we have

$$\begin{aligned}(\mathcal{L}_V \Omega)(fW) &= V(\Omega(fW)) - \Omega[V, fW] = V(f\Omega W) - \Omega(f[V, W] + (Vf)W) \\ &= (Vf)(\Omega W) + fV(\Omega W) - f\Omega[V, W] - (Vf)\Omega W \\ &= f(V(\Omega W) - \Omega[V, W]) = f(\mathcal{L}_V \Omega)(W).\end{aligned}$$

Exercise 6.30. Let M be a smooth manifold, and $V \in T_0^1 M$.
For all $f \in C^\infty(M)$, and $\Omega, \Theta \in T_1^0 M$, prove that

$$\mathcal{L}_V(f\Omega + \Theta) = (\mathcal{L}_V f)\Omega + f\mathcal{L}_V \Omega + \mathcal{L}_V \Theta = (Vf)\Omega + f\mathcal{L}_V \Omega + \mathcal{L}_V \Theta.$$

Analogous to Definition 6.4, we now give the following.

Definition 6.6. (Lie derivative of (r, s)-tensor fields.)
Let M be a smooth manifold.
Let $V \in T_0^1 M$ and $r \geqslant 2$ or $s \geqslant 1$.
If $T \in T_s^r M$, then the *Lie derivative* $\mathcal{L}_V T \in T_s^r M$ is defined by

$$\begin{aligned}&(\mathcal{L}_V T)(\Omega^1, \cdots, \Omega^r, W_1, \cdots, W_s) \\ &= V(T(\Omega^1, \cdots, \Omega^r, W_1, \cdots, W_s)) \\ &\quad - \sum_{i=1}^{r} T(\Omega^1, \cdots, \Omega^{i-1}, \mathcal{L}_V \Omega^i, \Omega^{i+1}, \cdots, \Omega^r, W_1, \cdots, W_s) \\ &\quad - \sum_{j=1}^{s} T(\Omega^1, \cdots, \Omega^r, W_1, \cdots, W_{j-1}, \mathcal{L}_V W_j, W_{j+1}, \cdots, W_s),\end{aligned}$$

for all $\Omega^1, \cdots, \Omega^r \in T_1^0 M$ and $W_1, \cdots, W_s \in T_0^1 M$.

Again, we must check $(\Omega^1, \cdots, \Omega^r, W_1, \cdots, W_s) \mapsto$ [the RHS of the above], is $C^\infty(M)$-multilinear. We check this for $(1,1)$-tensor fields T. The proof in the general case is analogous. For $f \in C^\infty(M)$, $\Theta, \Omega \in T_1^0 M$, $W \in T_0^1 M$,

$$\begin{aligned}
&(\mathcal{L}_V T)(f\Omega + \Theta, W) \\
&= V(T(f\Omega + \Theta, W)) - T(\mathcal{L}_V(f\Omega + \Theta), W) - T(f\Omega + \Theta, \mathcal{L}_V W) \\
&= V(fT(\Omega, W) + T(\Theta, W)) \\
&\quad -T((Vf)\Omega + f\mathcal{L}_V\Omega + \mathcal{L}_V\Theta, W) - fT(\Omega, \mathcal{L}_V W) - T(\Theta, \mathcal{L}_V W) \\
&= \cancel{(Vf)T(\Omega, W)} + fV(T(\Omega, W)) + V(T(\Theta, W)) \\
&\quad -\cancel{(Vf)T(\Omega, W)} - fT(\mathcal{L}_V\Omega, W) - T(\mathcal{L}_V\Theta, W) \\
&\quad -fT(\Omega, \mathcal{L}_V W) - T(\Theta, \mathcal{L}_V W) \\
&= f(V(T(\Omega, W)) - T(\mathcal{L}_V\Omega, W) - T(\Omega, \mathcal{L}_V W)) \\
&\quad +V(T(\Theta, W)) - T(\mathcal{L}_V\Theta, W) - T(\Theta, \mathcal{L}_V W) \\
&= f(\mathcal{L}_V T)(\Omega, W) + (\mathcal{L}_V T)(\Theta, W).
\end{aligned}$$

The $C^\infty(M)$-linearity in the second slot follows similarly.

Exercise 6.31. (Killing vector field and Killing's equation.)
Let $(M, \mathbf{g})$ be a semi-Riemannian manifold with the Levi-Civita connection ∇ induced by the metric $\mathbf{g}$. A vector field $V \in T_0^1 M$ is called a *Killing*[5] *vector field* if $\mathcal{L}_V \mathbf{g} = 0$. Show that if V is a Killing vector field, then for all $X, Y \in T_0^1 M$,

- $\mathbf{g}([V, X], Y) + \mathbf{g}(X, [V, Y]) = V(\mathbf{g}(X, Y))$, and
- (Killing's equation) $\mathbf{g}(\nabla_X V, Y) + \mathbf{g}(X, \nabla_Y V) = 0$.

Exercise 6.32. Let $M = \mathbb{R}^3$ be equipped with the standard smooth structure and the metric $\mathbf{g}$ described in Exercise 5.14 (p.83). Show that the vector fields L_x, L_y, L_z given in Exercise 3.24 (p.55) are Killing vector fields for $\mathbf{g}$.

Exercise 6.33. Let $(M, \mathbf{g})$ be an m-dimensional semi-Riemannian manifold, and $(U, \mathbf{x})$ be an admissible chart. Suppose that for all $1 \leqslant i, j \leqslant m$, the functions $g_{ij} \circ \mathbf{x}^{-1}$ are independent of u^k, that is,

$$\frac{\partial(g_{ij} \circ \mathbf{x}^{-1})}{\partial u^k} \equiv 0 \text{ on } \mathbf{x}(U).$$

Show that $\partial_{x^k} \in T_0^1 U$ is a Killing vector field for $\mathbf{g} \in T_2^0 U$.

Recall that a complete vector field V on a smooth manifold M is one whose integral curves are defined on all of $\mathbb{R}$. Then its flow is a family of diffeomorphisms $\psi_t : M \to M$, $t \in \mathbb{R}$. Suppose that M is a semi-Riemannian manifold with a metric $\mathbf{g}$. Recall that isometries on $(M, \mathbf{g})$ are special types of diffeomorphisms. It is natural to ask:

[5]After the German mathematician Wilhelm Killing (1847 – 1923) who made contributions to Lie theory.

For what vector fields V are the flow maps isometries?

Answer: Killing vector fields.

We prove this below. First, we have the following analogue of Propositions 3.1 (p.55) and 3.2 (p.56).

Proposition 6.6. *Let M be a smooth manifold, and $V \in T_0^1 M$ be a complete vector field with the flow $\{\psi_t : M \to M,\ t \in \mathbb{R}\}$. Let $T \in T_2^0 M$. Then*

$$\mathcal{L}_V T = \lim_{t \to 0} \frac{(\psi_t)^* T - T}{t}.$$

Here we mean the pointwise limit, that is, the right-hand side means the map which sends $X, Y \in T_0^1 M$ to the function given by

$$M \ni p \mapsto \lim_{t \to 0} \frac{(((\psi_t)^* T - T)(X, Y))(p)}{t}.$$

That this then defines an element in $T_2^0 M$ will follow from the proof below, since it matches the left-hand side $\mathcal{L}_V T \in T_2^0 M$.

Proof. To ease the notation, we will use

$$\mathbb{L}(\cdot) := \lim_{t \to 0} \frac{1}{t}(\cdot).$$

For $X, Y \in T_0^1 M$, and $p \in M$,

$$\begin{aligned}
&\mathbb{L}(((\psi_t)^* T - T)(X, Y)(p)) \\
&= \mathbb{L}(T(\psi_t p)((d\psi_t)_p X_p, (d\psi_t)_p Y_p) - T(p)(X_p, Y_p)) \\
&= \mathbb{L}(T(\psi_t p)((d\psi_t)_p X_p, (d\psi_t)_p Y_p) - T(\psi_t p)(X_{\psi_t p}, Y_{\psi_t p})) \\
&\quad + \mathbb{L}(T(\psi_t p)(X_{\psi_t p}, Y_{\psi_t p}) - T(p)(X_p, Y_p)). \qquad (6.2)
\end{aligned}$$

We will consider the two summands in the last two lines above separately. First, with $\gamma_p(t) := \psi_t p$, $t \in \mathbb{R}$, we have $v_{\gamma_p, p} = V_p$ as γ_p is an integral curve of V. With $f := T(X, Y)$, we have for the second summand that

$$\begin{aligned}
S_2 &:= \mathbb{L}(T(\psi_t p)(X_{\psi_t p}, Y_{\psi_t p}) - T(p)(X_p, Y_p)) \\
&= \frac{d(T(\gamma_p(t))(X_{\gamma_p(t)}, Y_{\gamma_p(t)}))}{dt}(0) \\
&= \frac{d(f \circ \gamma_p)}{dt}(0) = v_{\gamma_p, p} f = V_p f = V_p(T(X, Y)).
\end{aligned}$$

For the first summand in (6.2), we add and subtract

$$\mathbb{L}(T(\psi_t p)(X_{\psi_t p}, (d\psi_t)_p Y_p)).$$

We will also use $\psi_t \circ \psi_{-t} = \psi_{t-t} = \psi_0 = \mathrm{id}_M$, giving

$$(d\psi_t)_p (d\psi_{-t})_{\psi_t p} = (d(\mathrm{id}_M))_{\psi_t p} = \mathrm{id}_{T_{\psi_t p}M}.$$

So

$$\begin{aligned}
S_1 :=& \, \mathbb{L}\big(T(\psi_t p)((d\psi_t)_p X_p, (d\psi_t)_p Y_p) - T(\psi_t p)(X_{\psi_t p}, Y_{\psi_t p})\big) \\
=& \, \mathbb{L}\big(T(\psi_t p)((d\psi_t)_p X_p, (d\psi_t)_p Y_p)\big) - \mathbb{L}\big(T(\psi_t p)(X_{\psi_t p}, (d\psi_t)_p Y_p)\big) \\
&+ \mathbb{L}\big(T(\psi_t p)(X_{\psi_t p}, (d\psi_t)_p Y_p)\big) - \mathbb{L}\big(T(\psi_t p)(X_{\psi_t p}, Y_{\psi_t p})\big) \\
=& \, \mathbb{L}\big(T(\psi_t p)((d\psi_t)_p X_p, (d\psi_t)_p Y_p) - T(\psi_t p)(X_{\psi_t p}, (d\psi_t)_p Y_p)\big) \\
&+ \mathbb{L}\big(T(\psi_t p)(X_{\psi_t p}, (d\psi_t)_p Y_p) - T(\psi_t p)(X_{\psi_t p}, Y_{\psi_t p})\big) \\
=& \, \mathbb{L}\big(T(\psi_t p)((d\psi_t)_p X_p - X_{\psi_t p}, (d\psi_t)_p Y_p)\big) \\
&+ \mathbb{L}\big(T(\psi_t p)(X_{\psi_t p}, (d\psi_t)_p Y_p - Y_{\psi_t p})\big) \\
=& \, \mathbb{L}\big(T(\psi_t p)((d\psi_t)_p (X_p - (d\psi_{-t})_{\psi_t p} X_{\psi_t p}), (d\psi_t)_p Y_p)\big) \\
&+ \mathbb{L}\big(T(\psi_t p)(X_{\psi_t p}, (d\psi_t)_p (Y_p - (d\psi_{-t})_{\psi_t p} Y_{\psi_t p}))\big).
\end{aligned}$$

Proposition 3.2 (p.56) yields[6]

$$S_1 = T(p)(-[V,X]_p, Y_p) + T(p)(X_p, -[V,Y]_p).$$

So

$$\begin{aligned}
S_1 + S_2 &= V_p(T(X,Y)) - T(p)([V,X]_p, Y_p) - T(p)(X_p, [V,Y]_p) \\
&= \big(V(T(X,Y)) - T([V,X],Y) - T(X,[V,Y])\big)(p) \\
&= ((\mathcal{L}_V T)(X,Y))(p).
\end{aligned}$$

Thus $\lim\limits_{t\to 0} \dfrac{(((\psi_t)^*T - T)(X,Y))(p)}{t} = ((\mathcal{L}_V T)(X,Y))(p)$. □

Finally, we will prove the following important result.

Theorem 6.3. *Let $(M, \mathbf{g})$ be a semi-Riemannian manifold, and $V \in T_0^1 M$ be a complete vector field with the flow $\{\psi_t : M \to M, t \in \mathbb{R}\}$. Then V is a Killing vector field if and only if for all $t \in \mathbb{R}$, ψ_t is an isometry.*

Proof.
'If part': Suppose that each ψ_t is an isometry. Then $(\psi_t)^*\mathbf{g} = \mathbf{g}$, so that Proposition 6.6 immediately yields $\mathcal{L}_V \mathbf{g} = 0$, that is, V is a Killing vector field.

[6] At least formally, this is clear, but we do not include a detailed proof here (which can be carried out by considering a chart containing p and all $\psi_t p$ for t close enough to 0, and using the smoothness of $\mathbf{g}_{ij}, \gamma_p$, and using the chart representative Jacobian matrix of $(d\psi_t)_p$).

'Only if part': Let $\mathcal{L}_V \mathbf{g} = 0$. Suppose that $X, Y \in T_0^1 M$. For a fixed $s \in \mathbb{R}$, set $\widetilde{X} = (\psi_s)_* X$ and $\widetilde{Y} = (\psi_s)_* Y$. For any $p \in M$, $((\mathcal{L}_V \mathbf{g})(\widetilde{X}, \widetilde{Y}))(\psi_s p) = 0$, and so by Proposition 6.6,

$$\begin{aligned}
0 &= \lim_{t \to 0} \frac{\mathbf{g}(\psi_t \psi_s p)((d\psi_t)_{\psi_s p} \widetilde{X}_{\psi_s p}, (d\psi_t)_{\psi_s p} \widetilde{Y}_{\psi_s p}) - \mathbf{g}(\psi_s p)(\widetilde{X}_{\psi_s p}, \widetilde{Y}_{\psi_s p})}{t} \\
&= \lim_{t \to 0} \frac{\mathbf{g}(\psi_{t+s} p)((d\psi_t)_{\psi_s p}(d\psi_s)_p X_p, (d\psi_t)_{\psi_s p}(d\psi_s)_p Y_p) - \mathbf{g}(\psi_s p)((d\psi_s)_p X_p, (d\psi_s)_p Y_p)}{t} \\
&= \lim_{t \to 0} \frac{\mathbf{g}(\psi_{t+s} p)((d(\psi_t \circ \psi_s))_p X_p, (d(\psi_t \circ \psi_s))_p Y_p) - \mathbf{g}(\psi_s p)((d\psi_s)_p X_p, (d\psi_s)_p Y_p)}{t} \\
&= \lim_{t \to 0} \frac{\mathbf{g}(\psi_{t+s} p)((d\psi_{t+s})_p X_p, (d\psi_{t+s})_p Y_p) - \mathbf{g}(\psi_s p)((d\psi_s)_p X_p, (d\psi_s)_p Y_p)}{t} \\
&= \frac{d(\mathbf{g}(\psi_\sigma p)((d\psi_\sigma)_p X_p, (d\psi_\sigma)_p Y_p))}{d\sigma}(s).
\end{aligned}$$

As $s \in \mathbb{R}$ was arbitrary, we conclude that the map

$$\mathbb{R} \ni \sigma \mapsto \mathbf{g}(\psi_\sigma p)((d\psi_\sigma)_p X_p, (d\psi_\sigma)_p Y_p)$$

is constant, so that (considering $\sigma = 0$)

$$\mathbf{g}(\psi_\sigma p)((d\psi_\sigma)_p X_p, (d\psi_\sigma)_p Y_p) = \mathbf{g}(p)(X_p, Y_p) \text{ for all } \sigma \in \mathbb{R}.$$

Hence $(\psi_\sigma)^* \mathbf{g} = \mathbf{g}$ for all $\sigma \in \mathbb{R}$. □

Example 6.10. Consider $\mathbb{R}^3$ with the standard smooth structure, and the Riemannian metric $\mathbf{g}$, given in the global chart $(\mathbb{R}^3, (x, y, z) \mapsto (x, y, z))$ by $\mathbf{g} = dx \otimes dx + dy \otimes dy + dz \otimes dz$ (Exercise 5.14, p.83). Recall (from Exercise 3.24, p.55) that the flow maps for the vector fields

$$\begin{aligned}
L_x &= y\,\partial_z - z\,\partial_y \\
L_y &= z\,\partial_x - x\,\partial_z \\
L_z &= x\,\partial_y - y\,\partial_x
\end{aligned}$$

are rotations about the x-, y-, z-axis, respectively, which are isometries (see Exercise 5.14, p.83). In Exercise 6.32 (p.129), we had verified that L_x, L_y, L_z are Killing vector fields. ◇

Exercise 6.34. Let $(M, \mathbf{g})$ be a semi-Riemannian manifold.
Show that if $X, Y \in T_0^1 M$, then $\mathcal{L}_{[X,Y]} \mathbf{g} = \mathcal{L}_X \mathcal{L}_Y \mathbf{g} - \mathcal{L}_Y \mathcal{L}_X \mathbf{g}$.
Conclude that if $X, Y \in T_0^1 M$ are Killing vector fields, then so is $[X, Y]$.

Chapter 7

Parallel transport

We will now learn that a connection induces a way of 'parallel transporting' vectors along a curve γ. So if $p = \gamma(a)$ and $q = \gamma(b)$ are points along the curve γ, then $P_{ab}^{\gamma} : T_pM \to T_qM$ will be a linear map which will transport vectors from T_pM to T_qM 'parallelly' along the curve. We will explain the meaning of this below. In this manner tangent spaces at different points can now be 'connected', and this is the motivation behind the name 'connection'. We will also see that if M is a semi-Riemannian manifold, with ∇ taken as the Levi-Civita connection, then the parallel transport map is an isometry.

7.1 Vector fields along curves

The tangent vector along a curve exists only along the points of the curve. In order to talk of the acceleration, we would like to differentiate this 'vector field' that 'lives along the curve'. This prompts the following definition.

Definition 7.1. (Vector field along a smooth curve.)
Let M be a smooth manifold, and $\gamma : I \to M$ be a smooth curve, where I is an open interval in $\mathbb{R}$. A *vector field along* γ is a smooth map $V : I \to TM$ such that $\pi \circ V = \gamma$, where $\pi : TM \to M$ is the canonical projection.
We denote the set of all vector fields along γ by $T_0^1\gamma$.

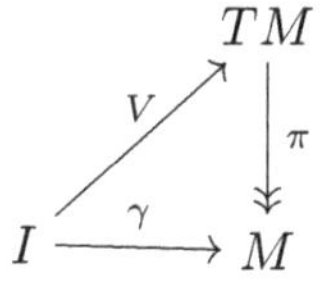

Example 7.1. (Restriction of a vector field to a curve is a vector field along the curve.) Let M be a smooth manifold, and let $\gamma : I \to M$ be a smooth curve, where I is an open interval. Suppose that $W \in T_0^1 M$. Set $V(t) = (\gamma(t), W_{\gamma(t)})$ for $t \in I$. Then $V = W \circ \gamma$. As $W : M \to TM$, $\gamma : I \to M$ are smooth, so is V. Moreover, $\pi \circ V = \pi \circ (W \circ \gamma) = (\pi \circ W) \circ \gamma = \mathrm{id}_M \circ \gamma = \gamma$. So V is a vector field along γ. ◇

A vector field along a curve is not necessarily the restriction of a vector field on M to $\gamma(I)$. Indeed, the curve may intersect itself, say at $p = \gamma(a) = \gamma(b)$, where $a, b \in I$ with $a \neq b$, and if the vector field V along γ is such that $V_{\gamma(a)} \neq V_{\gamma(b)}$, then we cannot create a global extension W of the V to a vector field on M, since there is a problem of specifying what W_p ought to be. The picture below illustrates the problem when we take the tangent vectors along γ (which we will see is a vector field along γ). However, a 'local' extension is possible for a part of the curve in the neighbourhood of $\gamma(t_0)$ if $v_{\gamma,\gamma(t_0)} \neq 0$; see Remark 7.1 (p.140).

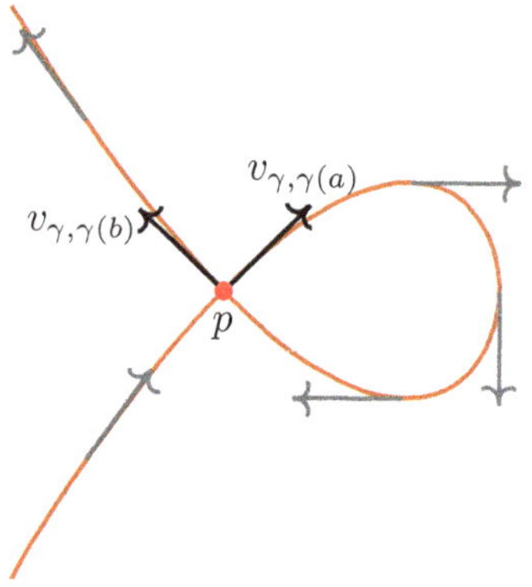

Lemma 7.1. *Let M be a smooth manifold, and $\gamma : I \to M$ be a smooth curve such that $\gamma(I) \subset U$, where I is an open interval in $\mathbb{R}$ and (U, φ) is an admissible chart for M. Suppose that $V : I \to TM$ is such that $\pi \circ V = \gamma$. Write $V(t) = (\gamma(t), V^i(t)\partial_{x^i,\gamma(t)})$, $t \in I$. Then V is smooth if and only if $V^i \in C^\infty(I)$ for all $i \in \{1, \cdots, m\}$.*

Proof. We recall that the admissible chart (U, φ) for M induces an admissible chart $(\mathbf{U}, \boldsymbol{\varphi})$ for TM, where $\mathbf{U} = \{(p, v) : p \in U,\ v \in T_pM\}$ and $\boldsymbol{\varphi}(p, v) = (\varphi(p), v(x^1), \cdots, v(x^m))$. Then $\boldsymbol{\varphi} : \mathbf{U} \to \boldsymbol{\varphi}(\mathbf{U}) \subset \mathbb{R}^{2m}$ is a diffeomorphism. Now suppose that V is smooth. Then $\boldsymbol{\varphi} \circ V$ is smooth, that is, $I \ni t \mapsto (\boldsymbol{\varphi} \circ V)(t) = (\varphi(\gamma(t)), V^1(t), \cdots, V^m(t))$ is smooth. In particular, $V^i \in C^\infty(I)$.

Conversely, if each of the V^i is smooth, then (since $\varphi \circ \gamma$ is already smooth) it follows from the above that $\varphi \circ V$ is smooth. But then $V = \varphi^{-1} \circ (\varphi \circ V)$ is smooth too. $\square$

Proposition 7.1. *Let M be a smooth manifold defined by the atlas $\mathcal{A}$, and $\gamma : I \to M$ be a smooth curve, where I is an open interval in $\mathbb{R}$. For any chart $(U, \mathbf{x})$ in $\mathcal{A}$, define the open subset $I_U = \gamma^{-1}U$ of I. Suppose that $V : I \to TM$ is such that $\pi \circ V = \gamma$. Then the following are equivalent:*

(1) *V is smooth.*

(2) *For any chart $(U, \mathbf{x})$ in $\mathcal{A}$ such that $I_U \neq \varnothing$, we have $V^i \in C^\infty(I_U)$ for all $i \in \{1, \cdots, m\}$, where $V(t) = (\gamma(t), V^i(t)\partial_{x^i,\gamma(t)})$, $t \in I_U$.*

Proof.
(1)⇒(2): Let $(U, \mathbf{x}) \in \mathcal{A}$ and $t \in I_U$. Then $\gamma(t) \in U$. By the continuity of γ, there exists a small enough open interval $I_t \subset I_U$ containing t. Then $\gamma(I_t) \subset U$. By Lemma 7.1 above, $V^i|_{I_t}$ is C^∞. As the choice of $t \in I_U$ was arbitrary, $V^i \in C^\infty(I_U)$.

(2)⇒(1): Let $t \in I$. Let $(U, \mathbf{x}) \in \mathcal{A}$ be such that $\gamma(t) \in U$. Then $t \in I_U \neq \varnothing$. As I_U is open, there exists an open interval $I_t \subset I_U$ containing t. By the hypothesis, $V^i \in C^\infty(I_U)$, and so $V^i \in C^\infty(I_t)$ as well. By Lemma 7.1, $V|_{I_t}$ is smooth. As $t \in I$ was arbitrary, V is smooth. $\square$

For a vector field $V \in T_0^1\gamma$ along a curve $\gamma : I \to M$, just as we have been doing with vector fields on M, we will identify $V(t) = (\gamma(t), v(t))$ $(t \in I)$, where $v(t) \in T_{\gamma(t)}M$, with $v(t)$.

Example 7.2. (Tangent/Velocity vector field along a curve.)
Let M be a smooth manifold and $\gamma : I \to M$ be a smooth curve, where I is an open interval in $\mathbb{R}$. For $t \in I$, define $V_\gamma(t) = v_{\gamma,\gamma(t)} \in T_{\gamma(t)}M$. Then V_γ is a vector field along γ. In any admissible chart $(U, \mathbf{x})$,

$$v_{\gamma,\gamma(t)} = \frac{d(x^i \circ \gamma)}{dt}(t)\, \partial_{x^i,\gamma(t)}, \quad t \in I_U,$$

so that $V_\gamma^i = \dfrac{d(x^i \circ \gamma)}{dt}$, and these are smooth on I_U (as γ is smooth). $\diamond$

Let M be a smooth manifold and $\gamma : I \to M$ be a smooth curve, where I is an open interval in $\mathbb{R}$. For $V, W \in T_0^1\gamma$, we define $V + W \in T_0^1\gamma$ by

$$(V + W)(t) = V(t) + W(t) \text{ for all } t \in I.$$

Also, if $f \in C^\infty(I)$, then we define $f \cdot V \in T_0^1\gamma$ by

$$(f \cdot V)(t) = f(t)V(t) \text{ for all } t \in I.$$

Then $T_0^1\gamma$ is a $C^\infty(I)$-module (just like T_0^1M is a $C^\infty(M)$-module).

7.2 Covariant derivative along a curve

Now we will learn that if ∇ is a connection on a smooth manifold M, then it induces for every smooth curve $\gamma : I \to M$, a 'covariant derivative' $\nabla_{V_\gamma} : T_0^1\gamma \to T_0^1\gamma$. Roughly speaking, it allows us to differentiate a vector field along a curve in the direction of the tangent/velocity vector field V_γ along the curve γ. We begin with the following result, where we first work with a chart. Below, we consider U as a smooth manifold with the atlas given by the single chart $(U, \mathbf{x})$, and the connection ∇^U (Prop. 6.1, p.112). For each smooth curve $\gamma : I \to U$, we denote by $T_0^1\gamma^U$ the vector fields along γ in the smooth manifold U, that is, they are smooth maps $V : I \to TU$ such that $\pi \circ V = \gamma$, where $\pi : TU \to U$ is the canonical projection.

Lemma 7.2. *Let M be a smooth manifold with a connection ∇, and let $(U, \mathbf{x})$ be an admissible chart for M. For each smooth curve $\gamma : I \to U$, where I is an open interval in $\mathbb{R}$, there is a unique map $\nabla^U_{V_\gamma}: T_0^1\gamma^U \to T_0^1\gamma^U$ such that the following hold for all $W, X \in T_0^1\gamma^U$, $c \in \mathbb{R}$ and $f \in C^\infty(I)$:*

- ($\mathbb{R}$-linearity) $\nabla^U_{V_\gamma}(cW + X) = c\nabla^U_{V_\gamma} W + \nabla^U_{V_\gamma} X$.
- (Leibniz) $\nabla^U_{V_\gamma}(fW) = \dfrac{df}{dt} W + f\nabla^U_{V_\gamma} W$.
- *If $\widetilde{W} \in T_0^1 U$ and $\widetilde{W}_{\gamma(t)} = W(t)$ for all $t \in I$, then we have $(\nabla^U_{V_\gamma} W)(t) = (\nabla_{v_{\gamma,\gamma(t)}} \widetilde{W})_{\gamma(t)}$ for all $t \in I$.*

We note that by Exercise 6.10 (p.114), the notation $\nabla_{v_{\gamma,\gamma(t)}} \widetilde{W}$ is legitimate, since $\nabla_X Y$ at a point p depends only on X_p (and on the values of Y in a neighbourhood of p). Thus, here we mean that if $t_0 \in I$ and $\widetilde{V} \in T_0^1 U$ is any vector field that coincides with the vector $v_{\gamma,\gamma(t_0)}$ at the point $\gamma(t_0) \in M$ (and such a vector field can be constructed using a bump function; see Lemma 3.1, p.41), then $\nabla_{v_{\gamma,\gamma(t)}} \widetilde{W} := (\nabla_{\widetilde{V}} \widetilde{W})_{\gamma(t)}$.

Proof.
Existence: Given $W \in T_0^1\gamma^U$, write $W = W^i \partial_{x^i}$. Also,

$$V_\gamma(t) = v_{\gamma,\gamma(t)} = \dot{\gamma}^i(t)\, \partial_{x^i,\gamma(t)}, \text{ where } \dot{\gamma}^i(t) := \frac{d(x^i \circ \gamma)}{dt}(t).$$

Set $\dot{W}^k = \dfrac{dW^k}{dt}$, and define

$$(\nabla^U_{V_\gamma} W)(t) = (\dot{W}^k(t) + \Gamma^k_{ij}(\gamma(t))\, \dot{\gamma}^j(t)\, W^i(t))\, \partial_{x^k,\gamma(t)}, \quad t \in I. \qquad (\star)$$

As $\dot{W}^k + (\Gamma^k_{ij} \circ \gamma)\, \dot{\gamma}^j\, W^i \in C^\infty(I)$, we have $\nabla^U_{V_\gamma} W \in T_0^1\gamma^U$ by Proposition 7.1. We now verify that $\nabla^U_{V_\gamma}$ has the claimed properties. The $\mathbb{R}$-linearity follows immediately since $(cW + X)^i = cW^i + X^i$.

Similarly, the Leibniz rule holds since

$$\nabla^{U}_{V_\gamma}(fW) = \Big(\frac{df}{dt}W^k + f\dot{W}^k + (\Gamma^k_{ij}\circ\gamma)\dot{\gamma}^j fW^i\Big)\partial_{x^k} = \frac{df}{dt}W + f\nabla^{U}_{V_\gamma}W.$$

Let $t_0 \in I$ and $\widetilde{V} \in T^1_0 U$ satisfy $\widetilde{V}_{\gamma(t_0)} = V_\gamma(t_0) = v_{\gamma,\gamma(t_0)} = V^i(t_0)\partial_{x^i,\gamma(t_0)}$. In particular, if $\widetilde{V} = \widetilde{V}^i\partial_{x^i}$, then $\widetilde{V}^i(\gamma(t_0)) = V^i(t_0)$. Also, writing $\widetilde{W} = \widetilde{W}^i\partial_{x^i}$, we have $\widetilde{W}^i(\gamma(t)) = W^i(t)$ for all $t \in I$. We have

$$\begin{aligned}
&(\nabla_{v_{\gamma,\gamma(t_0)}}\widetilde{W})_{\gamma(t_0)}\\
&= (\nabla_{\widetilde{V}}(\widetilde{W}^i\partial_{x^i}))_{\gamma(t_0)} = (\widetilde{V}^j\nabla_{\partial_{x^j}}(\widetilde{W}^i\partial_{x^i}))_{\gamma(t_0)}\\
&= \big(\widetilde{V}^j(\partial_{x^j}\widetilde{W}^k)\partial_{x^k} + \widetilde{V}^j\widetilde{W}^i\Gamma^k_{ij}\partial_{x^k}\big)_{\gamma(t_0)}\\
&= \Big((\widetilde{V}^j(\gamma(t_0))\partial_{x^j,\gamma(t_0)})\widetilde{W}^k + \widetilde{V}^j(\gamma(t_0))\widetilde{W}^i(\gamma(t_0))\Gamma^k_{ij}(\gamma(t_0))\Big)\partial_{x^k,\gamma(t_0)}\\
&= \Big(v_{\gamma,\gamma(t_0)}(\widetilde{W}^k) + V^j(t_0)W^i(t_0)\Gamma^k_{ij}(\gamma(t_0))\Big)\partial_{x^k,\gamma(t_0)}\\
&= \Big(\frac{d(\widetilde{W}^k\circ\gamma)}{dt}(t_0) + V^j(t_0)W^i(t_0)\Gamma^k_{ij}(\gamma(t_0))\Big)\partial_{x^k,\gamma(t_0)}\\
&= \Big(\frac{dW^k}{dt}(t_0) + V^j(t_0)W^i(t_0)\Gamma^k_{ij}(\gamma(t_0))\Big)\partial_{x^k,\gamma(t_0)} = (\nabla^{U}_{V_\gamma}W)(t_0).
\end{aligned}$$

This completes the proof of the 'existence' part.

Uniqueness: Suppose that there exists a map $D_\gamma : T^1_0\gamma^U \to T^1_0\gamma^U$ satisfying the given properties. Given $W \in T^1_0\gamma^U$, write $W(t) = W^i(t)\,\partial_{x^i,\gamma(t)}$, $t \in I$. Then we have, using the Leibniz rule, that

$$(D_\gamma W)(t) = (D_\gamma(W^i\partial_{x^i,\gamma(\cdot)}))(t) = \frac{dW^i}{dt}(t)\partial_{x^i,\gamma(t)} + W^i(t)(D_\gamma\partial_{x^i,\gamma(\cdot)})(t). \quad (*)$$

Let $t \in I$. To evaluate $(D_\gamma\partial_{x^i,\gamma(\cdot)})(t)$, we take any $\widetilde{V} \in T^1_0 U$ such that at this particular fixed t, we have $\widetilde{V}_{\gamma(t)} = V_\gamma(t) = v_{\gamma,\gamma(t)}$. If $\widetilde{V}^j := \widetilde{V}(x^j)$, then

$$\widetilde{V}_{\gamma(t)} = \widetilde{V}^i(\gamma(t))\,\partial_{x^i,\gamma(t)} = v_{\gamma,\gamma(t)} = \dot{\gamma}^i(t)\,\partial_{x^i,\gamma(t)}.$$

Then $\widetilde{V}^j(\gamma(t)) = \dot{\gamma}^j(t)$. Since $\partial_{x^i} \in T^1_0 U$ extends $\partial_{x^i,\gamma(\cdot)} \in T^1_0\gamma^U$,

$$\begin{aligned}
(D_\gamma\partial_{x^i,\gamma(\cdot)})(t) &= (\nabla_{v_{\gamma,\gamma(t)}}\partial_{x^i})_{\gamma(t)}\\
&= (\nabla_{\widetilde{V}}\partial_{x^i})_{\gamma(t)}\\
&= (\nabla_{\widetilde{V}^j\partial_{x^j}}\partial_{x^i})_{\gamma(t)}\\
&= (\widetilde{V}^j\nabla_{\partial_{x^j}}\partial_{x^i})_{\gamma(t)}\\
&= (\widetilde{V}^j\,\Gamma^k_{ij}\,\partial_{x^k})_{\gamma(t)} = \widetilde{V}^j(\gamma(t))\,\Gamma^k_{ij}(\gamma(t))\,\partial_{x^k,\gamma(t)}\\
&= \dot{\gamma}^j(t)\,\Gamma^k_{ij}(\gamma(t))\,\partial_{x^k,\gamma(t)}.
\end{aligned}$$

Substitution in $(*)$ yields $(\star)$ from the proof of the existence part, showing the claimed uniqueness. □

Theorem 7.1. *Let M be a smooth manifold with a connection ∇. For each smooth curve $\gamma : I \to M$, where I is an open interval in $\mathbb{R}$, there exists a unique map $\nabla_{V_\gamma} : T_0^1\gamma \to T_0^1\gamma$ such that the following hold for all $W, X \in T_0^1\gamma$, $c \in \mathbb{R}$ and $f \in C^\infty(I)$:*

- ($\mathbb{R}$-linearity) $\nabla_{V_\gamma}(cW+X) = c\nabla_{V_\gamma}W + \nabla_{V_\gamma}X$.
- (Leibniz) $\nabla_{V_\gamma}(fW) = \dfrac{df}{dt}W + f\nabla_{V_\gamma}W$.
- *If $\widetilde{W} \in T_0^1 M$ and $\widetilde{W}_{\gamma(t)} = W(t)$ for all $t \in I$, then we have* $(\nabla_{V_\gamma}W)(t) = (\nabla_{v_{\gamma,\gamma(t)}}\widetilde{W})_{\gamma(t)}$ *for all $t \in I$.*

If $(U, \mathbf{x})$ is an admissible chart for M, then

$$(\nabla_{V_\gamma}W)(t) = \big(\dot{W}^k(t) + \Gamma^k_{ij}(\gamma(t))\,\dot{\gamma}^j(t)\,W^i(t)\big)\,\partial_{x^k,\gamma(t)} \quad (t \in I_U),$$

where $W(t) = W^i(t)\,\partial_{x^i,\gamma(t)}$, for $t \in I_U := \gamma^{-1}U$.

Proof. Uniqueness: Let there exist a map $D_\gamma : T_0^1\gamma \to T_0^1\gamma$ satisfying the given properties. Let $t_0 \in I$, and $(U, \mathbf{x})$ be an admissible chart containing $\gamma(t_0)$. Let $\chi \in C^\infty(M)$ be a bump function which is identically equal to 1 in a neighbourhood $U_0 \subset U$ of $\gamma(t_0)$ and equal to 0 outside U_1, where $\overline{U_1} \subset U$. Let $J \subset I$ be an open interval containing t_0 such that $\gamma(J) \subset U_0$. Given $W \in T_0^1\gamma$, write $W(t) = W^i(t)\,\partial_{x^i,\gamma(t)}$, $t \in J$. Define $X \in T_0^1\gamma$ by $X = \chi(\gamma(\cdot))W^i\chi(\gamma(\cdot))\partial_{x^i,\gamma(\cdot)}$. We claim $(D_\gamma W)(t_0) = (D_\gamma X)(t_0)$. Using linearity and the Leibniz Rule,

$$\begin{aligned}
&(D_\gamma W)(t_0) - (D_\gamma X)(t_0) = (D_\gamma(W-X))(t_0) = (D_\gamma((1-(\chi(\gamma(\cdot))^2))W))(t_0)\\
&= \frac{d(1-(\chi(\gamma(\cdot))^2))}{dt}\Big|_{t=t_0} W(t_0) + (1-(\chi(\gamma(t_0))^2))(D_\gamma W)(t_0)\\
&= 0W(t_0) + 0(D_\gamma W)(t_0) = 0.
\end{aligned}$$

Using the Leibniz rule again,

$$\begin{aligned}
(D_\gamma W)(t_0) = (D_\gamma X)(t_0) &= (D_\gamma(\chi(\gamma(\cdot))W^i\chi(\gamma(\cdot))\partial_{x^i,\gamma(\cdot)}))(t_0)\\
&= \frac{dW^i}{dt}(t_0)\partial_{x^i,\gamma(t_0)} + W^i(t_0)(D_\gamma(\chi(\gamma(\cdot))\partial_{x^i,\gamma(\cdot)}))(t_0). \qquad (\star)
\end{aligned}$$

Now we find $(D_\gamma(\chi(\gamma(\cdot))\partial_{x^i,\gamma(\cdot)}))(t_0)$. Take any $\widetilde{V} \in T_0^1 M$ extending $V_\gamma(t_0)$, i.e., $\widetilde{V}_{\gamma(t_0)} = v_{\gamma,\gamma(t_0)}$. If $\widetilde{V}^j = \widetilde{V}(x^j)$, then $\widetilde{V}^i(\gamma(t_0))\partial_{x^i,\gamma(t_0)} = \dot{\gamma}^i(t_0)\partial_{x^i,\gamma(t_0)}$. So $\widetilde{V}^j(\gamma(t_0)) = \dot{\gamma}^j(t_0)$. As $\chi\partial_{x^i} \in T_0^1 M$ satisfies $(\chi\partial_{x^i})|_{\gamma(t)} = \chi(\gamma(t))\partial_{x^i,\gamma(t)}$ for all $t \in I$, we have (using Exercise 6.10, p.114)

$$\begin{aligned}
(D_\gamma(\chi(\gamma(\cdot))\partial_{x^i,\gamma(\cdot)}))(t_0) &= (\nabla_{v_{\gamma,\gamma(t_0)}}(\chi\partial_{x^i}))_{\gamma(t_0)} = (\nabla_{\widetilde{V}}\partial_{x^i})_{\gamma(t_0)}\\
&= (\nabla_{\widetilde{V}^j\partial_{x^j}}\partial_{x^i})_{\gamma(t_0)} = (\widetilde{V}^j\nabla_{\partial_{x^j}}\partial_{x^i})_{\gamma(t_0)}\\
&= (\widetilde{V}^j\Gamma^k_{ij}\,\partial_{x^k})_{\gamma(t_0)} = \widetilde{V}^j(\gamma(t_0))\,\Gamma^k_{ij}(\gamma(t_0))\,\partial_{x^k,\gamma(t_0)}\\
&= \dot{\gamma}^j(t_0)\,\Gamma^k_{ij}(\gamma(t_0))\,\partial_{x^k,\gamma(t_0)}.
\end{aligned}$$

Substitution in $(\star)$ gives $(D_\gamma W)(t_0) = (\nabla^U_{V_\gamma}W)(t_0)$, showing uniqueness.

Existence: We will define ∇_{V_γ} by using the charts from an atlas $\mathcal{A}$ defining the smooth structure on M. (Then by uniqueness, it follows that the construction of ∇_{V_γ} does not depend on the atlas $\mathcal{A}$.) Let $t_0 \in I$, and $(U, \mathbf{x})$ be a chart from $\mathcal{A}$ containing the point $\gamma(t_0)$. Let $J \subset I$ be an interval around t_0 such that $\gamma(J) \subset U$. For $t \in J$, $(\nabla_{V_\gamma} W)(t) := (\nabla^U_{V_\gamma} W)(t)$, i.e.,

$$(\nabla_{V_\gamma} W)(t) = \big(\dot{W}^k(t) + \Gamma^k_{ij}(\gamma(t))\,\dot{\gamma}^j(t)\,W^i(t)\big)\,\partial_{x^k,\gamma(t)},$$

where $W(t) = W^i(t)\,\partial_{x^i,\gamma(t)}$, $V_\gamma(t) = v_{\gamma,\gamma(t)} = \dot{\gamma}^i(t)\,\partial_{x^i,\gamma(t)}$, and

$$\dot{W}^i = \frac{dW^i}{dt}, \quad \dot{\gamma}^i(t) = \frac{d(x^i \circ \gamma)}{dt}(t).$$

Then ∇_{V_γ} as defined above has the claimed properties by the proof of the existence part of Lemma 7.2. Also, we note that if $(\widetilde{U}, \widetilde{\mathbf{x}})$ is another chart containing the point $\gamma(t)$, then the uniqueness part of Lemma 7.2 applied to $U \cap \widetilde{U}$ shows that

$$(\nabla^U_{V_\gamma} W)(t) = (\nabla^{U \cap \widetilde{U}}_{V_\gamma} W)(t) = (\nabla^{\widetilde{U}}_{V_\gamma} W)(t).$$

That $I \ni t \mapsto (\nabla_{V_\gamma} W)(t)$ is an element of $T^1_0\gamma$ follows from Lemma 7.1.

The last claim on the expression for $\nabla_{V_\gamma} W$ in a chart follows from the existence part combined with the uniqueness part. □

Exercise 7.1. Let M be a smooth manifold with a connection ∇. Let I, J be open subsets of $\mathbb{R}$, and $J \ni u \mapsto h(u) \in I$ be a C^∞ function. Let $\gamma : I \to M$ be a smooth curve. From Exercise 2.4 (p.23), the velocity vector field of $\gamma \circ h : J \to M$ is given by $V_{\gamma \circ h}(s) = \dot{h}(s) V_\gamma(h(s))$ for $s \in J$. For a given $Z \in T^1_0\gamma$, define $W \in T^1_0(\gamma \circ h)$ by $W(s) = Z(h(s))$, $s \in J$. Show that $(\nabla_{V_{\gamma \circ h}} W)(s) = \dot{h}(s)(\nabla_{V_\gamma} Z)(h(s))$, $s \in J$.

Exercise 7.2. The aim of this exercise is show a further strengthening of the results in Lemma 6.1 (p.112) and Exercise 6.10 (p.114), namely that $\nabla_V W$ is determined at a point p by just knowing

- the value V_p of V at p and
- the values of W along a *curve* γ passing through p whose velocity vector at p is V_p (instead of knowing W in an arbitrarily small *neighbourhood* of p as in Exercise 6.10, p.114).

Let M be a smooth manifold with a connection ∇, $p \in M$ and $v \in T_pM$.
Let $V, \widetilde{V} \in T^1_0 M$ be any two vector fields such that $V_p = v = \widetilde{V}_p$.
Let $\gamma : I \to M$ be a smooth curve, where I is an open interval in $\mathbb{R}$ containing 0, such that $\gamma(0) = p$ and $v_{\gamma,\gamma(0)} = v$. Suppose that $W, \widetilde{W} \in T^1_0 M$ are such that

$$W_{\gamma(t)} = \widetilde{W}_{\gamma(t)} \quad (t \in I).$$

Prove that

$$(\nabla_V W)_p = (\nabla_{\widetilde{V}} \widetilde{W})_p.$$

Hint: Define the vector field $X \in T^1_0\gamma$ by $X(t) = W_{\gamma(t)} = \widetilde{W}_{\gamma(t)}$ and consider $(\nabla_{V_\gamma} X)(0)$.

Remark 7.1. (Extending a vector field along a curve to a locally defined vector field.) We now show the following technical fact, which will be used later on. We had seen earlier that given a point p in a smooth manifold M, and a vector $v \in T_pM$, there exists a vector field $V \in T_0^1 M$ such that $V_p = v$ (Lemma 3.1, p.41). Now we show that if we have a smooth curve γ passing through a point p with a nonzero velocity at p, and a smooth vector field V along γ, then in the vicinity of p, V is the restriction of a vector field on M. The precise statement is given below.

Fact: Let M be an m-dimensional smooth manifold. Suppose $\gamma : I \to M$ is a smooth curve, where I is an open interval in $\mathbb{R}$ containing 0, and $v_{\gamma,\gamma(0)} \neq 0$. Let $V \in T_0^1\gamma$. Then there exists an open interval $J \subset I$ with $0 \in J$, an admissible chart $(U, \mathbf{x})$ containing $\gamma(J)$, and a vector field $\widetilde{V} \in T_0^1 U$ such that $\widetilde{V}_{\gamma(t)} = V(t)$ for all $t \in J$.

This can be shown as follows. By Remark 2.2 (p.32), there exists an open interval $J \subset I$ containing 0, and an admissible chart $(\widetilde{U}, \mathbf{x})$ containing $\gamma(J)$, such that $(\mathbf{x} \circ \gamma)(t) = (t, 0, \cdots, 0)$ for all $t \in J$. Write $V(t) = V^i(t)\partial_{x^i,\gamma(t)}$ for $t \in J$. Let S be the 'strip'

$$S := \{\mathbf{y} = (y^1, \cdots, y^m) \in \mathbf{x}(\widetilde{U}) : y^1 \in J\} \subset \mathbf{x}(\widetilde{U}).$$

For $1 \leqslant i \leqslant m$, define the smooth functions $\widetilde{V}^i : S \to \mathbb{R}$ by $\widetilde{V}^i(\mathbf{y}) = V^i(y^1)$ for $\mathbf{y} = (y^1, \cdots, y^m) \in S \subset \mathbf{x}(\widetilde{U})$. Set

$$\widetilde{V}(q) = (\widetilde{V}^i \circ \mathbf{x})(q)\, \partial_{x^i,q} \text{ for } q \in U := \mathbf{x}^{-1}S \subset \widetilde{U}.$$

Then the chart $(U, \mathbf{x}|_U)$ is admissible, and $\widetilde{V} \in T_0^1 U$ (as its components are smooth). For all $t \in J$, $\gamma(t) \in U$, and so

$$\widetilde{V}(\gamma(t)) = (\widetilde{V}^i \circ \mathbf{x})(\gamma(t))\, \partial_{x^i,\gamma(t)} = \widetilde{V}^i(t, 0, \cdots, 0)\partial_{x^i,\gamma(t)} = V^i(t)\partial_{x^i,\gamma(t)} = V(t).$$

This completes the justification of the fact. *

We now give the following natural definition.

Definition 7.2. (Parallel vector field along a smooth curve.)
Let M be a smooth manifold with a connection ∇, I be an open interval in $\mathbb{R}$, $\gamma : I \to M$ be a smooth curve, and $W \in T_0^1\gamma$ be a vector field along γ. We say W is *parallel along* γ if $(\nabla_{V_\gamma} W)(t) = 0$ for all $t \in I$.

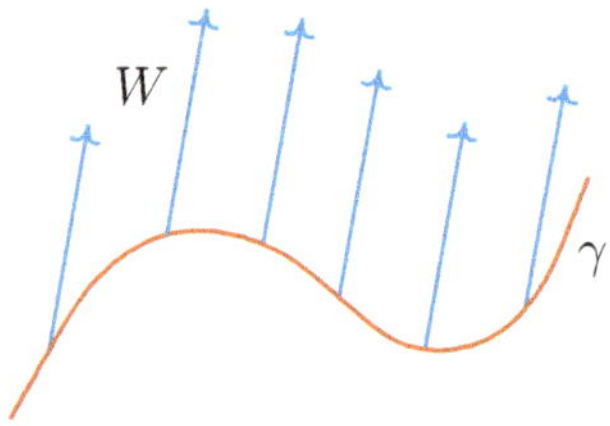

Example 7.3. (Sphere.) Recall Example 5.7 (p.85), where we considered the sphere S^2 as a Riemannian manifold using the metric $\mathbf{g}$ induced from the Euclidean inner product on $\mathbb{R}^3$. Let ∇ denote the Levi-Civita connection induced by $\mathbf{g}$. Consider the meridian γ given by

$$\gamma(t) = \big((\cos\phi_0)\sin t, (\sin\phi_0)\sin t, \cos t\big) \text{ for all } t \in (0,\pi),$$

with a fixed longitude $\phi_0 \in (0, 2\pi)$ in the chart (U, φ) described in Example 5.7 (p.85). We claim that the vector field W along γ given by the latitudinal vector field ∂_ϕ, is not parallel along γ. Firstly, we have $W(t) = \partial_{\phi,\gamma(t)}$ for all $t \in (0,\pi)$, and so $W^\phi \equiv 1$, $W^\theta \equiv 0$, giving $\dot{W}^\phi \equiv 0 \equiv \dot{W}^\theta$. Next, $v_{\gamma,\gamma(t)} = \dot{\gamma}^\theta \, \partial_{\theta,\gamma(t)} + \dot{\gamma}^\phi \, \partial_{\phi,\gamma(t)}$, where

$$\dot{\gamma}^\theta(t) = \frac{d(\theta\circ\gamma)}{dt}(t) = \frac{dt}{dt}(t) = 1, \text{ and}$$
$$\dot{\gamma}^\phi(t) = \frac{d(\phi\circ\gamma)}{dt}(t) = \frac{d\phi_0}{dt}(t) = 0.$$

Thus, keeping track of the nonzero terms, we have

$$\begin{aligned}(\nabla_{V_\gamma} W)(t) &= \big(\dot{W}^\theta(t) + \Gamma^\theta_{ij}(\gamma(t))\,\dot{\gamma}^j(t)\,W^i(t)\big)\,\partial_{\theta,\gamma(t)} \\ &\quad + \big(\dot{W}^\phi(t) + \Gamma^\phi_{ij}(\gamma(t))\,\dot{\gamma}^j(t)\,W^i(t)\big)\,\partial_{\phi,\gamma(t)} \\ &= \Gamma^\theta_{\phi\theta}(\gamma(t))\,\dot{\gamma}^\theta(t)\,W^\phi(t)\,\partial_{\theta,\gamma(t)} + \Gamma^\phi_{\phi\theta}(\gamma(t))\,\dot{\gamma}^\theta(t)\,W^\phi(t)\,\partial_{\phi,\gamma(t)}.\end{aligned}$$

But from Example 6.4 (p.120), $\Gamma^\theta_{\phi\theta} = 0$ and $\Gamma^\phi_{\phi\theta}(\gamma(t)) = \dfrac{\cos t}{\sin t}$, so that

$$(\nabla_{V_\gamma} W)(t) = \frac{\cos t}{\sin t}\cdot 1\cdot 1\,\partial_{\phi,\gamma(t)} = \frac{\cos t}{\sin t}\,\partial_{\phi,\gamma(t)}.$$

So W is not parallel along γ.

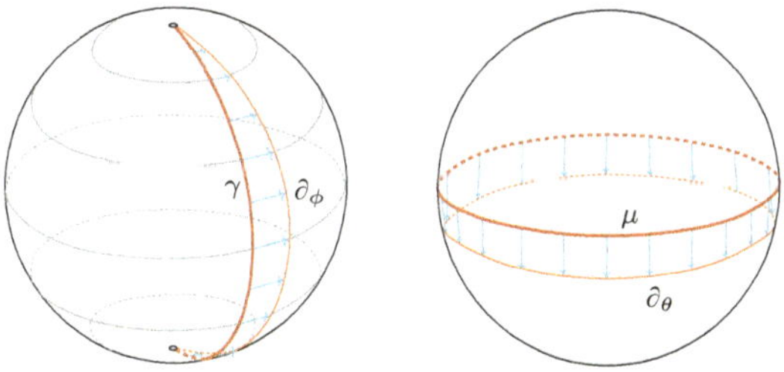

Consider the equator μ, given by $\mu(t) = (\cos t, \sin t, 0)$, for all $t \in (0, 2\pi)$. We claim that the vector field X along μ given by the longitudinal vector field ∂_θ, is parallel along μ. Firstly, $X(t) = \partial_{\theta,\mu(t)}$ for all $t \in (0, 2\pi)$, and so $X^\theta \equiv 1$, $X^\phi \equiv 0$, giving $\dot{X}^\theta \equiv 0 \equiv \dot{X}^\phi$. Next, $v_{\mu,\mu(t)} = \dot{\mu}^\theta \, \partial_{\theta,\mu(t)} + \dot{\mu}^\phi \, \partial_{\phi,\mu(t)}$, where

$$\dot{\mu}^\theta(t) = \frac{d(\theta\circ\mu)}{dt}(t) = \frac{d(\frac{\pi}{2})}{dt}(t) = 0, \text{ and}$$
$$\dot{\mu}^\phi(t) = \frac{d(\phi\circ\mu)}{dt}(t) = \frac{dt}{dt}(t) = 1.$$

Thus, keeping track of the nonzero terms, we have

$$\begin{aligned}(\nabla_{V_\mu} X)(t) &= (\dot{X}^\theta(t) + \Gamma^\theta_{ij}(\mu(t))\,\dot{\mu}^j(t)\,X^i(t))\,\partial_{\theta,\mu(t)} \\ &\quad + (\dot{X}^\phi(t) + \Gamma^\phi_{ij}(\mu(t))\,\dot{\mu}^j(t)\,X^i(t))\,\partial_{\phi,\mu(t)} \\ &= \Gamma^\theta_{\theta\phi}(\mu(t))\,\dot{\mu}^\phi(t)\,X^\theta(t)\,\partial_{\theta,\mu(t)} + \Gamma^\phi_{\theta\phi}(\mu(t))\,\dot{\mu}^\phi(t)\,X^\theta(t)\,\partial_{\phi,\mu(t)} \\ &= 0\cdot 1\cdot 1\,\partial_{\theta,\mu(t)} + \frac{\cos\theta}{\sin\theta}\Big|_{\theta=\frac{\pi}{2}}\cdot 1\cdot 1\,\partial_{\phi,\mu(t)} = 0+0=0.\end{aligned}$$

Here we used $\theta(\mu(t)) = \frac{\pi}{2}$ for all $t \in (0, 2\pi)$. So X is parallel along μ. ◇

Exercise 7.3. Consider the same curve μ from Example 7.3: $\mu(t) = (\cos t, \sin t, 0)$, $t \in (0, 2\pi)$. Take $Y(t) = \partial_{\phi,\mu(t)}$, $t \in (0, 2\pi)$. Prove that Y is parallel along μ. Conclude that for $\alpha, \beta \in \mathbb{R}$, $\alpha X + \beta Y$ is also parallel along μ, where X is the vector field along μ given in Example 7.3, namely $X(t) = \partial_{\theta,\mu(t)}$, $t \in (0, 2\pi)$.

Theorem 6.2 (p.125) manifested the metric-compatibility property (C5) of the Levi-Civita connection on a smooth manifold with a metric $\mathbf{g}$, namely $\nabla.\mathbf{g} = 0$. We now meet another avatar of this.

Proposition 7.2.
Let

- $(M, \mathbf{g})$ *be a semi-Riemannian manifold with the metric* $\mathbf{g}$,
- $\gamma : I \to M$ *be a smooth curve, where* I *is an open interval in* $\mathbb{R}$,
- $W, X \in T^1_0\gamma$, *and*
- ∇_{V_γ} *denote the covariant derivative along* γ *induced by the Levi-Civita connection* ∇ *corresponding to the metric* $\mathbf{g}$.

Then

$$\frac{d(\mathbf{g}(W, X))}{dt} = \mathbf{g}(\nabla_{V_\gamma} W, X) + \mathbf{g}(W, \nabla_{V_\gamma} X).$$

Here t arguments are suppressed: $\dfrac{d(\mathbf{g}(W, X))}{dt} = \dfrac{d(\mathbf{g}(\gamma(t))(W(t), X(t)))}{dt}$ etc.

Proof. It is enough to show this in admissible charts that meet $\gamma(I)$. Let $(U, \mathbf{x})$ be such a chart. Then we have

$$\frac{d(\mathbf{g}(W, X))}{dt} = (\dot{W}^i X^j + W^i \dot{X}^j)\,\mathbf{g}(\partial_{x^i}, \partial_{x^j}) + W^i X^j \frac{d(\mathbf{g}(\partial_{x^i}, \partial_{x^j}))}{dt}.$$

Let us first note that

$$\begin{aligned}\frac{d(\mathbf{g}(\partial_{x^i}, \partial_{x^j}))}{dt}(t) &= \frac{d(\mathbf{g}(\gamma(t))(\partial_{x^i,\gamma(t)}, \partial_{x^j,\gamma(t)}))}{dt}(t) \\ &= v_{\gamma,\gamma(t)}(U \ni p \mapsto \mathbf{g}(p)(\partial_{x^i,p}, \partial_{x^j,p})).\end{aligned}$$

Now fix $t \in I$, and let V be any vector field such that $V_{\gamma(t)} = v_{\gamma,\gamma(t)}$.

Then from the above,

$$\frac{d(\mathbf{g}(\partial_{x^i}, \partial_{x^j}))}{dt}(t) = (V(\mathbf{g}(\partial_{x^i}, \partial_{x^j})))(\gamma(t)).$$

Now we use the metric compatibility property (C5) of the Levi-Civita connection to expand the right-hand side $(V(\mathbf{g}(\partial_{x^i}, \partial_{x^j})))(\gamma(t))$, obtaining

$$\begin{aligned}
&\frac{d(\mathbf{g}(\partial_{x^i}, \partial_{x^j}))}{dt}(t) \\
&= (\mathbf{g}(\nabla_V \partial_{x^i}, \partial_{x^j}))(\gamma(t)) + (\mathbf{g}(\partial_{x^i}, \nabla_V \partial_{x^j}))(\gamma(t)) \\
&= \mathbf{g}(\gamma(t))((\nabla_V \partial_{x^i})_{\gamma(t)}, \partial_{x^j,\gamma(t)}) + \mathbf{g}(\gamma(t))(\partial_{x^i,\gamma(t)}, (\nabla_V \partial_{x^j})_{\gamma(t)}) \\
&= \mathbf{g}(\gamma(t))((\nabla_{V_\gamma} \partial_{x^i})(t), \partial_{x^j,\gamma(t)}) + \mathbf{g}(\gamma(t))(\partial_{x^i,\gamma(t)}, (\nabla_{V_\gamma} \partial_{x^j})(t)).
\end{aligned}$$

Thus, again suppressing the argument t, we have

$$\begin{aligned}
&\frac{d(\mathbf{g}(W, X))}{dt} \\
&= (\dot{W}^i X^j + W^i \dot{X}^j)\mathbf{g}(\partial_{x^i}, \partial_{x^j}) + W^i X^j \frac{d(\mathbf{g}(\partial_{x^i}, \partial_{x^j}))}{dt} \\
&= \mathbf{g}(\dot{W}^i \partial_{x^i}, X) + \mathbf{g}(W, \dot{X}^j \partial_{x^j}) + W^i X^j (\mathbf{g}(\nabla_{V_\gamma} \partial_{x^i}, \partial_{x^j}) + \mathbf{g}(\partial_{x^i}, \nabla_{V_\gamma} \partial_{x^j})) \\
&= \mathbf{g}(\dot{W}^i \partial_{x^i}, X) + \mathbf{g}(W, \dot{X}^j \partial_{x^j}) + \mathbf{g}(W^i \nabla_{V_\gamma} \partial_{x^i}, X) + \mathbf{g}(W, X^j \nabla_{V_\gamma} \partial_{x^j}) \\
&= \underline{\mathbf{g}(\dot{W}^i \partial_{x^i}, X)} + \underline{\mathbf{g}(W, \dot{X}^j \partial_{x^j})} + \underline{\mathbf{g}(W^i \nabla_{V_\gamma} \partial_{x^i}, X)} + \underline{\mathbf{g}(W, X^j \nabla_{V_\gamma} \partial_{x^j})} \\
&= \underline{\mathbf{g}(\nabla_{V_\gamma} W, X)} + \underline{\mathbf{g}(W, \nabla_{V_\gamma} X)} \qquad \text{(linearity and Leibniz rule for } \nabla_{V_\gamma}\text{)}.
\end{aligned}$$

This completes the proof. □

Corollary 7.1. *Let $(M, \mathbf{g})$ be a semi-Riemannian manifold with the metric $\mathbf{g}$ and the Levi-Civita connection ∇. Let $\gamma : I \to M$ be a smooth curve, where I is an open interval in $\mathbb{R}$. Let $W, X \in T_0^1\gamma$ be parallel along γ. Then*

$$\frac{d(\mathbf{g}(W, X))}{dt} = 0,$$

that is, there exists $c \in \mathbb{R}$ such that for all $t \in I$, $\mathbf{g}(\gamma(t))(W(t), X(t)) = c$.

Hence, in a Riemannian manifold, for pointwise 'unit length vector fields' W, X along a smooth curve that are parallel along γ, the 'angle' between $W(t)$ and $X(t)$ stays the same.

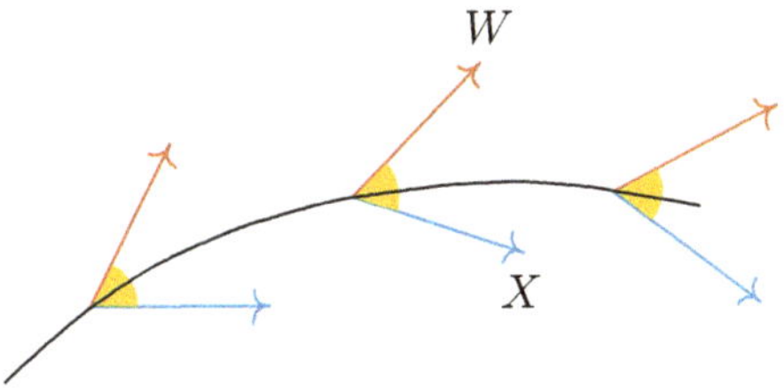

7.3 Parallel transport

In this section, we will meet the parallel transport map. Given a smooth curve $\gamma : I \to M$, where M is a smooth manifold with a connection ∇, and I is an open interval in $\mathbb{R}$ containing the points a and b the parallel transport map

$$P^{\gamma}_{ab} : T_{\gamma(a)}M \to T_{\gamma(b)}M$$

will transport tangent vectors from $T_{\gamma(a)}M$ to tangent vectors in $T_{\gamma(b)}M$, parallelly along the curve γ. In this manner, it 'connects' the two 'distant' tangent spaces using the connection ∇. The transportation itself will be done by working locally in charts, setting up a system of ordinary differential equations for components, and using a known existence and uniqueness result from the theory of differential equations.

Proposition 7.3.
Let I be an open interval in $\mathbb{R}$ and $A : I \to \mathbb{R}^{m\times m}$ have C^∞ components. Then given $a \in I$ and $B \in \mathbb{R}^m$, there exists a unique $Y : I \to \mathbb{R}^m$ such that the components of Y are C^∞, and

$$\begin{cases} \dot{Y}(t) = A(t)Y(t), & t \in I, \\ Y(a) = B. \end{cases}$$

Proof. This follows from a C^1-version[1] of the result (where the A is just assumed to be continuous), by the following argument: First view A as C^0, from which we deduce that there is a unique C^1 solution Y. But then look at the right-hand side AY, which is C^1, since A is C^1 and so is Y. Hence we conclude that $\dot{Y} = AY$ is C^1, which is the same as saying that Y is C^2. Again $\dot{Y} = AY$ is now C^2, since the right-hand side is the product of $A, Y \in C^2$, and so $Y \in C^3$, and so on. □

Theorem 7.2. *Let M be a smooth manifold with a connection ∇. Let $\gamma : I \to M$ be a smooth curve, where I is an open interval in $\mathbb{R}$. If $a \in I$ and $w \in T_{\gamma(a)}M$, then there exists a unique $W \in T^1_0\gamma$ such that W is parallel along γ and $W(a) = w$.*

Proof. First suppose that $\gamma(I)$ lies entirely within an admissible chart $(U, \mathbf{x})$. We write $W = W^i \partial_{x^i}$. Then W is parallel along γ if and only if for all $t \in I$, $(\nabla_{V_\gamma} W)(t) = 0$, i.e.,

$$\dot{W}^k(t) + \dot{\gamma}^j(t)\Gamma^k_{ij}(\gamma(t))W^i(t) = 0, \quad 1 \leqslant k \leqslant m.$$

[1]See for example, [Apostol (1969), §7.21].

Moreover, we have $W(a)=w$ if and only if $W^i(a) = b^i$, $1 \leqslant i \leqslant m$, where $w =: b^i\, \partial_{x^i,\gamma(a)}$. Define the vector $B=(b^1,\cdots,b^m)$. Then the claimed W exists if and only if the following initial value problem for a system of first order linear differential equations, in the unknown $Y=(W^1,\cdots,W^m)$, has a solution:

$$\begin{cases} \dot{Y}(t) = A(t)Y(t), & t \in I, \\ Y(a) = B. \end{cases}$$

Here $A(\cdot)$ is the matrix whose entry in the ith row and j^{th} column is given by the function $-\dot{\gamma}^k(\cdot)\Gamma^i_{jk}(\gamma(\cdot)) \in C^\infty(I)$. By Proposition 7.3, there exists a unique C^∞ solution Y. Consequently, there exists a unique $W \in T^1_0\gamma$ parallel to γ satisfying $W(a)=w$.

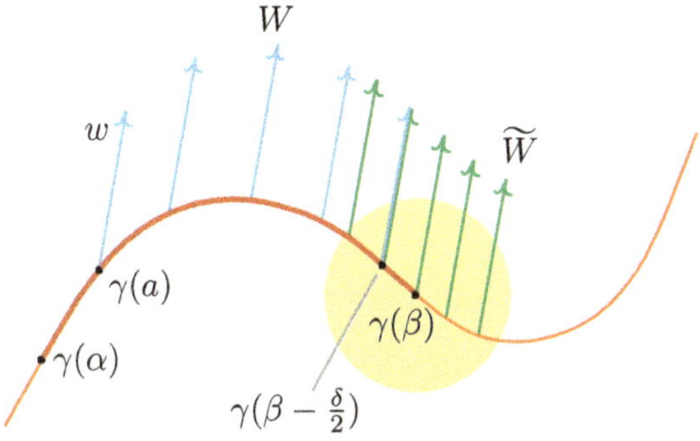

Now[2] suppose that there does not exist a single admissible chart covering all of $\gamma(I)$. Let (α,β) be the maximal interval containing a on which a $W \in T^1_0\gamma|_{(\alpha,\beta)}$ exists which is parallel along $\gamma|_{(\alpha,\beta)}$ and $W(a) = w$. If $\beta=\sup I$ and $\alpha=\inf I$, then we are done. Suppose that $\beta < \sup I$. (The proof for $\alpha > \inf I$ is analogous.) There is an admissible chart $(V,\mathbf{y})$ such that $\gamma(\beta)\in V$. We can then choose a $\delta>0$ such that $\gamma((\beta-\delta,\beta+\delta)) \in V$. We know that there exists a vector $W(\beta-\frac{\delta}{2})$. Taking this as an initial condition, there exists a unique vector field $\widetilde{W}$ on $(\beta-\delta,\beta+\delta)$ such that

$$\widetilde{W}\Big(\beta-\frac{\delta}{2}\Big)=W\Big(\beta-\frac{\delta}{2}\Big)$$

and $\widetilde{W}$ is parallel to $\gamma|_{(\beta-\delta,\beta+\delta)}$. But by the uniqueness, this $\widetilde{W}$ coincides with W on $(\beta-\delta,\beta)$. However, as $\widetilde{W}$ itself is defined on $(\beta-\delta,\beta+\delta)$, this means that W could have been extended beyond β, a contradiction to the maximality of (α,β). □

[2]We follow [Lee (2018), Theorem 4.32].

Definition 7.3. (Parallel transport of a vector along a smooth curve.) Let M be a smooth manifold with a connection ∇. Suppose that $\gamma : I \to M$ is a smooth curve, where I is an open interval in $\mathbb{R}$. Let $a \in I$. For a vector $w \in T_{\gamma(a)}M$, let $P_a^\gamma w \in T_0^1\gamma$ be the unique vector field along γ such that $P_a^\gamma w$ is parallel along γ and $(P_a^\gamma w)(a) = w$. Then $P_a^\gamma w$ is called the *parallel transport of w along γ*. Define the maps $P_{at}^\gamma : T_{\gamma(a)}M \to T_{\gamma(t)}M$ $(t \in I)$, by $P_{at}^\gamma w = (P_a^\gamma w)(t)$ for all $w \in T_{\gamma(a)}M$.

Proposition 7.4. *Let M be a smooth manifold with a connection ∇. Let $\gamma : I \to M$ be a smooth curve, where I is an open interval in $\mathbb{R}$. Let $a, b \in I$. Then $P_{ab}^\gamma : T_{\gamma(a)}M \to T_{\gamma(b)}M$ is a linear isomorphism.*

Proof. Let $w, x \in T_{\gamma(a)}M$, and denote by W, X, respectively, their parallel transports along γ: That is, $W, X \in T_0^1\gamma$, $\nabla_{V_\gamma} W = 0$, $W(a) = w$, and $\nabla_{V_\gamma} X = 0$, $X(a) = x$. Let $c \in \mathbb{R}$. Then by the $\mathbb{R}$-linearity of ∇_{V_γ}, we have

$$\nabla_{V_\gamma}(W + cX) = \nabla_{V_\gamma} W + c\nabla_{V_\gamma} X = 0 + 0 = 0,$$

and by the definition of addition in $T_0^1\gamma$,

$$(W + cX)(a) = W(a) + cX(a) = w + cx.$$

Hence, $W + cX$ is the parallel transport of $w + cx$ along γ. In particular,

$$P_{ab}^\gamma(w + cx) = (W + cX)(b) = W(b) + cX(b) = P_{ab}^\gamma w + cP_{ab}^\gamma x.$$

Thus, P_{ab}^γ is a linear map. Next, we show that $P_{ab}^\gamma : T_{\gamma(a)}M \to T_{\gamma(b)}M$ is invertible. If M is m-dimensional, then since $\dim T_{\gamma(a)}M = m = \dim T_{\gamma(b)}M$, it is enough to show injectivity. Let $w \in T_{\gamma(a)}M$, and W denote the corresponding parallel transport of w along γ. Then $P_{ab}^\gamma(w) = W(b)$. But now suppose we want the parallel transport of the vector $x := W(b) \in T_{\gamma(b)}M$ along the curve γ. Clearly, W already satisfies $\nabla_{V_\gamma} W = 0$ and $W(b) = x$. By the uniqueness established in Theorem 7.2, we conclude that the parallel transport of x along γ is W. In particular, $P_{ba}^\gamma x = W(a) = w$. Hence

$$(P_{ba}^\gamma \circ P_{ab}^\gamma) w = P_{ba}^\gamma (P_{ab}^\gamma w) = P_{ba}^\gamma x = w = \mathrm{id}_{T_{\gamma(a)}M} w.$$

As $w \in T_{\gamma(a)}M$ was arbitrary, we have $P_{ba}^\gamma \circ P_{ab}^\gamma = \mathrm{id}_{T_{\gamma(a)}M}$. Thus P_{ab}^γ is injective. So $P_{ab}^\gamma : T_{\gamma(a)}M \to T_{\gamma(b)}M$ is a linear isomorphism. $\square$

Theorem 7.3. *Let M be a semi-Riemannian manifold with metric* $\mathbf{g}$ *and the Levi-Civita connection ∇. Suppose that $\gamma : I \to M$ is a smooth curve, where I is an open interval in $\mathbb{R}$. Let $a, b \in I$. Then the parallel transport map $P_{ab}^\gamma : T_{\gamma(a)}M \to T_{\gamma(b)}M$ is an isometry, that is, it is an invertible linear map preserving the scalar product.*

Proof. We have already seen that P^γ_{ab} is linear and invertible. It remains to show that it preserves the scalar product. Let $w, x \in T_{\gamma(a)}M$, and denote by $W, X \in T^1_0\gamma$, respectively, the corresponding parallel transports along γ. Corollary 7.1 (p.143) implies that

$$\begin{aligned}\mathbf{g}(\gamma(b))(P^\gamma_{ab}w, P^\gamma_{ab}x) &= \mathbf{g}(\gamma(b))(W(b), X(b)) = \mathbf{g}(\gamma(a))(W(a), X(a))\\ &= \mathbf{g}(\gamma(a))(w, x).\end{aligned}$$

Hence $P^\gamma_{ab} : T_{\gamma(a)}M \to T_{\gamma(b)}M$ is an isometry. □

Remark 7.2. In this chapter, we have seen that once we have a connection ∇ on a smooth manifold, then it induces a notion of a covariant derivative ∇_{V_γ} along each smooth curve γ, and also a notion of parallel transport along each smooth curve γ. One can also show[3] that the parallel transport maps determine the covariant differentiation operators along curves via

$$(\nabla_{V_\gamma} W)(t) = \lim_{\tau\to t} \frac{P^\gamma_{\tau t}(W(\tau)) - W(t)}{\tau - t}, \quad W \in T^1_0\gamma, \qquad (\star)$$

and also the manifold's connection by

$$(\nabla_V W)_p = \lim_{h\to 0} \frac{P^\gamma_{h0}(W_{\gamma(h)}) - W_p}{h}, \quad p \in M,\ V, W \in T^1_0 M,$$

where γ is any smooth curve such that $\gamma(0) = p$ and $v_{\gamma,p} = V_p$. We sketch an argument for $(\star)$. Let $(U, \mathbf{x})$ be an admissible chart containing $\gamma(t)$. Write $W(\tau) = W^i(\tau)\partial_{x^i,\gamma(\tau)}$ for all $\tau \in I_U$, where I_U is an open interval containing t such that $\gamma(I_U) \subset U$. Then $P^t_{\tau t}(W(\tau)) = Y^k(t,\tau)\partial_{x^k,\gamma(t)}$, where $Y^k(\cdot,\cdot) : I_U \times I_U \to \mathbb{R}$, $1 \leqslant k \leqslant m$, satisfy for each $\tau \in I_U$,

$$\begin{aligned}\frac{\partial Y^k}{\partial s}(s,\tau) + \Gamma^k_{ij}(\gamma(s))\dot\gamma^j(s)Y^i(s,\tau) &= 0 \qquad (1 \leqslant k \leqslant m,\ s \in I_U),\\ Y^k(\tau,\tau) &= W^k(\tau) \quad (1 \leqslant k \leqslant m).\end{aligned}$$

For a fixed t, taking variable $\tau > t$, and applying the mean value theorem to $Y^k(\cdot,\tau)$ on the closed interval $[t,\tau]$, gives a $\theta^k_\tau \in (t,\tau)$ such that

$$\begin{aligned}Y^k(t,\tau) &= Y^k(\tau,\tau) + (t-\tau)\frac{\partial Y^k}{\partial s}(\theta^k_\tau,\tau)\\ &= W^k(\tau) + (\tau - t)\Gamma^k_{ij}(\gamma(\theta^k_\tau))\dot\gamma^j(\theta^k_\tau)Y^i(\theta^k_\tau,\tau).\end{aligned}$$

Thus

$$\begin{aligned}\lim_{\tau\to t}\frac{Y^k(t,\tau) - W^k(t)}{\tau - t} &= \lim_{\tau\to t}\Big(\frac{W^k(\tau) - W^k(t)}{\tau - t} + \Gamma^k_{ij}(\gamma(\theta^k_\tau))\dot\gamma^j(\theta^k_\tau)Y^i(\theta^k_\tau,\tau)\Big)\\ &= \dot W^k(t) + \Gamma^k_{ij}(\gamma(t))\dot\gamma^j(t)Y^i(t,t)\\ &= \dot W^k(t) + \Gamma^k_{ij}(\gamma(t))\dot\gamma^j(t)W^i(t),\end{aligned}$$

which coincides with $((\nabla_{V_\gamma} W)(t))(x^k)$. ∗

[3]See e.g. [Lee (2018), Theorem 4.34, Corollary 4.35].

Example 7.4. (Parallel transport under the flat connection on $\mathbb{R}^m$.) Consider $\mathbb{R}^m$ with the standard smooth structure and the flat connection $\overline{\nabla}$. We use the global admissible chart $(\mathbb{R}^m, \mathrm{id}_{\mathbb{R}^m})$. Then for any $p \in M$, we have a linear isomorphism

$$\mathbb{R}^m \ni \mathbf{w} = (w^1, \cdots, w^m) \overset{\iota}{\mapsto} w^i \partial_{x^i,p} \in T_p\mathbb{R}^m.$$

Let $\gamma : I \to \mathbb{R}^m$ be any smooth curve, where I is an open interval in $\mathbb{R}$ that passes through two points $p, q \in \mathbb{R}^m$. Let $a, b \in I$ be such that $\gamma(a) = p$ and $\gamma(b) = q$.

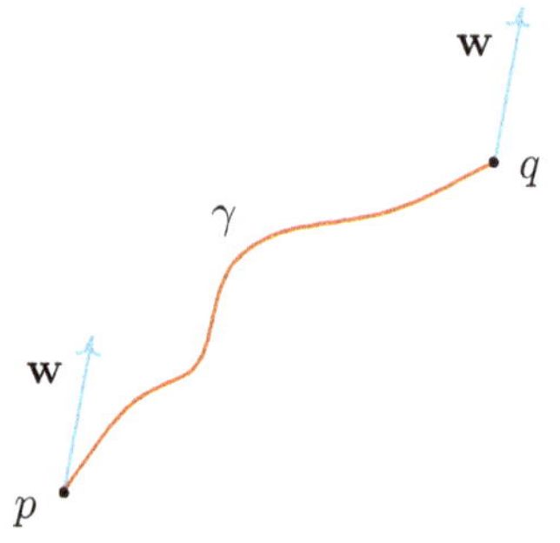

Claim: $P^\gamma_{ab}(\iota\mathbf{w}) = \iota\mathbf{w}$ for all $\mathbf{w} \in \mathbb{R}^m$, i.e.,

$$P^\gamma_{ab}(w^i\partial_{x^i,p}) = w^i\partial_{x^i,q} \text{ for all } w = w^i\partial_{x^i,p} \in T_p\mathbb{R}^m.$$

To see this, let W be the parallel transport of w along γ. Decompose $W(t) = W^i(t)\,\partial_{x^i,\gamma(t)}$, $t \in I$. Then for all $k \in \{1, \cdots, m\}$,

$$\dot{W}^k(t) + \Gamma^k_{ij}(\gamma(t))\,\dot{\gamma}^j(t)\,W^i(t) = 0, \quad t \in I.$$

But in the chart $(\mathbb{R}^m, \mathrm{id}_{\mathbb{R}^m})$, all the Γ-symbols are identically zero, and so $\dot{W}^k \equiv 0$, giving $W^k \equiv c^k$, a constant. As $W^k(a) = w^k$, where $w = w^k\partial_{x^k,p}$, we obtain $c^k = w^k$, $1 \leqslant k \leqslant m$. Hence

$$W(t) = W^k(t)\partial_{x^k,\gamma(t)} = w^k\partial_{x^k,\gamma(t)}, \quad t \in I.$$

In particular, $P^\gamma_{ab}w = W(b) = w^k\partial_{x^k,q}$. ◇

In the example above, the parallel transport that maps vectors in T_pM to vectors in T_qM was independent of the path γ connecting the two points $p, q \in M$. However, this was because the space was 'flat' (and in the calculation above, in fact the terms containing $\dot{\gamma}^j$ vanished in our chosen chart as the Γ-symbols happened to be zeroes). The parallel transport between two tangent spaces at $p, q \in M$ in general *does* depend on the curve used. In fact, it is a manifestation of the 'curvature' of the manifold. The following is an illustration of this in the case of the sphere.

Example 7.5. (Sphere.) Consider the sphere S^2 as a Riemannian manifold with the metric induced from the Euclidean inner product on $\mathbb{R}^3$, and the corresponding Levi-Civita connection. In Example 7.3 (p.141), we had considered the smooth curve μ, given by $\mu(t) = (\cos t, \sin t, 0)$, $t \in (0, 2\pi)$. In the admissible chart (U, φ) considered there, the chart map gives the usual spherical polar coordinates (θ^U, ϕ^U). In this chart (U, φ), let the vector field X along μ be given by $X(t) = \partial_{\theta^U, \mu(t)}$, $t \in (0, 2\pi)$. Then we had seen that X is parallel along μ (the equator in the chart U). Let $a := \frac{\pi}{2}$ and $b := 3\frac{\pi}{2}$, so that $p := \mu(a) = (0, 1, 0)$, and $q := \mu(b) = (0, -1, 0)$. Then with $x = \partial_{\theta^U, p} = X(a)$, we have $P^{\mu}_{ab} x = X(b) = \partial_{\theta^U, q}$.

Now consider another admissible chart $(\widetilde{U}, \widetilde{\varphi})$, where we use new spherical coordinates $(\theta^{\widetilde{U}}, \phi^{\widetilde{U}})$, measuring the polar $\theta^{\widetilde{U}}$ angle of a point from the positive x-axis (instead of the usual positive z-axis), and measuring the azimuthal angle $\phi^{\widetilde{U}}$ from the negative z-axis (instead of the usual positive x-axis). Consider the curve $\widetilde{\mu}$ given by $\widetilde{\mu}(t) = (0, \sin t, -\cos t)$, $t \in (0, 2\pi)$, and take $Y(t) = -\partial_{\phi^{\widetilde{U}}, \widetilde{\mu}(t)}$, $t \in (0, 2\pi)$.

By Exercise 7.3 (or by a direct calculation), Y is parallel along $\widetilde{\mu}$. With $a = \frac{\pi}{2}$ and $b = 3\frac{\pi}{2}$, we have $\widetilde{\mu}(a) = (0, 1, 0) = p$, and $\widetilde{\mu}(b) = (0, -1, 0) = q$, and

$$Y(a) = -\partial_{\phi^{\widetilde{U}}, p} \overset{(*)}{=} \partial_{\theta^U, p} = x.$$

That $(*)$ holds is convincing based on the following picture, but a justification is given at the end of this example.

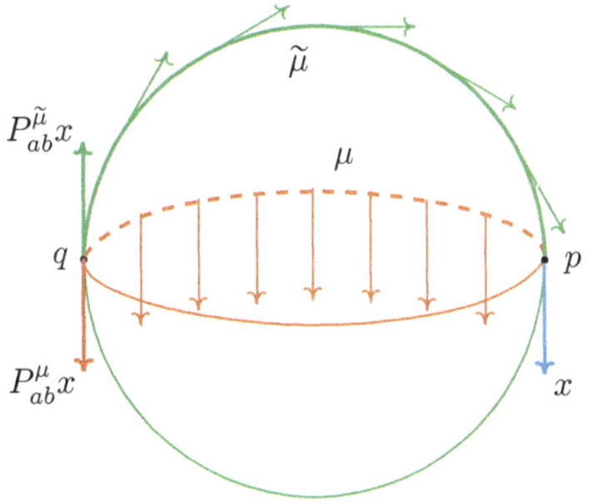

Referring to the picture above,

$$P^{\widetilde{\mu}}_{ab} x = Y(b) = -\partial_{\phi^{\widetilde{U}}, q} \overset{(\star)}{=} -\partial_{\theta^U, q} = -P^{\mu}_{ab} x.$$

Thus the result of parallel transporting x from p to q is different along the two curves μ and $\widetilde{\mu}$.

Justification of $(*)$ and $(\star)$:
Equation $(\star\star)$ on page 21 implies that the map

$$T_pS^2 \ni \mathrm{v} \mapsto \mathbf{v} = (\mathrm{v}x, \mathrm{v}y, \mathrm{v}z) \in \mathbb{R}^3$$

is injective. Thus it suffices to show

$$(\partial_{\theta^U,p}x, \partial_{\theta^U,p}y, \partial_{\theta^U,p}z) = -(\partial_{\phi^{\widetilde{U}},p}x, \partial_{\phi^{\widetilde{U}},p}y, \partial_{\phi^{\widetilde{U}},p}z)$$

in order to prove $\partial_{\theta^U,p} = -\partial_{\phi^V,p}$. We have

$$\begin{bmatrix} \partial_{\theta^U,p}x \\ \partial_{\theta^U,p}y \\ \partial_{\theta^U,p}z \end{bmatrix} = \begin{bmatrix} \dfrac{\partial((\sin\theta)\cos\phi)}{\partial\theta} \\ \dfrac{\partial((\sin\theta)\sin\phi)}{\partial\theta} \\ \dfrac{\partial(\cos\theta)}{\partial\theta} \end{bmatrix}_{\substack{\theta=\frac{\pi}{2} \\ \phi=t=\frac{\pi}{2}}} = \begin{bmatrix} 0 \\ 0 \\ -1 \end{bmatrix}.$$

Also,

$$\begin{bmatrix} \partial_{\phi^{\widetilde{U}},p}x \\ \partial_{\phi^{\widetilde{U}},p}y \\ \partial_{\phi^{\widetilde{U}},p}z \end{bmatrix} = \begin{bmatrix} \dfrac{\partial(\cos\theta)}{\partial\phi} \\ \dfrac{\partial((\sin\theta)\sin\phi)}{\partial\phi} \\ \dfrac{\partial(-(\sin\theta)\cos\phi)}{\partial\phi} \end{bmatrix}_{\substack{\theta=\frac{\pi}{2} \\ \phi=t=\frac{\pi}{2}}} = \begin{bmatrix} 0 \\ 0 \\ 1 \end{bmatrix}.$$

Thus $-\partial_{\phi^{\widetilde{U}},p} = \partial_{\theta^U,p}$, that is, $(*)$ holds. Similarly,

$$\begin{bmatrix} \partial_{\theta^U,q}x \\ \partial_{\theta^U,q}y \\ \partial_{\theta^U,q}z \end{bmatrix} = \begin{bmatrix} \dfrac{\partial((\sin\theta)\cos\phi)}{\partial\theta} \\ \dfrac{\partial((\sin\theta)\sin\phi)}{\partial\theta} \\ \dfrac{\partial(\cos\theta)}{\partial\theta} \end{bmatrix}_{\substack{\theta=\frac{\pi}{2} \\ \phi=\frac{3\pi}{2}}} = \begin{bmatrix} 0 \\ 0 \\ -1 \end{bmatrix} = \begin{bmatrix} \dfrac{\partial(\cos\theta)}{\partial\phi} \\ \dfrac{\partial((\sin\theta)\sin\phi)}{\partial\phi} \\ \dfrac{\partial(-(\sin\theta)\cos\phi)}{\partial\phi} \end{bmatrix}_{\substack{\theta=\frac{\pi}{2} \\ \phi=\frac{3\pi}{2}}} = \begin{bmatrix} \partial_{\phi^{\widetilde{U}},q}x \\ \partial_{\phi^{\widetilde{U}},q}y \\ \partial_{\phi^{\widetilde{U}},q}z \end{bmatrix},$$

justifying $\partial_{\phi^{\widetilde{U}},q} = \partial_{\theta^U,q}$, that is, $(\star)$ holds. ◇

Chapter 8

Geodesics

In the previous chapter, we learnt that a connection ∇ on a smooth manifold M allows us to differentiate a vector field W along a curve γ, using the induced operator ∇_{V_γ}, where V_γ is the velocity vector field along the curve. Thus, we can now talk about the acceleration along a curve by taking $W = V_\gamma$ itself, so that $\nabla_{V_\gamma} V_\gamma$ measures how fast V_γ changes as we move in the direction of V_γ, which exactly coincides with our intuitive notion of acceleration. If the acceleration everywhere along the curve is zero, then we call the curve a 'geodesic'. In the spacetime context, geodesics are the worldlines of 'freely falling observers'.

Geometrically speaking, geodesics are the 'straightest possible curves' in a smooth manifold M with a connection ∇, and are the generalisation of straight lines from the flat Euclidean plane case. However, we will also see that we should not think of geodesics $\gamma : I \to M$ merely in terms of the image $\gamma(I)$ of points it describes, but the parametrisation is crucial, and a reparametrisation of a geodesic may no longer be a geodesic. Roughly speaking, only affine linear maps of the arc-length parametrisation are admissible for maintaining the geodesic nature of a curve.

8.1 Geodesic equation

Definition 8.1. (Geodesic.)
Let M be a smooth manifold with a connection ∇. Let $I \subset \mathbb{R}$ be an open interval in $\mathbb{R}$. A smooth curve $\gamma : I \to M$ is called a *geodesic* if

$$\mathfrak{a}(t) := (\nabla_{V_\gamma} V_\gamma)(t) = 0 \text{ for all } t \in I.$$

We call $\mathfrak{a}(t) \in T_{\gamma(t)}M$ the *instantaneous acceleration at the point* $\gamma(t) \in M$. We refer to $\nabla_{V_\gamma} V_\gamma = 0$ as the *geodesic equation.*

Trivially, any constant curve, $\gamma \equiv p \in M$, satisfies the geodesic equation, and so is a geodesic. In a chart, the geodesic equation $\nabla_{V_\gamma} V_\gamma = 0$ is a *second* order ordinary differential equation for the curve coordinate functions.

Proposition 8.1. *Let M be an m-dimensional smooth manifold with a connection ∇. Let $\gamma : I \to M$ be a smooth curve, where $I \subset \mathbb{R}$ is an open interval. Then γ is a geodesic if and only if for each admissible chart $(U, \mathbf{x})$ such that $\gamma(I) \cap U \neq \varnothing$, we have for all $t \in I_U := \gamma^{-1}U$, and all $1 \leqslant k \leqslant m$,*

$$\ddot{\gamma}^k(t) + \Gamma_{ij}^k(\gamma(t))\,\dot{\gamma}^i(t)\,\dot{\gamma}^j(t) = 0.$$

Here $\gamma^i = x^i \circ \gamma$, $\dot{\gamma}^i = \dfrac{d\gamma^i}{dt}$, $\ddot{\gamma}^i = \dfrac{d\dot{\gamma}^i}{dt}$, and Γ_{ij}^k are the Christoffel symbols.

Proof. This follows from Theorem 7.1 (p.138), as $V_\gamma(t) = \gamma^i(t)\,\partial_{x^i,\gamma(t)}$. □

The following example shows that when we think of a geodesic $\gamma : I \to M$, we should not only think of the range $\gamma(I)$, as the parameterisation matters.

Example 8.1. Consider $\mathbb{R}^2$ with the standard smooth structure and the flat connection $\overline{\nabla}$. In the chart $(\mathbb{R}^2, \mathrm{id})$, all the connection coefficients vanish. If $I \subset \mathbb{R}$ is an open interval, then a curve $I \ni t \mapsto \gamma(t) = (x(t), y(t))$ is a geodesic if and only if we have $\ddot{x}(t) = 0 = \ddot{y}(t)$ for all $t \in I$, that is, $(x(t), y(t)) = (x_0, y_0) + t\mathbf{v}$ for some $(x_0, y_0), \mathbf{v} \in \mathbb{R}^2$. So γ is a straight line if $\mathbf{v} \neq (0, 0)$, and otherwise it is just the constant geodesic, given by $\gamma(t) = (x_0, y_0)$, $t \in I$.

However, not all curves whose image lies along a straight line will be a geodesic. As an example, consider $\mu : \mathbb{R} \to \mathbb{R}^2$, $\mu(t) = (t^3, t^3)$ for all $t \in \mathbb{R}$. Then $V_\mu(t) = 3t^2\,\partial_{x,\mu(t)} + 3t^2\,\partial_{y,\mu(t)}$, and so we have $\dot{\gamma}^x(t) = 3t^2 = \dot{\gamma}^y(t)$, giving $\ddot{\gamma}^x(t) = 6t = \ddot{\gamma}^y(t)$. Thus the geodesic equation is not satisfied, as $\ddot{\gamma}^k(t) + \Gamma_{ij}^k(\gamma(t))\,\dot{\gamma}^i(t)\,\dot{\gamma}^j(t) = 6t \neq 0$ for $t \neq 0$. So μ is not a geodesic. Hence, geodesics 'lie along' straight lines, but not all curves lying along a straight line are geodesics. ◇

Exercise 8.1. (Sphere.) Consider the sphere S^2 as a Riemannian manifold with metric $\mathbf{g}$ induced from the Euclidean inner product on $\mathbb{R}^3$. Let S^2 be given the Levi-Civita connection obtained from $\mathbf{g}$. Show that the geodesic equations in the chart (U, φ) given in Example 7.3 (p.141) are:

$$\begin{aligned} \ddot{\theta} - (\sin\theta)(\cos\theta)\dot{\phi}^2 &= 0, \\ \ddot{\phi} + 2(\cot\theta)\,\dot{\theta}\,\dot{\phi} &= 0. \end{aligned}$$

Suppose the geodesic is traversed at unit speed, i.e., $\mathbf{g}(\gamma(t))(v_{\gamma,\gamma(t)}, v_{\gamma,\gamma(t)}) = 1$ for all $t \in I$ (see Proposition 8.2, where it will be shown that the squared length of the tangent vector is constant along a geodesic). Show this gives $(\sin\theta)^2\dot{\phi}^2 + \dot{\theta}^2 = 1$. Writing the ϕ-component of the geodesic equation as

$$\frac{d}{dt}\log\dot{\phi} = \frac{\ddot{\phi}}{\dot{\phi}} = -2(\cot\theta)\,\dot{\theta} = -2\frac{d}{dt}\log(\sin\theta),$$

prove that $\dot{\phi} = \dfrac{c}{\sin\theta}$ for a constant c.

Using the unit speed relation, show that $\dot{\theta}=\sqrt{1-\frac{c^2}{(\sin\theta)^2}}$, if $\dot{\theta}>0$.

Dividing the above expressions for $\dot{\phi}$ and $\dot{\theta}$, prove that $\dfrac{d\phi}{d\theta}=\dfrac{c(\operatorname{cosec}\theta)^2}{\sqrt{1-c^2-c^2(\cot\theta)^2}}$.

Substitute $t=\frac{c}{\sqrt{1-c^2}}\cot\theta$, to show that for a constant α, $\sin(\alpha-\phi)=\frac{c}{\sqrt{1-c^2}}\cot\theta$.

Using the trigonometric angle-addition formula for $\sin(\alpha-\phi)$, show that this equation can be rearranged to read $\langle\gamma(t),\mathbf{n}\rangle_{\mathbb{R}^3}=0$ for a suitable fixed nonzero vector $\mathbf{n}\in\mathbb{R}^3$. Thus, $\gamma(t)$ lies in the intersection with S^2 of a plane passing through the origin with normal $\mathbf{n}$, that is, γ is a 'great circle' on the sphere.

Proposition 8.2. *Let M be a semi-Riemannian manifold with a metric $\mathbf{g}$ and the Levi-Civita connection ∇. Let $I\subset\mathbb{R}$ be an open interval and $\gamma: I\to M$ be a geodesic. Then there exists a constant $c\in\mathbb{R}$ such that for all $t\in I$, $\mathbf{g}(\gamma(t))(v_{\gamma,\gamma(t)},v_{\gamma,\gamma(t)})=c$.*

Proof. This is an immediate consequence of Proposition 7.2 (p.142), since

$$\begin{aligned}\frac{d(\mathbf{g}(\gamma(\cdot))(v_{\gamma,\gamma(\cdot)},v_{\gamma,\gamma(\cdot)}))}{dt}&=\frac{d(\mathbf{g}(\gamma(\cdot))(V_\gamma(\cdot),V_\gamma(\cdot)))}{dt}\\&=\mathbf{g}(\gamma(\cdot))((\nabla_{V_\gamma}V_\gamma,V_\gamma)+\mathbf{g}(\gamma(\cdot))(V_\gamma,\nabla_{V_\gamma}V_\gamma)\\&=\mathbf{g}(\gamma(\cdot))(0,V_\gamma)+\mathbf{g}(\gamma(\cdot))(V_\gamma,0)=0.\end{aligned}$$

□

Exercise 8.2. Let $(M,\mathbf{g})$ be a semi-Riemannian manifold, and let $X\in T_0^1M$ be a Killing vector field. Suppose that the curve $\gamma: I\to M$ is a geodesic, where $I\subset\mathbb{R}$ is an open interval, and $v_{\gamma,\gamma(t)}\neq 0$ for all $t\in I$. Show that

$$\frac{d\mathbf{g}(\gamma(\cdot))(v_{\gamma,\gamma(\cdot)},X_{\gamma(\cdot)})}{dt}(t)=0\quad(t\in I).$$

Exercise 8.3. (Isometries preserve geodesy.)
Let $(M,\mathbf{g})$, $(N,\mathbf{h})$ be semi-Riemannian manifolds, and $f: M\to N$ be an isometry. Show that if $\gamma: I\to M$ is a geodesic in M, then $f\circ\gamma: I\to N$ is a geodesic in N.
Hint: Use Exercise 6.15 (p.119).

Definition 8.2. (Spacelike, timelike, lightlike/null curves.)
Let $(M,\mathbf{g})$ be a Lorentzian manifold with the metric $\mathbf{g}$.
A smooth curve $\gamma: I\to M$, where I is an open interval in $\mathbb{R}$, is called

- *timelike* if for all $t\in I$, $v_{\gamma,\gamma(t)}$ is timelike.
- *null/lightlike* if for all $t\in I$, $v_{\gamma,\gamma(t)}$ is lightlike.
- *spacelike* if for all $t\in I$, $v_{\gamma,\gamma(t)}$ is spacelike.

An arbitrary smooth curve in a Lorentzian manifold may not necessarily have one of these 'causal characters'. However, a geodesic with respect to the Levi-Civita connection necessarily does. In particular, a timelike geodesic cannot change its causal character.

Exercise 8.4. Show that each geodesic in a Lorentzian manifold with the Levi-Civita connection, possesses a causal character, i.e., it is either timelike, spacelike or lightlike. *Hint:* Proposition 8.2.

Exercise 8.5. Let $(M, \mathbf{g})$ be the Schwarzschild spacetime (Example 5.10, p.87), $\mathbf{p} \in S^2$, $r \in (2m, \infty)$. Consider a stationary timelike observer $\gamma : \mathbb{R} \to M$ given by $\gamma(t) = (t, r, \mathbf{p})$, $t \in \mathbb{R}$. Show that γ is not a geodesic (that is, stationary observers are not freely falling).

8.2 Existence and uniqueness

Now we'll show that given any point $p \in M$ and a direction $v \in T_pM$, there exists a geodesic $\gamma : I \to M$ defined on some open interval $I \subset \mathbb{R}$ containing 0, such that $\gamma(0) = p$ and $v_{\gamma,\gamma(0)} = v$, called the 'geodesic passing through p in the direction of v'. To show this, we will use the following result, which follows from a well-known existence and uniqueness result from the theory of ordinary differential equations; see e.g. [Apostol (1969), Thm. 7.19, p.229].

Proposition 8.3. *Let $V \subset \mathbb{R}^m$ be an open set, $F_{ij}^k \in C^\infty(V)$, $1 \leqslant i, j, k \leqslant m$, $P \in V$, $\mathbf{v} \in \mathbb{R}^m$. Then there exists an open interval $I \subset \mathbb{R}$ containing 0, and a map $I \ni t \mapsto Y(t) = (Y^1(t), \cdots, Y^m(t)) \in V$, which is a solution to the following initial value problem:*

$$\left.\begin{aligned} \ddot{Y}^k(t) + F_{ij}^k(Y(t))\dot{Y}^i(t)\dot{Y}^j(t) &= 0 \quad (1 \leqslant k \leqslant m,\ t \in I) \\ Y(0) &= P \\ \dot{Y}(0) &= \mathbf{v}. \end{aligned}\right\} \tag{8.1}$$

Moreover, any two solutions $Y : I \to V$ and $\widetilde{Y} : \widetilde{I} \to V$ coincide on $I \cap \widetilde{I}$.

Proof. For all $1 \leqslant k \leqslant m$, let us define $A^k(t) = Y^k(t)$ and $B^k(t) = \dot{Y}^k(t)$. Set $A = (A^1, \cdots, A^m)$ and $B = (B^1, \cdots, B^m)$. Then Y satisfies the given (second order) initial value problem if and only if (A, B) is a solution to the following (first order!) initial value problem:

$$\left.\begin{aligned} \dot{A}^k &= B^k & (1 \leqslant k \leqslant m) \\ \dot{B}^k &= -F_{ij}^k(A)\,B^i\,B^j & (1 \leqslant k \leqslant m) \\ A(0) &= P & \\ B(0) &= \mathbf{v}. & \end{aligned}\right\} \tag{8.2}$$

Let $r > 0$ be such that the closed ball $\overline{B(P, r)}$ with center P and radius r is contained in V. Then the function

$$\begin{aligned} K := \overline{B(P, r)} \times \overline{B(\mathbf{v}, r)} &\to \mathbb{R}^{2m} \\ (\underbrace{a^1, \cdots, a^m}_{=:\boldsymbol{a}}, b^1, \cdots, b^m) &\mapsto (b^1, \cdots, b^m, -F_{ij}^1(\boldsymbol{a})\,b^i\,b^j, \cdots, -F_{ij}^m(\boldsymbol{a})\,b^i\,b^j) \end{aligned}$$

is a Lipschitz function (since its first order partial derivatives are all bounded on the compact set K). So by [Apostol (1969), Thm. 7.19, p.229], there exists a C^1 solution $(A, B) : I \to \mathbb{R}^{2m}$ on some interval I containing 0, satisfying the initial value problem (8.2). But as the functions F^k_{ij} are all smooth, the right-hand sides of the differential equations in (8.2) are C^1. So $(\dot{A}, \dot{B})$ is C^1, i.e., (A, B) is C^2. Again as the F^k_{ij} are all smooth, the right-hand sides of the differential equations in (8.2) are C^2, implying that $(\dot{A}, \dot{B})$ is C^2, i.e., (A, B) is C^3. Continuing in this manner, $(A, B) \in C^\infty(I)$. So there exists a $C^\infty(I)$ solution $Y := A$ to (8.1).

By the local uniqueness of solutions [Apostol (1969), Thm. 7.19, p.229], $Y, \widetilde{Y}$ coincide in a small interval $(-\epsilon, \epsilon)$ around 0. Set

$$T_{\max} = \sup\{\tau \in I \cap \widetilde{I} : Y(t) = \widetilde{Y}(t) \text{ for all } t \leqslant \tau\}.$$

We want $T_{\max} = \sup(I \cap \widetilde{I})$. Let $T_{\max} < \sup(I \cap \widetilde{I})$. As $Y, \widetilde{Y}$ are C^1,

$$Y(T_{\max}) = \widetilde{Y}(T_{\max}) =: P_{\max} \quad \text{and} \quad \dot{Y}(T_{\max}) = \dot{\widetilde{Y}}(T_{\max}) =: \mathbf{v}_{\max}.$$

Considering a new initial value problem with the initial conditions $Y(T_{\max}) = P_{\max}$ and $\dot{Y}(T_{\max}) = \mathbf{v}_{\max}$, we conclude that Y and $\widetilde{Y}$ must coincide in a neighbourhood of $T_{\max}$, contradicting the choice of $T_{\max}$. Hence $Y(t) = \widetilde{Y}(t)$ for all nonnegative $t \in I \cap \widetilde{I}$. A similar argument shows the agreement also for nonpositive values of $t \in I \cap \widetilde{I}$. So $Y(t) = \widetilde{Y}(t)$ for all $t \in I \cap \widetilde{I}$. □

Remark 8.1. (Smooth dependence on **v**.)
We remark that with the same notation as in the theorem above, it can be shown that there exists a neighbourhood N of $\mathbf{v}$, an $\epsilon > 0$, and a C^∞ map $\mathcal{Y} : N \times (-\epsilon, \epsilon) \to V$, such that for each $\mathbf{w} \in N$, $Y(t) := \mathcal{Y}(\mathbf{w}, t)$, $t \in (-\epsilon, \epsilon)$ solves the initial value problem

$$\begin{cases} \ddot{Y}^k(t) + \Gamma^k_{ij}(Y(t))\dot{Y}^i(t)\dot{Y}^j(t) = 0 \quad (1 \leqslant k \leqslant m) \\ Y(0) = P \\ \dot{Y}(0) = \mathbf{w}. \end{cases} \tag{8.3}$$

This is not contained in [Apostol (1969), Thm. 7.19, p.229] mentioned above. Instead, we refer the reader to [Hartman (2002), Thm. 4.1, p.100]. We will need this result in order to obtain the smoothness of the 'exponential map', to be studied later in this chapter. *

Theorem 8.1. *Let M be a smooth manifold with a connection ∇.*
For any $p \in M$ and any $v \in T_pM$, there exists an open interval $I \subset \mathbb{R}$ containing 0, and a geodesic $\gamma : I \to M$ such that $\gamma(0) = p$ and $v_{\gamma,p} = v$.
Moreover, any two such geodesics $\gamma : I \to M$ and $\widetilde{\gamma} : \widetilde{I} \to M$ coincide on $I \cap \widetilde{I}$.

Proof. Take any admissible chart $(U, \mathbf{x})$ containing p. Set $V = \mathbf{x}(U)$, $P = \mathbf{x}(p)$ and $\mathbf{v} = (v^1, \cdots, v^m)$, where $v = v^i \partial_{x^i,p}$. By Proposition 8.3, and taking $F_{ij}^k = \Gamma_{ij}^k \circ \mathbf{x}^{-1}$, there exists an interval I containing 0, and a $Y : I \to V$, having smooth components, satisfying (8.1). Define the smooth curve $\gamma : I \to M$ by $\gamma(t) = (\mathbf{x}^{-1} \circ Y)(t)$ for $t \in I$. Then

$$\gamma^i = x^i \circ \gamma = x^i \circ (\mathbf{x}^{-1} \circ Y) = (\mathbf{x} \circ \mathbf{x}^{-1} \circ Y)^i = Y^i.$$

So the differential equation in (8.1) implies that γ satisfies the geodesic equation. Moreover, we have that $\gamma(0) = \mathbf{x}^{-1}(Y(0)) = \mathbf{x}^{-1}(P) = p$, and $v_{\gamma,p} = \dot{\gamma}^i(0)\, \partial_{x^i,\gamma(0)} = \dot{Y}^i(0)\, \partial_{x^i,\gamma(0)} = v^i \partial_{x^i,p} = v$.

The uniqueness part is immediate from Proposition 8.3, as long as one is in a single chart. The argument in the general case is easily adapted from the uniqueness part of the proof of Proposition 8.3. First, it follows from Proposition 8.3 that in a small interval $(-\epsilon, \epsilon)$ around 0, the curves $\gamma, \widetilde{\gamma}$ do coincide. Set $T_{\max} = \sup\{\tau \in I \cap \widetilde{I} : \gamma(t) = \widetilde{\gamma}(t) \text{ for all } t \leqslant \tau\}$. Suppose that $T_{\max} < \sup(I \cap \widetilde{I})$. By continuity, $\gamma(T_{\max}) = \widetilde{\gamma}(T_{\max}) =: p_{\max}$. Moreover, $v_{\gamma,p_{\max}} = v_{\widetilde{\gamma},p_{\max}} =: v_{p_{\max}}$ (since their components can be seen to be the same by considering the limit from below $T_{\max}$). By the uniqueness in a single chart (containing the point $p_{\max}$), it follows that a geodesic passing through $p_{\max}$ with velocity $v_{\max}$ is unique, and so γ and $\widetilde{\gamma}$ must coincide in a neighbourhood of $T_{\max}$, contradicting the choice of $T_{\max}$. Hence $\gamma(t) = \widetilde{\gamma}(t)$ for all nonnegative $t \in I \cap \widetilde{I}$. A similar argument shows the agreement also for nonpositive $t \in I \cap \widetilde{I}$. Consequently, $\gamma(t) = \widetilde{\gamma}(t)$ for all $t \in I \cap \widetilde{I}$. □

Definition 8.3. (Maximal geodesic, geodesically complete manifold.)
Let M be a smooth manifold with a connection ∇.
A geodesic $\gamma : I \to M$, where $I \subset \mathbb{R}$ is an open interval, is called *maximal* if it cannot be extended to a geodesic on a larger open interval, that is, there does not exist a geodesic $\widetilde{\gamma} : \widetilde{I} \to M$, where $\widetilde{I} \subset \mathbb{R}$ is an open interval such that $I \subsetneq \widetilde{I}$ and $\widetilde{\gamma}|_I = \gamma$. We call M *geodesically complete* if every maximal geodesic is defined on the entire real line.

Corollary 8.1. *Let M be a smooth manifold with a connection ∇. For any $p \in M$ and any $v \in T_pM$, there exists a unique maximal geodesic $\gamma : I \to M$, where I is an open interval containing* 0, *such that $\gamma(0) = p$ and $v_{\gamma,p} = v$.*

Proof. By Theorem 8.1, we know that there exists a geodesic $\mu : J \to M$, where J is an open interval containing 0, such that $\mu(0) = p$ and $v_{\mu,p} = v$. Let I be the union of all such intervals J. Then J is open and contains 0. Moreover, since any two geodesics $\mu, \widetilde{\mu}$ defined on two such intervals $J, \widetilde{J}$

coincide on their overlap $J \cap \widetilde{J}$, we can unambiguously define $\gamma : I \to M$ by $\gamma(t) := \mu(t)$, where $\mu : J \to M$ is any such geodesic for which $t \in J$. We claim that γ is itself a geodesic. If $t \in I$, then there must exist a J which contains t, and then $\gamma|_J = \mu$ is a geodesic, and so it satisfies the geodesic equation. The maximality follows from the definition of I. □

Example 8.2. (Cylindrical spacetime.) Consider the cylindrical spacetime M in Example 5.8 (p.86), and the charts (U, φ), (V, ψ) given there. Suppose $I \subset \mathbb{R}$ is an open interval, and $\gamma : I \to U$ is a geodesic. The components θ, z are given by $(\theta(t), z(t)) := \varphi(\gamma(t)) \in (0, 2\pi) \times \mathbb{R}$. Then we have that $\gamma(t) = \big(\cos(\theta(t)), \sin(\theta(t)), z(t)\big)$, $t \in I$. From Example 6.6 (p.121), we know that the connection coefficients in the chart (U, φ) are all 0. So the geodesic equation in U is given, component-wise, by $\ddot{\theta} = 0$ and $\ddot{z} = 0$. Hence $\theta(t) = \alpha t + \theta_0$ and $z(t) = \beta t + z_0$ for some constants α, β, θ_0 and z_0. So the geodesic in U is given by $\gamma(t) = \big(\cos(\alpha t + \theta_0), \sin(\alpha t + \theta_0), \beta t + z_0\big)$, $t \in I$. These describe:

1° $\alpha = 0$ and $\beta = 0$: a point
2° $\alpha = 0$ and $\beta \neq 0$: a vertical line parallel to the z-axis
3° $\alpha \neq 0$ and $\beta = 0$: a circular arc
4° $\alpha \neq 0$ and $\beta \neq 0$: a helix.

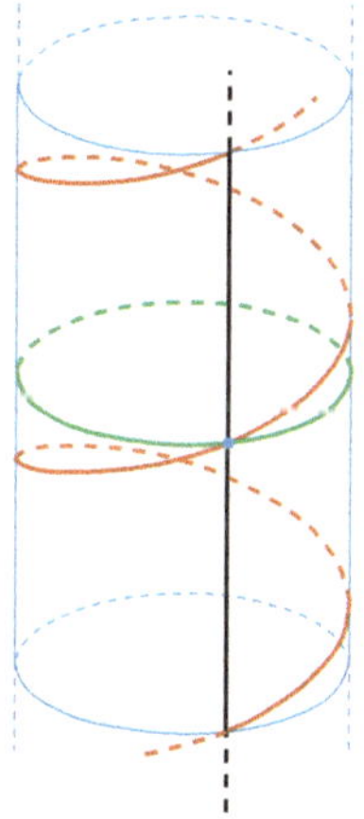

But now take $I = \mathbb{R}$, and $\gamma(t)$ defined by the same expression as above. Then $\gamma(t) \in M$, and γ also satisfies the geodesic equation in V (unless $\alpha = 0$, when it does not enter V at all, and $\gamma(\mathbb{R}) \subset U$). So γ is a maximal geodesic defined on $\mathbb{R}$. Similarly, we can show that every geodesic with image in V can be extended to a maximal geodesic defined on $\mathbb{R}$. Consequently, M is geodesically complete. ◇

Remark 8.2. (Parallel postulate.) A version of the *parallel postulate* of Euclid states that given a straight line γ in the flat plane and a point p outside γ, there exists a unique straight line passing through p which does not intersect γ. In Example 8.1 for $M = \mathbb{R}^2$ with the flat connection $\overline{\nabla}$, if $\gamma : \mathbb{R} \to \mathbb{R}^2$ is a nonconstant geodesic, then $\gamma(t) = \mathbf{a} + t\mathbf{v}$ for some $\mathbf{a} \in \mathbb{R}^2$ and $\mathbf{v} \in \mathbb{R}^2 \backslash \{(0,0)\}$. Then for a point $\mathbf{p}$ not belonging to the image of γ, define the geodesic $\widetilde{\gamma} : \mathbb{R} \to \mathbb{R}^2$ by $\widetilde{\gamma}(t) = \mathbf{p} + t\mathbf{v}$, $t \in \mathbb{R}$. Now if there exist a $t_1, t_2 \in \mathbb{R}$ such that $\widetilde{\gamma}(t_1) = \gamma(t_2)$, then we get $\mathbf{p} = \mathbf{a} + (t_2 - t_1)\mathbf{v}$, a contradiction to the fact that $\mathbf{p}$ does not belong to the image of γ. To show uniqueness of $\widetilde{\gamma}$, suppose that $\mu : \mathbb{R} \to \mathbb{R}^2$ is a nonconstant geodesic passing through $\mathbf{p}$ at $t = 0$. Then $\mu(t) = \mathbf{p} + t\mathbf{w}$ for some $\mathbf{w} \in \mathbb{R}^2 \backslash \{(0,0)\}$. If $\mathbf{w} \notin \operatorname{span} \mathbf{v}$, then $\{\mathbf{v}, \mathbf{w}\}$ form a basis for $\mathbb{R}^2$, and so there exist $t_1, t_2 \in \mathbb{R}$ such that $\mathbf{p} - \mathbf{a} = t_1 \mathbf{w} + t_2 \mathbf{v}$, i.e., $\mu(-t_1) = \gamma(t_2)$, and so μ and γ intersect. So the parallel postulate holds for the 'flat' plane $(\mathbb{R}^2, \overline{\nabla})$.

In Exercise 8.1, we considered the sphere S^2 as a Riemannian manifold with the metric $\mathbf{g}$ induced from the Euclidean inner product on $\mathbb{R}^3$, and showed that great circles are geodesics. Thus the parallel postulate fails for this geometry of the sphere. Indeed, if we take a point $p \in S^2$ outside a great circle γ, then any nonconstant geodesic $\widetilde{\gamma}$ passing through p will again be a great circle, and hence will intersect γ at exactly two diametrically opposite points. Here, the parallel postulate fails because there is *no* nonconstant geodesic passing through p which does not intersect γ.

On the other hand, in Exercise 8.6 below, we consider a Riemannian manifold, called the *Poincaré half-plane*, where the parallel postulate fails because there are *infinitely many* geodesics $\widetilde{\gamma}$ which pass through a point p outside a geodesic γ, such that the images of $\gamma, \widetilde{\gamma}$ do not intersect. $\ast$

Exercise 8.6. (Poincaré half-plane.) Recall from Exercise 5.24 (p.93) the smooth manifold $M = \{(x,y) \in \mathbb{R}^2 : y > 0\}$ with the smooth structure $[\{(M, \mathrm{id})\}]$, and the Riemannian metric $\mathbf{g}$ given in the global chart (M, id) by

$$\mathbf{g}(\mathbf{p})(v^x \partial_{x,\mathbf{p}} + v^y \partial_{y,\mathbf{p}}, w^x \partial_{x,\mathbf{p}} + w^y \partial_{y,\mathbf{p}}) = \frac{v^x w^x + v^y w^y}{y^2} \text{ for all } \mathbf{p} = (x,y) \in M.$$

It can be shown that the nonzero connection coefficients in the chart (M, id) are given by $\Gamma^y_{yy} = \Gamma^x_{xy} = \Gamma^x_{yx} = -\frac{1}{y}$ and $\Gamma^y_{xx} = \frac{1}{y}$. Define $\lambda, \sigma : \mathbb{R} \to M$ by $\lambda(t) := (0, e^t)$ and $\sigma(t) := (\tanh t, 1/\cosh t)$, $t \in \mathbb{R}$. Prove that λ, μ are (maximal) geodesics. The image of λ is the positive y-axis, while the image of σ is the semicircle in the upper half plane, with centre at $(0,0)$ and radius equal to 1. Similar calculations also show that for $a \in \mathbb{R}$ and $r > 0$, $\lambda_a(t) := (a, e^t)$ and $\sigma_{a,r}(t) := (r \tanh t - a, r \cosh t)$, $t \in \mathbb{R}$, describe maximal geodesics λ_a, $\sigma_{a,r}$. Draw a picture showing that if p is a point outside λ, then there are infinitely many geodesics that pass through p which do not intersect λ.

Exercise 8.7. Let M be a smooth manifold with a connection ∇.
Let $p \in M$, and $\gamma : I \to M$, where I is an open interval in $\mathbb{R}$ containing 0, be a geodesic such that $\gamma(0) = p$ and $v_{\gamma,p} = 0$. Prove that $\gamma(t) = p$ for all $t \in I$.
Hint: Consider $\widetilde{\gamma}(t) := p$ for all $t \in \mathbb{R}$, and show that it is a maximal geodesic passing through p with velocity 0.

Exercise 8.8. Consider $M = \mathbb{R}^2$ with the standard smooth structure, and a connection ∇ described via its connection coefficients in the global chart $(\mathbb{R}^2, \mathrm{id})$ (see Exercise 6.3, p.108), given by $\Gamma^x_{xy} = \Gamma^x_{yx} = 1$, and all other Γ-symbols identically 0. Determine the geodesics. Given any two points p, q in the plane, is there a geodesic that starts at p and passes through q? Plot, using the computer, geodesics starting at $p = (0,0)$ in the direction $v = (\cos\theta)\partial_{x,p} + (\sin\theta)\partial_{y,p}$, for various values of θ, say $\theta = k \cdot \frac{2\pi}{24}$, for $k = 0, \cdots, 23$.

Exercise 8.9. Consider Example 6.3 (p.116). In the polar coordinate chart $(V, \mathbf{y})$ given there, show that the chart-induced components r, θ of a geodesic satisfy $\ddot{r} - r\dot{\theta}^2 = 0$ and $\ddot{\theta} + \frac{2}{r}\dot{r}\dot{\theta} = 0$. Let $\cos^{-1} : (-1,1) \to (0,\pi)$ be the inverse cosine function, and $y_0 > 0$. Show that $r(t) := \sqrt{y_0^2 + t^2}$ and $\theta(t) := \cos^{-1}(t/\sqrt{y_0^2+t^2})$, $t \in \mathbb{R}$, satisfy the equations above. Define $\gamma : \mathbb{R} \to V$ by $\gamma(t) = \mathbf{y}^{-1}(r(t), \theta(t))$, $t \in \mathbb{R}$. Then γ is maximal. Show that $\gamma(t) = (t, y_0)$, $t \in \mathbb{R}$, and so the image of γ is the straight line parallel to the x-axis, passing through the point $(0, y_0) \in V$.

8.3 Affine linear reparametrisation

We had earlier remarked that a geodesic should not merely be thought of geometrically as the set of points in the range of γ, since its parametrisation does play a role in its being a geodesic. In other words, a reparametrisation may render a geodesic a non-geodesic. What are the allowed parametrisations? We will show that the parametrisation is unique upto an affine linear map, that is, affine linear maps are allowed and these are the only ones.

Lemma 8.1. *Let M be a smooth manifold with a connection ∇.*
Let I, J be open subsets of $\mathbb{R}$, and $h : J \to I$ be a C^∞ function.
Let $\gamma : I \to M$ be a smooth curve. Then for all $s \in J$,

$$(\nabla_{V_{\gamma\circ h}} V_{\gamma\circ h})(s) = \ddot{h}(s)\, V_\gamma(h(s)) + (\dot{h}(s))^2 (\nabla_{V_\gamma} V_\gamma)(h(s)).$$

Proof. From Exercise 2.4 (p.23), $V_{\gamma\circ h}(s) = \dot{h}(s)\, V_\gamma(h(s))$. Also, it follows from Exercise 7.1 (p.139) that $\big(\nabla_{V_{\gamma\circ h}} V_\gamma(h(\cdot))\big)(s) = \dot{h}(s)\ (\nabla_{V_\gamma} V_\gamma)(h(s))$. Consequently, using the facts above, and the Leibniz rule for the covariant derivative along the curve $\gamma \circ h$, we obtain, for all $s \in J$

$$\begin{aligned}(\nabla_{V_{\gamma\circ h}} V_{\gamma\circ h})(s) &= \big(\nabla_{V_{\gamma\circ h}} (\dot{h}\, V_\gamma(h(\cdot)))\big)(s)\\ &= \tfrac{d(\dot{h})}{du}(s)\, V_\gamma(h(s)) + \dot{h}(s)\big(\nabla_{V_{\gamma\circ h}} V_\gamma(h(\cdot))\big)(s)\\ &= \ddot{h}(s)\, V_\gamma(h(s)) + \dot{h}(s)\,\dot{h}(s)\,(\nabla_{V_\gamma} V_\gamma)(h(s))\\ &= \ddot{h}(s)\, V_\gamma(h(s)) + (\dot{h}(s))^2 (\nabla_{V_\gamma} V_\gamma)(h(s)).\end{aligned}$$

□

Exercise 8.10. Let M be a smooth manifold with a connection and $\gamma:(a,b)\to M$ be a geodesic. Show that if $v_{\gamma,\gamma(c)}=0$ for some $c\in(a,b)$, then γ is constant. (In the contrapositive form, the above says that if γ is a not the constant geodesic, then its velocity vector field is nowhere zero.)
Hint: Let $J=(a-c,b-c)$ and $h(s)=s+c$, $s\in J$. Show that $\gamma\circ h$ is constant using Exercise 8.7 (p.159).

Let $\gamma: I\to M$ be a geodesic in a smooth manifold with a connection ∇, where I is an open interval of $\mathbb{R}$. The 'variable' $t\in I$, parametrising the range $\{\gamma(t): t\in I\}$ of γ, is loosely called the 'affine parameter' of γ. The justification for the name 'affine' comes from the following result, combined with Example 8.3 below.

Theorem 8.2. *Let M be a smooth manifold with a connection ∇. Let $\gamma: I\to M$ be a nonconstant geodesic, where I is an open interval in $\mathbb{R}$. Let J be an open interval of $\mathbb{R}$, and $h: J\to I$ be a C^∞ function. Then $\gamma\circ h: J\to M$ is a geodesic if and only if there exist constants $a,b\in\mathbb{R}$ such that $h(s)=as+b$ for all $s\in J$.*

Proof. We first note that for all $s\in J$,

$$\begin{aligned}(\nabla_{V_{\gamma\circ h}}V_{\gamma\circ h})(s) &= \ddot{h}(s)\,V_\gamma(h(s))+(\dot{h}(s))^2(\nabla_{V_\gamma}V_\gamma)(h(s))\\ &= \ddot{h}(s)\,V_\gamma(h(s))+(\dot{h}(s))^2\,0=\ddot{h}(s)\,V_\gamma(h(s)),\end{aligned}$$

since $(\nabla_{V_\gamma}V_\gamma)(h(s))=0$ by virtue of γ being a geodesic.

If $\gamma\circ h$ is a geodesic, then $0=(\nabla_{V_{\gamma\circ h}}V_{\gamma\circ h})(s)=\ddot{h}(s)\,V_\gamma(h(s))$ for all $s\in J$. But as γ is nonconstant, $V_\gamma(h(s))\neq 0$ for all $s\in J$ (Exercise 8.10). So $\ddot{h}\equiv 0$ in J, and thus there exist constants $a,b\in\mathbb{R}$ such that $h(s)=as+b$ for all $s\in J$.

Vice versa, if $h(s)=as+b$, then $\ddot{h}\equiv 0$ in J, and so we have that $(\nabla_{V_{\gamma\circ h}}V_{\gamma\circ h})(s)=\ddot{h}(s)V_\gamma(h(s))=0$ for all $s\in J$. Thus $\gamma\circ h$ is a geodesic. □

Example 8.3. (Arclength parametrisation.)
Let $(M,\mathbf{g})$ be a Lorentzian manifold, and let $\gamma: I\to M$ be a geodesic, where $I\subset\mathbb{R}$ is an open interval. Suppose that γ is timelike or spacelike (but not lightlike). Let $t_0\in I$. Define the *arclength parameter along* γ by

$$\tau(t)=\int_{t_0}^{t}\sqrt{|\mathbf{g}(\gamma(s))(v_{\gamma,\gamma(s)},v_{\gamma,\gamma(s)})|}\,ds,\quad t\in I.$$

If $(M,\mathbf{g})$ has a time-orientation, and if γ is the worldline of an observer (a future-pointing timelike smooth curve), then the arclength parameter measures the proper time elapsed from the event at $\gamma(t_0)$.

As $I\ni s\mapsto \mathbf{g}(\gamma(s))(v_{\gamma,\gamma(s)},v_{\gamma,\gamma(s)})$ is constant along γ (Proposition 8.2), say taking value $c\in\mathbb{R}\backslash\{0\}$, we have $\tau(t)=\sqrt{|c|}\,(t-t_0)$ for $t\in I$. We remark that $c \neq 0$, as γ is assumed to be spacelike or timelike. Let $a := 1/\sqrt{|c|}$, $b := t_0$, and $h(\tau) := a\tau + b$ for all $\tau \in J := \{\sqrt{|c|}(t - t_0) : t \in I\}$. Since h is affine linear, $\gamma \circ h$ is also a geodesic, which we refer to as the geodesic γ 'parametrised by arclength'. As $v_{\gamma\circ h,\gamma\circ h(\tau)} = \frac{dh}{d\tau}(\tau)\, v_{\gamma,\gamma(h(\tau))} = a\, v_{\gamma,\gamma(h(\tau))}$, the tangent vector of $\gamma \circ h$ has 'unit length' everywhere:

$$\begin{aligned} &|\mathbf{g}((\gamma \circ h)(\tau))(v_{\gamma\circ h,\gamma\circ h(\tau)}, v_{\gamma\circ h,\gamma\circ h(\tau)})| \\ &= a^2|\mathbf{g}(\gamma(h(\tau)))(v_{\gamma,\gamma(h(\tau))}, v_{\gamma,\gamma(h(\tau))})| = a^2|c| = 1. \end{aligned}$$

For lightlike geodesics $\gamma : I \to M$, we have that $\tau(\cdot) \equiv 0$, that is, the 'proper time' experienced by a lightlike particle (photon) between any two events along its worldline is zero. Thus to label each point along the lightlike geodesic, the arclength parameter is to no avail. On the other hand, by Theorem 8.2, we know that any affine linear transformation of the interval I gives an alternative parameterisation. ◇

Example 8.4. (Minkowski spacetime, and Newton's first law in inertial frames.) Recall the Minkowski spacetime $(M, \mathbf{g})$ from Example 5.6 (p.83), and let ∇ be the Levi-Civita connection induced by $\mathbf{g}$ on M. Consider an inertial frame $(M, \mathbf{x})$ induced by the choice of a point $p \in M$ and an orthonormal basis $\{\mathbf{f}_0, \mathbf{f}_1, \mathbf{f}_2, \mathbf{f}_3\}$ for V with respect to g (the Minkowski scalar product on V), where we suppose that $\mathbf{f}_0$ is timelike. We will show that the timelike geodesics are straight lines parametrised affine-linearly in the 'time-coordinate' (of the inertial frame) along the geodesic. Also, the motion of a force-free particle (that is, a geodesic) is recorded in the inertial frame as 'uniform-in-time motion, along a straight-line-in-space', which is *Newton's first law of motion.* Suppose $\gamma : I \to M$ is a geodesic, where I is an open interval in $\mathbb{R}$ containing 0. Write $\gamma(\tau) = p + \gamma^i(\tau)\,\mathbf{f}_i$. Because the connection coefficients in the inertial frame are identically zero (Example 6.5), the geodesic equation gives $\ddot{\gamma}^i(\tau) = 0$, and so $\gamma^i(\tau) = a^i + b^i\tau$, $i = 0, 1, 2, 3$. Then $-(b^0)^2 + (b^1)^2 + (b^2)^2 + (b^3)^2 < 0$ since γ is timelike. In particular, $b^0 \neq 0$. The inertial frame 'time-coordinate' of a point $q := \gamma(\tau)$ ($\tau \in I$) on the geodesic is $t(q) := \gamma^0(\tau) = a^0 + b^0\tau$. The affine parameter of the geodesic at $\gamma(\tau)$ is then given in terms of $t(q)$ by $\tau = (t(q) - a^0)/b^0$, which is affine-linear in the time-coordinate $t(q)$ recorded by the inertial frame for q. The 'spatial position' of q recorded in the inertial frame is

$$\begin{aligned} (\gamma^1(\tau), \gamma^2(\tau), \gamma^3(\tau)) &= (a^1, a^2, a^3) + \tau(b^1, b^2, b^3) \\ &= (a^1, a^2, a^3) + \frac{(t(q) - a^0)}{b^0}(b^1, b^2, b^3), \end{aligned}$$

which is affine-linear in $t(q)$ (i.e., in the inertial frame, 'the motion is along a straight line in space uniformly in time'). ◇

Exercise 8.11. We use the same notation as in Example 8.4 (p.161).
Let $v > 0$ and $\alpha \in (-\frac{\pi}{2}, \frac{\pi}{2}]$. Define the curve γ by $\gamma(t) = p + \gamma^i(t)\mathbf{f}_i$, $t \geqslant 0$, where

$$\gamma^0(t) = t, \quad \gamma^1(t) = v(\cos\alpha)t, \quad \gamma^2(t) = v(\sin\alpha)t - \frac{g}{2}t^2, \quad \gamma^3(t) = 0.$$

Determine the acceleration $\nabla_{V_\gamma} V_\gamma$ along γ. Is γ a geodesic?

Exercise 8.12. (Circular timelike geodesic in Schwarzschild spacetime and GPS.)
We consider the spacetime around the Earth as given by Schwarzschild spacetime (Example 5.10, p.87), where $m = M_\oplus \approx 0.45\,\text{cm}$ is the mass of Earth. Let (U, φ) be the spherical polar coordinate chart, given in Example 1.8 (p.7). Consider a timelike geodesic $\gamma : I \to M$ in the equatorial plane ($\theta = \frac{\pi}{2}$) having a constant r-coordinate r, and parameterised by proper time: $\gamma(\tau) = \big(t(\tau), r, \varphi^{-1}(\frac{\pi}{2}, \phi(\tau))\big)$ for all $\tau \in I$. With $\cdot' = \frac{d}{d\tau}$, show that $t'' = 0$, $\phi'' = 0$, $(\phi')^2 = \frac{m}{r^3}t'^2$, $(1 - \frac{3m}{r})t'^2 = 1$. Let T_{orbit} be the circular orbit's period as measured by the observer γ, and T_{Earth} be the period as measured by an observer at rest on the surface of the Earth. The radius of Earth is $R_\oplus \approx 6400\,\text{km}$. We ignore the rotational motion of the Earth.
Show that $\dfrac{T_{\text{orbit}}}{T_{\text{Earth}}} \approx 1 - \dfrac{3m}{2r} + \dfrac{m}{R_\oplus}$.
Hint: Assume the light signal from the events at the start and end of the circular orbit reaches the observer on Earth radially, and revisit Example 5.16 (p.96).
For Global Positioning System (GPS) satellites, $r \approx 27000\,\text{km}$, giving

$$\frac{T_{\text{orbit}} - T_{\text{Earth}}}{T_{\text{Earth}}} \approx 4.5 \times 10^{-10},$$

and this is taken into account in the design of the GPS.

Exercise 8.13. (Conformally related metrics and lightlike geodesics.)
Let M be a smooth manifold with metrics $\mathbf{g}, \widetilde{\mathbf{g}}$ such that $\widetilde{\mathbf{g}} = f\widetilde{\mathbf{g}}$ for some pointwise positive function $f \in C^\infty(M)$. (Such metrics are said to be *conformally related.*) Let $\nabla, \widetilde{\nabla}$ denote Levi-Civita connections corresponding to $\mathbf{g}, \widetilde{\mathbf{g}}$, respectively. See Exercise 6.17, (p.119). Let $p \in M$, and $v \in T_pM$ be a lightlike vector. Let $I \subset \mathbb{R}$ be an open interval containing 0, and $\gamma : I \to M$ be a lightlike geodesic with respect to ∇ such that $\gamma(0) = p$ and $v_{\gamma,p} = v$. Show there exists an open interval $J \subset \mathbb{R}$ containing 0 and a smooth map $h : J \to I$ such that $\gamma \circ h : J \to M$ is lightlike, $(\gamma \circ h)(0) = p$, $v_{\gamma\circ h,p} = v$, and $\gamma \circ h$ is a geodesic with respect to $\widetilde{\nabla}$. Loosely speaking, a lightlike geodesic in $(M, \mathbf{g})$ can be reparametrised to become a lightlike geodesic in $(M, \widetilde{\mathbf{g}})$.

Remark 8.3. A Penrose diagram is a 2-dimensional picture representing spacetime, with two dimensions suppressed, in which lightlike geodesics have slopes ± 1. Such a diagram is useful to analyse global causality relations. For any spherically symmetric spacetime, ignoring the angular coordinates, one obtains a 2-dimensional Lorentzian manifold. Any 2-dimensional Lorentzian manifold is conformally flat (see Exercise 9.12 on p.188 and Remark 9.1), and so it can be conformally embedded into 2-dimensional Minkowski spacetime as a bounded open set (the Penrose diagram). This includes Minkowski spacetime itself, whose Penrose diagram is a triangle, as explained below. For the spherically symmetric

Minkowski spacetime M (identified with $\mathbb{R}^4$ after the choice of $p \in M$ and an orthonormal basis for V), the spherical angular coordinates θ and ϕ are suppressed, and the 'null coordinates' $u = t + r$ and $v = t - r$ are introduced, for which $\partial_{u,p}$ and $\partial_{v,p}$ are lightlike for all $p \in M$. We compactify the null coordinates by setting $U = \tan^{-1} u$ and $V = \tan^{-1} v$. With $T = U + V$ and $X = U - V$, the metric acquires the form $f(-dT \otimes dT + dX \otimes dX + (\sin X)^2 (d\theta \otimes d\theta + (\sin\theta)^2 d\phi \otimes d\phi))$, where $f = f(T, X) > 0$ pointwise. By the previous exercise, the radial lightlike geodesics run along 45° lines in the picture in the XT-plane. Minkowski spacetime looks like a triangle, as shown in the following picture.

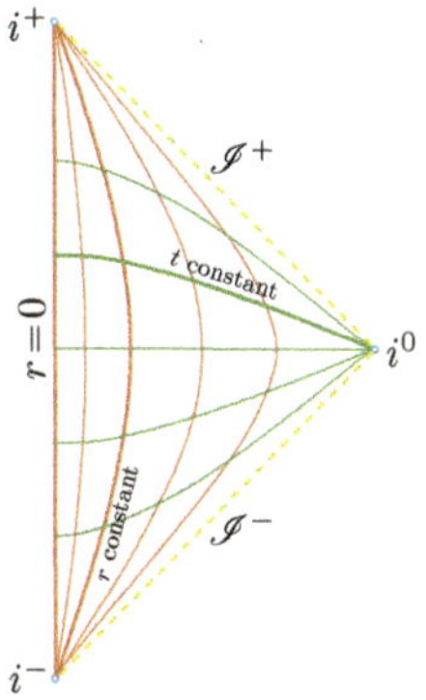

The vertical side on the left represents all points in M with the radial coordinate $r = 0$ at all moments of time. The lower vertex i^- is the infinitely remote past of all spatial points, and the upper vertex i^+ is the infinitely remote future of these points. The vertex i^0 on the right is spacelike spatial infinity at all finite moments of time. The slanted sides of the triangle are 'lightlike infinities' of the past (below, $\mathscr{I}^-$) and of the future (above, $\mathscr{I}^+$): $\mathscr{I}^-$ represents an infinitely remote 'region' of the spacetime from which light can go through all finite points of spacetime, and similarly $\mathscr{I}^+$ represents an infinitely remote region of the spacetime to which light can go from all finite events. The boundaries $\mathscr{I}^-$, $\mathscr{I}^+$ and the points i^+, i^-, i^0 do not belong to the spacetime. For a detailed discussion of Penrose diagrams for spacetimes, see e.g., [Natário (2021), Ch. 2], or [d'Inverno (2022), Ch. 18]. $\ast$

8.4 Lightlike geodesics in Schwarzschild spacetime

We now give the calculation behind one of the first observational verifications of the spacetime viewpoint of gravity. In the Newtonian viewpoint, the path $\mathbb{R} \ni t \mapsto (t, \mathbf{x}(t)) \in \mathbb{R} \times \mathbb{R}^3$ of light is a straight line, or 'light travels in a straight line'. However, a light ray, from our spacetime viewpoint, is just a lightlike/null geodesic. The region of spacetime in the vicinity of a star is described reasonably well by the Schwarzschild spacetime[1]. We will now consider the path of a light ray, and show that its path is 'bent' by the star. One can first photograph a certain celestial region, recording the light

[1]Or we imagine that the universe contains only the star and nothing else.

rays emanating from distant fixed stars, measured at night, and when the Sun is far away from the path of the light rays reaching the observer (so that the bending effect on light due to the Sun's gravity can be neglected). See the following picture.

One photographs the same celestial zone during a solar eclipse[2] and a different time of the year:

A comparison the two photographs should reveal a change in the apparent position of the stars due to the bending of the light ray path by the Sun's gravity. This angular deflection can be calculated by using the geodesic equation, and the predicted value was actually measured during the 1919 solar eclipse.

Proposition 8.4. (Schwarzschild lightlike geodesics.)
Let $(M = \mathbb{R}\times(2m,\infty)\times S^2, \mathbf{g})$ *be the Schwarzschild spacetime with the Levi-Civita connection. Let* (U,φ) *be the chart for* S^2 *from* Example 5.7 (p.85). *Let* $(\mathbf{U},\Phi)$ *be the chart for* M *given by* $\mathbf{U} = \mathbb{R}\times(2m,\infty)\times U$, *and the chart map* $\Phi = \mathrm{id}_{\mathbb{R}}\times \mathrm{id}_{(2m,\infty)}\times\varphi$. *Let* $\gamma : I \to \mathbf{U}$ *be a lightlike geodesic, where* I *is an open interval in* $\mathbb{R}$, *and with coordinates defined by*

$$(t(\tau), r(\tau), \theta(\tau), \phi(\tau)) = \Phi(\gamma(\tau)) \quad (\tau \in I).$$

Then with $\dot{} := \frac{d}{d\tau}$,

$$\ddot{t} = -\frac{2m}{r^2}\Big(1-\frac{2m}{r}\Big)^{-1}\dot{t}\,\dot{r}$$

$$\ddot{r} = -\frac{m}{r^2}\Big(1-\frac{2m}{r}\Big)\dot{t}^2 + \frac{m}{r^2}\Big(1-\frac{2m}{r}\Big)^{-1}\dot{r}^2 + r\Big(1-\frac{2m}{r}\Big)\dot{\theta}^2 + r\Big(1-\frac{2m}{r}\Big)(\sin\theta)^2\dot{\phi}^2$$

$$\ddot{\theta} = (\sin\theta)(\cos\theta)\dot{\phi}^2 - \frac{2}{r}\dot{r}\,\dot{\theta}$$

$$\ddot{\phi} = -\frac{2}{r}\dot{r}\,\dot{\phi} - 2(\cot\theta)\,\dot{\theta}\,\dot{\phi}$$

together with the constraint

$$\Big(1-\frac{2m}{r}\Big)\dot{t}^2 - \Big(1-\frac{2m}{r}\Big)^{-1}\dot{r}^2 - r^2\dot{\theta}^2 - r^2(\sin\theta)^2\dot{\phi}^2 = 0.$$

---

[2]The solar eclipse is relevant here only because during a solar eclipse, the stars are visible, and can be photographed.

Proof. The first four equations follow immediately from the geodesic equation expressed in the chart $(\mathbf{U}, \Phi)$ by using the connection coefficients listed in Example 6.8 (p.121). The constraint equation is the consequence of the causal character of γ being lightlike, that is, for all $\tau \in I$, $\mathbf{g}(\gamma(\tau))(v_{\gamma,\gamma(\tau)}, v_{\gamma,\gamma(\tau)}) = 0$, expressed in coordinates, and using the component matrix for the metric given in Example 5.10 (p.87). □

If a light ray satisfies, for a $\tau_0 \in I$, $\theta(\tau_0) = \frac{\pi}{2}$ and $\dot{\theta}(\tau_0) = 0$, then by looking at the differential equation involving $\ddot{\theta}$, and the uniqueness of solutions, we conclude that $\theta \equiv \frac{\pi}{2}$, that is, the geodesic then stays in the equatorial plane. Thus for such a light ray, setting $\theta \equiv \frac{\pi}{2}$ and $\dot{\theta} \equiv 0$ in the equations for $\ddot{\phi}$ and $\ddot{r}$, we obtain

$$\ddot{\phi} = -\frac{2}{r}\dot{r}\dot{\phi} \tag{8.4}$$

$$\ddot{r} = -\frac{m}{r^2}\Big(\big(1-\frac{2m}{r^2}\big)\dot{t}^2 - \big(1-\frac{2m}{r}\big)^{-1}\dot{r}^2\Big) + r\big(1-\frac{2m}{r}\big)\dot{\phi}^2. \tag{8.5}$$

With $\theta \equiv \frac{\pi}{2}$ and $\dot{\theta} \equiv 0$, the constraint equation becomes

$$\big(1-\frac{2m}{r}\big)\dot{t}^2 - \big(1-\frac{2m}{r}\big)^{-1}\dot{r}^2 - r^2\dot{\phi}^2 = 0.$$

We use this in equation (8.5) for $\ddot{r}$, to obtain

$$\ddot{r} = (r - 3m)\dot{\phi}^2. \tag{8.6}$$

Suppose that $\tau \mapsto \phi(\tau)$ is strictly monotone, so that ϕ is invertible, with the inverse map $h = \phi^{-1}$. Define thc function u by $\phi \mapsto u(\phi) := \frac{1}{(r \circ h)(\phi)}$. Using the chain rule and the inverse function theorem,

$$\frac{du}{d\phi}(\cdot) = -\frac{\dot{r}(h(\cdot))}{(r(h(\cdot)))^2}\frac{1}{\dot{\phi}(h(\cdot))}.$$

Also (with argument $h(\cdot)$ suppressed everywhere on the right-hand side),

$$\begin{aligned}
\frac{d^2u}{d\phi^2}(\cdot) &= -\frac{\ddot{r}\frac{1}{\dot{\phi}}r^2\dot{\phi} - \big(2r\dot{r}\frac{1}{\dot{\phi}}\dot{\phi} + r^2\ddot{\phi}\frac{1}{\dot{\phi}}\big)\dot{r}}{r^4\dot{\phi}^2} = \frac{\big(2r\dot{r} + r^2\frac{\ddot{\phi}}{\dot{\phi}}\big)\dot{r} - r^2\ddot{r}}{r^4\dot{\phi}^2} \\
&= \frac{\big(2r\dot{r} + r^2\big(-\frac{2}{r}\dot{r}\big)\big)\dot{r} - r^2\ddot{r}}{r^4\dot{\phi}^2} \qquad \text{(using (8.4))} \\
&= -\frac{\ddot{r}}{r^2\dot{\phi}^2} = -\frac{r-3m}{r^2} \qquad \text{(using (8.6))} \\
&= -\frac{1}{r} + 3m\frac{1}{r^2}.
\end{aligned}$$

Thus we obtain the differential equation

$$\frac{d^2u}{d\phi^2}(\phi) = -u(\phi) + 3m(u(\phi))^2. \tag{8.7}$$

We now determine much the light ray bends due to the gravity of the Sun, for which $m = \mathrm{M}_\odot = 1.5\mathrm{km}$ (the mass of the Sun in units where $c = G_\mathrm{N} = 1$), and we consider the situation described in the following picture.

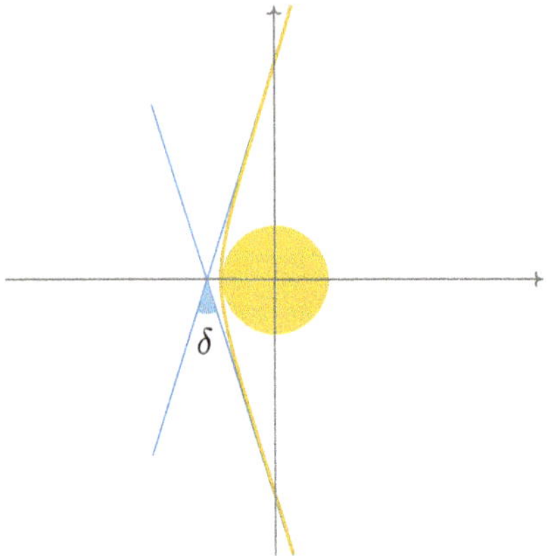

Here, a light ray starts in the equatorial plane, from a fixed distant star. So, we assume that the light ray comes in from $r = +\infty$ and leaves with $r = +\infty$ eventually. We assume that the point of closest approach of the light ray to the centre of the Sun is exactly equal to the radius $\mathrm{R}_\odot$ of the Sun, so that the ray just grazes its surface. We choose this point of closest approach as $(-\mathrm{R}_\odot, 0)$, lying in the equatorial xy-plane, as shown above. In order to compute the deflection angle, we need to compute the angles with the x-axis of the two asymptotes, that is, the azimuthal angle ϕ at $r = +\infty$ when coming in, and the angle ϕ at $r = +\infty$ when going out. In both cases, $r = +\infty$, which corresponds to $u = 0$. Moreover, $u(\pi) = \frac{1}{\mathrm{R}_\odot}$ and $\frac{du}{d\phi}(\pi) = 0$ (r-minimisation/u-maximisation condition).

Equation (8.7), namely $u'' + u = 3mu^2$ (where $\cdot' = \frac{d}{d\phi}$), is a nonlinear differential equation. Note that

$$u = \frac{1}{r} \leqslant \frac{1}{\mathrm{R}_\odot},$$

where $\mathrm{R}_\odot \approx 7 \times 10^5\,\mathrm{km}$. As $3m = 3\mathrm{M}_\odot \approx 4.5\,\mathrm{km}$, we have $\frac{3mu^2}{u} = 3mu \ll 1$. In light of this, we use a perturbative method to solve the differential equation approximately. With $w := \mathrm{R}_\odot u$, $w'' + w = \varepsilon w^2$, where $\varepsilon := \frac{3m}{\mathrm{R}_\odot} \ll 1$. Write $w = w_0 + \varepsilon w_1 + \varepsilon^2 w_2 + \cdots$, where w_0 is the solution to

$$w_0'' + w_0 = 0, \text{ with } w_0(\pi) = 1 \text{ and } w_0'(\pi) = 0.$$

The general solution is $w_0(\phi) = A\cos\phi + B\sin\phi$. Using the above 'initial' conditions, we get $w_0(\phi) = -\cos\phi$ (and so $u_0(\phi) := \frac{w_0}{\mathrm{R}_\odot} = -\frac{\cos\phi}{\mathrm{R}_\odot}$, corresponding to the vertical straight line, parallel to the y-axis, lying to its left at a distance of $R_\odot$). Substituting $w = w_0 + \varepsilon w_1 + \varepsilon^2 w_2 + \cdots$ in $w'' + w = \varepsilon w^2$, and comparing terms in ε,

$$w_1'' + w_1 = w_0^2 = (\cos\phi)^2 = \frac{1+\cos(2\phi)}{2}.$$

The initial conditions for w_1 are $w_1(\pi)=0$ and $w_1'(\pi)=0$. To find a solution, we first try to find a particular solution assuming $w_{1,p} = a + b\cos(2\phi)$, so that $w_{1,p}' = -2b\sin(2\phi)$, and $w_{1,p}'' = -4b\cos(2\phi)$. Substituting this in the differential equation, and comparing the coefficients of 1 and $\cos(2\phi)$, we get $a = \frac{1}{2}$ and $b = -\frac{1}{6}$. Thus $w_{1,p} = \frac{3-\cos(2\phi)}{6}$. We add a homogeneous solution to $w_{1,p}$ in order to satisfy the initial conditions $w_1(\pi) = 0$ and $w_1'(\pi) = 0$, that is, $w_1(\phi) = A\cos\phi + B\sin\phi + \frac{3-\cos(2\phi)}{6}$. Then using $w_1(\pi) = 0$ and $w_1'(\pi) = 0$, we get $A = \frac{1}{3}$ and $B = 0$. So

$$w_1(\phi) = \frac{\cos\phi}{3} + \frac{3-\cos(2\phi)}{6}.$$

Hence $w(\phi) = -\cos\phi + \varepsilon\left(\frac{\cos\phi}{3} + \frac{3-\cos(2\phi)}{6}\right) + \varepsilon^2 w_2 + \cdots$.

Neglecting the terms containing ε^2, we get

$$u(\phi) = \frac{w(\phi)}{R_\odot} \approx -\frac{\cos\phi}{R_\odot} + \frac{3m}{R_\odot^2}\left(\frac{\cos\phi}{3} + \frac{3-\cos(2\phi)}{6}\right).$$

To find the approximate value of the incoming azimuthal angle ϕ_{in} of the light ray, we set $u(\phi_{\text{in}})=0$, giving the following quadratic equation in $\cos\phi_{\text{in}}$:

$$-\frac{\cos\phi_{\text{in}}}{R_\odot} + \frac{3m}{R_\odot^2}\left(\frac{\cos\phi_{\text{in}}}{3} + \frac{3-(2(\cos\phi_{\text{in}})^2-1)}{6}\right) = 0.$$

Define $\delta_{\text{in}} = \frac{\pi}{2} - \phi_{\text{in}} > 0$. Then since[3] the bending takes place near the vicinity of the Sun, we may assume that δ_{in} is small. Then we obtain

$$\begin{aligned}\delta_{\text{in}} \approx \sin\delta_{\text{in}} = \cos\phi_{\text{in}} &= \frac{1}{2}\left(1 - \frac{R_\odot}{m} + \sqrt{\left(\frac{R_\odot}{m} - 1\right)^2 + 8}\,\right)\\ &= 4\left(\frac{R_\odot}{m} - 1 + \sqrt{\left(\frac{R_\odot}{m} - 1\right)^2 + 8}\,\right)^{-1}.\end{aligned}$$

As $m \ll R_\odot$,

$$\delta_{\text{in}} \approx 4\left(\frac{R_\odot}{m} + \frac{R_\odot}{m}\right)^{-1} = \frac{2m}{R_\odot}$$

Similarly, the angular deviation δ_{out} from the vertical is given by the same expression, so that the total angular deviation

$$\delta = \delta_{\text{in}} + \delta_{\text{out}} \approx \frac{4m}{R_\odot} = \frac{4M_\odot}{R_\odot} = \frac{4\times 1.5}{7\times 10^5} \text{ radians} \approx 1.77''.$$

This prediction was confirmed by observation during the 1919 solar eclipse.

[3] The Schwarzschild metric resembles the Minkowski metric for $r \gg 2m$, and in the case of Minkowski spacetime, (lightlike) geodesics are along straight lines.

8.5 The exponential map

Given a smooth manifold M with a connection ∇, and a point $p \in M$, we know that for each tangent vector $v \in T_pM$, there is a unique maximal geodesic $\gamma_v : I_v \to M$, defined on some open interval $I_v \subset \mathbb{R}$ containing 0, that passes through p in the direction of v, i.e., $\gamma_v(0) = p$ and $v_{\gamma_v,p}(0) = v$. The exponential map $\exp_p$ describes the dependence of γ_v on v, and sends straight lines $\ell_v := \{tv : t \in \mathbb{R}\}$ in T_pM to geodesics passing through p in M. We will see in the next section that in the context of a Lorentzian manifold modelling spacetime, the exponential map will give rise to 'inertial coordinates' at a spacetime point. We begin with the following definition.

Definition 8.4. (Exponential map.)
Let M be a smooth manifold with a connection ∇, and let $p \in M$.
Given a tangent vector $v \in T_pM$, denote by $\gamma_v : I_v \to M$, where $I_v \subset \mathbb{R}$ is an open interval containing 0, the unique maximal geodesic such that $\gamma_v(0) = p$ and $v_{\gamma_v,p}(0) = v$. Let $\mathcal{D}_p := \{v \in T_pM : 1 \in I_v\} \subset T_pM$. We define the *exponential map* $\exp_p : \mathcal{D}_p \to M$ *at* p, by $\exp_p v = \gamma_v(1)$, $v \in \mathcal{D}_p$.

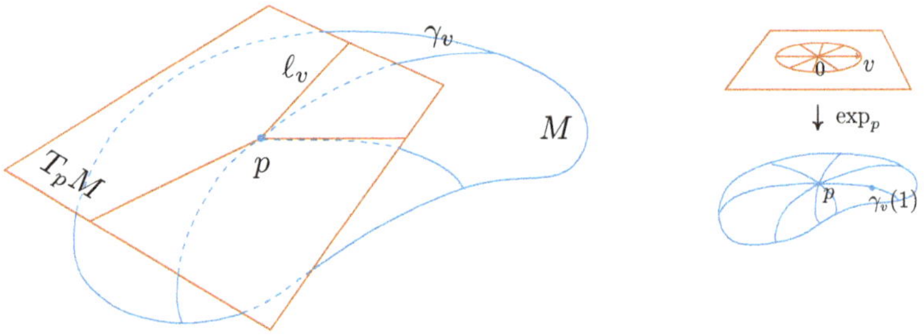

Note that $\exp_p 0 = p$. The following example gives some motivation for the terminology.

Example 8.5. (Exponential map for the unit circle.) In the Euclidean plane $\mathbb{R}^2$, consider the unit circle $S^1 = \{(x, y) \in \mathbb{R}^2 : x^2 + y^2 = 1\}$. Let U denote the circular arc which is the image of the injective map given by $(-\pi, \pi) \ni \theta \mapsto (\cos\theta, \sin\theta) \in S^1$, and denote its inverse by ϕ. Similarly, let V be the image of the injective map $(-\pi, \pi) \ni \theta \mapsto (-\cos\theta, \sin\theta) \in S^1$, and denote its inverse by ψ. Then it can be seen that $\mathcal{A} = \{(U, \phi), (V, \psi)\}$ forms an atlas for S^1. We consider S^1 as a Riemannian manifold with metric induced from the Euclidean inner product on $\mathbb{R}^2$, that is, for $p \in S^1$ and $v, w \in T_pS^1$, $\mathbf{g}(p)(v, w) = \langle (vx, vy), (wx, wy) \rangle_{\mathbb{R}^2}$, where x, y denote the coordinate maps given by $S^1 \ni (a,b) \mapsto x(a,b) := a$, $y(a,b) := b$. In (U,ϕ), $\mathbf{g}(p)_{\theta\theta} = \mathbf{g}(p)(\partial_{\theta,p}, \partial_{\theta,p}) = (\partial_{\theta,p}x)^2 + (\partial_{\theta,p}y)^2 = (-\sin\theta)^2 + (\cos\theta)^2 = 1$, for $p = (\cos\theta, \sin\theta) \in U$. Let ∇ be the corresponding Levi-Civita connection. Using Proposition 6.3 (p.119), the connection coefficient in (U,ϕ) is $\Gamma^\theta_{\theta\theta} = 0$. For $v \in B := (-1,1) \subset \mathbb{R}$, consider the curve $\gamma_v : (-\pi,\pi) \to S^1$ defined by $\gamma_v(t) = (\cos(vt), \sin(vt))$ for $t \in (-\pi,\pi)$. Then $\gamma_v(0) = (1,0) =: p \in U \subset S^1$. Moreover, we have $\gamma_v^\theta(t) = vt$, so that $\dot{\gamma}_v^\theta(t) = v$, and $v_{\gamma_v,p} = v\partial_{\theta,p}$. The curve γ_v is a geodesic because

$$\ddot{\gamma}_v^\theta + \Gamma^\theta_{\theta\theta}\dot{\gamma}_v^\theta\dot{\gamma}_v^\theta = 0 + 0 = 0.$$

Thus $\{v\partial_{\theta,p} : v \in B\} \subset \mathcal{D}_p \subset T_pS^1$ and the exponential map at $p = (1,0)$ sends $v\partial_{\theta,p}$, $v \in B$, to $\gamma_v(1) = (\cos v, \sin v) \in S^1$.

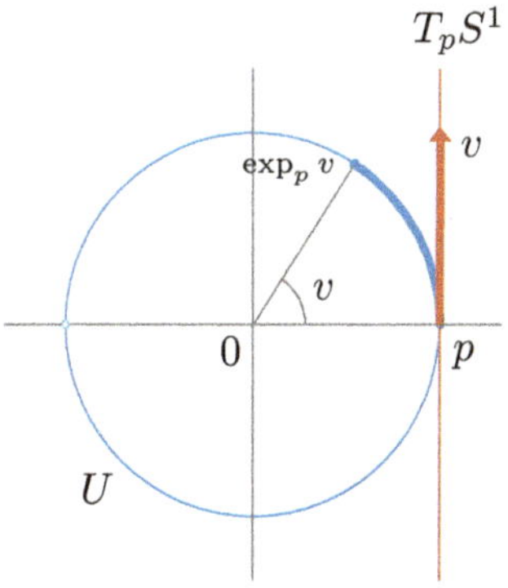

Identifying $\mathbb{R}^2$ with $\mathbb{C}$, and T_pS^1 with $\mathbb{R}$ (by sending $\mathbb{R} \ni v \mapsto v\partial_{\theta,p} \in T_pM$), we write $\exp_p v = (\cos v, \sin v) = \cos v + i\sin v = e^{iv}$. ◇

We had seen that an affine linear parametric change of variables of a curve preserves its geodesic nature. The following result relates the initial speeds in the case of dilations.

Lemma 8.2. (Scaling lemma.)

Let • *M be a smooth manifold with a connection ∇,*

• *$p \in M$, $v \in T_pM$, and $c > 0$,*

• *$\gamma : I \to M$ be a geodesic, where $I \subset \mathbb{R}$ is an open interval containing 0, and such that $\gamma(0) = p$ and $v_{\gamma,p} = v$.*

Then $\mu : \frac{1}{c}I \to M$ given by $\mu(t) = \gamma(ct)$, $t \in \frac{1}{c}I = \{\frac{\tau}{c} : \tau \in I\}$, is a geodesic such that $\mu(0) = p$ and $v_{\mu,p} = cv$.

Proof. If $I = (a,b)$, then $\frac{1}{c}I = (\frac{a}{c}, \frac{b}{c})$. With $h(t) := ct$ for $t \in \frac{1}{c}I$, $\mu = \gamma \circ h$. As γ is a geodesic, and h is linear, Thm. 8.2 (p.160) implies that μ is a geodesic too. Moreover, $\mu(0) = \gamma(c0) = \gamma(0) = p$ and $v_{\mu,p} = \dot{h}(0)\,v_{\gamma,p} = cv$. □

Corollary 8.2. *Let*

- *M be a smooth manifold with a connection ∇,*
- *$p \in M$, $v \in T_pM$, $c > 0$, and I_v, I_{cv} be open intervals in $\mathbb{R}$ containing 0,*
- *$\gamma_v : I_v \to M$ be the unique maximal geodesic such that $\gamma_v(0) = p$ and $v_{\gamma_v,p} = v$,*
- *$\gamma_{cv} : I_{cv} \to M$ be the unique maximal geodesic such that $\gamma_{cv}(0) = p$, $v_{\gamma_{cv},p} = cv$.*

Then $cI_{cv} = I_v$ and for all $t \in I_{cv}$, $\gamma_{cv}(t) = \gamma_v(ct)$.

Proof. We use the Scaling lemma, applied twice with the roles of γ_v and γ_{cv} swapped, and the maximality of the two geodesics. First, by the Scaling lemma, $\gamma_v(c\,\cdot) : \frac{1}{c}I_v \to M$ is a geodesic passing through p at $t = 0$ with velocity at p equal to $c\,v$. The maximality of $\gamma_{cv} : I_{cv} \to M$ implies $\frac{1}{c}I_v \subset I_{cv}$ and $\gamma_{cv}|_{\frac{1}{c}I_v} = \gamma_v(c\,\cdot)$. Also, starting from γ_{cv}, it follows by the Scaling lemma that $\gamma_{cv}(c^{-1}\cdot) : \frac{1}{c^{-1}}I_{cv} \to M$ is a geodesic passing through p at $t = 0$, with velocity at p equal to $c^{-1}cv = v$. As $\gamma_v : I_v \to M$ is maximal, $cI_{cv} \subset I_v$. So $cI_{cv} = I_v$ and $\gamma_{cv} = \gamma_v(c\,\cdot)$ on $I_{cv} = \frac{1}{c}I_v$. □

In particular, if $v \in \mathcal{D}_p$, that is, if $1 \in I_v$, then for all $c \in (0, 1]$, we have that $1 \in I_{cv} = \frac{1}{c}I_v$ and $\exp_p(cv) = \gamma_{cv}(1) = \gamma_v(c)$. Thus the exponential map takes the straight line segment $\{tv : t \in [0, 1]\}$ in T_pM, and maps it to the 'arc' $\{\gamma_v(t) : t \in [0, 1]\}$ of the geodesic γ_v.

Lemma 8.3. *Let M be a smooth manifold with a connection ∇, and $p \in M$. Then $\mathcal{D}_p$ contains a neighbourhood W of $0 \in T_pM$ and $\exp_p|_W : W \to M$ is smooth.*

We note that $T_pM \simeq \mathbb{R}^m$ via the isomorphism ι, $v = v^i \partial_{x^i,p} \overset{\iota}{\mapsto} (v^1, \cdots, v^m)$, where $(U, \mathbf{x})$ is an admissible chart containing p. Thus we may endow $T_pM \simeq \mathbb{R}^m$ with the usual Euclidean metric topology induced by the Euclidean norm $\|\cdot\|$. Recall from Example 2.3 (p.27) that the matrix relating the components of a vector in T_pM, with respect to induced bases from two charts, is invertible. It follows that this topology on T_pM is independent of the admissible chart $(U, \mathbf{x})$ containing p used to define the isomorphism ι.

Proof. Consider an admissible chart $(U, \mathbf{x})$ for M containing p. If γ is a geodesic through p in the direction $w = w^i \partial_{x^i,p}$, then $Y(t) := \mathbf{x}(\gamma(t))$ satisfies

$$\left\{\begin{aligned} \ddot{Y}^k(t) + \Gamma^k_{ij}(Y(t))\dot{Y}^i(t)\dot{Y}^j(t) &= 0 \qquad (1 \leqslant k \leqslant m) \\ Y(0) &= \mathbf{x}(p) \\ \dot{Y}(0) &= \mathbf{w} \end{aligned}\right.$$

where $\mathbf{w} = (w^1, \cdots, w^m)$. By Remark 8.1 (p.155), there exists a neighbourhood $N \subset \mathbb{R}^m$ of $\mathbf{0}$, an $\epsilon > 0$, and a C^∞ map $\mathcal{Y} : N \times (-\epsilon, \epsilon) \to \mathbf{x}(U)$ such that for all $\mathbf{w} \in N$, $Y(t) := \mathcal{Y}(\mathbf{w}, t)$ solves the above initial value problem.

But then $\widetilde{N} := \iota^{-1}N$ is a neighbourhood of $0 \in T_pM$, and for each $w \in \widetilde{N}$, we have that $\gamma_w(t) = \mathbf{x}^{-1}(\mathcal{Y}(\iota w, t))$, $t \in (-\epsilon, \epsilon) \subset I_w$, where I_w is the open interval of $\mathbb{R}$ containing 0 such that $\gamma_w : I_w \to M$ is the maximal geodesic such that $\gamma_w(0) = p$ and $v_{\gamma_w,p} = w$. Now we apply Corollary 8.2. Let $c := \epsilon/2$. For each $\widetilde{w} = cw \in c\widetilde{N}$, where $w \in \widetilde{N}$, we have

$$1 \in (-2,2) = c^{-1}(-\epsilon,\epsilon) \subset c^{-1}I_w = c^{-1}I_{c^{-1}cw} = c^{-1}I_{c^{-1}\widetilde{w}} = I_{\widetilde{w}}.$$

Thus the open neighbourhood $W := c\widetilde{N}$ of $0 \in T_pM$ is contained in $\mathcal{D}_p$. Also, by the smoothness of $\mathcal{Y}$, $c\widetilde{N} = W \ni \widetilde{w} \mapsto \exp_p \widetilde{w} = \gamma_{\widetilde{w}}(1) = \mathbf{x}^{-1}(\mathcal{Y}(\iota \frac{1}{c}\widetilde{w}, \frac{\epsilon}{2}))$ is smooth too. □

Theorem 8.3. *Let M be a smooth manifold with a connection ∇, and let $p \in M$. Then $\mathcal{D}_p$ contains a neighbourhood V of $0 \in T_pM$, and there is a neighbourhood U of $p \in M$, such that $\exp_p|_V : V \to U$ is a diffeomorphism.*

Proof. By Lemma 8.3, there exists a neighbourhood W of $0 \in T_pM$ such that $W \subset \mathcal{D}_p$ and $\exp_p|_W : W \to M$ is smooth. We'll show the derivative of this smooth map at $0 \in T_pM$ is invertible, and use the inverse function theorem (Thm. 2.3, p.31). The derivative map at $0 \in T_pM$ is a linear transformation $d(\exp_p)_0 : T_0(T_pM) \to T_pM$. As $\dim T_0(T_pM) = \dim T_pM$, it is enough to show surjectivity of $d(\exp_p)_0$. Let $v \in T_pM$. Consider the line $\ell_v : \mathbb{R} \to T_pM$ given by $\ell_v(t) = tv$. For small enough $t > 0$, $tv \in W$, and $\exp_p(\ell_v(t)) = \exp_p(tv) = \gamma_{tv}(1) = \gamma_v(t)$. (Here $\gamma_w : I_w \to M$, where $I_w \subset \mathbb{R}$ is an open interval containing 0, denotes the maximal geodesic in M such that $\gamma_w(0) = p$ and $v_{\gamma_w,p} = w$.) Then for any $f \in C^\infty(M)$,

$$\begin{aligned}\big((d(\exp_p))_0(v_{\ell_v,0})\big)(f) &= v_{\ell_v,0}(f \circ \exp_p) = \frac{d(f \circ \exp_p \circ \ell_v)}{dt}(0) = \frac{d(f \circ \gamma_v)}{dt}(0)\\ &= v_{\gamma_v,p}f = vf.\end{aligned}$$

Hence, $(d(\exp_p))_0(v_{\ell_v,0}) = v \in T_pM$. As the chosen $v \in T_pM$ at the outset was arbitrary, we have shown that $d(\exp_p)_0$ is surjective (and hence invertible), as wanted. So by the inverse function theorem (Theorem 2.3), there exists a neighbourhood $V \subset W$ of $0 \in T_pM$ and a neighbourhood U of $p \in M$ such that $\exp_p|_V : V \to U$ is a diffeomorphism. □

Exercise 8.14. (Exponential map is the exponential map.) Consider $\mathrm{GL}_2(\mathbb{R})$ as a 4-dimensional Lorentzian manifold as in Exercise 5.21 (p.88) with the trace metric $\mathbf{g}$, and equip $\mathrm{GL}_2(\mathbb{R})$ with the Levi-Civita connection corresponding to $\mathbf{g}$. It can be shown that $P : \mathbb{R} \to \mathrm{GL}_2(\mathbb{R})$ satisfies the geodesic equation if and only if $\ddot{P} - P^{-1}\dot{P}P^{-1} = 0$ (see Exercises 8.16, 8.17, p.179). For $A \in \mathbb{R}^{2\times 2}$ and $t \in \mathbb{R}$, let $P(t) := e^{tA}$ (matrix exponential of tA). Show that P is a geodesic in $\mathrm{GL}_2(\mathbb{R})$ satisfying $P(0) = I_2$ and $v_{P,I_2} = A$. Show that $\exp_{I_2} : T_I(\mathrm{GL}_2(\mathbb{R})) \to \mathrm{GL}_2(\mathbb{R})$ is the exponential map, i.e., $\exp_{I_2} A = e^A$ for all $A \in \mathbb{R}^{2\times 2} \simeq T_I(\mathrm{GL}_2(\mathbb{R}))$.

8.6 Normal coordinates

Given a smooth manifold M with a connection ∇, the tangent space at a point $p \in M$ can be viewed as a kind of 'flattening' of the manifold around p, since we have seen that radial straight rays from the origin in T_pM are mapped to geodesics emanating from the point p in M. In fact, we will see in this section that given any basis $\{e_i : 1 \leqslant i \leqslant m\}$ for T_pM, exponential map sets up a coordinate system in a neighbourhood of p, called the 'normal coordinate system', such that in this coordinate system, all the connection coefficients vanish at p (assuming ∇ is torsion-free). As the connection coefficients may be thought of as the 'components' of the connection in a chart (see equation (6.1), p.114), the normal coordinates as those which make (M, ∇) appear 'flat' (but only) at p.

Definition 8.5. (Normal chart/neighbourhood/coordinates.)
Let

- M be an m-dimensional smooth manifold with a connection ∇,
- $p \in M$, U be a neighbourhood of p, V be a neighbourhood of $0 \in T_pM$, such that $\exp_p : V \to U$ is a diffeomorphism,
- $B = \{e_i : 1 \leqslant i \leqslant m\} \subset V$ be a basis for T_pM, and $\iota : \mathbb{R}^m \to T_pM$ be the isomorphism $\mathbb{R}^m \ni (v^1, \cdots, v^m) \mapsto v^i e_i \in T_pM$,
- $\mathbf{x} : U \to \iota^{-1}V$ be given by $\mathbf{x} = \iota^{-1} \circ \exp_p^{-1}$.

Then the admissible[4] chart $(U, \mathbf{x})$ for M is called the *normal chart* $(U, \mathbf{x})$ *with the normal neighbourhood* U*, and normal coordinates* x^i, $1 \leqslant i \leqslant m$.

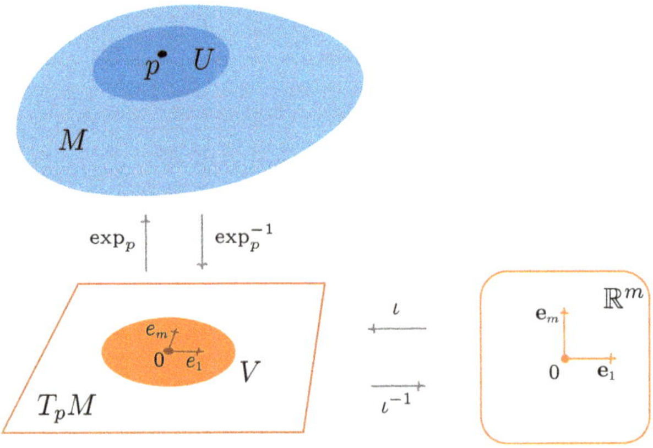

[4]The admissibility claim follows from the fact that $\mathbf{x}$ is a diffeomorphism (being the composition of diffeomorphisms).

Lemma 8.4. *With the notation*[5] *from* Definition 8.5, *we have*:

(1) $x^j(\exp_p(e_i))=\delta_i^j$, $1\leqslant i,j\leqslant m$.
(2) $\partial_{x^i,p}=e_i$, $1\leqslant i\leqslant m$.
(3) *For each* $q\in U$, $\exp_p^{-1} q=x^i(q)\partial_{x^i,p}$.

Proof.

(1) This follows immediately from the definition, since

$$\begin{aligned}(x^j\circ\exp_p)(e_i) &= (\mathbf{x}\circ\exp_p)^j(e_i) = (\iota^{-1}\circ\exp_p^{-1}\circ\exp_p)^j(e_i)\\ &= (\iota^{-1})^j(e_i) = \delta_i^j.\end{aligned}$$

(2) Let $\gamma_{e_i}: I_i \to M$, where $I_i\subset\mathbb{R}$ is an open interval containing 0, be the maximal geodesic such that $\gamma_{e_i}(0)=p$ and $v_{\gamma_{e_i},p}=e_i$. We had seen that for $t\in[0,1]$, $\gamma_{e_i}(t)=\exp_p(te_i)$. So

$$\begin{aligned}(dx^j)_p(e_i) &= e_i(x^j) = v_{\gamma_{e_i},p}(x^j) = \frac{d(x^j\circ\gamma_{e_i})}{dt}(0)\\ &= \frac{d(x^j\circ\exp_p(te_i))}{dt}(0) = \frac{d((\iota^{-1})^j\circ(te_i))}{dt}(0)\\ &= \frac{d(t\delta_i^j)}{dt}(0) = \delta_i^j = (dx^j)_p(\partial_{x^i,p}).\end{aligned}$$

As j was arbitrary, this proves the claim in (2).

(3) If $\mathbf{e}_i := \iota^{-1}e_i \in\mathbb{R}^m$, then it follows from the definition of ι that $\mathbf{e}_1,\cdots,\mathbf{e}_m$ are the standard basis vectors for $\mathbb{R}^m$. Thus for $q\in U$,

$$\begin{aligned}x^i(q)\partial_{x^i,p} &= x^i(q)e_i = x^i(q)\iota(\mathbf{e}_i) = \iota(x^i(q)\mathbf{e}_i) = \iota(x^1(q),\cdots,x^m(q))\\ &= \iota(\mathbf{x}(q)) = \iota((\iota^{-1}\circ\exp_p^{-1})(q)) = \exp_p^{-1} q.\end{aligned}$$

This completes the proof. □

Theorem 8.4. *Let M be a smooth manifold with a torsion-free connection ∇ (that is, ∇ satisfies* (C4) : $\nabla_X Y - \nabla_Y X = [X,Y]$ *for all* $X,Y\in T_0^1 M$*). Let $p\in M$. Then in a normal chart for p, $\Gamma_{ij}^k(p)=0$ for all $1\leqslant i,j,k\leqslant m$.*

Proof. We will use the notation from Definition 8.5. Consider a vector $\mathbf{v}=(v^1,\cdots,v^m)\in\iota^{-1}V\subset\mathbb{R}^m$. Set $v=v^i\partial_{x^i,p}=v^ie_i$. Let $\gamma_v: I_v\to M$, where $I_v\subset\mathbb{R}$ is an open interval containing 0, be the maximal geodesic such that $\gamma_v(0)=p$ and $v_{\gamma_v,p}=v$. For $t\in[0,1]$, we have

$$\begin{aligned}\gamma_v^i(t) &= (x^i\circ\gamma_v)(t) = x^i(\exp_p(tv)) = (\mathbf{x}\circ\exp_p)^i(tv)\\ &= (\iota^{-1})^i(tv) = (\iota^{-1})^i(tv^je_j) = tv^i.\end{aligned}$$

[5]Note in particular that the basis $B=\{e_1,\cdots,e_m\}$ for T_pM is chosen from within the domain V of the map $\exp_p: V\to U$, so that $\exp_p(e_i)$ is well-defined. In general, we can always scale the tangent vectors in the basis to achieve this.

γ_v satisfies the geodesic equation, and so $\ddot{\gamma}_v^k + \Gamma_{ij}^k(\gamma_v(t))\,\dot{\gamma}_v^i(t)\,\dot{\gamma}_v^j(t) = 0$ for all $t \in [0, 1]$. As

$$\ddot{\gamma}_v^k = \frac{d^2(tv^k)}{dt^2} = 0,$$

we obtain $\Gamma_{ij}^k(\gamma_v(t))\,v^i v^j = 0$. In particular, for $t = 0$, $\Gamma_{ij}^k(p)\,v^i v^j = 0$ for all $1 \leqslant k \leqslant m$. Fix an i, and take $\mathbf{v} = \mathbf{e}_i = \iota^{-1}(e_i)$, to obtain

$$\Gamma_{ii}^k(p) = 0.$$

Next, fix distinct i, j, and take $\mathbf{v} = \mathbf{e}_i + \mathbf{e}_j = \iota^{-1}(e_i + e_j)$ to obtain

$$\Gamma_{ii}^k(p) + \Gamma_{ij}^k(p) + \Gamma_{ji}^k(p) + \Gamma_{jj}^k(p) = 0.$$

But $\Gamma_{ii}^k(p) = 0 = \Gamma_{jj}^k(p)$. Thus the torsion freeness of ∇ (implying $\Gamma_{ij}^k = \Gamma_{ji}^k$, see Exercise 6.12, p.116) now yields $2\Gamma_{ij}^k(p) = 0$, i.e., $\Gamma_{ij}^k(p) = 0$. □

In the context of a spacetime $(M, \mathbf{g})$, that is, a Lorentzian manifold with the Levi-Civita connection ∇, we know that ∇ is torsion free. Moreover, if the normal neighbourhood U is such that $\exp_p^{-1} U$ contains an orthonormal basis $B = \{e_1, \cdots, e_m\}$ for T_pM with respect to the scalar product $\mathbf{g}(p)$, then we have the following result.

Corollary 8.3. *Let $(M, \mathbf{g})$ be an m-dimensional Lorentzian manifold, with the Levi-Civita connection ∇. Let $p \in M$. In the notation from* Def. 8.5, *suppose that the basis $B = \{e_1, \cdots, e_m\}$ for T_pM chosen there is orthonormal with respect to the scalar product $\mathbf{g}(p)$ for T_pM:*

$$\mathbf{g}(p)(e_i, e_j) = \eta_{ij} := \begin{cases} -1 & \text{if } i = j = 1, \\ 1 & \text{if } i = j \neq 1, \\ 0 & \text{if } i \neq j. \end{cases}$$

Then in the normal chart $(U, \mathbf{x})$ for p, we have $\mathbf{g}_{ij}(p) = \eta_{ij}$, and

$$\Gamma_{ij}^k(p) = 0 \text{ for all } 1 \leqslant i, j, k \leqslant m.$$

Roughly speaking this means that locally spacetime at a point looks like Minkowski spacetime. If γ is a freely falling observer parametrised by proper time, whose velocity vector $v_{\gamma,p}$ at an event p is taken as e_0, and we choose orthonormal vectors $e_1, e_2, e_3 \in v_{\gamma,p}^{\perp}$, then[6] the trajectory of another freely falling particle will (approximately) appear to us to be uniform-in-time, along a straight line in our perception of space.

[6]In the physics literature, this is sometimes expressed by saying that 'gravitational forces can be transformed away locally/infinitesimally'. If we happen to be in a freely falling elevator, then we will observe the motion of a freely falling dropped coin to be along a straight line, and uniform in our proper time.

Using the result of the following exercise, it follows that in Corollary 8.3, even if the orthonormal basis $B = \{e_1, \cdots, e_m\}$ for T_pM with respect to the scalar product $\mathbf{g}(p)$ does not lie inside $V := \exp_p^{-1} U$, where U is a normal neighbourhood of p, we can select appropriate small $k_1, \cdots, k_m \in \mathbb{R}\backslash\{0\}$ so that $\widetilde{B} := \{k_1 e_1, \cdots, k_m e_m\}$ lies inside V, and define a new chart map $\widetilde{\mathbf{x}}$ using the normal coordinates $\mathbf{x}$ obtained by starting with $\widetilde{B}$, such that with the new chart map $\widetilde{x}$ we have $\partial_{\widetilde{x}^i,p} = e_i$ for all $i \in \{1, \cdots, m\}$.

Exercise 8.15. Let M be a smooth manifold, and let $p \in M$. Suppose that $e_1, \cdots, e_m \in T_pM$, $k_1, \cdots, k_m \in \mathbb{R}\backslash\{0\}$, and that there exists an admissible chart $(U, \mathbf{x})$ for M, such that $\partial_{x^i,p} = k_i e_i$ for all $i \in \{1, \cdots, m\}$. Let $D: \mathbb{R}^m \to \mathbb{R}^m$ be the multiplication map $D(v^1, \cdots, v^m) = (k^1 v^2, \cdots, k^m v^m)$ for all $(v^1, \cdots, v^m) \in \mathbb{R}^m$. Define the chart map $\widetilde{\mathbf{x}} : U \to \mathbb{R}^m$ by $\widetilde{\mathbf{x}} = D \circ \mathbf{x}$. Show that the admissible chart $(U, \widetilde{\mathbf{x}})$ is such that $\partial_{\widetilde{x}^i,p} = e_i$ for all $i \in \{1, \cdots, m\}$, and the connection coefficients $\widetilde{\Gamma}^k_{ij}$ in this chart satisfy $\widetilde{\Gamma}^k_{ij}(p) = 0$, $1 \leqslant i, j, k, \leqslant m$.

8.7 Geodesics as extremal paths

In Euclidean space $\mathbb{R}^3$ (i.e., $\mathbb{R}^3$ endowed with the Euclidean inner product), we know that the *straight line segment* between two points $\mathbf{p}$, $\mathbf{q}$, is also the *shortest path* from $\mathbf{p}$ to $\mathbf{q}$. Likewise, on the sphere in $\mathbb{R}^3$ considered as a Riemannian manifold, the 'short arcs' of great circles give the shortest path between the endpoints of the arc. In this section, we will give an interpretation of timelike geodesics as being stationary for the proper time 'functional'. More precisely, we will show the following.

Theorem 8.5. (Timelike proper time maximisers are geodesics.)
Suppose $(M, \mathbf{g})$ is a time-oriented Lorentzian manifold with Levi-Civita connection ∇. Let $(U, \mathbf{x})$ be any admissible chart, $p, q \subset U$, and $a, b \in \mathbb{R}$, $a < b$, be such that the set

$$\mathcal{O}_{pq} := \{\mu : [a, b] \to U \mid \mu \text{ is an observer such that } \mu(a) = p,\ \mu(b) = q\}$$

is not empty. Define $\tau : \mathcal{O}_{pq} \to \mathbb{R}$ by

$$\tau(\mu) = \int_a^b \sqrt{-\mathbf{g}(\mu(t))(v_{\mu,\mu(t)}, v_{\mu,\mu(t)})}\, dt, \quad \mu \in \mathcal{O}_{pq}.$$

If the curve $\mu_* \in \mathcal{O}_{pq}$ maximises τ, then there exists a diffeomorphism $t : (0, \tau(\mu_*)) \to (a, b)$, such that $\mu_* \circ t : (0, \tau(\mu_*)) \to U$ is a geodesic.

The fact that μ_* is a geodesic *after* a reparametrisation is not a surprise, since we know that arclength/proper time is invariant under a reparametrisation (see Proposition 5.4, p.91), while the property of being a geodesic is not preserved under reparametrisations that are not affine linear.

In order to prove this result, we will prove the necessity of the Euler-Lagrange equations for maximising curves.

Proposition 8.5. (Necessity of Euler-Lagrange equations.)
Suppose that $a, b \in \mathbb{R}$, $a<b$, $\mathbf{U} \subset \mathbb{R}^m$ is open, and $\boldsymbol{\Omega} \subset \mathbf{U} \times \mathbb{R}^m$ is open. Let $\mathcal{L} : \boldsymbol{\Omega} \to \mathbb{R}$, $\mathbf{U} \times \mathbb{R}^m \supset \boldsymbol{\Omega} \ni (\boldsymbol{\alpha}, \boldsymbol{\beta}) \mapsto \mathcal{L}(\boldsymbol{\alpha}, \boldsymbol{\beta}) \in \mathbb{R}$, be a C^∞ function. Define the open set $\mathbf{U}_{\boldsymbol{\Omega}}$ of $\mathbb{R}^m$ given by

$$\mathbf{U}_{\boldsymbol{\Omega}} = \{\boldsymbol{\alpha} \in \mathbb{R}^m : \textit{there exists a } \boldsymbol{\beta} \in \mathbb{R}^m \textit{ such that } \mathcal{L}(\boldsymbol{\alpha}, \boldsymbol{\beta}) \in \boldsymbol{\Omega}\}.$$

Let $\mathbf{p}, \mathbf{q} \in \mathbf{U}_{\boldsymbol{\Omega}}$, and $\mathcal{O}_{\mathbf{pq}} := \{\boldsymbol{\mu} : [a, b] \to \mathbf{U}_{\boldsymbol{\Omega}} \mid \boldsymbol{\mu}(a)=\mathbf{p},\ \boldsymbol{\mu}(b)=\mathbf{q}\}$. Define $f : \mathcal{O}_{\mathbf{pq}} \to \mathbb{R}$ by

$$f(\boldsymbol{\mu}) = \int_a^b \mathcal{L}(\boldsymbol{\mu}(t), \dot{\boldsymbol{\mu}}(t))\, dt, \quad \boldsymbol{\mu} \in \mathcal{O}_{\mathbf{pq}}.$$

If $\boldsymbol{\mu}_*$ maximises f, then $\boldsymbol{\mu}_*$ satisfies the Euler-Lagrange equations:

$$\frac{\partial \mathcal{L}}{\partial \alpha^i}(\boldsymbol{\mu}(t), \dot{\boldsymbol{\mu}}(t)) - \frac{d}{dt}\Big(\frac{\partial \mathcal{L}}{\partial \beta^i}(\boldsymbol{\mu}(t), \dot{\boldsymbol{\mu}}(t))\Big) = 0, \quad t \in [a, b], \quad 1 \leqslant i \leqslant m.$$

(Here we use the notation $\dot{\boldsymbol{\mu}} = \dfrac{d\boldsymbol{\mu}}{dt}$ and α^i for the components of $\boldsymbol{\alpha}$ etc.)

Proof. Let $\mathbf{h} : [a, b] \to \mathbf{U}_{\boldsymbol{\Omega}}$ be any C^∞ map, such that $\mathbf{h}(a) = \mathbf{0} = \mathbf{h}(b)$. Then for any $\epsilon \in \mathbb{R}$, $(\boldsymbol{\mu}_* + \epsilon \mathbf{h})(a) = \mathbf{p}$ and $(\boldsymbol{\mu}_* + \epsilon \mathbf{h})(b) = \mathbf{q}$. Moreover, as $\mathbf{h}([a, b])$ is a compact subset of the open set $\mathbf{U}_{\boldsymbol{\Omega}}$, it follows that for all $|\epsilon|$ small enough, $\boldsymbol{\mu}_* + \epsilon \mathbf{h}$ assumes values in $\mathbf{U}_{\boldsymbol{\Omega}}$, and so $\boldsymbol{\mu}_* + \epsilon \mathbf{h} \in \mathcal{O}_{\mathbf{pq}}$. As $\boldsymbol{\mu}_*$ is a maximiser for f, it follows that g, given by $g(\epsilon) := f(\boldsymbol{\mu}_* + \epsilon \mathbf{h})$, has a maximum at $\epsilon = 0$, and so by elementary calculus, $g'(0) = 0$. Thus

$$0 = g'(0) = \frac{df(\boldsymbol{\mu}_* + \epsilon \mathbf{h})}{d\epsilon}\Big|_{\epsilon=0} = \frac{d}{d\epsilon} \int_a^b \mathcal{L}(\boldsymbol{\mu}_*(t) + \epsilon \mathbf{h}(t), \dot{\boldsymbol{\mu}}_*(t) + \epsilon \dot{\mathbf{h}}(t))\, dt\Big|_{\epsilon=0}$$

$$= \int_a^b \frac{\partial}{\partial \epsilon} \mathcal{L}(\boldsymbol{\mu}_*(t) + \epsilon \mathbf{h}(t), \dot{\boldsymbol{\mu}}_*(t) + \epsilon \dot{\mathbf{h}}(t))\Big|_{\epsilon=0} dt$$

$$= \int_a^b \Big(\frac{\partial \mathcal{L}}{\partial \alpha^i}(\boldsymbol{\mu}_*(t), \dot{\boldsymbol{\mu}}_*(t))\, \mathbf{h}^i(t) + \frac{\partial \mathcal{L}}{\partial \beta^i}(\boldsymbol{\mu}_*(t), \dot{\boldsymbol{\mu}}_*(t))\, \dot{\mathbf{h}}^i(t)\Big)\, dt. \tag{8.8}$$

Integration by parts on the second summand yields

$$\int_a^b \frac{\partial \mathcal{L}}{\partial \beta^i}(\boldsymbol{\mu}_*(t), \dot{\boldsymbol{\mu}}_*(t))\, \dot{\mathbf{h}}^i(t)\, dt$$

$$= -\int_a^b \frac{d}{dt}\Big(\frac{\partial \mathcal{L}}{\partial \beta^i}(\boldsymbol{\mu}_*(t), \dot{\boldsymbol{\mu}}_*(t))\Big)\, \mathbf{h}^i(t)\, dt + \frac{\partial \mathcal{L}}{\partial \beta^i}(\boldsymbol{\mu}_*(t), \dot{\boldsymbol{\mu}}_*(t))\, \mathbf{h}^i(t)\Big|_{t=a}^{t=b}$$

$$= -\int_a^b \frac{d}{dt}\Big(\frac{\partial \mathcal{L}}{\partial \beta^i}(\boldsymbol{\mu}_*(t), \dot{\boldsymbol{\mu}}_*(t))\Big)\, \mathbf{h}^i(t)\, dt + 0 - 0,$$

where we used the boundary conditions $\mathbf{h}(a) = \mathbf{0} = \mathbf{h}(b)$. Putting this in (8.8), we get that for all smooth $\mathbf{h} : [a, b] \to \mathbb{R}^m$ satisfying $\mathbf{h}(a) = \mathbf{0} = \mathbf{h}(b)$,

$$\int_a^b \Big(\frac{\partial \mathcal{L}}{\partial \alpha^i}(\boldsymbol{\mu}_*(t), \dot{\boldsymbol{\mu}}_*(t)) - \frac{d}{dt}\Big(\frac{\partial \mathcal{L}}{\partial \beta^i}(\boldsymbol{\mu}_*(t), \dot{\boldsymbol{\mu}}_*(t))\Big)\Big)\, \mathbf{h}^i(t)\, dt = 0.$$

Now consider $\mathbf{h}$ as a column vector in which all entries are equal to the zero function, except for the ith one, which is any smooth function $h : [a, b] \to \mathbb{R}$ such that $h(a)=0=h(b)$. Thus we have, for all such h, that

$$\int_a^b \Big(\frac{\partial \mathcal{L}}{\partial \alpha^i}(\boldsymbol{\mu}_*(t), \dot{\boldsymbol{\mu}}_*(t)) - \frac{d}{dt}\Big(\frac{\partial \mathcal{L}}{\partial \beta^i}(\boldsymbol{\mu}_*(t), \dot{\boldsymbol{\mu}}_*(t))\Big)\Big)\, h(t)\, dt = 0.$$

We will show that this implies the ith Euler-Lagrange equation. Suppose on the contrary that there is some $t_0 \in [a, b]$ such that $\varphi(t_0) \neq 0$, where

$$\varphi(t) := \frac{\partial \mathcal{L}}{\partial \alpha^i}(\boldsymbol{\mu}_*(t), \dot{\boldsymbol{\mu}}_*(t)) - \frac{d}{dt}\Big(\frac{\partial \mathcal{L}}{\partial \beta^i}(\boldsymbol{\mu}_*(t), \dot{\boldsymbol{\mu}}_*(t))\Big), \quad t \in [a, b].$$

Without loss of generality, suppose that $\varphi(t_0) > 0$. Also, by the continuity of φ, we may choose t_0 in the *open* interval (a, b). Let $\delta > 0$ be such that $\varphi(t) > \varphi(t_0)/2$ for $t \in (t_0 - \delta, t_0 + \delta)$. Take any nonnegative bump function h such that $h \equiv 1$ on $(t_0 - \delta/2, t_0 + \delta/2)$ and $h \equiv 0$ outside $(t_0 - \delta, t_0 + \delta)$. Then we get the contradiction that

$$0 = \int_a^b \varphi(t)\, h(t)\, dt = \int_{t_0-\delta}^{t_0+\delta} \varphi(t)\, h(t)\, dt > \int_{t_0-\frac{\delta}{2}}^{t_0+\frac{\delta}{2}} \frac{\varphi(t_0)}{2} 1\, dt > \frac{\varphi(t_0)}{2}\delta > 0.$$

□

Proof. (Of Theorem 8.5.)
Let $\mathbf{U} := \mathbf{x}(U)$. Then $\boldsymbol{\Omega} = \{(\boldsymbol{\alpha}, \boldsymbol{\beta}) \in \mathbf{U} \times \mathbb{R}^m : \mathbf{g}_{ij}(\mathbf{x}^{-1}(\boldsymbol{\alpha}))\beta^i\beta^j < 0\}$ (contained in $\mathbf{U} \times \mathbb{R}^m$) is an open set. Define $g_{ij} = \mathbf{g}_{ij} \circ \mathbf{x}^{-1}$. Then

$$\partial_{x^k,p}\, \mathbf{g}_{ij} = \frac{\partial(\mathbf{g}_{ij} \circ \mathbf{x}^{-1})}{\partial \alpha^k}(\mathbf{x}(p)) = \frac{\partial g_{ij}}{\partial \alpha^k}(\mathbf{x}(p)). \tag{8.9}$$

Define $\mathcal{L}(\boldsymbol{\alpha}, \boldsymbol{\beta}) = \sqrt{-g_{ij}(\boldsymbol{\alpha})\beta^i\beta^j}$ for $(\boldsymbol{\alpha}, \boldsymbol{\beta}) \in \boldsymbol{\Omega} \subset \mathbf{U} \times (\mathbb{R}^m \setminus \{\mathbf{0}\})$. Then $\mathcal{L}$ is smooth (using the fact that $x \mapsto \sqrt{x}$ is C^∞ on $(0, \infty)$). For $\mu \in \mathcal{O}_{pq}$, we define $\boldsymbol{\mu}(t) = \mathbf{x}(\mu(t))$ for $t \in [a, b]$. We have $v_{\mu,\mu(t)} = \dot{\mu}^i(t)\partial_{x^i,\mu(t)}$, where the chart-induced curve components $\mu^i := (\mathbf{x} \circ \mu)^i = \boldsymbol{\mu}^i$, $1 \leqslant i \leqslant m$. It follows that $\mathbf{g}(\mu(t))(v_{\mu,\mu(t)}, v_{\mu,\mu(t)}) = g_{ij}(\boldsymbol{\mu}(t))\dot{\boldsymbol{\mu}}^i(t)\dot{\boldsymbol{\mu}}^j(t)$, and

$$\tau(\mu) = \int_a^b \mathcal{L}(\boldsymbol{\mu}(t), \dot{\boldsymbol{\mu}}(t))\, dt =: f(\boldsymbol{\mu}),$$

where f is defined on $\{\mathbf{x} \circ \mu : \mu \in \mathcal{O}_{pq}\}$. The set $\{\mathbf{x} \circ \mu : \mu \in \mathcal{O}_{pq}\}$ is the same as $\mathcal{O}_{\mathbf{pq}}$ defined in Proposition 8.5 with $\mathbf{p} := \mathbf{x}(p)$ and $\mathbf{q} := \mathbf{x}(q)$. By Proposition 8.5, the Euler-Lagrange equations hold. We note that

$$\frac{\partial \mathcal{L}}{\partial \alpha^i} = \frac{-\dfrac{\partial g_{k\ell}}{\partial \alpha^i}\beta^k\beta^\ell}{2D}, \text{ where } D = D(\boldsymbol{\alpha}, \boldsymbol{\beta}) := \sqrt{-g_{ij}(\boldsymbol{\alpha})\beta^i\beta^j}, \text{ and}$$

$$\frac{\partial \mathcal{L}}{\partial \beta^i} = -\frac{1}{2D}(g_{k\ell}\,\delta_i^k\beta^\ell + g_{k\ell}\beta^k\delta_i^\ell) = -\frac{1}{D}g_{ki}\beta^k.$$

Thus the Euler-Lagrange equation is

$$\frac{-\frac{1}{2}\frac{\partial g_{k\ell}}{\partial \alpha^i}\frac{d\mu_*^k}{dt}\frac{d\mu_*^\ell}{dt}}{D} + \frac{d}{dt}\frac{g_{ki}\frac{d\mu_*^k}{dt}}{D} = 0. \tag{8.10}$$

Define $s : [a,b] \to \mathbb{R}$ by

$$s(t) = \int_a^t \sqrt{-g_{ij}(\boldsymbol{\mu}_*(t))\frac{d\mu_*^i}{dt}(t)\frac{d\mu_*^j}{dt}(t)}\, dt, \quad t \in [a,b].$$

Then $s(a)=0$, $s(b)=\tau(\mu_*)$, and since μ_* is timelike,

$$\frac{ds}{dt}(t) = D(\boldsymbol{\mu}_*(t), \dot{\boldsymbol{\mu}}_*(t)) > 0 \quad (t \in [a,b]).$$

Thus the smooth function s is strictly increasing, and $s([a,b]) = [0, \tau(\mu_*)]$. In particular, s is a bijection, and has an inverse map $t : [0,\tau(\mu_*)] \to [a,b]$. Also, since the derivative of s is everywhere positive on (a,b), it follows that the map $s : (a,b) \to (0,\tau(\mu_*))$ is a diffeomorphism. Thus its inverse t is smooth. So by the chain rule and the inverse function theorem, we have that for any smooth $h : (a,b) \to \mathbb{R}$,

$$\frac{d(h\circ t)}{ds}(s) = \frac{1}{\frac{ds}{dt}(t(s))}\frac{dh}{dt}(t(s)) = \frac{1}{D(\boldsymbol{\mu}_*(t(s)), \dot{\boldsymbol{\mu}}_*(t(s)))}\frac{dh}{dt}(t(s)).$$

In particular, defining $\boldsymbol{\gamma}_* : (0,\tau(\mu_*)) \to \mathbf{U}$ by $\boldsymbol{\gamma}_*(s) = \boldsymbol{\mu}_*(t(s))$, we have

$$\frac{d\boldsymbol{\gamma}_*}{ds}(s) = \frac{1}{D(\boldsymbol{\mu}_*(t(s)), \dot{\boldsymbol{\mu}}_*(t(s)))}\frac{d\boldsymbol{\mu}_*}{dt}(t(s)).$$

Thus (8.10) becomes (after dividing throughout by D, and using the above)

$$-\frac{1}{2}\frac{\partial g_{k\ell}}{\partial \alpha^i}\frac{d\gamma_*^k}{ds}\frac{d\gamma_*^\ell}{ds} + \frac{d}{ds}\Big(g_{ki}\frac{d\gamma_*^k}{ds}\Big) = 0.$$

Expanding the second summand, we obtain

$$\frac{d}{ds}\Big(g_{ki}(\boldsymbol{\gamma}_*(s))\frac{d\gamma_*^k}{ds}(s)\Big) - \frac{\partial g_{ki}}{\partial \alpha^\ell}(\boldsymbol{\gamma}_*(s))\frac{d\gamma_*^\ell}{ds}(s)\frac{d\gamma_*^k}{ds}(s) + g_{ki}(\boldsymbol{\gamma}_*(s))\frac{d^2\gamma_*^k}{ds^2}(s).$$

Thus

$$\begin{aligned} 0 &= g_{ki}\frac{d^2\gamma_*^k}{ds^2} + \frac{\partial g_{ki}}{\partial \alpha^\ell}\frac{d\gamma_*^\ell}{ds}\frac{d\gamma_*^k}{ds} - \frac{1}{2}\frac{\partial g_{k\ell}}{\partial \alpha^i}\frac{d\gamma_*^k}{ds}\frac{d\gamma_*^\ell}{ds} \\ &= g_{ki}\frac{d^2\gamma_*^k}{ds^2} + \frac{1}{2}\Big(\frac{\partial g_{ki}}{\partial \alpha^\ell}\frac{d\gamma_*^\ell}{ds}\frac{d\gamma_*^k}{ds} + \frac{\partial g_{\ell i}}{\partial \alpha^k}\frac{d\gamma_*^k}{ds}\frac{d\gamma_*^\ell}{ds}\Big) - \frac{1}{2}\frac{\partial g_{k\ell}}{\partial \alpha^i}\frac{d\gamma_*^k}{ds}\frac{d\gamma_*^\ell}{ds} \\ &= g_{ki}\frac{d^2\gamma_*^k}{ds^2} + \frac{1}{2}\Big(\frac{\partial g_{ki}}{\partial \alpha^\ell} + \frac{\partial g_{\ell i}}{\partial \alpha^k} - \frac{\partial g_{k\ell}}{\partial \alpha^i}\Big)\frac{d\gamma_*^k}{ds}\frac{d\gamma_*^\ell}{ds}. \end{aligned}$$

(To get the second equality, we double the coloured middle summand by swapping the k, ℓ dummy indices.)

Set $g^{ij} := \mathbf{g}^{ij} \circ \mathbf{x}^{-1}$. Then multiplying by the evaluation of g^{ij} on $\gamma_*(s)$, and summing over i yields, in light of (8.9) and Proposition 6.3 (p.119),

$$\begin{aligned} 0 &= g^{ij} g_{ki} \frac{d^2\gamma_*^k}{ds^2} + \frac{g^{ij}}{2}\Big(\frac{\partial g_{ki}}{\partial \alpha^\ell} + \frac{\partial g_{\ell i}}{\partial \alpha^k} - \frac{\partial g_{k\ell}}{\partial \alpha^i}\Big)\frac{d\gamma_*^k}{ds}\frac{d\gamma_*^\ell}{ds} \\ &= \frac{d^2\gamma_*^j}{ds^2} + \Gamma_{k\ell}^j(\mathbf{x}^{-1}(\gamma_*(s)))\frac{d\gamma_*^k}{ds}\frac{d\gamma_*^\ell}{ds}. \end{aligned}$$

Consequently, $\gamma_* := \mathbf{x}^{-1}(\boldsymbol{\gamma}_*) : (0, \tau(\mu_*)) \to U$ satisfies the geodesic equation in $(U, \mathbf{x})$, and hence it is a geodesic. We note that

$$\gamma_*(s) = \mathbf{x}^{-1}(\boldsymbol{\gamma}_*(s)) = \mathbf{x}^{-1}(\boldsymbol{\mu}_*(t(s))) = (\mu_* \circ t)(s) \text{ for all } s \in (0, \tau(\mu_*)).$$

This completes the proof of Theorem 8.5. □

Essentially the same proof can be used to show that length minimisers in Riemannian manifolds are geodesics.

Remark 8.4. (Hopf-Rinow theorem.) Let $(M, \mathbf{g})$ be a connected Riemannian manifold. For $p, q \in M$, we say that a smooth curve $\gamma : [a, b] \to M$ *connects p and q* if $\gamma(a) = p$ and $\gamma(b) = q$, and then we set

$$L(\gamma) = \int_a^b \sqrt{\mathbf{g}(\gamma(t))(v_{\gamma,\gamma(t)}, v_{\gamma,\gamma(t)})}\, dt.$$

The *distance* between p and q is defined as

$$d(p, q) = \inf\{L(\gamma) : \gamma \text{ connects } p \text{ and } q\}.$$

Then (M, d) is a metric space[7]. The Hopf-Rinow theorem states that (M, d) is a complete metric space if and only if $(M, \mathbf{g})$ is geodesically complete[8]. However, for general Lorentzian manifolds, metric completeness and geodesic completeness are unrelated[9]. ∗

Exercise 8.16. Let $(M, \mathbf{g})$ be a semi-Riemannian manifold with the metric $\mathbf{g}$. For a smooth curve $\gamma : [a, b] \to M$, define the *energy of* γ by

$$\mathcal{E}(\gamma) = \int_a^b \mathbf{g}(\gamma(t))(v_{\gamma,\gamma(t)}, v_{\gamma,\gamma(t)})\, dt.$$

Suppose that the range of γ lies in an admissible chart $(U, \mathbf{x})$.
Show that the Euler-Lagrange equation for $\mathcal{E}$ is the geodesic equation.

Exercise 8.17. Consider $\mathrm{GL}_2(\mathbb{R})$ as a 4-dimensional Lorentzian manifold as in Exercise 5.21 (p.88) with the trace metric. Show that $P : \mathbb{R} \to M$ satisfies the Euler-Lagrange equation corresponding to the energy functional $\mathcal{E}$ from Exercise 8.16 above if and only if $\ddot{P} - \dot{P}P^{-1}\dot{P} = 0$.

---

[7]See, e.g. [O'Neill (1983), §18, Chapter 5].
[8]See, e.g., [O'Neill (1983), §21, Chapter 5].
[9]See [Beem, Ehrlich and Easley (1996)].

Chapter 9

Curvature

The extent to which the commutator of covariant derivatives differs from the covariant derivative along the commutator of the vector field is used as a measure of the 'curvedness' of a manifold with a connection. In this chapter, we will learn about this object, namely a (1, 3)-tensor field $\mathbf{R}$, called the curvature tensor field defined as follows: for vector fields $X, Y, Z \in T_0^1 M$ and a 1-form field $\Omega \in T_1^0 M$,

$$R(X,Y)Z := [\nabla_X, \nabla_Y]Z - \nabla_{[X,Y]}Z := \nabla_X \nabla_Y Z - \nabla_Y \nabla_X Z - \nabla_{[X,Y]}Z$$
$$\mathbf{R}(\Omega, X, Y, Z) := \Omega(R(X,Y)Z).$$

We had met the curvature tensor field $\mathbf{R}$ in Exercise 6.20 (p.122), where we had seen that for a locally flat semi-Riemannian manifold $(M, \mathbf{g})$, the curvature tensor field $\mathbf{R}$ is zero. For example, $\mathbf{R}$ is zero for the Minkowski[1] and cylindrical[2] spacetimes, as these are locally flat. But $\mathbf{R} \neq \mathbf{0}$ for S^2 considered as a Riemannian manifold with the metric induced on its tangent spaces from the Euclidean inner product on $\mathbb{R}^3$ (Exercise 6.20, p.122). In the appendix to this chapter we will also learn that if $\mathbf{R} = \mathbf{0}$, then the semi-Riemannian manifold is locally flat. This provides the motivation for considering $\mathbf{R}$ to be a quantification of curvature. We begin with the following definition.

Definition 9.1. (Curvature tensor field.)
Let M be a smooth manifold with a connection ∇.
Define the *curvature operator* $R : T_0^1 M \times T_0^1 M \times T_0^1 M \to T_0^1 M$ by

$$R(X,Y)Z = \nabla_X \nabla_Y Z - \nabla_Y \nabla_X Z - \nabla_{[X,Y]}Z \text{ for all } X, Y, Z \in T_0^1 M.$$

Define the *curvature tensor field* $\mathbf{R} \in T_3^1 M$ by

$$\mathbf{R}(\Omega, X, Y, Z) = \Omega(R(X,Y)Z) = \Omega(\nabla_X \nabla_Y Z - \nabla_Y \nabla_X Z - \nabla_{[X,Y]}Z)$$

for all $X, Y, Z \in T_0^1 M$, $\Omega \in T_1^0 M$.
If $(M, \mathbf{g})$ is a semi-Riemannian manifold with the Levi-Civita connection ∇, then we call $\mathbf{R}$ the *Riemann curvature tensor field.*

[1]Example 5.6, p.83.
[2]Example 5.8, p.86.

We had seen in Exercise 6.20 (p.122) that $\mathbf{R}$ is indeed a $(1,3)$-tensor field.

Exercise 9.1. (Commutator of Lie derivatives.)
Let M be a smooth manifold. By Definition 3.14 (p.55), the Lie derivative of $Y \in T_0^1 M$ with respect to $X \in T_0^1 M$ is $\mathcal{L}_X Y = [X,Y] \in T_0^1 M$. Show that for all $X, Y, Z \in T_0^1 M$, $\mathcal{L}_X \mathcal{L}_Y Z - \mathcal{L}_Y \mathcal{L}_X Z - \mathcal{L}_{[X,Y]} Z = 0$. *Hint:* Jacobi identity.

Exercise 9.2. Let $(G, \mathbf{g})$ be a Lie group with a metric $\mathbf{g}$ which is left-invariant. Also suppose that for all $X, Y, Z \in \mathfrak{g}$, $\mathbf{g}([Z,X],Y) = -\mathbf{g}(X,[Z,Y])$. Show that $R(X,Y)Z = -\frac{1}{4}[[X,Y],Z]$ for all $X, Y, Z \in \mathfrak{g}$. *Hint:* Exercise 6.16 (p.119).

We can express the chart-induced components of $\mathbf{R}$ in terms of the connection coefficients.

Proposition 9.1. *Let M be a smooth manifold with a connection ∇, and let $(U, \mathbf{x})$ be an admissible chart. Then*

$$\mathbf{R}^{\ell}_{ijk} = \partial_{x^i}\Gamma^{\ell}_{kj} - \partial_{x^j}\Gamma^{\ell}_{ki} + \Gamma^{r}_{kj}\Gamma^{\ell}_{ri} - \Gamma^{r}_{ki}\Gamma^{\ell}_{rj}.$$

Proof. We have

$$\begin{aligned}
\mathbf{R}^{\ell}_{ijk} &= dx^{\ell}(\nabla_{\partial_{x^i}}\nabla_{\partial_{x^j}}\partial_{x^k} - \nabla_{\partial_{x^j}}\nabla_{\partial_{x^i}}\partial_{x^k} - \nabla_{[\partial_{x^i},\partial_{x^j}]}\partial_{x^k}) \\
&= dx^{\ell}(\nabla_{\partial_{x^i}}(\Gamma^{r}_{kj}\partial_{x^r}) - \nabla_{\partial_{x^j}}(\Gamma^{r}_{ki}\partial_{x^r}) - \nabla_0\partial_{x^k}) \\
&= dx^{\ell}((\partial_{x^i}\Gamma^{r}_{kj})\partial_{x^r} + \Gamma^{r}_{kj}\nabla_{\partial_{x^i}}\partial_{x^r} - (\partial_{x^j}\Gamma^{r}_{ki})\partial_{x^r} - \Gamma^{r}_{ki}\nabla_{\partial_{x^j}}\partial_{x^r} - 0) \\
&= (\partial_{x^i}\Gamma^{r}_{kj})\delta^{\ell}_{r} + dx^{\ell}(\Gamma^{r}_{kj}\Gamma^{s}_{ri}\partial_{x^s}) - (\partial_{x^j}\Gamma^{r}_{ki})\delta^{\ell}_{r} - dx^{\ell}(\Gamma^{r}_{ki}\Gamma^{s}_{rj}\partial_{x^s}) \\
&= \partial_{x^i}\Gamma^{\ell}_{kj} + \Gamma^{r}_{kj}\Gamma^{\ell}_{ri} - \partial_{x^j}\Gamma^{\ell}_{ki} - \Gamma^{r}_{ki}\Gamma^{\ell}_{rj}.
\end{aligned}$$

□

Exercise 9.3. Let M be a smooth manifold with a connection ∇, and let $(U, \mathbf{x})$ be an admissible chart. Show that $R(\partial_{x^i}, \partial_{x^j})\partial_{x^k} = \mathbf{R}^{\ell}_{ijk}\partial_{x^\ell}$.
Hint: Apply dx^r on both sides.

For a 4-dimensional manifold, there are $4^4 = 256$ components of $\mathbf{R}$. Fortunately, one does not have to do as many computations because of presence of various symmetries. We begin with the following.

Lemma 9.1. *Let M be a smooth manifold with a connection ∇. Then for all vector fields $X, Y, Z \in T_0^1 M$, we have $R(X,Y)Z = -R(Y,X)Z$.*

In terms of components of $\mathbf{R}$ in an admissible chart $(U, \mathbf{x})$, $\mathbf{R}^{\ell}_{ijk} = -\mathbf{R}^{\ell}_{jik}$.

Proof. We have

$$\begin{aligned}
R(X,Y)Z &= \nabla_X\nabla_Y Z - \nabla_Y\nabla_X Z - \nabla_{[X,Y]}Z \\
&= -(-\nabla_X\nabla_Y Z + \nabla_Y\nabla_X Z + \nabla_{-[Y,X]}Z) \\
&= -(\nabla_Y\nabla_X Z - \nabla_X\nabla_Y Z - \nabla_{[Y,X]}Z) \\
&= -R(Y,X)Z.
\end{aligned}$$

□

Proposition 9.2. (First Bianchi[3] identity.)
Let $(M, \mathbf{g})$ be a semi-Riemannian manifold with the Levi-Civita connection. Then for all $X, Y, Z \in T_0^1 M$, $R(X,Y)Z + R(Y,Z)X + R(Z,X)Y = 0$.

In terms of components of $\mathbf{R}$ in an admissible chart $(U, \mathbf{x})$, we have that $\mathbf{R}_{ijk}^{\ell} + \mathbf{R}_{jki}^{\ell} + \mathbf{R}_{kij}^{\ell} = 0$.

Proof. We will use the torsion-freeness of the Levi-Civita connection (for all $V, W \in T_0^1 M$, $\nabla_V W - \nabla_W V = [V, W]$), and the Jacobi identity. We have

$$\begin{aligned}
&R(X,Y)Z + R(Y,Z)X + R(Z,X)Y \\
= \;& \nabla_X \nabla_Y Z - \nabla_Y \nabla_X Z - \nabla_{[X,Y]} Z \\
&+ \nabla_Y \nabla_Z X - \nabla_Z \nabla_Y X - \nabla_{[Y,Z]} X \\
&+ \nabla_Z \nabla_X Y - \nabla_X \nabla_Z Y - \nabla_{[Z,X]} Y \\
\overset{\text{(C4)}}{=}\;& \nabla_X [Y,Z] - \nabla_{[Y,Z]} X + \nabla_Y [Z,X] - \nabla_{[Z,X]} Y + \nabla_Z [X,Y] - \nabla_{[X,Y]} Z \\
\overset{\text{(C4)}}{=}\;& [X,[Y,Z]] + [Y,[Z,X]] + [Z,[X,Y]] \overset{\text{(Jacobi)}}{=} 0.
\end{aligned}$$

□

In a semi-Riemannian manifold $(M, \mathbf{g})$, we can use the musical isomorphism to construct, from the Riemann curvature tensor field, a totally covariant $(0,4)$-tensor field $\widetilde{\mathbf{R}}$.

Definition 9.2. (Covariant curvature tensor field.)
Let $(M, \mathbf{g})$ be a semi-Riemannian manifold.
Define *covariant curvature tensor field* $\widetilde{\mathbf{R}} \in T_4^0 M$ by

$$\widetilde{\mathbf{R}}(X,Y,Z,W) = \mathbf{R}(W^\flat, X, Y, Z) = \mathbf{g}(W, R(X,Y)Z)$$

for all $W, X, Y, Z \in T_0^1 M$.
In a chart $(U, \mathbf{x})$, the covariant curvature tensor field has components $\widetilde{\mathbf{R}}_{ijk\ell} = \mathbf{g}_{\ell r} \mathbf{R}_{ijk}^{r}$, as

$$\begin{aligned}
\widetilde{\mathbf{R}}_{ijk\ell} &= \widetilde{\mathbf{R}}(\partial_{x^i}, \partial_{x^j}, \partial_{x^k}, \partial_{x^\ell}) = \mathbf{g}(\partial_{x^\ell}, R(\partial_{x^i}, \partial_{x^j})\partial_{x^k}) \\
&= \mathbf{g}(\partial_{x^\ell}, \mathbf{R}_{ijk}^{r} \partial_{x^r}) = \mathbf{R}_{ijk}^{r} \mathbf{g}(\partial_{x^\ell}, \partial_{x^r}) = \mathbf{R}_{ijk}^{r} \mathbf{g}_{\ell r} = \mathbf{g}_{\ell r} \mathbf{R}_{ijk}^{r}.
\end{aligned}$$

Exercise 9.4. (First Bianchi identity for the covariant curvature tensor field.) Let $(M, \mathbf{g})$ be a semi-Riemannian manifold with the Levi-Civita connection. Let $W, X, Y, Z \in T_0^1 M$. Show that $\widetilde{\mathbf{R}}(X,Y,Z,W) + \widetilde{\mathbf{R}}(Y,Z,X,W) + \widetilde{\mathbf{R}}(Z,X,Y,W) = 0$.

Exercise 9.5. Let $(M, \mathbf{g})$ be a semi-Riemannian manifold with the Levi-Civita connection ∇. Show that for all $\Omega \in T_1^0 M$ and all $X, Y, Z \in T_0^1 M$,

$$\mathbf{R}(\Omega, X, Y, Z) = \widetilde{\mathbf{R}}(X, Y, Z, \Omega^\sharp).$$

Hence or otherwise, conclude that in a chart $(U, \mathbf{x})$, $\mathbf{R}_{ijk}^{\ell} = \widetilde{\mathbf{R}}_{ijkr} \mathbf{g}^{\ell r}$.

Exercise 9.6. Let $(M, \mathbf{g})$ be a semi-Riemannian manifold. Prove $\widetilde{\mathbf{R}} = \mathbf{C}_5^1(\mathbf{R} \otimes \mathbf{g})$.

[3]After Luigi Bianchi (1856–1928), an Italian mathematician. We will meet the *second* Bianchi identity in Prop. 9.4 (p.185), which will involve the covariant derivatives of $\mathbf{R}$.

We have the following symmetries for $\widetilde{\mathbf{R}}$.

Proposition 9.3. *Let $(M, \mathbf{g})$ be a semi-Riemannian manifold with the Levi-Civita connection ∇. Then:*

For all $X, Y, Z, W \in T_0^1 M$,	*In any admissible chart $(U, \mathbf{x})$,*
$\widetilde{\mathbf{R}}(X,Y,Z,W) = -\widetilde{\mathbf{R}}(Y,X,Z,W)$	$\widetilde{\mathbf{R}}_{ijk\ell} = -\widetilde{\mathbf{R}}_{jik\ell}$
$\widetilde{\mathbf{R}}(X,Y,Z,W) = -\widetilde{\mathbf{R}}(X,Y,W,Z)$	$\widetilde{\mathbf{R}}_{ijk\ell} = -\widetilde{\mathbf{R}}_{ij\ell k}$
$\widetilde{\mathbf{R}}(X,Y,Z,W) = \widetilde{\mathbf{R}}(Z,W,X,Y)$	$\widetilde{\mathbf{R}}_{ijk\ell} = \widetilde{\mathbf{R}}_{k\ell ij}$

In the right column above, the i, j, k, ℓ are arbitrary in $\{1, \cdots, m\}$.

Proof. The first claim follows from Lemma 9.1 (p.182), since

$$\widetilde{\mathbf{R}}(X,Y,Z,W) = \mathbf{g}(W, R(X,Y)Z) = \mathbf{g}(W, -R(Y,X)Z) = -\widetilde{\mathbf{R}}(Y,X,Z,W).$$

For the second claim, we use the metric-compatibility of the Levi-Civita connection. First,

$$\begin{aligned}
&\mathbf{g}(W, \nabla_X \nabla_Y Z) \\
&= -\mathbf{g}(\nabla_X W, \nabla_Y Z) + X\mathbf{g}(W, \nabla_Y Z) \\
&= \mathbf{g}(\nabla_Y \nabla_X W, Z) - Y\mathbf{g}(\nabla_X W, Z) + XY\mathbf{g}(W, Z) - X\mathbf{g}(\nabla_Y W, Z). \quad (9.1)
\end{aligned}$$

Swapping X and Y, we also have

$$\begin{aligned}
&\mathbf{g}(W, \nabla_Y \nabla_X Z) \\
&= \mathbf{g}(\nabla_X \nabla_Y W, Z) - X\mathbf{g}(\nabla_Y W, Z) + YX\mathbf{g}(W, Z) - Y\mathbf{g}(\nabla_X W, Z).
\end{aligned}$$

Moreover, $\mathbf{g}(W, \nabla_{[X,Y]} Z) = -\mathbf{g}(\nabla_{[X,Y]} W, Z) + [X,Y]\mathbf{g}(W, Z)$. Subtracting the last two equations from (9.1), we obtain

$$\begin{aligned}
\mathbf{g}(W, R(X,Y)Z) &= \mathbf{g}(\nabla_Y \nabla_X W, Z) - \mathbf{g}(\nabla_X \nabla_Y W, Z) + \mathbf{g}(\nabla_{[X,Y]} W, Z) \\
&= -\mathbf{g}(R(X,Y)W, Z),
\end{aligned}$$

and so $\widetilde{\mathbf{R}}(X,Y,Z,W) = -\widetilde{\mathbf{R}}(X,Y,W,Z)$, proving the second claim.

Finally, using the anti-symmetry in the last two components, and the first Bianchi identity, we have

$$\widetilde{\mathbf{R}}(X,Y,Z,W) = -\widetilde{\mathbf{R}}(X,Y,W,Z) = \widetilde{\mathbf{R}}(Y,W,X,Z) + \widetilde{\mathbf{R}}(W,X,Y,Z).$$

Also, $\widetilde{\mathbf{R}}(X,Y,Z,W) = -\widetilde{\mathbf{R}}(Y,X,Z,W) = \widetilde{\mathbf{R}}(X,Z,Y,W) + \widetilde{\mathbf{R}}(Z,Y,X,W)$. Adding these yields

$$\begin{aligned}
&2\widetilde{\mathbf{R}}(X,Y,Z,W) \\
&= \underline{\widetilde{\mathbf{R}}(Y,W,X,Z)} + \underline{\widetilde{\mathbf{R}}(W,X,Y,Z)} + \underline{\widetilde{\mathbf{R}}(X,Z,Y,W)} + \widetilde{\mathbf{R}}(Z,Y,X,W). \quad (9.2)
\end{aligned}$$

By replacing (X, Y, Z, W) by (Z, W, X, Y), we obtain also

$$\begin{aligned}
&2\widetilde{\mathbf{R}}(Z, W, X, Y) \\
&= \widetilde{\mathbf{R}}(W, Y, Z, X) + \widetilde{\mathbf{R}}(Y, Z, W, X) + \widetilde{\mathbf{R}}(Z, X, W, Y) + \underline{\widetilde{\mathbf{R}}(X, W, Z, Y)}. \quad (9.3)
\end{aligned}$$

Since a simultaneous swap of the first two and of the last two arguments of $\widetilde{\mathbf{R}}$ does not change its value, a comparison of the right-hand sides of (9.2) and (9.3) reveals that the left-hand sides are also equal. □

Exercise 9.7. Let $(M, \mathbf{g})$ be a semi-Riemannian manifold with the Levi-Civita connection ∇. Show that in any admissible chart $(U, \mathbf{x})$, we have $\mathbf{R}^k_{kij} = \mathbf{R}^k_{kji}$.

Exercise 9.8. Recall Example 5.7 (p.85), where we considered the sphere S^2 as a Riemannian manifold using the metric $\mathbf{g}$ induced on the tangent spaces from the Euclidean inner product on $\mathbb{R}^3$. In the chart (U, φ) described there, determine the component $\mathbf{R}^\theta_{\phi\theta\phi}$ using Proposition 9.1. Also determine $\widetilde{\mathbf{R}}_{\phi\theta\phi\theta}$, and hence determine all the 16 components of $\widetilde{\mathbf{R}}$ using the symmetries of $\widetilde{\mathbf{R}}$. Find all the components of $\mathbf{R}$ as well.

We now show the second Bianchi identity, which involves derivatives, and bears some resemblance to the first Bianchi identity seen earlier. First we note that as $\mathbf{R}$ is a $(1,3)$-tensor field, given any $V \in T^1_0 M$, its covariant derivative $\nabla_V \mathbf{R}$ is a $(1,3)$-tensor field. For $\Omega \in T^0_1 M$ and $X, Y, Z \in T^1_0 M$,

$$\begin{aligned}
&(\nabla_V \mathbf{R})(\Omega, X, Y, Z) \\
&= V(\mathbf{R}(\Omega, X, Y, Z)) - \mathbf{R}(\nabla_V \Omega, X, Y, Z) \\
&\quad - \mathbf{R}(\Omega, \nabla_V X, Y, Z) - \mathbf{R}(\Omega, X, \nabla_V Y, Z) - \mathbf{R}(\Omega, X, Y, \nabla_V Z) \\
&= V(\Omega(R(X, Y)Z)) - (\nabla_V \Omega)(R(X, Y)Z) \\
&\quad - \Omega(R(\nabla_V X, Y)Z) - \Omega(R(X, \nabla_V Y)Z) - \Omega(R(X, Y)\nabla_V Z) \\
&= \Omega\big(\nabla_V(R(X, Y)Z) - R(\nabla_V X, Y)Z - R(X, \nabla_V Y)Z - R(X, Y)\nabla_V Z\big).
\end{aligned}$$

So defining the map $(\nabla_V R)(X, Y) : T^1_0 M \to T^1_0 M$ by

$$\begin{aligned}
&((\nabla_V R)(X, Y))Z \\
&:= \nabla_V(R(X, Y)Z) - R(\nabla_V X, Y)Z - R(X, \nabla_V Y)Z - R(X, Y)\nabla_V Z, \quad (9.4)
\end{aligned}$$

we see that $(\nabla_V \mathbf{R})(\Omega, X, Y, Z) = \Omega((\nabla_V R)(X, Y)Z)$. We are now ready to state the result on the second Bianchi identity.

Proposition 9.4. (Second Bianchi identity.)
Let $(M, \mathbf{g})$ be a semi-Riemannian manifold with the Levi-Civita connection ∇. Then with the notation in (9.4), we have for all $X, Y, Z \in T^1_0 M$ that

$$(\nabla_X R)(Y, Z) + (\nabla_Y R)(Z, X) + (\nabla_Z R)(X, Y) = 0. \quad (9.5)$$

In terms of the components of the Riemann curvature tensor field in an admissible chart $(U, \mathbf{x})$, $(\nabla_{\partial_{x^h}} \mathbf{R})^i_{k\ell j} + (\nabla_{\partial_{x^k}} \mathbf{R})^i_{\ell hj} + (\nabla_{\partial_{x^\ell}} \mathbf{R})^i_{hkj} = 0$.

By abuse of notation, the last equation is sometimes written in the literature as $\mathbf{R}^i_{k\ell j;\,h} + \mathbf{R}^i_{\ell hj;\,k} + \mathbf{R}^i_{hkj;\,\ell} = 0$.

Proof. The left-hand side of (9.5), henceforth denoted by LHS, is the (pointwise defined) sum of maps from $T^1_0 M$ to $T^1_0 M$, and we want to show that this is the zero map, i.e., its action on each $V \in T^1_0 M$ is the zero vector field $0 \in T^1_0 M$. By using the definition in (9.4),

- $((\nabla_Z R)(X,Y))V$
  $= \nabla_Z(R(X,Y)V) - \underline{R(\nabla_Z X, Y)V} - \underline{R(X, \nabla_Z Y)V} - R(X,Y)\nabla_Z V,$
- $((\nabla_X R)(Y,Z))V$
  $= \nabla_X(R(Y,Z)V) - \underline{R(\nabla_X Y, Z)V} - \underline{R(Y, \nabla_X Z)V} - R(Y,Z)\nabla_X V,$
- $((\nabla_Y R)(Z,X))V$
  $= \nabla_Y(R(Z,X)V) - \underline{R(\nabla_Y Z, X)V} - \underline{R(Z, \nabla_Y X)V} - R(Z,X)\nabla_Y V.$

We will now add these equations. But before doing so, we note that using the skew-symmetry of R (Lemma 9.1, p.182) with the torsion-freeness of the Levi-Civita connection, we can combine the respective underlined terms:

$$\begin{aligned}
-R(\nabla_Z X, Y)V - R(Y, \nabla_X Z)V &= R(Y, \nabla_Z X - \nabla_X Z)V = R(Y,[Z,X])V \\
-R(X, \nabla_Z Y)V - R(\nabla_Y Z, X)V &= R(X, \nabla_Y Z - \nabla_Z Y)V = R(X,[Y,Z])V \\
-R(\nabla_X Y, Z)V - R(Z, \nabla_Y X)V &= R(Z, \nabla_X Y - \nabla_Y X)V = R(Z,[X,Y])V.
\end{aligned}$$

Then we obtain the following:

$$\begin{aligned}
\text{LHS} = {} & \nabla_Z(R(X,Y)V) + \underline{R(Y,[Z,X])V} - R(X,Y)\nabla_Z V \\
& + \nabla_X(R(Y,Z)V) + \underline{R(X,[Y,Z])V} - R(Y,Z)\nabla_X V \\
& + \nabla_Y(R(Z,X)V) + \underline{R(Z,[X,Y])V} - R(Z,X)\nabla_Y V \\
= {} & \nabla_Z\nabla_X\nabla_Y V - \nabla_Z\nabla_Y\nabla_X V - \nabla_Z\nabla_{[X,Y]}V \\
& + \nabla_Y\nabla_{[Z,X]}V - \nabla_{[Z,X]}\nabla_Y V - \nabla_{[Y,[Z,X]]}V \\
& - \nabla_X\nabla_Y\nabla_Z V + \nabla_Y\nabla_X\nabla_Z V + \nabla_{[X,Y]}\nabla_Z V \\
& + \nabla_X\nabla_Y\nabla_Z V - \nabla_X\nabla_Z\nabla_Y V - \nabla_X\nabla_{[Y,Z]}V \\
& + \nabla_X\nabla_{[Y,Z]}V - \nabla_{[Y,Z]}\nabla_X V - \nabla_{[X,[Y,Z]]}V \\
& - \nabla_Y\nabla_Z\nabla_X V + \nabla_Z\nabla_Y\nabla_X V + \nabla_{[Y,Z]}\nabla_X V \\
& + \nabla_Y\nabla_Z\nabla_X V - \nabla_Y\nabla_X\nabla_Z V - \nabla_Y\nabla_{[Z,X]}V \\
& + \nabla_Z\nabla_{[X,Y]}V - \nabla_{[X,Y]}\nabla_Z V - \nabla_{[Z,[X,Y]]}V \\
& - \nabla_Z\nabla_X\nabla_Y V + \nabla_X\nabla_Z\nabla_Y V + \nabla_{[Z,X]}\nabla_Y V.
\end{aligned}$$

The terms on the right-hand side all cancel except for the following:

$$\begin{aligned}\text{LHS} &= -\nabla_{[Y,[Z,X]]}V - \nabla_{[X,[Y,Z]]}V - \nabla_{[Z,[X,Y]]}V \\ &= -\nabla_{[Y,[Z,X]]+[X,[Y,Z]]+[Z,[X,Y]]}V \overset{\text{(Jacobi)}}{=} -\nabla_0 V = 0,\end{aligned}$$

proving the second Bianchi identity.

To obtain the identity for components in a chart, we first note that the second Bianchi identity above implies that for all $\Omega \in T_1^0 M$ and all $X, Y, Z, V \in T_0^1 M$,

$$\begin{aligned}0 &= \Omega\big(((\nabla_X R)(Y,Z))V + ((\nabla_Y R)(Z,X))V + ((\nabla_Z R)(X,Y))V\big) \\ &= (\nabla_X \mathbf{R})(\Omega, Y, Z, V) + (\nabla_Y \mathbf{R})(\Omega, Z, X, V) + (\nabla_Z \mathbf{R})(\Omega, X, Y, V).\end{aligned}$$

With $\Omega = dx^i \in T_1^0 U$, $X = \partial_{x^h}$, $Y = \partial_{x^k}$, $Z = \partial_{x^\ell}$, $V = \partial_{x^j}$ in $T_0^1 U$, we get

$$\begin{aligned}0 = &\ (\nabla_{\partial_{x^h}} \mathbf{R})(dx^i, \partial_{x^k}, \partial_{x^\ell}, \partial_{x^j}) \\ &+ (\nabla_{\partial_{x^k}} \mathbf{R})(dx^i, \partial_{x^\ell}, \partial_{x^h}, \partial_{x^j}) \\ &+ (\nabla_{\partial_{x^\ell}} \mathbf{R})(dx^i, \partial_{x^h}, \partial_{x^k}, \partial_{x^j}) \\ = &\ (\nabla_{\partial_{x^h}} \mathbf{R})^i_{k\ell j} + (\nabla_{\partial_{x^k}} \mathbf{R})^i_{\ell hj} + (\nabla_{\partial_{x^\ell}} \mathbf{R})^i_{hkj}.\end{aligned}$$

□

9.1 Ricci and scalar curvatures

The curvature tensor field $\mathbf{R}$ is a complicated $(1,3)$-tensor field, but we will now see that by contracting it, simpler tensor fields can be obtained.

Definition 9.3. (Ricci curvature tensor field.)
Let $(M, \mathbf{g})$ be a semi-Riemannian manifold with the Levi-Civita connection ∇. The *Ricci curvature tensor field*, $\mathbf{Ric} \in T_2^0 M$ is the contraction $\mathbf{C}_1^1 \mathbf{R}$.

More explicitly, for $Y, Z \in T_0^1 M$,

$$\mathbf{Ric}(Y,Z) = (\mathbf{C}_1^1 \mathbf{R})(Y,Z) = \mathbf{C}\big(T_1^0 M \times T_0^1 M \ni (\Omega, X) \mapsto \mathbf{R}(\Omega, X, Y, Z)\big),$$

and so in an admissible chart $(U, \mathbf{x})$, $\mathbf{Ric}(Y,Z) = \mathbf{R}(dx^k, \partial_{x^k}, Y, Z)$, giving

$$\mathbf{Ric}_{ij} = \mathbf{R}(dx^k, \partial_{x^k}, \partial_{x^i}, \partial_{x^j}) = \mathbf{R}^k_{kij}.$$

We now show that $\mathbf{Ric}$ is symmetric.

Proposition 9.5. *Let $(M, \mathbf{g})$ be a semi-Riemannian manifold with the Levi-Civita connection ∇. For all $Y, Z \in T_0^1 M$, $\mathbf{Ric}(Y,Z) = \mathbf{Ric}(Z,Y)$.*

Proof. It is enough to show that in any admissible chart, $\mathbf{Ric}_{ij} = \mathbf{Ric}_{ji}$, which follows from the symmetry properties of $\widetilde{\mathbf{R}}$ (Exercise 9.7, p.185):

$$\begin{aligned}\mathbf{R}^k_{kij} &= \mathbf{g}^{k\ell}\widetilde{\mathbf{R}}_{kij\ell} = \mathbf{g}^{k\ell}\widetilde{\mathbf{R}}_{j\ell ki} = (-1)(-1)\mathbf{g}^{k\ell}\widetilde{\mathbf{R}}_{\ell jik} \\ &= \mathbf{g}^{\ell k}\widetilde{\mathbf{R}}_{\ell jik} = \mathbf{R}^\ell_{\ell ji} = \mathbf{R}^k_{kji}.\end{aligned}$$

□

Exercise 9.9. Let $(M, \mathbf{g})$ be a semi-Riemannian manifold with the Levi-Civita connection ∇. Show that $\mathbf{Ric} = -\mathbf{C}^1_2\mathbf{R}$. *Hint:* Use chart components and the symmetries of $\widetilde{\mathbf{R}}$.

We note that $\mathbf{Ric} \in T^0_2 M$ can be considered as a $(1,1)$-tensor field $\mathbf{Ric}^\sharp$ by 'pulling up a covariant index' using the musical map, say the first one:

$$T^0_1 M \times T^1_0 M \ni (\Omega, Z) \mapsto \mathbf{Ric}^\sharp(\Omega, Z) := \mathbf{Ric}(\Omega^\sharp, Z) \in C^\infty(M).$$

We can further take the 'trace' of $\mathbf{Ric}^\sharp$, i.e., operate by $\mathbf{C}$, and obtain a $C^\infty(M)$-function.

Definition 9.4. (Scalar curvature function.)
Let $(M, \mathbf{g})$ be a semi-Riemannian manifold with the Levi-Civita connection ∇. The *scalar curvature (function)*, $\mathbf{S} \in C^\infty(M)$ is the contraction $\mathbf{C}(\mathbf{Ric}^\sharp) = \mathbf{C}(\mathbf{Ric}(\cdot^\sharp, \cdot))$.

Thus if $(U, \mathbf{x})$ is any admissible chart, then

$$\begin{aligned} \mathbf{S} &= \mathbf{Ric}((dx^i)^\sharp, \partial_{x^i}) = \mathbf{Ric}(\mathbf{g}^{ij}\partial_{x^j}, \partial_{x^i}) \\ &= \mathbf{g}^{ij}\mathbf{Ric}(\partial_{x^j}, \partial_{x^i}) = \mathbf{g}^{ij}\mathbf{Ric}_{ji} \\ &= \mathbf{g}^{ij}\mathbf{Ric}_{ij} \\ &= \mathbf{g}^{ij}\mathbf{R}^k_{kij}. \end{aligned}$$

Exercise 9.10. In continuation to Exercise 9.8 (p.185), calculate the components of the Ricci curvature tensor field in the chart (U, φ) for the Riemannian manifold $(S^2, \mathbf{g})$. Also determine the scalar curvature function on S^2.

For 2-dimensional semi-Riemannian manifolds, one has the following result.

Exercise 9.11. Let $(M, \mathbf{g})$ be a 2-dimensional semi-Riemannian manifold.
Show that there exists a function $k \in C^\infty(M)$ such that for all $X, Y, Z, W \in T^1_0 M$,

$$\begin{aligned} \widetilde{\mathbf{R}}(X, Y, Z, W) &= k(\mathbf{g}(X, W)\mathbf{g}(Y, Z) - \mathbf{g}(p)(X, Z)\mathbf{g}(p)(Y, W)), \\ \mathbf{Ric} &= k\mathbf{g}, \\ \mathbf{S} &= 2k. \end{aligned}$$

The function k is called the *Gaussian curvature* of M. Calculate k for the sphere from Exercise 9.10, and for the Poincaré half-plane from Exercise 8.6 (p.158).

Exercise 9.12. (Weyl/conformal curvature tensor field.)
Let $(M, \mathbf{g})$ be a semi-Riemannian manifold of dimension $m > 2$.
Define the *Weyl/conformal curvature tensor field* $\boldsymbol{C} \in T^0_4 M$ by

$$\begin{aligned} \boldsymbol{C}(X, Y, Z, W) = \widetilde{\mathbf{R}}(X, Y, Z, W) & \\ -\frac{1}{m-2}&(\mathbf{g}(X, W)\mathbf{Ric}(Y, Z) + \mathbf{g}(Y, Z)\mathbf{Ric}(X, W) \\ &\quad - \mathbf{g}(X, Z)\mathbf{Ric}(Y, W) - \mathbf{g}(Y, W)\mathbf{Ric}(X, Z)) \\ -\frac{1}{(m-1)(m-2)}&(\mathbf{g}(X, Z)\mathbf{g}(Y, W) - \mathbf{g}(X, W)\mathbf{g}(Y, Z))\mathbf{S}. \end{aligned}$$

One can check that $\boldsymbol{C}$ shares the following three symmetry properties with $\widetilde{\mathbf{R}}$. For all $X, Y, Z, W \in T_0^1 M$, we have

(S1) $\boldsymbol{C}(X, Y, Z, W) = -\boldsymbol{C}(Y, X, Z, W)$

(S2) $\boldsymbol{C}(X, Y, Z, W) = -\boldsymbol{C}(X, Y, W, Z)$

(S3) $\boldsymbol{C}(X, Y, Z, W) = \boldsymbol{C}(Z, W, X, Y)$.

Show that $\boldsymbol{C}$ has an additional property, namely,

(S4) In any admissible chart, $\mathbf{g}^{ik}\boldsymbol{C}_{ijk\ell} = 0$.

Use this and the above symmetries to conclude that $\boldsymbol{C} = \mathbf{0}$ if $m = 3$.

Remark 9.1. Let $(M, \mathbf{g})$ be an m-dimensional semi-Riemannian manifold with the Levi-Civita connection.

- If $m = 1$, then $\widetilde{\mathbf{R}} = \mathbf{0}$.
- If $m = 2$, then Exercise 9.11 shows that $\widetilde{\mathbf{R}}$ is essentially determined by the scalar curvature $\mathbf{S}$.
- If $m = 3$, then in Exercise 9.12 we have seen that the Weyl curvature vanishes, so that $\widetilde{\mathbf{R}}$ is essentially determined by the Ricci curvature $\mathbf{Ric}$.
- If $m \geqslant 4$, then even when $\mathbf{Ric} = 0$, we have that $\widetilde{\mathbf{R}}$ may not be zero, and in this case $\widetilde{\mathbf{R}}$ coincides with the Weyl curvature $\boldsymbol{C}$.

The property (S4) of the Weyl curvature means that $\boldsymbol{C}$ captures the 'traceless' part of $\widetilde{\mathbf{R}}$, and $\widetilde{\mathbf{R}}$ (or equivalently $\mathbf{R}$) can be reconstructed using the trace $\mathbf{Ric}$ and the traceless part $\boldsymbol{C}$.

If the cosmological constant $\Lambda = 0$, then it follows from the field equation that in a vacuum region of spacetime (where $\mathbf{T}$ is zero), $\mathbf{Ric}$ vanishes. So the Weyl curvature describes a gravity in a vacuum region of spacetime. An example is that of gravitational waves (see for instance Exercise 14.3, p.306). Since $\boldsymbol{C} = 0$ when $m = 3$, there can be no gravity in a vacuum region.

Lorentzian metrics $\mathbf{g}, \widetilde{\mathbf{g}}$ on a smooth manifold M are said to be *conformally related* if there exists a pointwise positive function $f \in C^\infty(M)$ such that $\widetilde{\mathbf{g}} = f\mathbf{g}$. Loosely speaking, 'angles' between vectors in each tangent space are preserved under multiplication by such a pointwise positive f, and so the causal structure is identical. It can be shown[4] that the Weyl curvature $\widetilde{\boldsymbol{C}}$ for $(M, \widetilde{\mathbf{g}})$ acquires the same factor f, that is, $\widetilde{\boldsymbol{C}} = f\boldsymbol{C}$, where $\boldsymbol{C}$ denotes the Weyl curvature for $(M, \mathbf{g})$.

The Weyl curvature $\boldsymbol{C}$ can be algebraically classified at each spacetime point, and the resulting classification types have interesting physical interpretations. The classification can be helpful in the search for exact solutions of the field equation, since simplifications are produced if one focuses on a single type. The classification[5] is done by considering the eigenvalues and eigenvectors of a symmetric matrix associated with $\boldsymbol{C}$.[6] $\ast$

[4] See, e.g., [Lee (2018), Theorem 7.30].

[5] *Petrov classification*, after the Russian mathematician, Aleksei Petrov (1910–72).

[6] See, e.g., [Hall (2004)].

Exercise 9.13. (Kretschmann scalar.)[7]
Let $(M, \mathbf{g})$ be a semi-Riemannian manifold. Define the $(4,0)$-tensor field $\widehat{\mathbf{R}}$ by

$$\widehat{\mathbf{R}} = \widetilde{\mathbf{R}}(\cdot^\sharp, \cdot^\sharp, \cdot^\sharp, \cdot^\sharp).$$

The *Kretschmann invariant/scalar* is defined by $\mathbf{K} = \mathbf{C}(\mathbf{C}_1^1(\mathbf{C}_1^1(\mathbf{C}_1^1(\widehat{\mathbf{R}} \otimes \widetilde{\mathbf{R}}))))$. Thus, in an admissible chart $(U, \mathbf{x})$, $\mathbf{K} = \widehat{\mathbf{R}}^{ijk\ell}\widetilde{\mathbf{R}}_{ijk\ell}$.

In continuation to Exercises 9.8 and 9.10, show that the Kretschmann scalar $\mathbf{K} \equiv 4$ for the Riemannian manifold $(S^2, \mathbf{g})$.

In Exercise 6.11 (p.116), we had defined the divergence of a vector field on a smooth manifold equipped with a connection. In a semi-Riemannian manifold $(M, \mathbf{g})$, we can use the musical map $\cdot^\sharp$ to define the divergence of $(0, s)$-tensor fields. We give the following definition.

Definition 9.5. (Divergence of a $(0, s)$-tensor field.)
Let $(M, \mathbf{g})$ be a semi-Riemannian manifold with the Levi-Civita connection ∇. If $T \in T_s^0 M$ is a $(0, s)$-tensor field, where $s \geqslant 1$, then its *divergence* $\operatorname{div} T \in T_{s-1}^0 M$ is the $(0, s-1)$-tensor field defined by
$(\operatorname{div} T)(V_1, \cdots, V_{s-1}) = \mathbf{C}(T_1^0 M \times T_0^1 M \ni (\Omega, V) \mapsto (\nabla_V T)(\Omega^\sharp, V_1, \cdots, V_{s-1}))$
for $V_1, \cdots V_{s-1} \in T_0^1 M$.

We note that for fixed $V_1, \cdots, V_{s-1} \in T_0^1 M$, the map $\widetilde{T}$ given by

$$T_1^0 M \times T_0^1 M \ni (\Omega, V) \stackrel{\widetilde{T}}{\longmapsto} (\nabla_V T)(\Omega^\sharp, V_1, \cdots, V_{s-1})$$

is a $(1,1)$-tensor field (the linearity in V follows from property (C3) of the connection, and the linearity in Ω is a consequence of the $C^\infty(M)$-linearity of $\cdot^\sharp$ and of $\nabla_V T$ in its first argument). So action of the contraction map $\mathbf{C}$ on $\widetilde{T}$ makes sense.

Exercise 9.14. Let $(M, \mathbf{g})$ be a semi-Riemannian manifold with the Levi-Civita connection ∇. Suppose that $T, S \in T_s^0 M$ are $(0, s)$-tensor fields, $s \geqslant 1$, and $c \in \mathbb{R}$. Show that $\operatorname{div}(T + cS) = (\operatorname{div} T) + c \operatorname{div} S$.

Exercise 9.15. Let $(M, \mathbf{g})$ be a semi-Riemannian manifold with the Levi-Civita connection ∇. Let $V \in T_0^1 M$. Prove that $\operatorname{div} V = \operatorname{div}(V^\flat) \in C^\infty(M)$.

Example 9.1. ($\operatorname{div} \mathbf{g} = 0$.) By Theorem 6.2 (p.125), $\nabla . \mathbf{g} = 0$. So in any admissible chart $(U, \mathbf{x})$, we have $(\operatorname{div} \mathbf{g})_i = (\nabla_{\partial_{x^j}} \mathbf{g})((dx^j)^\sharp, \partial_{x^i}) = 0$. ◇

---

[7]After Erich Kretschmann (1887 – 1973), a German physicist. In Chapter 14 on the field equation, Exercise 14.1 (p.306) shows that 'vacuum spacetimes' (spacetimes for which the energy-momentum tensor field $\mathbf{T} = 0$ and the cosmological constant is 0) are 'Ricci-flat' ($\mathbf{Ric} = 0$). So the scalar curvature $\mathbf{S} \equiv 0$ for vacuum spacetimes. But the Kretschmann scalar can then be useful since it may be still nonzero. For example, the Schwarzschild spacetime is Ricci-flat, but the Kretchmann scalar is nonzero; see equation (15.4), p.335.

Proposition 9.6. *Let* $(M, \mathbf{g})$ *be a semi-Riemannian manifold with the Levi-Civita connection* ∇. *Then* $\operatorname{div}\mathbf{Ric} = \frac{1}{2}d\mathbf{S}$.

Proof. Let $(U, \mathbf{x})$ be any admissible chart. Then by the second Bianchi identity, we have (after replacing h by i, and summing over i)

$$0 = (\nabla_{\partial_{x^i}}\mathbf{R})^i_{k\ell j} + (\nabla_{\partial_{x^k}}\mathbf{R})^i_{\ell ij} + (\nabla_{\partial_{x^\ell}}\mathbf{R})^i_{ikj}. \tag{9.6}$$

As contraction commutes with the covariant derivative, we obtain using the symmetry property of R (Lemma 9.1, p.182) that

$$\mathbf{C}^1_2(\nabla_{\partial_{x^k}}\mathbf{R}) = \nabla_{\partial_{x^k}}(\mathbf{C}^1_2\mathbf{R}) = -\nabla_{\partial_{x^k}}(\mathbf{C}^1_1\mathbf{R}) = -\mathbf{C}^1_1(\nabla_{\partial_{x^k}}\mathbf{R}),$$

and so

$$\begin{aligned}(\nabla_{\partial_{x^k}}\mathbf{Ric})_{\ell j} &= (\nabla_{\partial_{x^k}}(\mathbf{C}^1_1\mathbf{R}))_{\ell j} = (\mathbf{C}^1_1(\nabla_{\partial_{x^k}}\mathbf{R}))_{\ell j} \\ &= -(\mathbf{C}^1_2(\nabla_{\partial_{x^k}}\mathbf{R}))_{\ell j} = -(\nabla_{\partial_{x^k}}\mathbf{R})^i_{\ell ij}.\end{aligned}$$

Moreover, $(\nabla_{\partial_{x^\ell}}\mathbf{R})^i_{ikj} = (\mathbf{C}^1_1(\nabla_{\partial_{x^\ell}}\mathbf{R}))_{kj} = (\nabla_{\partial_{x^\ell}}(\mathbf{C}^1_1\mathbf{R}))_{kj} = (\nabla_{\partial_{x^\ell}}\mathbf{Ric})_{kj}$. Using these in (9.6), and multiplying by $\mathbf{g}^{kj}$ (and summing over j), we obtain

$$0 = \mathbf{g}^{ki}(\nabla_{\partial_{x^i}}\mathbf{R})^i_{k\ell j} - \mathbf{g}^{kj}(\nabla_{\partial_{x^k}}\mathbf{Ric})_{\ell j} + \mathbf{g}^{kj}(\nabla_{\partial_{x^\ell}}\mathbf{Ric})_{kj}.$$

Using $(\nabla.\mathbf{Ric})^\sharp = \nabla.(\mathbf{Ric}^\sharp)$ (Exercise 6.29, p.127), we have

$$\begin{aligned}\mathbf{g}^{kj}(\nabla_{\partial_{x^\ell}}\mathbf{Ric})_{kj} &= \mathbf{g}^{kj}(\nabla_{\partial_{x^\ell}}\mathbf{Ric})(\partial_{x^k}, \partial_{x^j}) = (\nabla_{\partial_{x^\ell}}\mathbf{Ric})((dx^j)^\sharp, \partial_{x^j}) \\ &= (\nabla_{\partial_{x^\ell}}\mathbf{Ric})^\sharp(dx^j, \partial_{x^j}) = (\nabla_{\partial_{x^\ell}}(\mathbf{Ric}^\sharp))(dx^j, \partial_{x^j}) \\ &= (\nabla_{\partial_{x^\ell}}(\mathbf{Ric}^\sharp))^j_j.\end{aligned}$$

Using also the symmetry of $\mathbf{Ric}$, $(\nabla_{\partial_{x^k}}\mathbf{Ric})_{\ell j} = (\nabla_{\partial_{x^k}}\mathbf{Ric})_{j\ell}$, and so

$$\begin{aligned}(\operatorname{div}\mathbf{Ric})_\ell &= (\operatorname{div}\mathbf{Ric})(\partial_{x^\ell}) = (\nabla_{\partial_{x^k}}\mathbf{Ric})((dx^k)^\sharp, \partial_{x^\ell}) \\ &= (\nabla_{\partial_{x^k}}\mathbf{Ric})(\mathbf{g}^{kj}\partial_{x^j}, \partial_{x^\ell}) \\ &= \mathbf{g}^{kj}(\nabla_{\partial_{x^k}}\mathbf{Ric})(\partial_{x^j}, \partial_{x^\ell}) = \mathbf{g}^{kj}(\nabla_{\partial_{x^k}}\mathbf{Ric})(\partial_{x^\ell}, \partial_{x^j}) \\ &= \mathbf{g}^{kj}(\nabla_{\partial_{x^k}}\mathbf{Ric})_{\ell j}.\end{aligned} \tag{9.7}$$

Thus

$$\begin{aligned}0 &= \mathbf{g}^{kj}(\nabla_{\partial_{x^i}}\mathbf{R})^i_{k\ell j} - \mathbf{g}^{kj}(\nabla_{\partial_{x^k}}\mathbf{Ric})_{\ell j} + \mathbf{g}^{kj}(\nabla_{\partial_{x^\ell}}\mathbf{Ric})_{kj} \\ &= \mathbf{g}^{kj}(\nabla_{\partial_{x^i}}\mathbf{R})^i_{k\ell j} - (\operatorname{div}\mathbf{Ric})_\ell + (\nabla_{\partial_{x^\ell}}(\mathbf{Ric}^\sharp))^j_j \\ &= \mathbf{g}^{kj}(\nabla_{\partial_{x^i}}\mathbf{R})^i_{k\ell j} - (\operatorname{div}\mathbf{Ric})_\ell + \mathbf{C}\nabla_{\partial_{x^\ell}}(\mathbf{Ric}^\sharp) \\ &= \mathbf{g}^{kj}(\nabla_{\partial_{x^i}}\mathbf{R})^i_{k\ell j} - (\operatorname{div}\mathbf{Ric})_\ell + \nabla_{\partial_{x^\ell}}\mathbf{C}(\mathbf{Ric}^\sharp) \\ &= \mathbf{g}^{kj}(\nabla_{\partial_{x^i}}\mathbf{R})^i_{k\ell j} - (\operatorname{div}\mathbf{Ric})_\ell + \nabla_{\partial_{x^\ell}}\mathbf{S} \\ &= \mathbf{g}^{kj}(\nabla_{\partial_{x^i}}\mathbf{R})^i_{k\ell j} - (\operatorname{div}\mathbf{Ric})_\ell + \partial_{x^\ell}\mathbf{S} \\ &= \mathbf{g}^{kj}(\nabla_{\partial_{x^i}}\mathbf{R})^i_{k\ell j} - (\operatorname{div}\mathbf{Ric})_\ell + (d\mathbf{S})_\ell.\end{aligned}$$

Finally, we will show that the first summand on the right-hand side of the last equality is $-(\mathrm{div}\,\mathbf{Ric})_\ell$, completing the proof. Since $\widetilde{\mathbf{R}} = \mathbf{C}_5^1(\mathbf{R}\otimes\mathbf{g})$ (Exercise 9.6, p.183), $\nabla.\widetilde{\mathbf{R}} = \mathbf{C}_5^1((\nabla.\mathbf{R})\otimes\mathbf{g})$, we get

$$(\nabla_{\partial_{x^i}}\widetilde{\mathbf{R}})_{pqrs} = (\nabla_{\partial_{x^i}}\mathbf{R})^t_{pqr}\mathbf{g}_{st}, \quad \text{and} \quad \mathbf{g}^{su}(\nabla_{\partial_{x^i}}\widetilde{\mathbf{R}})_{pqrs} = (\nabla_{\partial_{x^i}}\mathbf{R})^u_{pqr}.$$

Using the skew-symmetry of $\widetilde{\mathbf{R}}$ in its last two entries, it follows (from the definition of $\nabla_{\partial_{x^i}}\widetilde{\mathbf{R}}$) that $(\nabla_{\partial_{x^i}}\widetilde{\mathbf{R}})_{k\ell js} = -(\nabla_{\partial_{x^i}}\widetilde{\mathbf{R}})_{k\ell sj}$, and so

$$\begin{aligned}
\mathbf{g}^{kj}(\nabla_{\partial_{x^i}}\mathbf{R})^i_{k\ell j} &= \mathbf{g}^{kj}\mathbf{g}^{is}(\nabla_{\partial_{x^i}}\widetilde{\mathbf{R}})_{k\ell js} = -\mathbf{g}^{kj}\mathbf{g}^{is}(\nabla_{\partial_{x^i}}\widetilde{\mathbf{R}})_{k\ell sj}\\
&= -\mathbf{g}^{kj}\mathbf{g}^{is}\mathbf{g}_{jr}(\nabla_{\partial_{x^i}}\mathbf{R})^r_{k\ell s} = -\mathbf{g}^{is}(\nabla_{\partial_{x^i}}\mathbf{R})^k_{k\ell s}\\
&= -\mathbf{g}^{is}(\mathbf{C}_1^1\nabla_{\partial_{x^i}}\mathbf{R})_{\ell s} = -\mathbf{g}^{is}(\nabla_{\partial_{x^i}}(\mathbf{C}_1^1\mathbf{R}))_{\ell s}\\
&= -\mathbf{g}^{is}(\nabla_{\partial_{x^i}}\mathbf{Ric})_{\ell s} \overset{(9.7)}{=} -(\mathrm{div}\,\mathbf{Ric})_\ell. \qquad \square
\end{aligned}$$

Corollary 9.1. *Let* $(M, \mathbf{g})$ *be a semi-Riemannian manifold with the Levi-Civita connection* ∇. *Let* $\Lambda \in \mathbb{R}$. *Then* $\mathrm{div}\,(\mathbf{Ric} - \frac{1}{2}\mathbf{S}\mathbf{g} + \Lambda\mathbf{g}) = 0$.

Proof. Let $(U, \mathbf{x})$ be any admissible chart. Then

$$\begin{aligned}
(\mathrm{div}(\mathbf{S}\mathbf{g}))_i &= \mathbf{C}\big(T_1^0 M \times T_0^1 M \ni (\Omega, V) \mapsto (\nabla_V(\mathbf{S}\mathbf{g}))(\Omega^\sharp, \partial_{x^i})\big)\\
&= (\nabla_{\partial_{x^k}}(\mathbf{S}\mathbf{g}))((dx^k)^\sharp, \partial_{x^i})\\
&= (\partial_{x^k}\mathbf{S})\mathbf{g}(\mathbf{g}^{k\ell}\partial_{x^\ell}, \partial_{x^i}) + \mathbf{S}(\nabla_{\partial_{x^k}}\mathbf{g})((dx^k)^\sharp, \partial_{x^i})\\
&= (d\mathbf{S})_k\mathbf{g}^{k\ell}\mathbf{g}(\partial_{x^\ell}, \partial_{x^i}) + \mathbf{S}\cdot 0 \qquad (\text{since } \nabla.\mathbf{g} = 0)\\
&= (d\mathbf{S})_k\mathbf{g}^{k\ell}\mathbf{g}_{\ell i} = (d\mathbf{S})_k\delta_i^k = (d\mathbf{S})_i.
\end{aligned}$$

So $\mathrm{div}(\mathbf{Ric} - \frac{1}{2}\mathbf{S}\mathbf{g}) = 0$. As $\mathrm{div}(\Lambda\mathbf{g}) = \Lambda\,\mathrm{div}\,\mathbf{g} = \Lambda(0) = 0$, we obtain that $\mathrm{div}\,(\mathbf{Ric} - \frac{1}{2}\mathbf{S}\mathbf{g} + \Lambda\mathbf{g}) = 0$. $\square$

9.2 Cosmological time and cosmological red-shift in FLRW

In this section, we will consider the FLRW spacetime $(M, \mathbf{g})$, as described in Example 5.9 (p.87), where $M = I \times \mathbb{R}^3$, and $I = (0, \infty)$. In the chart $(M, \mathrm{id}_{I\times\mathbb{R}^3})$, the chart-induced component matrix $G = [\mathbf{g}_{ij}]$ of the metric is the diagonal matrix with diagonal entries $-1, a^2, a^2, a^2$. We will mostly use a general a, but sometimes we will take a special[8] $a : I \to I$, namely $a(t) = t^{\frac{2}{3}}$, $t \in (0, \infty)$. We will first calculate the scalar curvature, and note that the chart-coordinate t has a geometric meaning. Then we will derive a 'cosmological version' of the red-shift.

[8]This choice corresponds to the so-called *Einstein-de Sitter universe*, which was for some time believed to be the correct model for our universe. It assumes a matter-only universe with a zero cosmological constant. The Einstein-de Sitter model is a good approximation to our universe in the period after the radiation-dominated era but before the dominance of the positive cosmological constant.

Scalar curvature and cosmological time when $a = t^{\frac{2}{3}}$

We had seen that in an admissible chart $(U, \mathbf{x})$, the curvature tensor field $\mathbf{R}$ has components given in terms of the connection coefficients by

$$\mathbf{R}^{\ell}_{ijk} = \partial_{x^i}\Gamma^{\ell}_{kj} - \partial_{x^j}\Gamma^{\ell}_{ki} + \Gamma^{r}_{kj}\Gamma^{\ell}_{ri} - \Gamma^{r}_{ki}\Gamma^{\ell}_{rj}.$$

So $\mathbf{Ric}_{ij} = \mathbf{R}^{k}_{kij} = \partial_{x^k}\Gamma^{k}_{ji} - \partial_{x^i}\Gamma^{k}_{jk} + \Gamma^{r}_{ji}\Gamma^{k}_{rk} - \Gamma^{r}_{jk}\Gamma^{k}_{ri}$. The connection coefficients for the FLRW spacetime were given in Example 6.7 (p.121). Using these, for example we can calculate, with $(t, x, y, z) \equiv (x^0, x^1, x^2, x^3)$,

$$\begin{aligned}
\mathbf{Ric}_{00} &= \partial_{x^k}\Gamma^{k}_{00} - \partial_t\Gamma^{k}_{0k} + \Gamma^{r}_{00}\Gamma^{k}_{rk} - \Gamma^{r}_{0k}\Gamma^{k}_{r0} \\
&= -\partial_t\Gamma^{1}_{01} - \partial_t\Gamma^{2}_{02} - \partial_t\Gamma^{3}_{03} - \Gamma^{r}_{01}\Gamma^{1}_{r0} - \Gamma^{r}_{02}\Gamma^{2}_{r0} - \Gamma^{r}_{03}\Gamma^{3}_{r0} \\
&= -3\,\partial_t\frac{\dot{a}}{a} - 3\frac{\dot{a}^2}{a^2} = -3\frac{\ddot{a}a - \dot{a}^2}{a^2} - 3\frac{\dot{a}^2}{a^2} = -3\frac{\ddot{a}}{a}.
\end{aligned}$$

Similarly,

$$\begin{aligned}
\mathbf{Ric}_{11} &= \partial_{x^k}\Gamma^{k}_{11} - \partial_{x^1}\Gamma^{k}_{1k} + \Gamma^{r}_{11}\Gamma^{k}_{rk} - \Gamma^{r}_{1k}\Gamma^{k}_{r1} \\
&= \partial_t\Gamma^{0}_{11} - \partial_{x^1}\Gamma^{0}_{10} + \Gamma^{0}_{11}\Gamma^{k}_{0k} - \Gamma^{0}_{11}\Gamma^{1}_{01} - \Gamma^{1}_{10}\Gamma^{0}_{11} \\
&= \partial_t(a\dot{a}) + 3a\dot{a}\frac{\dot{a}}{a} - 2a\dot{a}\frac{\dot{a}}{a} = a\ddot{a} + \dot{a}^2 + 3\dot{a}^2 - 2\dot{a}^2 = a\ddot{a} + 2\dot{a}^2.
\end{aligned}$$

By symmetric role played by the x^1, x^2, x^3 variables, $\mathbf{Ric}_{11} = \mathbf{Ric}_{22} = \mathbf{Ric}_{33}$. The other components of $\mathbf{Ric}$ are all zeroes. Since $[\mathbf{g}_{ij}]$ is a diagonal matrix,

$$(\mathbf{Ric}^{\sharp})^{0}_{0} = -\mathbf{Ric}_{00} = 3\frac{\ddot{a}}{a}, \quad (\mathbf{Ric}^{\sharp})^{1}_{1} = (\mathbf{Ric}^{\sharp})^{2}_{2} = (\mathbf{Ric}^{\sharp})^{3}_{3} = \frac{1}{a^2}\mathbf{Ric}_{11} = \frac{a\ddot{a} + 2\dot{a}^2}{a^2}.$$

Consequently, the scalar curvature is given by

$$\mathbf{S} = (\mathbf{Ric}^{\sharp})^{k}_{k} = 3\frac{\ddot{a}}{a} + 3\frac{a\ddot{a} + 2\dot{a}^2}{a^2} = 6\frac{a\ddot{a} + \dot{a}^2}{a^2}.$$

If $a = t^{\frac{2}{3}}$, then we have $\dot{a} = \frac{2}{3}t^{-\frac{1}{3}}$, $\ddot{a} = -\frac{2}{9}t^{-\frac{4}{3}}$, and so

$$\mathbf{S} = 6\frac{-\frac{2}{9}t^{-\frac{2}{3}} + \frac{4}{9}t^{-\frac{2}{3}}}{t^{\frac{4}{3}}} = \frac{4}{3t^2}.$$

Thus the t-coordinate in the global chart $(M, \mathrm{id}_{I\times\mathbb{R}^3})$ can be expressed in terms of the geometric quantity $\mathbf{S}$. We call this the 'cosmological time'.

Definition 9.6. (Cosmological time for FLRW.)
Let $(M, \mathbf{g})$ be the FLRW spacetime with $a(t) = t^{\frac{2}{3}}$. The function $t: M \to \mathbb{R}$,

$$t(p) = \sqrt{\frac{4}{3\,\mathbf{S}(p)}} \quad \text{for all } p \in M,$$

is called the[9] *cosmological time.*

[9] A similar definition of cosmological time can be made for many other choices of $a(t)$. For example, if $a(t) = t^p$, where $p > \frac{1}{2}$, then $\mathbf{S} = 6p(2p-1)/t^2$.

We note that as $t(p) \to 0$, $\mathbf{S}(p) \to \infty$, and so the scalar curvature blows up as we go back in cosmological time (and we think of this as approaching the 'big-bang singularity').

Geodesics in FLRW

Let $J \subset \mathbb{R}$ be an open interval, and suppose that $\gamma : J \to M$ is a lightlike geodesic. In the global chart $(I \times \mathbb{R}^3, \text{id})$, write $\gamma(s) = (t(s), x(s), y(s), z(s))$, $s \in J$. As γ is lightlike, $\mathbf{g}(\gamma(s))(v_{\gamma,\gamma(s)}, v_{\gamma,\gamma(s)}) = 0$ for all $s \in J$, giving

$$0 = -t'^2 + (a(t))^2(x'^2 + y'^2 + z'^2), \tag{9.8}$$

where $\cdot' = \frac{d}{ds}$. The geodesic equation gives in particular (using the connection coefficient expressions from Example 6.7, p.121, and noting that Γ^0_{ij} is nonzero only when $0 \neq i = j$, and then $\Gamma^0_{ii} = a\dot{a}$, $i = 1, 2, 3$):

$$t'' + a(t)\,\dot{a}(t)(x'^2 + y'^2 + z'^2) = 0. \tag{9.9}$$

Here $\dot{a}(t) = \frac{da}{dt}(t(s))$, $s \in J$. Equations (9.8), (9.9) imply that $Q := a(t)\,t'$ is conserved along the lightlike geodesic, since

$$\begin{aligned} Q' &= (a(t)\,t')' = \dot{a}(t)\,t'\,t' + a(t)\,(t')' \\ &= \dot{a}(t)\,(a(t))^2(x'^2 + y'^2 + z'^2) + a(t)\,t'' \\ &= a(t)(t'' + a(t)\,\dot{a}(t)(x'^2 + y'^2 + z'^2)) \\ &= a(t)\,(0) = 0. \end{aligned}$$

We will use this fact below to derive the so-called 'cosmological red-shift'. Now instead of a lightlike geodesic, consider for $\mathbf{p}, \mathbf{q} \in \mathbb{R}^3$, the curves $\mu_{\mathbf{p}}, \mu_{\mathbf{q}} : I \to M$, given as follows, in the chart $(I \times \mathbb{R}^3, \text{id})$:

$$\mu_{\mathbf{p}}(t) = (t, \mathbf{p}), \quad \mu_{\mathbf{q}}(t) = (t, \mathbf{q}), \quad t \in I = (0, \infty).$$

Both are geodesics: As $v_{\mu_{\mathbf{p}},\mu_{\mathbf{p}}(t)} = \partial_{t,\mu_{\mathbf{p}}(t)}$, we have for all $t \in I$ that

$$(\nabla_{V_{\mu_{\mathbf{p}}}} V_{\mu_{\mathbf{p}}})(t) = (\nabla_{\partial_t} \partial_t)_{\mu_{\mathbf{p}}(t)} = \Gamma^k_{00}(\mu_{\mathbf{p}}(t))\partial_{x^k,\mu_{\mathbf{p}}(t)} = 0.$$

Similarly also $\nabla_{V_{\mu_{\mathbf{q}}}} V_{\mu_{\mathbf{q}}} = 0$. These geodesics can be thought of as the worldlines of two freely-falling galaxies.

Cosmological red-shift when $a = t^{\frac{2}{3}}$

With notation as above, now consider a light ray, $\gamma : J \to M$, emanating at the event $A = (t, \mathbf{q}) = \gamma(s_A)$ on the worldline of $\mu_{\mathbf{q}}$, which reaches the worldline of $\mu_{\mathbf{p}}$ at the event $B = (T, \mathbf{p}) = \gamma(s_B)$, as shown in the following picture.

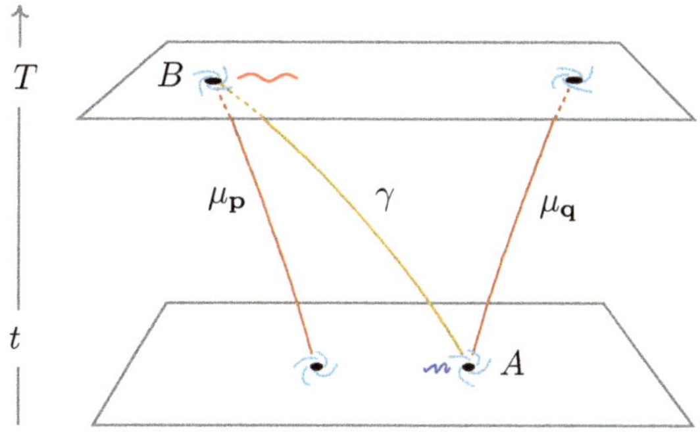

The energy/frequency E_A of the light signal transmitted/perceived by the galaxy $\mu_{\mathbf{q}}$ at A is given by

$$E_A = \frac{-\mathbf{g}(A)(v_{\mu_{\mathbf{q}},A}, v_{\gamma,A})}{\sqrt{-\mathbf{g}(A)(v_{\mu_{\mathbf{q}},A}, v_{\mu_{\mathbf{q}},A})}}$$
$$= \frac{-\mathbf{g}(A)(\partial_{t,A}, t'(s_A)\partial_{t,A}+x'(s_A)\partial_{x,A}+y'(s_A)\partial_{y,A}+z'(s_A)\partial_{z,A})}{\sqrt{-\mathbf{g}(A)(\partial_{t,A}, \partial_{t,A})}} = t'(s_A).$$

Similarly, the energy/frequency E_B of the light signal received/perceived by the galaxy $\mu_{\mathbf{p}}$ at the event B is $E_B = t'(s_B)$. Since $Q = a(t)\,t'$ is conserved along the lightlike geodesic γ, we have $a(t(s_A))\,t'(s_A) = a(t(s_B))\,t'(s_B)$. Consequently,

$$\frac{E_A}{E_B} = \frac{t'(s_A)}{t'(s_B)} = \frac{a(t(s_B))}{a(t(s_A))} = \frac{a(T)}{a(t)}.$$

If $a(t) = t^{\frac{2}{3}}$, then

$$\frac{E_A}{E_B} = \Big(\frac{T}{t}\Big)^{\frac{2}{3}} > 1.$$

Thus the frequency, when received by a 'comoving' observer along the galaxy worldline $\mu_{\mathbf{p}}$ (imagine this to be us), is shifted towards the red end of the spectrum (i.e., lower frequency), as compared to what is measured by a comoving observer with galaxy $\mu_{\mathbf{q}}$. This reduction in frequency, described by the quantity

$$z = \frac{E_A}{E_B} - 1 = \Big(\frac{T}{t}\Big)^{\frac{2}{3}} - 1,$$

is called the *cosmological red-shift*.

Exercise 9.16. The age of the universe is estimated to be $T = 13.8$ billion years. If the light from a distant galaxy, reaching us now, has the cosmological red-shift $z = \frac{1}{5}$, then determine how long ago the light was emitted. (This exercise assumes that the Einstein-de Sitter model is the correct model for our universe. As mentioned earlier, this is approximately correct from the Big Bang until today, but not at all in the far future.)

9.3 Geodesic deviation and curvature

Intuitively, we expect that the manner in which geodesics emanating from a point 'diverge' should tell us something about the curvature.

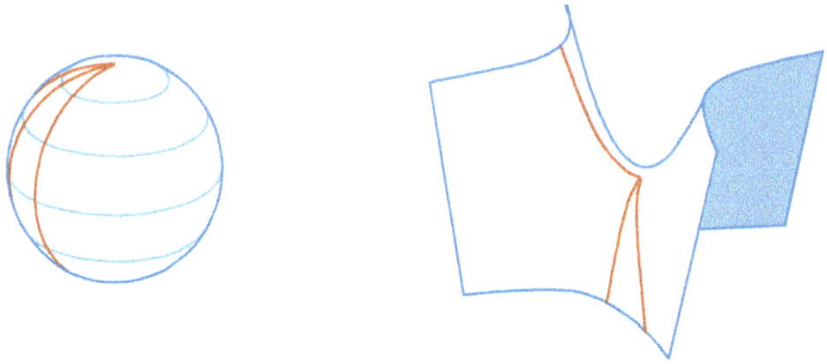

If we imagine the geodesic as a thread on a surface passing through a point, and we slide it across this surface through the point in such a way that each intermediate position of the thread is still a geodesic, then the velocity with which the points on the geodesics travel transversely, defines a vector field along the original geodesic.

Definition 9.7. (Family of geodesics; main and transverse curves.)
Let M be a smooth manifold with a connection ∇, $I \subset \mathbb{R}$ be an open interval, and $\epsilon > 0$. A *family of geodesics* is a smooth map $\Gamma : (-\epsilon, \epsilon) \times I \to M$ such that for each $s \in (-\epsilon, \epsilon)$, the smooth curve $\Gamma_s : I \to M$, given by $\Gamma_s(t) = \Gamma(s,t)$ for all $t \in I$, is a geodesic. The geodesics Γ_s, $s \in (-\epsilon, \epsilon)$, are called *main curves*. The smooth curves $\widetilde{\Gamma}_t : (-\epsilon, \epsilon) \to M$, $t \in I$, given by $\widetilde{\Gamma}_t(s) = \Gamma(s,t)$ for all $s \in (-\epsilon, \epsilon)$, are called *transverse curves*.
For $s \in (-\epsilon, \epsilon)$ and $t \in I$, define

$$\begin{aligned} V(s,t) &= v_{\Gamma_s, \Gamma_s(t)} \in T_{\Gamma(s,t)}M, \\ J(t,s) &= v_{\widetilde{\Gamma}_t, \widetilde{\Gamma}_t(s)} \in T_{\Gamma(s,t)}M. \end{aligned}$$

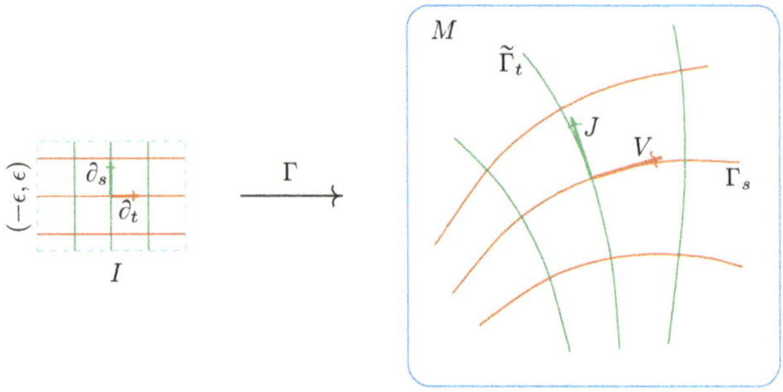

We already know that the velocity vector field $V(s, \cdot)$ along a main curve Γ_s is a vector field along Γ_s, that is, $V(s, \cdot) \in T_0^1\Gamma_s$ for each $s \in (-\epsilon, \epsilon)$.

Similarly, any transverse curve $\widetilde{\Gamma}_t$ has its velocity vector field $J(\cdot, t) \in T_0^1\widetilde{\Gamma}_t$ for each $t \in I$. Moreover, we have the following.

Lemma 9.2. *We use the same notation as in* Definition 9.7. *We have*:
(1) $V(\cdot, t) \in T_0^1\widetilde{\Gamma}_t$ *for each* $t \in I$.
(2) $J(s, \cdot) \in T_0^1\Gamma_s$ *for each* $s \in (-\epsilon, \epsilon)$.
(3) $\nabla_{J(\cdot,t)}V(\cdot, t) \in T_0^1\widetilde{\Gamma}_t$.
(4) $\nabla_{V(s,\cdot)}J(s, \cdot) \in T_0^1\Gamma_s$.
(5) $\nabla_{V(s,\cdot)}V(s, \cdot) = 0$ *for all* $s \in (-\epsilon, \epsilon)$.

Proof. Below, $(U, \mathbf{x})$ is any admissible chart containing the point $\Gamma(t, s)$.
(1) We have $v_{\Gamma_s,\Gamma_s(t)} = \dfrac{\partial(\mathbf{x} \circ \Gamma)^i}{\partial t}(s, t)\ \partial_{x^i,\widetilde{\Gamma}_t(s)}$. As $\dfrac{\partial(\mathbf{x} \circ \Gamma)^i}{\partial t}(\cdot, t)$ is smooth, on $\Gamma(\cdot, t)^{-1}U \subset (-\epsilon, \epsilon)$. By Proposition 7.1 (p.135), $V(\cdot, t) \in T_0^1\widetilde{\Gamma}_t$.
(2) Analogous to the above, by observing $\dfrac{\partial(\mathbf{x} \circ \Gamma)^i}{\partial s}(s, \cdot) \in C^\infty(I)$.
(3) Follows from (1).
(4) Follows from (2).
(5) Since Γ_s is a geodesic, we have $\nabla_{V(s,\cdot)}V(s, \cdot) = 0$ for all $s \in (-\epsilon, \epsilon)$. □

Exercise 9.17. With the notation used in Definition 9.7, show that
(1) $(I \ni t \mapsto (\nabla_{J(\cdot,t)}V(\cdot, t))(s)) \in T_0^1\Gamma_s$ for each $s \in (-\epsilon, \epsilon)$.
(2) $((-\epsilon, \epsilon) \ni s \mapsto (\nabla_{V(s,\cdot)}J(s, \cdot))(t)) \in T_0^1\widetilde{\Gamma}_t$ for each $t \in I$.

We now establish the promised link between the behaviour of a family of geodesics and the curvature operator R. To do so, we will assume that we have a semi-Riemannian manifold with its Levi-Civita connection (since as we shall see in the proof, we will use the torsion-freeness of the connection).

Theorem 9.1.
Let $(M, \mathbf{g})$ *be a semi-Riemannian manifold with the Levi-Civita connection* ∇, $I \subset \mathbb{R}$ *be an open interval,* $\epsilon > 0$, *and* $\Gamma : (-\epsilon, \epsilon) \times I \to M$ *be a family of geodesics. Suppose that there exist vector fields* $\widetilde{J}, \widetilde{V} \in T_0^1M$ *such that* $\widetilde{J}_{\Gamma(s,t)} = J(s, t)$ *and* $\widetilde{V}_{\Gamma(s,t)} = V(s, t)$, *for all* $(s, t) \in (-\epsilon, \epsilon) \times I$. *Then we have*[10] $\nabla_V\nabla_V J = -(R(\widetilde{J}, \widetilde{V})\widetilde{V}) \circ \Gamma$ *on* $(-\epsilon, \epsilon) \times I$.

Proof. Let $\mathbb{R}^2$ be given the standard smooth structure. We consider $N = (-\epsilon, \epsilon) \times I$ as a smooth manifold with the induced smooth structure from $\mathbb{R}^2$. Then $[\partial_t, \partial_s] = 0$ in the global chart (N, id). Using this, we will show that also $[\widetilde{J}, \widetilde{V}] \circ \Gamma = 0$ on $(-\epsilon, \epsilon) \times I$.

[10] We call this the *Jacobi equation*.

Firstly, $V(s,t) = (d\Gamma)_{(s,t)}(\partial_{t,(s,t)})$, because for all $f \in C^\infty(M)$, we have

$$\begin{aligned}((d\Gamma)_{(s,t)}(\partial_{t,(s,t)}))f &= \partial_{t,(t,s)}(f \circ \Gamma) \\ &= \frac{\partial(f \circ \Gamma)}{\partial t}(s,t) \\ &= \frac{d(f \circ \Gamma_s)}{dt}(t) = v_{\Gamma_s,\Gamma_s(t)} f = V(s,t)f.\end{aligned}$$

Similarly, $J(s,t) = (d\Gamma)_{(s,t)}(\partial_{s,(s,t)})$.

Secondly, for $(s,t) \in (-\epsilon,\epsilon) \times I$, and $f \in C^\infty(M)$,

$$\begin{aligned}((\widetilde{V} f) \circ \Gamma)(s,t) &= (\widetilde{V} f)(\Gamma(s,t)) = \widetilde{V}_{\Gamma(s,t)} f = V(s,t) f \\ &= ((d\Gamma)_{(s,t)}(\partial_{t,(s,t)}))(f) = \partial_{t,(s,t)}(f \circ \Gamma) = (\partial_t(f \circ \Gamma))(s,t).\end{aligned}$$

Similarly,

$$((\widetilde{J} f) \circ \Gamma)(s,t) = \partial_{s,(s,t)}(f \circ \Gamma) = (\partial_s(f \circ \Gamma))(s,t).$$

Hence for all $(s,t) \in (-\epsilon,\epsilon) \times I$ and all $f \in C^\infty(M)$,

$$\begin{aligned}[\widetilde{J},\widetilde{V}]_{\Gamma(s,t)} f &= \widetilde{J}_{\Gamma(s,t)}(\widetilde{V} f) - \widetilde{V}_{\Gamma(s,t)}(\widetilde{J} f) = J(s,t)(\widetilde{V} f) - V(s,t)(\widetilde{J} f) \\ &= ((d\Gamma)_{(s,t)}(\partial_{s,(s,t)}))(\widetilde{V} f) - ((d\Gamma)_{(s,t)}(\partial_{t,(s,t)}))(\widetilde{J} f) \\ &= \partial_{s,(s,t)}((\widetilde{V} f) \circ \Gamma) - \partial_{t,(s,t)}((\widetilde{J} f) \circ \Gamma) \\ &= \partial_{s,(s,t)}(\partial_t(f \circ \Gamma)) - \partial_{t,(s,t)}(\partial_s(f \circ \Gamma)) \\ &= \frac{\partial^2(f \circ \Gamma)}{\partial s \partial t}(s,t) - \frac{\partial^2(f \circ \Gamma)}{\partial t \partial s}(s,t) = 0.\end{aligned}$$

Consequently, $[\widetilde{J},\widetilde{V}] \circ \Gamma = 0$. Now we are ready to show the equation given in the theorem statement. By the torsion-freeness of the Levi-Civita connection, $\nabla_{\widetilde{J}} \widetilde{V} - \nabla_{\widetilde{V}} \widetilde{J} = [\widetilde{J},\widetilde{V}]$. Thus $(\nabla_{\widetilde{J}} \widetilde{V}) \circ \Gamma = (\nabla_{\widetilde{V}} \widetilde{J}) \circ \Gamma$. By Exercise 7.2 (p.139), and the fact that the Γ_s are all geodesics, we have $(\nabla_{\widetilde{V}} \widetilde{V}) \circ \Gamma = \nabla_V V = 0$. Recalling the definition of the operator R, we have

$$\begin{aligned}(R(\widetilde{J},\widetilde{V})\widetilde{V}) \circ \Gamma &= (\nabla_{\widetilde{J}} \nabla_{\widetilde{V}} \widetilde{V} - \nabla_{\widetilde{V}} \nabla_{\widetilde{J}} \widetilde{V} - \nabla_{[\widetilde{J},\widetilde{V}]} \widetilde{V}) \circ \Gamma \\ &= (\nabla_{\widetilde{J}} 0 - \nabla_{\widetilde{V}} \nabla_{\widetilde{V}} \widetilde{J} - \nabla_0 \widetilde{V}) \circ \Gamma \\ &= 0 - \nabla_V \nabla_V J - 0 \\ &= -\nabla_V \nabla_V J.\end{aligned}$$

□

Exercise 9.18. We use the notation from Theorem 9.1. Show that

$$\frac{d^2}{dt^2}(\mathbf{g}(\Gamma_s(t))(J(s,t), V(s,t))) = 0, \quad s \in (-\epsilon,\epsilon),\ t \in I.$$

Hint: Use Proposition 7.2 (p.142) and the Jacobi equation.

Let $t_0 \in I$, and $J(s,t_0)$, $(\nabla_{V(s,\cdot)} J(s,\cdot))(t_0)$ be orthogonal to $v_{\Gamma_s,\Gamma_s(t_0)}$. Conclude that $J(s,t)$ and $(\nabla_{V(s,\cdot)} J(s,\cdot))(t)$ are orthogonal to $v_{\Gamma_s,\Gamma_s(t)}$ for all $t \in I$.

9.4 Tidal forces

Theorem 9.1, relating the geodesic deviation with the curvature operator R, can be used to explain the notion of tidal forces in the spacetime context. We first adopt the classical Newtonian perspective. Consider any two constituent particles of an extended body in an inhomogeneous gravitational field. The resulting different accelerations lead to internal stresses in the body, which are called tidal forces. Imagine for example a long rod falling in the gravitational field of the Earth as shown in the picture below.

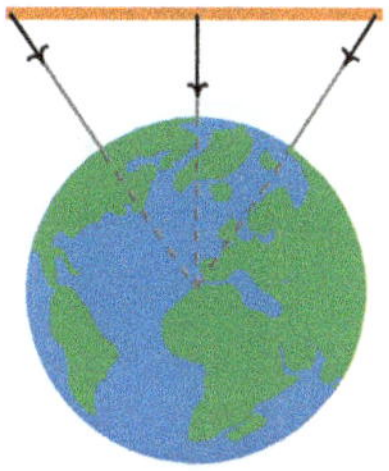

In contrast to the classical understanding, in the spacetime viewpoint the trajectories of two particles are just geodesics, and the curvature of spacetime produces geodesic deviation, resulting in relative acceleration amongst the particles, which is perceived by each one of the particles as a force. To make this precise, we first define an instantaneous observer.

Definition 9.8. (Instantaneous observer, space perceived.)
Let $(M, \mathbf{g})$ be a time-oriented Lorentzian manifold. An *instantaneous observer at* $p \in M$ is a vector[11] $v \in T_pM$ which is timelike and future-pointing. The *space perceived* by an instantaneous observer $v \in T_pM$ is defined to be $v^\perp \subset T_pM$, where $v^\perp := (\text{span}\{v\})^\perp = \{w \in T_pM : \mathbf{g}(p)(v, w) = 0\}$.

We imagine the worldline γ of a 'freely-falling' observer (that is, a geodesic) passing through the event $p = \gamma(0)$, where it has the velocity $v_{\gamma,p} = v$. Note that $U := \text{span}\{v\} \in \mathcal{N}_{\mathbf{g}(p)}$, as $\mathbf{g}(p)|_U$ is negative definite. So by Thm. 5.2 (p.76), $\mathbf{g}(p)|_{v^\perp}$ is positive definite. Suppose that there is a family of geodesics Γ such that $\Gamma(0, \cdot) = \gamma$, and for each[12] t, $J(0,t) \in v^\perp_{\gamma,\gamma(t)}$. By Taylor's formula, for small t, s, in any admissible chart $(U, \mathbf{x})$ containing p,

$$(\mathbf{x} \circ \Gamma)^i(s,t) - (\mathbf{x} \circ \Gamma)^i(0,t) \approx s\frac{\partial(\mathbf{x} \circ \Gamma)^i}{\partial s}(0,t) = s(J(0,t))^i.$$

[11] As worldlines can be reparametrised, a more accurate definition would be that an instantaneous observer is the ray $\{\alpha v : \alpha > 0\}$. It can be checked that each physical quantity derived from the instantaneous observer v (e.g. the perceived space $v^\perp$ or the relative velocity expression ($\star$) on p.266) is invariant under a scaling by $\alpha > 0$.
[12] This is guaranteed if $J(0,0) \in v^\perp$ and $(\nabla_{V(0,\cdot)} J(0,\cdot))(0) \in v^\perp$; see Exercise 9.18.

Hence, formally, we can think of the the spacelike vector $sJ(0,t)$ as describing the vector separation, perceived at time t in $v^{\perp}_{\gamma,\gamma(t)}$, by the instantaneous observer $v_{\gamma,\gamma(t)}$, of another particle at $\Gamma(s,t)$. The proper time elapsed between $\gamma(0)$ and $\gamma(t)$ for γ is $\approx \sqrt{-\mathbf{g}(\gamma(t))(v_{\gamma,\gamma(t)}, v_{\gamma,\gamma(t)})}\,dt$. So for small t, the rate of change of this separation with the elapsed proper time is

$$\frac{\nabla_{V_\gamma}(\text{separation}) \cdot dt}{\text{proper time elapsed}} = \frac{(\nabla_{V_\gamma} sJ(0,\cdot))(t)}{\sqrt{-\mathbf{g}(\gamma(t))(v_{\gamma,\gamma(t)}, v_{\gamma,\gamma(t)})}},$$

see ($\star$) in Remark 7.2 (p. 147). Thus the instantaneous 'acceleration' of the second particle reckoned by the observer v is

$$\frac{(\nabla_{V_\gamma}\nabla_{V_\gamma} sJ(0,\cdot))(0)}{-\,\mathbf{g}(p)(v,v)} = \frac{R(p)(x,v)v}{\mathbf{g}(p)(v,v)},$$

where $x := sJ(0,0)$. Here $R(p)(x,v)v$ means $(R(X,\widetilde{V})\widetilde{V})_p$; $X, \widetilde{V} \in T^1_0 M$ are any vector fields such that $X_p = x$ and $\widetilde{V}_p = v$; see Exercise 6.20 (p.122). This motivates the following definition.

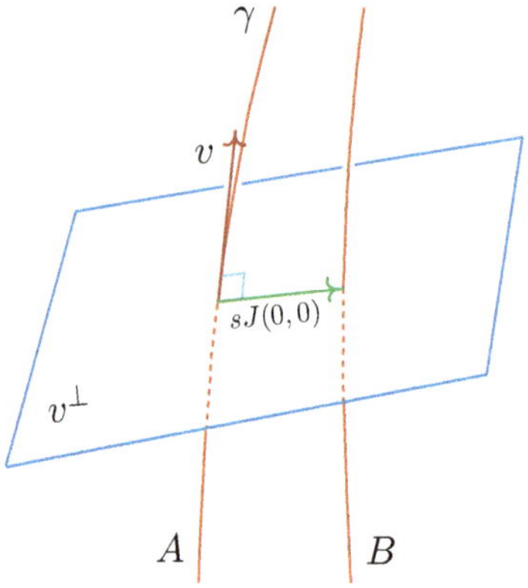

Definition 9.9. (Tidal force[13].)
Let $(M, \mathbf{g})$ be a time-oriented Lorentzian spacetime.
Let $v \in T_pM$ be an instantaneous observer at $p \in M$, and $v^{\perp} := (\text{span}\{v\})^{\perp}$.
Then the *tidal force* is a map $F_v : v^{\perp} \to v^{\perp}$ given by

$$F_v(x) = \frac{R(p)(x,v)v}{\mathbf{g}(p)(v,v)} \text{ for all } x \in v^{\perp}.$$

F_v is well-defined, i.e., $F_v(w)$ does belong to $v^{\perp}$ for each $x \in v^{\perp}$, because

$$\mathbf{g}(p)(v, F_v x) = \frac{\mathbf{g}(p)(v, R(p)(x,v)v)}{\mathbf{g}(p)(v,v)} = \frac{\widetilde{\mathbf{R}}(p)(x,v,v,v)}{\mathbf{g}(p)(v,v)} = \frac{0}{\mathbf{g}(p)(v,v)} = 0,$$

where we have used the skew-symmetry of $\widetilde{\mathbf{R}}$ in its last two slots. We also note that F_v is a *linear* map (from the subspace $v^{\perp}$ of T_pM to $v^{\perp}$).

[13]Although we call it a force, it is more appropriately an acceleration.

Example 9.2. (Schwarzschild spacetime.) Recall Schwarzschild spacetime $(M, \mathbf{g})$ with the time-orientation from Example 5.14 (p.90). Consider the chart $\mathbb{R} \times (2m, \infty) \times U$ given there, where (U, φ) is the chart for S^2 from Example 5.7 (p.85). Let $p \in \mathbb{R} \times (2m, \infty) \times U$ have the coordinates (t, r, θ, ϕ). Set $v = \partial_{t,p}$. Then the subspace $v^\perp$ of T_pM has a basis B comprising three vectors, $B = \{0 \oplus \partial_{r,p} \oplus 0,\ 0 \oplus 0 \oplus \partial_{\theta,p},\ 0 \oplus 0 \oplus \partial_{\phi,p}\}$; see Example 5.10 (p.87), and we will henceforth denote these three vectors simply by $\partial_{r,p}, \partial_{\theta,p}, \partial_{\phi,p}$. To determine the matrix of the linear map $F_v : v^\perp \to v^\perp$ with respect to this basis, we first compute

$$F_v \partial_{r,p} = \frac{R(p)(\partial_{r,p}, \partial_{t,p})\partial_{t,p}}{\mathbf{g}(p)(\partial_{t,p}, \partial_{t,p})} = \frac{(\nabla_{\partial_r} \nabla_{\partial_t} \partial_t - \nabla_{\partial_t} \nabla_{\partial_r} \partial_t - 0)_p}{-(1 - \frac{2m}{r})},$$

where we have used $[\partial_r, \partial_t] = 0$. Using the connection coefficients noted in Example 6.8 (p.121), we have $\nabla_{\partial_t} \partial_t = \Gamma^r_{tt} \partial_r = \frac{m}{r^2}(1 - \frac{2m}{r})\partial_r$, and so

$$\begin{aligned}\nabla_{\partial_r} \nabla_{\partial_t} \partial_t &= \nabla_{\partial_r}\Big(\frac{m}{r^2}\Big(1 - \frac{2m}{r}\Big)\partial_r\Big) = \partial_r\Big(\frac{m}{r^2}\Big(1 - \frac{2m}{r}\Big)\Big)\partial_r + \frac{m}{r^2}\Big(1 - \frac{2m}{r}\Big)\nabla_{\partial_r}\partial_r \\ &= \Big(-\frac{2m}{r^3} + \frac{6m^2}{r^4}\Big)\partial_r + \frac{m}{r^2}\Big(1 - \frac{2m}{r}\Big)\Gamma^r_{rr}\partial_r = \Big(-\frac{2m}{r^3} + \frac{5m^2}{r^4}\Big)\partial_r.\end{aligned}$$

Also,

$$\nabla_{\partial_t} \nabla_{\partial_r} \partial_t = \nabla_{\partial_t}(\Gamma^t_{tr}\partial_t) = \nabla_{\partial_t}\Big(\frac{m}{r^2(1 - \frac{2m}{r})}\partial_t\Big) = \frac{m}{r^2(1 - \frac{2m}{r})}\Gamma^r_{tt}\partial_r = \frac{m^2}{r^4}\partial_r.$$

Consequently, $F_v \partial_{r,p} = \dfrac{(-\frac{2m}{r^3} + \frac{5m^2}{r^4} - \frac{m^2}{r^4})\partial_{r,p}}{-(1 - \frac{2m}{r})} = \dfrac{2m}{r^3}\partial_{r,p}$.

Similarly, $F_v \partial_{\theta,p} = -\frac{m}{r^3}\partial_{\theta,p}$ and $F_v \partial_{\phi,p} = -\frac{m}{r^3}\partial_{\phi,p}$. Thus

$$[F_v]_B = \begin{bmatrix} \frac{2m}{r^3} & & \\ & -\frac{m}{r^3} & \\ & & -\frac{m}{r^3} \end{bmatrix}$$

is the matrix of the linear map $F_v : v^\perp \to v^\perp$ with respect to the ordered basis B consisting of the vectors $\partial_{r,p}, \partial_{\theta,p}, \partial_{\phi,p}$ (taken in this order).

The matrix of the map F_v shows that the tidal forces of spherically symmetric bodies stretch along the radial direction and compress along the tangential directions; a consequence of this is that the Moon deforms the oceans by creating two bulges along the Moon-Earth line, which gives rise to the two daily tides as the Earth rotates. ◇

Exercise 9.19. Show that the tidal force map F_v from Definition 9.9 is self-adjoint, that is, for all $x, y \in v^\perp$, $\mathbf{g}(p)(F_v(x), y) = \mathbf{g}(p)(x, F_v(y))$.

Exercise 9.20. Let $(M, \mathbf{g})$ be a time-oriented Lorentzian spacetime. Let $v \in T_pM$ be an instantaneous observer at $p \in M$. Show that

$$\mathrm{trace}(F_v) = \frac{\mathbf{Ric}(p)(v,v)}{\mathbf{g}(p)(v,v)}.$$

Exercise 9.21. Suppose that an astronaut can withstand an acceleration of $100\,\mathrm{s}^{-2}$ per unit length along his/her height. Consider the Schwarzschild spacetime, which in this exercise, is assumed to model the spacetime in the vicinity of a black hole. Determine the minimum mass m of a black hole, in terms of multiples of solar masses, that a radially aligned astronaut can tolerate arbitrarily close to the 'event horizon', i.e., $r = 2m+\delta$, $\delta > 0$ arbitrary. Suppose that the astronaut is an instantaneous observer with $v = \partial_{t,p} \in T_pM$ as in Example 9.2. The mass $\mathrm{M}_\odot$ of the sun is $\mathrm{M}_\odot \approx 5\,\mu\mathrm{s} = 5\times 10^{-6}\,\mathrm{s}$ (in units such that $G_\mathrm{N} = c = 1$).

Appendix: 'Locally flat' means $\mathbf{R} = 0$

In this final section, we will show that for a semi-Riemannian manifold $(M, \mathbf{g})$ with Levi-Civita connection ∇, the curvature tensor field $\mathbf{R}$ vanishes if and only if M is locally flat. Recall from Exercise 6.20 (p.122) that an admissible chart $(U, \mathbf{x})$ is called an *affine chart* if the connection coefficients with respect to the chart are all identically zero, and the manifold M is called *locally flat* if for every point $p \in M$, there exists an affine chart $(U, \mathbf{x})$ such that $p \in U$. We will show Theorem 9.2, for which we will need the following result, called the Frobenius theorem[14].

Proposition 9.7. (Frobenius.)
Let $V \subset \mathbb{R}^m$ be an open set, $\mathbf{p} \in V$, and $f_{ij} : V \times \mathbb{R}^n \to \mathbb{R}$ be smooth maps for $1 \leqslant i \leqslant m$, and $1 \leqslant j \leqslant n$. Consider the following 'initial value' problem for the system of partial differential equations in the unknown real-valued functions $Y^j = Y^j(x^1, \cdots, x^m)$, $1 \leqslant j \leqslant n$ (with $\mathbf{x} = (x^1, \cdots, x^m) \in V$):

$$\begin{cases} \dfrac{\partial Y^j}{\partial x^i}(\mathbf{x}) = f_{ij}(\mathbf{x}, Y^1, \cdots, Y^n), & 1 \leqslant i \leqslant m, \; 1 \leqslant j \leqslant n, \\ Y^j(\mathbf{p}) = y^j, & 1 \leqslant j \leqslant n. \end{cases} \tag{9.10}$$

If the maps $(x^1, \cdots, x^m, X^1, \cdots, X^n) \mapsto f_{ij} = f_{ij}(x^1, \cdots, x^m, X^1, \cdots, X^n)$ satisfy the following consistency condition in $V \times \mathbb{R}^n$:

$$\frac{\partial f_{ij}}{\partial x^k} + \frac{\partial f_{ij}}{\partial X^\ell} f_{k\ell} = \frac{\partial f_{kj}}{\partial x^i} + \frac{\partial f_{kj}}{\partial X^\ell} f_{i\ell}, \tag{$*$}$$

then there exists an open set $U \subset V$ containing $\mathbf{p}$ such that (9.10) has a unique C^∞ solution $Y^1, \cdots, Y^n : U \to \mathbb{R}$.

[14] See for example, [Frobenius (1877)] or [Shilov (1975), Chapter 2, §2.5], but we include a proof here based on [Hakopian and Tonoyan (2004), Theorem 1.1].

Before giving the proof, we note that if there were a solution, then clearly the Schwarz theorem on the exchangeability of the order of differentiation for smooth functions implies

$$\frac{\partial}{\partial x^k}\frac{\partial Y^j}{\partial x^i} = \frac{\partial}{\partial x^i}\frac{\partial Y^j}{\partial x^k},$$

which motivates the consistency condition $(*)$. The content of the theorem is that the failure of $(*)$ is the only possible obstruction for the solvability of (9.10).

Proof. We will use induction on m. In the base case $m=1$, we just have an ordinary differential equation system with an initial condition at $p \in \mathbb{R}$:

$$\begin{cases} \dfrac{dY^j}{dx}(x) = f_j(x, Y^1, \cdots, Y^n), & 1 \leqslant j \leqslant n, \\ Y^j(p) = y^j, & 1 \leqslant j \leqslant n, \end{cases}$$

and the consistency condition is trivially satisfied. Then the usual result[15] on existence and uniqueness of a smooth solution guarantees the existence of a neighbourhood of p where the above system has a unique solution.

Now suppose that the statement is true for $m = k-1$. Let $m = k$. We write $\mathbf{p} = (p^1, \cdots, p^k)$. Consider the following $(k-1)$-variate problem in the hyperplane $H = \{(x^1, \cdots, x^k) \in \mathbb{R}^k : x^k = p^k\}$:

$$\frac{\partial Y_H^j}{\partial x^i}(x^1, \cdots, x^{k-1}) = f_{ij}(x^1, \cdots, x^{k-1}, p^k, Y_H^1, \cdots, Y_H^n) \qquad (1 \leqslant i \leqslant k-1, \ \ 1 \leqslant j \leqslant n) \tag{9.11}$$

$$Y_H^j(p^1, \cdots, p^{k-1}) = y^j \qquad (1 \leqslant j \leqslant n). \tag{9.12}$$

By the induction hypothesis, this has a unique smooth solution in the neighbourhood of a compact subset of H given by

$$K_1 = \{(x^1, \cdots, x^{k-1}, p^k) : |x^i - p^i| \leqslant \delta_1, \ 1 \leqslant i \leqslant k-1\},$$

for some $\delta_1 > 0$.

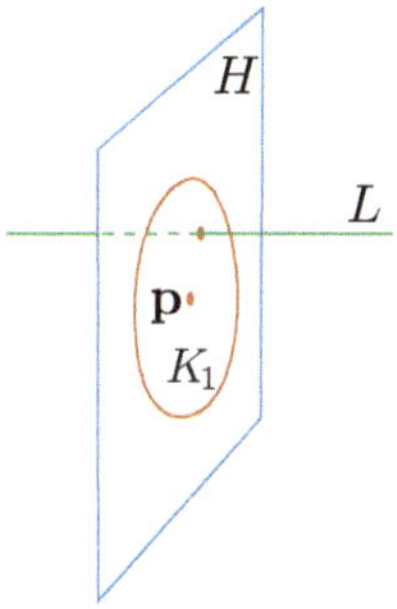

[15]See [Apostol (1969), Thm. 7.19, p.229] for the continuously differentiable (C^1) version. The C^∞ version can be obtained by successively differentiating the ODE and applying the C^1 result.

Next, for each $(\widetilde{x}^1, \cdots, \widetilde{x}^{k-1}, p^k) \in K_1 \subset H$, consider the univariate initial value problem on the line $L = \{(x^1, \cdots, x^k) : x^1 = \widetilde{x}^1, \cdots, x^{k-1} = \widetilde{x}^{k-1}\}$:

$$\frac{dY_L^j}{dx^k}(x^k) = f_{kj}(\widetilde{x}^1, \cdots, \widetilde{x}^{k-1}, x^k, Y_L^1, \cdots, Y_L^n) \qquad (1 \leqslant j \leqslant n) \tag{9.13}$$

$$Y_L^j(p^k) = Y_H^i(\widetilde{x}^1, \cdots, \widetilde{x}^{k-1}) \qquad (1 \leqslant j \leqslant n). \tag{9.14}$$

By the ODE result on existence and uniqueness, this problem also has a unique solution in a neighbourhood (in the subspace topology of L) of

$$K_2 := \{(\widetilde{x}^1, \cdots, \widetilde{x}^{k-1}, x^k) : |x^k - p^k| \leqslant \delta_2\},$$

for some $\delta_2 > 0$. The smooth dependence on initial conditions[16] implies that $Y_L^1, \cdots, Y_L^n$ depend smoothly on the parameters $\widetilde{x}^1, \cdots, \widetilde{x}^k$. A compactness argument can be given to arrange a *uniform* $\delta_2 > 0$ which works for all lines emanating from all initial conditions $(\widetilde{x}^1, \cdots, \widetilde{x}^{k-1}, p^k) \in K_1$. Let

$$\widetilde{U} := \{(x^1, \cdots, x^k) \in \mathbb{R}^m : |x^k - p^k| < \delta_2,\ |x^i - p^i| < \delta_1 \text{ for } 1 \leqslant i \leqslant k-1\} \ (\ni \mathbf{p}).$$

If a solution of (9.10) exists on $\widetilde{U}$, then by the uniqueness part of the induction hypothesis, this solution must coincide with $Y_H^1, \cdots, Y_H^k$ on an open subset $U_1 \subset K_1$ of H with $\mathbf{p} \in U_1$. Thus in view of the uniqueness result for ODEs, it also coincides with $Y_L^1, \cdots, Y_L^k$ on each line L emanating from a point in $U_1 \subset K_1$. So the solution is unique on the open set

$$U := \widetilde{U} \cap \{(x^1, \cdots, x^{k-1}, x^k) \in \mathbb{R}^m : (x^1, \cdots, x^{k-1}, p^k) \in U_1\}.$$

Denote the line emanating from $(x^1, \cdots, x^{k-1}, p^k) \in K_1$ by $L(x^1, \cdots, x^{k-1})$, i.e. $L(x^1, \cdots, x^{k-1}) = \{(x^1, \cdots, x^{k-1}, x^k) : x^k \in \mathbb{R}\}$. It remains to show

$$Y^j(x^1, \cdots, x^k) := Y^j_{L(x^1, \cdots, x^{k-1})}(x^k), \quad (x^1, \cdots, x^k) \in K, \quad 1 \leqslant j \leqslant n,$$

is a solution of (9.10) on $\widetilde{U}$. The initial condition is satisfied, as for $1 \leqslant j \leqslant n$,

$$\begin{aligned} Y^j(p^1, \cdots, p^{k-1}, p^k) &= Y^i_{L(p^1, \cdots, p^{k-1})}(p^k) \\ &= Y_H^i(p^1, \cdots, p^{k-1}) \\ &= y^i. \end{aligned}$$

As Y_L^j is already a solution of (9.13), we only need to check that

$$\eta_i^j := \frac{\partial Y^j}{\partial x^i} - f_{ij}(x^1, \cdots, x^k, Y^1, \cdots, Y^n) = 0 \quad (1 \leqslant i \leqslant k-1,\ 1 \leqslant j \leqslant n).$$

[16] See e.g. [Hartman (2002), Theorem 4.1, p.100].

We have

$$\begin{aligned}
\frac{\partial}{\partial x^k}\eta_i^j &= \frac{\partial}{\partial x^k}\frac{\partial Y^j}{\partial x^i} - \frac{\partial f_{ij}}{\partial x^k} - \frac{\partial f_{ij}}{\partial X^\ell}\frac{\partial Y^\ell}{\partial x^k} \\
&= \frac{\partial}{\partial x^i}\frac{\partial Y^j}{\partial x^k} - \frac{\partial f_{ij}}{\partial x^k} - \frac{\partial f_{ij}}{\partial X^\ell}\frac{\partial Y^\ell}{\partial x^k} \\
&\overset{(9.13)}{=} \frac{\partial}{\partial x^i}(f_{kj}(x^1, \cdots, x^k, Y^1, \cdots, Y^n)) - \frac{\partial f_{ij}}{\partial x^k} - \frac{\partial f_{ij}}{\partial X^\ell}\cdot f_{k\ell} \\
&= \frac{\partial f_{kj}}{\partial x^i} + \frac{\partial f_{kj}}{\partial X^\ell}\frac{\partial Y^\ell}{\partial x^i} - \frac{\partial f_{ij}}{\partial x^k} - f_{k\ell}\frac{\partial f_{ij}}{\partial X^\ell} \\
&= \cancel{\frac{\partial f_{kj}}{\partial x^i}} + \frac{\partial f_{kj}}{\partial X^\ell}\frac{\partial Y^\ell}{\partial x^i} - \cancel{\frac{\partial f_{kj}}{\partial x^i}} - f_{i\ell}\frac{\partial f_{kj}}{\partial X^\ell} \quad \text{(consistency condition)} \\
&= \frac{\partial f_{kj}}{\partial X^\ell}\Big(\frac{\partial Y^\ell}{\partial x^i} - f_{i\ell}\Big) = \frac{\partial f_{kj}}{\partial X^\ell}\eta_i^\ell.
\end{aligned}$$

So the vector $\boldsymbol{\eta}_i = (\eta_i^1, \cdots, \eta_i^n)$ satisfies a homogeneous linear system of ODEs given by

$$\frac{d}{dx^k}\boldsymbol{\eta}_i = M\boldsymbol{\eta}_i, \tag{9.15}$$

where $M = [M_\ell^j]$ is the $n\times n$ matrix with entries $M_\ell^j = \frac{\partial f_{kj}}{\partial X^\ell}$. But with $\mathbf{x} := (x^1, \cdots, x^{k-1})$,

$$\begin{aligned}
\eta_i^j(\mathbf{x}, p^k) &= \frac{\partial Y^j}{\partial x^i}(\mathbf{x}, p^k) - f_{ij}(\mathbf{x}, p^k, Y^1(\mathbf{x}, p^k), \cdots, Y^n(\mathbf{x}, p^k)) \\
&\overset{(9.14)}{=} \frac{\partial Y_H^j}{\partial x^i}(\mathbf{x}) - f_{ij}(\mathbf{x}, p^k, Y_H^1(\mathbf{x}), \cdots, Y_H^n(\mathbf{x})) \overset{(9.11)}{=} 0.
\end{aligned}$$

It now follows from (9.15), and the uniqueness of solutions to the initial value problem for ODEs, that $\boldsymbol{\eta}_i$ must be the trivial solution, that is, $\boldsymbol{\eta}_i = 0$, as wanted. □

Exercise 9.22. Consider the smooth manifold $M = \mathbb{R}^3$ with the standard smooth structure. Show that a 1-form field $\Omega \in T_1^0 M$ is the gradient df of some smooth function $f \in C^\infty(M)$ if and only if the 'curl of $\Omega^\sharp$' is zero, that is,

$$\partial_y\Omega_x = \partial_x\Omega_y, \quad \partial_z\Omega_y = \partial_y\Omega_z, \quad \partial_x\Omega_z = \partial_z\Omega_x,$$

where Ω is decomposed as $\Omega = \Omega_x dx + \Omega_y dy + \Omega_z dz$ in the global admissible chart $(\mathbb{R}^3, (x, y, z) \mapsto (x, y, z))$.

Theorem 9.2. *Let $(M, \mathbf{g})$ be a semi-Riemannian manifold with Levi-Civita connection ∇. Then M is locally flat if and only if $\mathbf{R} = 0$.*

Proof. The 'only if' part is trivial, and was noted in Exercise 6.20 (p.122). We now prove the 'if' part. Let $p \in M$, and $v_1, \cdots, v_m$ be a basis of T_pM.

Step 1. We first show that there exist vector fields $V_1, \cdots, V_m$ defined in a neighbourhood U of p such that for all $1 \leqslant i \leqslant m$, we have $(V_i)_p = v_i$, $\nabla.V_i = 0$, and for all $q \in U$, $\{(V_1)_q, \cdots, (V_m)_q\}$ is a basis for T_qM. To do this, we first take an admissible chart $(V, \mathbf{x})$ containing p, and use it to write the as-yet-undetermined vector fields in terms of their components: $V_r = V_r^j \partial_{x^j}$. Fix an $r \in \{1, \cdots, m\}$. Then the conditions $\nabla.V_r = 0$ and $(V_r)_p = v_r$ give the following initial value problem for the component functions $V_r^1, \cdots, V_r^m$:

$$\begin{cases} \partial_{x^i} V_r^j = -\Gamma^j_{si} V_r^s, & 1 \leqslant i, j \leqslant m, \\ (V_r^i)_p = v_r(x^i), & 1 \leqslant i \leqslant m. \end{cases} \tag{9.16}$$

These can be rewritten as an initial value problem for a system of differential equations in the region $\mathbf{x}(V) \times \mathbb{R}^m$ for unknown functions $V_r^j \circ \mathbf{x}^{-1}$, and using the Frobenius theorem, it can be seen that (9.16) is guaranteed a smooth solution in some open set $U \subset V$ containing p if the following consistency conditions are satisfied:

$$-(\partial_{x^k}\Gamma^j_{si})X^s - \Gamma^j_{si}\delta^s_\ell(-\Gamma^\ell_{tk}X^t) = -(\partial_{x^i}\Gamma^j_{sk})X^s - \Gamma^j_{sk}\delta^s_\ell(-\Gamma^\ell_{ti}X^t)$$

for all $X^1, \cdots, X^m \in \mathbb{R}$. By rearranging and relabelling dummy indices, this can be rewritten as $(\partial_{x^k}\Gamma^j_{si} - \partial_{x^i}\Gamma^j_{sk} + \Gamma^j_{\ell k}\Gamma^\ell_{si} - \Gamma^j_{\ell i}\Gamma^\ell_{sk})X^s = 0$, that is, $\mathbf{R}^j_{kis}X^s = 0$, which is true because $\mathbf{R} = 0$. Hence there is a subset $U \subset V$ containing p and C^∞ functions $V_r^1, \cdots, V_r^m$ such that with $V_r := V_r^i \partial_{x^i}$, we have that $\nabla.V_r = 0$ in U, and $(V_r)_p = v_r$. Repeating this with the other $r \in \{1, \cdots, m\}$ gives m vector fields $V_1, \cdots, V_m$ defined on some open set $U \subset V$ (taken as the intersection of each of the sets $U = U_r$ obtained for the various r) that contains p. Next we will show that for each $q \in U$, $(V_1)_q, \cdots, (V_m)_q$ form a basis for T_qM. It is enough to show that they are linearly independent. Let $c^1, \cdots, c^m \in \mathbb{R}$ be such that $c^i(V_i)_q = 0$. Consider the vector field $W = c^i V_i$. This solves the equation $\nabla.W = 0$ with the initial condition $W_q = 0$. But from the calculation done above, this is equivalent to an initial value problem

$$\begin{cases} \partial_{x^i} W^j = -\Gamma^j_{si} W^s, & 1 \leqslant i, j \leqslant m, \\ W^i_q = 0, & 1 \leqslant i \leqslant m. \end{cases}$$

And just as done above, again the consistency conditions are satisfied due to the vanishing of $\mathbf{R}$. The Frobenius theorem then implies that the solution is unique, implying that $W \equiv 0$ in a small neighbourhood of q. In fact, the subset $\widetilde{U} := \{\widetilde{q} \in U : W_{\widetilde{q}} = 0\}$ is nonempty ($q \in \widetilde{U}$), and (using the above argument) open. Also the set $U \setminus \widetilde{U}$ consisting of the set of points where the evaluation of W is a nonzero tangent vector is open (this is true for any smooth vector field, since the component function are smooth, and in particular, continuous). Thus we have a decomposition $U = \widetilde{U} \cup (U \setminus \widetilde{U})$ as a disjoint union of open sets. By shrinking U at the outset of necessary, we could have assumed that U is connected, and thus we are able to conclude that the nonempty set $\widetilde{U}$ must be all of U. In particular, $c^i(V_i)_p = 0$, that is, $c^i v_i = 0$, and so $c^1 = \cdots = c^m = 0$ (since $v_1, \cdots, v_m$ was a basis for T_pM).

Step 2. In this step, we will construct a new chart $(U, \mathbf{y})$ containing p, which will be the desired affine chart. To do this, we will first construct 1-form fields Ω^i from the V_i of Step 1, and then solve a system of differential equations involving the components of the Ω^i in order to find the coordinate functions y^i. For each $q \in U$, let $(\Omega^1)_q, \cdots, (\Omega^m)_q$ be the basis for $(T_qM)^*$ which is dual to the basis $(V_1)_q, \cdots, (V_m)_q$ for T_qM.

Claim: For $1 \leqslant k \leqslant m$, $U \ni q \mapsto (\Omega^k)_q \in (T_qM)^*$ defines a 1-form field $\Omega^i \in T_1^0 U$.

Firstly, since the vectors $(V_1)_q, \cdots, (V_m)_q$ form a basis for T_qM, the matrix $M(q) := [((V_j)_q)^\ell]$ is invertible for all $q \in U$ Also, as $q \mapsto ((V_j)_q)^\ell = (V_j^\ell)(q)$ are smooth maps, the entries of $M(\cdot)^{-1}$ belong to $C^\infty(U)$. Finally, we have $\delta_j^k = (\Omega^k)_q(V_j)_q = (\Omega^k)_q(((V_j)_q)^\ell \partial_{x^\ell,q}) = ((V_j)_q)^\ell (\Omega^k)_{q,\ell}$, showing that $(\Omega^k)_{q,\ell} = (M(q)^{-1})^k_\ell$. So $q \mapsto (\Omega^k)_{q,\ell}$ belongs to $C^\infty(U)$, and Ω^k is smooth.

Also, $\nabla . \Omega^k = 0$. To see this, we note that for all $W \in T_0^1 U$,

$$\begin{aligned} 0 = W(\delta_i^j) &= W(\Omega^j(V_i)) = (\nabla_W \Omega^j) V_i + \Omega^j(\nabla_W V_i) \\ &= (\nabla_W \Omega^j) V_i + \Omega^j(0) = (\nabla_W \Omega^j) V_i. \end{aligned}$$

But for each q, $(V_1)_q, \cdots, (V_m)_q$ forms a basis for T_qM. Consequently, $\nabla_W \Omega^j = 0$, $1 \leqslant j \leqslant m$.

Now we seek real-valued smooth functions y^i, $1 \leqslant i \leqslant m$ such that

$$\partial_{x^j} y^i = \Omega^i_j, \quad 1 \leqslant i, j \leqslant m. \tag{9.17}$$

(We do not specify an initial condition for $y^i(p)$, $1 \leqslant i \leqslant m$, which can be taken as an arbitrary m-tuple.) Again, using the Frobenius theorem, this is possible if $\partial_{x^k} \Omega^i_j = \partial_{x^j} \Omega^i_k$, $1 \leqslant i, j, k \leqslant m$. But this is true, and can be seen as follows. Firstly,

$$\begin{aligned} \partial_{x^k} \Omega^i_j &= \partial_{x^k}(\Omega^i(\partial_{x^j})) = (\nabla_{\partial_{x^k}} \Omega^i)(\partial_{x^j}) + \Omega^i \nabla_{\partial_{x^k}} \partial_{x^j} \\ &= 0(\partial_{x^j}) + \Omega^i(\Gamma^\ell_{jk} \partial_{x^\ell}) = \Gamma^\ell_{jk} \Omega^i_\ell. \end{aligned}$$

Similarly, $\partial_{x^j} \Omega^i_k = \Gamma^\ell_{kj} \Omega^i_\ell$. By the torsion-freeness of the Levi-Civita connection, $\Gamma^\ell_{kj} = \Gamma^\ell_{jk}$. Hence the consistency condition in Frobenius theorem is satisfied for all i, j, k. We will continue to denote the possibly smaller subset of U containing p on which the functions y^i exist, still by U. Define the matrix

$$N := [\partial_{x^i} y^j] \overset{(9.17)}{=} [\Omega^j_i].$$

Then N is pointwise invertible in U since $(\Omega^1)_q, \cdots, (\Omega^m)_q$ form a basis for $(T_q M)^*$ for each $q \in U$. Hence the Jacobian of the 'transition function' $\mathbf{y} \circ \mathbf{x}^{-1}$ is pointwise invertible on $\mathbf{x}(U)$. Also, as N is smooth on U, it follows that $\mathbf{y} \circ \mathbf{x}^{-1}$ is smooth on $\mathbf{x}(U)$. By the inverse function theorem, and by shrinking the neighbourhood U of p if necessary, it follows that $\mathbf{y} \circ \mathbf{x}^{-1} : \mathbf{x}(U) \to (\mathbf{y} \circ \mathbf{x}^{-1})(\mathbf{x}(U)) = \mathbf{y}(U)$ is a diffeomorphism. Thus $\mathbf{y} = \mathbf{y} \circ \mathbf{x}^{-1} \circ \mathbf{x} : U \to \mathbb{R}^m$ is a diffeomorphism onto the open set $\mathbf{y}(U)$. Hence $(U, \mathbf{y})$ is an admissible chart for M. We have

$$\delta^i_j = \Omega^i(V_j) = \Omega^i_k dx^k(V^\ell_j \partial_{x^\ell}) = \Omega^i_k V^\ell_j \delta^k_\ell = \Omega^i_k V^k_j = V^k_j \partial_{x^i} y^j,$$

i.e., $[V^k_j][\partial_{x^i} y^j] = I$. But then also $[\partial_{x^i} y^j][V^i_k] = I$, that is, $(\partial_{x^i} y^j) V^i_k \overset{(\star)}{=} \delta^j_k$. Thus

$$V_j = V_j(y^k)\partial_{y^k} = ((V^\ell_j \partial_{x^\ell}) y^k)\partial_{y^k} = V^\ell_j (\partial_{x^\ell} y^k)\partial_{y^k} \overset{(\star)}{=} \delta^k_j \partial_{y^k} = \partial_{y^j}.$$

In the chart $(U, \mathbf{y})$, we have that

$$\Gamma^k_{ji} = dy^k(\nabla_{\partial_{y^i}} \partial_{y^j}) = dy^k(\nabla_{\partial_{y^i}} V_j) = dy^k(0) = 0.$$

Consequently, the admissible chart $(U, \mathbf{y})$ containing p is an affine chart, as wanted. □

Chapter 10

Form fields

In this chapter, we will discuss special types of tensor fields on M, called 'form fields' (a special but trivial example of which is a 1-form field), and develop a useful calculus of these, after introducing the 'exterior derivative'. This will enable us to write Maxwell's equations of electromagnetism succinctly in a later chapter. The exterior derivative generalises the grad-div-curl vector calculus language from $\mathbb{R}^3$ to smooth manifolds.

The language of form fields will also allow us to talk in particular about a 'volume form field' on a Lorentzian manifold, enabling a coordinate-free manner of integrating functions on a manifold. The simplest form field is a 1-form field. A one form field can be integrated along a curve: at any point of the curve, the point evaluation of the 1-form field on the tangent vector to the curve produces a number, and we add these contributions along the curve. The higher k-forms can be used to integrate along higher dimensional 'submanifolds', but in this book we will only consider the full manifold case and integrate m-forms on m-dimensional manifolds.

10.1 k-forms

An 'alternating' $(0,k)$-tensor on a vector space V is called a k-form.

Definition 10.1. (k-form.)
Let V be a vector space and $k \geqslant 2$. Then a *k-form on V* is a $(0,k)$-tensor $\omega : V \times \cdots \times V \to \mathbb{R}$ which is *alternating*, that is, for all $1 \leqslant i < j \leqslant k$ and all $v_1, \cdots, v_k \in V$, swapping the ith and j^{th} entries produces a minus sign, that is, $\omega(\cdots, v_i, \cdots, v_j, \cdots) = -\omega(\cdots, v_j, \cdots, v_i, \cdots)$. The set of all k-forms on V is denoted by $\bigwedge^k V^*$. We set $\bigwedge^1 V = V^*$ and $\bigwedge^0 V = \mathbb{R}$.

Clearly the action of a k-form with $k \geqslant 2$ on k vectors, any two of which are identical, is 0. This also true if the k vectors are linearly dependent. In particular, if V is m-dimensional, then if $k > m$, every k-form ω is 0, and so $\bigwedge^k V^* = \{0\}$ for all $k > m$.

We now develop notation and collect preliminary facts on permutations, which will be needed in the proofs.

We recall that a bijective map $\pi : \{1, \cdots, k\} \to \{1, \cdots, k\}$ is called a *permutation*. We specify a permutation π by writing the image $\pi(i)$ below $i \in \{1, \cdots, n\}$ in a row as follows:

$$\begin{pmatrix} 1 & \cdots & n \\ \pi(1) & \cdots & \pi(n) \end{pmatrix}. \qquad (\star)$$

The set of all permutations on $\{1, \cdots, k\}$ is denoted by S_k, and it forms a group under composition. The group is non-Abelian if $k \geqslant 3$:

$$\begin{pmatrix} 1\,2\,3 \\ 1\,3\,2 \end{pmatrix} \circ \begin{pmatrix} 1\,2\,3 \\ 2\,1\,3 \end{pmatrix} = \begin{pmatrix} 1\,2\,3 \\ 3\,1\,2 \end{pmatrix} \neq \begin{pmatrix} 1\,2\,3 \\ 2\,3\,1 \end{pmatrix} = \begin{pmatrix} 1\,2\,3 \\ 2\,1\,3 \end{pmatrix} \circ \begin{pmatrix} 1\,2\,3 \\ 1\,3\,2 \end{pmatrix}.$$

The identity element in S_k is the identity map, $\{1, \cdots, k\} \ni i \mapsto i$. *Transpositions* are special permutations that just swap two different elements of the set $\{1, \cdots, k\}$ while keeping the others fixed. It can be shown[1] that every permutation can be written as a composition of transpositions. While this decomposition is not unique, the parity of the number of factors is[2]. That is, no matter how a permutation π is factored into transpositions, the number of factors is either

- always even (and then we define $\operatorname{sign} \pi = +1$), or else it is
- always odd (and then we define $\operatorname{sign} \pi = -1$).

$\operatorname{sign} \pi$ is called the *sign/signature of the permutation* π. For all $\sigma, \pi \in S_k$, $\operatorname{sign}(\sigma \circ \pi) = (\operatorname{sign} \sigma)(\operatorname{sign} \pi)$. If $\{-1, 1\}$ is considered as a group, with the usual multiplication, then $\operatorname{sign} : S_k \to \{-1, +1\}$ is a group homomorphism. Let the *alternating set* A_k be the set of all even permutations of $\{1, \cdots, k\}$. The kernel of sign is A_k. We can find the sign of a permutation π by determining the number of swaps K needed on the second row in $(\star)$ above to make it $1, \cdots, k$, and then $\operatorname{sign} \pi = (-1)^K$.

---

[1] See e.g. [Halmos (1987), §27, Theorem 3]. The idea is to decompose the permutation π into a product of disjoint 'cycles' by operating repeatedly by π starting from an element, and then writing each cycle as a product of transpositions.

[2] The inverse of a transposition is itself. The claim then boils down to showing that the identity permutation cannot be written as a product of an odd number of transpositions. Denote the transposition swapping distinct i, j by (i, j). Then for distinct a, b, c, d, we have $(a, b) \circ (a, b) = \text{id}$, $(c, d) \circ (a, b) = (a, b) \circ (c, d)$, $(b, c) \circ (a, b) = (c, a) \circ (c, b)$ and $(a, c) \circ (a, b) = (b, a) \circ (b, c)$. In each of the last three cases, we can move the a to the left. We prove the result by induction on the number m of factors in a factorisation of id by transpositions. A transposition cannot be the identity and so $m > 1$. If $m = 2$, then we are done. Let a be the largest number appearing among all the transpositions in a factorisation of id by transpositions. Consider the right-most occurrence of a and the transposition just before it. If we are in the '$(a, b) \circ (a, b)$' case, then this collapses to id, reducing the number of factors by 2, and we are done by the induction hypothesis. In any of the other three cases, move a to the left. As we move all of the occurrences of a to the left, and if at no stage we have the '$(a, b) \circ (a, b)$' case, then eventually $\text{id} = (a, b) \circ \pi$ with no a appearing in the product π of transpositions. But this is a contradiction since the LHS sends a to a, while the RHS sends a to b.

For a k-form ω on V, $\omega(v_1, \cdots, v_k) = (\text{sign } \pi)\ \omega(v_{\pi(1)}, \cdots, v_{\pi(k)})$, or equivalently, $(\text{sign } \pi)\ \omega(v_1, \cdots, v_k) = \omega(v_{\pi(1)}, \cdots, v_{\pi(k)})$, for all vectors $v_1, \cdots, v_k \in V$. Vice versa, a $(0, k)$-tensor on V having this property is seen to be an k-form by specialising the permutations to be transpositions.

Definition 10.2. (Permutation action on a $(0, k)$-tensor.)
Let V be a vector space, $k \geqslant 1$, $\tau \in T_k^0 V$, and $\pi \in S_k$. Then $\pi(\tau) \in T_k^0 V$ is defined by $(\pi(\tau))(v_1, \cdots, v_k) = \tau(v_{\pi(1)}, \cdots, v_{\pi(k)})$ for all $v_1, \cdots, v_k \in V$.

Exercise 10.1. Suppose that V is a vector space, $k \geqslant 1$, $\tau \in T_k^0 V$, and $\pi, \sigma \in S_k$. Show that $\pi(\sigma\tau) = (\pi \circ \sigma)\tau$.

Any $(0, k)$-tensor τ on V can be converted into a k-form $\text{Alt}(\omega)$ by an operation called alternation.

Definition 10.3. (Alternation.)
Let V be a vector space. For $\tau \in T_k^0 V$, the *alternation* $\text{Alt}\,\tau \in \bigwedge^k V^*$ *of* τ is

$$\text{Alt}\,\tau = \frac{1}{k!} \sum_{\pi \in S_k} (\text{sign}\,\pi)\,\pi\tau.$$

Let us check that $\text{Alt}\,\tau \in \bigwedge^k V^*$. Indeed, for $v_1, \cdots, v_k \in V$, we have

$$\begin{aligned}
(\text{sign}\,\pi)(\text{Alt}\,\tau)(v_{\pi(1)}, \cdots, v_{\pi(k)}) &= \frac{\text{sign}\,\pi}{k!} \sum_{\sigma \in S_k} (\text{sign}\,\sigma)\,\tau(v_{\pi(\sigma(1))}, \cdots, v_{\pi(\sigma(k))}) \\
&= \frac{\text{sign}\,\pi}{k!} \sum_{\sigma \in S_k} (\text{sign}\,\sigma)\,\tau(v_{\pi\circ\sigma(1)}, \cdots, v_{\pi\circ\sigma(k)}) \\
&= \frac{1}{k!} \sum_{\sigma \in S_k} (\text{sign}(\pi \circ \sigma))\tau(v_{\pi\circ\sigma(1)}, \cdots, v_{\pi\circ\sigma(k)}) \\
&= \frac{1}{k!} \sum_{\mu \in S_k} (\text{sign}\,\mu)\,\tau(v_{\mu(1)}, \cdots, v_{\mu(k)}) \\
&= (\text{Alt}\,\tau)(v_1, \cdots, v_k),
\end{aligned}$$

since for each $\mu \in S_k$, there is a unique $\sigma \in S_k$ such that $\mu = \pi \circ \sigma$, namely, $\sigma = \pi^{-1} \circ \mu$.

Example 10.1. Let V be a vector space. If $\omega \in V^* = T_1^0 V$, then $\text{Alt}\,\omega = \omega$. If $\tau \in T_2^0 V$, then for all $v, w \in V$, $(\text{Alt}\,\tau)(v, w) = \frac{1}{2!}(\tau(v, w) - \tau(w, v))$. $\Diamond$

Exercise 10.2. Let V be a vector space, and $\tau \in T_3^0 V$. Determine the action of $\text{Alt}\,\tau$ on (x, y, z), where $x, y, z \in V$, in terms of that of τ.

Exercise 10.3. Let V be a vector space, and $\omega \in \bigwedge^k V$. Show that $\text{Alt}\,\omega = \omega$.

Lemma 10.1. *Let V be a vector space, $\omega \in \bigwedge^k V^*$, $\theta \in \bigwedge^\ell V^*$. Then:*

(1) $\text{Alt}\,((\text{Alt}\,\omega) \otimes \theta) = \text{Alt}\,(\omega \otimes \theta)$.
(2) $\text{Alt}\,(\omega \otimes \theta) = (-1)^{k\ell}\text{Alt}\,(\theta \otimes \omega)$
(3) $\text{Alt}\,(\omega \otimes (\text{Alt}\,\theta)) = \text{Alt}\,(\omega \otimes \theta)$.

Proof.

(1) We have, using the distributivity of $\otimes$ over addition, that

$$\operatorname{Alt}((\operatorname{Alt}\omega)\otimes\theta)=\frac{1}{(k+\ell)!}\sum_{\pi\in S_{k+\ell}}(\operatorname{sign}\pi)\,\pi\Big(\frac{1}{k!}\sum_{\sigma\in S_k}(\operatorname{sign}\sigma)\,(\sigma\omega)\otimes\theta\Big).$$

Let us consider $\sigma\in S_k$ as an element (denoted again by σ) of $S_{k+\ell}$, where we imagine that σ fixes $k+1,\cdots,k+\ell$. Then $(\sigma\omega)\otimes\theta=\sigma(\omega\otimes\theta)$ and the sign of σ, considered as an element of $S_{k+\ell}$, is the sign of $\sigma\in S_k$. Thus the above gives

$$\operatorname{Alt}((\operatorname{Alt}\omega)\otimes\theta)=\frac{1}{(k+\ell)!\,k!}\sum_{\pi\in S_{k+\ell}}\sum_{\sigma\in S_k}\operatorname{sign}(\pi\circ\sigma)\,(\pi\circ\sigma)(\omega\otimes\theta).$$

But for each $\sigma\in S_k$, the map $S_{k+\ell}\ni\pi\mapsto\pi\circ\sigma\in S_{k+\ell}$ is a bijection, and since S_k has $k!$ elements, we obtain

$$\operatorname{Alt}((\operatorname{Alt}\omega)\otimes\theta)=\frac{1}{(k+\ell)!\,k!}k!\sum_{\mu\in S_{k+\ell}}(\operatorname{sign}\mu)\,\mu(\omega\otimes\theta)=\operatorname{Alt}(\omega\otimes\theta).$$

(2) Suppose that μ denotes the permutation $\begin{pmatrix}1&\cdots&\ell&\ell+1\cdots\ell+k\\ k+1\cdots&k+\ell&1&\cdots\ \ k\end{pmatrix}$. To bring the 1 in the second row to the first position, we need ℓ transpositions (and then $k+1,\cdots k+\ell$ are shifted to the right by one space). Next, to bring the 2 in the second row to the second position, we again need ℓ transpositions, and so on. Thus, starting from μ, in order to move $1,\cdots,k$ in the second row to occupy the first k places (and in the process shift the $k+1,\cdots,k+\ell$ rightwards to occupy the last ℓ places), the number of transpositions needed is $\ell+\cdots+\ell$ (k times) $=k\ell$, and then the resulting new permutation sends each i to itself (i.e., it is the identity map). Hence $\operatorname{sign}\mu=(-1)^{k\ell}$. Now for $v_1,\cdots,v_{k+\ell}\in V$, we have

$$\begin{aligned}&(\operatorname{Alt}(\omega\otimes\theta))(v_1,\cdots,v_{k+\ell})\\&=\frac{1}{(k+\ell)!}\sum_{\pi\in S_{k+\ell}}(\operatorname{sign}\pi)\,\omega(v_{\pi(1)},\cdots,v_{\pi(k)})\,\theta(v_{\pi(k+1)},\cdots,v_{\pi(k+\ell)})\\&=\frac{1}{(k+\ell)!}\sum_{\pi\in S_{k+\ell}}(\operatorname{sign}\pi)\,\omega(v_{\pi(\mu(\ell+1))},\cdots,v_{\pi(\mu(\ell+k))})\,\theta(v_{\pi(\mu(1))},\cdots,v_{\pi(\mu(\ell))})\\&=\frac{(\operatorname{sign}\mu)}{(k+\ell)!}\sum_{\pi\in S_{k+\ell}}\operatorname{sign}(\pi\circ\mu)\,\theta(v_{(\pi\circ\mu)(1)},\cdots,v_{(\pi\circ\mu)(\ell)})\cdot\\&\qquad\qquad\omega(v_{(\pi\circ\mu)(\ell+1)},\cdots,v_{(\pi\circ\mu)(\ell+k)}).\end{aligned}$$

The map $S_{k+\ell}\ni\pi\mapsto\pi\circ\mu$ is a bijection, and so we obtain

$$\begin{aligned}&\operatorname{Alt}(\omega\otimes\theta)(v_1,\cdots,v_{k+\ell})\\&=\frac{\operatorname{sign}\mu}{(k+\ell)!}\sum_{\sigma\in S_{k+\ell}}(\operatorname{sign}\sigma)\,\theta(v_{\sigma(1)},\cdots,v_{\sigma(\ell)})\,\omega(v_{\sigma(\ell+1)},\cdots,v_{\sigma(\ell+k)})\\&=(-1)^{k\ell}\operatorname{Alt}(\theta\otimes\omega)(v_1,\cdots,v_{k+\ell}).\end{aligned}$$

(3) $\text{Alt}\,(\omega \otimes (\text{Alt}\,\theta)) \overset{(2)}{=} (-1)^{k\ell}\text{Alt}\,((\text{Alt}\,\theta) \otimes \omega) \overset{(1)}{=} (-1)^{k\ell}\text{Alt}\,(\theta \otimes \omega)$
$\overset{(2)}{=} \text{Alt}\,(\omega \otimes \theta).$ □

The tensor product of forms is a tensor, but it may fail to be a form, since it may not be alternating. To get an alternating form, we apply the alternation operation. The resulting operation is the exterior product.

Definition 10.4. (Exterior product of forms.)
Suppose that V is a vector space. Let the integers $k, \ell \geqslant 1$.
The *exterior product* $\wedge : \bigwedge^k V^* \times \bigwedge^\ell V^* \to \bigwedge^{k+\ell} V^*$ is defined as follows:
For $\omega \in \bigwedge^k V^*$ and $\theta \in \bigwedge^\ell V^*$, $\omega \wedge \theta := \dfrac{(k+\ell)!}{k!\,\ell!}\text{Alt}(\omega \otimes \theta)$.
For $a, b \in \mathbb{R} = \bigwedge^0 V^*$, $a \wedge b := ab$.
For $a \in \mathbb{R} = \bigwedge^0 V^*$ and $\omega \in \bigwedge^k V^*$, $a \wedge \omega = \omega \wedge a := a\omega$.

More explicitly, for $v_1, \cdots, v_{k+\ell} \in V$,

$$(\omega \wedge \theta)(v_1, \cdots, v_{k+\ell}) = \frac{1}{k!\,\ell!} \sum_{\pi \in S_{k+\ell}} (\text{sign}\,\pi)\,(\omega \otimes \theta)(v_{\pi(1)}, \cdots, v_{\pi(k+\ell)}).$$

It is straightforward to verify that for $a \in \mathbb{R}$, $\omega \in \bigwedge^k V^*$ and $\theta \in \bigwedge^\ell V^*$,

$$a(\omega \wedge \theta) = (a\omega) \wedge \theta = \omega \wedge (a\theta).$$

Also, for $\omega, \widetilde{\omega} \in \bigwedge^k V^*$ and $\theta \in \bigwedge^\ell V^*$, we have

$$(\omega + \widetilde{\omega}) \wedge \theta = \omega \wedge \theta + \widetilde{\omega} \wedge \theta \quad \text{and} \quad \theta \wedge (\omega + \widetilde{\omega}) = \theta \wedge \omega + \theta \wedge \widetilde{\omega}.$$

We summarise by saying that the wedge product is $\mathbb{R}$-bilinear. We now show that the wedge product is associative.

Proposition 10.1. *Let V be a vector space, and $\omega^i \in \bigwedge^{k_i} V^*$, $i = 1, 2, 3$. Then $\omega^1 \wedge (\omega^2 \wedge \omega^3) = (\omega^1 \wedge \omega^2) \wedge \omega^3$.*

Proof. We have

$$\begin{aligned}
\omega^1 \wedge (\omega^2 \wedge \omega^3) &= \frac{(k_1+k_2+k_3)!}{k_1!\,(k_2+k_3)!}\text{Alt}\,(\omega^1 \otimes (\omega^2 \wedge \omega^3)) \\
&= \frac{(k_1+k_2+k_3)!}{k_1!\,(k_2+k_3)!}\,\frac{(k_2+k_3)!}{k_2!\,k_3!}\text{Alt}\,(\omega^1 \otimes \text{Alt}\,(\omega^2 \otimes \omega^3)) \\
&= \frac{(k_1+k_2+k_3)!}{k_1!\,k_2!\,k_3!}\text{Alt}\,(\omega^1 \otimes (\omega^2 \otimes \omega^3)) \quad \text{(Lemma 10.1)} \\
&= \frac{(k_1+k_2+k_3)!}{k_1!\,k_2!\,k_3!}\text{Alt}\,((\omega^1 \otimes \omega^2) \otimes \omega^3) \quad (\otimes \text{ associative}) \\
&= \frac{(k_1+k_2+k_3)!}{(k_1+k_2)!\,k_3!}\,\frac{(k_1+k_2)!}{k_1!\,k_2!}\text{Alt}\,(\text{Alt}\,(\omega^1 \otimes \omega^2) \otimes \omega^3)) \\
&= \frac{(k_1+k_2+k_3)!}{(k_1+k_2)!\,k_3!}\text{Alt}\,((\omega^1 \wedge \omega^2) \otimes \omega^3)) = (\omega^1 \wedge \omega^2) \wedge \omega^3.
\end{aligned}$$

□

In light of the associativity of $\wedge$, there is no ambiguity in using the notation $\omega^1 \wedge \cdots \wedge \omega^n$, where $\omega^i \in \bigwedge^{k_i} V^*$, $i=1,\cdots,n$.

Lemma 10.2. *Let V be a vector space, and $\omega^i \in \bigwedge^{k_i} V^*$, $i=1,\cdots,n$. Then $\omega^1 \wedge \cdots \wedge \omega^n = \dfrac{(k_1+\cdots+k_n)!}{k_1!\cdots k_n!}\mathrm{Alt}\,(\omega^1 \otimes \cdots \otimes \omega^n)$.*

Proof. We give an inductive argument on the number of factors in the wedge product. For $n=1$, there is nothing to prove since $\mathrm{Alt}\,\omega^1=\omega^1$, and for $n=2$, this is just the definition of the wedge product. The case $n=3$ was shown in the proof of Proposition 10.1. Suppose that $n>3$, and that the result has been shown when the number of factors is $\leqslant n-1$. Then

$$\begin{aligned}
&\omega^1 \wedge \omega^2 \wedge \cdots \wedge \omega^n \\
&= \frac{(k_1+(k_2+\cdots+k_n))!}{k_1!\,(k_2+\cdots+k_n)!}\mathrm{Alt}\,(\omega^1 \otimes (\omega^2 \wedge \cdots \wedge \omega^n)) \\
&= \frac{(k_1+k_2+\cdots+k_n)!}{k_1!\,(k_2+\cdots+k_n)!}\,\frac{(k_2+\cdots+k_n)!}{k_2!\cdots k_n!}\mathrm{Alt}\,(\omega^1 \otimes \mathrm{Alt}\,(\omega^2 \otimes \cdots \otimes \omega^n)) \\
&= \frac{(k_1+k_2+\cdots+k_n)!}{k_1!\,k_2!\cdots k_n!}\mathrm{Alt}\,(\omega^1 \otimes (\omega^2 \otimes \cdots \otimes \omega^n)) \\
&= \frac{(k_1+\cdots+k_n)!}{k_1!\cdots k_n!}\mathrm{Alt}\,(\omega^1 \otimes \cdots \otimes \omega^n).
\end{aligned}$$

□

Exercise 10.4. Let V be a vector space and $\epsilon^i \in \bigwedge^1 V^* = V^*$, $i=1,\cdots,n$. Show that for all $v_i \in V$, $i=1,\cdots,n$, $(\epsilon^1 \wedge \cdots \wedge \epsilon^n)(v_1,\cdots,v_n) = \det[\epsilon^i(v_j)]$.

Exercise 10.5. Let V be a vector space, $\epsilon^1,\cdots,\epsilon^k \in V^*$, and $\pi \in S_k$. Show that $\epsilon^{\pi(1)} \wedge \cdots \wedge \epsilon^{\pi(k)} = (\mathrm{sign}\,\pi)\,\epsilon^1 \wedge \cdots \wedge \epsilon^k$.

We have the following consequence of Lemma 10.1 (2).

Proposition 10.2. *Let V be a vector space, $\omega \in \bigwedge^k V^*$, and $\theta \in \bigwedge^\ell V^*$. Then $\omega \wedge \theta = (-1)^{k\ell}(\theta \wedge \omega)$.*

Proof. We have

$$\omega \wedge \theta = \frac{(k+\ell)!}{k!\,\ell!}\mathrm{Alt}\,(\omega \otimes \theta) = \frac{(k+\ell)!}{k!\,\ell!}(-1)^{k\ell}\mathrm{Alt}\,(\theta \otimes \omega) = (-1)^{k\ell}\theta \wedge \omega.$$

□

Proposition 10.3. *Let V be an m-dimensional vector space, with a basis $\{e_1,\cdots,e_m\}$, and let $\{\epsilon^1,\cdots,\epsilon^m\}$ be the corresponding dual basis for the dual space V^*. Then*

$$B = \{\epsilon^{i_1} \wedge \cdots \wedge \epsilon^{i_k} : 1 \leqslant i_1 < \cdots < i_k \leqslant m\}$$

is a basis for $\bigwedge^k V^*$, and $\dim \bigwedge^k V^* = \binom{m}{k}$.

Proof. It is clear that the elements in B are in one-to-one correspondence with all k-tuples $(i_1, \cdots, i_k)$ chosen from $\{1, \cdots, m\}$ that are strictly increasing, which is just the number of ways of choosing k distinct numbers from $\{1, \cdots, m\}$ (since once a choice is made, they can then be arranged in an increasing order). Thus B has $\binom{m}{k}$ elements, and once we show it is a basis, the claim on the dimension follows.

$\operatorname{span} B = \bigwedge^k V^*$: Let $\omega \in \bigwedge^k V^* \subset T_k^0 V$, and so

$$\omega = \omega(e_{i_1}, \cdots, e_{i_k})\, \epsilon^{i_1} \otimes \cdots \otimes \epsilon^{i_k}.$$

As ω is alternating, we have

$$\begin{aligned}\omega &= \operatorname{Alt} \omega = \omega(e_{i_1}, \cdots, e_{i_k}) \operatorname{Alt}(\epsilon^{i_1} \otimes \cdots \otimes \epsilon^{i_k})\\ &= \omega(e_{i_1}, \cdots, e_{i_k}) \frac{1}{k!} \epsilon^{i_1} \wedge \cdots \wedge \epsilon^{i_k}.\end{aligned}$$

It can be seen, using Exercise 10.5 (with π an appropriate transposition), that the summands with some i_r equal to some i_s, with $r \neq s$, will be zero. In a summand where the $i_1, \cdots, i_k$ are all distinct, we can write $\epsilon^{i_1} \wedge \cdots \wedge \epsilon^{i_k} = (-1)^\ell \epsilon^{j_1} \wedge \cdots \wedge \epsilon^{j_k}$ for some $\ell \in \mathbb{N}$, with $j_1, \cdots, j_k$ a permutation of $i_1, \cdots, i_k$ and such that $j_1 < \cdots < j_k$. This shows that B spans $\bigwedge^k V^*$.

B is independent: Suppose that we have $c_{i_1 \cdots i_k} \epsilon^{i_1} \wedge \cdots \wedge \epsilon^{i_k} = 0$. Then acting both sides on the tuple $(e_{j_1}, \cdots, e_{j_k})$, with $1 \leqslant j_1 < \cdots < j_k \leqslant m$, yields

$$\begin{aligned}0 &= c_{i_1 \cdots i_k} \epsilon^{i_1} \wedge \cdots \wedge \epsilon^{i_k}(e_{j_1}, \cdots, e_{j_k}) = c_{i_1 \cdots i_k} \det \begin{bmatrix} \epsilon^{i_1}(e_{j_1}) & \cdots & \epsilon^{i_1}(e_{j_k}) \\ \vdots & \ddots & \vdots \\ \epsilon^{i_k}(e_{j_1}) & \cdots & \epsilon^{i_k}(e_{j_k}) \end{bmatrix}\\ &= c_{i_1 \cdots i_k} \det \underbrace{\begin{bmatrix} \delta^{i_1}_{j_1} & \cdots & \delta^{i_1}_{j_k} \\ \vdots & \ddots & \vdots \\ \delta^{i_k}_{j_1} & \cdots & \delta^{i_k}_{j_k} \end{bmatrix}}_{=:\Delta^I_J}.\end{aligned}$$

We give an inductive argument (on k) to show that the determinant $\det \Delta^I_J$ in a summand is nonzero only if $I := (i_1, \cdots, i_k) = (j_1, \cdots, j_k) =: J$. If $k = 1$, then clearly $\det \Delta^I_J \neq 0$ implies $I = J$. Suppose that the claim is true for some k. Then for a matrix of size $k+1$, the top row must be nonzero, so that $i_1 = j_\ell$ for some ℓ. Expanding the determinant along the top row, $0 \neq \det \Delta^I_J = (-1)^\ell \det \Delta^{I'}_{J'}$, where I', J' are defined by $I' := (i_2, \cdots, i_{k+1})$, $J' := (j_1, \cdots, j_{\ell-1}, j_{\ell+1}, \cdots, j_{k+1})$. Thus $I' = J'$ by the induction hypothesis. Now if $\ell > 1$, then we get the contradiction that $i_1 = j_\ell > j_1 = i_2$. So $i_1 = j_1$, and then together with $I' = J'$, we obtain $I = J$. Moreover, if $I = J$, then clearly Δ^I_J is the identity matrix, and $\det \Delta^I_J = 1$. So we obtain $0 = c_{i_1 \cdots i_k} \epsilon^{i_1} \wedge \cdots \wedge \epsilon^{i_k}(e_{j_1}, \cdots, e_{j_k}) = c_{j_1 \cdots j_k}$. Thus B is independent. □

10.2 k-form fields

A k-form field is just a $(0, k)$-tensor field Ω on a smooth manifold M whose pointwise evaluations at each point $p \in M$ is a k-form $\Omega(p) \in \bigwedge^k (T_pM)^*$ on T_pM. But then for all $p \in M$, all permutations π of $\{1, \cdots, k\}$ and all vector fields $V_1, \cdots, V_k \in T_0^1 M$,

$$\begin{aligned}(\operatorname{sign}\pi)(\Omega(V_{\pi(1)}, \cdots, V_{\pi(k)}))(p) &= (\operatorname{sign}\pi)\,\Omega(p)((V_{\pi(1)})_p, \cdots, (V_{\pi(k)})_p) \\ &= \Omega(p)((V_1)_p, \cdots, (V_k)_p) \\ &= (\Omega(V_1, \cdots, V_k))(p),\end{aligned}$$

and so we can equivalently define a k-form field as follows.

Definition 10.5. (k-form field.)
Let M be a smooth manifold. For $\pi \in S_k$ and $T \in T_k^0 M$, define $\pi T \in T_k^0 M$ by $(\pi T)(V_1, \cdots, V_k) = T(V_{\pi(1)}, \cdots, V_{\pi(k)})$ for all $V_1, \cdots, V_k \in T_0^1 M$.
Let $k \geqslant 2$. A *k-form field* is a $(0, k)$-tensor field $\Omega \in T_k^0 M$ which is *alternating*, that is, for all permutations $\pi \in S_k$, $\Omega = (\operatorname{sign} \pi)\, \pi\, \Omega$.
The set of all k-form fields on M is denoted by $\mathbf{\Omega}^k M$. We set $\mathbf{\Omega}^1 M = T_1^0 M$ and $\mathbf{\Omega}^0 M = C^\infty(M)$. If $\Omega \in \mathbf{\Omega}^k M$, then k is called the *degree of* Ω.

$\mathbf{\Omega}^k M$ is a $C^\infty(M)$-module with the usual operations for tensor fields. Recall that for a vector space V, $\bigwedge^k V^* = \{0\}$ for $k > \dim V$. If M is m-dimensional (so $\dim T_pM = m$ for all $p \in M$), then $\mathbf{\Omega}^k M = \{0\}$ for all $k > m$.

Exercise 10.6. (Lie differentiation of a form field gives a form field.) Let M be a smooth manifold, and $V \in T_0^1 M$. Show that if $\Omega \in \mathbf{\Omega}^k M$, then $\mathcal{L}_V \Omega \in \mathbf{\Omega}^k M$.

Alternation can be applied to the pointwise evaluation of any $(0, k)$-tensor field T to obtain a k-form field: $(\operatorname{Alt} T)(p) = \operatorname{Alt}(T(p))$, $p \in M$. By Proposition 4.1, $\operatorname{Alt} T \in T_k^0 M$, and since for each $p \in M$, $\operatorname{Alt}(T(p)) \in \bigwedge^k (T_pM)^*$, we have $\operatorname{Alt} T \in \mathbf{\Omega}(M)$. Or equivalently:

Definition 10.6. (Alternation.)
Let M be a smooth manifold and $T \in T_k^0 M$. Then the *alternation of* T is the k-form field $\operatorname{Alt} T \in \mathbf{\Omega}(M)$ defined by

$$\operatorname{Alt} T = \frac{1}{k!} \sum_{\pi \in S_k} (\operatorname{sign}\pi)\, \pi T.$$

If $\Omega \in \mathbf{\Omega}^k M$, then

$$\operatorname{Alt}\Omega = \frac{1}{k!} \sum_{\pi \in S_k} (\operatorname{sign}\pi)\, \pi\, \Omega = \frac{1}{k!} \sum_{\pi \in S_k} (\operatorname{sign}\pi)^2 \Omega = \frac{1}{k!} \sum_{\pi \in S_k} \Omega = \frac{1}{k!} k!\, \Omega = \Omega.$$

Exercise 10.7. Let M be a smooth manifold, $f \in C^\infty(M)$, and $T \in T_k^0 M$. Show that $\operatorname{Alt}(fT) = f \operatorname{Alt} T$.

We have the following analogue of Lemma 10.1, which can be obtained as a consequence, or directly in the same manner.

Lemma 10.3. *Let M be a smooth manifold, $\Omega \in \mathbf{\Omega}^k M$, $\Theta \in \mathbf{\Omega}^\ell M$. Then:*
(1) $\operatorname{Alt}((\operatorname{Alt}\Omega) \otimes \Theta) = \operatorname{Alt}(\Omega \otimes \Theta)$.
(2) $\operatorname{Alt}(\Omega \otimes \Theta) = (-1)^{k\ell} \operatorname{Alt}(\Theta \otimes \Omega)$.
(3) $\operatorname{Alt}(\Omega \otimes (\operatorname{Alt}\Theta)) = \operatorname{Alt}(\Omega \otimes \Theta)$.

Proof.
(1) We have for all $p \in M$ that

$$\begin{aligned}(\operatorname{Alt}((\operatorname{Alt}\Omega) \otimes \Theta))(p) &= \operatorname{Alt}(((\operatorname{Alt}\Omega) \otimes \Theta)(p)) \\ &= \operatorname{Alt}(((\operatorname{Alt}\Omega)(p)) \otimes \Theta(p)) \\ &= \operatorname{Alt}((\operatorname{Alt}(\Omega(p))) \otimes \Theta(p)) = \operatorname{Alt}(\Omega(p) \otimes \Theta(p)) \\ &= \operatorname{Alt}((\Omega \otimes \Theta)(p)) = (\operatorname{Alt}(\Omega \otimes \Theta))(p).\end{aligned}$$

So $\operatorname{Alt}(\operatorname{Alt}(\Omega) \otimes \Theta) = \operatorname{Alt}(\Omega \otimes \Theta)$.
(2) For all $p \in M$,

$$\begin{aligned}(\operatorname{Alt}(\Omega \otimes \Theta))(p) &= \operatorname{Alt}((\Omega \otimes \Theta)(p)) = \operatorname{Alt}(\Omega(p) \otimes \Theta(p)) \\ &= (-1)^{k\ell} \operatorname{Alt}(\Theta(p) \otimes \Omega(p)) = (-1)^{k\ell} \operatorname{Alt}((\Theta \otimes \Omega)(p)) \\ &= ((-1)^{k\ell} \operatorname{Alt}(\Theta \otimes \Omega))(p).\end{aligned}$$

Thus $\operatorname{Alt}(\Omega \otimes \Theta) = (-1)^{k\ell} \operatorname{Alt}(\Theta \otimes \Omega)$.
(3) Analogous to (1). □

The tensor product of form fields is a tensor field, but it may fail to be a form field since it may not be alternating. To get an alternating form field, we apply the alternation operation. The resulting operation is called the exterior product.

Definition 10.7. (Exterior product.)
Let M be a smooth manifold, and let the integers $k, \ell \geqslant 1$.
The *exterior product* $\wedge : \mathbf{\Omega}^k M \times \mathbf{\Omega}^\ell M \to \mathbf{\Omega}^{k+\ell} M$ is defined as follows:
For $\Omega \in \mathbf{\Omega}^k M$ and $\Theta \in \mathbf{\Omega}^\ell M$, $\Omega \wedge \Theta := \dfrac{(k+\ell)!}{k!\,\ell!} \operatorname{Alt}(\Omega \otimes \Theta)$.
For $f, g \in C^\infty(M) = \mathbf{\Omega}^0(M)$, $f \wedge g := fg$.
For $f \in C^\infty(M) = \mathbf{\Omega}^0(M)$ and $\Omega \in \mathbf{\Omega}^k(M)$, $f \wedge \Omega = \Omega \wedge f := f\Omega$.

Clearly, for all $p \in M$, we have $(\Omega \wedge \Theta)(p) = (\Omega(p)) \wedge (\Theta(p)) \in \bigwedge^{k+\ell}(T_pM)^*$.
More explicitly, for $V_1, \cdots, V_{k+\ell} \in T_0^1 M$,

$$(\Omega \wedge \Theta)(V_1, \cdots, V_{k+\ell}) = \frac{1}{k!\,\ell!} \sum_{\pi \in S_{k+\ell}} (\operatorname{sign}\pi)\,(\Omega \otimes \Theta)(V_{\pi(1)}, \cdots, V_{\pi(k+\ell)}).$$

Example 10.2. Let M be a smooth manifold and $\Omega, \Theta \in T_1^0 M$. Then for $V, W \in T_0^1 M$,

$$\begin{aligned}(\Omega \wedge \Theta)(V, W) &= (\Omega \otimes \Theta)(V, W) - (\Omega \otimes \Theta)(W, V)\\ &= \Omega(V)\Theta(W) - \Theta(V)\Omega(W)\\ &= (\Omega \otimes \Theta - \Theta \otimes \Omega)(V, W).\end{aligned}$$

Thus $\Omega \wedge \Theta = \Omega \otimes \Theta - \Theta \otimes \Omega$. ◇

It is straightforward to verify that for $f \in C^\infty(M)$, $\Omega \in \mathbf{\Omega}^k M$ and $\Theta \in \mathbf{\Omega}^\ell M$,

$$f(\Omega \wedge \Theta) = (f\Omega) \wedge \Theta = \Omega \wedge (f\Theta).$$

Also, for $\Omega, \widetilde{\Omega} \in \mathbf{\Omega}^k M$ and $\Theta \in \mathbf{\Omega}^\ell M$, we have

$$(\Omega + \widetilde{\Omega}) \wedge \Theta = \Omega \wedge \Theta + \widetilde{\Omega} \wedge \Theta \text{ and } \Theta \wedge (\Omega + \widetilde{\Omega}) = \Theta \wedge \Omega + \Theta \wedge \widetilde{\Omega}.$$

Thus the wedge product is bilinear over $C^\infty(M)$. We now show that the wedge product is associative.

Proposition 10.4. *Suppose that M is a smooth manifold, and $\Omega^i \in \mathbf{\Omega}^{k_i} M$, $i = 1, 2, 3$. Then $\Omega^1 \wedge (\Omega^2 \wedge \Omega^3) = (\Omega^1 \wedge \Omega^2) \wedge \Omega^3$.*

Proof. For all $p \in M$ we have

$$\begin{aligned}(\Omega^1 \wedge (\Omega^2 \wedge \Omega^3))(p) &= \Omega^1(p) \wedge ((\Omega^2 \wedge \Omega^3)(p))\\ &= \Omega^1(p) \wedge (\Omega^2(p) \wedge \Omega^3(p))\\ &= (\Omega^1(p) \wedge \Omega^2(p)) \wedge \Omega^3(p)\\ &= ((\Omega^1 \wedge \Omega^2)(p)) \wedge \Omega^3(p)\\ &= ((\Omega^1 \wedge \Omega^2) \wedge \Omega^3)(p).\end{aligned}$$

Thus $\Omega^1 \wedge (\Omega^2 \wedge \Omega^3) = (\Omega^1 \wedge \Omega^2) \wedge \Omega^3$. □

As $\wedge$ is associative, there is no ambiguity in writing $\Omega^1 \wedge \cdots \wedge \Omega^n$ for $\Omega^i \in \mathbf{\Omega}^{k_i} M$, $i = 1, \cdots, n$.

Exercise 10.8. Let M be a smooth manifold and let $\Omega^i \in \mathbf{\Omega}^{k_i} M$ for $i = 1, \cdots, n$. Show that $\Omega^1 \wedge \cdots \wedge \Omega^n = \dfrac{(k_1 + \cdots + k_n)!}{k_1! \cdots k_n!} \operatorname{Alt}(\Omega^1 \otimes \cdots \otimes \Omega^n)$.

An application of this exercise is the following result on the wedge product of 1-form fields.

Proposition 10.5. *Let M be a smooth manifold and $\Omega^i \in \mathbf{\Omega}^1 M = T_1^0 M$, $i = 1, \cdots, n$. Then for all $V_i \in T_0^1 M$, $i = 1, \cdots, n$, we have*

$$(\Omega^1 \wedge \cdots \wedge \Omega^n)(V_1, \cdots, V_n) = \det[\Omega^i(V_j)].$$

Proof. We have

$$\begin{aligned}(\Omega^1 \wedge \cdots \wedge \Omega^n)(V_1, \cdots, V_n) &= n!\,\mathrm{Alt}\,(\Omega^1 \otimes \cdots \otimes \Omega^n)(V_1, \cdots, V_n)\\ &= \sum_{\pi \in S_n} (\operatorname{sign}\pi)\,(\Omega^1 \otimes \cdots \otimes \Omega^n)(V_{\pi(1)}, \cdots, V_{\pi(n)})\\ &= \sum_{\pi \in S_n} (\operatorname{sign}\pi)\,\Omega^1(V_{\pi(1)}) \cdots \Omega^n(V_{\pi(n)})\\ &= \det[\Omega^i(V_j)].\end{aligned}$$

□

We have the following consequence of Lemma 10.3 (2).

Proposition 10.6. *Let M be a smooth manifold, $\Omega \in \mathbf{\Omega}^k M$ and $\Theta \in \mathbf{\Omega}^\ell M$. Then $\Omega \wedge \Theta = (-1)^{k\ell}(\Theta \wedge \Omega)$.*

Proof. We have

$$\Omega \wedge \Theta = \frac{(k+\ell)!}{k!\,\ell!}\mathrm{Alt}\,(\Omega \otimes \Theta) = \frac{(k+\ell)!}{k!\,\ell!}(-1)^{k\ell}\mathrm{Alt}\,(\Theta \otimes \Omega) = (-1)^{k\ell}\Theta \wedge \Omega.$$

□

Exercise 10.9. Suppose that M is a smooth manifold, k is odd, and $\Omega \in \mathbf{\Omega}^k M$. Show that $\Omega \wedge \Omega = 0$.

Recall that in an admissible chart $(U, \mathbf{x})$ for a smooth manifold M, we can decompose a given $T \in T^0_k M$ as $T = T_{i_1 \cdots i_k} dx^{i_1} \otimes \cdots \otimes dx^{i_k}$, for some functions $T_{i_1 \cdots i_k} \in C^\infty(U)$. Now if $\Omega \in \mathbf{\Omega}^k M$, then we have in particular $\Omega = \Omega_{i_1 \cdots i_k} dx^{i_1} \otimes \cdots \otimes dx^{i_k}$, for some functions $\Omega_{i_1 \cdots i_k} \in C^\infty(U)$, but

$$\Omega = \mathrm{Alt}\,\Omega = \Omega_{i_1 \cdots i_k} \mathrm{Alt}\,(dx^{i_1} \otimes \cdots \otimes dx^{i_k}) = \frac{\Omega_{i_1 \cdots i_k}}{k!} dx^{i_1} \wedge \cdots \wedge dx^{i_k}.$$

Using Proposition 10.6, we can rearrange the factors of the wedge product within a summand, and combine terms to obtain

$$\Omega = \sum_{1 \leqslant i_1 < \cdots < i_k \leqslant m} \omega_{i_1 \cdots i_k} dx^{i_1} \wedge \cdots \wedge dx^{i_k}$$

for some functions $\omega_{i_1 \cdots i_k} \in C^\infty(U)$. The functions $\omega_{i_1 \cdots i_k}$ are actually $\Omega_{i_1 \cdots i_k}$. Indeed, recall first the definition of $\Omega_{i_1 \cdots i_k}$ and note that a rearrangement of indices produces the sign of the permutation used, as Ω is alternating. The same factor is picked while rearranging the indices in $dx^{i_1} \wedge \cdots \wedge dx^{i_k}$. Finally, each term in the rearranged sum collects $k!$ terms in the original one. In particular, if M is m-dimensional, then every m-form field $\Omega \in \mathbf{\Omega}^m M$ can be decomposed in an admissible chart $(U, \mathbf{x})$ as $\Omega = \omega\, dx^1 \wedge \cdots \wedge dx^m$, where $\omega = \Omega(\partial_{x^1}, \cdots, \partial_{x^m}) \in C^\infty(U)$.

Exercise 10.10. Consider the smooth manifold $M = \mathbb{R}^3$ with the standard smooth structure, and the global admissible chart $(\mathbb{R}^3, (x, y, z) \mapsto (x, y, z))$. Suppose that $\Omega := x\,dx + y\,dy + z\,dz \in \mathbf{\Omega}^1 M$, $\Theta := y\,dx + z\,dy + x\,dz \in \mathbf{\Omega}^1 M$. Determine $\Omega \wedge \Theta$.

Exercise 10.11. Let M be a smooth manifold. In an admissible chart $(U, \mathbf{x})$, suppose that $\Omega := f\, dx^{i_1} \wedge \cdots \wedge dx^{i_k}$, for a function $f \in C^\infty(U)$ and a fixed k-tuple $I = (i_1, \cdots, i_k)$, where $1 \leqslant i_1 < \cdots < i_k \leqslant m$. Show that if $J = (j_1, \cdots, j_k)$, with $1 \leqslant j_1 < \cdots < j_k \leqslant m$, then

$$\Omega_{j_1 \cdots j_k} = \begin{cases} f & \text{if } J = I, \\ 0 & \text{if } J \neq I. \end{cases}$$

Pull-back

We had seen that if $f : M \to N$ is a smooth map between smooth manifolds M and N, then a $(0, k)$-tensor field $T \in T_k^0 N$ can be pulled back under f to a $(0, k)$-tensor field $f^* T$ on M. If $\Omega \in \mathbf{\Omega}^k N$, then $f^* \Omega \in \mathbf{\Omega}^k M$. Indeed, for all $V_1, \cdots, V_k \in T_0^1 M$, any permutation $\pi \in S_k$, and all $p \in M$, we have

$$\begin{aligned} ((f^*\Omega)(V_{\pi(1)}, \cdots, V_{\pi(k)}))(p) &= \Omega(f(p))(df_p((V_{\pi(1)})_p), \cdots, df_p((V_{\pi(k)})_p)) \\ &= (\operatorname{sign} \pi)\, \Omega(f(p))(df_p((V_1)_p), \cdots, df_p((V_k)_p)) \\ &= (\operatorname{sign} \pi)\, ((f^*\Omega)(V_1, \cdots, V_k))(p), \end{aligned}$$

and so $(f^*\Omega)(V_{\pi(1)}, \cdots, V_{\pi(k)}) = (\operatorname{sign} \pi)\, (f^*\Omega)(V_1, \cdots, V_k)$. Consequently, $f^*\Omega \in \mathbf{\Omega}^k M$.

Exercise 10.12. Let M, N be smooth manifolds, $f : M \to N$ be a smooth map, $\Omega \in \mathbf{\Omega}^k N$, and $\Theta \in \mathbf{\Omega}^\ell N$. Prove that $f^*(\Omega \wedge \Theta) = (f^*\Omega) \wedge (f^*\Theta)$.

Exercise 10.13. Let $N = \mathbb{R}^3$ with the standard smooth structure, and M be the open subset of N given by $M = (0, \infty) \times (0, \pi) \times (0, 2\pi)$ with the induced smooth structure. In the admissible chart $(V = N, (x, y, z) \mapsto (x, y, z))$ for N, consider the 3-form field $\Omega = dx \wedge dy \wedge dz$. Let $f : M \to N$ be the smooth map given by $f(r, \theta, \phi) = (r(\sin\theta)(\cos\phi), r(\sin\theta)(\sin\phi), r\cos\theta)$, for $(r, \theta, \phi) \in M$. Determine the pull-back $f^*\Omega \in \mathbf{\Omega}^3 M$ of Ω.

10.3 Exterior derivative

For a 0-form on a smooth manifold M, that is, for a smooth function $f \in C^\infty(M)$, we had defined the gradient df of f as a 1-form field. Thus $d : \mathbf{\Omega}^0 M \to \mathbf{\Omega}^1 M$. We now give an extension of this operation, which is called the 'exterior derivative'.

Definition 10.8. (Exterior derivative of a k-form field.)
Let M be a smooth manifold and let $\Omega \in \mathbf{\Omega}^k M$ be a k-form field.
The *exterior derivative* $d\Omega \in \mathbf{\Omega}^{k+1} M$ *of* Ω is defined by

$$d\Omega = \frac{\partial_{x^i} \Omega_{i_1 \cdots i_k}}{k!} dx^i \wedge dx^{i_1} \wedge \cdots \wedge dx^{i_k} = \sum_{1 \leqslant i_1 < \cdots < i_k \leqslant m} \partial_{x^i} \Omega_{i_1 \cdots i_k}\, dx^i \wedge dx^{i_1} \wedge \cdots \wedge dx^{i_k}$$

in any admissible chart $(U, \mathbf{x})$ for M.

We check this notion is well-defined. Firstly, the above gives a $(k+1)$-form field in any admissible chart $(U, \mathbf{x})$. So we only need to check that there is no conflict of definition in overlapping charts. Then by Proposition 4.1 (p.65), $d\,\Omega$ is a $(0,k)$-tensor field. Moreover, the alternating property holds everywhere, as it holds in every chart. Let $(U, \mathbf{x})$ and $(V, \mathbf{y})$ be overlapping admissible charts for M. Then $\Omega = \frac{1}{k!}\Omega(\partial_{y^{j_1}}, \cdots, \partial_{y^{j_k}})\, dy^{j_1} \wedge \cdots \wedge dy^{j_k}$, and

$$\begin{aligned}
&\partial_{y^j}\Omega(\partial_{y^{j_1}}, \cdots, \partial_{y^{j_k}})\, dy^j \wedge dy^{j_1} \wedge \cdots \wedge dy^{j_k} \\
&= (\partial_{y^j} x^i)\, \partial_{x^i}\Omega(\partial_{y^{j_1}} x^{i_1}\, \partial_{x^{i_1}}, \cdots, \partial_{y^{j_k}} x^{i_k}\, \partial_{x^{i_k}}) \cdot \\
&\quad \partial_{x^\ell} y^j\, dx^\ell \wedge \partial_{x^{\ell_1}} y^{j_1}\, dx^{\ell_1} \wedge \cdots \wedge \partial_{x^{\ell_k}} y^{j_k}\, dx^{\ell_k} \\
&= (\partial_{y^j} x^i)(\partial_{x^\ell} y^j)\partial_{x^i}\big((\partial_{y^{j_1}} x^{i_1}) \cdots (\partial_{y^{j_k}} x^{i_k})\,\Omega(\partial_{x^{i_1}}, \cdots, \partial_{x^{i_k}})\big) \cdot \\
&\quad (\partial_{x^{\ell_1}} y^{j_1}) \cdots (\partial_{x^{\ell_k}} y^{j_k}) dx^\ell \wedge dx^{\ell_1} \wedge \cdots \wedge dx^{\ell_k} \\
&= \partial_{x^i}\big((\partial_{y^{j_1}} x^{i_1}) \cdots (\partial_{y^{j_k}} x^{i_k})\,\Omega(\partial_{x^{i_1}}, \cdots, \partial_{x^{i_k}})\big) \cdot \\
&\quad (\partial_{x^{\ell_1}} y^{j_1}) \cdots (\partial_{x^{\ell_k}} y^{j_k})\, dx^i \wedge dx^{\ell_1} \wedge \cdots \wedge dx^{\ell_k} \\
&= \partial_{x^i}(\Omega(\partial_{x^{i_1}}, \cdots, \partial_{x^{i_k}}))\, dx^i \wedge dx^{i_1} \wedge \cdots \wedge dx^{i_k} + \Omega(\partial_{x^{i_1}}, \cdots, \partial_{x^{i_k}}) \cdot S,
\end{aligned}$$

where the product rule gives the last equality, and S is given by

$$S = \sum_{r=1}^{k} (\partial_{x^{\ell_r}} y^{j_r})(\partial_{x^i}\partial_{y^{j_r}} x^{i_r})\, dx^i \wedge dx^{i_1} \wedge \cdots \wedge dx^{i_{r-1}} \wedge dx^{\ell_r} \wedge dx^{i_{r+1}} \wedge \cdots \wedge dx^{i_k},$$

which we will now show is 0. We have

$$(\partial_{x^{\ell_r}} y^{j_r})(\partial_{x^i}\partial_{y^{j_r}} x^{i_r}) = \partial_{x^i}((\partial_{x^{\ell_r}} y^{j_r})(\partial_{y^{j_r}} x^{i_r})) - (\partial_{x^i}\partial_{x^{\ell_r}} y^{j_r})(\partial_{y^{j_r}} x^{i_r})$$

The first summand is $\partial_{x^i}((\partial_{x^{\ell_r}} y^{j_r})(\partial_{y^{j_r}} x^{i_r})) = \partial_{x^i}\delta^{i_r}_{\ell_r} = 0$. So

$$S = -\sum_{r=1}^{k} (\partial_{y^{j_r}} x^{i_r})(\partial_{x^i}\partial_{x^{\ell_r}} y^{j_r})\, dx^i \wedge dx^{i_1} \wedge \cdots \wedge dx^{i_{r-1}} \wedge dx^{\ell_r} \wedge dx^{i_{r+1}} \wedge \cdots \wedge dx^{i_k}.$$

A swapping $i \leftrightarrow \ell^r$ in $dx^i \wedge dx^{i_1} \wedge \cdots \wedge dx^{i_{r-1}} \wedge dx^{\ell_r} \wedge dx^{i_{r+1}} \wedge \cdots \wedge dx^{i_k}$ yields an overall sign $(-1)^r(-1)^{r-1} = -1$, but the term $\partial_{x^i}\partial_{x^{\ell_r}} y^{j_r}$ does not change under such a swapping of these dummy indices. Thus $S = 0$. This shows the well-definition.

The map $d : \mathbf{\Omega}^k M \to \mathbf{\Omega}^{k+1} M$ is a linear transformation, considering $\mathbf{\Omega}^k M, \mathbf{\Omega}^{k+1} M$ as vector spaces.

Exercise 10.14. Let M be a smooth manifold and $(U, \mathbf{x})$ be an admissible chart. Let $f \in C^\infty(U)$, and $(i_1, \cdots, i_k)$ be a fixed k-tuple from $\{1, \cdots, m\}$ with distinct entries. Let $\Omega = f dx^{i_1} \wedge \cdots \wedge dx^{i_k}$. Show that $d\Omega = (\partial_{x^i} f)\, dx^i \wedge dx^{i_1} \wedge \cdots \wedge dx^{i_k}$.

By the linearity of d, it follows by the above exercise that if $\Omega \in \mathbf{\Omega}^k M$ has a decomposition in a chart given by $\Omega = \omega_{i_1 \cdots i_k} dx^{i_1} \wedge \cdots \wedge dx^{i_k}$, for some functions $\omega_{i_1 \cdots i_k} \in C^\infty(U)$, then

$$d\Omega = (\partial_{x^i}\omega_{i_1 \cdots i_k})\, dx^i \wedge dx^{i_1} \wedge \cdots \wedge dx^{i_k}.$$

The exterior derivative map also satisfies a type of Leibniz rule with respect to the wedge product (Prop. 10.7), which justifies calling it a derivative.

Notation 10.1. Let $(U, \mathbf{x})$ be an admissible chart in a smooth manifold M, and $\Omega \in \mathbf{\Omega}^k M$. For a k-tuple $(i_1, \cdots, i_k)$ from $\{1, \cdots, m\}$, we introduce the abbreviations $I = (i_1, \cdots, i_k)$, $dx^I = dx^{i_1} \wedge \cdots \wedge dx^{i_k}$, and

$$\Omega_I = \frac{\Omega_{i_1 \cdots i_k}}{k!}.$$

Then $\Omega = \Omega_I dx^I$.

Proposition 10.7. *Suppose that M is a smooth manifold. Let $\Omega \in \mathbf{\Omega}^k M$ and $\Theta \in \mathbf{\Omega}^\ell M$. Then $d(\Omega \wedge \Theta) = (d\Omega) \wedge \Theta + (-1)^k \Omega \wedge d\Theta$.*

Proof. In any admissible chart $(U, \mathbf{x})$, we have

$$\begin{aligned} d(\Omega \wedge \Theta) &= d(\Omega_I dx^I \wedge \Theta_J dx^J) \\ &= d(\Omega_I \Theta_J dx^I \wedge dx^J) \\ &= \partial_{x^i}(\Omega_I \Theta_J)\, dx^i \wedge dx^I \wedge dx^J \\ &= ((\partial_{x^i} \Omega_I)\Theta_J + \Omega_I \partial_{x^i} \Theta_J)\, dx^i \wedge dx^I \wedge dx^J \\ &= (\partial_{x^i} \Omega_I)\, dx^i \wedge dx^I \wedge \Theta_J dx^J + (-1)^k \Omega_I dx^I \wedge \partial_{x^i} \Theta_J dx^i \wedge dx^J \\ &= (d\Omega) \wedge \Theta + (-1)^k \Omega \wedge d\Theta, \end{aligned}$$

where the second summand acquires the factor $(-1)^k$ since dx^i has to pass through the k-form $dx^I = dx^{i_1} \wedge \cdots \wedge dx^{i_k}$. □

In particular, we have $d(f\Omega) = d(f \wedge \Omega) = (df) \wedge \Omega + f \wedge d\Omega$.

Exercise 10.15. Consider the 1-form fields Ω and Θ from Exercise 10.10 again. Show that $d\Omega = 0$, and $d\Theta = -(dx \wedge dy + dy \wedge dz + dz \wedge dx)$.

In the previous exercise, we note that for some form fields the exterior derivative vanishes, while it does not for others. This prompts the first of following definitions.

Definition 10.9. (Closed and exact form fields.)
Let M be a smooth manifold. A k-form field $\Omega \in \mathbf{\Omega}^k M$ is called *closed* if $d\Omega = 0$. If $k \geqslant 1$, then a k-form field $\Omega \in \mathbf{\Omega}^k M$ is called *exact* if there exists a $\Theta \in \mathbf{\Omega}^{k-1} M$ such that $\Omega = d\Theta$.

We now show $d^2 = d \circ d = 0$. This implies every exact form field is closed, since for an exact form field Ω, we have $\Omega = d\Theta$, and so $d\Omega = d(d\Theta) = 0$.

Proposition 10.8. *Suppose that M is a smooth manifold, and $\Omega \in \mathbf{\Omega}^k M$. Then $d(d\Omega) = 0$.*

Proof. Let $(U, \mathbf{x})$ be any admissible chart for M. Using Notation 10.1,

$$\begin{aligned} d(d\Omega) &= d((\partial_{x^i}\Omega_I)\, dx^i \wedge dx^I) \\ &= (\partial_{x^j}\partial_{x^i}\Omega_I)\, dx^j \wedge dx^i \wedge dx^I \\ &= (\partial_{x^i}\partial_{x^j}\Omega_I)\, dx^j \wedge dx^i \wedge dx^I \\ &= -(\partial_{x^i}\partial_{x^j}\Omega_I)\, dx^i \wedge dx^j \wedge dx^I \\ &= -d(d\Omega), \end{aligned}$$

where we used $[\partial_{x^i}, \partial_{x_j}]=0$, and $dx^i \wedge dx^j = -dx^j \wedge dx^i$. So $d(d\Omega)=0$. □

Exact form fields are closed, but the converse may not always hold.

Example 10.3. (Closed $\not\Rightarrow$ exact.)
Let $\mathbb{R}^2$ be equipped with the standard smooth structure, and consider the open set $M := \mathbb{R}^2 \backslash \{(0,0)\}$ with the induced smooth structure from $\mathbb{R}^2$. In the global admissible chart $(\mathbb{R}^2 \backslash \{(0,0)\}, (x,y) \mapsto (x,y))$ for M, let the 1-form field $\Omega \in T_1^0 M$ be given by

$$\Omega = \frac{-y}{x^2+y^2}\, dx + \frac{x}{x^2+y^2}\, dy.$$

Then Ω is closed, since

$$\begin{aligned} d\Omega &= \partial_y \frac{-y}{x^2+y^2}\, dy \wedge dx + \partial_x \frac{x}{x^2+y^2}\, dx \wedge dy \\ &= \frac{y^2-x^2}{(x^2+y^2)^2}\, dy \wedge dx + \frac{y^2-x^2}{(x^2+y^2)^2}\, dx \wedge dy = 0. \end{aligned}$$

However, we now argue that Ω is not exact. Suppose, on the contrary, that there exists an $f \in C^\infty(M)$, such that $\Omega = df = \partial_x f\, dx + \partial_y dy$. Then

$$\frac{\partial f}{\partial x} = \frac{-y}{x^2+y^2} \quad \text{and} \quad \frac{\partial f}{\partial y} = \frac{x}{x^2+y^2}.$$

Consider the smooth curve $[0, 2\pi] \ni t \overset{C}{\mapsto} (\cos t, \sin t) \in M$. We have

$$\begin{aligned} 0 &= f((1,0)) - f((1,0)) = f(C(2\pi)) - f(C(0)) \\ &= \int_0^{2\pi} \frac{d}{dt} f(C(t))\, dt = \int_0^{2\pi} \frac{d}{dt} f(\cos t, \sin t)\, dt \\ &= \int_0^{2\pi} \Big(\frac{\partial f}{\partial x}(\cos t, \sin t)(-\sin t) + \frac{\partial f}{\partial y}(\cos t, \sin t)(\cos t)\Big)\, dt \\ &= \int_0^{2\pi} \Big(\frac{(-\sin t)}{(\cos t)^2+(\sin t)^2}(-\sin t) + \frac{(\cos t)}{(\cos t)^2+(\sin t)^2}(\cos t)\Big)\, dt \\ &= \int_0^{2\pi} 1\, dt = 2\pi, \end{aligned}$$

a contradiction. So Ω is not exact. ◇

If $(U, \mathbf{x})$ is an admissible chart for a smooth manifold M, and Ω is a k-form field of the type $\Omega = dx^{i_1} \wedge \cdots \wedge dx^{i_k} \in \mathbf{\Omega}^k U$, then it follows from Propositions 10.7 and 10.8 that $d\Omega = 0$.

Exercise 10.16. Consider the smooth manifold $N = \mathbb{R}^2$ with the standard smooth structure. Show that every closed 1-form field $\Omega \in \mathbf{\Omega}^1 M$ is exact.

Remark 10.1. (de Rham cohomology.)
The failure of the converse (closed $\Rightarrow$ exact) is linked to the global topology of the manifold M. (For instance, in Example 10.3, $M = \mathbb{R}^2 \setminus \{(0,0)\}$ has a 'hole' at the origin, while in Exercise 10.16, $N = \mathbb{R}^2$ does not.) We have the following sequence of vector spaces and maps

$$0 \xrightarrow{d} \mathbf{\Omega}^0 M \xrightarrow{d} \mathbf{\Omega}^1 M \xrightarrow{d} \mathbf{\Omega}^2 M \xrightarrow{d} \cdots \xrightarrow{d} \mathbf{\Omega}^m M \xrightarrow{d} 0.$$

(Note that the first map on the left is just the zero map, and we have set $\mathbf{\Omega}^{-1} M = \{0\}$.) Also, this explains the terminology 'exact' from Definition 10.9: recall that a sequence of linear maps

$$A \xrightarrow{f} B \xrightarrow{g} C$$

of vector spaces is *exact at* B if $\operatorname{ran} f = \ker g$. In order to specify which map d we mean, we use d^k for the map $d : \mathbf{\Omega}^k M \to \mathbf{\Omega}^{k+1} M$. Then $d^2 = 0$ implies that $\operatorname{ran} d^{k-1} \subset \ker d^k$. The quotient vector space

$$H^k_{\mathrm{dR}} := (\ker d^k)/(\operatorname{ran} d^{k-1})$$

is called the kth *de Rham cohomology vector space* H^k_{dR} *of* M. The association of these algebraic objects (vector spaces) to a manifold gives rise to a dictionary for translating topological properties of M into algebraic properties of the de Rham cohomological vector spaces. For example,

$$\begin{aligned} H^0_{\mathrm{dR}} &= (\ker d : C^\infty(M) \to \mathbf{\Omega}^1 M)/\{0\} \\ &\simeq (\ker d : C^\infty(M) \to \mathbf{\Omega}^1 M) \\ &= \{f \in C^\infty(M) : df = 0\}. \end{aligned}$$

But if M is connected, then it can be shown that $df = 0$ implies that f must be constant on M, and in general, $\{f \in C^\infty(M) : df = 0\}$ is the collection of all functions that are constant on each connected component of M. Thus $H^0_{\mathrm{dR}} \simeq \mathbb{R}^{b_0(M)}$, where $b_0(M)$ is the number of connected components of M. (The integer $b_0(M)$ is called the 0th *Betti number of* M; see for instance [Bishop and Goldberg (1980), p.186].) $\ast$

Vector calculus in $\mathbb{R}^3$

In multivariable calculus, a *vector-valued function* $\vec{V}$ on $\mathbb{R}^3$ is a smooth map $\vec{V} : \mathbb{R}^3 \to \mathbb{R}^3$ that associates to each point $p \in \mathbb{R}^3$ an element $\vec{V}(p) \in \mathbb{R}^3$. Considering $\mathbb{R}^3$ as a smooth manifold equipped with the standard smooth structure, we have $\mathbb{R}^3 \simeq T_p\mathbb{R}^3$ using the global chart-induced basis. So a vector-valued function in the multivariable-calculus-sense can be considered a vector field in the manifold language:

$$\vec{V} = \begin{bmatrix} V^x \\ V^y \\ V^z \end{bmatrix} \quad \leftrightarrow \quad V = V^x\partial_x + V^y\partial_y + V^z\partial_z \in T_0^1\mathbb{R}^3.$$

We also know that we can create a vector field from a smooth function $f : \mathbb{R}^3 \to \mathbb{R}$ by taking its 'gradient'

$$\vec{\nabla} f := \begin{bmatrix} \frac{\partial f}{\partial x} \\ \frac{\partial f}{\partial y} \\ \frac{\partial f}{\partial z} \end{bmatrix}.$$

The exterior derivative of $f \in \boldsymbol{\Omega}^0 M$ is the 1-form field $df = \frac{\partial f}{\partial x}dx + \frac{\partial f}{\partial y}dy + \frac{\partial f}{\partial z}dz$, which has the same components as $\vec{\nabla} f$.

Let $\vec{V}$ be a vector-valued function with the component functions given by (V^x, V^y, V^z). Recall from multivariable calculus that the differential operator $\vec{\nabla}\times$, the *curl*, acts on $\vec{V}$ to produce the vector-valued function

$$\vec{\nabla} \times \vec{V} = \begin{bmatrix} \frac{\partial V^z}{\partial y} - \frac{\partial V^y}{\partial z} \\ \frac{\partial V^x}{\partial z} - \frac{\partial V^z}{\partial x} \\ \frac{\partial V^y}{\partial x} - \frac{\partial V^x}{\partial y} \end{bmatrix}.$$

If we consider the 1-form field $\Omega := V^x dx + V^y dy + V^z dz$ using the components of $\vec{V}$, then its exterior derivative $d\Omega$ is the 2-form field given by

$$\begin{aligned} d\Omega &= d(V^x dx + V^y dy + V^z dz) \\ &= \frac{\partial V^x}{\partial y} dy \wedge dx + \frac{\partial V^x}{\partial z} dz \wedge dx + \frac{\partial V^y}{\partial x} dx \wedge dy + \frac{\partial V^y}{\partial z} dz \wedge dy \\ &\quad + \frac{\partial V^z}{\partial x} dx \wedge dz + \frac{\partial V^z}{\partial y} dy \wedge dz \\ &= \left(\frac{\partial V^z}{\partial y} - \frac{\partial V^y}{\partial z}\right) dy \wedge dz + \left(\frac{\partial V^x}{\partial z} - \frac{\partial V^z}{\partial x}\right) dz \wedge dx + \left(\frac{\partial V^y}{\partial x} - \frac{\partial V^x}{\partial y}\right) dx \wedge dy, \end{aligned}$$

which has the same components as $\vec{\nabla} \times \vec{V}$. We recall that the divergence map $\vec{\nabla}\cdot$ in multivariable calculus acts on a vector-valued function $\vec{V}$, and produces a smooth function,

$$\vec{\nabla}\cdot\vec{V} = \frac{\partial V^x}{\partial x} + \frac{\partial V^y}{\partial y} + \frac{\partial V^z}{\partial z}.$$

If we consider the 2-form field

$$\Omega = V^x dy \wedge dz + V^y dz \wedge dx + V^z dx \wedge dy$$

constructed using the components of $\vec{V}$, then the exterior derivative $d\Omega$ is the 3-form field given by

$$\begin{aligned} d\Omega &= d(V_x\, dy \wedge dz + V_y\, dz \wedge dx + V_z\, dx \wedge dy) \\ &= \frac{\partial V_x}{\partial x} dx \wedge dy \wedge dz + \frac{\partial V_y}{\partial y} dy \wedge dz \wedge dx + \frac{\partial V_z}{\partial z} dz \wedge dx \wedge dy \\ &= (\vec{\nabla}\cdot\vec{V})\, dx \wedge dy \wedge dz. \end{aligned}$$

We know that $C^\infty(M) = \mathbf{\Omega}^0 M$ by definition, and since all 3-form fields Ω on $M = \mathbb{R}^3$ can be decomposed as $\Omega = f\ dx \wedge dy \wedge dz$ for some function $f \in C^\infty(M)$, we can identify $\mathbf{\Omega}^3 M$ with $C^\infty(M)$ too. Furthermore, 1-form fields Ω admit a decomposition $\Omega = f\ dx + g\ dy + h\ dz$, and 2-form fields admit a decomposition

$$\Omega = f\, dy \wedge dz + g\, dz \wedge dx + h\, dx \wedge dy,$$

for some functions $f, g, h \in C^\infty(M)$, and so they can be identified with vector-valued functions with the three component functions given by (f, g, h). Thus if $\mathbf{V}(\mathbb{R}^3)$ denotes the $C^\infty(\mathbb{R}^3)$-module of all vector-valued functions on $\mathbb{R}^3$ (with the module operations defined component-wise in the obvious way), then both of $\mathbf{\Omega}^1\mathbb{R}^3$ and $\mathbf{\Omega}^2\mathbb{R}^3$ are isomorphic as $C^\infty(\mathbb{R}^3)$-modules to $\mathbf{V}(\mathbb{R}^3)$ under the identifications described above. Hence we obtain the following commutative diagram.

$$\begin{array}{ccccccc} \mathbf{\Omega}^0\mathbb{R}^3 & \xrightarrow{d} & \mathbf{\Omega}^1\mathbb{R}^3 & \xrightarrow{d} & \mathbf{\Omega}^2\mathbb{R}^3 & \xrightarrow{d} & \mathbf{\Omega}^3\mathbb{R}^3 \\ \downarrow\wr & & \downarrow\wr & & \downarrow\wr & & \downarrow\wr \\ C^\infty(\mathbb{R}^3) & \xrightarrow[\vec{\nabla}]{} & \mathbf{V}(\mathbb{R}^3) & \xrightarrow[\vec{\nabla}\times]{} & \mathbf{V}(\mathbb{R}^3) & \xrightarrow[\vec{\nabla}\cdot]{} & C^\infty(\mathbb{R}^3) \end{array}$$

The familiar facts from multivariable calculus saying that

- the 'curl of the gradient vanishes', $\vec{\nabla}\times(\vec{\nabla}f) = 0$ for smooth functions f
- the 'divergence of the curl vanishes', $\vec{\nabla}\cdot(\vec{\nabla}\times\vec{V}) = 0$ for vector-valued functions $\vec{V}$,

are just manifestations of the property $d \circ d = 0$ for the exterior derivative.

Pull-back

Pulling back commutes with exterior differentiation.

Proposition 10.9. *Suppose M, N are smooth manifolds, and $f : M \to N$ is a smooth map. If $\Omega \in \mathbf{\Omega}^k N$, then $f^*(d\Omega) = d(f^*\Omega)$.*

Proof. We work locally. For a point $p \in M$, let $(U, \mathbf{x})$ be an admissible chart for M containing p, and $(V, \mathbf{y})$ be an admissible chart for N containing $f(p)$ such that $f(U) \subset V$. Using Notation 10.1, we write

$$\Omega = \frac{\Omega_{i_1\cdots i_k}}{k!} dy^{i_1} \wedge \cdots \wedge dy^{i_k} = \Omega_I dy^{i_1} \wedge \cdots \wedge dy^{i_k}.$$

Then it follows from Exercises 10.12 (p.220), 3.20 (p.51) and 3.21 that

$$f^*\Omega = f^*(\Omega_I dy^{i_1} \wedge \cdots \wedge dy^{i_k}) = (\Omega_I \circ f) d(y^{i_1} \circ f) \wedge \cdots \wedge d(y^{i_k} \circ f).$$

Using $d^2 = 0$, it follows by the Leibniz rule (Proposition 10.7, p.222) that

$$d(f^*\Omega) = d(\Omega_I \circ f) \wedge d(y^{i_1} \circ f) \wedge \cdots \wedge d(y^{i_k} \circ f). \tag{10.1}$$

On the other hand, $d\Omega = d(\Omega_I dy^{i_1} \wedge \cdots \wedge dy^{i_k}) = d\Omega_I \wedge dy^{i_1} \wedge \cdots \wedge dy^{i_k}$, and so by Exercise 10.12,

$$\begin{aligned} f^*(d\Omega) &= f^*(d\Omega_I \wedge dy^{i_1} \wedge \cdots \wedge dy^{i_k}) \\ &= f^*(d\Omega_I) \wedge f^*(dy^{i_1}) \wedge \cdots \wedge f^*(dy^{i_k}) \\ &= d(\Omega_I \circ f) \wedge d(y^{i_1} \circ f) \wedge \cdots \wedge d(y^{i_k} \circ f) \overset{(10.1)}{=} d(f^*\Omega). \qquad \square \end{aligned}$$

An intrinsic expression for the exterior derivative

We had defined the exterior derivative by using charts, and checked that the choice of chart did not matter. We now give an explicit intrinsic definition for the exterior derivative. This alternative definition is given in the form of a formula, and this is the content of the following result.

Theorem 10.1. *Suppose that M is a smooth manifold, and $\Omega \in \mathbf{\Omega}^k M$. For all $V_1, \cdots, V_k, V_{k+1}$,*

$$\begin{aligned} (d\Omega)(V_1, \cdots, V_{k+1}) &= \sum_{i=1}^{k+1} (-1)^{i-1} V_i(\Omega(V_1, \cdots, \widehat{V_i}, \cdots, V_{k+1})) \\ &\quad + \sum_{1 \leqslant i < j \leqslant k+1} (-1)^{i+j} \Omega([V_i, V_j], V_1, \cdots, \widehat{V_i}, \cdots, \widehat{V_j}, \cdots, V_{k+1}), \end{aligned}$$

where $\widehat{\cdot}$ indicates an omission of the corresponding argument.

Proof. Let us call the right-hand side $\Theta(V_1, \cdots, V_{k+1})$, and the two sums appearing there as $\Theta_1(V_1, \cdots, V_{k+1})$ and $\Theta_2(V_1, \cdots, V_{k+1})$. We note that $\Theta : T_0^1 M \times \cdots \times T_0^1 M (k+1 \text{ times}) \to C^\infty(M)$ is $\mathbb{R}$-multilinear. We will show that it is in fact also $C^\infty(M)$-multilinear.

For $1 \leqslant \ell \leqslant k+1$, and $f \in C^\infty(M)$,

$$\begin{aligned}
&\Theta_1(V_1, \cdots, fV_\ell, \cdots, V_{k+1}) \\
&= (-1)^{\ell-1}(fV_\ell)(\Omega(V_1, \cdots, \widehat{V_\ell}, \cdots, V_{k+1})) \\
&\quad + \sum_{\ell<i\leqslant k+1} (-1)^{i-1}V_i(\Omega(V_1, \cdots, fV_\ell, \cdots, \widehat{V_i}, \cdots, V_{k+1})) \\
&\quad + \sum_{1\leqslant i<\ell} (-1)^{i-1}V_i(\Omega(V_1, \cdots, \widehat{V_i}, \cdots, fV_\ell, \cdots, V_{k+1})) \\
&= f\Theta_1(V_1, \cdots, V_\ell, \cdots, V_{k+1}) + A + B,
\end{aligned}$$

where

$$\begin{aligned}
A :=& \sum_{\ell<i\leqslant k+1} (-1)^{i-1}(V_i f)\,\Omega(V_1, \cdots, V_\ell, \cdots, \widehat{V_i}, \cdots, V_{k+1}) \\
=& \sum_{\ell<i\leqslant k+1} (-1)^{i-1+\ell-1}(V_i f)\,\Omega(V_\ell, V_1, \cdots, \widehat{V_\ell}, \cdots, \widehat{V_i}, \cdots, V_{k+1}) \\
=& \sum_{\ell<j\leqslant k+1} (-1)^{j+\ell}(V_j f)\,\Omega(V_\ell, V_1, \cdots, \widehat{V_\ell}, \cdots, \widehat{V_j}, \cdots, V_{k+1}),
\end{aligned}$$

(the dummy i was replaced by j to obtain the last line), and

$$\begin{aligned}
B :=& \sum_{1\leqslant i<\ell} (-1)^{i-1}(V_i f)\,\Omega(V_1, \cdots, \widehat{V_i}, \cdots, V_\ell, \cdots, V_{k+1}) \\
=& \sum_{1\leqslant i<\ell} (-1)^{i-1+\ell-2}(V_i f)\,\Omega(V_\ell, V_1, \cdots, \widehat{V_i}, \cdots, \widehat{V_\ell}, \cdots, V_{k+1}) \\
=& -\sum_{1\leqslant i<\ell} (-1)^{i+\ell}(V_i f)\,\Omega(V_\ell, V_1, \cdots, \widehat{V_i}, \cdots, \widehat{V_\ell}, \cdots, V_{k+1}).
\end{aligned}$$

Similarly, to compute $\Theta_2(V_1, \cdots, fV_\ell, \cdots, V_{k+1})$, we split the sum into the parts when $i = \ell$, when $j = \ell$, and the rest (and the f factors out from this last part). When $i = \ell$ or $j = \ell$, we use $[fV_\ell, V_j] = f[V_\ell, V_j] - (V_j f)V_\ell$ and $[V_i, fV_\ell] = f[V_i, V_\ell] + (V_i f)V_\ell$. Thus

$$\begin{aligned}
&\Theta_2(V_1, \cdots, fV_\ell, \cdots, V_{k+1}) \\
&= f\,\Theta_2(V_1, \cdots, V_{k+1}) - \sum_{\ell<j\leqslant k+1} (-1)^{\ell+j}(V_j f)\,\Omega(V_\ell, V_1, \cdots, \widehat{V_\ell}, \cdots, \widehat{V_j}, \cdots, V_{k+1}) \\
&\quad + \sum_{1\leqslant i<\ell} (-1)^{\ell+i}(V_i f)\,\Omega(V_\ell, V_1, \cdots, \widehat{V_i}, \cdots, \widehat{V_\ell}, \cdots, V_{k+1}) \\
&= f\,\Theta_2(V_1, \cdots, V_{k+1}) - A - B.
\end{aligned}$$

Hence the A, B terms cancel in the sum $\Theta = \Theta_1 + \Theta_2$, showing that Θ is $C^\infty(M)$-linear in the ℓth slot, for all $1 \leqslant \ell \leqslant k+1$. Now that both $d\Omega$ and Θ are $(0, k+1)$-tensor fields, it is enough to check that in any admissible chart $(U, \mathbf{x})$, their action on $(k+1)$-tuples of chart-induced vector fields ∂_{x^i} coincide. We note that the Lie bracket $[\partial_{x^i}, \partial_{x^j}]$ is zero, and so the Θ_2 summand disappears. We can decompose $\Omega = \Omega_I dx^I$. Since both sides depend additively on Ω, it is enough to check that both sides coincide for an Ω of the form $f dx^I$ with a fixed $I = (i_1, \cdots, i_k)$, and a smooth $f \in C^\infty(M)$. Then $d\Omega = \partial_{x^i} f\, dx^i \wedge dx^I$.

So we have

$$\begin{aligned}(d\Omega)(\partial_{x^{j_1}},\cdots,\partial_{x^{j_{k+1}}}) &= (\partial_{x^i}f)(dx^i\wedge dx^{i_1}\wedge\cdots\wedge dx^{i_k})(\partial_{x^{j_1}},\cdots,\partial_{x^{j_{k+1}}})\\ &= (\partial_{x^i}f)\det\begin{bmatrix}\delta^i_{j_1} & \cdots & \delta^i_{j_{k+1}}\\ \delta^{i_1}_{j_1} & \cdots & \delta^{i_1}_{j_{k+1}}\\ \vdots & & \vdots\\ \delta^{i_k}_{j_1} & \cdots & \delta^{i_k}_{j_{k+1}}\end{bmatrix}.\end{aligned}$$

Looking at the top row, we see that the result will be zero unless i equals some j_ℓ. So only such terms (in the sum over $i\in\{1,\cdots,m\}$, $m:=\dim M$) will survive, giving (by expansion along the top row where all the entries are zeroes except for the 1 which appears in the ℓth place if $i=j_\ell$):

$$\begin{aligned}(d\Omega)(\partial_{x^{j_1}},\cdots,\partial_{x^{j_{k+1}}}) &= \sum_{\ell=1}^{k+1}(-1)^{\ell-1}(\partial_{x^{j_\ell}}f)dx^I(\partial_{x^{j_1}},\cdots,\widehat{\partial_{x^{j_\ell}}},\cdots,\partial_{x^{j_{k+1}}})\\ &= \sum_{i=1}^{k+1}(-1)^{i-1}\partial_{x^{j_i}}((f\,dx^I)(\partial_{x^{j_1}},\cdots,\widehat{\partial_{x^{j_i}}},\cdots,\partial_{x^{j_{k+1}}}))\\ &= \Theta_1(\partial_{x^{j_1}},\cdots,\partial_{x^{j_{k+1}}}) = \Theta(\partial_{x^{j_1}},\cdots,\partial_{x^{j_{k+1}}}),\end{aligned}$$

where the second equality follows by the Leibniz rule and by noting that we have $dx^I(\partial_{x^{j_1}},\cdots,\partial_{x^{j_{k+1}}})=$ a constant. □

So if $\Omega\in\mathbf{\Omega}^1(M)$, then $(d\Omega)(X,Y)=X(\Omega(Y))-Y(\Omega(X))-\Omega([X,Y])$, for all $X,Y\in T^1_0M$.

10.4 Interior multiplication

While the exterior derivative increases the degree of the form field, we now learn about the operation of interior multiplication by a vector field V which decreases the degree. Sometimes interior multiplication by V is referred to as the interior derivative with respect to V, because it also satisfies a type of Leibniz rule akin to the one for the exterior derivative.

Definition 10.10. (Interior multiplication.)
Let M be a smooth manifold, and $V\in T^1_0M$. If $k>1$ and $\Omega\in\mathbf{\Omega}^kM$, then the *interior multiplication of* Ω *by* V, denoted $i_V\Omega\in\mathbf{\Omega}^{k-1}M$, is defined by

$$(i_V(\Omega))(W_1,\cdots,W_{k-1})=\Omega(V,W_1,\cdots,W_{k-1}),$$

for all vector fields $W_1,\cdots,W_{k-1}\in T^1_0M$. If $k=1$, then for a 1-form field $\Omega\in T^0_1M$, we set $i_V(\Omega)=\Omega V\in C^\infty(M)=\mathbf{\Omega}^0M$.

In particular, for any $f\in C^\infty(M)$, $i_V(df)=df(V)=Vf$. So it is natural to think of i_V as a differentiation process.

Exercise 10.17. Suppose that M is a smooth manifold, and $V \in T_0^1 M$. Show that $i_V : \mathbf{\Omega}^k M \to \mathbf{\Omega}^{k-1} M$ is a $C^\infty(M)$-linear map between the $C^\infty(M)$-modules $\mathbf{\Omega}^k M$ and $\mathbf{\Omega}^{k-1} M$.

An application of Proposition 10.5 yields the following local description of the interior multiplication by V.

Lemma 10.4. *Let M be a smooth manifold, and $V \in T_0^1 M$. If $(U, \mathbf{x})$ is an admissible chart for M, then in the chart $(U, \mathbf{x})$ (with $\widehat{\cdot}$ meaning omission of the corresponding term),*

$$\begin{aligned} &i_V(dx^{i_1} \wedge \cdots \wedge dx^{i_k}) \\ &= \sum_{r=1}^{k} (-1)^{r-1} dx^{i_r}(V)\, dx^{i_1} \wedge \cdots \wedge dx^{i_{r-1}} \wedge \widehat{dx^{i_r}} \wedge dx^{i_{r+1}} \wedge \cdots \wedge dx^{i_k}. \end{aligned}$$

Proof. For vector fields $W_1, \cdots, W_{k-1} \in T_0^1 M$, by Proposition 10.5,

$$\begin{aligned} &(i_V(dx^{i_1} \wedge \cdots \wedge dx^{i_k}))(W_1, \cdots, W_{k-1}) \\ &= (dx^{i_1} \wedge \cdots \wedge dx^{i_k})(V, W_1, \cdots, W_{k-1}) \\ &= \begin{bmatrix} dx^{i_1}(V) & dx^{i_1}(W_1) & \cdots & dx^{i_1}(W_{k-1}) \\ \vdots & \vdots & & \vdots \\ dx^{i_k}(V) & dx^{i_k}(W_1) & \cdots & dx^{i_k}(W_{k-1}) \end{bmatrix}. \end{aligned}$$

We expand the determinant down the first column, and use Proposition 10.5 again, now to convert the determinant of the submatrix obtained by deleting the first column and the rth row, into the action of the $(k-1)$-form field $dx^{i_1} \wedge \cdots \wedge \widehat{dx^{i_r}} \wedge \cdots \wedge dx^{i_k}$ on $(W_1, \cdots, W_{k-1})$. Thus we have

$$\begin{aligned} &(i_V(dx^{i_1} \wedge \cdots \wedge dx^{i_k}))(W_1, \cdots, W_{k-1}) \\ &= \sum_{r=1}^{k} (-1)^{r-1} dx^{i_r}(V)\, (dx^{i_1} \wedge \cdots \wedge \widehat{dx^{i_r}} \wedge \cdots \wedge dx^{i_k})(W_1, \cdots, W_{k-1}). \end{aligned}$$ □

We will now show the following Leibniz type of rule for i_V.

Proposition 10.10. *Let M be a smooth manifold, $V \in T_0^1 M$, $\Omega \in \mathbf{\Omega}^k M$ and $\Theta \in \mathbf{\Omega}^\ell M$. Then $i_V(\Omega \wedge \Theta) = (i_V \Omega) \wedge \Theta + (-1)^k \Omega \wedge (i_V \Theta)$.*

Proof. We show this locally, using an admissible chart $(U, \mathbf{x})$. Decompose $\Omega = \Omega_I dx^I = \Omega_I dx^{i_1} \wedge \cdots \wedge dx^{i_k}$, $\Theta = \Theta_J dx^J = \Theta_J dx^{j_1} \wedge \cdots \wedge dx^{j_\ell}$. Then

$$\begin{aligned} i_V(\Omega \wedge \Theta) &= \Omega_I \Theta_J\, i_V(dx^{i_1} \wedge \cdots \wedge dx^{i_k} \wedge dx^{j_1} \wedge \cdots \wedge dx^{j_\ell}) \\ &= \Omega_I \Theta_J \sum_{r=1}^{k} (-1)^{r-1} dx^{i_r}(V) dx^{i_1} \wedge \cdots \wedge \widehat{dx^{i_r}} \wedge \cdots \wedge dx^{i_k} \wedge dx^J \\ &\quad + \Omega_I \Theta_J \sum_{r=1}^{\ell} (-1)^{k+r-1} dx^{j_r}(V) dx^I \wedge dx^{j_1} \wedge \cdots \wedge \widehat{dx^{j_r}} \wedge \cdots \wedge dx^{j_\ell} \\ &= (i_V \Omega) \wedge \Theta + (-1)^k \Omega \wedge (i_V \Theta). \end{aligned}$$

Note that after the second equality, in the second summand the extra $(-1)^k$ is accounted by the fact that the summation index runs from $r=1$ to $r=\ell$, instead of from $r=k+1$ to $r=k+\ell$. □

Just as $d\circ d=0$, we have $i_V\circ i_V=0$. This is a consequence of the following (taking $V=W$).

Proposition 10.11. *Let M be a smooth manifold, $V,W\in T_0^1M$, $k\geqslant 2$, and $\Omega\in\mathbf{\Omega}^kM$. Then $i_V(i_W\Omega)=-i_W(i_V\Omega)$.*

Proof. First, let $k>2$. For all $W_1,\cdots,W_{k-2}\in T_0^1M$, we have

$$\begin{aligned}(i_V(i_W\Omega))(W_1,\cdots,W_{k-2})&=(i_W\Omega)(V,W_1,\cdots,W_{k-2})\\&=\Omega(W,V,W_1,\cdots,W_{k-2})\\&=-\Omega(V,W,W_1,\cdots,W_{k-2})\\&=-i_V\Omega(W,W_1,\cdots,W_{k-2})\\&=-(i_W(i_V\Omega))(W_1,\cdots,W_{k-2})\end{aligned}$$

and so $i_V(i_W\Omega)=-i_W(i_V\Omega)$. If $k=2$, then $i_W\Omega, i_V\Omega\in\mathbf{\Omega}^1M$, and thus

$$i_V(i_W\Omega)=(i_W\Omega)(V)=\Omega(W,V)=-\Omega(V,W)=-(i_V\Omega)(W)=-i_W(i_V\Omega).$$

This completes the proof. □

Using only the smooth structure of a smooth manifold, we have learnt two notions of differentiation of form fields: the exterior derivative and the Lie derivative (p.128). We will now show that although they were defined in quite different ways, they are related via interior multiplication.

Theorem 10.2. (Cartan's magic formula[3].)
Let M be a smooth manifold, and $V\in T_0^1M$. Then $\mathcal{L}_V=d\circ i_V+i_V\circ d$.

More explicitly, for all k-form fields $\Omega\in\mathbf{\Omega}^kM$, $\mathcal{L}_V\Omega=d(i_V\Omega)+i_V(d\Omega)$.

Proof. Let $\Omega\in\mathbf{\Omega}^kM$. Let $V_1,\cdots,V_k\in T_0^1M$ be arbitrary vector fields. Set $V_0=V$. Then

$$\begin{aligned}&(i_V(d\Omega))(V_1,\cdots,V_k)=(d\Omega)(V,V_1,\cdots,V_k)=(d\Omega)(V_0,V_1,\cdots,V_k)\\&=\sum_{i=0}^{k}(-1)^iV_i(\Omega(V_0,\cdots,\widehat{V_i},\cdots,V_k))\\&\quad+\sum_{0\leqslant i<j\leqslant k}(-1)^{i+j}\Omega([V_i,V_j],V_0,\cdots,\widehat{V_i},\cdots,\widehat{V_j},\cdots,V_k).\end{aligned}$$

[3]After Élie Cartan (1869 – 1951), French mathematician, who created the theory of form fields.

Note that in the first sum we get an extra (-1) since the i runs from $i=0$ to k (instead of from $i=1$ to $k+1$ as in Theorem 10.1), but in the second sum, now as both i and j have been reduced by 1, the factor $(-1)^{i+j}$ stays the same. We have

$$\begin{aligned}
&(d(i_V\Omega))(V_1,\cdots,V_k)\\
&=\sum_{i=1}^{k}(-1)^{i-1}V_i((i_V\Omega)(V_1,\cdots,\widehat{V_i},\cdots,V_k))\\
&\quad+\sum_{1\leqslant i<j\leqslant k}(-1)^{i+j}(i_V\Omega)([V_i,V_j],V_1,\cdots,\widehat{V_i},\cdots,\widehat{V_j},\cdots,V_k),\\
&=\sum_{i=1}^{k}(-1)^{i-1}V_i(\Omega(V_0,V_1,\cdots,\widehat{V_i},\cdots,V_k))\\
&\quad+\sum_{1\leqslant i<j\leqslant k}(-1)^{i+j}\Omega(V_0,[V_i,V_j],V_1,\cdots,\widehat{V_i},\cdots,\widehat{V_j},\cdots,V_k).
\end{aligned}$$

Adding $(i_V(d\Omega))(V_1,\cdots,V_k)$ and $(d(i_V\Omega))(V_1,\cdots,V_k)$, we note that only the $i=0$ term survives from both summands. Here we also use

$$\Omega(V_0,[V_i,V_j],\cdots)=-\Omega([V_i,V_j],V_0,\cdots).$$

Consequently,

$$\begin{aligned}
&(d(i_V\Omega)+i_V(d\Omega))(V_1,\cdots,V_k)\\
&=(-1)^0V_0(\Omega(V_1,\cdots,V_k))+\sum_{0=i<j\leqslant k}(-1)^{0+j}\Omega([V_0,V_j],V_1,\cdots,\widehat{V_j},\cdots,V_k)\\
&=V(\Omega(V_1,\cdots,V_k))-\sum_{j=1}^{k}\Omega(V_1,\cdots,\mathcal{L}_VV_j,\cdots,V_k)=(\mathcal{L}_V\Omega)(V_1,\cdots,V_k).
\end{aligned}$$

Thus $d(i_V\Omega)+i_V(d\Omega)=\mathcal{L}_V\Omega$. □

Exercise 10.18. (Lie differentiation commutes with exterior differentiation.) Let M be a smooth manifold, and $V\in T_0^1M$. Show that $\mathcal{L}_V\circ d=d\circ\mathcal{L}_V$. Thus for all k-form fields $\Omega\in\mathbf{\Omega}^kM$, $\mathcal{L}_V(d\Omega)=d(\mathcal{L}_V\Omega)$.
Hint: Use Cartan's formula and $d^2=0$.

Chapter 11

Integration

In the previous chapter, we learnt about k-form fields on smooth manifolds. We will learn in this chapter that these are the objects which can be integrated over 'oriented' manifolds. If we want an intrinsic notion of integration, we will need the notion of 'orientation'. The root cause of this is that when performing the usual Riemann integral on $\mathbb{R}$, there is a subconscious choice made of orientation, namely

$$\int_{(a,b)} f(x)dx := \int_a^b f(x)dx,$$

where $(a,b) \subset \mathbb{R}$ is an interval, and when integrating, we go from the smaller number a to the bigger number b. Thus the integral is set up so that for an everywhere positive function f, this gives a positive area under its graph. In the case of a manifold, on the other hand, there is no such fixed coordinate system, and so an orientation has to be set up by hand. It turns out that for arbitrary smooth manifolds, this may not always be possible, but the ones where it is are called 'orientable', and then we have an intrinsic notion of integration available on such manifolds.

So far, we defined a spacetime as a Lorentzian manifold $(M, \mathbf{g})$ with a time-orientation. This was reasonable based on our everyday experience that we can only 'go forward in time'. Certain particle physics experiments involving weak interactions show that even[1] at the fundamental level of elementary particle processes, the universe possesses chirality, that is, 'handed-ness', and has a preference for one type over the other[2]. Thus it makes sense to also talk about spacetime being an 'oriented' manifold.

[1]Macroscopic chirality is familiar; e.g. human hearts lie on the left.

[2]The weak force operating in the nucleus is what governs the production of beta rays during radioactive decay. Beta rays are energetic rays of electrons. Electrons possess an intrinsic spin, and hence they can be classified as right- or left-handed depending on whether they are moving along or against their spin axis. A famous 1957 experiment demonstrated that beta particles produced during radioactive decay have a definite chiral asymmetry in that left-handed electrons outnumber right-handed ones.

11.1 Orientation

We begin with orientations in the vector space setting. Let V be an m-dimensional real vector space, and let $\omega \in (\bigwedge^m V^*)\backslash\{0\}$, i.e., ω is a nonzero m-form. An m-form, where $m = \dim V$, is sometimes referred to as a 'top form'. The action of ω on an m-tuple formed by the elements of any basis must be nonzero. Suppose on the contrary, $\omega(e_1, \cdots, e_m) = 0$ for a basis $B = \{e_1, \cdots, e_m\}$. Then since $\omega \neq 0$, there are vectors $v_1, \cdots, v_m \in V$ such that $\omega(v_1, \cdots, v_m) \neq 0$. Writing $v_j = c_j^i e_i$, we have

$$\omega(v_1, \cdots, v_m) = \omega(c_1^{i_1} e_{i_1}, \cdots, c_m^{i_m} e_{i_m}) = c_1^{i_1} \cdots c_m^{i_m} \omega(e_{i_1}, \cdots, e_{i_m}).$$

If in a summand corresponding to the index tuple $I = (i_1, \cdots, i_m)$, we have $i_j = i_\ell$, then $\omega(e_{i_1}, \cdots, e_{i_m}) = 0$, and that term vanishes, thanks to the alternating nature of ω. So only the summands $I = (i_1, \cdots, i_m)$ where $(i_1, \cdots, i_m)$ is a permutation of $(1, \cdots, m)$ will survive. But then in such a term, if $\pi \in S_m$ is the permutation $\pi(1) = i_1, \cdots, \pi(m) = i_m$, then $\omega(e_{i_1}, \cdots, e_{i_m}) = \omega(e_{\pi(1)}, \cdots, e_{\pi(m)}) = (\operatorname{sign} \pi)\,\omega(e_1, \cdots, e_m) = 0$, and so the term becomes zero. So we obtain $\omega(v_1, \cdots, v_m) = 0$, a contradiction.

So in light of the above, we can use the action of ω to give an orientation to V by declaring an ordered basis $B = (e_1, \cdots, e_m)$ to be 'positively oriented' if the action satisfies $\omega(e_1, \cdots, e_m) > 0$, and negatively oriented if this action is < 0. Thus it would seem that there are as many orientations as there are nonzero 'top-forms'. But the choice is not as large as it may seem, since any two nonzero top forms are a nonzero multiple of each other. Suppose that $\omega, \theta \in (\bigwedge^m V^*)\backslash\{0\}$. Let $B = (e_1, \cdots, e_m)$ be any ordered basis for V. As the action of ω and of θ on B is nonzero, there is some $k \neq 0$ such that $\omega(e_1, \cdots, e_m) = k \cdot \theta(e_1, \cdots, e_m)$. Thus we also obtain $\omega(e_{i_1}, \cdots, e_{i_m}) = k \cdot \theta(e_{i_1}, \cdots, e_{i_m})$. For any vectors v_j, $1 \leqslant j \leqslant m$, in V, by writing $v_j = c_j^i e_i$, we obtain that

$$\begin{aligned}\omega(v_1, \cdots, v_m) &= \omega(c_1^{i_1} e_{i_1}, \cdots, c_m^{i_m} e_{i_m}) = c_1^{i_1} \cdots c_m^{i_m} \omega(e_{i_1}, \cdots, e_{i_m}) \\ &= c_1^{i_1} \cdots c_m^{i_m}\, k \cdot \theta(e_{i_1}, \cdots, e_{i_m}) = k \cdot \theta(v_1, \cdots, v_m).\end{aligned}$$

So any two nonzero top forms are a nonzero multiple of each other. Define the relation $\sim$ on the set $(\bigwedge^m V^*)\backslash\{0\}$, by setting $\omega \sim \theta$ if $\omega = k\,\theta$ for a positive k. Then $\sim$ is an equivalence relation. Under $\sim$, there are clearly only two equivalence classes: $[\omega]$ and $[-\omega]$, where ω is any nonzero top form. Indeed, for any equivalence class $[\theta]$, we have either

- $\omega = k\theta$ with $k > 0$, and so $[\omega] = [\theta]$, or
- $\omega = k\theta$ with $k < 0$, i.e., $\theta = (-k^{-1})(-\omega)$, and so $\theta \sim -\omega$, i.e., $[\theta] = [-\omega]$.

This prompts the following.

Definition 11.1. (Oriented space, positively/negatively oriented bases.) Let V be an m-dimensional vector space. The equivalence relation $\sim$ on $(\wedge^m V^*)\backslash\{0\}$ is defined by setting $\omega \sim \theta$ if $\omega = k\theta$ for a positive k. An *orientation on* V is an equivalence class $[\omega]$ under $\sim$, where $\omega \in (\wedge^m V^*)\backslash\{0\}$. After choosing an $\omega \in (\wedge^m V^*)\backslash\{0\}$, V is said to be *oriented* with the orientation $[\omega]$. Then an ordered basis $B = (e_1, \cdots, e_m)$ is called

- *positively oriented* if $\omega(e_1, \cdots, e_m) > 0$
- *negatively oriented* if $\omega(e_1, \cdots, e_m) < 0$.

Example 11.1. (Standard orientation on $\mathbb{R}^m$.)
Consider the vector space $V = \mathbb{R}^m$. Define the top form $\omega \in \wedge^m V^*$ by $\omega(\mathbf{v}_1, \cdots, \mathbf{v}_m) = \det[\mathbf{v}_1 \ \cdots \ \mathbf{v}_m] = \det[v_j^i]$, for all vectors $\mathbf{v}_1, \cdots, \mathbf{v}_m \in \mathbb{R}^m$, where $\mathbf{v}_j = v_j^i \mathbf{e}_i$. Then ω is nonzero (e.g. $\omega(\mathbf{e}_1, \cdots, \mathbf{e}_m) = \det I = 1 > 0$). The orientation $[\omega]$ on $\mathbb{R}^m$ is called the *standard orientation* on $\mathbb{R}^m$. $\diamond$

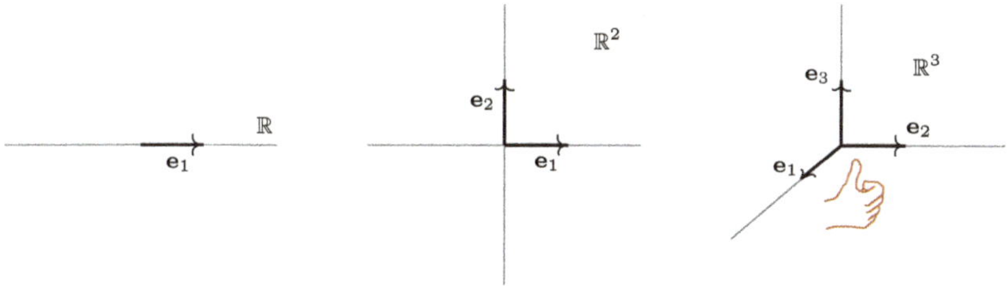

Exercise 11.1. Let V be an m-dimensional vector space. Then each ordered basis $B = (e_1, \cdots, e_m)$ induces an orientation $[\omega_B]$ on V, called the *induced orientation on V by the ordered basis B* by setting $\omega_B = \epsilon^1 \wedge \cdots \wedge \epsilon^m$, where $(\epsilon^1, \cdots, \epsilon^m)$ is the dual basis for V^* corresponding to B. Show that $\omega_B \neq 0$, and $(e_1, \cdots, e_m)$ is positively oriented with respect to $[\omega_B]$. Prove that for another ordered basis $B' = (e_1', \cdots, e_m')$, the following are equivalent:

(1) B' positively oriented with respect to $[\omega_B]$.

(2) If $e_j' = a_j^i e_i$, then the change of basis matrix $[a_j^i]$ has a positive determinant.

(3) $\omega_B \sim \omega_{B'}$.

In the case of an m-dimensional smooth manifold M, we would like the orientations on the tangent spaces T_pM, $p \in M$, to vary smoothly. So, in order to fix an 'orientation on the manifold', we want an m-form field (also referred to as a top-form field) $\Omega \in \mathbf{\Omega}^m M$, which vanishes nowhere, so that the evaluation $\Omega(p) \in \wedge^m (T_pM)^*$ is nonzero, and can be used to give an orientation to T_pM.

Not every smooth manifold is 'orientable'. An example is the Möbius strip (with the edge removed, so that it is an 'open strip', and hence a legitimate 2-dimensional manifold inside $\mathbb{R}^3$), and the nonorientability can be intuitively seen as follows. We refer to the picture below. Suppose M is orientable, so that there exists a nowhere vanishing 2-form field Ω. Let V_1 be the globally defined horizontal vector field along the length of the rectangular strip making the Möbius strip. Take a point p on the 'equator' C. Consider the locally defined vector field V_2 which points upwards, defined in a little width determined by a small arc $A \subset C$ around p. Then $\Omega(p)((V_1)_p, (V_2)_p) \neq 0$, and without loss of generality, suppose that it is $2\delta > 0$. As $A \ni q \mapsto \Omega(q)((V_1)_q, (V_2)_q)$ is smooth and in particular continuous, we can find a point $q \in A$ near p such that $\Omega(q)((V_1)_q, (V_2)_q) > \delta$. But now if r is a point on the arc A between p and q, then we imagine a vertical cut at r, and on the resulting open set U in the Möbius strip (i.e., U is the cut Möbius strip), we consider two vector fields: V_1 as before, and $\widetilde{V}_2$, which starts at p pointing upwards just as V_2 was, but by the time it comes back to q, we have $(\widetilde{V}_2)_q = -(V_2)_q$. But $\Omega|_U(V_1, \widetilde{V}_2)$ is smooth on U, and in particular continuous along the curve $C \cap U$. As it is also nonzero[3], it should not change sign. However, $\Omega(p)((V_1)_p, (\widetilde{V}_2)_p) = \Omega(p)((V_1)_p, (V_2)_p) = 2\delta > 0$, and $\Omega(q)((V_1)_q, (\widetilde{V}_2)_q) = \Omega(q)((V_1)_q, -(V_2)_q) < -\delta < 0$, a contradiction.

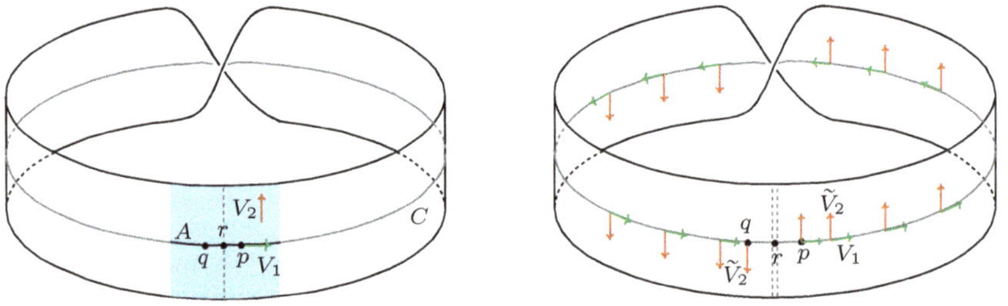

If a manifold is orientable, then does every nowhere vanishing m-form field define a different orientation? Just as with vector spaces, the choice is smaller than it first appears. First, we note that multiplying a nowhere vanishing m-form field by an everywhere positive smooth function, does not change the orientation. We introduce the relation $\sim$ on the set of all nowhere vanishing m-form fields on M by setting $\Omega \sim \Theta$ if $\Omega = f\Theta$ for some function $f : M \to \mathbb{R}$ such that $f(p) > 0$ for all $p \in M$. Then it is easily

[3] Since at any point x along the equator, $(\widetilde{V}_2)_x$ is perpendicular to $(V_1)_x$, so that $\{(V_1)_x, (\widetilde{V}_2)_x\}$ forms a basis for the tangent space T_xU.

seen that $\sim$ is an equivalence relation on the set of all nowhere vanishing m-form fields on M. We remark that for a connected[4] orientable manifold M, it can be shown that[5] there are only two equivalence classes under $\sim$, namely $[\Omega]$ and $[-\Omega]$, where Ω is any nowhere vanishing top form.

Definition 11.2. (Orientation on a manifold, induced orientation.)
A smooth manifold M is called *orientable* if there exists an $\Omega \in \boldsymbol{\Omega}^m M$ such that for each $p \in M$, $\Omega(p) \in (\bigwedge^m (T_pM)^*)\backslash\{0\}$. Let $\sim$ be the equivalence relation defined by setting $\Omega \sim \Theta$ if $\Omega = f\Theta$ for some function $f > 0$ everywhere on M. If M is orientable, then an *orientation* on M is an equivalence class $[\Omega]$, where $\Omega \in \boldsymbol{\Omega}^m M$ is nowhere zero. After choosing an orientation $[\Omega]$ for an orientable manifold M, we say M is *oriented with the orientation* $[\Omega]$. For $p \in T_pM$, the *induced orientation* is given by equivalence class $[\Omega(p)]$ (under the equivalence relation $\sim_p$ on $(\bigwedge^m (T_pM)^*)\backslash\{0\}$ defined by $\omega \sim_p \theta$ if $\omega = k\theta$ for a positive k).

Having defined orientability, we formally give the definition of a spacetime.

Definition 11.3. (Spacetime.)
A *spacetime* $(M, \mathbf{g})$ is a connected, 4-dimensional, oriented, time-oriented Lorentzian manifold with the metric $\mathbf{g}$, equipped with the Levi-Civita connection induced by $\mathbf{g}$.

Example 11.2. (FLRW spacetime.)
Consider the FLRW spacetime $M = I \times \mathbb{R}^3$, where the interval $I = (0, \infty)$, with the smooth structure given by the atlas comprising the single chart $(M, \mathrm{id}_{I\times\mathbb{R}^3})$, and metric $\mathbf{g}$ given in Example 5.9 (p.87). We give M the orientation $[\Omega]$, where $\Omega = dx \wedge dy \wedge dz \wedge dt \in \boldsymbol{\Omega}^4 M$, t, x, y, z being the components of $\mathrm{id}_{I\times\mathbb{R}^3}$. ◇

Orientability of a manifold is needed for integration since oriented manifolds possess 'oriented atlases', as we shall see. First, we explain what we mean by an oriented atlas.

[4]Recall that a topological space X of *connected* if there do not exist two disjoint, nonempty, open subsets of X whose union is X. Also, X is *path-connected* if every pair of points $x, y \in X$ can be joined by a path in X. For a smooth manifold M, M is connected if and only if it is path-connected; see e.g. [Lee (2013), Proposition 1.11, p.8].

[5]We will not prove this here, as we will not use this result, but here is a sketch: If Ω and Θ are two top form fields that are nonvanishing everywhere, then by looking at the action on chart-induced vector fields, for admissible charts, it can be seen that there must exist a smooth function f which is nowhere zero, such that $\Omega = f\Theta$. As M is connected, this f must be either everywhere positive or everywhere negative. This means that either $\Theta \in [\Omega]$ or $\Theta \in [-\Omega]$.

Definition 11.4. (Oriented atlas.)
An atlas $\mathcal{A}$ on a set M is said to be *oriented* if for all charts $(U, \mathbf{x}), (V, \mathbf{y}) \in \mathcal{A}$ for which $U \cap V \neq \varnothing$, the determinant $\det[(\mathbf{y} \circ \mathbf{x}^{-1})']$ of the Jacobi matrix $(\mathbf{y} \circ \mathbf{x}^{-1})'$ of the smooth map $\mathbf{y} \circ \mathbf{x}^{-1} : \mathbf{x}(U \cap V) \to \mathbf{y}(U \cap V)$ is positive everywhere on $\mathbf{x}(U \cap V)$.

Example 11.3. (Minkowski spacetime.)
Consider the Minkowski spacetime $(M, \mathbf{g})$ from Example 5.6. Let $p \in M$ and $B = \{\mathbf{e}_1, \mathbf{e}_2, \mathbf{e}_3, \mathbf{e}_4\}$ be an orthonormal basis[6] for V such that $g(\mathbf{e}_i, \mathbf{e}_i) = 1$ for $i = 1, 2, 3$ and $g(\mathbf{e}_4, \mathbf{e}_4) = -1$. For $q \in M$, writing $q = p + x^i(q)\, \mathbf{e}_i$, gives a global admissible chart $(M, \mathbf{x}_{p,B})$, where the chart map $\mathbf{x}_{p,B}$ is given by $\mathbf{x}_{p,B}(q) = (x^1(q), x^2(q), x^3(q), x^4(q))$ for all $q \in M$. The atlas $\mathcal{A} = \{(M, \mathbf{x}_{p,B})\}$ is trivially oriented.

If we take a different point $p' \in M$ and a different orthonormal basis $B' = \{\mathbf{e}_1', \mathbf{e}_2', \mathbf{e}_3', \mathbf{e}_4'\}$ for V, then $\mathcal{A} \cup \{(M, \mathbf{x}_{p',B'})\}$ is oriented if and only if the Lorentz transformation taking B to B' has determinant $+1$ (see Proposition 5.3, p.82). This follows from the chart transition map expression derived in Example 1.7 (p.5). Also, from the solution to Exercise 5.8, we know that the determinant of any Lorentz transformation is ± 1. ◇

Exercise 11.2. Consider the atlas $\mathcal{A} = \{(U_{\mathbf{n}}, \varphi_{\mathbf{n}}), (U_{\mathbf{s}}, \varphi_{\mathbf{s}})\}$ from Example 1.5 (p.5), defining the smooth structure of the sphere S^2. Show that $\mathcal{A}$ is not oriented. Keeping the chart $(U_{\mathbf{s}}, \varphi_{\mathbf{s}})$, modify the chart map in $(U_{\mathbf{n}}, \varphi_{\mathbf{n}})$ by now setting $\widetilde{\varphi}_{\mathbf{n}} = R \circ \varphi_{\mathbf{n}}$, where R is a reflection in the u-axis, and check that the resulting atlas $\widetilde{\mathcal{A}} = \{(U_{\mathbf{s}}, \varphi_{\mathbf{s}}), (U_{\mathbf{n}}, \widetilde{\varphi}_{\mathbf{n}})\}$ is oriented. Show that $\widetilde{\mathcal{A}} \cup \{(U, \varphi)\}$ is oriented, where (U, φ) is the 'spherical coordinate' chart given in Example 1.8 (p.7).

Proposition 11.1.
Suppose that M is a smooth manifold with the smooth structure $[\mathcal{A}]$. Then M is orientable if and only if there exists an oriented atlas $\widetilde{\mathcal{A}} \in [\mathcal{A}]$.

To prove this, we will need a result on the existence of a 'partition of unity'.

11.2 Partitions of unity

If M is a smooth manifold, and $S \subset M$, then we will denote by $\overline{S}$ the closure of S in M, namely the intersection of all closed sets containing S. For $f \in C^\infty(M)$, the *support of f*, denoted by $\operatorname{supp} f$, is the closure of the set of points where f is nonzero:

$$\operatorname{supp} f := \overline{\{p \in M : f(p) \neq 0\}}.$$

[6]Thus, in contrast to Example 5.6, we will now label $\mathbf{e}_0$ by $\mathbf{e}_4$ instead. This will naturally put the timelike vector ∂_t after spacelike vectors, and this is the usual orientation convention used for the metric with index 1.

A collection $\{F_i : i \in I\}$ of subsets of M is said to be *locally finite* if for each $p \in M$, there exists a neighbourhood $U(p)$ of p which intersects only finitely many sets $F_{i_1}, \cdots, F_{i_{n(p)}}$. A collection $\{U_i : i \in I\}$ of open sets in M such that $\bigcup_{i\in I} U_i = M$ is called an *open cover* of M.

Definition 11.5. (Partition of unity; subordinate to a cover.)
Let M be a smooth manifold. A *partition of unity on* M is a collection $\{\varphi_i : i \in I\}$ of smooth functions $\varphi_i \in C^\infty(M)$ such that

- $\varphi_i(p) \geqslant 0$ for all $p \in M$ and $i \in I$,
- $\{\operatorname{supp} \varphi_i : i \in I\}$ is locally finite,
- $\sum_{i\in I} \varphi_i = 1$. (Note that at each point, this is a finite sum.)

Given an open cover $\mathcal{C} = \{U_j : j \in J\}$, a partition of unity $\{\varphi_i : i \in I\}$ is said to be *subordinate to* $\mathcal{C}$ if for each $i \in I$, there exists an $j(i) \in J$ such that $\operatorname{supp} \varphi_i \subset U_{j(i)}$.

Lemma 11.1. *Let M be a smooth manifold. Then M has a countable basis all of whose elements have a compact closure.*

Proof. Start with a countable basis $\mathcal{B}$. Let $\mathcal{B}_c \subset \mathcal{B}$ be the subcollection of elements from $\mathcal{B}$ which have a compact closure. We will prove that $\mathcal{B}_c$ is itself a basis. Let $U \subset M$ be open, and let $p \in U$. Choose a neighbourhood V of p such that $V \subset U$ and V has a compact closure. (To see this we can take a small neighbourhood of p, that is contained in U, which is homeomorphic under a (chart) map $\mathbf{x}$ to an open ball $B(\mathbf{0}, r)$ in $\mathbb{R}^m$. Now take V to be the image $\mathbf{x}^{-1}B(\mathbf{0}, \frac{r}{2})$. Then $\overline{V}$ is compact.) As $\mathcal{B}$ is a basis, there exists an open set $B \in \mathcal{B}$ such that $p \in B \subset V \subset U$. But then $\overline{B} \subset \overline{V}$ is compact too. Thus $B \in \mathcal{B}_c$. So given any open U and any $p \in U$, we found a $B \in \mathcal{B}_c$ such that $p \in B \subset U$. Hence $\mathcal{B}_c$ is a basis, and being a subset of the countable collection $\mathcal{B}$, is also countable. $\square$

Proposition 11.2. *Let M be a manifold.*
Then there exists a sequence $(V_n)_{n\in\mathbb{N}}$ such that

- *V_n is open and $\overline{V_n}$ is compact for all $n \in \mathbb{N}$*
- *$V_1 \subset \overline{V_1} \subset V_2 \subset \overline{V_2} \subset V_3 \subset \overline{V_3} \subset \cdots$*
- *$\bigcup_{n=1}^{\infty} V_n = M$.*

Proof. By Lemma 11.1, there exists a countable basis $\{B_n : n \in \mathbb{N}\}$ with each $\overline{B_n}$ compact. Set $V_1 = B_1$. By compactness, there exists an $n_1 > 1$ such that $\overline{V_1} \subset B_1 \cup \cdots \cup B_{n_1} =: V_2$. Now[7] $\overline{V_2} = \overline{B_1 \cup \cdots \cup B_{n_1}} = \overline{B_1} \cup \cdots \cup \overline{B_{n_1}}$ is compact, and so there exists an $n_2 > n_1$ such that $\overline{V_2} \subset B_1 \cup \cdots \cup B_{n_2} =: V_3$, and proceed inductively in this manner. Finally,

$$\begin{aligned} M &= \bigcup_{n=1}^{\infty} B_n = B_1 \cup \cdots \cup B_{n_1} \cup \cdots \cup B_{n_2} \cup \cdots \\ &= \bigcup_{k=1}^{\infty} (B_1 \cup \cdots \cup B_{n_k}) = \bigcup_{k=1}^{\infty} V_k. \end{aligned}$$

□

In the above construction, we note that as $\overline{V_{n+1}} \setminus V_n$ is a closed subset of the compact set $\overline{V_{n+1}}$, it is itself compact. Moreover, since $\overline{V_{n-1}} \subset V_n$, we have $V_{n+2} \setminus V_n \subset V_{n+2} \setminus \overline{V_{n-1}}$. Now since $\overline{V_{n+1}} \subset V_{n+2}$, we also obtain $\overline{V_{n+1}} \setminus V_n \subset V_{n+2} \setminus V_n$. Putting these together, we have:

$$_{\text{(compact)}}\ \overline{V_{n+1}} \setminus V_n \subset V_{n+2} \setminus \overline{V_{n-1}}\ _{\text{(open)}}.$$

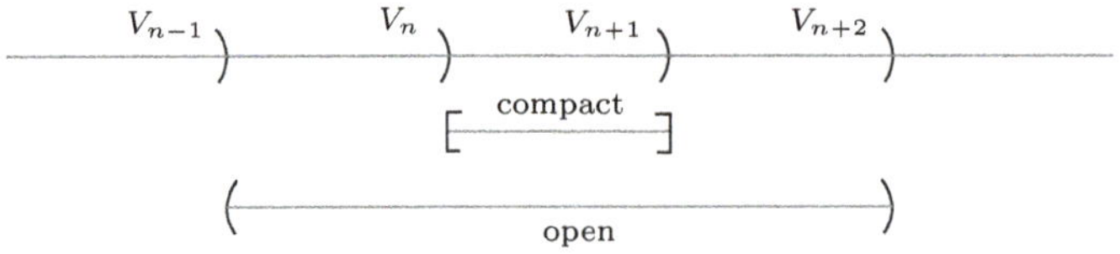

Exercise 11.3. Set $V_0 = \varnothing$. Show that for $\{V_n : n \in \mathbb{N}\}$ from Proposition 11.2, $\bigcup_{n=0}^{\infty} \overline{V_{n+1}} \setminus V_n = M$.

Theorem 11.1. *Let M be a smooth manifold, and let $\mathcal{C} = \{U_j : j \in J\}$ be an open cover of M. Then there exists a partition of unity subordinate to $\mathcal{C}$.*

Proof. Set $V_0 = \varnothing$. Let $\{V_n : n \in \mathbb{N}\}$ be a sequence of open sets as in Proposition 11.2. Fix $n \in \mathbb{N}$. For each p in the compact set $\overline{V_{n+1}} \setminus V_n$, let $j(p) \in J$ be an index such that $p \in U_{j(p)}$. Then p belongs to the open set $(V_{n+2} \setminus \overline{V_{n-1}}) \cap U_{j(p)}$.

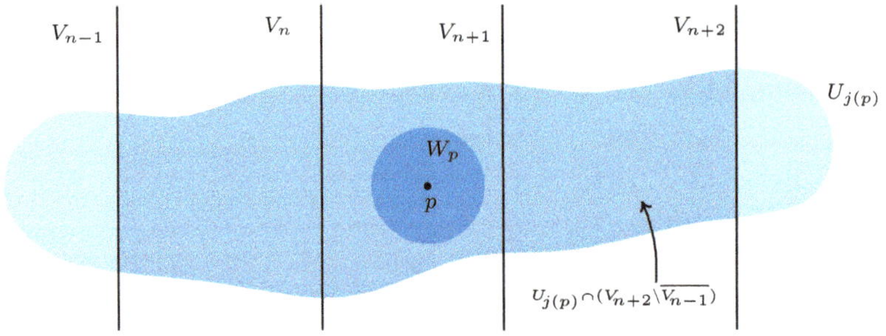

[7]Here we use the fact that the closure of a finite union is the union of the closures; see e.g. [Bishop and Goldberg (1980), p.9].

Let $\psi_p \in C^\infty(M)$ be a bump function such that $\psi_p \equiv 1$ in an open neighbourhood W_p of p, and with $\text{supp}\,\psi_p \subset (V_{n+2}\backslash\overline{V_{n-1}}) \cap U_{j(p)}$. Then $\{W_p : p \in \overline{V_{n+1}} \backslash V_n\}$ is an open cover of the compact set $\overline{V_{n+1}} \backslash V_n$. So there exist finitely many points $p_1^n, \cdots, p_{k(n)}^n$ supplying a finite subcover $\{W_{p_1^n}, \cdots, W_{p_{k(n)}^n}\}$, with associated bump functions $\psi_{p_1^n}, \cdots, \psi_{p_{k(n)}^n}$ satisfying

$$\text{supp}\,\psi_{p_\ell^n} \subset (V_{n+2}\backslash\overline{V_{n-1}}) \cap U_{j(p_\ell^n)}, \quad 1 \leqslant \ell \leqslant k(n). \tag{11.1}$$

As the $n \in \mathbb{N}$ changes, we get countably many bump functions:

$$\{\psi_{p_\ell^n} : 1 \leqslant \ell \leqslant k(n),\, n \in \mathbb{N}\}.$$

The collection $\{\text{supp}\,\psi_{p_\ell^n} : 1 \leqslant \ell \leqslant k(n),\, n \in \mathbb{N}\}$ of their supports is locally finite, since firstly, using Exercise 11.3, we see that the $W_{p_\ell^n}$ (which is contained in $\text{supp}\,\psi_{p_\ell^n}$, and which together cover $\overline{V_{n+1}}\backslash V_n$) together cover M, and secondly, given any n, only finitely many of the supports will intersect $W_{p_\ell^n}$: $(\text{supp}\,\psi_{p_\ell^N}) \cap W_{p_\ell^n} = \varnothing$ whenever $N \geqslant n+3$ by (11.1).

Any point $p \in M$ is contained in $\overline{V_{n+1}} \backslash V_n$ for some n, and so $p \in W_{p_\ell^n}$ for some ℓ (because $\{W_{p_1^n}, \cdots, W_{p_{k(n)}^n}\}$ forms a cover for $\overline{V_{n+1}} \backslash V_n$). Then $\psi_{p_\ell^n}(p) = 1 > 0$. So

$$\psi := \sum_{n=1}^{\infty} \sum_{\ell=1}^{k(n)} \psi_{p_\ell^n}$$

is pointwise > 0 on M. We remark that the above sum is well-defined thanks to the local finiteness of $\{\text{supp}\,\psi_{p_\ell^n} : 1 \leqslant \ell \leqslant k(n),\, n \in \mathbb{N}\}$. Now relabel the summands occurring in the double sum as $\psi_1, \psi_2, \psi_3, \cdots$, and set $\varphi_k := \psi_k/\psi$. Then clearly $\sum_k \varphi_k = 1$, and $\varphi_k \geqslant 0$ pointwise for all $k \in \mathbb{N}$. Finally, as $\varphi_k = \psi_{p_\ell^n}/\psi$ for some n and some ℓ, we have

$$\text{supp}\,\varphi_k = \text{supp}\,\psi_{p_\ell^n} \subset U_{j(p_\ell^n)}$$

showing the desired subordination property. □

Oriented atlas ↔ Orientability

We now return to the task left out at the end of the previous section on orientation, i.e., to prove Proposition 11.1, stating that a smooth manifold is orientable if and only if it possesses an oriented atlas in its smooth structure.

Proof. (Of Proposition 11.1.) Suppose that M is orientable, and that $\Omega \in \boldsymbol{\Omega}^m M$ is a nowhere vanishing top form field. We take any atlas $\mathcal{A}$ from the smooth structure such that for each chart $(U, \mathbf{x}) \in \mathcal{A}$, U is connected[8].

[8]Atlases with connected charts always exist, since a nonconnected chart can be split into connected components.

We will alter $\mathcal{A}$ to make it into an oriented atlas $\widetilde{\mathcal{A}}$ by changing, if necessary, the chart maps, but not the chart domains. Consider any chart $(U, \mathbf{x}) \in \mathcal{A}$, then we can write $\Omega = f\, dx^1 \wedge \cdots \wedge dx^m$ for some $f \in C^\infty(U)$. Since the function f vanishes nowhere in the connected set U, it is either everywhere >0 or everywhere <0. If $f>0$ on U, then we keep the chart $(U, \mathbf{x}) \in \widetilde{\mathcal{A}}$. But if $f<0$ on U, then include in $\widetilde{\mathcal{A}}$, instead of $(U, \mathbf{x})$, the new admissible chart $(U, p \mapsto (-x^1(p), x^2(p), \cdots, x^m(p)) =: \mathbf{y}(p))$. Then

$$\begin{aligned}\Omega &= f\, dx^1 \wedge \cdots \wedge dx^m = f\,((\partial_{y^{i_1}} x^1) dy^{i_1}) \wedge \cdots \wedge ((\partial_{y^{i_m}} x^m) dy^{i_m}) \\ &= f\,(\partial_{y^{i_1}} x^1) \cdots (\partial_{y^{i_m}} x^m)\, dy^{i_1} \wedge \cdots \wedge dy^{i_m} \\ &= f \sum_{\pi \in S_m} (\partial_{y^{\pi(1)}} x^1) \cdots (\partial_{y^{\pi(m)}} x^m)\, dy^{\pi(1)} \wedge \cdots \wedge dy^{\pi(m)} \\ &= f \det[\partial_{y^j} x^i]\, dy^1 \wedge \cdots \wedge dy^m = g\, dy^1 \wedge \cdots \wedge dy^m,\end{aligned}$$

where $g := f \det[\partial_{y^j} x^i]$. As $\det[\partial_{y^j} x^i] = -1$ and $f<0$ on U, $g>0$ on U.

Now we have a new atlas $\widetilde{\mathcal{A}}$, with the property that for any chart $(U, \mathbf{x})$ belonging to $\widetilde{\mathcal{A}}$, if we decompose $\Omega = f\, dx^1 \wedge \cdots \wedge dx^m$ in U, then $f>0$ in U. We claim that this new atlas is oriented. To see this, let $(U, \mathbf{x})$ and $(V, \mathbf{y})$ be two overlapping charts in $\widetilde{\mathcal{A}}$. Then in $U \cap V$,

$$\begin{aligned}\Omega &= f\, dx^1 \wedge \cdots \wedge dx^m = f\,((\partial_{y^{i_1}} x^1) dy^{i_1}) \wedge \cdots \wedge ((\partial_{y^{i_m}} x^m) dy^{i_m}) \\ &= f \det[\partial_{y^j} x^i]\, dy^1 \wedge \cdots \wedge dy^m.\end{aligned}$$

Since we also know that $\Omega = g\, dy^1 \wedge \cdots \wedge dy^m$, we obtain $g = f \det[\partial_{y^j} x^i]$. But by the aforementioned property of the atlas, we know that $f>0$ in U, and $g>0$ in V, so that they are in particular also >0 in $U \cap V$. The relation $g = f \det[\partial_{y^j} x^i]$ now implies that $\det[\partial_{y^j} x^i] > 0$ in $U \cap V$. This completes the proof of the 'only if' part, showing that the orientable manifold possesses the oriented atlas $\widetilde{\mathcal{A}}$ in its smooth structure $[\mathcal{A}] = [\widetilde{\mathcal{A}}]$.

It remains to show that if M admits an oriented atlas $\mathcal{A}$ in its smooth structure, then M is orientable, that is, there exists a nowhere vanishing top form field $\Omega \in \mathbf{\Omega}^m M$. Let $\mathcal{A} = \{(U_\alpha, \mathbf{x}_\alpha) : \alpha \in I\}$. Take a partition of unity $\{\varphi_\alpha : \alpha \in I\}$ subordinate to $\{U_\alpha : \alpha \in I\}$. Set

$$\Omega = \sum_{\alpha \in I} \varphi_\alpha\, dx_\alpha^1 \wedge \cdots \wedge dx_\alpha^m.$$

Let $p \in M$. Since $\sum_\alpha \varphi_\alpha = 1$, and all the φ_α are pointwise nonnegative, there exists an index $\beta \in I$ such that $\varphi_\beta(p) > 0$. Then

$$\begin{aligned}\Omega|_{U_\beta} &= \sum_{\alpha \in I : U_\alpha \cap U_\beta \neq \varnothing} \varphi_\alpha\, dx_\alpha^1 \wedge \cdots \wedge dx_\alpha^m \\ &= \sum_{\alpha \in I : U_\alpha \cap U_\beta \neq \varnothing} \varphi_\alpha \det[\partial_{x_\beta^j} x_\alpha^i]\, dx_\beta^1 \wedge \cdots \wedge dx_\beta^m,\end{aligned}$$

and so the action of the top form $\Omega(p) \in \bigwedge^m (T_pM)^*$ on the ordered basis $(\partial_{x^1_\beta,p}, \cdots, \partial_{x^m_\beta,p})$ for T_pM is

$$\begin{aligned}\Omega(p)(\partial_{x^1_\beta,p}, \cdots, \partial_{x^m_\beta,p}) &= \sum_{\alpha \in I: U_\alpha \cap U_\beta \neq \varnothing} \varphi_\alpha(p)(\det[\partial_{x^j_\beta} x^i_\alpha])(p) \det I \\ &\geqslant \varphi_\beta(p)(\det[\partial_{x^j_\beta} x^i_\beta])(p) > 0,\end{aligned}$$

where we have used $\det[\partial_{x^j_\beta} x^i_\alpha] > 0$ on $U_\beta \cap U_\alpha$. Thus $\Omega(p) \neq 0$. As $p \in M$ was arbitrary, Ω is nonvanishing everywhere, and so M is orientable. □

Definition 11.6. (Orientation-compatible oriented atlas.)
Let M be a smooth manifold with an orientation $[\Omega]$ for a nowhere vanishing top-form field $\Omega \in \mathbf{\Omega}^m M$. An oriented atlas $\mathcal{A}$ from the smooth structure of M is called $[\Omega]$-*compatible* if in each chart $(U, \mathbf{x}) \in \mathcal{A}$, the decomposition $\Omega = \Omega^U dx^1 \wedge \cdots \wedge dx^m$ yields $\Omega^U > 0$ everywhere on U.

Orientation-compatibility of an atlas is independent of the Ω chosen in the orientation, and so this notion, as specified above, is well-defined.

Exercise 11.4. Let M be a smooth manifold with an orientation $[\Omega]$, and let $\mathcal{A}, \mathcal{A}'$ be atlases in the smooth structure of M that are both $[\Omega]$-compatible. Show that $\mathcal{A} \cup \mathcal{A}'$ is also an oriented atlas which is $[\Omega]$-compatible.

11.3 Integration

We first provide some heuristic motivation for the definition that is to follow. The aim is to give intrinsic definition of the integral on a manifold.

We would like to 'integrate' a function $f : M \to \mathbb{R}$. Assume that f is compactly supported, and even smooth, hoping that there will not be any problems from the analysis point of view while defining such an object. We could use charts as long as at the end we have a chart-independent number. Using charts means that the problem boils down to the integral in $\mathbb{R}^m$, which is already familiar to us. We recall that if $f : \mathbb{R}^m \to \mathbb{R}$ has compact support, then we define

$$\int_{\mathbb{R}^m} f(u^1, \cdots, u^m)\, du^1 \cdots du^m := \int_{-\infty}^{\infty} \Big(\cdots \Big(\int_{-\infty}^{\infty} f(u^1, \cdots, u^m)\, du^1 \Big) \cdots \Big)\, du^m,$$

i.e., the integral is defined via iterations of 1-dimensional integrations. Also, recall the change of variables formula: If $\Phi : \mathbb{R}^m \to \mathbb{R}^m$ is a diffeomorphism, then

$$\begin{aligned}&\int_{\mathbb{R}^m} f(u^1, \cdots, u^m)\, du^1 \cdots du^m \\ &= \int_{\mathbb{R}^m} (f \circ \Phi^{-1})(u^1, \cdots, u^m) \Big| \det \Big[\frac{\partial (\Phi^{-1})^i}{\partial u^j}(u^1, \cdots, u^m) \Big] \Big|\, du^1 \cdots du^m.\end{aligned}$$

An attempt at a definition of the integral of a smooth, compactly supported function f on a smooth manifold M could be made as follows. Assume for the moment that a smooth manifold M has global charts $(M, \mathbf{x})$ and $(M, \mathbf{y})$ with bijections $\mathbf{x}, \mathbf{y} : M \to \mathbb{R}^m$. Then we could try

$$\int_M f := \int_{\mathbb{R}^m} (f \circ \mathbf{x}^{-1})(u^1, \cdots, u^m)\, du^1 \cdots du^m,$$

and hope that if we used $\mathbf{y}$ instead, then the result would be the same. But this is clearly not true, since the above change of variables formula shows that if $M = \mathbb{R}^m$, with the standard smooth structure, and we consider the admissible charts $(\mathbb{R}^m, \mathbf{x} = \mathrm{id})$, and $(\mathbb{R}^m, \mathbf{y} = \Phi)$, where Φ is any diffeomorphism as in the change of variables formula, then

$$\begin{aligned}
&\int_{\mathbb{R}^m} (f \circ \mathbf{x}^{-1})(u^1, \cdots, u^m)\, du^1 \cdots du^m = \int_{\mathbb{R}^m} f(u^1, \cdots, u^m)\, du^1 \cdots du^m \\
&= \int_{\mathbb{R}^m} (f \circ \Phi^{-1})(u^1, \cdots, u^m) \left| \det \left[\frac{\partial (\Phi^{-1})^i}{\partial u^j} (u^1, \cdots, u^m) \right] \right| du^1 \cdots du^m \\
&= \int_{\mathbb{R}^m} (f \circ \mathbf{y}^{-1})(u^1, \cdots, u^m)\, |(\det[(\partial_{y^j} x^i)])(\mathbf{y}^{-1}(u^1, \cdots, u^m))|\, du^1 \cdots du^m.
\end{aligned}$$

So, in general,

$$\int_{\mathbb{R}^m} (f \circ \mathbf{x}^{-1})(u^1, \cdots, u^m)\, du^1 \cdots du^m \neq \int_{\mathbb{R}^m} (f \circ \mathbf{y}^{-1})(u^1, \cdots, u^m)\, du^1 \cdots du^m.$$

We note that, firstly, the absolute value on the determinant would disappear if we used an oriented atlas. Secondly, if we formally replace $dx^1 \cdots dx^m$ by $dx^1 \wedge \cdots \wedge dx^m$, then the definition of the integral of a top form field Ω by setting

$$\int_M \Omega = \int_{\mathbb{R}^m} (f \circ \mathbf{x}^{-1})(u^1, \cdots, u^m)\, du^1 \cdots du^m$$

where $\Omega = f\, dx^1 \wedge \cdots \wedge dx^m$ in a global chart $(U = M, \mathbf{x})$, where $\mathbf{x} : M \to \mathbb{R}^m$ is a bijection, seems to be an intrinsic definition. Indeed, if $(V = M, \mathbf{y})$ is another global chart, where $\mathbf{y} : M \to \mathbb{R}^m$ is a bijection, then writing $\Omega = f\, dx^1 \wedge \cdots \wedge dx^m = g\, dy^1 \wedge \cdots \wedge dy^m$, we have $g = f \det[\partial_{y^j} x^i]$, and so

$$\begin{aligned}
&\int_{\mathbb{R}^m} (g \circ \mathbf{y}^{-1})(u^1, \cdots, u^m)\, du^1 \cdots du^m \\
&= \int_{\mathbb{R}^m} (f \circ \mathbf{y}^{-1})(u^1, \cdots, u^m)(\det[\partial_{y^j} x^i])(\mathbf{y}^{-1}(u^1, \cdots, u^m))\, du^1 \cdots du^m \\
&= \int_{\mathbb{R}^m} (f \circ \mathbf{y}^{-1})(u^1, \cdots, u^m) |\det[\partial_{y^j} x^i])(\mathbf{y}^{-1}(u^1, \cdots, u^m)|\, du^1 \cdots du^m \\
&= \int_{\mathbb{R}^m} (f \circ \mathbf{x}^{-1} \circ \mathbf{x} \circ \mathbf{y}^{-1})(u^1, \cdots, u^m) \left| \det \left[\frac{\partial (\mathbf{x} \circ \mathbf{y}^{-1})^i}{\partial u^i} (u^1, \cdots, u^m) \right] \right| du^1 \cdots du^m \\
&= \int_{\mathbb{R}^m} (f \circ \mathbf{x}^{-1})(u^1, \cdots, u^m)\, du^1 \cdots du^m.
\end{aligned}$$

When global charts are not available, we use a partition of unity to patch local integrals.

Definition 11.7. (Integral of top form fields.)
Suppose that M is a smooth manifold with an orientation $[\Theta]$.
Let $\mathcal{A}=\{(U_\alpha,\mathbf{x}_\alpha):\alpha\in A\}$ be a $[\Theta]$-compatible oriented atlas in the smooth structure of M, and $\{\varphi_i : i\in I\}$ be a partition of unity subordinate to $\{U_\alpha : \alpha\in A\}$. For each $i\in I$, let $\alpha(i)\in A$ be such that $\operatorname{supp}\varphi_i\subset U_{\alpha(i)}$. Let Ω be any m-form field with a compact support, that is, there exists a compact set $K\subset M$ such that $\Omega|_{M\setminus K}\equiv 0$. Then we define

$$\begin{aligned}\int_M \Omega &= \sum_{i\in I}\int_{U_{\alpha(i)}}\varphi_i\,\Omega\\ &= \sum_{i\in I}\int_{\mathbf{x}_{\alpha(i)}U_{\alpha(i)}}(\varphi_i\circ\mathbf{x}_{\alpha(i)}^{-1})(u^1,\cdots,u^m)\,\Omega^{U_{\alpha(i)}}(\mathbf{x}_{\alpha(i)}^{-1}(u^1,\cdots,u^m))\,du^1\cdots du^m,\end{aligned}$$

where $\Omega^{U_{\alpha(i)}}\in C^\infty(U_{\alpha(i)})$ are the functions such that

$$\Omega=\Omega^{U_{\alpha(i)}}dx^1_{\alpha(i)}\wedge\cdots\wedge dx^m_{\alpha(i)}$$

in the chart $(U_{\alpha(i)},\mathbf{x}_{\alpha(i)})$.

If we accept well-posedness, it is clear that the above gives a $\mathbb{R}$-linear map

$$\int_M : \{\Omega\in\mathbf{\Omega}^m M\,|\,\Omega \text{ has compact support}\}\to\mathbb{R}.$$

We need to check the well-definedness at various levels. First, we note that even with a particular $[\Theta]$-compatible atlas, charts may overlap, and we should check that in the overlap region, the right-hand side above is the same. In the intersection region of two charts $(U,\mathbf{x}),(V,\mathbf{y})\in\mathcal{A}$, if $\Omega=\Omega^U dx^1\wedge\cdots\wedge dx^m=\Omega^V dy^1\wedge\cdots\wedge dy^m$, then $\Omega^U=\Omega^V\det[\partial_{x^j}y^i]$. With $\Phi:=\mathbf{y}\circ\mathbf{x}^{-1}$, for $\mathbf{u}\in\mathbf{x}(U\cap V)$, we have

$$\begin{aligned}\Omega^U(\mathbf{x}^{-1}(\mathbf{u})) &= \Omega^V(\mathbf{x}^{-1}(\mathbf{u}))(\det[\partial_{x^j}y^i])(\mathbf{x}^{-1}(\mathbf{u}))\\ &= \Omega^V((\mathbf{y}^{-1}\circ\Phi)(\mathbf{u}))\det\Big[\frac{\partial(\mathbf{y}\circ\mathbf{x}^{-1})^i}{\partial u^j}(\mathbf{u})\Big]\\ &= \Omega^V((\mathbf{y}^{-1}\circ\Phi)(\mathbf{u}))\det\Big[\frac{\partial\Phi^i}{\partial u^j}(\mathbf{u})\Big].\end{aligned}$$

Thus for any $f\in C^\infty(U\cap V)$, we have

$$\begin{aligned}&\int_{\mathbf{x}(U\cap V)}(f\circ\mathbf{x}^{-1})(\mathbf{u})\,\Omega^U(\mathbf{x}^{-1}\mathbf{u})\,du^1\cdots du^m\\ &=\int_{(\mathbf{x}\circ\mathbf{y}^{-1}\circ\mathbf{y})(U\cap V)}(f\circ\mathbf{y}^{-1}\circ\Phi)(\mathbf{u})\,\Omega^V((\mathbf{y}^{-1}\circ\Phi)(\mathbf{u}))\det\Big[\frac{\partial\Phi^i}{\partial u^j}(\mathbf{u})\Big]du^1\cdots du^m\\ &=\int_{(\Phi^{-1}\circ\mathbf{y})(U\cap V)}(f\circ\mathbf{y}^{-1}\circ\Phi)(\mathbf{u})\,\Omega^V((\mathbf{y}^{-1}\circ\Phi)(\mathbf{u}))\,\Big|\det\Big[\frac{\partial\Phi^i}{\partial u^j}(\mathbf{u})\Big]\Big|\,du^1\cdots du^m\\ &=\int_{\mathbf{y}(U\cap V)}(f\circ\mathbf{y}^{-1})(\mathbf{u})\,\Omega^V(\mathbf{y}^{-1}\mathbf{u})\,du^1\cdots du^m.\end{aligned}$$

Next, we need to check that the integral depends neither on the chosen $[\Theta]$-compatible oriented atlas from the smooth structure of M, nor on the partition of unity. Suppose that $\mathcal{A}=\{(U_\alpha, \mathbf{x}_\alpha): \alpha \in A\}$ and $\mathcal{A}'=\{(V_\beta, \mathbf{y}_\beta): \beta \in B\}$ are two oriented $[\Theta]$-compatible atlases, with corresponding subordinate partitions of unity $\{\varphi_i : i \in I\}$ and $\{\psi_j : j \in J\}$, respectively. We use the same notation $\alpha(i)$ as before, and for each $j \in J$, let $\beta(j) \in B$ be such that $\operatorname{supp}\psi_j \subset V_{\beta(j)}$. We have

$$\begin{aligned}
\sum_i \int_{U_{\alpha(i)}} \varphi_i \Omega &= \sum_i \int_{U_{\alpha(i)}} \varphi_i \sum_j \psi_j \Omega && (\text{as } \textstyle\sum_j \psi_j = 1) \\
&= \sum_i \sum_j \int_{U_{\alpha(i)}} \varphi_i \psi_j \Omega && \begin{array}{l}(\text{these are finite sums,}\\ \text{since } \Omega \text{ has compact support})\end{array} \\
&= \sum_i \sum_j \int_{U_{\alpha(i)} \cap V_{\beta(j)}} \varphi_i \psi_j \Omega && (\text{since } \operatorname{supp}(\varphi_i \psi_j) \subset U_{\alpha(i)} \cap V_{\beta(j)})
\end{aligned}$$

and by symmetry, we get the same last expression also starting from $\sum_j \int_{V_{\beta(j)}} \psi_j \Omega$. This completes the justification of the well-posedness.

Exercise 11.5. (Baby[9] Stokes' theorem.)
The aim of this exercise is to prove a basic version of Stokes theorem:
If M is an m-dimensional oriented smooth manifold, and $\Omega \in \mathbf{\Omega}^{m-1}M$ is compactly supported, then

$$\int_M d\Omega = 0.$$

Proceed as follows. Let $[\Theta]$ be the orientation on M, $\mathcal{A}=\{(U_\alpha, \mathbf{x}_\alpha): \alpha \in A\}$ be a $[\Theta]$-compatible oriented atlas, and $\{\varphi_i : i \in I\}$ be a partition of unity subordinate to $\mathcal{A}$. Then $\Omega = \sum_i \varphi_i \Omega$ since $\sum_i \varphi_i = 1$. If for $i \in I$, $\operatorname{supp}\varphi_i \subset U_{\alpha(i)}$, then in $U_{\alpha(i)}$, justify the decomposition

$$\varphi_i \Omega = (-1)^{k-1} f_k \, dx^1 \wedge \cdots \wedge \widehat{dx^k} \wedge \cdots \wedge dx^m,$$

for some smooth functions $f_k \in C^\infty(U_{\alpha(i)})$ having compact supports in $U_{\alpha(i)}$. Using this, show that $d(\varphi_i \Omega) = (\partial_{x^1} f_1 + \cdots + \partial_{x^m} f_m) \, dx^1 \wedge \cdots \wedge dx^m$. Thus

$$\int_{U_{\alpha(i)}} d(\varphi_i \Omega) = \int_{\mathbf{x}_{\alpha(i)} U_{\alpha(i)}} (\partial_{x^1} f_1 + \cdots + \partial_{x^m} f_m)(\mathbf{x}_{\alpha(i)}^{-1} \mathbf{u}) \, du^1 \cdots du^m.$$

Show that this is 0 using Fubini's theorem. Conclude that $\int_M d\Omega = 0$.

[9] A more general version of Stokes' theorem for smooth manifolds M possessing a 'boundary' ∂M exists, which roughly speaking, says that

$$\int_M d\Omega = \int_{\partial M} \Omega$$

for compactly supported $(m-1)$-form fields Ω on M. Thus in our baby version, there is no boundary: $\partial M = \varnothing$. However, we will not need this more general result in the sequel, and as it will be too much of a detour (with a necessary discussion of manifolds with a boundary, submanifolds, and so on), we will not include this important result in this book. We refer the interested reader to [Lee (2013), pp.411–415].

11.4 Volume form field in semi-Riemannian manifolds

In the case of a semi-Riemannian manifold $(M, \mathbf{g})$ which admits an oriented atlas $\mathcal{A}$ in its smooth structure, there is a natural top form field $\mathrm{vol}_{\mathbf{g}}$ which is everywhere nonvanishing, and also such that $\mathcal{A}$ is $[\mathrm{vol}_{\mathbf{g}}]$-compatible. Besides integrating compactly supported m-form fields using this orientation, we can also use it to compute the integrals of compactly supported smooth functions f on M canonically[10] by setting

$$\int_M f = \int_M f \cdot \mathrm{vol}_{\mathbf{g}}.$$

In particular, for a compact manifold, the volume of M is defined to be $\int_M 1 = \int_M \mathrm{vol}_{\mathbf{g}}$.

Definition 11.8. (Volume form field.)
Let $(M, \mathbf{g})$ be an m-dimensional semi-Riemannian manifold with metric $\mathbf{g}$, and suppose that M admits an oriented atlas $\mathcal{A}$ in its smooth structure. The *volume form field* $\mathrm{vol}_{\mathbf{g}} \in \mathbf{\Omega}^m M$ is defined as follows: In any chart $(U, \mathbf{x}) \in \mathcal{A}$,

$$\mathrm{vol}_{\mathbf{g}}|_U = \sqrt{|\det[\mathbf{g}(\partial_{x^i}, \partial_{x^j})]|}\, dx^1 \wedge \cdots \wedge dx^m.$$

Let us first check the chart-independence of the above definition. If $(U, \mathbf{x})$ and $(V, \mathbf{y})$ are overlapping charts in $\mathcal{A}$, then in $U \cap V$, we have

$$\begin{aligned}
&\sqrt{|\det[\mathbf{g}(\partial_{y^i}, \partial_{y^j})]|}\, dy^1 \wedge \cdots \wedge dy^m \\
&= \sqrt{|\det[\mathbf{g}((\partial_{y^i} x^k)\partial_{x^k}, (\partial_{y^j} x^\ell)\partial_{x^\ell})]|}\, (\partial_{x^{i_1}} y^1) dx^{i_1} \wedge \cdots \wedge (\partial_{x^{i_m}} y^m) dx^{i_m} \\
&= \sqrt{|\det[(\partial_{y^i} x^k)(\partial_{y^j} x^\ell)\, \mathbf{g}(\partial_{x^k}, \partial_{x^\ell})]|}\, \det[\partial_{x^s} y^r]\, dx^1 \wedge \cdots \wedge dx^m \\
&= \sqrt{|\det([\partial_{y^i} x^k]^{\mathrm{t}} [\mathbf{g}(\partial_{x^k}, \partial_{x^\ell})][\partial_{y^j} x^\ell])|}\, \det[\partial_{x^s} y^r]\, dx^1 \wedge \cdots \wedge dx^m \\
&= \sqrt{(\det[\partial_{y^i} x^j])^2 |\det[\mathbf{g}(\partial_{x^k}, \partial_{x^\ell})]|}\, \det[\partial_{x^s} y^r]\, dx^1 \wedge \cdots \wedge dx^m \\
&= \sqrt{|\det[\mathbf{g}(\partial_{x^k}, \partial_{x^\ell})]|}\, \det[\partial_{y^i} x^j]\, \det[\partial_{x^s} y^r]\, dx^1 \wedge \cdots \wedge dx^m \\
&= \sqrt{|\det[\mathbf{g}(\partial_{x^k}, \partial_{x^\ell})]|}\, \det[(\partial_{y^r} x^j)(\partial_{x^s} y^r)]\, dx^1 \wedge \cdots \wedge dx^m \\
&= \sqrt{|\det[\mathbf{g}(\partial_{x^k}, \partial_{x^\ell})]|}\, \det[\delta^j_s]\, dx^1 \wedge \cdots \wedge dx^m \\
&= \sqrt{|\det[\mathbf{g}(\partial_{x^k}, \partial_{x^\ell})]|}\, (1)\, dx^1 \wedge \cdots \wedge dx^m.
\end{aligned}$$

[10]We could have also done this when we only had an orientation $[\Omega]$, but then we would have to make an arbitrary choice of the top form field in $[\Omega]$. In a sense, given an oriented atlas, now the metric *chooses* a natural orientation from among the many available.

(In the above, while extracting $\det[\partial_{y^i} x^k]$ outside the square root sign, we have used its positivity, thanks to the fact that $\mathcal{A}$ is oriented.) This shows that the definition of $\mathrm{vol}_{\mathbf{g}}$ does not depend on the chosen chart from $\mathcal{A}$.

So for each $p \in M$, we have a well-defined top form $\mathrm{vol}_{\mathbf{g}}(p) \in \bigwedge^m (T_pM)^*$. For vector fields $V_1, \cdots, V_m \in T_0^1 M$, the map

$$\begin{aligned} p \mapsto\ & \mathrm{vol}_{\mathbf{g}}(p)((V_1)_p, \cdots, (V_m)_p) \\ &= \sqrt{|\det[\mathbf{g}_{ij}(p)]|}\,((dx^1)_p \wedge \cdots \wedge (dx^m)_p)((V_1)_p, \cdots, (V_m)_p) \\ &= \sqrt{|\det[\mathbf{g}_{ij}(p)]|}\,\det[V_j(x^i)](p) \end{aligned}$$

is smooth in each chart $(U, \mathbf{x})$, and hence is smooth on M. It follows from Proposition 4.1 that $\mathrm{vol}_{\mathbf{g}}$ is a $(0, m)$-tensor field. As it is clearly alternating in each chart, it belongs to $\mathbf{\Omega}^m M$. We also note that for any $p \in (U, \mathbf{x})$, the action of $\mathrm{vol}_{\mathbf{g}}(p) \in \bigwedge^m (T_pM)^*$ on the ordered basis $(\partial_{x^1,p}, \cdots, \partial_{x^m,p})$ for T_pM is given by

$$(\mathrm{vol}_{\mathbf{g}}(p))(\partial_{x^1,p}, \cdots, \partial_{x^m,p}) = \sqrt{|\det[\mathbf{g}(\partial_{x^k,p}, \partial_{x^\ell,p})]|}\,\det I > 0.$$

So $\mathrm{vol}_{\mathbf{g}}$ is nowhere vanishing. Moreover, since $\sqrt{|\det[\mathbf{g}(\partial_{x^k}, \partial_{x^\ell})]|} > 0$ on $(U, \mathbf{x})$, it follows that $\mathcal{A}$ is $[\mathrm{vol}_{\mathbf{g}}]$-compatible.

Definition 11.9. (Integration of functions, volume.)
Let $(M, \mathbf{g})$ be an m-dimensional semi-Riemannian manifold with metric $\mathbf{g}$, and suppose that M admits an oriented atlas $\mathcal{A}$ in its smooth structure. Let M be given the orientation $[\mathrm{vol}_{\mathbf{g}}]$. If $f \in C^\infty(M)$ has compact support, then its *integral on M* is defined by

$$\int_M f = \int_M f \cdot \mathrm{vol}_{\mathbf{g}},$$

where the integral of the right-hand side is that of the compactly supported top form field $f \cdot \mathrm{vol}_{\mathbf{g}} \in \mathbf{\Omega}^m M$ in the sense of Definition 11.7. If M is compact, then its *volume* is defined by

$$\mathrm{Vol}_{\mathbf{g}} M = \int_M 1 = \int_M \mathrm{vol}_{\mathbf{g}}.$$

Exercise 11.6. Consider the sphere $(S^2, \mathbf{g})$ as a Riemannian manifold with the metric $\mathbf{g}$ induced on its tangent spaces from the Euclidean inner product on $\mathbb{R}^3$, as described in Example 5.7 (p.85). Determine its volume $\mathrm{Vol}_{\mathbf{g}} S^2$ with the orientation induced by the oriented atlas $\widetilde{\mathcal{A}}$ from Exercise 11.2 (p.238).

Example 11.4. (Schwarzschild spacetime.)
Consider the Schwarzschild spacetime $(M, \mathbf{g})$, where $M = \mathbb{R} \times (2m, \infty) \times S^2$ has the smooth structure and the metric $\mathbf{g}$ described in Example 5.10 (p.87). Consider the sphere $(S^2, \mathbf{g})$ as a Riemannian manifold with the metric $\mathbf{g}_{S^2}$ induced on its tangent spaces from the Euclidean inner product on $\mathbb{R}^3$, as described in Example 5.7 (p.85), and the orientation induced by the oriented atlas $\widetilde{\mathcal{A}}$ from Exercise 11.2 (p.238). Denote the volume form field on $(S^2, \mathbf{g}_{S^2})$ by $\mathrm{vol}_{\mathbf{g}_{S^2}}$. Then

$$\Omega := r^2 dr \wedge \mathrm{vol}_{\mathbf{g}_{S^2}} \wedge dt \in \mathbf{\Omega}^4 M \tag{11.2}$$

is nowhere vanishing, and we equip $(M, \mathbf{g})$ with the orientation $[\Omega]$. Here the functions $t, r : M \to \mathbb{R}$ are given by

$$\begin{aligned} t(p) &= \tau, \quad \text{and} \\ r(p) &= \rho, \end{aligned}$$

for $p = (\tau, \rho, \mathbf{p}) \in \mathbb{R} \times (2m, \infty) \times S^2 = M$. ◇

Exercise 11.7. Given the oriented atlas $\mathcal{A}_* := \widetilde{\mathcal{A}} \cup \{(U, \varphi)\}$ for S^2 in Example 11.2 (p.238), consider the oriented atlas

$$\mathcal{A}_\star = \{(\mathbb{R} \times (2m, \infty) \times W,\ p = (\tau, \rho, \mathbf{p}) \mapsto (\rho, \sigma(\mathbf{p}), \tau)),\ (W, \sigma) \in \mathcal{A}_*\}$$

for the Schwarzschild spacetime $(M, \mathbf{g})$. Show that the volume form field $\mathrm{vol}_{\mathbf{g}}$ is the same as (11.2) above.

11.5 Hodge star

If V is an m-dimensional real vector space, then $\dim \wedge^k V^* = \binom{m}{k}$ for all $0 \leqslant k \leqslant m$. So

$$\dim \wedge^{m-k} V^* = \binom{m}{m-k} = \binom{m}{k} = \dim \wedge^k V^*.$$

So the two spaces $\wedge^k V^*$ and $\wedge^{m-k} V^*$ are isomorphic as vector spaces. We will see in this section that in the context of an oriented real vector space with a scalar product, the Hodge star operator[11] is such an isomorphism. This operator will be used to define the magnetic field perceived by an instantaneous observer in spacetime.

Lemma 11.2. *Let V be an m-dimensional real vector space with a scalar product g, and an orientation $[\omega]$, for an $\omega \in \wedge^m V^* \backslash \{0\}$.*
Let $(e_1, \cdots, e_m), (f_1, \cdots, f_m)$ be ordered bases for V that are positively oriented, and are also orthonormal with respect to g.
If $f_j = a_j^i e_i$, $1 \leqslant j \leqslant m$, then $\det[a_j^i] = 1$.

[11] Named after W. Hodge (1903–1975), a British geometer.

Proof. We have (no sum over j):

$$g(f_j, f_j) = g(a^i_j e_i, a^k_j e_k) = a^i_j a^k_j g(e_i, e_k) = \sum_{k=1}^{m} a^k_j a^k_j g(e_k, e_k).$$

For $i \neq j$, $0 = g(f_i, f_j) = g(a^k_i e_k, a^\ell_j e_\ell) = a^k_i a^\ell_j g(e_k, e_\ell) = \sum_{k=1}^{m} a^k_i a^k_j g(e_k, e_k)$. Thus,

$$\begin{bmatrix} g(f_1,f_1) & & \\ & \ddots & \\ & & g(f_m,f_m) \end{bmatrix} = \begin{bmatrix} a^1_1 & \cdots & a^m_1 \\ \vdots & & \vdots \\ a^1_m & \cdots & a^m_m \end{bmatrix} \begin{bmatrix} g(e_1,e_1) & & \\ & \ddots & \\ & & g(e_m,e_m) \end{bmatrix} \begin{bmatrix} a^1_1 & \cdots & a^1_m \\ \vdots & & \vdots \\ a^m_1 & \cdots & a^m_m \end{bmatrix}.$$

Taking determinants, $(-1)^{\mathfrak{i}_g} = (\det[a^i_j])^2 (-1)^{\mathfrak{i}_g}$. So $\det[a^i_j] \in \{-1, 1\}$. But

$$\begin{aligned} 0 < \omega(f_1, \cdots, f_m) &= \omega(a^{i_1}_1 e_{i_1}, \cdots, a^{i_m}_m e_{i_m}) = a^{i_1}_1 \cdots a^{i_m}_m \,\omega(e_{i_1}, \cdots, e_{i_m}) \\ &= \sum_{\pi \in S_m} a^{\pi(1)}_1 \cdots a^{\pi(m)}_m (\operatorname{sign} \pi)\, \omega(e_1, \cdots, e_m) \\ &= (\det[a^i_j])\, \omega(e_1, \cdots, e_m), \end{aligned}$$

and since $\omega(e_1, \cdots, e_m) > 0$ too, it follows that $\det[a^i_j] = 1$. □

Definition 11.10. (Volume form on a vector space with a scalar product.) Let V be an m-dimensional real vector space with a scalar product g of index $\mathfrak{i}_g$, and an orientation $[\omega]$, for an $\omega \in \bigwedge^m V^* \backslash \{0\}$. Let $(e_1, \cdots, e_m)$ be an ordered basis which is positively oriented, and orthonormal with respect to g. Then the top form $\operatorname{vol}_g := (-1)^{\mathfrak{i}_g} (e_1)^\flat \wedge \cdots \wedge (e_m)^\flat \in \bigwedge^m V^*$ is called the *volume form* on V with respect to g and $[\omega]$.

It might seem that vol_g depends on the chosen basis, but it does not, since if $(f_1, \cdots, f_m)$ is also an ordered basis which is positively oriented with respect to $[\omega]$, and orthonormal with respect to g, then with $f_j = a^i_j e_i$,

$$\begin{aligned} (f_1)^\flat \wedge \cdots \wedge (f_m)^\flat &= (a^{i_1}_1 e_{i_1})^\flat \wedge \cdots \wedge (a^{i_m}_m e_{i_m})^\flat \\ &= a^{i_1}_1 \cdots a^{i_m}_m \,(e_{i_1})^\flat \wedge \cdots \wedge (e_{i_m})^\flat \\ &= \sum_{\pi \in S_m} a^{\pi(1)}_1 \cdots a^{\pi(m)}_m (\operatorname{sign} \pi)\, (e_1)^\flat \wedge \cdots \wedge (e_m)^\flat \\ &= \det[a^i_j]\, (e_1)^\flat \wedge \cdots \wedge (e_m)^\flat = 1\, (e_1)^\flat \wedge \cdots \wedge (e_m)^\flat. \end{aligned}$$

This shows that vol_g is independent of the chosen positively oriented orthonormal basis. Also, $\operatorname{vol}_g \in [\omega]$: Indeed, $\omega(e_1, \cdots, e_m) > 0$ and

$$\begin{aligned} \operatorname{vol}_g(e_1, \cdots, e_m) &= (-1)^{\mathfrak{i}_g} ((e_1)^\flat \wedge \cdots \wedge (e_m)^\flat)(e_1, \cdots, e_m) \\ &= (-1)^{\mathfrak{i}_g} \det[g(e_i, e_j)] = (-1)^{\mathfrak{i}_g} (-1)^{\mathfrak{i}_g} = 1 > 0. \end{aligned}$$

Thus $\omega = c \cdot \operatorname{vol}_g$, where $c := \omega(e_1, \cdots, e_m) > 0$, and so $\operatorname{vol}_g \sim \omega$.

Exercise 11.8. Let V be an oriented m-dimensional real vector space with a scalar product g of index $\mathfrak{i}_g$, and the volume form vol_g. Let $(e_1, \cdots, e_m)$ be a positively oriented ordered basis such that it is also orthonormal with respect to g, and let $(\epsilon^1, \cdots, \epsilon^m)$ be the corresponding dual basis for V^*. Let $\iota(k) := g(e_k, e_k)$, $1 \leqslant k \leqslant m$. Show that $(e_k)^\flat = \iota(k)\epsilon^k$, $1 \leqslant k \leqslant m$. Conclude that $\mathrm{vol}_g = \epsilon^1 \wedge \cdots \wedge \epsilon^m$.

Exercise 11.9. Let V be an oriented m-dimensional real vector space with a scalar product$g$ of index $\mathfrak{i}_g$ and volume form vol_g. Let $(v_1, \cdots, v_m)$ be an arbitrary (not necessarily orthonormal) positively oriented basis for V, and let $(\omega^1, \cdots, \omega^m)$ be the corresponding dual basis for V^*. Show $\mathrm{vol}_g = \sqrt{|\det[g(v_i, v_j)]|}\,\omega^1 \wedge \cdots \wedge \omega^m$.

Let V be an oriented m-dimensional real vector space with a scalar product g, and the volume form vol_g. Let $\omega \in \bigwedge^k V^*$, and $v_1, \cdots, v_{m-k} \in V$. Then $\omega \wedge (v_1)^\flat \wedge \cdots \wedge (v_{m-k})^\flat \in \bigwedge^m V^*$, and so using the basis $\{\mathrm{vol}_g\}$ for $\bigwedge^m V^*$, there is a unique $f_\omega(v_1, \cdots, v_{m-k}) \in \mathbb{R}$ such that

$$\omega \wedge (v_1)^\flat \wedge \cdots \wedge (v_{m-k})^\flat = f_\omega(v_1, \cdots, v_{m-k})\,\mathrm{vol}_g. \qquad (*)$$

By the $\mathbb{R}$-linearity of the wedge product in each of its factors, the map $f_\omega : V^{m-k} \to \mathbb{R}$, given by $V^{m-k} \ni (v_1, \cdots, v_{m-k}) \mapsto f_\omega(v_1, \cdots, v_{m-k})$ is multilinear. The LHS of $(*)$, being a wedge product, is alternating in $v_1, \cdots, v_{m-k}$. So f_ω is alternating. Thus $f_\omega \in \bigwedge^{m-k} V^*$. By the linearity of the wedge product, the map $\bigwedge^k V^* \ni \omega \mapsto f_\omega \in \bigwedge^{m-k} V^*$ is linear. This map is called the Hodge star operator (one for each fixed $0 \leqslant k < m$).

If $k = m$, then we set $\star(\omega) = \omega(e_1, \cdots, e_m) \in \mathbb{R}$, for $\omega \in \bigwedge^m V^*$, where $(e_1, \cdots, e_m)$ is any positively oriented orthonormal basis. This gives a linear map $\star : \bigwedge^m V^* \to \mathbb{R} = \bigwedge^0 V^*$. To show that the map does not depend on the chosen ordered orthonormal basis, we note that since $\{\mathrm{vol}_g\}$ is a basis for $\bigwedge^m V^*$, for $\omega \in \bigwedge^m V^*$, there exists a $c \in \mathbb{R}$ such that $\omega = c\,\mathrm{vol}_g$, and if $(e_1, \cdots, e_m)$, $(f_1, \cdots, f_m)$ are positively oriented orthonormal bases, then $\omega(e_1, \cdots, e_m) = c\,\mathrm{vol}_g(e_1, \cdots, e_m) = c1 = c\,\mathrm{vol}_g(f_1, \cdots, f_m) = \omega(f_1, \cdots, f_m)$.

Definition 11.11. (Hodge star operator.)
Let V be an oriented m-dimensional real vector space with a scalar product g, and volume form vol_g. The *Hodge star operator* $\star : \bigwedge^k V^* \to \bigwedge^{m-k} V^*$ is defined to be the linear map that satisfies, for all $\omega \in \bigwedge^k V^*$, and all $v_1, \cdots, v_{m-k} \in V$, that $\omega \wedge (v_1)^\flat \wedge \cdots \wedge (v_{m-k})^\flat = (\star\omega)(v_1, \cdots, v_{m-k})\,\mathrm{vol}_g$.

Lemma 11.3. *Let V be an oriented m-dimensional real vector space with a scalar product g of index $\mathfrak{i}_g$, and the volume form vol_g. Let $\omega \in \bigwedge^k V^*$, $v, v_1, \cdots, v_{m-k} \in V$. Then:*

(1) $\star\,\mathrm{vol}_g = 1$.

(2) $\star 1 = (-1)^{\mathfrak{i}_g}\mathrm{vol}_g$.

(3) $(\star\omega)(v_1, \cdots, v_{m-k}) = \star(\omega \wedge (v_1)^\flat \wedge \cdots \wedge (v_{m-k})^\flat)$.

Proof.

(1) $\star \text{vol}_g = \text{vol}_g(e_1, \cdots, e_m) = 1$. Here $(e_1, \cdots, e_m)$ is any ordered positively oriented orthonormal basis.

(2) Let $w_1, \cdots, w_m \in V$. The top form $\omega := (w_1)^\flat \wedge \cdots \wedge (w_m)^\flat$ is equal to $\omega(e_1, \cdots, e_m)\text{vol}_g$, as $\omega = c\,\text{vol}_g$, and we had seen that $c = \omega(e_1, \cdots, e_m)$ for any positively oriented orthonormal basis $(e_1, \cdots, e_m)$. We have

$$\begin{aligned}\omega(e_1, \cdots, e_m) &= ((w_1)^\flat \wedge \cdots \wedge (w_m)^\flat)(e_1, \cdots, e_m)\\ &= \det[(w_i)^\flat e_j] = \det[g(w_i, e_j)] = \det[(e_j)^\flat w_i]\\ &= ((e_1)^\flat \wedge \cdots \wedge (e_m)^\flat)(w_1, \cdots, w_m)\\ &= (-1)^{\mathfrak{i}_g}\text{vol}_g(w_1, \cdots, w_m).\end{aligned}$$

Thus

$$\begin{aligned}(\star 1)(w_1, \cdots, w_m)\text{vol}_g &= 1 \wedge (w_1)^\flat \wedge \cdots \wedge (w_m)^\flat\\ &= (-1)^{\mathfrak{i}_g}\text{vol}_g(w_1, \cdots, w_m)\text{vol}_g.\end{aligned}$$

Consequently, that $\star 1 = (-1)^{\mathfrak{i}_g}\text{vol}_g$.

(3) It follows by the $\mathbb{R}$-homogeneity of $\star : \bigwedge^m V^* \to \mathbb{R} = \bigwedge^0 V^*$ that

$$\begin{aligned}\star(\omega \wedge (v_1)^\flat \wedge \cdots \wedge (v_{m-k})^\flat) &= \star((\star\omega)(v_1, \cdots, v_{m-k})\text{vol}_g)\\ &= (\star\omega)(v_1, \cdots, v_{m-k})\cdot(\star\text{vol}_g)\\ &= (\star\omega)(v_1, \cdots, v_{m-k})\cdot 1.\end{aligned}$$

□

Analogous to the interior multiplication of form fields, we introduce the following 'pointwise' version for forms.

Definition 11.12. (Interior multiplication.)
Let V be a real vector space, and $v \in V$. If $k \geqslant 1$ and $\omega \in \bigwedge^k V^*$, then the *interior multiplication* $i_v\omega \in \bigwedge^{k-1} V^*$ *of* ω *by* v is defined by

$$(i_v\omega)(v_1, \cdots, v_{k-1}) = \omega(v, v_1, \cdots, v_{k-1}), \quad \text{for all } v_1, \cdots, v_{k-1} \in V.$$

If $k = 1$, then for a 1-form $\omega \in V^*$, we set $i_v\omega = \omega(v) \in \mathbb{R} = \bigwedge^0 V^*$.

Exercise 11.10. Let V be a vector space. For all $v \in V$, $\omega^1, \cdots, \omega^k \in V^* = \bigwedge^1 V^*$, show that $i_v(\omega^1 \wedge \cdots \wedge \omega^k) = \sum_{r=1}^{k} (-1)^{r-1}\omega^r(v)\omega^1 \wedge \cdots \wedge \widehat{\omega^r} \wedge \cdots \wedge \omega^k$.

Exercise 11.11. Let V be a vector space and $v \in V$. For all $\omega \in \bigwedge^k V^*$, $\theta \in \bigwedge^\ell V^*$, show that $i_v(\omega \wedge \theta) = (i_v\omega) \wedge \theta + (-1)^k\omega \wedge i_v\theta$.

Lemma 11.4. *Let V be an m-dimensional real vector space with a scalar product g. Let $\omega \in \bigwedge^k V^*$ and $v, v_1, \cdots, v_{m-k+1} \in V$. Then*

$$\begin{aligned}&(i_v\omega) \wedge (v_1)^\flat \wedge \cdots \wedge (v_{m-k+1})^\flat\\ &= \sum_{r=1}^{m-k+1} (-1)^{k+r} v^\flat(v_r)\,\omega \wedge (v_1)^\flat \wedge \cdots \wedge \widehat{(v_r)^\flat} \wedge \cdots \wedge (v_{m-k+1})^\flat.\end{aligned}$$

Proof. $\omega \wedge (v_1)^\flat \wedge \cdots \wedge (v_{m-k+1})^\flat \in \bigwedge^{m+1} V^* = \{0\}$, and Exercises 11.11 and 11.10 imply

$$\begin{aligned} 0 &= i_v(0) = i_v(\omega \wedge (v_1)^\flat \wedge \cdots \wedge (v_{m-k+1})^\flat) \\ &= (i_v\omega) \wedge (v_1)^\flat \wedge \cdots \wedge (v_{m-k+1})^\flat \\ &\quad + (-1)^k \sum_{r=1}^{m-k+1} (-1)^{r-1}((v_r)^\flat v)\, \omega \wedge (v_1)^\flat \wedge \cdots \wedge \widehat{(v_r)^\flat} \wedge \cdots \wedge (v_{m-k+1})^\flat. \end{aligned}$$

The proof is completed by observing that $(v_r)^\flat v = g(v, v_r) = v^\flat(v_r)$ and rearranging. □

Lemma 11.5. *Let V be an oriented m-dimensional real vector space with a scalar product g and the volume form vol_g. Let $v \in V$, and let $\omega \in \bigwedge^k V^*$. Then $\star(\omega \wedge v^\flat) = i_v(\star\omega)$.*

Proof. For $v_1, \cdots, v_{m-k-1} \in V$, using Lemma 11.3.(3), we have

$$\begin{aligned} (\star(\omega \wedge v^\flat))(v_1, \cdots, v_{m-k-1}) &= \star((\omega \wedge v^\flat) \wedge (v_1)^\flat \wedge \cdots \wedge (v_{m-k-1})^\flat) \\ &= \star(\omega \wedge (v^\flat \wedge (v_1)^\flat \wedge \cdots \wedge (v_{m-k-1})^\flat)) \\ &= (\star\omega)(v, v_1, \cdots, v_{m-k-1}) \\ &= (i_v(\star\omega))(v_1, \cdots, v_{m-k-1}). \end{aligned}$$ □

Lemma 11.6. *Let V be an oriented m-dimensional real vector space with a scalar product g and the volume form vol_g. Let $\omega \in \bigwedge^k V^*$, and let $v \in V$. Then $\star(i_v\omega) = (-1)^{m-1}(\star\omega) \wedge v^\flat$.*

Proof. We have

$$(-1)^{m-1}(\star\omega) \wedge v^\flat = (-1)^{m-1}(-1)^{m-k} v^\flat \wedge \star\omega = (-1)^{k+1} v^\flat \wedge \star\omega.$$

Let $\star\omega = f_I \epsilon^I$, where for a multi-index $I = (i_1, \cdots, i_{m-k})$, $\epsilon^I := \epsilon^{i_1} \wedge \cdots \wedge \epsilon^{i_{m-k}}$, and $\{\epsilon^1, \cdots, \epsilon^m\}$ is a basis for V^*. For all $v_1, \cdots, v_{m-k+1} \subset V$,

$$\begin{aligned} ((-1)^{m-1}(\star\omega) \wedge v^\flat)(v_1, \cdots, v_{m-k+1}) &= (-1)^{k+1}(v^\flat \wedge (\star\omega))(v_1, \cdots, v_{m-k+1}) \\ &= (-1)^{k+1} f_I \det \begin{bmatrix} v^\flat v_1 & \cdots & v^\flat v_{m-k+1} \\ \epsilon^{i_1} v_1 & \cdots & \epsilon^{i_1} v_{m-k+1} \\ \vdots & & \vdots \\ \epsilon^{i_{m-k}} v_1 & \cdots & \epsilon^{i_{m-k}} v_{m-k+1} \end{bmatrix}. \end{aligned}$$

Expanding the determinant along the top row, and using Lemma 11.4

$$\begin{aligned} &((-1)^{m-1}(\star\omega) \wedge v^\flat)(v_1, \cdots, v_{m-k+1}) \\ &= (-1)^{k+1} \sum_{r=1}^{m-k+1} (-1)^{r-1}(v^\flat v_r)\,(\star\omega)(v_1, \cdots, \widehat{v_r}, \cdots, v_{m-k+1}) \\ &= \sum_{r=1}^{m-k+1} (-1)^{k+r}(v^\flat v_r)\, \star(\omega \wedge (v_1)^\flat \wedge \cdots \wedge \widehat{(v_r)^\flat} \wedge \cdots \wedge (v_{m-k+1})^\flat) \\ &= \star((i_v\omega) \wedge (v_1)^\flat \wedge \cdots \wedge (v_{m-k+1})^\flat) = (\star(i_v\omega))(v_1, \cdots, v_{m-k+1}). \end{aligned}$$ □

Exercise 11.12. Let V be an oriented m-dimensional real vector space with a scalar product g of index $\mathfrak{i}_g$, and the volume form vol_g. Let $\omega \in \bigwedge^k V^*$ and $v_1, \cdots, v_k \in V$. Prove that

$$(-1)^{k(m-1)}(\star\omega) \wedge v_1^\flat \wedge \cdots \wedge v_k^\flat = \star(i_{v_k}(\cdots(i_{v_1}\omega)\cdots)) = (-1)^{\mathfrak{i}_g}\omega(v_1, \cdots, v_k)\,\text{vol}_g.$$

Lemma 11.7. *Let V be an oriented m-dimensional real vector space with a scalar product g of index $\mathfrak{i}_g$, and the volume form vol_g. Let $\omega \in \bigwedge^k V^*$. Then $\star\star\omega = (-1)^{k(m-1)+\mathfrak{i}_g}\omega$.*

Proof. Using Lemma 11.3 and Exercise 11.12, for all $v_1, \cdots, v_k \in V$,

$$\begin{aligned}(\star(\star\omega))(v_1, \cdots, v_k) &= \star((\star\omega) \wedge (v_1)^\flat \wedge \cdots \wedge (v_{m-k})^\flat) \\ &= (-1)^{k(m-1)} \star(\omega(v_1, \cdots, v_k)(-1)^{\mathfrak{i}_g}\,\text{vol}_g) \\ &= (-1)^{k(m-1)}(-1)^{\mathfrak{i}_g}\omega(v_1, \cdots, v_k) \star \text{vol}_g \\ &= (-1)^{k(m-1)+\mathfrak{i}_g}\omega(v_1, \cdots, v_k)1.\end{aligned}$$

Thus $\star\star\omega = (-1)^{k(m-1)+\mathfrak{i}_g}\omega$. □

In particular, $\star : \bigwedge^k V^* \to \bigwedge^{m-k} V^*$ is an injective linear transformation, and since $\dim \bigwedge^k V^* = \binom{m}{k} = \binom{m}{m-k} = \dim \bigwedge^{m-k} V^*$, it follows that $\star$ is an isomorphism.

Lemma 11.8. *Let V be an oriented m-dimensional real vector space with a scalar product g and the volume form vol_g. Suppose that $\omega, \nu \in \bigwedge^k V^*$. Then $\omega \wedge \star\nu = \nu \wedge \star\omega$.*

Proof. Thanks to the linearity in each of the arguments ω, ν on both sides, it is enough to prove the result for $\omega = (w_1)^\flat \wedge \cdots \wedge (w_k)^\flat$ and for $\nu = (v_1)^\flat \wedge \cdots \wedge (v_k)^\flat$, where $w_1, \cdots w_k, v_1, \cdots, v_k \in V$. We have

$$\begin{aligned}\omega \wedge \star\nu &= (-1)^{k(m-k)}(\star\nu) \wedge \omega = (-1)^{k(m-k)}(\star\nu) \wedge ((w_1)^\flat \wedge \cdots \wedge (w_k)^\flat) \\ &= (-1)^{k(m-k)}(-1)^{k(m-1)+\mathfrak{i}_g}\nu(w_1, \cdots, w_k)\,\text{vol}_g \\ &= (-1)^{\mathfrak{i}_g}((v_1)^\flat \wedge \cdots \wedge (v_k)^\flat)(w_1, \cdots, w_k)\,\text{vol}_g \\ &= (-1)^{\mathfrak{i}_g} \det[(v_i)^\flat w_j]\,\text{vol}_g.\end{aligned}$$

Swapping ω, ν, $\nu \wedge \star\omega = (-1)^{\mathfrak{i}_g}\det[(w_i)^\flat v_j]\,\text{vol}_g$. Since taking the transpose preserves the determinant, and as $(v_i)^\flat w_j = g(v_i, w_j) = (w_j)^\flat v_i$, we conclude that $\omega \wedge \star\nu = \nu \wedge \star\omega$. □

The following result (Proposition 11.3) tells us how to compute the $\star$ of a k-form, as it describes the action on the basis vectors for $\bigwedge^k V^*$.

Definition 11.13. (Complement of a tuple, index of a tuple.)
Let $I = (i_1, \cdots, i_k)$ be a k-tuple obtained from $\{1, \cdots, m\}$ with $i_1 < \cdots < i_k$. Then the unique $(m-k)$-tuple $I_c := (j_1, \cdots, j_{m-k})$, with $j_1 < \cdots < j_{m-k}$ and $I \cup I_c = \{1, \cdots, m\}$, is called the *complement of I*.

Now let V be an m-dimensional real vector space with a scalar product g, and let $\{e_1, \cdots, e_m\}$ be an orthonormal basis with respect to g. For the k-tuple $I=(i_1, \cdots, i_k)$ of elements from $\{1, \cdots, m\}$, we define the *g-index of* I by $\iota(I):=g(e_{i_1}, e_{i_1})\cdots g(e_{i_k}, e_{i_k})=\iota(i_1)\cdots\iota(i_k)$.

Below, for an r-tuple $L=(\ell_1, \cdots, \ell_r)$, we set $\epsilon^L:=\epsilon^{\ell_1}\wedge\cdots\wedge\epsilon^{\ell_r}$.

Proposition 11.3. *Let V be an oriented m-dimensional real vector space with a scalar product g and the volume form vol_g. Let $(e_1, \cdots, e_m)$ be a positively oriented ordered basis such that it is also orthonormal with respect to g, and let $(\epsilon^1, \cdots, \epsilon^m)$ be the corresponding dual basis for V^*. Let $I=(i_1, \cdots, i_k)$ with $i_1<\cdots<i_k$, I_c be its complement, and let $\pi_{II^c}\in S_m$ be the permutation taking (I, I_c) to $(1, \cdots, m)$. Then $\star\epsilon^I = (\operatorname{sign}\pi_{II_c})\,\iota(I_c)\,\epsilon^{I_c}$.*

Proof. Let $J=(j_1, \cdots, j_{m-k})$ be any $(m-k)$-tuple of distinct elements of $\{1, \cdots, m\}$. By Lemma 11.3(3), and using Exercise 11.8 (p.251), we have

$$\begin{aligned}(\star\epsilon^I)(e_{j_1}, \cdots, e_{j_{m-k}}) &= \star(\epsilon^I\wedge(e_{j_1})^\flat\wedge\cdots\wedge(e_{j_{m-k}})^\flat)\\ &= \iota(j_1)\cdots\iota(j_{m-k})\star(\epsilon^I\wedge\epsilon^J),\end{aligned}$$

which is zero if $I\cap J\neq\varnothing$. On the other hand, if $J=I_c$, then

$$\star(\epsilon^I\wedge\epsilon^J) = \star(\epsilon^I\wedge\epsilon^{I_c}) = \star((\operatorname{sign}\pi_{II_c})\mathrm{vol}_g) = (\operatorname{sign}\pi_{II_c})\star\mathrm{vol}_g = (\operatorname{sign}\pi_{II_c})\,1.$$

We used Exercise 10.5 to get the second equality. So if $I_c =: (\ell_1, \cdots, \ell_{m-k})$, then $(\star\epsilon^I)(e_{\ell_1}, \cdots, e_{\ell_{m-k}})=(\operatorname{sign}\pi_{II_c})\,\iota(I_c)$. Summarising, we have shown

$$(\star\epsilon^I)(e_{j_1}, \cdots, e_{j_{m-k}}) = \left\{\begin{array}{l}0 \text{ if } I\cap J\neq\varnothing,\\ (\operatorname{sign}\pi_{II_c})\,\iota(I_c) \text{ if } J=I_c.\end{array}\right\} \tag{11.3}$$

We will show that this coincides with the action of $(\operatorname{sign}\pi_{II_c})\,\iota(I_c)\,\epsilon^{I_c}$ on $(e_{j_1}, \cdots, e_{j_{m-k}})$. Let $I_c=(\ell_1, \cdots, \ell_{m-k})$. Then

$$\epsilon^{I_c}(e_{j_1}, \cdots, e_{j_{m-k}}) = (\epsilon^{\ell_1}\wedge\cdots\wedge\epsilon^{\ell_{m-k}})(e_{j_1}, \cdots, e_{j_{m-k}}) = \det[\epsilon^{\ell_r}(e_{j_s})].$$

If $I\cap J\neq\varnothing$, then at least one column of the matrix $[\epsilon^{\ell_r}(e_{j_s})]$ is zero, so that the determinant above vanishes. On the other hand, if $J=I_c$, then

$$\det[\epsilon^{\ell_r}(e_{j_s})] = \det[\epsilon^{\ell_r}(e_{\ell_s})] = \det[\delta^r_s] = 1,$$

and so $\epsilon^{I_c}(e_{j_1}, \cdots, e_{j_{m-k}})=1$. Summarising, we have shown that

$$(\operatorname{sign}\pi_{II_c})\,\iota(I_c)\,\epsilon^{I_c}(e_{j_1}, \cdots, e_{j_{m-k}}) = \left\{\begin{array}{l}0 \text{ if } I\cap J\neq\varnothing,\\ (\operatorname{sign}\pi_{II_c})\,\iota(I_c) \text{ if } J=I_c.\end{array}\right\} \tag{11.4}$$

The result now follows from (11.3) and (11.4). □

Now suppose that $(M, \mathbf{g})$ is an m-dimensional semi-Riemannian manifold with metric $\mathbf{g}$, which admits an oriented atlas $\mathcal{A}$ in its smooth structure. Let M be given the orientation $[\mathrm{vol}_{\mathbf{g}}]$. For each $p \in M$, the tangent space T_pM is given the orientation $[\mathrm{vol}_{\mathbf{g}}(p)]$. Then we obtain the Hodge star operator $\star : \wedge^k(T_pM)^* \to \wedge^{m-k}(T_pM)^*$ corresponding to the volume form $\mathrm{vol}_{\mathbf{g}(p)}$ belonging to the orientation $[\mathrm{vol}_{\mathbf{g}}(p)]$ on T_pM. In fact, for $\Omega \in \mathbf{\Omega}^k M$, the pointwise definition $\star\,\Omega := (M \ni p \mapsto \star(\Omega(p)) \in \wedge^{m-k}(T_pM)^*)$ gives an element[12] $\star\Omega \in \mathbf{\Omega}^{m-k}M$. Thus we have Hodge star operators

$$\star : \mathbf{\Omega}^k M \to \mathbf{\Omega}^{m-k} M$$

for $0 \leqslant k \leqslant m$, mapping k-form fields to $(m-k)$-form fields.

Exercise 11.13. Let $(M, \mathbf{g})$ be an m-dimensional semi-Riemannian manifold with metric $\mathbf{g}$, which admits an oriented atlas $\mathcal{A}$ in its smooth structure. Let M be given the orientation $[\mathrm{vol}_{\mathbf{g}}]$. Show that $\star : \mathbf{\Omega}^k M \to \mathbf{\Omega}^{m-k} M$ is a $C^\infty(M)$-linear map. Also show that $\star \star \Omega = (-1)^{k(m-1)+\iota(\mathbf{g})}\Omega$ for all $\Omega \in \mathbf{\Omega}^k M$.

Exercise 11.14. Let $(M, \mathbf{g})$ be a semi-Riemannian manifold with metric $\mathbf{g}$, which admits an oriented atlas $\mathcal{A}$ in its smooth structure. Let M be given the orientation $[\mathrm{vol}_{\mathbf{g}}]$, and for $p \in M$, T_pM be given the orientation $[\mathrm{vol}_{\mathbf{g}}(p)]$. Show that $\mathrm{vol}_{\mathbf{g}(p)} = \mathrm{vol}_{\mathbf{g}}(p)$.

Example 11.5. (Minkowski spacetime.)
Consider the Minkowski spacetime $(M, \mathbf{g})$ from Example 11.3 with the oriented atlas $\mathcal{A} = \{(M, \mathbf{x}_{p,B})\}$ induced by a choice of a $p \in M$ and an orthonormal basis $B = \{\mathbf{e}_1, \mathbf{e}_2, \mathbf{e}_3, \mathbf{e}_4\}$ for V with respect to g. We will use the two notations $\mathbf{x}_{p,B} = (x^1, x^2, x^3, x^4)$ and $\mathbf{x}_{p,B} = (x, y, z, t)$ for the chart map components of $\mathbf{x}_{p,B}$ interchangeably. The metric $\mathbf{g}$ is given by specifying its component matrix in the chart $(M, \mathbf{x}_{p,B})$ by $\mathbf{g}(\partial_i, \partial_j) = 1$ if $i = j \in \{1, 2, 3\}$, $\mathbf{g}(\partial_i, \partial_j) = -1$ if $i = j = 4$, and $\mathbf{g}(\partial_i, \partial_j) = 0$ if $i \neq j$. The volume form field $\mathrm{vol}_{\mathbf{g}} \in \mathbf{\Omega}^4 M$ is given by $\mathrm{vol}_{\mathbf{g}} = dx \wedge dy \wedge dz \wedge dt$. Let M be given the orientation $[\mathrm{vol}_{\mathbf{g}}]$. This amounts to deeming the ordered basis $(\partial_{x,p}, \partial_{y,p}, \partial_{z,p}, \partial_{t,p})$ as positively oriented in T_pM, at each $p \in M$. We then say the Minkowski space $(M, \mathbf{g})$ has the *standard orientation induced by* (p, B).

We will now determine $\star$ on $\mathbf{\Omega}^1 M$, $\mathbf{\Omega}^2 M$ and $\mathbf{\Omega}^3 M$.
Any $\Omega \in \mathbf{\Omega}^1 M$ has the decomposition $\Omega = \Omega_i\, dx^i$ for $\Omega_i \in C^\infty(M)$, and so $\star\Omega = \Omega_i \star (dx^i)$. So to describe $\star$ on $\mathbf{\Omega}^1 M$, it is enough to determine $\star(dx^i)$.

[12] Using an analogue of the formula given in Proposition 11.3, but for not necessarily orthonormal basis, see e.g. [Marsden and Ratiu (2007), Example 7.2.14.B–D], the smoothness of $p \mapsto (\star(\Omega(p)))(\partial_{x^{i_1},p}, \cdots, \partial_{x^{i_{m-k}},p})$ can be established in any admissible chart $(U, \mathbf{x})$. By Proposition 4.1, $\star\Omega$ is a $(0, m-k)$-tensor field on M. The alternating property of $\star\Omega$ follows from the pointwise alternating property.

We use pointwise Proposition 11.3, giving

$$\begin{aligned}
\star dx &= \star dx^1 = \operatorname{sign}\begin{pmatrix}1\,2\,3\,4\\1\,2\,3\,4\end{pmatrix}\mathbf{g}(\partial_y,\partial_y)\,\mathbf{g}(\partial_z,\partial_z)\,\mathbf{g}(\partial_t,\partial_t)\,dy\wedge dz\wedge dt\\
&= (1)(1)(1)(-1)\,dy\wedge dz\wedge dt = -dy\wedge dz\wedge dt,\\
\star dy &= \star dx^2 = \operatorname{sign}\begin{pmatrix}1\,2\,3\,4\\2\,1\,3\,4\end{pmatrix}\mathbf{g}(\partial_x,\partial_x)\,\mathbf{g}(\partial_z,\partial_z)\,\mathbf{g}(\partial_t,\partial_t)\,dx\wedge dz\wedge dt\\
&= (-1)(1)(1)(-1)\,dx\wedge dz\wedge dt = dx\wedge dz\wedge dt,\\
\star dz &= \star dx^3 = \operatorname{sign}\begin{pmatrix}1\,2\,3\,4\\3\,1\,2\,4\end{pmatrix}\mathbf{g}(\partial_x,\partial_x)\,\mathbf{g}(\partial_y,\partial_y)\,\mathbf{g}(\partial_t,\partial_t)\,dx\wedge dy\wedge dt\\
&= (1)(1)(1)(-1)\,dx\wedge dy\wedge dt = -dx\wedge dy\wedge dt,\\
\star dt &= \star dx^4 = \operatorname{sign}\begin{pmatrix}1\,2\,3\,4\\4\,1\,2\,3\end{pmatrix}\mathbf{g}(\partial_x,\partial_x)\,\mathbf{g}(\partial_y,\partial_y)\,\mathbf{g}(\partial_z,\partial_z)\,dx\wedge dy\wedge dz\\
&= (-1)(1)(1)(1)\,dx\wedge dy\wedge dz = -dx\wedge dy\wedge dz.
\end{aligned}$$

To determine $\star$ on $\mathbf{\Omega}^2M$, we compute for example

$$\begin{aligned}
\star(dx\wedge dt) &= \star(dx^1\wedge dx^4) = \operatorname{sign}\begin{pmatrix}1\,2\,3\,4\\1\,4\,2\,3\end{pmatrix}\mathbf{g}(\partial_y,\partial_y)\,\mathbf{g}(\partial_z,\partial_z)\,dy\wedge dz\\
&= (1)(1)(1)\,dy\wedge dz = dy\wedge dz.
\end{aligned}$$

Analogously, one obtains

$$\begin{aligned}
&\star(dx\wedge dy) = -dz\wedge dt, \quad \star(dx\wedge dz) = dy\wedge dt, \quad \star(dy\wedge dz) = -dx\wedge dt,\\
&\star(dy\wedge dt) = -dx\wedge dz, \quad \star(dz\wedge dt) = dx\wedge dy.
\end{aligned}$$

For $\Omega\in\mathbf{\Omega}^2M$, $\star\Omega$ can be found by decomposing

$$\Omega = \sum_{1\leqslant i<j\leqslant 4}\Omega_{ij}\,dx^i\wedge dx^j,$$

where $\Omega_{ij}\in C^\infty(M)$. Finally, we compute for example

$$\begin{aligned}
\star(dx\wedge dy\wedge dt) &= \star(dx^1\wedge dx^2\wedge dx^4) = \operatorname{sign}\begin{pmatrix}1\,2\,3\,4\\1\,2\,4\,3\end{pmatrix}\mathbf{g}(\partial_z,\partial_z)\,dz\\
&= (-1)(1)\,dz = -dz.
\end{aligned}$$

Similarly $\star(dx\wedge dy\wedge dz) = -dt$, $\star(dx\wedge dz\wedge dt) = dy$, $\star(dy\wedge dz\wedge dt) = -dx$. These determine $\star$ on $\mathbf{\Omega}^3M$. We also have $\star(dx\wedge dy\wedge dz\wedge dt) = \star\operatorname{vol}_{\mathbf{g}} = 1$, the constant function equal to 1 everywhere. ◇

Exercise 11.15. In the same set up as Example 11.5, verify directly that

$$\star\star\Omega = (-1)^{k(m-1)+\iota(\mathbf{g})}\Omega$$

from Exercise 11.13 for all $\Omega\in\mathbf{\Omega}^3M$.

Chapter 12

Minkowski spacetime physics

In this chapter, we will study Minkowski spacetime physics which served as a precursor to the development of the geometric theory of gravitation. At the end of the 19th century, it was realised that the classical (Newtonian/Galilean) viewpoint of space and time was inadequate, as it could not explain experimental observations involving the measurement of the 'speed of light' c. The revision of the notions of space and time with the 1905 'special relativity theory' by Einstein gave rise to the simplest spacetime, namely Minkowski[1] spacetime, which we will discuss in this chapter. But before doing so, we quickly recall the familiar classical viewpoint, in order to be able to contrast it relatively easily with the Minkowski spacetime.

We remark that most of the definitions and results in this chapter pertain to objects in the tangent space T_pM of a general time-oriented Lorentzian manifold M at a point p. Even in the case of Lorentzian manifolds, the local region around a point resembles Minkowski spacetime, and the exponential map $\exp_p$ maps a neighbourhood $U \subset T_pM$ of 0 onto a neighbourhood $\exp_p U$ of p in M.

12.1 Classical spacetime

Definition 12.1. (Classical spacetime.)
Classical spacetime is a triple $(M, \mathbf{t}, \eta)$, where

- M is an affine space over a 4-dimensional real vector space V
- $\mathbf{t} : V \to \mathbb{R}$ is a nonzero linear map
- η is an inner product on the vector space $\ker \mathbf{t}$.

M is thought of as the collection of all events, and $\mathbf{t}$ is the *absolute time.* For $p \in M$, define the set S_p *of events that are simultaneous with* p by $S_p = p + \ker \mathbf{t} = \{q \in M : \mathbf{t}(\mathbf{v}_{pq}) = 0\}$.

[1]After Hermann Minkowski (1864–1909), who realised that space and time should be considered as a single whole (a 4-dimensional geometric object called spacetime), rather than two separate entities, and that special relativity, introduced by Einstein (who was Minkowski's former student) is best understood in this mathematical setting.

The inner product η provides the angles and distances in each spatial section S_p, $p \in M$. By the rank-nullity theorem, $\dim(\ker \mathbf{t}) = 3$. For $\mathbf{v} \in V$, define the smooth curve $\gamma_{\mathbf{v}}(t) = p + t\mathbf{v}$ for all $t \in \mathbb{R}$. We had seen that the map $\iota : V \to T_pM$ given by $\iota(\mathbf{v}) = v_{\gamma_{\mathbf{v}},p}$ for all $\mathbf{v} \in V$, is an isomorphism of vector spaces. Henceforth we do not write down ι, ι^{-1} explicitly.

Definition 12.2. (Instantaneous observer, space perceived.)
We use the same notation as in Definition 12.1. An *instantaneous observer at* $p \in M$ *in classical spacetime* is a vector $v \in T_pM$ such that $\mathbf{t}(v) > 0$. The *space perceived* by an instantaneous observer $v \in T_pM$ is $S_p = p + \ker \mathbf{t}$.

We note that S_p is the space perceived by *any* instantaneous observer at p, that is, it is independent of $v \in T_pM$, and depends just on p. This 'affine subspace' S_p divides the affine space M into two parts, namely

- the 'future' of events in S_p, consisting of all events q such that $\mathbf{t}(\mathbf{v}_{pq}) > 0$,
- the 'past' of events in S_p, consisting of all events q such that $\mathbf{t}(\mathbf{v}_{pq}) < 0$.

Definition 12.3. (Particle.)
We use notation from Definition 12.1. A *particle in classical spacetime* is a triple (γ, m, e), where $m > 0$, $e \in \mathbb{R}$, and $\gamma : I \to M$ is a smooth curve defined on an interval $I \subset \mathbb{R}$, such that $\mathbf{t}(v_{\gamma,\gamma(t)}) > 0$ for all $t \in I$. The number m is called the *mass* of the particle, and e the *charge* of the particle.

If (γ, m, e) is a particle passing through a point $p \in M$ with $v_{\gamma,p} = w \in T_pM$, or if w is an instantaneous observer at p, then an instantaneous observer $v \in T_pM$ will assign to w a certain relative velocity u, by observing the motion of the particle or the other observer in its perceived space. We give the following definition.

Definition 12.4. (Relative velocity.)
We use notation from Definition 12.1. Let $v \in T_pM$ be an instantaneous observer in classical spacetime. Let $w \in T_pM$ be such that $\mathbf{t}(w) > 0$. Then the *relative velocity* $u \in \ker \mathbf{t}$ *of* w *with respect to an instantaneous observer* v is defined by

$$u = \frac{w}{\mathbf{t}(w)} - \frac{v}{\mathbf{t}(v)}.$$

Note that $\mathbf{t}(u) = \mathbf{t}\big(\frac{w}{\mathbf{t}(w)} - \frac{v}{\mathbf{t}(v)}\big) = 1 - 1 = 0$. So the relative velocity $u \in \ker \mathbf{t}$.

Also if $w = v$, then the relative velocity $u = 0$, as expected. In the relative velocity expression, the seemingly strange factors $\mathbf{t}(w)$ and $\mathbf{t}(v)$ in the denominators are necessary. These arise since the 'proper' time elapsed between an event p and a later event q is given by $\mathbf{t}(\mathbf{v}_{pq})$. Roughly speaking, the curve parametrisation only keeps track of what order the events along curve are visited, but what is physical, and is 'felt', is the proper time elapsed between events.

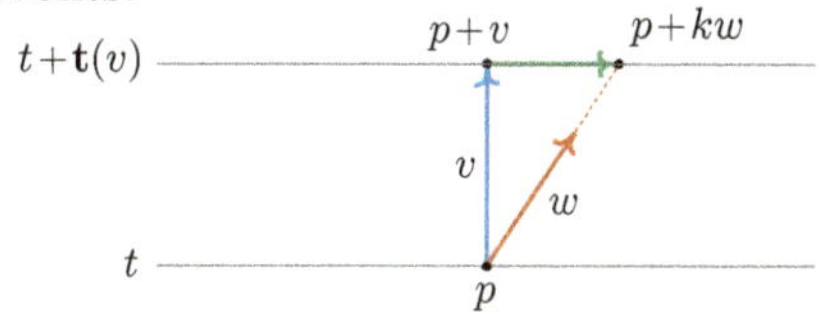

So formally, referring to the picture, if the observer 'evolves' from event p to $p+v$, then it does so in the time $\mathbf{t}(v)$. In the meantime, the particle starting at p with $v_{\gamma,p}=w$ will go to some point q belonging to the spatial section S_{p+v} simultaneous with $p+v$. To determine q, we write $q=p+kw$, and solve for k using $\mathbf{t}(kw)=\mathbf{t}(v)$ to obtain $k=\frac{\mathbf{t}(v)}{\mathbf{t}(w)}$. So it makes sense that the relative velocity of the particle as reckoned by the instantaneous observer v is defined by $\frac{kw-v}{\mathbf{t}(v)}=\frac{w}{\mathbf{t}(w)}-\frac{v}{\mathbf{t}(v)}$. As the inner product η on $\ker \mathbf{t}$ provides the 'spatial length' $\|u\|$ of a vector $u \in \ker \mathbf{t}$ via $\|u\|:=\sqrt{\eta(u,u)}$, if u is the relative velocity, then it is reasonable to call $\|u\|$ the relative speed of w with respect to v.

Proposition 12.1. (Galilean addition of relative velocities.)
Let $(M,\mathbf{t},\eta)$ be a classical spacetime, and $p \in M$. Suppose that $v, \widetilde{v} \in T_pM$ are instantaneous observers, and $w \in T_pM$ is such that $\mathbf{t}(w)>0$. Then the relative velocities $u, \widetilde{u}$ of w with respect to v and $\widetilde{v}$, respectively, are related by $\widetilde{u}=u-x$, where

$$x=\frac{\widetilde{v}}{\mathbf{t}(\widetilde{v})}-\frac{v}{\mathbf{t}(v)}$$

is the relative velocity of $\widetilde{v}$ with respect to v.

Proof. Subtract $u=\frac{w}{\mathbf{t}(w)}-\frac{v}{\mathbf{t}(v)}$ from $\widetilde{u}=\frac{w}{\mathbf{t}(w)}-\frac{\widetilde{v}}{\mathbf{t}(\widetilde{v})}$. □

Thus according to the Galilean viewpoint, if $v, \widetilde{v}$ are observers such that[2] $\mathbf{t}(v)=\mathbf{t}(\widetilde{v})=1$, and the relative velocity of light with respect to the observer v is c, then the relative velocity of light with respect to the observer $\widetilde{v}$ ought to be $c+v-\widetilde{v}$. However, this was not observed experimentally, notably in a famous experiment performed by Michelson and Morley in 1887. The speed of light is the *same* relative to all observers, and is a universal constant, namely $c=3\times 10^8 \mathrm{m\,s}^{-1}$. We will see later in this chapter that in Minkowski spacetime, we obtain a different addition law of relative velocities, which will automatically render c as a constant.

[2]There is no loss of generality, since the relative velocity is the difference of terms of the form $\frac{v}{\mathbf{t}(v)}$, $v \in T_pM$, and this is also equal to $\frac{\alpha v}{\mathbf{t}(\alpha v)}$ for any real scalar $\alpha \neq 0$.

Exercise 12.1. (Doppler effect in classical spacetime.) Let $\mathbf{e}_0 \in V$ be such that $\mathbf{t}(\mathbf{e}_0) = 1$, and $\mathbf{e}_1 \in \ker \mathbf{t}$ be such that $\eta(\mathbf{e}_1, \mathbf{e}_1) = 1$. Define $\gamma : \mathbb{R} \to M$ by $\gamma(t) = p + t\mathbf{e}_0$ for all $t \in \mathbb{R}$. Let $\theta > 0$, and consider the worldline $\widetilde{\gamma} : \mathbb{R} \to M$ given by $\widetilde{\gamma}(t) = p + t(\mathbf{e}_0 + (\tanh\theta)\mathbf{e}_1)$ for all $t \in \mathbb{R}$. The two worldlines meet at $\gamma(0) = p = \widetilde{\gamma}(0)$, and the relative velocity of $v_{\widetilde{\gamma},p}$ with respect to $v_{\gamma,p}$ at p is $x = (\tanh\theta)I(\mathbf{e}_1)$, where $\mathrm{I}: V \to T_pM$ is the isomorphism from Example 2.2 (p.27). Suppose that particles are emitted from $\widetilde{\gamma}$ to γ at the events $\widetilde{\gamma}(\widetilde{t_n})$, $\widetilde{t_n} = \frac{n}{\nu}$, $n \in \mathbb{Z}$. Here the fixed number $\nu > 0$ is the frequency of emission of particles. The nth particle has the worldline $\lambda_n : \mathbb{R} \to M$ given by $\lambda_n(t) = \widetilde{\gamma}(\widetilde{t_n}) + (t - \widetilde{t_n})(\mathbf{e}_0 - \mathbf{e}_1)$, $t \in \mathbb{R}$. Thus the particles have relative velocities $-\mathrm{I}(\mathbf{e}_1)$ with respect to $v_{\gamma,\gamma(t_n)}$, when they are received by γ at the events $\gamma(t_n)$, $n \in \mathbb{Z}$. Show that there exists a $\widetilde{\nu} > 0$ such that $t_n = \frac{n}{\widetilde{\nu}}$, $n \in \mathbb{Z}$, and that $\frac{\widetilde{\nu}}{\nu} = (1 + \tanh\theta)^{-1}$.

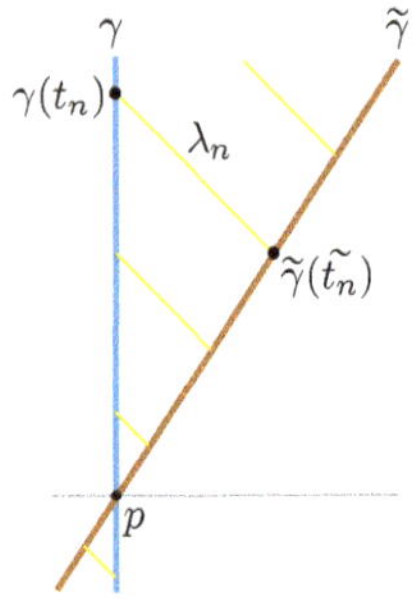

12.2 Minkowski space

Recall that Minkowski spacetime is a pair $(M, \mathbf{g})$, where

- M is an affine space over a 4-dimensional real vector space V,
- M is considered[3] as a smooth manifold, and
- V has a Minkowski scalar product g, inducing[4] the metric $\mathbf{g}$ on M.

We had introduced the notion of an instantaneous observer in a time-oriented Lorentzian manifold in Definition 9.8: it is a timelike future-pointing tangent vector $v \in T_pM$. Set $v^\perp = \{w \in T_pM : \mathbf{g}(p)(v, w) = 0\}$. Then the space perceived by an instantaneous observer $v \in T_pM$ is $p + v^\perp = \{p + w \in T_pM : \mathbf{g}(p)(v, w) = 0\}$.

In classical spacetime, S_p, the set of all events which are simultaneous with p, is the 3-dimensional affine space $p + \ker \mathbf{t}$, which divides M into half-spaces, namely, the past of S_p, and the future of S_p. That S_p does not depend on the instantaneous observer $v \in T_pM$ in classical spacetime, is a key difference to what happens in Minkowski spacetime. In classical spacetime, *all* instantaneous observers at an event $p \in M$ agree on the set $S_p = p + \ker \mathbf{t}$ of events that are simultaneous with p.

---

[3]See Example 1.9, p.8.
[4]See Example 5.6, p.83.

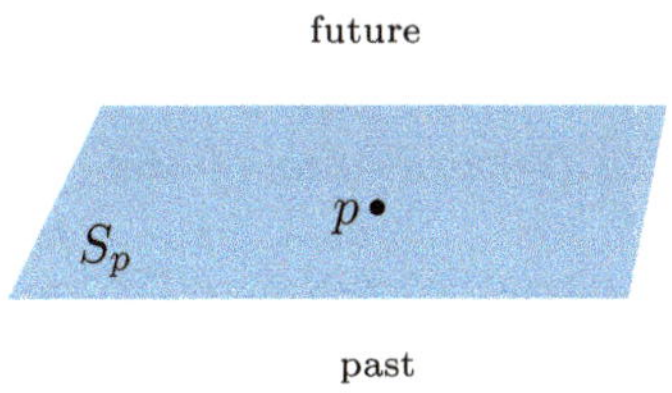

Classical spacetime

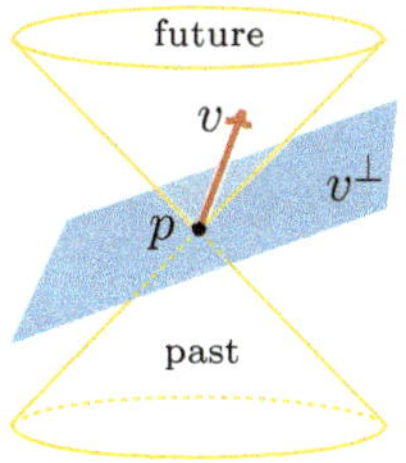

Minkowski spacetime

On the other hand, in Minkowski spacetime, given an instantaneous observer $v \in T_pM$, the set of events that are simultaneous with p is $p + v^{\perp}$, and depends very much on the observer v. Hence what are the simultaneous events to one instantaneous observer at p, may not be perceived as being simultaneous to another observer. This is not really a surprise, since in Minkowski spacetime, there is no universal 'absolute time' that allows a universal 'foliation' of the spacetime into 'space sections'. Instead in Minkowski spacetime, we will see that while all instantaneous observers agree on what is the future of p, and what is the past of p, they will dispute on what are events simultaneous with p, as it will depend on their own 'motion' through spacetime.

Example 12.1. (Time dilation.) Let $(M, \mathbf{g})$ be Minkowski spacetime. Suppose that $p \in M$, and that $\mathbf{e}_0 \in V$ satisfies $g(\mathbf{e}_0, \mathbf{e}_0) = -1$. Let $\mathbf{e}_1 \in \mathbf{e}_0^{\perp}$ be such that $g(\mathbf{e}_1, \mathbf{e}_1) = 1$. We endow M with the time-orientation induced by $\mathbf{e}_0$. Consider the two observers $\gamma, \widetilde{\gamma} : \mathbb{R} \to M$,

$$\begin{aligned}\gamma(t) &= p + t\mathbf{e}_0,\\ \widetilde{\gamma}(t) &= p + t(\alpha\mathbf{e}_0 + \beta\mathbf{e}_1),\end{aligned}$$

where $t \in \mathbb{R}$, and $\alpha, \beta \in \mathbb{R}$ are fixed, and satisfy $-\alpha^2 + \beta^2 = -1$ and $\beta \neq 0$.

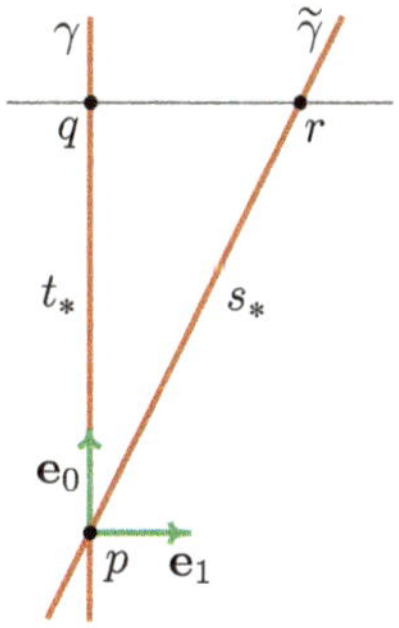

We note that $\alpha > 0$, since $v_{\widetilde{\gamma},\widetilde{\gamma}(t)}$ is future-pointing for all $t \in \mathbb{R}$ (as $\widetilde{\gamma}$ is an observer). For some $t_* > 0$, consider the event $q = \gamma(t_*) = p + t_*\mathbf{e}_0$. Then the instantaneous observer $v_{\gamma,q}$ perceives space as $q + \mathbf{e}_0^\perp$. In this space, there is an event r belonging to the worldline of the observer $\widetilde{\gamma}$, that is, $r = \widetilde{\gamma}(s_*)$. To determine s_*, we use $s_*(\alpha\,\mathbf{e}_0 + \beta\,\mathbf{e}_1) - t_*\mathbf{e}_0 \in \mathbf{e}_0^\perp$, so that $t_* = s_*\alpha = s_*\sqrt{1+\beta^2} > s_*$. Now the proper time elapsed between p and q for the observer γ is given by $\int_0^{t_*}\sqrt{-g(\mathbf{e}_0,\mathbf{e}_0)}\,dt = \int_0^{t_*} 1\,dt = t_*$. But the proper time elapsed between p and r for the observer $\widetilde{\gamma}$ is $\int_0^{s_*}\sqrt{-g(\alpha\mathbf{e}_0+\beta\mathbf{e}_1,\alpha\mathbf{e}_0+\beta\mathbf{e}_1)}\,dt = \int_0^{s_*}\sqrt{-(-\alpha^2+\beta^2)}\,dt = \int_0^{s_*} 1\,dt = s_*$. So, although the events q and r are deemed simultaneous by $v_{\gamma,q}$, we note that γ has experienced more proper time to go from p to q than the proper time experienced by $\widetilde{\gamma}$ in going from p to r. This is strange only if we have not freed ourselves from the shackles of classical-spacetime-thinking. $\diamond$

For an instantaneous observer $v \in T_pM$ at a point p in Minkowski spacetime, the space perceived is $p + v^\perp$. To 'measure' lengths in the space perceived, the observer uses the length provided by the inner product $\mathbf{g}(p)|_{v^\perp}$ (which is the analogue of the inner product η on $\ker \mathbf{t}$ in classical spacetime): The length between p and $p + w$, where $w \in v^\perp$, is thus $\sqrt{\mathbf{g}(p)(w,w)}$.

Example 12.2. (Length contraction.) Let $(M, \mathbf{g})$ be Minkowski spacetime. Let $p \in M$, and $\mathbf{e}_0 \in V$ be such that $g(\mathbf{e}_0,\mathbf{e}_0) = -1$ and let $\mathbf{e}_1 \in \mathbf{e}_0^\perp$ be such that $g(\mathbf{e}_1,\mathbf{e}_1) = 1$. We endow M with the time-orientation induced by $\mathbf{e}_0$. Consider the strip $\mathcal{S} = \{p + a\,\mathbf{e}_0 + b\,\mathbf{e}_1 : a \in \mathbb{R} \text{ and } |b| \leqslant L\}$. (Imagine a 'rigid rod' whose atoms have worldlines in M, and the union of these worldlines constitutes the set $\mathcal{S}$.) Consider the two observers $\gamma, \widetilde{\gamma} : \mathbb{R} \to M$ given by $\gamma(t) = p + t\,\mathbf{e}_0$ and $\widetilde{\gamma}(t) = p + t(\alpha\,\mathbf{e}_0 + \beta\,\mathbf{e}_1)$, where $t \in \mathbb{R}$, and the fixed $\alpha, \beta \in \mathbb{R}$ satisfy $-\alpha^2 + \beta^2 = -1$, $\alpha > 0$, $\beta \neq 0$.

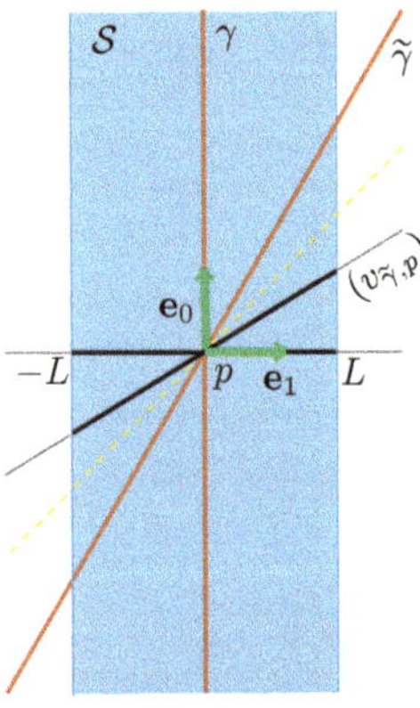

At $p = \widetilde{\gamma}(0)$, the instantaneous observer $v_{\widetilde{\gamma},p} = \alpha\,\mathbf{e}_0 + \beta\,\mathbf{e}_1$ perceives space as the set $p + (\alpha\,\mathbf{e}_0 + \beta\,\mathbf{e}_1)^\perp = \{p + \mathbf{w} : g(\mathbf{w}, \alpha\,\mathbf{e}_0 + \beta\,\mathbf{e}_1) = 0\} \subset M$, and deems the rigid rod to be along the spatial section

$$\begin{aligned}\mathcal{S} \cap (p + (\alpha\mathbf{e}_0 + \beta\mathbf{e}_1)^\perp) &= \{p + a\,\mathbf{e}_0 + b\,\mathbf{e}_1 : a \in \mathbb{R},\ |b| \leqslant L,\ -a\alpha + b\beta = 0\} \\ &= \{p + b(\alpha^{-1}\beta\,\mathbf{e}_0 + \mathbf{e}_1) : |b| \leqslant L\}.\end{aligned}$$

So the observer $\widetilde{\gamma}$ at p reckons the rod's length to be

$$2\widetilde{L} := 2L\sqrt{g\big(\tfrac{\beta}{\alpha}\mathbf{e}_0 + \mathbf{e}_1, \tfrac{\beta}{\alpha}\mathbf{e}_0 + \mathbf{e}_1\big)} = 2L\sqrt{-\frac{\beta^2}{\alpha^2} + 1} = 2L\frac{1}{\alpha} = 2L\frac{1}{\sqrt{1+\beta^2}} < 2L.$$

The instantaneous observer $v_{\gamma,p}$ at p, perceives the rod to lie along the spatial section $\{p + b\,\mathbf{e}_1 : |b| \leqslant L\}$, and reckons the rod's length to be $2L$. The inequality $2\widetilde{L} < 2L$ says that the length of the rod 'measured' by γ at p is strictly greater than that measured by $\widetilde{\gamma}$ at p. ◇

Exercise 12.2. Consider the Minkowski spacetime $(M, \mathbf{g})$. Let $p \in M$, and let B be an orthonormal basis for V with respect to g, such that $\mathbf{e}_0 \in B$ satisfies $g(\mathbf{e}_0, \mathbf{e}_0) = -1$. We endow M with the time-orientation induced by $\mathbf{e}_0$. Let $v \in T_pM$ be a timelike future-pointing vector, and let $\gamma : \mathbb{R} \to M$ be given by $\gamma(t) = p + tv$, $t \in \mathbb{R}$. Let $w \in v^\perp$ be such that $\mathbf{g}(p)(w, w) = 1$. Then $n_+ = v + w$ and $n_- = v - w$ are lightlike future-pointing vectors. Let $k \in \mathbb{R}$. Consider light signals $\lambda_+, \lambda_- : \mathbb{R} \to M$ meeting the event $p + kw$, given by $\lambda_+(s) = p + kw + sn_+$ and $\lambda_-(t) = p + kw + sn_-$, $s \in \mathbb{R}$. Let λ_+, λ_- meet the worldline γ at events $\gamma(t_+)$, $\gamma(t_-)$ for some $t_+, t_- \in \mathbb{R}$. Show that the proper time along γ elapsed between $\gamma(t_-)$ and p equals that elapsed between $\gamma(t_+)$ and p. This justifies the fact that for an instantaneous observer $v \in T_pM$, the perceived space is $v^\perp$.

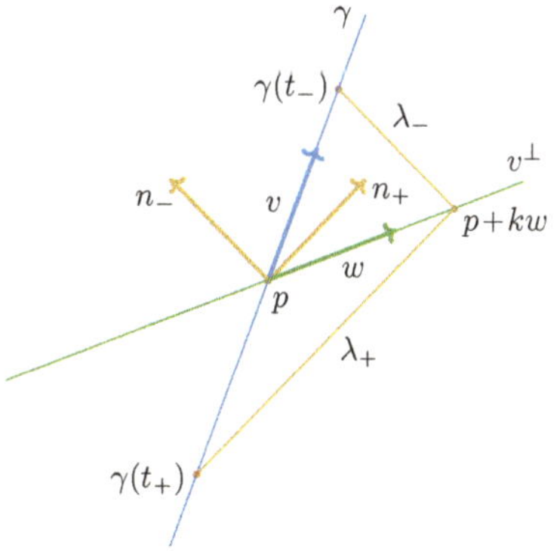

Definition 12.5. (Particle, mass, charge.)
Let $(M, \mathbf{g})$ be a time-oriented Lorentzian manifold. A *particle* is a triple (γ, m, e), where $m > 0$, $e \in \mathbb{R}$, and $\gamma : I \to M$ is a timelike, future-pointing, smooth curve, defined on an interval $I \subset \mathbb{R}$. The number m is the *mass* of the particle, and the number e is the *charge* of the particle.

For an instantaneous observer v, we now give the definition of the relative velocity of a particle/another instantaneous observer, as perceived by v.

Definition 12.6. (Relative velocity.)
Let $(M, \mathbf{g})$ be a time-oriented Lorentzian manifold, $p \in M$, and $v \in T_pM$ be an instantaneous observer at p.
If $w \in T_pM$ is timelike and future-pointing, then the *relative velocity* $u \in v^\perp$ *of* w *with respect to* v is defined as

$$u = -\frac{\sqrt{-\mathbf{g}(p)(v,v)}}{\mathbf{g}(p)(w,v)}w - \frac{v}{\sqrt{-\mathbf{g}(p)(v,v)}}.$$

Note that it follows by the reversed Cauchy-Schwarz inequality (p.81) that since v, w are timelike, $(\mathbf{g}(p)(v,w))^2 > 0$, and in particular nonzero, so that the above formula for u is well-defined.

Exercise 12.3. We use the same notation as in Definition 12.6 above. Show that $u \in v^\perp$. Also, prove that $u = 0$ if and only if $w = cv$ for some $c > 0$.

The rationale behind this definition is analogous to what was done in the previous section, although the time elapsed for the instantaneous observer in going from p to $p + v$ is now given by $\sqrt{-\mathbf{g}(p)(v,v)}$.

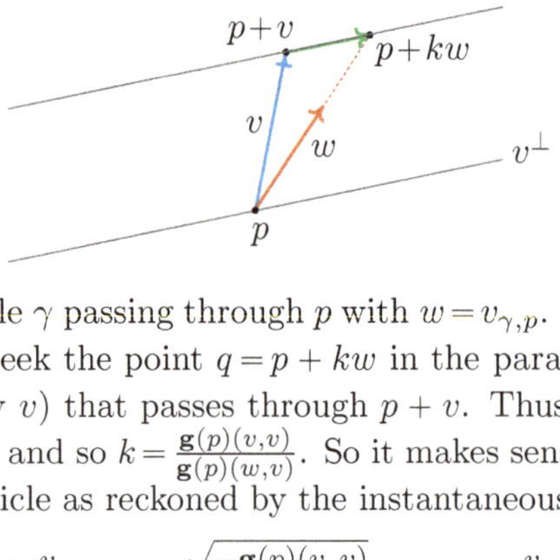

Consider the particle γ passing through p with $w = v_{\gamma,p}$. By referring to the picture above, we seek the point $q = p + kw$ in the parallel translate of $v^\perp$ (space perceived by v) that passes through $p + v$. Thus $kw - v \in v^\perp$, i.e., $\mathbf{g}(p)(kw - v, v) = 0$, and so $k = \frac{\mathbf{g}(p)(v,v)}{\mathbf{g}(p)(w,v)}$. So it makes sense that the relative velocity of the particle as reckoned by the instantaneous observer is

$$\frac{kw - v}{\sqrt{-\mathbf{g}(p)(v,v)}} = -\frac{\sqrt{-\mathbf{g}(p)(v,v)}}{\mathbf{g}(p)(w,v)}w - \frac{v}{\sqrt{-\mathbf{g}(p)(v,v)}}. \qquad (\star)$$

We now show there is a universal speed limit, $c := 1$, for relative speeds.

Proposition 12.2. *Suppose that* $(M, \mathbf{g})$ *is a time-oriented Lorentzian manifold,* $p \in M$*, and* $v \in T_pM$ *is an instantaneous observer. Let* $w \in T_pM$ *be timelike and future-pointing, and let* u *be the relative velocity of* w *with respect to* v*. Then* $0 \leqslant \mathbf{g}(p)(u,u) < 1$.

Proof. Let $k = \dfrac{\mathbf{g}(p)(v,v)}{\mathbf{g}(p)(w,v)}$. The relative velocity u of w with respect to v is

$$u = \frac{kw - v}{\sqrt{-\mathbf{g}(p)(v,v)}}.$$

Since $kw - v \in v^{\perp}$, we have $\mathbf{g}(p)(kw - v, v) = 0$. Using this, we obtain

$$\begin{aligned}\mathbf{g}(p)(u,u) &= \frac{\mathbf{g}(p)(kw - v, kw - v)}{-\mathbf{g}(p)(v,v)} = \frac{\mathbf{g}(p)(kw - v, kw)}{-\mathbf{g}(p)(v,v)} \\ &= -k^2 \frac{\mathbf{g}(w,w)}{\mathbf{g}(v,v)} + k\frac{\mathbf{g}(p)(v,w)}{\mathbf{g}(p)(v,v)} = -\frac{\mathbf{g}(p)(v,v)\cdot\mathbf{g}(p)(w,w)}{(\mathbf{g}(p)(v,w))^2} + 1 \\ &\left\{\begin{array}{l} < \ \ 0+1 = 1, \\ \geqslant -1+1 = 0 \end{array}\right.\end{aligned}$$

using the fact that as v, w are timelike vectors, the reversed Cauchy-Schwarz inequality (p.81) gives $(\mathbf{g}(p)(v,w))^2 \geqslant \mathbf{g}(p)(v,v)\cdot\mathbf{g}(p)(w,w) > 0$. □

Exercise 12.4. Recall Example 12.2 on length contraction. Determine the relative velocity u of $\widetilde{\gamma}$ at p with respect to γ at p. Find $|u| = \sqrt{\mathbf{g}(p)(u,u)}$, and express the ratio of lengths, $\dfrac{2L}{2\widetilde{L}}$, reckoned by the two observers at p, in terms of $|u|$.

12.3 Relative velocity addition

Without loss of generality, we will assume that instantaneous observers and particles $v \in T_pM$ are 'normalised', i.e., $\mathbf{g}(p)(v,v) = -1$, by the following.

Lemma 12.1. (Reparametrisation by proper time.)
Let $(M, \mathbf{g})$ be a Lorentzian manifold, and $\gamma : I \to M$ be a timelike smooth curve, where I is an open interval. Then there exists an open interval J, a smooth bijection $s : I \to J$ with a smooth inverse $s^{-1} : J \to I$, such that $\widetilde{\gamma} := \gamma \circ s^{-1} : J \to M$ satisfies $\mathbf{g}(\widetilde{\gamma}(\sigma))(v_{\widetilde{\gamma},\widetilde{\gamma}(\sigma)}, v_{\widetilde{\gamma},\widetilde{\gamma}(\sigma)}) = -1$ for all $\sigma \in J$.

Proof. Take any $t_0 \in I$. Set $s(t) = \int_{t_0}^{t} \sqrt{-\mathbf{g}(\gamma(t))(v_{\gamma,\gamma(t)}, v_{\gamma,\gamma(t)})}\, dt$, $t \in I$. So

$$\frac{ds}{dt}(t) = \sqrt{-\mathbf{g}(\gamma(t))(v_{\gamma,\gamma(t)}, v_{\gamma,\gamma(t)})} > 0, \quad t \in I.$$

Thus with $J := s(I)$, it follows that $s : I \to J$ is a smooth bijection with a smooth inverse. Defining $\widetilde{\gamma} : J \to M$ by $\widetilde{\gamma}(\sigma) = (\gamma \circ s^{-1})(\sigma)$ for $\sigma \in J$, we have by Exercise 2.4 (p.23) that

$$v_{\widetilde{\gamma},\widetilde{\gamma}(\sigma)} = \frac{d(s^{-1})}{d\sigma}(\sigma)\, v_{\gamma,\gamma(s^{-1}\sigma)} = \frac{1}{\frac{ds}{dt}(s^{-1}\sigma)}\, v_{\gamma,\gamma(s^{-1}\sigma)} = \frac{v_{\gamma,\widetilde{\gamma}(\sigma)}}{\sqrt{-\mathbf{g}(\widetilde{\gamma}(\sigma))(v_{\gamma,\widetilde{\gamma}(\sigma)}, v_{\gamma,\widetilde{\gamma}(\sigma)})}},$$

and so $\mathbf{g}((\widetilde{\gamma}(\sigma))(v_{\widetilde{\gamma},\widetilde{\gamma}(\sigma)}, v_{\widetilde{\gamma},\widetilde{\gamma}(\sigma)}) = -1$ for all $\sigma \in J$. □

For a *normalised* instantaneous observer, the relative velocity formula from Definition 12.6 becomes

$$u = -\frac{w}{\mathbf{g}(p)(w,v)} - v.$$

The following is a recovery formula for the normalised particle velocity, knowing its relative velocity.

Lemma 12.2. *Let $(M, \mathbf{g})$ be a time-oriented Lorentzian manifold, $p \in M$, $v \in T_pM$ be a normalised instantaneous observer, and $w \in T_pM$ be a timelike, future-pointing vector such that $\mathbf{g}(p)(w,w) = -1$. If the relative velocity of w with respect to v is u, then*

$$w = \frac{u+v}{\sqrt{1-\mathbf{g}(p)(u,u)}}.$$

Proof. We know that

$$u = -\frac{w}{\mathbf{g}(p)(w,v)} - v. \tag{12.1}$$

Taking scalar product with $u \in v^\perp$ gives

$$\mathbf{g}(p)(u,u) = -\frac{\mathbf{g}(p)(w,u)}{\mathbf{g}(p)(w,v)}. \tag{12.2}$$

Also taking scalar product in (12.1) with w yields

$$\mathbf{g}(p)(w,u) = \frac{1}{\mathbf{g}(p)(w,v)} - \mathbf{g}(p)(w,v). \tag{12.3}$$

Substituting $\mathbf{g}(p)(w,u)$ from (12.2) into (12.3) gives, after rearranging,

$$(\mathbf{g}(p)(w,v))^2 = \frac{1}{1-\mathbf{g}(p)(u,u)}.$$

Since w, v are both timelike future-pointing, we have[5] $\mathbf{g}(p)(w,v) < 0$. Thus,

$$\mathbf{g}(p)(w,v) = -\frac{1}{\sqrt{1-\mathbf{g}(p)(u,u)}}.$$

Substituting this in (12.1), and rearranging proves the result. □

An instantaneous observer v can choose an orthonormal basis $\{e_1, e_2, e_3\}$ for its perceived space $v^\perp$, so that if $v =: e_0$ is normalised, then $\{e_0, e_1, e_2, e_3\}$ forms an orthonormal basis for T_pM. A relative velocity $u \in v^\perp$ acquires components u^1, u^2, u^3 with respect to the basis $\{e_1, e_2, e_3\}$, obtained by decomposing $u = u^1e_1 + u^2e_2 + u^3e_3$.

Now suppose we have yet another normalised instantaneous observer $\widetilde{v}$ with a certain nonzero relative velocity $x \in v^\perp$ with respect to v. Suppose

[5] See Exercise 5.5, p.81.

that v chooses e_1 to be a positive multiple of x. So we then have $x=\beta e_1$ for some $\beta\in(0,1)$. Then by Lemma 12.2,

$$\widetilde{v}=\frac{v+\beta e_1}{\sqrt{1-\beta^2}}.$$

It is straightforward to verify that

$$\widetilde{e}_1:=\frac{\beta v+e_1}{\sqrt{1-\beta^2}},\quad \widetilde{e}_2=e_2,\quad \widetilde{e}_3=e_3$$

belong to $\widetilde{v}^\perp$ and form an orthonormal basis for it. We have the following result on the addition of relative velocities.

Proposition 12.3. *Suppose that* $(M,\mathbf{g})$ *is a time-oriented Lorentzian manifold,* $p\in M$, $v,\widetilde{v}\in T_pM$ *are normalised instantaneous observers such that the relative velocity of* $\widetilde{v}$ *with respect to* v *is equal to* βe_1, *where* $e_1\in T_pM$ *satisfies* $\mathbf{g}(p)(e_1,e_1)=1$ *and* $\beta\in(0,1)$. *Let* e_2,e_3 *be such that* $\{e_1,e_2,e_3\}$ *forms an orthonormal basis for* $v^\perp$. *Define*

$$\widetilde{e}_1:=\frac{\beta v+e_1}{\sqrt{1-\beta^2}},\quad \widetilde{e}_2=e_2,\quad \widetilde{e}_3=e_3.$$

Then $\{\widetilde{e}_1,\widetilde{e}_2,\widetilde{e}_3\}$ *forms an orthonormal basis for* $\widetilde{v}^\perp$.

Let $w\in T_pM$ *be timelike, future-pointing, and such that* $\mathbf{g}(p)(w,w)=-1$. *Suppose that the relative velocities of* w *with respect to* $v,\widetilde{v}$ *are, respectively,*

$$\begin{aligned} u &= u^1e_1+u^2e_2+u^3e_3, \text{ and}\\ \widetilde{u} &= \widetilde{u}^1\widetilde{e}_1+\widetilde{u}^2\widetilde{e}_2+\widetilde{u}^3\widetilde{e}_3.\end{aligned}$$

Then $\displaystyle \widetilde{u}^1=\frac{u^1-\beta}{1-\beta u^1},\quad \widetilde{u}^2=\frac{u^2\sqrt{1-\beta^2}}{1-\beta u^1},\quad \widetilde{u}^3=\frac{u^3\sqrt{1-\beta^2}}{1-\beta u^1}.$

Proof. By Lemma 12.2, $\widetilde{v}=\dfrac{v+\beta e_1}{\sqrt{1-\beta^2}}$ and $w=\dfrac{v+u^ie_i}{\sqrt{1-\mathbf{g}(p)(u,u)}}$. Thus

$$\begin{aligned}\widetilde{u} &= -\frac{w}{\mathbf{g}(p)(w,\widetilde{v})}-\widetilde{v}=-\frac{\dfrac{v+u^ie_i}{\sqrt{1-\mathbf{g}(p)(u,u)}}}{\mathbf{g}(p)\Big(\frac{v+u^ie_i}{\sqrt{1-\mathbf{g}(p)(u,u)}},\frac{v+\beta e_1}{\sqrt{1-\beta^2}}\Big)}-\frac{v+\beta e_1}{\sqrt{1-\beta^2}}\\
&=\frac{\sqrt{1-\beta^2}\,(v+u^1e_1)}{1-\beta u^1}-\frac{v+\beta e_1}{\sqrt{1-\beta^2}}+\frac{u^2\sqrt{1-\beta^2}}{1-\beta u^1}e_2+\frac{u^3\sqrt{1-\beta^2}}{1-\beta u^1}e_3\\
&=\frac{(1-\beta^2)(v+u^1e_1)-(1-\beta u^1)(v+\beta e_1)}{(1-\beta u^1)\sqrt{1-\beta^2}}+\frac{u^2\sqrt{1-\beta^2}}{1-\beta u^1}\widetilde{e}_2+\frac{u^3\sqrt{1-\beta^2}}{1-\beta u^1}\widetilde{e}_3\\
&=\frac{(u^1-\beta)(\beta v+e_1)}{(1-\beta u^1)\sqrt{1-\beta^2}}+\frac{u^2\sqrt{1-\beta^2}}{1-\beta u^1}\widetilde{e}_2+\frac{u^3\sqrt{1-\beta^2}}{1-\beta u^1}\widetilde{e}_3.\end{aligned}$$

Using the definition of $\widetilde{e}_1$, the components $\widetilde{u}^i$ in terms of the u^i, can now be read off. □

Now we examine the limiting case where the instantaneous particle velocity w approaches the light cone. With the same notation as in Proposition 12.3, consider $\widetilde{w} := e_0 + \lambda\, e_1$. Then $\mathbf{g}(p)(\widetilde{w}, \widetilde{w}) = -1 + \lambda^2$, and so to make this vector $\widetilde{w}$ approach the light cone, we take the limit $\lambda \to 1$ (from below, as $\widetilde{w}$ is meant to be timelike). To obtain a normalised w, we take

$$w := \frac{\widetilde{w}}{\sqrt{-\mathbf{g}(p)(\widetilde{w}, \widetilde{w})}} = \frac{e_0 + \lambda e_1}{\sqrt{1-\lambda^2}}.$$

The picture shows that the tips of the arrows representing the future-pointing vector w lie along the top branch of a hyperbola:

$$\Big\{(x, y) = \Big(\frac{\lambda}{\sqrt{1-\lambda^2}}, \frac{1}{\sqrt{1-\lambda^2}}\Big) : |\lambda| < 1\Big\}$$

(note that $y^2 - x^2 = 1$), and as $\lambda \to 1$, these vector tips approach the 45° asymptote (as $x/y = \lambda \to 1$).

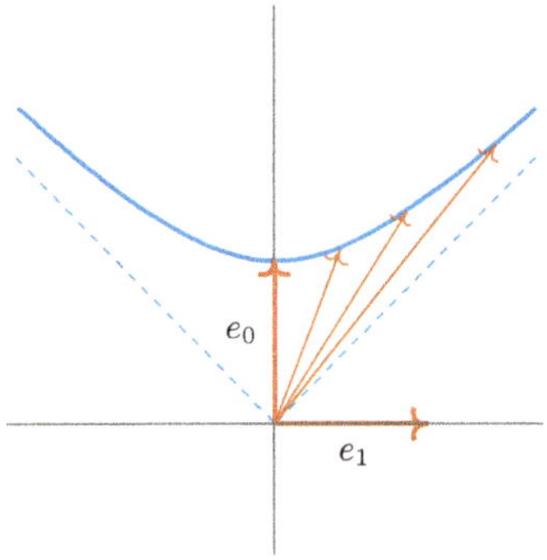

The relative velocity u of w with respect to $v = e_0$ is given by

$$u = -\frac{w}{\mathbf{g}(p)(w, v)} - v = -\frac{\widetilde{w}}{\mathbf{g}(p)(\widetilde{w}, v)} - v = -\frac{e_0 + \lambda e_1}{\mathbf{g}(p)(e_0 + \lambda e_1, e_0)} - e_0 = \lambda e_1.$$

Then $u^2 = u^3 = 0$ and so $\widetilde{u}^2 = \widetilde{u}^3 = 0$ as well. As $u^1 = \lambda$, we obtain

$$\widetilde{u}^1 = \frac{\lambda - \beta}{1 - \beta\lambda}.$$

In the limit $\lambda \to 1$, we see that $\widetilde{u}^1 \to 1$ too. So the second observer also reckons that the photon has relative speed 1, irrespective of what β is.

We also remark that if the relative speeds are much smaller than the speed of light, that is, if $\beta \ll 1$ and $\lambda \ll 1$, then

$$\widetilde{u}^1 = \frac{\lambda - \beta}{1 - \beta\lambda} = (\lambda - \beta)(1 + \lambda\beta + \lambda^2\beta^2 + \cdots) \approx \lambda - \beta,$$

and so $\widetilde{u} = \widetilde{u}^1 e_1 \approx (\lambda - \beta) e_1 = u - x$, where $x = \beta e_1$ is the relative velocity of $\widetilde{v}$ with respect to v. We note that the right-hand side, $u - x$, is the result of Galilean addition of relative velocities (Proposition 12.1).

Exercise 12.5. We use the same notation as in Proposition 12.3 (p.269). Let the relative velocity of w with respect to $\widetilde{v}$ be $\widetilde{u}=\alpha\widetilde{e}_1$, where $\alpha\in(0,1)$. Show that the relative velocity of w with respect to v is given by $(\alpha\circledast\beta)e_1$, where

$$\alpha\circledast\beta := \frac{\alpha+\beta}{1+\alpha\beta}.$$

(This corresponds to the situation that if a person sitting in a train moving with speed β with respect to the ground, sees a fly go past at a speed α, then a person on the ground reckons that the speed of the fly is $\alpha\circledast\beta$.)

Given $\alpha,\beta\in(0,1)$, the number $\alpha\circledast\beta$ can be interpreted geometrically as follows.

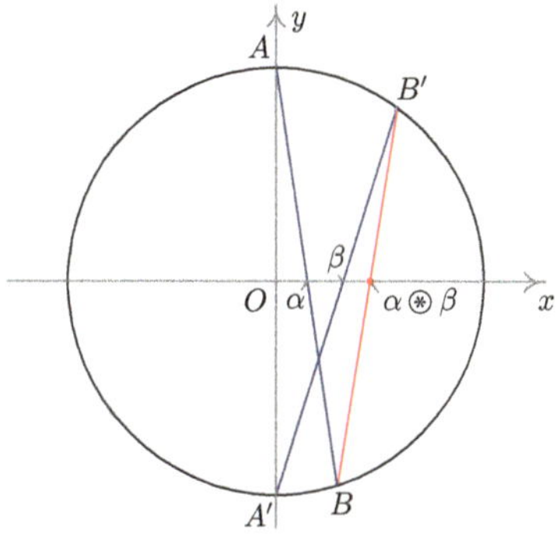

Draw a unit circle in $\mathbb{R}^2$ with center O at $(0,0)$ and vertical diameter AA', where $A=(0,1)$ and $A'=(0,-1)$. Join A to $(\alpha,0)$ and extend it to meet the circle at B. Similarly, join A' to $(\beta,0)$ and extend it to meet the circle at B'. Join BB' and let it meet the x-axis at P. Show that the x-coordinate of P is $\alpha\circledast\beta$.

Hint: This can be done using coordinates, but we could also use the work already done in establishing the 3D formula given Exercise 1.2 (p.4).

The set $G=(-1,1)$ with the operation $\circledast$ forms an Abelian group. Show, based on the above geometric construction, that for all $\alpha,\beta\in(-1,1)$ we have the properties that $\alpha\circledast\beta=\beta\circledast\alpha$, $\alpha\circledast 0=\alpha$, $\alpha\circledast(-\alpha)=0$. Also, show that if we allow $\beta=1$, then the geometric construction gives $\alpha\circledast 1=1$.

Exercise 12.6. Show that the operation $\circledast$ from Exercise 12.5 is the following: If $\alpha,\beta\in(-1,1)$ are written as $\alpha=\tanh a$ and $\beta=\tanh b$ for some $a,b\in\mathbb{R}$, then $\alpha\circledast\beta=\tanh(a+b)$, that is, $\circledast=\big((\alpha,\beta)\mapsto\tanh(\tanh^{-1}\alpha+\tanh^{-1}\beta)\big)$.

Exercise 12.7. A cart rolls on a table with relative speed $\beta\in(0,1)$. A smaller cart rolls on the first cart in the same direction with a relative speed β with respect to the first cart. A third cart rolls on the second cart in the same direction with a relative speed β with respect to the second cart, and so on, up to n carts. Determine the relative speed u_n of the nth cart with respect to the table. What is $\lim\limits_{n\to\infty} u_n$?

Exercise 12.8. With the notation of Proposition 12.3, set also $e_0:=v$ and $\widetilde{e}_0:=\widetilde{v}$. Decompose a $z\in T_pM$ as $z=z^ie_i=\widetilde{z}^i\widetilde{e}_i$. Let $\Lambda=[\Lambda^i_j]\in\mathbb{R}^{4\times 4}$ be such that $z^i=\Lambda^i_j\widetilde{z}^j$. Show that Λ is a 'boost' (see Exercise 5.10, p.82), where $\phi=\tanh^{-1}\beta$.

12.4 Energy and momentum

In classical spacetime, a particle of mass m and velocity $\mathbf{u}$ is said to have a momentum $m\mathbf{u}$. When calculating the work done, one takes the inner product of force (the rate of change in momentum) with the infinitesimal displacement, that is, $\frac{m\mathbf{u}(t+\Delta t)-m\mathbf{u}(t)}{\Delta t}\cdot(\Delta t)\mathbf{u}(t)$. Hence we could view this operation also as follows: Convert $\mathbf{u}$ to a 1-form using the flat operator $\cdot^\flat$, and act it on the velocity, that is, $(m\mathbf{u}(t+\Delta t)^\flat - m\mathbf{u}(t)^\flat)(\mathbf{u}(t))$. With this motivation[6], we give the following definition.

Definition 12.7. (Energy-momentum 1-form of a particle.)
Let $(M, \mathbf{g})$ be a time-oriented Lorentzian manifold, $p \in M$, and (γ, m, e) be a particle passing through p, with $\mathbf{g}(p)(w, w) = -1$, where $w = v_{\gamma,p}$.
The *energy-momentum 1-form* $\mathfrak{p}$ of the particle at p is the 1-form given by $\mathfrak{p} = mw^\flat = m\mathbf{g}(p)(w, \cdot) \in (T_pM)^*$.

For an instantaneous observer v, the instantaneous space is $v^\perp$, and so the momentum of the particle ought to be the restriction of $\mathfrak{p}$ to $v^\perp$. We give the following definition.

Definition 12.8. (Particle momentum/energy measured by an observer.)
Let $(M, \mathbf{g})$ be a time-oriented Lorentzian manifold, $p \in M$, (γ, m, e) be a particle passing through p, with $\mathbf{g}(p)(w, w) = -1$, where $w = v_{\gamma,p}$, and let $\mathfrak{p}$ denote the energy-momentum 1-form of the particle at p. Suppose $v \in T_pM$ is a normalised instantaneous observer at p, i.e., $\mathbf{g}(p)(v, v) = -1$. Then

- the *momentum of w measured by v* is $\mathfrak{p}|_{v^\perp}$, and
- the *energy of w measured by v* is $E = -\mathfrak{p}(v) = -m\mathbf{g}(p)(w, v)$.

If $u \in v^\perp$ denotes the relative velocity of w with respect to v, then we have

$$E = -m\mathbf{g}(p)(w, v) = -m\mathbf{g}(p)\Big(\frac{u+v}{\sqrt{1-\mathbf{g}(p)(u,u)}}, v\Big) = \frac{m}{\sqrt{1-\mathbf{g}(p)(u,u)}}.$$

The reason for calling this the energy can be motivated as follows. Imagine the scenario when u is much smaller than the speed of light (so that the classical spacetime picture is approximately valid). Consider the function $f : (-1, 1) \to \mathbb{R}$, given by

$$f(t) = \frac{1}{\sqrt{1-t^2}}, \quad t \in (-1, 1).$$

[6]There are also further reasons, from the Lagrangian and Hamiltonian approaches to classical mechanics, and also from quantum mechanics, for defining the momentum using 1-forms; see e.g. [Dodson and Poston (1997), Chapter IX, §2.05].

By a Taylor expansion of f around the point $t=0$, we obtain

$$\frac{1}{\sqrt{1-t^2}} = 1 + \frac{1}{2}t^2 + h(t),$$

where $\lim_{t\to 0} \frac{h(t)}{t^3} = 0$. So if $|u| := \sqrt{\mathbf{g}(p)(u,u)}$, then for $|u| \ll 1$, we have

$$E = \frac{m}{\sqrt{1-\mathbf{g}(p)(u,u)}} = \frac{m}{\sqrt{1-|u|^2}} \approx m + \frac{m}{2}|u|^2.$$

We identify the second summand as the kinetic energy possessed by the particle as measured by the instantaneous observer, and this motivates also calling E as an energy. We note that if $u=0$ (i.e., the particle is at rest with respect to the instantaneous observer), then $E=m$. So m is sometimes referred to as the 'rest energy'.

We remark that if we know $\mathfrak{p}|_{v^\perp}$ and E, then we can recover $\mathfrak{p}$. Choose an orthonormal basis $\{e_1, e_2, e_3\}$ for $v^\perp$. For $i=1,2,3$, we define

$$\begin{aligned} p_i := \mathfrak{p}|_{v^\perp}(e_i) &= m\,\mathbf{g}(p)(w, e_i) = m\,\mathbf{g}(p)\Big(\frac{u+v}{\sqrt{1-\mathbf{g}(p)(u,u)}}, e_i\Big) \\ &= \frac{m}{\sqrt{1-\mathbf{g}(p)(u,u)}}\,\mathbf{g}(p)(u,e_i) = \frac{m}{\sqrt{1-\mathbf{g}(p)(u,u)}}\,u^i, \end{aligned}$$

where $u=u^1e_1+u^2e_2+u^3e_3$. (If $\sqrt{\mathbf{g}(p)(u,u)} \ll 1$, then we have $p_i \approx mu^i$ are the three components of the classical momentum.) Consequently, for any $x := x^1e_1+x^2e_2+x^3e_3 \in v^\perp$, we get

$$\mathfrak{p}|_{v^\perp}(x) = x^1p_1 + x^2p_2 + x^3p_3.$$

With $e_0 := v$, $\{e_0, e_1, e_2, e_3\}$ forms an orthonormal basis for T_pM, so that any $z \in T_pM$ has a decomposition $z = z^0e_0 + z^1e_1 + z^2e_2 + z^3e_3$, and then

$$\mathfrak{p}(z) = \mathfrak{p}(z^0v + z^1e_1 + z^2e_2 + z^3e_3) = -z^0E + \mathfrak{p}|_{v^\perp}(z^1e_1 + z^2e_2 + z^3e_3).$$

We note that

$$\begin{aligned} &-E^2 + (p_1)^2 + (p_2)^2 + (p_3)^2 \\ &= -\frac{m^2}{1-\mathbf{g}(p)(u,u)} + \frac{m^2((u^1)^2+(u^2)^2+(u^3)^2)}{1-\mathbf{g}(p)(u,u)} \qquad (12.4) \\ &= -\frac{m^2}{1-\mathbf{g}(p)(u,u)} + \frac{m^2\mathbf{g}(p)(u,u)}{1-\mathbf{g}(p)(u,u)} = -m^2. \qquad (12.5) \end{aligned}$$

Also the calculation in (12.4) shows that

$$\frac{\sqrt{(p_1)^2+(p_2)^2+(p_3)^2}}{E} = \sqrt{(u^1)^2+(u^2)^2+(u^3)^2} = \sqrt{\mathbf{g}(u,u)}.$$

Thus in the limiting case that the particle relative velocity approaches the light cone, i.e., $|u| = \sqrt{\mathbf{g}(p)(u,u)}$ approaches the speed of light $c=1$, we note that $(p_1)^2 + (p_2)^2 + (p_3)^2 = E^2$. But then (12.5) shows that $m \to 0$, that is, the light 'particles', i.e., photons, must have zero mass.

Momentum of light

If we think of light as an electromagnetic wave, then the fact that it carries momentum can be loosely justified as follows. Consider a charged particle like an electron with a charge q encountering monochromatic light, assumed to be an electromagnetic wave, propagating the z-direction. With $\mathbf{e}_x, \mathbf{e}_y, \mathbf{e}_z$ forming an orthonormal basis for $\mathbb{R}^3$, $\mathbf{E}(t,x,y,z) = E_0 \sin(kz - \omega t)\mathbf{e}_x$ and $\mathbf{B}(t,x,y,z) = B_0 \sin(kz - \omega t)\mathbf{e}_y$, where $E_0 = cB_0$, and c is the speed of light (which is set to 1 with an appropriate choice of units). The electric field $\mathbf{E}$ will move the charge up and down, along the x direction. This moving charge (with an instantaneous velocity, say v_x, along the positive x-direction), is subject to a perpendicular magnetic field, namely $\mathbf{B}$, along the positive y-direction. By the Lorentz force law, the charge experiences a Lorentz force $F(t)$ due to the magnetic field:

$$F(t) = \frac{q v_x E_0 \sin(kz - \omega t)}{c} \mathbf{e}_z.$$

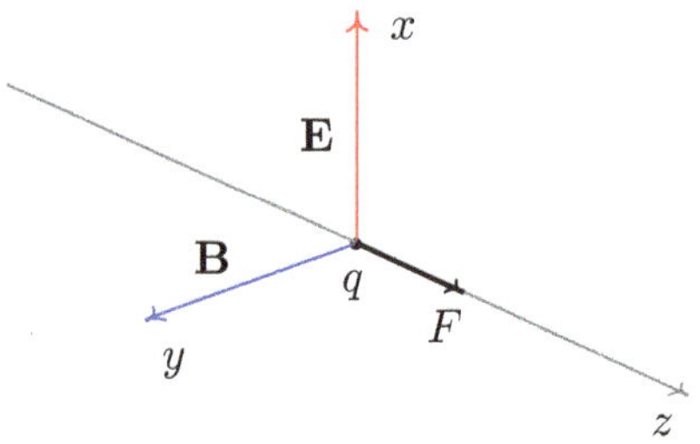

So when light shines on a charge, and the charge oscillates in response to the light's electromagnetic component, and then the light's magnetic component pushes the charge in the direction of propagation of the light beam. Thus light possesses momentum which it transfers to the electron.

We now give a heuristic argument to show that the pushing momentum delivered by light per unit time equals $\frac{1}{c}$ times the energy absorbed from light per unit time. As everything is oscillating, the time average $\langle F \rangle$ of the force is

$$\frac{q\langle v_x E_0 \sin(kz - \omega t)\rangle}{c}.$$

But the charge times the electric field is the force on the charge, and the force times the velocity of the charge is the work done per unit time on the charge. So the pushing momentum per unit time matches $\frac{1}{c}$ times the energy absorbed per unit time. This lends further credibility to E being equal (when $c = 1$) to the size of the relative momentum, which we had noted earlier.

In the quantum theory of light, one imagines light to behave like a particle, and the energy of the light-particle is

$$E = \hbar\omega,$$

where $\omega = 2\pi\nu$ is the 'angular frequency' of light, ν is the frequency of light (number of oscillations per unit time if imagined to be an electromagnetic wave), and $\hbar = h/2\pi$, h being Planck's constant. As we know that the momentum is then the energy divided by c, we have that the photons carry a momentum

$$\frac{\hbar\omega}{c} =: \hbar k.$$

One can then do kinematic computations, for example in the photoelectric effect, Compton effect, etc. Even the light coming from a mobile phone exerts pressure, but it is of course very weak and is thus not felt. The radiation pressure effect is visible in the case of a comet approaching the Sun, when the dust tail turns around considerably due to the pressure of the solar radiation.

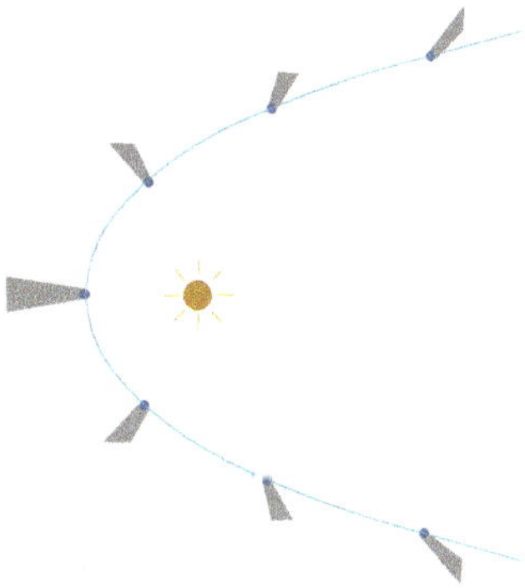

With these preliminaries in mind, we define the momentum of a light signal.

Definition 12.9. (Light signal, energy-momentum of a photon.)
Let $(M, \mathbf{g})$ be a time-oriented Lorentzian manifold.
A *light signal* is a geodesic[7] $\lambda : I \to M$, where $I \subset M$ is an interval and for all $t \in I$, $v_{\lambda,\lambda(t)}$ is lightlike and future-pointing.
Let $\lambda : I \to M$ be a light signal that passes through $p \in M$, and $w = v_{\lambda,p}$.
The *energy-momentum* of λ at p is the 1-form $\mathfrak{p} = w^\flat = \mathbf{g}(p)(w, \cdot) \in (T_pM)^*$.

[7] We know that an affine change of parameter will maintain the geodesic nature of λ, but here when we talk about a light signal, we mean a specific curve, that is, a specific parametrisation. This is related to what *some* observer records to be the frequency/wavelength/energy of the light signal, which 'decides/fixes' a particular parametrisation among all feasible parametrisations.

As opposed to particle energy-momentum at a point, where we multiply by $m > 0$, we do not have this factor now. The energy and momentum of a light signal at a point measured by a normalised instantaneous observer at that point are defined exactly as in Definition 12.8.

Definition 12.10. (Photon momentum/energy measured by an observer.) Let $(M, \mathbf{g})$ be a time-oriented Lorentzian manifold.
Let $p \in M$, and λ be a light signal passing through p, with $w := v_{\lambda,p} \in T_pM$. Let $\mathfrak{p} := w^\flat \in (T_pM)^*$. Let $v \in T_pM$ be a normalised instantaneous observer at p, that is, $\mathbf{g}(p)(v, v) = -1$. Then

- the *momentum of* w *measured by* v is $\mathfrak{p}|_{v^\perp}$, and
- the *energy of* w *measured by* v is $E = -\mathfrak{p}(v) = -\mathbf{g}(p)(w, v)$.

In the following two exercises, we motivate the definitions of the energy/frequency of a photon measured by an instantaneous observer (given above, and in Definition 5.10 from p.96).

Exercise 12.9. Let $(M, \mathbf{g})$ be a time-oriented Lorentzian manifold, $p \in M$, and $\lambda : I \to M$ be a light signal passing through p, with $w := v_{\lambda,p} \in T_pM$. Let $v \in T_pM$ be a normalised instantaneous observer, and suppose that v reckons the frequency of the light signal is ν. We think of T_pM as an approximate local picture of M. We imagine successive light 'wavefront crests' occurring at the event p and another event $q \in M$. One then expects that $1/\nu$ should be the proper time measured by v between p and q. Denoting the vector segment joining p and q by v_{pq}, we have $v_{pq} = (1/\nu)v$. In the infinitesimal picture in T_pM, we may suppose that the light signal λ is a straight line in the direction w. So we expect that another normalised instantaneous observer $\widetilde{v} \in T_pM$ observes the successive wavefront crests (that v had observed) at p and at $\widetilde{q}$, where the vector segment joining p to $\widetilde{q}$ is $v_{p\widetilde{q}} = \alpha w$ for some $\alpha \in \mathbb{R}$, see the picture below.

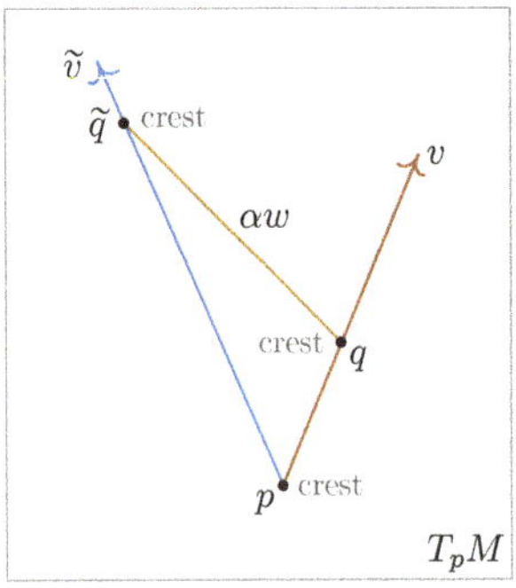

Show that if $\widetilde{\nu}$ denotes the perceived frequency of the light signal by $\widetilde{v}$, then

$$\frac{\widetilde{\nu}}{\nu} = \frac{\mathbf{g}(p)(w, \widetilde{v})}{\mathbf{g}(p)(w, v)}.$$

Exercise 12.10. Let $(M, \mathbf{g})$ be Minkowski spacetime, $p \in M$, $\mathbf{v}, \widetilde{\mathbf{v}}, \mathbf{w} \in V$ be such that $g(\mathbf{v}, \mathbf{v}) = -1 = g(\widetilde{\mathbf{v}}, \widetilde{\mathbf{v}})$, and $g(\mathbf{w}, \mathbf{w}) = 0$. Suppose that $\nu := -g(\mathbf{w}, \mathbf{v}) > 0$. Let $p, q \in M$. Define $\gamma, \widetilde{\gamma} : \mathbb{R} \to M$ by $\gamma(t) = p + t\mathbf{v}$ and $\widetilde{\gamma}(t) = q + t\widetilde{\mathbf{v}}$ for all $t \in \mathbb{R}$. For fixed $t \in \mathbb{R}$, define the lightlike geodesic $\lambda_t : \mathbb{R} \to M$ given by $\lambda_t(s) = \gamma(t) + s\mathbf{w}$ for all $s \in \mathbb{R}$. Show that if $\widetilde{\gamma}$ meets λ_t at the event $\widetilde{\gamma}(\widetilde{t})$, then

$$\widetilde{t} = -\frac{g(\mathbf{w}, \mathbf{v}_{pq})}{g(\mathbf{w}, \widetilde{\mathbf{v}})} + t\frac{g(\mathbf{w}, \mathbf{v})}{g(\mathbf{w}, \widetilde{\mathbf{v}})}.$$

Interpretation: Suppose that γ encounters wavefront crests at a frequency of $\nu = -g(\mathbf{w}, \mathbf{v})$, at its proper times (measured from p) n/ν, $n \in \mathbb{Z}$. Then these crests are perceived by $\widetilde{\gamma}$ at a frequency of $\widetilde{\nu} = -g(\mathbf{w}, \widetilde{\mathbf{v}})$, at its proper times (measured from q) $\widetilde{t}_0 + n/\widetilde{\nu}$, where $\widetilde{t}_0 = -g(\mathbf{w}, \mathbf{v}_{pq})/\widetilde{\nu}$ and $n \in \mathbb{Z}$. We note that

$$\frac{\widetilde{\nu}}{\nu} = \frac{g(\mathbf{w}, \widetilde{\mathbf{v}})}{g(\mathbf{w}, \mathbf{v})}.$$

in accordance with the previous exercise.

Exercise 12.11. (Doppler effect.) Although the speed of light is independent of the motion of the source, the frequency measured is not, and this phenomenon is called the *Doppler effect.* If the source moves towards us, we perceive a higher frequency than that measured by the source (the emitted frequency), while if it recedes, the frequency is lower. This effect is already expected classically (see Exercise 12.1, p.262), but in Minkowski spacetime, the change in the frequency is quantitatively different. Let $(M, \mathbf{g})$ be Minkowski spacetime. Let $p \in M$, and $\mathbf{e}_0 \in V$ be such that $g(\mathbf{e}_0, \mathbf{e}_0) = -1$ and let $\mathbf{e}_1 \in \mathbf{e}_0^\perp$ be such that $g(\mathbf{e}_1, \mathbf{e}_1) = 1$. We endow M with the time-orientation induced by $\mathbf{e}_0$. Let $\gamma(t) = p + t\mathbf{e}_0$, $t \in \mathbb{R}$. Let $\theta \in \mathbb{R}$, let $\widetilde{\gamma}(t) := p + t((\cosh\theta)\mathbf{e}_0 + (\sinh\theta)\mathbf{e}_1)$, $t \in \mathbb{R}$, be an emitter of light signals that are intercepted by γ. Suppose that the frequency of these light signals perceived by $v_{\widetilde{\gamma}, \widetilde{\gamma}(t)}$ at each $\widetilde{\gamma}(t)$, $t \in \mathbb{R}$, is constant, say ν. At p, the relative velocity of $v_{\widetilde{\gamma}, p}$ with respect to $v_{\gamma, p}$ is $u = (\tanh\theta)\mathrm{I}\mathbf{e}_1$, where $\mathrm{I} : V \to T_pM$ is the isomorphism from Example 2.2 (p.27). Show that at each $t \in \mathbb{R}$, the ratio of the frequency of the light signal perceived by $v_{\gamma, \gamma(t)}$ to ν is given by $e^{-\theta}$. Show that for $0 < \theta \ll 1$, this coincides with the ratio in the classical case found in Exercise 12.1. See the following picture.

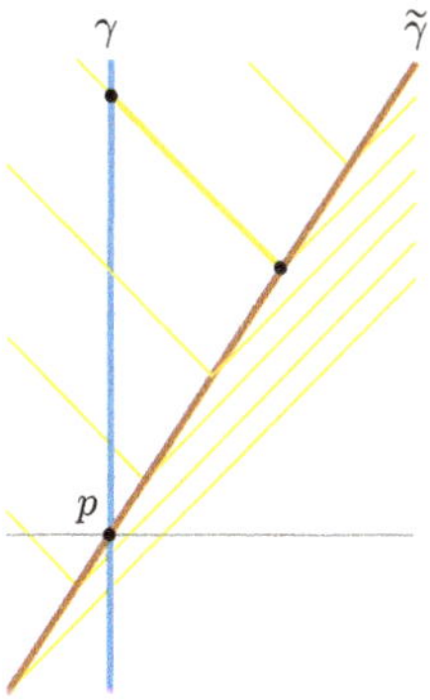

Particles and photons can 'collide' at an event, and can exchange energies and momenta. In such an event, the total energy-momentum is conserved. We make this precise now.

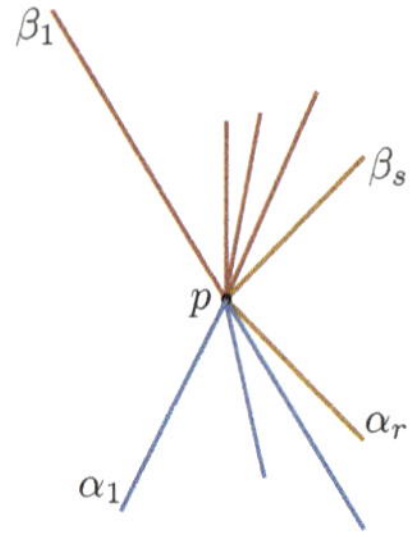

Definition 12.11. (Collision, law of conservation of energy-momentum.)
Let $(M, \mathbf{g})$ be a time-oriented Lorentzian manifold.
A *collision at* $p \in M$ is a collection of

- r *incoming* particles/photons $\alpha_i : [a_i, 0] \to M$, with $\alpha_i(0) = p$, $1 \leqslant i \leqslant r$,
- s *outgoing* particles/photons $\beta_j : [0, b_i] \to M$, with $\beta_j(0) = p$, $1 \leqslant j \leqslant s$.

Additionally, denote the the energy-momentum 1-form of α_i at p by $\mathfrak{p}^i$, $1 \leqslant i \leqslant r$, and the energy-momentum 1-form of β_j at p by $\mathfrak{q}^j$, $1 \leqslant j \leqslant s$.
An *allowed collision*[8] is one for which the following *law of conservation of energy-momentum* holds:

$$\sum_{i=1}^{r} \mathfrak{p}^i = \sum_{j=1}^{s} \mathfrak{q}^j.$$

Example 12.3. (Compton scattering.)
Compton scattering is a process where a photon interacts with an electron, which results in a change in the photon's frequency/wavelength. It was an important historical milestone because it implied that light has a particle nature. (On the other hand, Thomson scattering is the non-quantum, classical scattering of an electromagnetic wave by a charged particle, which is incapable of accounting for the change in wavelength. Compton scattering is the quantum relativistic description of the electron-photon collision.) We wish to derive Compton's formula for the scattered photon energy

$$\widetilde{E} = \frac{E}{1 + \dfrac{E}{m_e}(1 - \cos\theta)}, \qquad (\star)$$

where E is the energy of the incident photon and θ is the 'scattering angle'.

[8] We will only consider allowed collisions. This is a key tenet in particle physics. Ultimately, there is some theory describing fields and interactions, described by a Lagrangian, whose symmetries lead to the conservation of momentum (see Remark 13.1, p.294). No violation has been experimentally observed for the currently accepted standard model of particle physics.

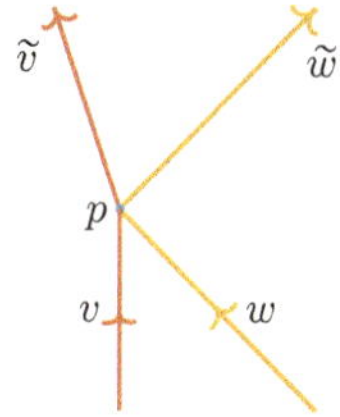

Let $(M, \mathbf{g})$ be a time-oriented Minkowski spacetime. At an event $p \in M$, suppose an incoming photon with energy-momentum $w^\flat$ scatters off an incoming electron having energy-momentum $m_e v^\flat$, and it produces an outgoing photon with energy-momentum $\widetilde{w}^\flat$ and an outgoing electron having energy-momentum $m_e \widetilde{v}^\flat$. We have $\mathbf{g}(p)(v,v) = -1 = \mathbf{g}(p)(\widetilde{v},\widetilde{v})$ and $\mathbf{g}(p)(w,w) = 0 = \mathbf{g}(p)(\widetilde{w},\widetilde{w})$.

The Compton scattering formula gives a relation between the 'scattering angle' of the photons as reckoned by the instantaneous observer $v \in T_pM$ in its perceived space $v^\perp$. In the observer's perceived space $v^\perp$, the incoming electron has relative velocity 0 (i.e., it appears stationary), while the photon is deflected through an angle θ.

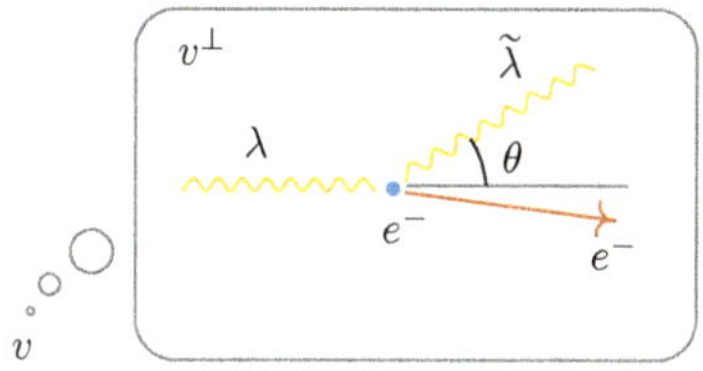

The 'projections' $x, \widetilde{x}$ of $w, \widetilde{w}$, respectively, onto $v^\perp$, are the spacelike vectors

$$\begin{aligned} x &= w + \mathbf{g}(p)(w,v)\,v = w - Ev \\ \widetilde{x} &= \widetilde{w} + \mathbf{g}(p)(\widetilde{w},v)\,v = \widetilde{w} - \widetilde{E}v, \end{aligned}$$

where $E, \widetilde{E}$ are the energies of $w, \widetilde{w}$, respectively, measured by v. Since $\mathbf{g}(p)(w,w)=0$ and $\mathbf{g}(p)(\widetilde{w},\widetilde{w})=0$, $-E^2 + \mathbf{g}(p)(x,x) = 0 = -\widetilde{E}^2 + \mathbf{g}(p)(\widetilde{x},\widetilde{x})$. Define the angle θ by

$$\cos\theta := \frac{\mathbf{g}(p)(x,\widetilde{x})}{\sqrt{\mathbf{g}(p)(x,x)}\sqrt{\mathbf{g}(p)(\widetilde{x},\widetilde{x})}}.$$

So

$$\begin{aligned} E\widetilde{E}\cos\theta &= (\cos\theta)\sqrt{\mathbf{g}(p)(x,x)}\sqrt{\mathbf{g}(p)(\widetilde{x},\widetilde{x})} = \mathbf{g}(p)(x,\widetilde{x}) \\ &= \mathbf{g}(p)(w - Ev, \widetilde{w} - \widetilde{E}v) \\ &= \mathbf{g}(p)(w,\widetilde{w}) - E\,\mathbf{g}(p)(v,\widetilde{w}) - \widetilde{E}\,\mathbf{g}(p)(w,v) + E\widetilde{E}\,\mathbf{g}(p)(v,v) \\ &= \mathbf{g}(p)(w,\widetilde{w}) + E\widetilde{E} + \widetilde{E}E - E\widetilde{E} = \mathbf{g}(p)(w,\widetilde{w}) + E\widetilde{E}. \qquad (12.6) \end{aligned}$$

We use the law of conservation of energy-momentum to derive an expression for $\mathbf{g}(p)(w,\widetilde{w})$: $m_e v^\flat + w^\flat = m_e\widetilde{v}^\flat + \widetilde{w}^\flat$ gives $m_e\widetilde{v} = m_e v + w - \widetilde{w}$. So

$$\begin{aligned}
-m_e^2 &= \mathbf{g}(p)(m_e\widetilde{v}, m_e\widetilde{v})\\
&= \mathbf{g}(p)(m_e v + w - \widetilde{w},\, m_e v + w - \widetilde{w})\\
&= -m_e^2 + 2m_e\,\mathbf{g}(p)(v,w) - 2m_e\,\mathbf{g}(p)(v,\widetilde{w}) + \mathbf{g}(p)(w-\widetilde{w}, w-\widetilde{w})\\
&= -m_e^2 - 2m_e(E-\widetilde{E}) + 0 + 0 - 2\mathbf{g}(p)(w,\widetilde{w}).
\end{aligned}$$

Hence we obtain $\mathbf{g}(p)(w,\widetilde{w}) = -m_e(E-\widetilde{E})$. Substitution in (12.6) yields

$$E\widetilde{E}\cos\theta = -m_e(E-\widetilde{E}) + E\widetilde{E}.$$

Solving for $\widetilde{E}$ yields ($\star$) from page 278. ◇

Exercise 12.12. A positron is a particle with the same mass as an electron, but with the opposite charge, i.e., if $-e$ is the charge of the electron, then that of the positron is $+e$. Show that in a time-oriented Lorentzian manifold $(M,\mathbf{g})$, the annihilation of an electron-positron pair into a photon is prohibited by the law of conservation of energy-momentum. *Hint:* For future-pointing timelike vectors, the scalar product is negative; see Exercise 5.5 (p.81).

12.5 Electromagnetism

The spacetime viewpoint clarifies how electricity and magnetism are really one and the same thing, described by a spacetime object called the 'Faraday tensor field' $\mathbf{F}$, which is a 2-form field on M. We will see that an instantaneous observer $v \in T_pM$ 'splits' $\mathbf{F}(p)$ into an electric field E and a magnetic field B (and so this splitting is not universal, but depends on the observer v). The equations determining $\mathbf{F}$ from the 'charge-current density' $\mathbf{J}$ in spacetime are the Maxwell equations. For us, now in the definitions below, the Faraday tensor field is just a 2-form field $\mathbf{F}$ on M.

Definition 12.12. (Electric field perceived by an observer.)
In a spacetime $(M,\mathbf{g})$, let $\mathbf{F} \in \mathbf{\Omega}^2 M$ be the Faraday tensor field, and $p \in M$. If $v \in T_pM$ be a normalised instantaneous observer, then the *electric field* E *perceived by* v is defined to be the unique vector $\mathrm{E} \in v^\perp$ such that $\mathbf{F}(p)(w,v) = \mathbf{g}(p)(w,\mathrm{E})$ for all $w \in v^\perp$.

Note that the map $v^\perp \ni w \mapsto \mathbf{F}(p)(w,v)$ is $\mathbb{R}$-linear. So $\mathbf{F}(p)(\cdot,v) \in (v^\perp)^*$. But then we can apply the sharp operator $\cdot^\sharp : (v^\perp)^* \to v^\perp$ (associated with the vector space $v^\perp$ with inner product $\mathbf{g}(p)|_{v^\perp}$) to get the vector E in $v^\perp$. Thus $\mathrm{E} = (\mathbf{F}(p)(\cdot,v)|_{v^\perp})^\sharp$ or equivalently, $\mathrm{E}^\flat = \mathbf{F}(p)(\cdot,v)|_{v^\perp}$.

To define the magnetic field perceived by an instantaneous observer v, we need an orientation on $v^\perp$.

Definition 12.13. (Induced orientation.)
Let $(M, \mathbf{g})$ be a spacetime, with an orientation $[\Omega]$, where $\Omega \in \mathbf{\Omega}^4 M$. For $p \in M$, T_pM is given the *induced orientation* $[\Omega(p)]$. Then for an instantaneous observer $v \in T_pM$, the *induced orientation* $[\omega_v]$ *on* $v^\perp$ is given by $\omega_v \in \bigwedge^3 (v^\perp)^*$, where $\omega_v(x,y,z) = \Omega(p)(x,y,z,v)$ for all $x,y,z \in v^\perp$.

ω_v is nonzero: As[9] $T_pM = \operatorname{span}\{v\} \oplus v^\perp$, we can find an orthonormal basis $\{\alpha v, e_1, e_2, e_3\}$, where $\alpha = 1/\sqrt{-\mathbf{g}(p)(v,v)} \neq 0$ and $\{e_1,e_2,e_3\}$ is an orthonormal basis for $v^\perp$. Since $\Omega(p) \in \bigwedge^4 (T_pM)^* \setminus \{0\}$, $\Omega(p)(e_1,e_2,e_3,\alpha v) \neq 0$. Hence also $\omega_v(e_1,e_2,e_3) \neq 0$.

Definition 12.14. (Magnetic field perceived by an observer.)
In a spacetime $(M, \mathbf{g})$, let $\mathbf{F} \in \mathbf{\Omega}^2 M$ be the Faraday tensor field, and $p \in M$. Let $v \in T_pM$ be a normalised instantaneous observer. Equip $v^\perp$ with the induced orientation. Then the *magnetic field* B *perceived by* v is defined to be the vector $\mathrm{B} \in v^\perp$ such that $\mathbf{F}(p)(x,y) = (\star(\mathrm{B}^\flat))(x,y)$ for all $x,y \in v^\perp$, where $\star : \bigwedge^1 (v^\perp)^* \to \bigwedge^2 (v^\perp)^*$ and $\cdot^\flat : v^\perp \to (v^\perp)^*$.

Let us see that the above definition makes sense. Firstly, $\mathbf{F}(p) \in \bigwedge^2 (T_pM)^*$, and so in particular, also the restriction of $\mathbf{F}(p)$ to $v^\perp \times v^\perp$ belongs to $\bigwedge^2 (v^\perp)^*$. We also know that the restriction of $\mathbf{g}(p)$ the vector space $v^\perp$ is an inner product by Theorem 5.2 (p.76).

Secondly, since the Hodge star operators are isomorphisms[10], for the oriented inner product space $(v^\perp, \mathbf{g}(p)|_{v^\perp}, [\omega_v])$, the Hodge star operator $\star : (v^\perp)^* = \bigwedge^1 (v^\perp)^* \to \bigwedge^2 (v^\perp)^*$ is an isomorphism. Finally, the map $\cdot^\flat : v^\perp \to (v^\perp)^*$ is also an isomorphism.

Exercise 12.13. In the setting of Definition 12.14, show that B is alternatively characterised by $\mathbf{F}(p)(x,y) = \mathrm{vol}_{\mathbf{g}(p)}(x,y,\mathrm{B},v)$, where $\mathrm{vol}_{\mathbf{g}(p)}$ is the volume form on the vector space T_pM with the orientation $[\Omega(p)]$ and the scalar product $\mathbf{g}(p)$.

Example 12.4. Let $(M, \mathbf{g})$ be the Minkowski spacetime with the standard orientation induced by (p, B) as in Example 11.5 (p.256). Let $f \in C^\infty(M)$. Suppose that $\mathbf{F} = f\, dz \wedge dx \in \mathbf{\Omega}^2 M$. For a $q \in M$, consider the normalised instantaneous observer $v = \partial_{t,q} \in T_qM$. The electric field E measured by v at q satisfies $\mathbf{F}(q)(v,w) = \mathbf{g}(q)(\mathrm{E}, w)$ for all $w \in v^\perp$. With the decomposition $w = w^x \partial_{x,q} + w^y \partial_{y,q} + w^z \partial_{z,q}$, we have

$$\begin{aligned}
\mathbf{F}(q)(w,v) &= \mathbf{F}(q)(w, \partial_{t,q}) = (f dz \wedge dx)(q)(w^x \partial_{x,q} + w^y \partial_{y,q} + w^z \partial_{z,q}, \partial_{t,q}) \\
&= f(q)\big(w^x dz \wedge dx(\partial_x, \partial_t) + w^y dz \wedge dx(\partial_y, \partial_t) + w^z dz \wedge dx(\partial_z, \partial_t)\big)(q) \\
&= f(q)(0+0+0) = 0.
\end{aligned}$$

So the electric field measured by v is $\mathrm{E} = 0$.

[9]See Theorem 5.1 (p.75), and take $U = \operatorname{span}\{v\}$ there.
[10]See the remark immediately following Lemma 11.7, p.254.

Let $v' = \frac{1}{2}(\sqrt{5}\,\partial_{t,q} + \partial_{x,q})$ be another normalised instantaneous observer. From the above, $\mathbf{F}(q)(w^x\partial_{x,q}+w^y\partial_{y,q}+w^z\partial_{z,q}, \partial_{t,q})=0$ for all $w^x, w^y, w^z \in \mathbb{R}$. Now if $w \in (v')^\perp$, and we decompose $w = w^t\partial_{t,q} + w^x\partial_{x,q}+w^y\partial_{y,q}+w^z\partial_{z,q}$ for some $w^t, w^x, w^y, w^z \in \mathbb{R}$, then we obtain

$$\begin{aligned}\mathbf{F}(q)(w, v') &= 0 + \tfrac{1}{2}\mathbf{F}(q)(w^x\partial_{x,q}+w^y\partial_{y,q}+w^z\partial_{z,q}, \partial_{x,q}) \\ &= \tfrac{1}{2}(f\,dz \wedge dx)(q)(w^x\partial_{x,q}+w^y\partial_{y,q}+w^z\partial_{z,q}, \partial_{x,q}) \\ &= \tfrac{1}{2}f(q)(0 + 0 + w_z dz \wedge dx(\partial_z, \partial_x))(q) \\ &= \tfrac{1}{2}f(q)\,w_z\,1 = \mathbf{g}(q)(\mathrm{E}', w),\end{aligned}$$

where $\mathrm{E}' = \frac{1}{2}f(q)\partial_{z,q}$. So the electric field measured by v' is $\mathrm{E}' \neq 0$ if $f(q) \neq 0$.

Consequently, we see explicitly from the two calculations above, that the electric field perceived at a spacetime point by an instantaneous observer depends on the observer's state of motion. A similar thing also happens with the magnetic field. ◇

Proposition 12.4. *Let $(M, \mathbf{g})$ be the Minkowski spacetime with the standard orientation induced by (p, B) as in* Example 11.5 (p.256), *and let $\mathbf{F} \in \mathbf{\Omega}^2 M$ be the Faraday tensor field. Suppose that $q \in M$, and that the normalised instantaneous observer $v = \partial_{t,q} \in T_qM$ perceives the electric field* E *and magnetic field* B, *where*

$$\begin{aligned}\mathrm{E} &= \mathrm{E}^x\partial_{x,q} + \mathrm{E}^y\partial_{y,q} + \mathrm{E}^z\partial_{z,q} \\ \mathrm{B} &= \mathrm{B}^x\partial_{x,q} + \mathrm{B}^y\partial_{y,q} + \mathrm{B}^z\partial_{z,q}.\end{aligned}$$

Then the Faraday tensor field at q is given by

$$\begin{aligned}\mathbf{F}(q) = \big(&\mathrm{E}^x dx \wedge dt + \mathrm{E}^y dy \wedge dt + \mathrm{E}^z dz \wedge dt \\ &+\mathrm{B}^z dx \wedge dy + \mathrm{B}^x dy \wedge dz + \mathrm{B}^y dz \wedge dx\big)(q).\end{aligned}$$

Proof. The 2-form field $\mathbf{F}(q)$ can be decomposed as

$$\begin{aligned}\mathbf{F}(q) = \big(&\mathrm{F}_{xy}dx \wedge dy+\mathrm{F}_{xz}dx \wedge dz+\mathrm{F}_{xt}dx \wedge dt \\ &+\mathrm{F}_{yz}dy \wedge dz +\mathrm{F}_{yt}dy \wedge dt +\mathrm{F}_{zt}dz \wedge dt\big)(q).\end{aligned}$$

First, we note that $v^\perp = \operatorname{span}\{\partial_{x,q}, \partial_{y,q}, \partial_{z,q}\}$, and the component matrix for the inner product $\mathbf{g}(q)|_{v^\perp}$ with respect to the ordered basis $(\partial_{x,q}, \partial_{y,q}, \partial_{z,q})$ for $v^\perp$ is the identity matrix. Thus, $\mathrm{B}^\flat = (\mathrm{B}^x dx + \mathrm{B}^y dy + \mathrm{B}^z dz)_q$. Using Proposition 11.3 (p.255), it follows that

$$\begin{aligned}\star(\mathrm{B}^\flat) &= (\mathrm{B}^x dy \wedge dz - \mathrm{B}^y dx \wedge dz + \mathrm{B}^z dx \wedge dy)(q) \\ &= (\mathrm{B}^x dy \wedge dz + \mathrm{B}^y dz \wedge dx + \mathrm{B}^z dx \wedge dy)(q).\end{aligned}$$

Since $\mathbf{F}(q)|_{v^\perp \times v^\perp} = \star(\mathrm{B}^\flat)$, by operating both sides on $(\partial_{x,q}, \partial_{y,q})$, and in light of the decompositions above, we obtain that $\mathrm{F}_{xy} = \mathrm{B}^z$. Similarly, operating on $(\partial_{y,q}, \partial_{z,q})$, and on $(\partial_{z,q}, \partial_{x,q})$ gives $\mathrm{F}_{yz} = \mathrm{B}^x$, respectively $-\mathrm{F}_{xz} = \mathrm{B}^y$. Next to find the $\mathrm{F}_{\cdot t}$ components, we will use $\mathbf{F}(q)(w,v) = \mathbf{g}(q)(w,\mathrm{E})$ for all $w \in v^\perp$. Taking $w = \partial_{x,q}$, we obtain $\mathrm{F}_{xt} = \mathrm{E}^x$. Similarly, setting w as $\partial_{y,q}$, and as $\partial_{z,q}$, gives $\mathrm{F}_{yt} = \mathrm{E}^y$, respectively $\mathrm{F}_{zt} = \mathrm{E}^z$. Substituting these components of $\mathbf{F}(q)$ in terms of the components of E and B, we obtain

$$\begin{aligned}\mathbf{F}(q) = (&\mathrm{B}^z dx \wedge dy - \mathrm{B}^y dx \wedge dz + \mathrm{E}^x dx \wedge dt \\ &+ \mathrm{B}^x dy \wedge dz + \mathrm{E}^y dy \wedge dt + \mathrm{E}^z dz \wedge dt)(q).\end{aligned}$$

□

Definition 12.15. (Instantaneous Lorentz force.)
Let $(M, \mathbf{g})$ be a spacetime, $p \in M$ and $\mathbf{F} \in \mathbf{\Omega}^2 M$ be the Faraday tensor field. Let (γ, m, e) be a particle passing through p, with $\mathbf{g}(p)(w,w) = -1$, where $w = v_{\gamma,p}$. The *instantaneous Lorentz force* experienced by the particle at p is the 1-form $e\mathbf{F}(p)(\cdot, w) \in (T_pM)^*$.

Proposition 12.5. *Let $(M, \mathbf{g})$ be a spacetime, $\mathbf{F} \in \mathbf{\Omega}^2 M$ be the Faraday tensor field, and $p \in M$. Let $v \in T_pM$ be a normalised instantaneous observer, perceiving the electric field* E *and magnetic field* B. *Let (γ, m, e) be a particle passing through p, with $\mathbf{g}(p)(w,w) = -1$, where $w = v_{\gamma,p}$. If the relative velocity of w with respect to v is u, then the instantaneous Lorentz force experienced by the particle at p is*

$$e\mathbf{F}(p)(\cdot, w)|_{v^\perp} = \frac{e\mathbf{g}(p)(\cdot, \mathrm{E} + (\star(u^\flat \wedge \mathrm{B}^\flat))^\sharp)}{\sqrt{1 - \mathbf{g}(p)(u,u)}}.$$

Proof. For $x \in v^\perp$, we have

$$\begin{aligned}\mathbf{F}(p)(x,w) &= \mathbf{F}(p)\Big(x, \frac{u+v}{\sqrt{1-\mathbf{g}(p)(u,u)}}\Big) \\ &= \frac{\mathbf{F}(p)(x,v) + \mathbf{F}(p)(x,u)}{\sqrt{1-\mathbf{g}(p)(u,u)}} \\ &= \frac{\mathbf{g}(p)(x,\mathrm{E}) + (\star(\mathrm{B}^\flat))(x,u)}{\sqrt{1-\mathbf{g}(p)(u,u)}}.\end{aligned}$$

By Lemma 11.3 (p.251),

$$\begin{aligned}(\star(\mathrm{B}^\flat))(x,u) &= \star(\mathrm{B}^\flat \wedge x^\flat \wedge u^\flat) \\ &= \star(u^\flat \wedge \mathrm{B}^\flat \wedge x^\flat) \\ &= (\star(u^\flat \wedge \mathrm{B}^\flat))(x) \\ &= \mathbf{g}(p)(x, (\star(u^\flat \wedge \mathrm{B}^\flat))^\sharp).\end{aligned}$$

Thus $e\mathbf{F}(p)(x,w) = \dfrac{e\mathbf{g}(p)(x, \mathrm{E} + (\star(u^\flat \wedge \mathrm{B}^\flat))^\sharp)}{\sqrt{1-\mathbf{g}(p)(u,u)}}$ for all $x \in v^\perp$. □

Example 12.5. (Minkowski spacetime.) Consider the Minkowski spacetime $(M, \mathbf{g})$ with the standard orientation[11] induced by (p, B). Let $q \in M$. Let $v \in T_qM$ be the normalised instantaneous observer $v = \partial_{t,q}$. Suppose that (γ, m, e) is a particle passing through q, $w = v_{\gamma,q}$, and $\mathbf{g}(q)(w,w) = -1$. Let $u \in v^\perp$ be the relative velocity of w with respect to v. Since we have $v^\perp = \text{span}\{\partial_{x,q}, \partial_{y,q}, \partial_{z,q}\}$, we decompose $u = u^x\partial_{x,q} + u^y\partial_{y,q} + u^z\partial_{z,q}$ and also $\mathrm{B} = \mathrm{B}^x\partial_{x,q} + \mathrm{B}^y\partial_{y,q} + \mathrm{B}^z\partial_{z,q}$. Then $u^\flat = u^x(dx)_q + u^y(dy)_q + u^z(dz)_q$ and $\mathrm{B}^\flat = \mathrm{B}^x(dx)_q + \mathrm{B}^y(dy)_q + \mathrm{B}^z(dz)_q$. Hence $u^\flat \wedge \mathrm{B}^\flat$ is

$$((u^x\mathrm{B}^y - u^y\mathrm{B}^x)(dx \wedge dy) + (u^y\mathrm{B}^z - u^z\mathrm{B}^y)(dy \wedge dz) + (u^z\mathrm{B}^x - u^x\mathrm{B}^z)(dz \wedge dx))(q).$$

Thus

$$\star(u^\flat \wedge \mathrm{B}^\flat) = (u^x\mathrm{B}^y - u^y\mathrm{B}^x)(dz)_q + (u^y\mathrm{B}^z - u^z\mathrm{B}^y)(dx)_q + (u^z\mathrm{B}^x - u^x\mathrm{B}^z)(dy)_q,$$

$$(\star(u^\flat \wedge \mathrm{B}^\flat))^\sharp = (u^x\mathrm{B}^y - u^y\mathrm{B}^x)\partial_{z,q} + (u^y\mathrm{B}^z - u^z\mathrm{B}^y)\partial_{x,q} + (u^z\mathrm{B}^x - u^x\mathrm{B}^z)\partial_{y,q}.$$

We recognise the components above those obtained by taking the usual cross product $\vec{\mathbf{u}} \times \vec{\mathbf{B}}$ in $\mathbb{R}^3$ of $\vec{\mathbf{u}} := (u^x, u^y, u^z)$ and $\vec{\mathbf{B}} := (\mathrm{B}^x, \mathrm{B}^y, \mathrm{B}^z)$. Let the electric field E perceived by v be decomposed as $\mathrm{E} = \mathrm{E}^x\partial_{x,q} + \mathrm{E}^y\partial_{y,q} + \mathrm{E}^z\partial_{z,q}$. Define $\vec{\mathbf{E}} := (\mathrm{E}^x, \mathrm{E}^y, \mathrm{E}^z)$. With respect to the basis $\{(dx)_q, (dy)_q, (dz)_q\}$ for $(v^\perp)^*$, the components of the instantaneous Lorentz force on the particle are those of

$$\frac{e(\vec{\mathbf{E}} + \vec{\mathbf{u}} \times \vec{\mathbf{B}})}{\sqrt{1 - \vec{\mathbf{u}} \cdot \vec{\mathbf{u}}}} \in \mathbb{R}^3. \qquad \diamond$$

Definition 12.16. (Charge-current density.)
Let $(M, \mathbf{g})$ be a spacetime. A vector field $\mathbf{J} \in T_0^1M$ is said to be a *charge-current density* if it satisfies the *continuity equation* $d(\star(\mathbf{J}^\flat)) = 0$, where $\star : \mathbf{\Omega}^1M (= T_1^0M) \to \mathbf{\Omega}^3M$.

Example 12.6. (Minkowski spacetime.) Let $(M, \mathbf{g})$ be Minkowski spacetime with the standard orientation induced by (p, B). Let $\mathbf{J} \in T_0^1M$ be a charge-current density. Let us write the continuity equation in the global chart $(M, \mathbf{x}_{p,B})$. We decompose $\mathbf{J} = \mathrm{j}^x\partial_x + \mathrm{j}^y\partial_y + \mathrm{j}^z\partial_z + \varrho\,\partial_t$, where the functions $\mathrm{j}^x, \mathrm{j}^y, \mathrm{j}^z, \varrho \in C^\infty(M)$. Then $\mathbf{J}^\flat = \mathrm{j}^x dx + \mathrm{j}^y dy + \mathrm{j}^z dz - \varrho dt$, and so

$$\star\,(\mathbf{J}^\flat) = -\mathrm{j}^x dy \wedge dz \wedge dt + \mathrm{j}^y dx \wedge dz \wedge dt - \mathrm{j}^z dx \wedge dy \wedge dt + \varrho dx \wedge dy \wedge dz.$$

Thus, the continuity equation becomes

$$0 = d(\star(\mathbf{J}^\flat)) = (-\partial_x\mathrm{j}^x - \partial_y\mathrm{j}^y - \partial_z\mathrm{j}^z - \partial_t\varrho)\, dx \wedge dy \wedge dz \wedge dt.$$

Set $\vec{\mathbf{j}} = (\mathrm{j}^x, \mathrm{j}^y, \mathrm{j}^z) \circ (\mathbf{x}_{p,B})^{-1} : \mathbb{R} \times \mathbb{R}^3 \to \mathbb{R}^3$ and $\rho = \varrho \circ (\mathbf{x}_{p,B})^{-1} : \mathbb{R} \times \mathbb{R}^3 \to \mathbb{R}$. Then we can rewrite the above continuity equation in $\mathbb{R} \times \mathbb{R}^3$ as

$$\vec{\nabla} \cdot \vec{\mathbf{j}} + \frac{\partial \rho}{\partial t} = 0. \qquad (\star) \qquad \diamond$$

---

[11] Example 11.5 (p.256).

One starts with a given $\mathbf{J}$, from which the Faraday tensor field is determined $\mathbf{F}$ by solving the Maxwell equations (assuming that the effect of $\mathbf{F}$ on the spacetime $(M, \mathbf{g})$ is negligible).

Definition 12.17. (Maxwell's equations.)
Let $(M, \mathbf{g})$ be a spacetime. Suppose that $\mathbf{J} \in T_0^1 M$ is a charge-current density. A $(0,2)$-form field $\mathbf{F} \in \mathbf{\Omega}^2 M$ is said to satisfy the[12] *Maxwell equations corresponding to* $\mathbf{J}$ if

- $d\mathbf{F} = 0$ (closed), and
- $d\star\mathbf{F} = \star(\mathbf{J}^\flat)$.

$\mathbf{F}$ is then called the *Faraday tensor field.*

The continuity equation is necessary for the second condition since $d^2 = 0$.

Example 12.7. (Minkowski space.) We continue with Example 12.6, and write down the Maxwell equations for $\mathbf{F}$ in terms of its components. In Proposition 12.4, the instantaneous observer $v = \partial_{t,q}$ measured, from $\mathbf{F}(q)$, the electric field $\mathrm{E}(q)$ and the magnetic field $\mathrm{B}(q)$ at q. If we vary q, and then the measurements by instantaneous observers $\partial_{t,q}$ give the maps $q \mapsto \mathrm{E}(q), \mathrm{B}(q)$. By Proposition 12.4 applied to each point $q \in M$, we have $\mathbf{F} = \mathrm{E}^x dx \wedge dt + \mathrm{E}^y dy \wedge dt + \mathrm{E}^z dz \wedge dt + \mathrm{B}^z dx \wedge dy + \mathrm{B}^x dy \wedge dz + \mathrm{B}^y dz \wedge dx$, where $\mathrm{E} = \mathrm{E}^x \partial_x + \mathrm{E}^y \partial_y + \mathrm{E}^z \partial_z$ and $\mathrm{B} = \mathrm{B}^x \partial_x + \mathrm{B}^y \partial_y + \mathrm{B}^z \partial_z$ for smooth functions $\mathrm{E}^x, \mathrm{E}^y, \mathrm{E}^z, \mathrm{B}^x, \mathrm{B}^y, \mathrm{B}^z \in C^\infty(M)$. Then

$$\star\mathbf{F} = \mathrm{E}^x dy \wedge dz - \mathrm{E}^y dx \wedge dz + \mathrm{E}^z dx \wedge dy - \mathrm{B}^z dz \wedge dt - \mathrm{B}^x dx \wedge dt - \mathrm{B}^y dy \wedge dt,$$

and so $d(\star\mathbf{F})$ is given by

$$\begin{aligned} &\partial_x \mathrm{E}^x dx \wedge dy \wedge dz + \partial_t \mathrm{E}^x dy \wedge dz \wedge dt + \partial_y \mathrm{E}^y dx \wedge dy \wedge dz - \partial_t \mathrm{E}^y dx \wedge dz \wedge dt \\ &+ \partial_z \mathrm{E}^z dx \wedge dy \wedge dz + \partial_t \mathrm{E}^z dx \wedge dy \wedge dt - \partial_x \mathrm{B}^z dx \wedge dz \wedge dt - \partial_y \mathrm{B}^z dy \wedge dz \wedge dt \\ &+ \partial_z \mathrm{B}^x dx \wedge dz \wedge dt + \partial_y \mathrm{B}^x dx \wedge dy \wedge dt + \partial_z \mathrm{B}^y dy \wedge dz \wedge dt - \partial_x \mathrm{B}^y dx \wedge dy \wedge dt. \end{aligned}$$

Equating this expression for $d\star\mathbf{F}$ with that for $\star(\mathbf{J}^\flat)$ found in Example 12.6,

$$\left.\begin{aligned} \partial_y \mathrm{B}^z - \partial_z \mathrm{B}^y - \partial_t \mathrm{E}^x &= \mathrm{j}^x \\ \partial_z \mathrm{B}^x - \partial_x \mathrm{B}^z - \partial_t \mathrm{E}^y &= \mathrm{j}^y \\ \partial_x \mathrm{B}^y - \partial_y \mathrm{B}^x - \partial_t \mathrm{E}^z &= \mathrm{j}^z \end{aligned}\right\}, \quad \text{and} \quad \partial_x \mathrm{E}^x + \partial_y \mathrm{E}^y + \partial_z \mathrm{E}^z = \varrho. \tag{12.7}$$

Set $\vec{\mathbf{E}} = (\mathrm{E}^x, \mathrm{E}^y, \mathrm{E}^z) \circ (\mathbf{x}_{p,B})^{-1}$, and $\vec{\mathbf{B}} = (\mathrm{B}^x, \mathrm{B}^y, \mathrm{B}^z) \circ (\mathbf{x}_{p,B})^{-1}$. Then we can rewrite (12.7) as

$$\vec{\nabla} \times \vec{\mathbf{B}} = \vec{\mathbf{j}} + \frac{\partial \vec{\mathbf{E}}}{\partial t}, \quad \text{and} \quad \vec{\nabla} \cdot \vec{\mathbf{E}} = \rho. \tag{12.8}$$

[12]We adopt 'geometrised units', meaning that the usual constants found in the classical version of the Maxwell PDEs have been absorbed in the charge-current density.

Next we unravel $d\mathbf{F} = 0$ using chart-induced components. Firstly, $d\mathbf{F}$ is

$$\begin{aligned}&-\partial_y \mathrm{E}^x dx \wedge dy \wedge dt - \partial_z \mathrm{E}^x dx \wedge dz \wedge dt + \partial_x \mathrm{E}^y dx \wedge dy \wedge dt - \partial_z \mathrm{E}^y dy \wedge dz \wedge dt\\ &+ \partial_x \mathrm{E}^z dx \wedge dz \wedge dt + \partial_y \mathrm{E}^z dy \wedge dz \wedge dt + \partial_z \mathrm{B}^z dx \wedge dy \wedge dz + \partial_t \mathrm{B}^z dx \wedge dy \wedge dt\\ &+ \partial_x \mathrm{B}^x dx \wedge dy \wedge dz + \partial_t \mathrm{B}^x dy \wedge dz \wedge dt + \partial_y \mathrm{B}^y dx \wedge dy \wedge dz - \partial_t \mathrm{B}^y dx \wedge dz \wedge dt.\end{aligned}$$

Thus $d\mathbf{F} = 0$ is equivalent to

$$\left.\begin{aligned}\partial_y \mathrm{E}^z - \partial_z \mathrm{E}^y + \partial_t \mathrm{B}^x &= 0\\ \partial_z \mathrm{E}^x - \partial_x \mathrm{E}^z + \partial_t \mathrm{B}^y &= 0\\ \partial_x \mathrm{E}^y - \partial_y \mathrm{E}^x + \partial_t \mathrm{B}^z &= 0\end{aligned}\right\}, \quad \text{and} \quad \partial_x \mathrm{B}^x + \partial_y \mathrm{B}^y + \partial_z \mathrm{B}^z = 0, \tag{12.9}$$

which can be rewritten as

$$\vec{\nabla} \times \vec{\mathbf{E}} + \frac{\partial \vec{\mathbf{B}}}{\partial t} = \mathbf{0} \quad \text{and} \quad \vec{\nabla} \cdot \vec{\mathbf{B}} = 0. \tag{12.10}$$

We recognise (12.8) and (12.10) as the usual Maxwell equations[13]. ◇

Exercise 12.14. Suppose that $\vec{\mathbf{E}}, \vec{\mathbf{B}}, \vec{\mathbf{j}} : \mathbb{R} \times \mathbb{R}^3 \to \mathbb{R}^3$ and $\rho : \mathbb{R} \times \mathbb{R}^3 \to \mathbb{R}$ satisfy the Maxwell equations (12.8), (12.10), and (hence necessarily) the continuity equation ($\star$) from Example 12.6 (p.284). Prove that the following wave equations are satisfied:

$$\frac{\partial^2 \vec{\mathbf{E}}}{\partial t^2} - \Delta \vec{\mathbf{E}} = -\frac{\partial \vec{\mathbf{j}}}{\partial t} - \vec{\nabla} \rho \quad \text{and} \quad \frac{\partial^2 \vec{\mathbf{B}}}{\partial t^2} - \Delta \vec{\mathbf{B}} = \vec{\nabla} \times \vec{\mathbf{j}}.$$

Here the Laplacian[14] $\Delta := \frac{\partial^2}{\partial x^2} + \frac{\partial^2}{\partial y^2} + \frac{\partial^2}{\partial z^2}$. If $\vec{\mathbf{j}} \equiv \mathbf{0}$ and $\rho \equiv 0$, then show:

$$\frac{\partial^2 \vec{\mathbf{E}}}{\partial t^2} - \Delta \vec{\mathbf{E}} = \mathbf{0} \quad \text{and} \quad \frac{\partial^2 \vec{\mathbf{B}}}{\partial t^2} - \Delta \vec{\mathbf{B}} = \mathbf{0}.$$

Show that $\vec{\mathbf{E}}(t,x,y,z) = (\sin(z-t), 0, 0)$ and $\vec{\mathbf{B}}(t,x,y,z) = (0, \sin(z-t), 0)$ satisfy the Maxwell equations when $\vec{\mathbf{j}} \equiv \mathbf{0}$ and $\rho \equiv 0$.

Remark 12.1. (Polarisation.)
A transverse wave (as opposed to a longitudinal wave) has the direction of propagation orthogonal to the oscillations. Polarisation refers to the orientation of the oscillations of a transverse wave in the plane orthogonal to the direction of propagation. Besides the solution considered above, more generally, a *monochromatic plane wave* solution of Maxwell's equation, propagating in the z-direction has an electric field $\vec{\mathbf{E}} = A\,\mathrm{Re}((\alpha \mathbf{e}_x + \beta \mathbf{e}_y) e^{i(kz - \omega t)})$, where $A, \omega, k > 0$, $\mathbf{e}_x = (1,0,0)$, $\mathbf{e}_y = (0,1,0)$, and α, β are complex numbers such that $|\alpha|^2 + |\beta|^2 = 1$. Here we take componentwise real parts.

[13]The Maxwell equations were known since the 1860s, but not in the vectorial form that we learn nowadays. The latter were given by Oliver Heaviside in 1884, reducing the more than 20 equations to the compactly written 4 equations familiar to us. The Maxwell equations played an important role in the genesis of the spacetime viewpoint. Encoded in them was the refutal of the earlier-fundamental '*instantaneous* action-at-a-distance viewpoint' of the operation of forces. Indeed, the Maxwell equations show that the electric field and the magnetic field behave like 'waves' (see Exercise 12.14), showing that there is a *finite* propagation speed, which is the speed of light!

[14]$\Delta \vec{\mathbf{E}}$ means Δ acting component-wise on $\vec{\mathbf{E}}$ (and a similar meaning for $\Delta \vec{\mathbf{B}}$).

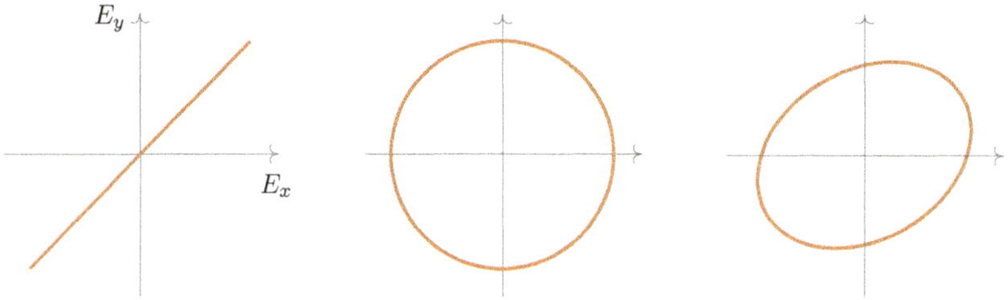

If α/β (or if undefined, β/α) is real, then the wave is *linearly polarised.* For example, if $\alpha = \beta = 1/\sqrt{2}$, and then writing $\overrightarrow{\mathbf{E}}(t,x,y,z) = (E_x(t), E_y(t), 0)$, the curve $t \mapsto (E_x(t), E_y(t))$ traces a line segment in the E_x, E_y plane, as shown in the picture on the left above. If $\alpha = \pm i\beta = \pm i(1/\sqrt{2})$, then we say the wave is *circularly polarised*, and in this case, $t \mapsto (E_x(t), E_y(t))$ is a circle, as shown in the middle picture above. In all other cases, we say the wave is *elliptically polarised*, see for example the picture on the right above. $\ast$

Exercise 12.15. (Reissner-Nordström spacetime.)
Akin to Schwarzschild spacetime, the Reissner-Nordström spacetime is the smooth manifold $M = \mathbb{R} \times (r_+, \infty) \times S^2$, where $r_+ \geqslant 0$. Let (U, φ) be the chart for S^2 from Example 1.8 (p.7). Let $e \geqslant 0$ and $m > 0$. In the chart $\mathcal{U} := \mathbb{R} \times (r_+, \infty) \times U$ with the chart map $\mathrm{id}_{\mathbb{R}} \times \mathrm{id}_{(r_+,\infty)} \times \varphi$, the metric $\mathbf{g}$ on M is given by

$$\mathbf{g} = -f(r)\, dt \otimes dt + \frac{1}{f(r)} dr \otimes dr + r^2 d\theta \otimes d\theta + r^2 (\sin\theta)^2 d\phi \otimes d\phi,$$

where $f(r) = 1 - \dfrac{2m}{r} + \dfrac{e^2}{r^2}$. Also, $r_+ := \begin{cases} 0 & \text{if } e^2 > m^2, \\ m + \sqrt{m^2 - e^2} & \text{if } e^2 \leqslant m^2. \end{cases}$

We use spacetime structures on M analogous to the Schwarzschild spacetime: the time-orientation is ∂_t and the orientation is described in Example 11.4 (p.249), with r_+ replacing $2m$. Then $(M, \mathbf{g})$ models the spacetime region outside a charged body of mass m and electric charge e. If $e = 0$, then $(M, \mathbf{g})$ is just the Schwarzschild spacetime. Consider the vector fields $V_i \in T_0^1 M$, given in $\mathcal{U}$ by

$$V_1 = \sqrt{f(r)}\, \partial_r, \quad V_2 = \frac{1}{r} \partial_\theta, \quad V_3 = \frac{1}{r \sin\theta} \partial_\phi, \quad V_4 = \frac{1}{\sqrt{f(r)}} \partial_t$$

and the 'dual' 1-form fields $\Omega^i \in T_1^0 M$, given in $\mathcal{U}$ by

$$\Omega^1 = \frac{1}{\sqrt{f(r)}} dr, \quad \Omega^2 = r\, d\theta, \quad \Omega^3 = r(\sin\theta)\, d\phi, \quad \Omega^4 = \sqrt{f(r)}\, dt.$$

Show that at each $p \in \mathcal{U}$, $((V_1)_p, (V_2)_p, (V_3)_p, (V_4)_p)$ is a positively oriented orthonormal basis for $T_p M$ with the induced orientation. Define $\mathbf{F} \in T_2^0 M$ by

$$\mathbf{F} = \frac{e}{r^2} \Omega^1 \wedge \Omega^4,$$

and let $\mathbf{J} := 0 \in T_0^1 M$. Prove that $\mathbf{F}$ satisfies the Maxwell equations.

Chapter 13

Matter

As we had mentioned in the preface, the field equation is

$$\text{(geometry)}\quad \mathbf{Ric} - \frac{1}{2}\mathbf{S}\mathbf{g} + \Lambda\mathbf{g} = 8\pi\mathbf{T}. \quad \text{(matter)}$$

We now focus on the right hand-side of this equation, and also provide some motivation for this equation. (The real belief in the equation as the governing equation for spacetime rests on its ability to predict physical phenomena, and we have already met two of these tests, namely the red-shift, and the deflection of light. We will also meet a third test, on the perihelion shift of the orbit of Mercury, where the observed discrepancy with the Newtonian prediction[1] was correctly accounted for within the spacetime geometric view of gravitation.) We begin by revisiting Newtonian gravity, and recalling the Poisson equation that shows how the matter sources determine the 'gravitational potential'. This serves as a precursor to the field equation above.

13.1 Newtonian gravity and the Poisson equation

We suppose that the 4-dimensional classical spacetime is described[2] by $\mathbb{R}\times\mathbb{R}^3$. The Newtonian gravitation law says that for point masses m, M located at $\mathbf{x}, \mathbf{y}$, respectively in $\mathbb{R}^3$, the instantaneous force of attraction that m feels due to the presence of M is[3]

$$\mathbf{F}(t,\mathbf{x}) = \frac{mM}{\|\mathbf{y}-\mathbf{x}\|^2}\frac{\mathbf{y}-\mathbf{x}}{\|\mathbf{y}-\mathbf{x}\|}.$$

[1]The Newtonian prediction was about 532 arcseconds of degree per century (due to perturbations by the other planets), whereas the observed value was 575 arcseconds per century, so that there was a discrepancy of 43 arcseconds per century. See §14.4.

[2]After making the choice of a point $p \in M$, a vector $\mathbf{e}_0 \in V$ such that $\mathbf{t}(\mathbf{e}_0) = 1$, and $(\mathbf{e}_x, \mathbf{e}_y, \mathbf{e}_z)$ forming an orthonormal basis for $\ker \mathbf{t}$ with respect to the inner product η.

[3]In units where the universal gravitational constant is set to 1.

Let $\rho : \mathbb{R}\times\mathbb{R}^3 \to \mathbb{R}$ describe the mass density of matter, which we assume, for the convenience of analysis below, to be a compactly supported smooth function. The instantaneous force on m is given by the improper integral

$$\mathbf{F}(t,\mathbf{x}) = m\int_{\mathbb{R}^3} \frac{\rho(t,\mathbf{y})}{\|\mathbf{y}-\mathbf{x}\|^2}\,\frac{\mathbf{y}-\mathbf{x}}{\|\mathbf{y}-\mathbf{x}\|}dV(\mathbf{y}).$$

Here $dV(\mathbf{y})=dxdydz$ is the usual volume element in $\mathbb{R}^3$. We note that the map $\mathbf{y}\mapsto \|\mathbf{y}-\mathbf{x}\|^{-2}$ is locally integrable on $\mathbb{R}^3$. If we think of m as a 'test mass', then the force on m is $m\boldsymbol{g}(t,\mathbf{x})$, where $\boldsymbol{g}$ is the *gravitational field*:

$$\boldsymbol{g}(t,\mathbf{x}) = \int_{\mathbb{R}^3} \frac{\rho(t,\mathbf{y})}{\|\mathbf{y}-\mathbf{x}\|^2}\,\frac{\mathbf{y}-\mathbf{x}}{\|\mathbf{y}-\mathbf{x}\|}dV(\mathbf{y}).$$

Let $\vec{\nabla}_{\mathbf{x}}$ denote the gradient operator in $\mathbb{R}^3$ acting on functions of the variable $\mathbf{x}\in\mathbb{R}^3$. With $\mathbf{y}\in\mathbb{R}^3$ fixed, for $\mathbf{x}\neq\mathbf{y}$, we have $\vec{\nabla}_{\mathbf{x}}\dfrac{1}{\|\mathbf{y}-\mathbf{x}\|} = \dfrac{\mathbf{y}-\mathbf{x}}{\|\mathbf{y}-\mathbf{x}\|^3}$. Thus

$$\begin{aligned}\boldsymbol{g}(t,\mathbf{x}) &= \int_{\mathbb{R}^3} \frac{\rho(t,\mathbf{y})}{\|\mathbf{y}-\mathbf{x}\|^2}\,\frac{\mathbf{y}-\mathbf{x}}{\|\mathbf{y}-\mathbf{x}\|}dV(\mathbf{y}) = \int_{\mathbb{R}^3} \rho(t,\mathbf{y})\,\vec{\nabla}_{\mathbf{x}}\frac{1}{\|\mathbf{y}-\mathbf{x}\|}dV(\mathbf{y})\\ &= \vec{\nabla}_{\mathbf{x}}\int_{\mathbb{R}^3}\frac{\rho(t,\mathbf{y})}{\|\mathbf{y}-\mathbf{x}\|}dV(\mathbf{y}) = -\vec{\nabla}_{\mathbf{x}}\Phi(t,\mathbf{x}),\end{aligned}$$

where Φ is the *gravitational potential*, defined by the improper integral

$$\Phi(t,\mathbf{x}) = -\int_{\mathbb{R}^3}\frac{\rho(t,\mathbf{y})}{\|\mathbf{y}-\mathbf{x}\|}dV(\mathbf{y}). \qquad (\star)$$

We claim that Φ satisfies the *Poisson equation*

$$\Delta\Phi(t,\mathbf{x}) = 4\pi\rho(t,\mathbf{x}),$$

where $\Delta=\Delta_{\mathbf{x}}$ is the Laplacian. Fix $\epsilon>0$, and let $B_\epsilon(\mathbf{0})$ be the ball with center $\mathbf{0}\in\mathbb{R}^3$ and radius ϵ, with the sphere $\partial B_\epsilon(\mathbf{0})$ as its boundary. First, using the substitution $\mathbf{y}'=\mathbf{y}-\mathbf{x}$,

$$\Phi(t,\mathbf{x}) = -\int_{\mathbb{R}^3}\frac{\rho(t,\mathbf{y})}{\|\mathbf{y}-\mathbf{x}\|}dV(\mathbf{y}) = -\int_{\mathbb{R}^3}\frac{\rho(t,\mathbf{y}+\mathbf{x})}{\|\mathbf{y}\|}dV(\mathbf{y}).$$

Thus

$$\begin{aligned}&\Delta_{\mathbf{x}}\Phi(t,\mathbf{x})\\ &= \underbrace{-\int_{B_\epsilon(\mathbf{0})}(\Delta_{\mathbf{x}}\rho(t,\mathbf{y}+\mathbf{x}))\frac{1}{\|\mathbf{y}\|}dV(\mathbf{y})}_{=:I_\epsilon} \underbrace{-\int_{\mathbb{R}^3\setminus B_\epsilon(\mathbf{0})}(\Delta_{\mathbf{x}}\rho(t,\mathbf{y}+\mathbf{x}))\frac{1}{\|\mathbf{y}\|}dV(\mathbf{y})}_{=:J_\epsilon}.\end{aligned} \qquad (13.1)$$

Since ρ is compactly supported and smooth, we have[4]

$$|I_\epsilon| \leqslant \|\Delta\rho(t,\cdot)\|_{L^\infty(\mathbb{R}^3)}\frac{\epsilon^2}{2}4\pi. \qquad (13.2)$$

[4]For a compactly supported smooth function $\mathbf{f}:\mathbb{R}^3\to\mathbb{R}^m$, $\|\mathbf{f}\|_{L^\infty(\mathbb{R}^3)} := \max\limits_{\mathbf{x}\in\mathbb{R}^3}\|\mathbf{f}(\mathbf{x})\|$.

For smooth f, g, $\vec{\nabla}\cdot(g\vec{\nabla}f)=(\vec{\nabla}g)\cdot\vec{\nabla}f+g\Delta f$. Using the Gauss divergence theorem[5], for smooth f, g, either of which is compactly supported,

$$\begin{aligned}\int_{\mathbb{R}^3\setminus B_\epsilon(\mathbf{0})} g\Delta f\, dV(\mathbf{y}) &= \int_{\mathbb{R}^3\setminus B_\epsilon(\mathbf{0})} -(\vec{\nabla}g)\cdot(\vec{\nabla}f)\,dV(\mathbf{y}) + \int_{\mathbb{R}^3\setminus B_\epsilon(\mathbf{0})} \vec{\nabla}\cdot(g\vec{\nabla}f)\,dV(\mathbf{y})\\ &= -\int_{\mathbb{R}^3\setminus B_\epsilon(\mathbf{0})} (\vec{\nabla}g)\cdot(\vec{\nabla}f)\,dV(\mathbf{y}) + \int_{\partial B_\epsilon(\mathbf{0})} (g\vec{\nabla}f)\cdot\widehat{\mathbf{n}}_{\mathbf{y}}\,dS(\mathbf{y}),\end{aligned}$$

where $\widehat{\mathbf{n}}_{\mathbf{y}}$ is the radially inward unit normal on the sphere $\partial B_\epsilon(\mathbf{0})$ at the point $\mathbf{y}$, and $dS(\mathbf{y})$ is the area element on the sphere. With $f(\mathbf{y})=\rho(\mathbf{y}+\mathbf{x})$ and $g(\mathbf{y})=\|\mathbf{y}\|^{-1}$, we get that J_ϵ equals

$$\begin{aligned}&-\int_{\mathbb{R}^3\setminus B_\epsilon(\mathbf{0})} (\Delta_{\mathbf{x}}\rho(t,\mathbf{y}+\mathbf{x}))\frac{1}{\|\mathbf{y}\|}dV(\mathbf{y}) = -\int_{\mathbb{R}^3\setminus B_\epsilon(\mathbf{0})} (\Delta_{\mathbf{y}}\rho(t,\mathbf{y}+\mathbf{x}))\frac{1}{\|\mathbf{y}\|}dV(\mathbf{y})\\ &= \int_{\mathbb{R}^3\setminus B_\epsilon(\mathbf{0})} (\vec{\nabla}_{\mathbf{y}}\rho(t,\mathbf{y}+\mathbf{x}))\cdot\vec{\nabla}_{\mathbf{y}}\frac{1}{\|\mathbf{y}\|}dV(\mathbf{y}) - \int_{\partial B_\epsilon(\mathbf{0})} \frac{1}{\|\mathbf{y}\|}(\vec{\nabla}_{\mathbf{y}}\rho(t,\mathbf{y}+\mathbf{x}))\cdot\widehat{\mathbf{n}}_{\mathbf{y}}\,dS(\mathbf{y}).\end{aligned}$$

Call these two last summand integrals as K_ϵ and L_ϵ, respectively. Firstly,

$$|L_\epsilon| \leqslant \|\vec{\nabla}\rho(t,\cdot)\|_{L^\infty(\mathbb{R}^3)}\int_{\partial B_\epsilon(\mathbf{0})}\frac{1}{\|\mathbf{y}\|}dS(\mathbf{y}) = \|\vec{\nabla}\rho(t,\cdot)\|_{L^\infty(\mathbb{R}^3)}4\pi\epsilon. \qquad (13.3)$$

For finding the limiting behaviour of K_ϵ as $\epsilon\to 0$, we integrate by parts:

$$\begin{aligned}K_\epsilon &= \int_{\mathbb{R}^3\setminus B_\epsilon(\mathbf{0})} (\vec{\nabla}_{\mathbf{y}}\rho(t,\mathbf{y}+\mathbf{x}))\cdot\vec{\nabla}_{\mathbf{y}}\frac{1}{\|\mathbf{y}\|}dV(\mathbf{y})\\ &= -\int_{\mathbb{R}^3\setminus B_\epsilon(\mathbf{0})} \rho(t,\mathbf{y}+\mathbf{x})\,\Delta_{\mathbf{y}}\frac{1}{\|\mathbf{y}\|}dV(\mathbf{y}) + \int_{\mathbb{R}^3\setminus B_\epsilon(\mathbf{0})} \vec{\nabla}_{\mathbf{y}}\cdot\Big(\rho(t,\mathbf{y}+\mathbf{x})\vec{\nabla}_{\mathbf{y}}\frac{1}{\|\mathbf{y}\|}\Big)dV(\mathbf{y})\\ &= -\int_{\mathbb{R}^3\setminus B_\epsilon(\mathbf{0})} \rho(t,\mathbf{y}+\mathbf{x})\Delta_{\mathbf{y}}\frac{1}{\|\mathbf{y}\|}dV(\mathbf{y}) + \int_{\partial B_\epsilon(\mathbf{0})} \rho(t,\mathbf{y}+\mathbf{x})\Big(\vec{\nabla}_{\mathbf{y}}\frac{1}{\|\mathbf{y}\|}\Big)\cdot\widehat{\mathbf{n}}_{\mathbf{y}}\,dS(\mathbf{y})\\ &= -0 + \int_{\partial B_\epsilon(\mathbf{0})} \rho(t,\mathbf{y}+\mathbf{x})\Big(\vec{\nabla}_{\mathbf{y}}\frac{1}{\|\mathbf{y}\|}\Big)\cdot\widehat{\mathbf{n}}_{\mathbf{y}}\,dS(\mathbf{y}).\end{aligned}$$

We used $\Delta_{\mathbf{y}}\frac{1}{\|\mathbf{y}\|}\equiv 0$ in $\mathbb{R}^3\setminus\{\mathbf{0}\}$ to get the last equality. As $\vec{\nabla}_{\mathbf{y}}\frac{1}{\|\mathbf{y}\|}=\frac{-\mathbf{y}}{\|\mathbf{y}\|^3}$,

$$\begin{aligned}K_\epsilon &= \int_{\partial B_\epsilon(\mathbf{0})} \rho(t,\mathbf{y}+\mathbf{x})\Big(\frac{-\mathbf{y}}{\|\mathbf{y}\|^3}\Big)\cdot\Big(\frac{-\mathbf{y}}{\|\mathbf{y}\|}\Big)dS(\mathbf{y})\\ &= \int_{\partial B_\epsilon(\mathbf{0})} \rho(t,\mathbf{y}+\mathbf{x})\frac{1}{\|\mathbf{y}\|^2}dS(\mathbf{y}) \stackrel{\epsilon\to 0}{\longrightarrow} 4\pi\rho(t,\mathbf{x}). \qquad (13.4)\end{aligned}$$

Taking the limit $\epsilon\to 0$ in (13.1), and using (13.2), (13.3), (13.4), we have

$$\Delta_{\mathbf{x}}\Phi(t,\mathbf{x}) = \lim_{\epsilon\to 0}(I_\epsilon+J_\epsilon) = \lim_{\epsilon\to 0}(I_\epsilon+K_\epsilon+L_\epsilon) = 0+4\pi\rho(t,\mathbf{x})+0 = 4\pi\rho(t,\mathbf{x}).$$

Exercise 13.1. Determine the gravitational potential at a distance $r_0 > R$ from the centre of a homogeneous ball of mass m and radius $R > 0$.

[5] See e.g. [Apostol (1969), Theorem 12.6, §12.19].

13.2 Classical dynamics from a geometric viewpoint

In the classical viewpoint, a particle of mass m at a point with coordinates $(t, \mathbf{x})$ in spacetime, in which there is a pervading gravitational potential Φ, will experience the gravitational field of strength $-\vec{\nabla}\Phi(t, \mathbf{x})$, and coupling to it with its ('gravitational') mass m, feel the force

$$m(-\vec{\nabla}\Phi(t, \mathbf{x})).$$

We use coordinates from the previous section. Parametrise the particle trajectory $\gamma : \mathbb{R} \to \mathbb{R} \times \mathbb{R}^3$ using coordinate time, i.e.,

$$\mathbb{R} \ni t \mapsto (t, \mathbf{x}(t)) \in \mathbb{R} \times \mathbb{R}^3.$$

The force, $-m\vec{\nabla}\Phi(t, \mathbf{x}(t))$, is equal to the ('inertial') mass m times the acceleration $\ddot{\mathbf{x}}(t)$, so that $\cancel{m}\ddot{\mathbf{x}}(t) = -\cancel{m}\vec{\nabla}\Phi(t, \mathbf{x}(t))$, that is,

$$\ddot{\mathbf{x}}(t) + \vec{\nabla}\Phi(t, \mathbf{x}(t)) = 0. \tag{13.5}$$

So all masses move along these curves, as famously demonstrated by Galileo's experiment in which two objects with different masses simultaneously hit the floor when let go from the same height.

An intriguing question then arises, namely, whether the above equation for the trajectory is in fact a geodesic equation in curved spacetime, where the curvature is created by the matter source ρ (determining the Φ which appears in the candidate for the geodesic equation). The answer to this question is 'yes', and it shows that classical spacetime is curved by mass sources within the framework of classical Newtonian gravity. This viewpoint also shows that Poisson's equation is really an equation relating geometry to matter, akin to the field equation, except that it is at odds with what we have learnt so far in physics: Namely that matter, as well as signals, ought to be described by future-pointing timelike or lightlike curves, and so there is a speed limit, the speed of light. But in the Poisson equation, there is no provision for a speed limit: if the source ρ changes suddenly, then it is clear from the definition of Φ that this effect is simultaneously felt everywhere, no matter how far the location we take. This 'flaw' in the geometric treatment of the Newtonian gravity will be rectified by the field equation. Below, we will proceed heuristically and formally, to identify (13.5) as part of the set of equations for a geodesic in the smooth manifold[6] $\mathbb{R} \times \mathbb{R}^3$, and then see what the Poisson equation tells us if we put on these 'geometric glasses'.

[6] Here we only use the smooth manifold structure of $\mathbb{R} \times \mathbb{R}^3$, and proceed formally; see [Misner, Thorne and Wheeler (2017), Chapter 12] for a more elaborate mathematical treatment.

With $\gamma^0(t) = t$, $\gamma^1(t) = x(t)$, $\gamma^2(t) = y(t)$, $\gamma^3(t) = z(t)$, we obtain $\dot{\gamma}^0(t) = 1$, and so

$$\begin{aligned}\ddot{\gamma}^0(t) &= 0\\ \ddot{\gamma}^1(t) + \frac{\partial \Phi}{\partial x}(\gamma(t))\dot{\gamma}^0(t)\dot{\gamma}^0(t) &= 0\\ \ddot{\gamma}^2(t) + \frac{\partial \Phi}{\partial y}(\gamma(t))\dot{\gamma}^0(t)\dot{\gamma}^0(t) &= 0\\ \ddot{\gamma}^3(t) + \frac{\partial \Phi}{\partial z}(\gamma(t))\dot{\gamma}^0(t)\dot{\gamma}^0(t) &= 0.\end{aligned}$$

A formal comparison with the equations obtained in Proposition 8.1 (p.152) suggests that the connection coefficients are all zero except the following, where $(t, x, y, z) = (x^0, x^1, x^2, x^3)$:

$$\Gamma^1_{00}(t, \mathbf{x}) = \frac{\partial \Phi}{\partial x}(t, \mathbf{x}), \quad \Gamma^2_{00}(t, \mathbf{x}) = \frac{\partial \Phi}{\partial y}(t, \mathbf{x}), \quad \Gamma^3_{00}(t, \mathbf{x}) = \frac{\partial \Phi}{\partial z}(t, \mathbf{x}).$$

Define a corresponding connection ∇ as in Exercise 6.3 (p.108) with these coefficient functions. Using Proposition 9.1 (p.182), it can be checked that for the curvature tensor field $\mathbf{R}$, we have

$$\mathbf{R}^1_{100} = \frac{\partial^2 \Phi}{\partial x^2}, \quad \mathbf{R}^2_{200} = \frac{\partial^2 \Phi}{\partial y^2}, \quad \mathbf{R}^3_{300} = \frac{\partial^2 \Phi}{\partial z^2}.$$

Also, $\mathbf{R}^0_{000} = 0$. Thus

$$\mathbf{Ric}_{00} = \mathbf{R}^0_{000} + \mathbf{R}^1_{100} + \mathbf{R}^2_{200} + \mathbf{R}^3_{300} = \Delta \Phi.$$

So the Poisson equation can be rewritten as

$$\mathbf{Ric}_{00} = 4\pi\rho.$$

As mentioned earlier, this equation is not appropriate in the Lorentzian manifold case, and an appropriate generalisation is sought of (to begin with) the right-hand side. Here only the mass appears, but we have already seen that the energy perceived by an instantaneous observer depends on the observer's state of motion, that is, his own normalised velocity. A wild guess is to replace the right-hand side by an appropriate $(0, 2)$-tensor field $\mathbf{T}$, so that we could maybe have the aesthetically pleasing equation

$$\mathbf{Ric} = 4\pi \mathbf{T}.$$

It turns out that this train of thought is in the right direction, but it does not work out exactly like this. We first investigate what an appropriate generalisation of the right-hand side is. We anticipate that it cannot be just built out of mass density (since we have seen that in Minkowski spacetime the fundamental notion for a particle at a point $p \in M$ is its momentum 1-form $\mathfrak{p}$, which is split into a perceived energy and a momentum by an instantaneous observer $v \in T_pM$), and so $\mathbf{T}$ should involve both energy and momentum.

13.3 Energy-momentum tensor field

We adopt a top-down approach, beginning with an abstract definition, and then seeing concrete physical examples.

Definition 13.1. (Energy-momentum tensor field.)
Let $(M, \mathbf{g})$ be a spacetime. An *energy-momentum tensor field* $\mathbf{T}$ is a $(0,2)$-tensor field $\mathbf{T} \in T_2^0 M$ that satisfies

- (symmetry) $\mathbf{T}(V, W) = \mathbf{T}(W, V)$ for all $V, W \in T_0^1 M$,
- (weak energy condition) $\mathbf{T}(V, V) \geqslant 0$ (pointwise) for all causal[7] $V \in T_0^1 M$,
- (divergence-free) $\operatorname{div} \mathbf{T} = 0 \in T_1^0 M$.

A trivial example of an energy-momentum tensor field is the $(0,2)$-tensor field that is identically zero. We refer to this as *vacuum*.

Remark 13.1. Where does $\mathbf{T}$ come from? In modern physics, the evolution equations for physical (matter) fields are the Euler-Lagrange equations for an 'action'. The action is an integral over spacetime of a Lagrangian. We have seen the Euler-Lagrange equation in a one-dimensional context in §8.7. A similar equation can also be derived in the multivariable setting. The variation of the Lagrangian with respect to the metric then yields the energy-momentum tensor field. The vanishing divergence condition is the energy-momentum conservation or the continuity equation, and in the Lagrangian viewpoint, it arises as a consequence of the invariance of the action under diffeomorphisms. See [Wald (1984), App. E] for these matters. $\ast$

Matter takes various complicated forms. For example, in the portion of spacetime where a galaxy evolves, one has

- a swarm of particles (stars constituting the galaxy),
- an inert dust cloud,
- even if matter were absent as massive stuff, there may be a field like the electromagnetic field, which contributes to the energy density.

The energy momentum tensor field $\mathbf{T}$ captures in one go[8] all of these.

Before looking at examples, we mention that if we imagine a continuous matter distribution in spacetime, and $v \in T_p M$ is a normalised instantaneous observer, then for this observer, $\mathbf{T}(p)(v, v)$ is meant to capture physically

[7] A vector field such that for each $p \in M$, V_p is causal, that is, either timelike or lightlike.

[8] In the Lagrangian viewpoint (Remark 13.1), the energy-momentum tensor field $\mathbf{T}$ is found from the overall matter-Lagrangian obtained by adding the Lagrangians of the various disparate matter sources.

the energy density (energy per unit 'spatial' volume in $v^\perp$). Similar physical interpretations[9] can be given for $\mathbf{T}(p)(u, w)$ for $u, w \in T_pM$.

Dust

We begin with a model of matter where we imagine non-colliding, non-interacting particles which are 'freely falling' (so that their worldlines are geodesics), and close-by particles have approximately parallel worldlines (so that we can imagine these worldlines as being the integral curves of a vector field). We give the precise definition below.

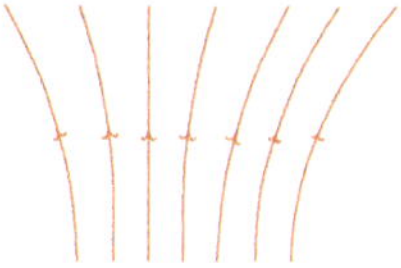

Definition 13.2. (Dust.)
Let $(M, \mathbf{g})$ be a spacetime. *Dust* is a triple (m, ν, V), where

- $m > 0$,
- $\nu : M \to \mathbb{R}$ is a smooth map, and $\nu(p) \geqslant 0$ for all $p \in M$,
- $V \in T_0^1 M$ satisfies $\nabla_V V = 0$, and for all $p \in M$, $\mathbf{g}(p)(V_p, V_p) = -1$ with V_p future-pointing,
- $\text{div}(\nu V) = 0$.

ν is called the *number density*, and V is called the *velocity field.*

To see what the number density ν describes, imagine an observer moving along an integral curve of V ('co-moving' observer), that passes through $p \in M$. Then the nearby particles appear stationary to the instantaneous observer V_p in the spatial section $(V_p)^\perp \subset T_pM$.

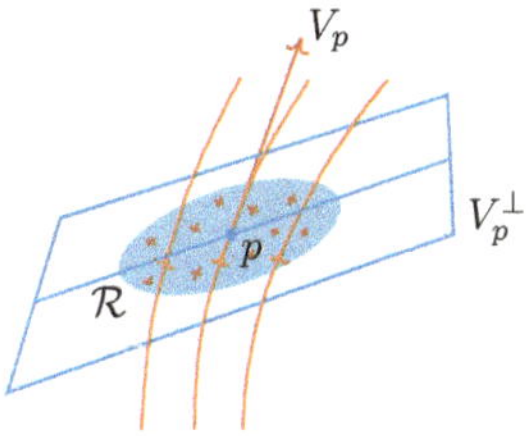

Then in an 'infinitesimal' region $\mathcal{R} \subset V_p^\perp$, the number of particles in $\mathcal{R}$ are given by $\int_{\mathcal{R}} \nu \, \text{vol}_{\mathbf{g}}(\cdot, \cdot, \cdot, V)$.

[9]E.g. $\mathbf{T}(p)(x, v)$ is the momentum density in the direction of the unit vector $x \in v^\perp$, and $\mathbf{T}(p)(x, y)$ is the stress (force component in the x-direction per unit area perpendicular to the y-direction), where x, y are unit vectors in $v^\perp$. For a detailed treatment, we refer the reader to [Misner, Thorne and Wheeler (2017), Chapter 5].

Definition 13.3. (Dust energy-momentum tensor field.)
Let $(M, \mathbf{g})$ be a spacetime, and (m, ν, V) be dust. Then the *dust energy-momentum tensor field* $\mathbf{T} \in T_2^0 M$ is given by $\mathbf{T} = m\nu V^\flat \otimes V^\flat$.

Before we check that this $\mathbf{T}$ satisfies all the properties demanded of the energy-momentum tensor field, let us provide some heuristic motivation for the above definition by checking that $\mathbf{T}(p)(v, v)$ does give the energy density measured by any normalised instantaneous observer $v \in T_pM$. Suppose that $v_1, v_2, v_3 \in v^\perp$ are linearly independent vectors with 'small' lengths $\|v_i\| = \sqrt{\mathbf{g}(p)(v_i, v_i)} \ll 1$, $i = 1, 2, 3$. Let $K \subset v^\perp$ be an infinitesimal parallelepiped formed by v_1, v_2, v_3.

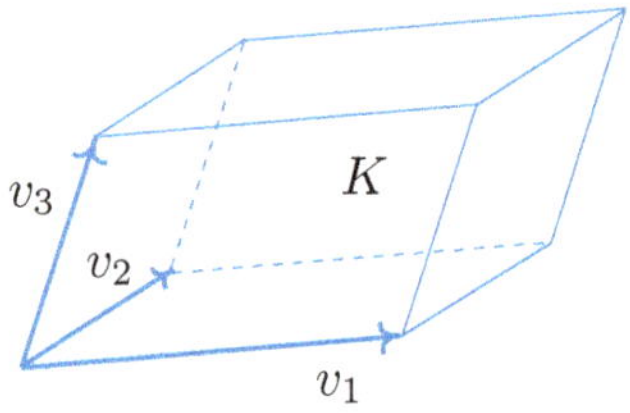

The 3-volume in $v^\perp$ of K is[10] defined by $\mu(K) = |\mathrm{vol}_\mathbf{g}(p)(v_1, v_2, v_3, v)|$. The number of particles in K is $N = \nu(p)|\mathrm{vol}_\mathbf{g}(p)(v_1, v_2, v_3, V_p)|$. The energy measured by v for each particle is $E = -\, m\mathbf{g}(p)(v, V_p)$. Thus the energy density measured by v is

$$\frac{NE}{\mu(K)}.$$

But $V_p = cv + w$, for some $c \in \mathbb{R}$ and some vector $w \in \mathrm{span}\{v_1, v_2, v_3\} = v^\perp$. Taking the scalar product with v, we obtain $\mathbf{g}(p)(V_p, v) = -\, c + 0$, so that $V_p = (E/m)v + w$. By the linearity and skew-symmetry of $\mathrm{vol}_\mathbf{g}$, it follows that $|\mathrm{vol}_\mathbf{g}(p)(v_1, v_2, v_3, V_p)| = (E/m)\, |\mathrm{vol}_\mathbf{g}(p)(v_1, v_2, v_3, v)|$. So the energy density measured by v is

$$\frac{NE}{\mu(K)} = \nu(p)\frac{E}{m}E = m\nu(p)\left(\mathbf{g}(p)(v, V_p)\right)^2.$$

We note that this is independent of the choice of the parallelepiped K. On the other hand, with $\mathbf{T} := m\nu V^\flat \otimes V^\flat$, we have

$$\mathbf{T}(p)(v, v) = m\nu(p)(V^\flat \otimes V^\flat)(p)(v, v) = m\nu(p)(\mathbf{g}(p)(V_p, v))^2,$$

which matches the energy density determined above.

Now we will check that $\mathbf{T}$ is an energy-momentum tensor field in the sense of Definition 13.1. First, we will show the following.

[10]Since $\mathrm{vol}_\mathbf{g}(p)(\cdot, \cdot, \cdot, v) = \mathrm{vol}_{\mathbf{g}(p)}(\cdot, \cdot, \cdot, v) \in \wedge^3(v^\perp)^*$; see Exercise 11.14 (p.256).

Lemma 13.1. *Let* $(M, \mathbf{g})$ *be a spacetime,* $f \in C^\infty(M)$*, and* $V \in T_0^1 M$*. Then* $\operatorname{div}(fV^\flat \otimes V^\flat) = (\operatorname{div}(fV))V^\flat + f(\nabla_V V)^\flat$.

Proof. For $W \in T_0^1 M$, we have

$$\begin{aligned}
&(\operatorname{div}(fV^\flat \otimes V^\flat))W \\
&= \mathbf{C}((\Omega, X) \mapsto (\nabla_X (fV^\flat \otimes V^\flat))(\Omega^\sharp, W)) \\
&= \mathbf{C}((\Omega, X) \mapsto ((\nabla_X (fV^\flat)) \otimes V^\flat)(\Omega^\sharp, W)) \\
&\quad + \mathbf{C}((\Omega, X) \mapsto (fV^\flat \otimes \nabla_X (V^\flat))(\Omega^\sharp, W)) \\
&= (V^\flat W)\mathbf{C}((\Omega, X) \mapsto (\nabla_X (fV^\flat))\Omega^\sharp) + f\,\mathbf{C}((\Omega, X) \mapsto V^\flat(\Omega^\sharp)(\nabla_X (V^\flat))W) \\
&= (V^\flat W)\operatorname{div}(fV^\flat) + f(\nabla_{\mathbf{C}((\Omega,X)\mapsto V^\flat(\Omega^\sharp)X)}(V^\flat))W.
\end{aligned}$$

By Exercise 9.15 (p.190), $\operatorname{div}(fV^\flat) = \operatorname{div}(fV)$. Also, in an admissible chart $(U, \mathbf{x})$, we have

$$\begin{aligned}
\mathbf{C}((\Omega, X) \mapsto V^\flat(\Omega^\sharp)X) &= \mathbf{g}(V, (dx^i)^\sharp)\,\partial_{x^i} = \mathbf{g}(V^k \partial_{x^k}, \mathbf{g}^{ij}\partial_{x^j})\,\partial_{x^i} \\
&= V^k \mathbf{g}^{ij}\mathbf{g}_{kj}\partial_{x^i} = V^k \delta^i_k \partial_{x^i} = V^i \partial_{x^i} = V.
\end{aligned}$$

So $\operatorname{div}(f\,V^\flat \otimes V^\flat) = (\operatorname{div}(f\,V))V^\flat + f\,\nabla_V(V^\flat) = (\operatorname{div}(f\,V))V^\flat + f\,(\nabla_V V)^\flat$, where the last equality follows using Exercise 6.22 (p.123). □

Proposition 13.1. *Let* $(M, \mathbf{g})$ *be a spacetime, and* (m, ν, V) *be dust. Then* $\mathbf{T} = m\nu V^\flat \otimes V^\flat$ *is an energy-momentum tensor field.*

Proof. Clearly $\mathbf{T} \in T_2^0 M$. To show symmetry, we note that for vector fields $X, Y \in T_0^1 M$,

$$\begin{aligned}
\mathbf{T}(X, Y) = m\nu(V^\flat \otimes V^\flat)(X, Y) &= m\nu V^\flat(X)V^\flat(Y) \\
&= m\nu V^\flat(Y)V^\flat(X) = m\nu(V^\flat \otimes V^\flat)(Y, X) = \mathbf{T}(Y, X).
\end{aligned}$$

Also, the weak energy condition is satisfied since for all vector fields (causal or not) $X \in T_0^1 M$, we have (pointwise)

$$\mathbf{T}(X, X) = m\nu(V^\flat \otimes V^\flat)(X, X) = m\nu(V^\flat(X))^2 \geqslant 0,$$

since $m > 0$ and $\nu \geqslant 0$ pointwise. The divergence freeness of $\mathbf{T}$ follows immediately from the previous lemma and the facts that $\nabla_V V = 0$ and $\operatorname{div}(\nu V) = 0$. □

Exercise 13.2. (Dust to no dust.)
Let $(M, \mathbf{g})$ be Minkowski spacetime. Let $V = \partial_t$ (in the notation described in Example 11.5, p.256), and $m > 0$, $\nu_* \geqslant 0$ be given real numbers. Denote by ν the constant function on M assuming the value ν_* everywhere. Verify that (m, ν, V) is dust, and determine the component matrix $[\mathbf{T}_{ij}]$ of the corresponding energy-momentum tensor field with respect to the ordered basis $(\partial_t, \partial_x, \partial_y, \partial_z)$. Show that the field equation is satisfied if and only if $\nu_* = 0$, that is, if there is no dust.

Exercise 13.3. Consider Newtonian/classical fluid flow in $\mathbb{R} \times \mathbb{R}^3$, such that at time t and position $\mathbf{x}$ the mass density is $\rho(t, \mathbf{x})$, and the velocity is $\mathbf{v}(t, \vec{\mathbf{x}})$. Below, we will denote the vector (spatial) differential operator $\vec{\nabla}_{\mathbf{x}}$ by $\vec{\nabla}$.

(1) Show the continuity equation $\frac{\partial \rho}{\partial t} + \vec{\nabla}\cdot(\rho \mathbf{v}) = 0$.
(Consider a spatial ball B_r with centre $\mathbf{x}$ and radius $r > 0$, and with spherical boundary ∂B_r having an outward normal $\mathbf{n_y}$ at $\mathbf{y} \in \partial B_r$, and use the Gauss divergence theorem to obtain $\int_{B_r} \vec{\nabla}\cdot(\rho \mathbf{v})\, dV = \int_{\partial B_r} \rho \mathbf{v}\cdot \mathbf{n_y}\, dA$. But the latter must equal the rate of change of mass in B_r.)

(2) Now take $\rho > 0$ to be a constant (incompressible flow). Show that $\vec{\nabla} \cdot \mathbf{v} = 0$. Also, we take the pressure to be 0 everywhere (dust), and so the acceleration of the fluid particles is zero. Let $\mathbf{v} =: (v^1, v^2, v^3)$. Derive the Euler equation
$$\frac{\partial v^i}{\partial t} + (\vec{\nabla} v^i)\cdot \mathbf{v} = 0 \quad (i = 1, 2, 3).$$
Consider a region Ω with a boundary $\partial\Omega$, having a normal $\mathbf{n_y}$ at $\mathbf{y} \in \partial\Omega$. Let $\{\mathbf{e}_1, \mathbf{e}_2, \mathbf{e}_3\}$ be the standard basis in $\mathbb{R}^3$. Define the momentum current density tensor field $P \in T_2^0\mathbb{R}^3$ by $P = \rho \mathbf{v}^\flat \otimes \mathbf{v}^\flat$. Show that the rate of change of classical momentum in Ω is given by
$$-\frac{d}{dt}\int_\Omega \rho v^i\, dV = \int_{\partial\Omega} P(\mathbf{e}_i, \mathbf{n_y})\, dA(\mathbf{y}) \quad (i = 1, 2, 3).$$

Perfect fluid

Yet another example of an energy-momentum tensor field is that of a perfect fluid, which is a bit more complicated than dust, because now we suppose that there is also a pressure p at each spacetime point. This pressure arises from the random thermal motion of the particles in the fluid, and is equal in all spatial directions for any instantaneous observer. The fluid is 'perfect' in the sense that there is no interparticle binding or viscosity.

Definition 13.4. (Perfect fluid.)
Let $(M, \mathbf{g})$ be a spacetime. A *perfect fluid* is a triple (ρ, p, V), where

- $\rho : M \to (0, \infty)$ and $\mathrm{p} : M \to [0, \infty)$ are smooth maps,
- $V \in T_0^1 M$ is such that for all $q \in M$, $\mathbf{g}(q)(V_q, V_q) = -1$, and V_q is future-pointing,
- $\mathrm{div}(\rho V) = -\mathrm{p}\, \mathrm{div} V$ and $\nabla_V V = -\frac{1}{\mathrm{p} + \rho}((V\mathrm{p})V + (d\mathrm{p})^\sharp)$.

p is called the *pressure*, ρ is called the *density*, and V is called the *velocity field*. The *energy-momentum tensor field of a perfect fluid* (ρ, p, V) is given by $\mathbf{T} = (\rho + \mathrm{p})V^\flat \otimes V^\flat + \mathrm{p}\mathbf{g}$.

We note that if $\mathrm{p} \equiv 0$, the energy-momentum tensor field of the perfect fluid is that of dust. Intuitively, the condition on $\nabla_V V$ makes sense, because

$\nabla_V V$ represents the acceleration of the fluid particles, and so the equation roughly says that mass density times acceleration is driven by the pressure gradient, akin to the wind flow on the surface of the Earth. The first equation, $\operatorname{div}(\rho V) = -\mathrm{p} \operatorname{div} V$ can likewise be roughly interpreted as saying that the change in energy density is driven by the work done by the pressure.

The symmetry of the $\mathbf{T}$ given above is clear.

Exercise 13.4. With the objects as in Definition 13.4, show that $\mathbf{T}$ satisfies the weak energy condition.

Exercise 13.5. With the objects as in Definition 13.4, show that $\operatorname{trace} \mathbf{T} = 3\mathrm{p} - \rho$.

We now check the divergence-freeness.

Lemma 13.2. *Let* $(M, \mathbf{g})$ *be a spacetime, and* (ρ, p, V) *be a perfect fluid. Then* $\operatorname{div} \mathbf{T} = 0$, *where* $\mathbf{T} = (\rho + \mathrm{p}) V^\flat \otimes V^\flat + \mathrm{p}\mathbf{g}$.

Proof. By Lemma 13.1, we have

$$\begin{aligned}
\operatorname{div} \mathbf{T} &= (\operatorname{div}((\rho + \mathrm{p})V))V^\flat + (\rho + \mathrm{p})(\nabla_V V)^\flat + \operatorname{div}(\mathrm{p}\mathbf{g}) \\
&= (-\mathrm{p} \operatorname{div} V + \operatorname{div}(\mathrm{p}V))V^\flat - (\rho + \mathrm{p})\frac{1}{\rho + \mathrm{p}}((V\mathrm{p})V + (d\mathrm{p})^\sharp)^\flat + \operatorname{div}(\mathrm{p}\mathbf{g}) \\
&= (-\mathrm{p} \operatorname{div} V + \operatorname{div}(\mathrm{p}V))V^\flat - (V\mathrm{p})V^\flat - d\mathrm{p} + \operatorname{div}(\mathrm{p}\mathbf{g}).
\end{aligned}$$

In any admissible chart $(U, \mathbf{x})$, we have

$$\begin{aligned}
\operatorname{div}(\mathrm{p}V) &= dx^i(\nabla_{\partial_{x^i}}(\mathrm{p}V)) = dx^i((\partial_{x^i}\mathrm{p})V + \mathrm{p}\nabla_{\partial_{x^i}} V) \\
&= (\partial_{x^i}\mathrm{p})V^i + \mathrm{p} \operatorname{div} V = V\mathrm{p} + \mathrm{p} \operatorname{div} V.
\end{aligned}$$

Also, for $W \in T_0^1 M$,

$$\begin{aligned}
(\operatorname{div}(\mathrm{p}\mathbf{g}))W &= (\nabla_{\partial_{x^i}}(\mathrm{p}\mathbf{g}))((dx^i)^\sharp, W) \\
&= (\partial_{x^i}\mathrm{p})\mathbf{g}(\mathbf{g}^{ij}\partial_{x^j}, W^k\partial_{x^k}) + 0 \qquad (\text{as } \nabla.\mathbf{g} = 0) \\
&= (\partial_{x^i}\mathrm{p})W^k\mathbf{g}^{ij}\mathbf{g}_{jk} = (\partial_{x^i}\mathrm{p})W^k\delta_k^i \\
&= (\partial_{x^i}\mathrm{p})W^i = W\mathrm{p} = (d\mathrm{p})W.
\end{aligned}$$

So $\operatorname{div}(\mathrm{p}\mathbf{g}) = d\mathrm{p}$. Consequently,

$$\begin{aligned}
\operatorname{div} \mathbf{T} &= (-\mathrm{p} \operatorname{div} V + \operatorname{div}(\mathrm{p}V))V^\flat - (V\mathrm{p})V^\flat - d\mathrm{p} + \operatorname{div}(\mathrm{p}\mathbf{g}) \\
&= (-\mathrm{p} \operatorname{div} V + V\mathrm{p} + \mathrm{p} \operatorname{div} V)V^\flat - (V\mathrm{p})V^\flat - d\mathrm{p} + d\mathrm{p} = 0. \qquad \square
\end{aligned}$$

Often, a relation between the pressure and density, of the form $\mathrm{p} = f(\rho)$, is also available. Such is relation is referred to[11] as an *equation of state*. For example, consider a perfect fluid with the equation of state $\mathrm{p} = w\rho$, where w is a smooth function on M. The case $w \equiv 0$ corresponds to dust, where the pressure is identically zero.

[11] This terminology arises from thermodynamics, where an equation of state relates 'state variables' such as pressure, volume, temperature, etc.

Remark 13.2. (Cosmological constant term arising from matter.) Consider the state equation $\mathrm{p} = w\rho$ with $w \equiv -1$. Strictly speaking, this is impossible if p and ρ are both supposed to be nonnegative. But we momentarily relax[12] the positivity condition on p in Definition 13.4. In the field equation $\mathbf{Ric} - \frac{1}{2}\mathbf{S}\mathbf{g} + \Lambda\mathbf{g} = 8\pi\mathbf{T}$, the term $\Lambda\mathbf{g}$ on the 'geometry side' can be brought over to the 'matter side', and it can be thought of as being the energy-momentum tensor of a perfect fluid with the equation of state as above with $w \equiv -1$, and the pressure $\mathrm{p} \equiv -\frac{\Lambda}{8\pi}$. *

Finally, we give a heuristic justification of the energy-momentum tensor field for a perfect fluid. We assume that at any spacetime point q, for any instantaneous observer $v \in T_qM$, the 'gas molecules' have uniformly distributed random speeds. Suppose that a normalised instantaneous observer $v \in T_qM$ chooses an orthonormal basis (e_1, e_2, e_3) for $v^\perp$, and observes a gas molecule having the relative speed $u_{\mathbf{n}} = u_0(q)(n^1 e_1 + n^2 e_2 + n^3 e_3)$, where $\mathbf{n} = (n^1, n^2, n^3)$ is a uniformly distributed random point on the unit sphere, which decides the direction of the relative velocity of the observed molecule. The uniform distribution assumption means that no spatial direction is preferred over any other, that is, there is 'isotropy' in the random motion of the gas molecules. The tangent vector $w_{\mathbf{n}}$ to the worldline of the gas molecule whose relative speed is perceived to be $u_{\mathbf{n}}$ by v is (see Lemma 12.2, p.268)

$$w_{\mathbf{n}} = \frac{v + u_{\mathbf{n}}}{\sqrt{1 - \mathbf{g}(q)(u_{\mathbf{n}}, u_{\mathbf{n}})}} = \frac{v + u_{\mathbf{n}}}{\sqrt{1 - (u_0(q))^2\, \mathbf{n}\cdot\mathbf{n}}} = \frac{v + u_{\mathbf{n}}}{\sqrt{1 - (u_0(q))^2}},$$

where $\mathbf{n}\cdot\mathbf{n}$, the Euclidean inner product in $\mathbb{R}^3$, is equal to 1. Now if we imagine each of these molecules as a dust particle (since there is no binding energy with other molecules), then the energy-momentum tensor (at q) contribution is given by $m\,\nu(q)\,(w_{\mathbf{n}})^\flat \otimes (w_{\mathbf{n}})^\flat$, whose component matrix in the basis (v, e_1, e_2, e_3) for T_qM is given by

$$[\mathbf{T}(q)_{\mathbf{n}}] := m\,\nu(q) \begin{bmatrix} -\dfrac{1}{\sqrt{1-(u_0(q))^2}} \\ \dfrac{n^1 u_0(q)}{\sqrt{1-(u_0(q))^2}} \\ \dfrac{n^2 u_0(q)}{\sqrt{1-(u_0(q))^2}} \\ \dfrac{n^3 u_0(q)}{\sqrt{1-(u_0(q))^2}} \end{bmatrix} \begin{bmatrix} -\dfrac{1}{\sqrt{1-(u_0(q))^2}} \\ \dfrac{n^1 u_0(q)}{\sqrt{1-(u_0(q))^2}} \\ \dfrac{n^2 u_0(q)}{\sqrt{1-(u_0(q))^2}} \\ \dfrac{n^3 u_0(q)}{\sqrt{1-(u_0(q))^2}} \end{bmatrix}^{\mathrm{t}}.$$

[12]Then the weak energy condition is not required to hold; see the solution to Exercise 13.4.

Intuitively, we expect that the 'effective' contribution of energy-momentum tensor at q will be the expected value of the above. Denoting the 'expectation operator' by $\mathbb{E}$, we have

$$\mathbb{E}\Big(\frac{n^i u_0(q)}{1-(u_0(q))^2}\Big)=0 \text{ for } i=1,2,3, \quad \text{and} \quad \mathbb{E}\Big(\frac{n^i n^j (u_0(q))^2}{1-(u_0(q))^2}\Big)=0 \text{ for } i\neq j.$$

But since $\mathbf{n}\cdot\mathbf{n}=1$, it follows that $\mathbb{E}((n^1)^2)=\mathbb{E}((n^2)^2)=\mathbb{E}((n^3)^2)=\frac{1}{3}$. Thus

$$[\mathbf{T}(q)] := \mathbb{E}[\mathbf{T}(q)_{\mathbf{n}}]$$

$$= m\nu(q)\begin{bmatrix} \frac{1}{1-(u_0(q))^2} & 0 & 0 & 0 \\ 0 & \frac{1}{3}\frac{(u_0(q))^2}{(1-(u_0(q))^2)} & 0 & 0 \\ 0 & 0 & \frac{1}{3}\frac{(u_0(q))^2}{(1-(u_0(q))^2)} & 0 \\ 0 & 0 & 0 & \frac{1}{3}\frac{(u_0(q))^2}{(1-(u_0(q))^2)} \end{bmatrix}.$$

Let us set

$$\rho(q) := m\nu(q)\frac{1}{1-(u_0(q))^2}, \quad \mathrm{p}(q) := m\nu(q)\frac{1}{3}\frac{(u_0(q))^2}{(1-(u_0(q))^2)} = \frac{1}{3}\rho(q)(u_0(q))^2.$$

Then

$$[\mathbf{T}(q)] = \begin{bmatrix} \rho(q) & 0 & 0 & 0 \\ 0 & \mathrm{p}(q) & 0 & 0 \\ 0 & 0 & \mathrm{p}(q) & 0 \\ 0 & 0 & 0 & \mathrm{p}(q) \end{bmatrix}.$$

On the other hand, we have that the components with respect to the basis $(V_q =: v, e_1, e_2, e_3)$, of the $(0,2)$-tensor $((\mathrm{p}+\rho)V^\flat \otimes V^\flat + \mathrm{p}\mathbf{g})(q) \in T_2^0 M(q)$, are given by the same matrix above. This motivates the Definition 13.4.

Remark 13.3. (Radiation or 'photon gas'.)
Let us consider the limiting case $u_0(q)\to 1$ in the above calculation, when the perfect fluid particles have the relative speed 1 for the instantaneous observer $v=V_q\in T_qM$, i.e., we think of the particles as photons. Suppose that ν behaves in such a way we have a finite limit

$$\rho(q) = m\nu(q)\frac{1}{1-(u_0(q))^2} \to \rho_0(q) \text{ as } (u_0(q))^2\to 1.$$

Then necessarily

$$\mathrm{p}(q) = m\nu(q)\frac{1}{3}\frac{(u_0(q))^2}{(1-(u_0(q))^2)} = \frac{1}{3}\rho(q)(u_0(q))^2 \to \frac{1}{3}\rho_0(q) \text{ as } (u_0(q))^2\to 1.$$

In light of this, we take the energy-momentum tensor field for 'radiation' as that of a perfect fluid with $\mathrm{p}=\frac{1}{3}\rho$ (so that $w\equiv\frac{1}{3}$ in $\mathrm{p}=w\rho$). *

Example 13.1. (FLRW spacetime.)
Consider the FLRW spacetime $M = I \times \mathbb{R}^3$, where $I = (0, \infty)$, and with the metric $\mathbf{g}$, time-orientation, and orientation as in Example 5.9 (p.87), Example 5.13 (p.90) and Example 11.2 (p.237). Recall also that in the chart $(M, \mathrm{id}_{I\times\mathbb{R}^3})$, we had determined the Ricci curvature tensor field and the scalar curvature function in Section 9.2. Suppose that the energy-momentum tensor field is given by a perfect fluid (ρ, p, V) with $V = \partial_t$. We will now obtain two differential equations involving a, ρ, p, called the *Friedmann equations*[13], which follow from the field equations, and play an important role in cosmology. Recall that in the chart basis $(\partial_t, \partial_x, \partial_y, \partial_z)$,

$$\mathbf{Ric}_{00} = -3\frac{\ddot{a}}{a}, \quad \text{and} \quad \mathbf{Ric}_{11} = \mathbf{Ric}_{22} = \mathbf{Ric}_{33} = a\ddot{a} + 2\dot{a}^2,$$

and the other components are all zeroes. Also, the scalar curvature is

$$\mathbf{S} = 6\,\frac{a\ddot{a} + \dot{a}^2}{a^2}.$$

The 00-component of the field equation $\mathbf{Ric} - \frac{\mathbf{S}\mathbf{g}}{2} + \Lambda\mathbf{g} = 8\pi\mathbf{T}$ then gives

$$-3\frac{\ddot{a}}{a} - \frac{1}{2}6\,\frac{a\ddot{a} + \dot{a}^2}{a^2}(-1) + \Lambda(-1) = 8\pi((\mathrm{p} + \rho)(-1)(-1) + \mathrm{p}(-1)),$$

and upon simplification, this gives the *first Friedmann equation*,

$$\frac{\dot{a}^2}{a^2} - \frac{\Lambda}{3} = \frac{8\pi\rho}{3}. \tag{13.6}$$

The 11-component (also 22-, 33-component) of the field equation yields

$$a\ddot{a} + 2\dot{a}^2 - \frac{1}{2}6\,\frac{a\ddot{a} + \dot{a}^2}{a^2}(a^2) + \Lambda(a^2) = 8\pi((\mathrm{p} + \rho)(0) + \mathrm{p}(a^2)).$$

Thus

$$-2\frac{\ddot{a}}{a} - \frac{\dot{a}^2}{a^2} + \Lambda = 8\pi\mathrm{p}. \tag{13.7}$$

Solving for $\frac{\dot{a}^2}{a^2}$, and substituting the resulting expression in the first Friedmann equation (13.6), results in the *second Friedmann equation*,

$$\frac{\ddot{a}}{a} - \frac{\Lambda}{3} = -\frac{4\pi}{3}(\rho + 3\mathrm{p}). \tag{13.8}$$

The other mixed components in the field equation just reduce to $0 = 0$. ◇

Exercise 13.6. (First law of thermodynamics for FLRW spacetime.)
We use notation from Example 13.1. Let p and ρ depend only on t. Show that

$$\frac{d}{dt}(\rho a^3) + \mathrm{p}\frac{d}{dt}(a^3) = 0.$$

---

[13] In the 'flat spatial' case.

We remark that setting
$dE = d(\rho\frac{4\pi}{3}a^3)$ (think mass-energy change in a sphere of radius $a(t)$), and
$dV = d(\frac{4\pi}{3}a^3)$ (think change in volume),
we can formally write the above as $dE + \mathrm{p}\,dV = 0$, reminiscent of the first law of thermodynamics (law of conservation of energy): The work done by pressure in the expansion accounts for the change in the mass-energy in the volume.

Remark 13.4. (Other examples of energy-momentum tensor fields.)
We mention two more commonly encountered examples of energy-momentum tensor fields, without details.

(a) Electromagnetic energy-momentum tensor field[14]: The Maxwell equations can be derived as the Euler-Lagrange equations for an 'electromagnetic action' involving a Lagrangian for the electromagnetic field, and using the procedure explained in Remark 13.1, the corresponding energy-momentum tensor field $\mathbf{T}$ can be derived, which is given in an admissible chart $(U, \mathbf{x})$ by $\mathbf{T}_{ij} = \frac{1}{4\pi}(\mathbf{F}_{im}\mathbf{F}_{jk}\mathbf{g}^{km} - \frac{1}{4}\mathbf{g}^{rm}\mathbf{g}^{sn}\mathbf{F}_{rs}\mathbf{F}_{mn}\,\mathbf{g}_{ij})$, where $\mathbf{F} \in \boldsymbol{\Omega}^2 M$ is the Faraday tensor field. We will consider an example in Exercise 13.7.

(b) Klein-Gordon scalar field[15]: The Klein-Gordon equation for $\phi : M \to \mathbb{R}$ is given by $\Box\,\phi - m^2\phi = 0$, where $\Box\,\phi = \operatorname{div}((d\phi)^\sharp)$. In an admissible chart $(U, \mathbf{x})$, this becomes $\mathbf{g}^{ij}\partial_{x^i}\partial_{x^j}\phi - \mathbf{g}^{ij}\Gamma^k_{ij}\partial_{x^k}\phi - m^2\phi = 0$. The Klein-Gordon equation is the Euler-Lagrange equation for the 'Klein-Gordon action'. The corresponding energy-momentum tensor field $\mathbf{T}$ is given by $\mathbf{T} = (d\phi) \otimes (d\phi) - \frac{1}{2}(\mathbf{g}((d\phi)^\sharp, (d\phi)^\sharp) + m^2\phi^2)\,\mathbf{g}$. *

Exercise 13.7. Recall from Exercise 12.15 (p.287) the Reissner-Nordström spacetime $(M, \mathbf{g})$, the 1-form fields $\Omega^i \in T^0_1 M$ $(i = 1, 2, 3, 4)$ defined there, and the Faraday tensor field $\mathbf{F} = \frac{e}{r^2}\Omega^1 \wedge \Omega^4$. The expression for the electromagnetic energy-momentum tensor field $\mathbf{T} \in T^0_2 M$ given in Remark 13.4(a), for this $\mathbf{F}$ and $\mathbf{g}$, is $\mathbf{T} = \frac{e^2}{8\pi r^4}(\Omega^4 \otimes \Omega^4 - \Omega^1 \otimes \Omega^1 + \Omega^2 \otimes \Omega^2 + \Omega^3 \otimes \Omega^3)$. Clearly $\mathbf{T}$ is symmetric. Show that $\mathbf{T}$ satisfies the weak energy condition. (In Exercise 14.14, p.322, we will check that, with the cosmological constant $\Lambda = 0$, the field equation $\mathbf{Ric} - \frac{1}{2}\mathbf{S}\mathbf{g} + \Lambda\mathbf{g} = \mathbf{T}$ is satisfied. By Corollary 9.1, p.192, we have $\operatorname{div}\mathbf{T} = \operatorname{div}(\mathbf{Ric} - \frac{1}{2}\mathbf{S}\mathbf{g}) = 0$. For a direct proof of $\operatorname{div}\mathbf{T} = 0$ using the Maxwell equations, where $\mathbf{T}$ is the general expression for electromagnetic energy-momentum tensor field $\mathbf{T}$ given in Remark 13.4(a), we refer the reader to [Kriele (2001), Lemma 5.2.4, p.268].)

---

[14]See [Sachs and Wu (1977), §3.7] and [Kriele (2001), §5.3.1] for a detailed treatment of the electromagnetic energy-momentum tensor field.
[15]See [Wald (1984), §4.3] for details.

Chapter 14

Field equation

In the last chapter, we saw that the energy-momentum tensor field $\mathbf{T}$ on a spacetime $(M, \mathbf{g})$ is assumed to satisfy

$$\operatorname{div} \mathbf{T} = 0.$$

If this is to match with a (geometric) $(0, 2)$-tensor field, then its divergence ought to be zero too, and we had seen[1] one such tensor field in Chapter 9:

$$\operatorname{div} \left(\mathbf{Ric} - \frac{1}{2} \mathbf{S} \mathbf{g} + \Lambda \mathbf{g} \right) = 0,$$

for any constant Λ. The field equation is postulated[2] to be

$$\mathbf{Ric} - \frac{1}{2} \mathbf{S} \mathbf{g} + \Lambda \mathbf{g} = 8\pi \mathbf{T}.$$

The constant Λ is called the *cosmological constant*. This equation is accepted as a postulate[3], in the sense of being the most powerful unfalsified model. Indeed, it accommodates all that the Newtonian theory has to say, but also supersedes it, since (for example) it provides an explanation of the perihelion precession of Mercury's orbit accurately.

Solving the field equation amounts to finding the metric $\mathbf{g}$ such that the field equation is satisfied. Thus if we start with a spacetime model as a certain smooth manifold, and we have also assumed knowledge of $\mathbf{T}$ (i.e., a matter model), then in an admissible chart $(U, \mathbf{x})$, the chart-induced components $\mathbf{g}_{ij}(\mathbf{x}^{-1}(\cdot))$ of $\mathbf{g}$ satisfy the 10 second order nonlinear partial differential equations obtained by expressing the field equation componentwise. As both sides of the field equation are symmetric tensor fields, there are not 16, but only 10 'independent' scalar PDEs.

[1] See Corollary 9.1 (p.192).

[2] This was published first by Albert Einstein with $\Lambda = 0$ in 1915. David Hilbert found the field equation almost simultaneously with Einstein based on a variational principle. Hilbert maintained that priority is due to Einstein, and that his investigation was founded on Einstein's previous work. The constant Λ was introduced later by Einstein in 1917. See [Mehra (1974)] for an account of the intellectual history behind the genesis of the field equation.

[3] However, starting from a Lagrangian formalism makes it possible to derive the field equation. This transfers the belief in field equation postulate over to the postulate on the Lagrangian we choose to consider. We refer the reader to [Wald (1984), App. E].

Determining the trajectories of test particles (e.g. a planet in a solar system) amounts to determining the geodesics. Although these are complicated problems, one usually has a lot of symmetry at hand, allowing much simplification, and the possibility of obtaining exact solutions, or very good approximations.

Exercise 14.1. Let the cosmological constant be $\Lambda = 0$. Recall the definition of the trace of a $(0,2)$-tensor field from Exercise 5.30 (p.103). By the definition of the scalar curvature, $\mathbf{S} = \operatorname{trace} \mathbf{Ric}$. Show that the field equation is equivalent to the equation $\mathbf{Ric} = 8\pi(\mathbf{T} - \frac{1}{2}(\operatorname{trace} \mathbf{T})\, \mathbf{g})$.

Exercise 14.2. Let $\mathbf{T}$ be the energy-momentum tensor field on a spacetime M. A vector field is called *lightlike* (respectively *timelike*) if at each $p \in M$, V_p is lightlike (respectively timelike). Using the weak energy condition on $\mathbf{T}$ and the field equation, show that $\mathbf{Ric}(V,V) \geqslant 0$ for all lightlike vector fields $V \in T_0^1 M$. This is called the *null convergence condition*, and is one of the hypotheses of Penrose's singularity theorem, which establishes geodesic incompleteness in any black hole for matter satisfying reasonable energy conditions. We say that $\mathbf{T}$ satisfies the *strong energy condition* if for all timelike vector fields $V \in T_0^1 M$, we have $\mathbf{T}(V,V) \geqslant \frac{1}{2}(\operatorname{trace} \mathbf{T})\, \mathbf{g}(V,V)$. Prove that if $\mathbf{T}$ satisfies the strong energy condition, and if the cosmological constant is $\Lambda = 0$, then $\mathbf{Ric}(V,V) \geqslant 0$ for all timelike vector fields $V \in T_0^1 M$. This is called the *timelike convergence condition*, and is one of the hypotheses of the Hawking singularity theorem, which, roughly speaking, says that any expanding spacetime satisfying physically reasonable conditions, is 'singular', that is, it is not geodesically complete. Prior to such singularity theorems of Penrose and Hawking from the 1960s, it was believed that spacetimes were singular only in highly symmetric situations, while more 'realistic' spacetimes would not be singular.

Exercise 14.3. (Plane gravitational waves.)
The aim of this exercise is to consider an idealised spacetime modelling gravitational radiation in a region far away from the source of the radiation (e.g. the source may be a binary system of stars rotating about their mutual centre of mass). We suppose that $M = \mathbb{R}^4$ with the standard smooth structure. Using the global chart $(\mathbb{R}^4, \mathrm{id})$, we give M the time-orientation $V = \partial_t$, the orientation $[dx \wedge dy \wedge dz \wedge dt]$, and the metric $\mathbf{g} = \boldsymbol{\eta} + 2H(x,y,t-z)\,(dt - dz) \otimes (dt - dz)$, where $H : \mathbb{R}^3 \to \mathbb{R}$ is a smooth function, and $\boldsymbol{\eta}$ denotes the Minkowski metric, $\boldsymbol{\eta} = -\,dt \otimes dt + dx \otimes dx + dy \otimes dy + dz \otimes dz$. Set $Y = \partial_t + \partial_z \in T_0^1 M$. Show that $\mathbf{g}(Y,Y) = 0$. Show that $\mathbf{g}$ is a Lorentzian metric by computing the component matrix $[\mathbf{g}(V_i, V_j)]$, where, denoting $h := H(x,y,t-z)$,

$$V_1 = \frac{2h+3}{2\sqrt{2}}\partial_t + \frac{2h+1}{2\sqrt{2}}\partial_z, \quad V_2 = \frac{-2h+1}{2\sqrt{2}}\partial_t + \frac{-2h+3}{2\sqrt{2}}\partial_z, \quad V_3 = \partial_x, \quad V_4 = \partial_y.$$

It can be shown[4] that the Riemann curvature tensor field is nonzero (so that the spacetime $(M, \mathbf{g})$ is 'not flat'), and that the Ricci curvature tensor field is given by $\mathbf{Ric} = -(\partial_x(\partial_x h) + \partial_y(\partial_y h))(dt - dz) \otimes (dt - dz)$.

[4]See, e.g., [Sachs and Wu (1977), §7.6] or [Beem, Ehrlich and Easley (1996), §13.1].

Show that the scalar curvature $\mathbf{S}=0$. Assuming that the cosmological constant Λ is 0, prove that the vacuum field equation is satisfied if and only if h is harmonic in the x, y variables, that is,

$$\frac{\partial^2 h}{\partial x^2}+\frac{\partial^2 h}{\partial y^2}=0.$$

Let $f, g: \mathbb{R} \to \mathbb{R}$ be smooth functions such that f^2+g^2 is not identically 0. Set

$$H(x,y,u)=\frac{f(u)}{2}(x^2-y^2)+g(u)xy.$$

Check that the resulting h is harmonic. The spacetime $(M, \mathbf{g})$ is called a *plane gravitational wave.*

Exercise 14.4. (Gödel spacetime.)
A first example of a spacetime solution $(M, \mathbf{g})$ to the field equation that allowed the existence of closed timelike curves was given by Kurt Gödel[5] in 1949. The Lorentzian manifold M is time-orientable, and the existence of timelike closed loops means that there are worldlines along which a particle, while always travelling into the future, ends up back in the past. The smooth manifold M is $\mathbb{R}^4$ with the standard smooth structure. The metric $\mathbf{g}$ in the global chart $(\mathbb{R}^4, \mathrm{id})$ is

$$\mathbf{g}=-dt\otimes dt+dx\otimes dx-\frac{e^{2x}}{2}dy\otimes dy+dz\otimes dz-e^x\,dt\otimes dy-e^x\,dy\otimes dt.$$

By 'completing squares' in dy, dt, show that $\mathbf{g}$ is Lorentzian. Show that $V:=\partial_t$ is timelike everywhere. We use the standard orientation $[dx\wedge dy\wedge dz\wedge dt]$ on $\mathbb{R}^4$ and the time-orientation is given by $V=\partial_t$. In the Gödel spacetime, the cosmological constant[6] is $\Lambda=-\frac{1}{2}<0$. The matter energy-momentum tensor field $\mathbf{T}$ is given by $\mathbf{T}=\frac{1}{8\pi}V^\flat\otimes V^\flat$. Check that this is dust. It can be shown[7] that the Ricci curvature tensor field is given by $\mathbf{Ric}=(\partial_t)^\flat\otimes(\partial_t)^\flat$. Determine the scalar curvature, and show that the field equation is satisfied. It can be shown[8] that the following curve γ is timelike: $\gamma(s)=(t(s),x(s),y(s),z(s))$, $s\in(-\pi,\pi)$, where, with $d:=\log(2+\sqrt{2}+\sqrt{5+4\sqrt{2}})$,

$$\begin{aligned}
t(s)&=2\sqrt{2}\tan^{-1}(e^{-d}\tan\tfrac{s}{2})-\sqrt{2}\,s,\\
x(s)&=\log((\cosh d)+(\cos s)\sinh d),\\
y(s)&=\frac{\sqrt{2}\,(\sin s)\sinh d}{(\cosh d)+(\cos s)\sinh d},\\
z(s)&=0.
\end{aligned}$$

γ has a continuous extension to $[-\pi,\pi]$. Check that γ is closed by showing $\gamma(\pi)=(0,d,0,0)=\gamma(-\pi)$. Suppressing the z-coordinate, use the computer to plot the curve $\mathbb{R}\ni s\mapsto(t(s),x(s),y(s))\in\mathbb{R}^3$.

[5] Logician and mathematician, 1906 – 1978, reputed for his 1931 incompleteness theorems in mathematical logic.
[6] Note that $\Lambda<0$, as opposed to $\Lambda>0$ in the current FLRW model of our spacetime.
[7] See for example [Kriele (2001), Proposition 9.4.1].
[8] See for example [Cooke (2017), p.319] and also [Momin (2002)].

Remark 14.1. (Field equation as an evolution equation.) When written in appropriate coordinates, the field equation in general relativity can be shown to be a set of evolution nonlinear wave equations together with constraint equations for the initial data. This is similar to Maxwell's equations of electromagnetism. In classical spacetime $\mathbb{R}^4$, the Maxwell equations in vacuum (charge density $\rho = 0$ and current density $\vec{\mathbf{j}} = \mathbf{0}$) for the unknowns $\vec{\mathbf{E}} : \mathbb{R} \times \mathbb{R}^3 \to \mathbb{R}^3$ and the magnetic field $\vec{\mathbf{B}} : \mathbb{R} \times \mathbb{R}^3 \to \mathbb{R}^3$ can be viewed as evolution equations (giving wave equations for $\vec{\mathbf{E}}$, $\vec{\mathbf{B}}$, see Ex. 12.14, p.286)

$$\left.\begin{aligned} \frac{\partial \vec{\mathbf{E}}}{\partial t}(t, \mathbf{x}) &= \vec{\nabla} \times \vec{\mathbf{B}}(t, \mathbf{x}) \\ \frac{\partial \vec{\mathbf{B}}}{\partial t}(t, \mathbf{x}) &= -\vec{\nabla} \times \vec{\mathbf{E}}(t, \mathbf{x}) \end{aligned}\right\} \quad (\mathbf{x} \in \mathbb{R}^3,\ t \geqslant 0)$$

together with the spatial initial conditions $\vec{\mathbf{E}}(\cdot, 0) : \mathbb{R} \times \mathbb{R}^3 \to \mathbb{R}^3$ and $\vec{\mathbf{B}}(\cdot, 0) : \mathbb{R} \times \mathbb{R}^3 \to \mathbb{R}^3$ that satisfy the constraints

$$\left.\begin{aligned} (\vec{\nabla} \cdot \vec{\mathbf{E}})(0, \mathbf{x}) &= 0 \\ (\vec{\nabla} \cdot \vec{\mathbf{B}})(0, \mathbf{x}) &= 0 \end{aligned}\right\} \quad (\mathbf{x} \in \mathbb{R}^3).$$

The problem is well-posed in the sense that for every pair of smooth initial conditions satisfying the constraint equations, there exists a unique smooth solution, and moreover the constraints propagate, that is,

$$\left.\begin{aligned} (\vec{\nabla} \cdot \vec{\mathbf{E}})(t, \mathbf{x}) &= 0 \\ (\vec{\nabla} \cdot \vec{\mathbf{B}})(t, \mathbf{x}) &= 0 \end{aligned}\right\} \quad (\mathbf{x} \in \mathbb{R}^3, t \geqslant 0).$$

Given smooth initial data obeying constraints on a 'spacelike hypersurface'[9], it can be shown that the gravitational field equation is well-posed, that is, there exists a unique local solution[10]. It is natural to ask how much into the future the solution can be extended, how much the smoothness of the initial data can be relaxed (and how much solution regularity is lost as a consequence), and stability questions (smooth dependence on the initial conditions). These are current active areas of research in the mathematical partial differential equation approach to general relativity[11]. *

[9] Roughly, a 3-dimensional submanifold of M such that every curve lying within it has spacelike tangent vectors at each point along it, with respect to the metric determined by the field equation. But one does not know in advance what the domain of dependence of the initial surface is. We just start with a 3-dimensional manifold S with the initial data, and it is required to find a 4-dimensional Lorentzian manifold $(M, \mathbf{g})$ such that $\mathbf{g}$ satisfies the field equation, and such that the surface S, embedded in M, is spacelike, with the solution agreeing with the initial values specified on S.

[10] A 1950s result due to Choquet-Bruhat. See [Choquet-Bruhat (2009)].

[11] A milestone result from 1993 due to Christodoulou and Klainerman proves the global stability of Minkowski spacetime. See e.g. [Klainerman and Nicolò (2003)].

Exercise 14.5. Consider a suitable open set $U \subset \mathbb{R}^4$ with the standard smooth structure. The smooth manifold U is endowed with the metric[12] given in the global chart (U, id) by $\mathbf{g} = -dt \otimes dt + (A(t,z))^2 dx \otimes dx + (B(t,z))^2 dy \otimes dy + dz \otimes dz$, where $A, B : \mathbb{R}^2 \to \mathbb{R}\backslash\{0\}$ are smooth functions. It can be shown that the nonzero Levi-Civita connection coefficients are given by (where $\dot{} := \partial_t$ and $' := \partial_z$):

$$\begin{aligned} \Gamma^t_{xx} &= A\dot{A} & \Gamma^t_{yy} &= B\dot{B} \\ \Gamma^x_{xt} &= \Gamma^x_{tx} = \tfrac{\dot{A}}{A} & \Gamma^x_{xz} &= \Gamma^x_{zx} = \tfrac{A'}{A} \\ \Gamma^y_{ty} &= \Gamma^y_{yt} = \tfrac{\dot{B}}{B} & \Gamma^y_{yz} &= \Gamma^{zy} = \tfrac{B'}{B} \\ \Gamma^z_{xx} &= -AA' & \Gamma^z_{yy} &= -BB'. \end{aligned}$$

Moreover, the nonzero components of the Ricci curvature are given as follows where i, j are taken in the order t, x, y, z:

$$[\mathbf{Ric}_{ij}] = \begin{bmatrix} \frac{\ddot{A}}{A} + \frac{\ddot{B}}{B} & & & \frac{\dot{A}'}{A} + \frac{\dot{B}'}{B} \\ & A(A'' - \ddot{A}) + \frac{A}{B}(A'B' - \dot{A}\dot{B}) & & \\ & & B(B'' - \ddot{B}) + \frac{B}{A}(A'B' - \dot{A}\dot{B}) & \\ \frac{\dot{A}'}{A} + \frac{\dot{B}'}{B} & & & \frac{A''}{A} + \frac{B''}{B} \end{bmatrix}.$$

(1) Show that the vacuum field equation is equivalent to the *evolution equations*

$$\text{(e1)} \qquad \partial_t^2 A - \partial_z^2 A = 0,$$
$$\text{(e2)} \qquad \partial_t^2 B - \partial_z^2 B = 0,$$

(which are wave equations for A and B), with the *constraint equations*

$$\text{(c1)} \qquad \frac{\partial_z^2 A}{A} + \frac{\partial_z^2 B}{B} = 0,$$
$$\text{(c2)} \qquad \frac{\partial_z \partial_t A}{A} + \frac{\partial_z \partial_t B}{B} = 0,$$
$$\text{(c3)} \qquad (\partial_t A)(\partial_t B) - (\partial_z A)(\partial_z B) = 0,$$

(constraining the initial data $A(t_0, z)$, $\partial_t A(t_0, z)$, $B(t_0, z)$ and $\partial_t B(t_0, z)$).

(2) Show that if $f, g \in C^\infty(\mathbb{R})$, then $A(t,z) = f(t-z)$ and $B(t,z) = g(t-z)$ are solutions of the evolution equations (e1), (e2).

Suppose that $\{z : \exists\, x, y \in \mathbb{R}$ such that $(t_0, x, y, z) \in U\} = \mathbb{R}$. Check that if the constraint equations (c1), (c2), (c3) hold at t_0, then $\frac{f''}{f} + \frac{g''}{g} = 0$.

Moreover, prove that if the constraint equations are satisfied at $t = 0$ then they are satisfied for all t.

(3) Fix $\mathbf{p} = (x_0, y_0, z_0) \in \mathbb{R}^3$. Define the curve $\gamma_\mathbf{p} : I_\mathbf{p} \to U$ by $\gamma_\mathbf{p}(t) := (t, \mathbf{p})$ for all $t \in I_\mathbf{p} := \{t \in \mathbb{R} : (t, \mathbf{p}) \in U\}$. Show that $\gamma_\mathbf{p}$ is a geodesic parametrised by proper time in each interval contained in $I_\mathbf{p}$.

(4) Consider the *expansion* $\theta := \frac{\partial_t A}{A} + \frac{\partial_t B}{B} = \partial_t \log(AB)$ of the family of geodesics, $\{\gamma_\mathbf{p} : (t_0, \mathbf{p}) \in U\}$. Show that $\partial_t \theta \leqslant -\frac{1}{2}\theta^2$. Conclude that if $\theta(t_0, z_0) < 0$ for some $z_0 \in \mathbb{R}$ then the spacetime is geodesically incomplete.

[12]This describes an exact + polarised plane gravitational wave propagating in the z-direction. See [d'Inverno (2022), Chapter 22] for the relation of this metric with that of Exercise 14.3 (p.306).

Remark 14.2. The geodesic incompleteness result in Exercise 14.5 can be seen as a baby version of the celebrated Hawking-Penrose singularity theorems, which guarantee, under very general conditions, that physically reasonable contracting solutions of the Einstein equations are geodesically incomplete. $\ast$

14.1 Newtonian limit

Recall from §13.1 that in classical Newtonian spacetime[13] $M=\mathbb{R}\times\mathbb{R}^3$, test matter moves according to the equation

$$\ddot{\mathbf{x}}(t)+\overrightarrow{\nabla}\Phi(t,\mathbf{x}(t))=0,$$

where

$$\Delta_{\mathbf{x}}\Phi(t,\mathbf{x})=4\pi\rho(t,\mathbf{x}),$$

where ρ describes the mass density distribution. We wish to show that for classical systems, so that the particles have velocities much smaller than the speed of light, and the objects are not too massive, the geodesic equation and the field equation reduce to the above (approximately).

We consider the spacetime $(M,\mathbf{g})$, where $M=\mathbb{R}\times\mathbb{R}^3$ is equipped with the standard smooth structure and a metric $\mathbf{g}$. We use the global chart $(M,\mathrm{id}=(t,x,y,z))$. We will also use the notation (x^0,x^1,x^2,x^3) for the component functions of this (identity) chart map. In our approximation, ∂_t can be regarded as a unit vector. Consider a simple matter distribution given by $\mathbf{T}=\rho V^\flat\otimes V^\flat$, where $V=\partial_t$. Then the chart-induced components of $\mathbf{T}$ are described by the entries of the matrix

$$[\mathbf{T}_{ij}]\approx\begin{bmatrix}\rho & & & \\ & 0 & & \\ & & 0 & \\ & & & 0\end{bmatrix}.$$

We expect that if the objects are not too massive, then the metric differs only slightly from the flat Minkowski spacetime metric described by η from Example 5.6 (p.83). So defining

$$\mathbf{h}_{ij}=\mathbf{g}_{ij}-\eta_{ij},$$

we assume that $|\mathbf{h}_{ij}|\ll 1$ and $|\partial_{x^i}\mathbf{h}_{jk}|\ll 1$. We also suppose that we have a 'stationary' spacetime metric where the components of $\mathbf{g}$ (and hence also those of $\mathbf{h}$) do not depend on t. If $G=[\mathbf{g}_{ij}]$, $Y=[\eta_{ij}]$, and $H=[\mathbf{h}_{ij}]$, then $Y=Y^{-1}$, and

$$\begin{aligned}G^{-1}&=(Y+H)^{-1}=(Y(I+YH))^{-1}=(I+YH)^{-1}Y\\&=(I-YH+(YH)^2-+\cdots)Y\approx Y-YHY.\end{aligned}$$

So $\mathbf{g}^{ij}=\eta_{ij}-\eta_{ik}\,\mathbf{h}_{k\ell}\,\eta_{\ell j}$. Note that $\operatorname{trace}\mathbf{T}=\mathbf{T}((dx^i)^\sharp,\partial_{x^i})=\mathbf{T}_{ji}\mathbf{g}^{ij}=\rho\,\mathbf{g}^{00}$.

[13] M can be identified with $\mathbb{R}\times\mathbb{R}^3$ as described in the footnote on page 289.

We will need this below (when we use the field equation in the form given in Exercise 14.1, p.306).

In the geodesic equation, which we will soon also use, we shall need the Christoffel symbols. The Christoffel symbols are given by

$$\Gamma_{ij}^k = \frac{\mathbf{g}^{\ell k}}{2}(\partial_{x^i}\mathbf{g}_{j\ell}+\partial_{x^j}\mathbf{g}_{\ell i}-\partial_{x^\ell}\mathbf{g}_{ij}) \approx \frac{\eta_{\ell k}}{2}(\partial_{x^i}\mathbf{h}_{j\ell}+\partial_{x^j}\mathbf{h}_{i\ell}-\partial_{x^\ell}\mathbf{h}_{ij}).$$

Also, we note that thanks to the stationarity assumption, for $i = 1, 2, 3$,

$$\Gamma_{i0}^k = \Gamma_{0i}^k \approx \frac{\eta_{k\ell}}{2}(\partial_{x^i}\mathbf{h}_{0\ell} - \partial_{x^\ell}\mathbf{h}_{i0}). \tag{14.1}$$

Neglecting the $\Gamma\Gamma$ terms in the expression from Proposition 9.1 (p.182) for the curvature tensor field components, we obtain

$$\mathbf{R}_{ijk}^{\ell} \approx \partial_{x^i}\Gamma_{kj}^{\ell} - \partial_{x^j}\Gamma_{ki}^{\ell}.$$

Hence $\mathbf{Ric}_{jk} = \mathbf{R}_{ijk}^{i} \approx \partial_{x^i}\Gamma_{kj}^{i} - \partial_{x^j}\Gamma_{ki}^{i}$, and in particular,

$$\begin{aligned}\mathbf{Ric}_{00} &\approx \partial_{x^i}\Gamma_{00}^{i} - \partial_t\Gamma_{0i}^{i}\\ &= \partial_{x^i}\Gamma_{00}^{i} - 0\\ &= \partial_{x^i}\Gamma_{00}^{i},\end{aligned}$$

where we used $\partial_t\Gamma_{0i}^{i}=0$ since the spacetime metric is stationary.

Newtonian equation of motion from the geodesic equation

Consider, in the Newtonian sense, the motion of a slow-moving particle, falling freely in the 'gravitational field' created by ρ. We can parametrise[14] the particle's worldline using the $x^0 = t$ coordinate function, and obtain a curve $\gamma : I \to M$, where $I \subset \mathbb{R}$ is an interval, and $\gamma(t) = (t, \mathbf{x}(t))$, $t \in I$. We have $\dot{\gamma}^0(t) = 1$, $\dot{\gamma}^1(t) = \dot{x}(t)$, $\dot{\gamma}^2(t) = \dot{y}(t)$, $\dot{\gamma}^3(t) = \dot{z}(t)$. By the Newtonian 'slow-moving' assumption, we mean $|\dot{x}(t)|, |\dot{y}(t)|, |\dot{z}(t)| \ll 1$ for all $t \in I$. Let τ denote the proper time recorded starting from $\gamma(c)$, for some $c \in I$:

$$\begin{aligned}\tau(\gamma(t)) &= \int_c^t \sqrt{-\mathbf{g}(\gamma(t))(v_{\gamma,\gamma(t)}, v_{\gamma,\gamma(t)})}\, dt\\ &\approx \int_c^t \sqrt{-(-1)}\, dt = t - c.\end{aligned}$$

Thus t is affine linearly related to the proper time. But as the particle is freely falling, there exists a reparametrisation of γ, namely a map $h : \widetilde{I} \to I$, such that $\widetilde{\gamma} := \gamma \circ h : \widetilde{I} \to M$ is a geodesic. From Example 8.3 (p.160), we may assume that $\widetilde{\gamma}$ is parametrised by proper time.

[14] If $\mu : J \to M$ is any parametrisation of the worldline of the particle, then since the map $J \ni \alpha \overset{k}{\mapsto} t(\mu(\alpha))$ is strictly increasing, it has an inverse $k^{-1} : I \to J$, where $I := k(J)$. Then we set $\gamma = \mu \circ k^{-1} : I \to M$.

We have (with $\cdot' := \frac{d}{d\tau}$)

$$\begin{aligned}
-1 &= \mathbf{g}(\tilde{\gamma}(\tau))(v_{\tilde{\gamma},\tilde{\gamma}(\tau)}, v_{\tilde{\gamma},\tilde{\gamma}(\tau)}) \\
&= \mathbf{g}(\gamma(h(\tau)))(h'(\tau)v_{\gamma,\gamma(h(\tau))}, h'(\tau)v_{\gamma,\gamma(h(\tau))}) \\
&= (h'(\tau))^2\, \mathbf{g}(\gamma(h(\tau)))(v_{\gamma,\gamma(h(\tau))}, v_{\gamma,\gamma(h(\tau))}) \\
&\approx (h'(\tau))^2(-1).
\end{aligned}$$

This implies (as h is smooth) that h' is constant, taking value everywhere either equal to $+1$ or equal to -1. Using Lemma 8.1 (p.159), it follows that

$$\begin{aligned}
0 &= (\nabla_{V_{\gamma\circ h}} V_{\gamma\circ h})(\tau) \\
&= h''(\tau)\, V_\gamma(h(\tau)) + (h'(\tau))^2 (\nabla_{V_\gamma} V_\gamma)(h(\tau)) \\
&= 0\, V_\gamma(h(\tau)) + 1\, (\nabla_{V_\gamma} V_\gamma)(h(\tau)) \\
&= (\nabla_{V_\gamma} V_\gamma)(h(\tau)).
\end{aligned}$$

So we conclude that γ itself is a geodesic. Thus γ satisfies the geodesic equation $\ddot{\gamma}^k(t) + \Gamma^k_{ij}(\gamma(t))\,\dot{\gamma}^i(t)\,\dot{\gamma}^j(t) \approx 0$. For $i, j = 1, 2, 3$, we neglect terms containing $\dot{\gamma}^i(t)\dot{\gamma}^j(t)$ or $\Gamma^k_{i0}\dot{\gamma}^i(t)$ by equation (14.1). So $\ddot{\gamma}^k(t) + \Gamma^k_{00}\cdot 1\cdot 1 \approx 0$. The x-component is $\ddot{x}(t) + \Gamma^1_{00} = 0$, where

$$\Gamma^1_{00} \approx \tfrac{1}{2}(\partial_t \mathbf{h}_{01} + \partial_t \mathbf{h}_{01} - \partial_x \mathbf{h}_{00}) \approx -\tfrac{1}{2}\,\partial_x \mathbf{h}_{00}.$$

Similarly, we obtain expressions for the y- and z-components. Altogether, $\ddot{\mathbf{x}} + \vec{\nabla}(-\frac{1}{2}\,\mathbf{h}_{00}) = 0$. This coincides with the Newtonian equation of motion $\ddot{\mathbf{x}} + \vec{\nabla}\Phi = 0$, if $\mathbf{h}_{00} = -2\,\Phi$ (up to an additive constant). So the Newtonian picture would be complete if we manage to show that $-\frac{1}{2}\mathbf{h}_{00}$ is in fact the solution to the Poisson equation $\Delta\Phi = 4\pi\rho$ (which, in Newtonian gravity, is the equation that describes how the matter source ρ produces a gravitational potential Φ, whose gradient $-\vec{\nabla}\Phi$ then gives the gravitational field). This is done below.

Newtonian gravitational potential from the field equation

By Exercise 14.1 (p.306), the field equation (assuming $\Lambda = 0$) is

$$\mathbf{Ric} = 8\pi\big(\mathbf{T} - \tfrac{1}{2}(\operatorname{trace}\mathbf{T})\,\mathbf{g}\big). \tag{14.2}$$

We had seen that $\operatorname{trace}\mathbf{T} = \rho\,\mathbf{g}^{00}$. It follows that the 00-component of the tensor field on the right-hand side is given approximately by

$$8\pi\big(\rho - \tfrac{1}{2}\rho(-1 - \mathbf{h}_{00})(-1 + \mathbf{h}_{00})\big) \approx 4\pi\rho.$$

On the other hand, the 00-component of the Ricci curvature tensor field is

$$\begin{aligned}
\mathbf{Ric}_{00} &\approx \partial_{x^i}\Gamma^i_{00} = \partial_t\Gamma^0_{00} + \partial_x\Gamma^1_{00} + \partial_y\Gamma^2_{00} + \partial_z\Gamma^3_{00} \\
&\approx 0 + \partial_x(-\tfrac{1}{2}\partial_x\mathbf{h}_{00}) + \partial_y(-\tfrac{1}{2}\partial_y\mathbf{h}_{00}) + \partial_z(-\tfrac{1}{2}\partial_z\mathbf{h}_{00}) = \Delta_{\mathbf{x}}(-\tfrac{1}{2}\,\mathbf{h}_{00}).
\end{aligned}$$

Here we used the fact that the time derivatives of the components of $\mathbf{h}$ are zero (our assumption that the spacetime metric is stationary).

Thus the 00-component of the field equation (14.2) above delivers the Poisson's equation in Newtonian gravity, $\Delta_{\mathbf{x}}(-\frac{1}{2}\,\mathbf{h}_{00}) = 4\pi\rho$. To show that $-\frac{1}{2}\,\mathbf{h}_{00} = \Phi$, we need to show the uniqueness of solutions in the Poisson equation, assuming appropriate decay, and we outline this argument below.

Suppose that the smooth functions $f, \Phi_1, \Phi_2 : \mathbb{R}^3 \to \mathbb{R}$ are such that $\Delta\Phi_i = f$, and Φ_i, $i = 1, 2$, decay fast enough to 0 in $\mathbb{R}^3$ in a way so that

$$\int_{\mathbb{R}^3} \|\nabla\Phi_i\|^2\, dV < \infty, \quad \lim_{r\to\infty} r \sup_{\|\mathbf{x}\|=r} |\Phi_i(\mathbf{x})| = 0, \text{ and } \lim_{r\to\infty} r \sup_{\|\mathbf{x}\|=r} \|\nabla\Phi_i(\mathbf{x})\| = 0.$$

Consider a ball B_r with centre at $\mathbf{0}$ and a radius $r > 0$, with boundary ∂B_r having an outward-pointing unit-length normal at $\mathbf{x} \in \partial B_r$ denoted by $\mathbf{n_x}$. Set $u = \Phi_1 - \Phi_2$. Then by the Gauss divergence theorem applied in the domain B_r, we have (using $\Delta u = \Delta\Phi_1 - \Delta\Phi_2 = f - f = 0$) that

$$\begin{aligned}\int_{B_r} \|\nabla u\|^2\, dV &= \int_{\partial B_r} u(\nabla u)\cdot \mathbf{n_x}\, dA - \int_{B_r} u\Delta u\, dV \\ &= \int_{\partial B_r} u(\nabla u)\cdot \mathbf{n_x}\, dA - \int_{B_r} u\,0\, dV = \int_{\partial B_r} u(\nabla u)\cdot \mathbf{n_x}\, dA.\end{aligned}$$

By the Cauchy-Schwarz inequality, $|u(\nabla u)\cdot\mathbf{n_x}| \leqslant |u|\|\nabla u\|$, and so from the above, we obtain

$$\int_{B_r} \|\nabla u\|^2 dV \leqslant 4\pi r^2 \sup_{\|\mathbf{x}\|=r} |u(\mathbf{x})| \sup_{\|\mathbf{x}\|=r} \|\nabla u(\mathbf{x})\|.$$

Passing to the limit as $r \to \infty$, we conclude that $\int_{\mathbb{R}^3} \|\nabla u\|^2 dV = 0$, and so $\|\nabla u\| = 0$ everywhere. Thus u is constant. But as

$$\lim_{r\to\infty} r \sup_{\|\mathbf{x}\|=r} |u(\mathbf{x})| = 0,$$

it follows that the constant value of u must be 0. So $\Phi_1 = \Phi_2$.

Exercise 14.6. (Linearised field equation.) Consider $M = \mathbb{R}^4$ as a smooth manifold with the standard smooth structure. Let $\{\mathbf{g}^{(s)} : s \in (a, b)\}$ be a family of Lorentzian metrics $\mathbf{g}^{(s)}$ such that $M \times (a, b) \ni (p, s) \mapsto g_{ij}(p, s) := (\mathbf{g}^{(s)})_{ij}(p)$ is smooth, where we use the global chart $(\mathbb{R}^4, \mathbf{x} = \mathrm{id})$. Let $0 \in (a, b)$, and let $\mathbf{g}^{(0)}$ be the Minkowski metric, i.e., $\mathbf{g}^{(0)}_{ij} = \eta_{ij}$; see Example 5.6 (p.83). Define $\mathbf{h} \in T^0_2 M$ by

$$\mathbf{h} := \frac{\partial g_{ij}(\cdot, s)}{\partial s}\Big|_{s=0} dx^i \otimes dx^j.$$

Thus $\mathbf{g}^{(s)} \approx \mathbf{g}^{(0)} + s\,\mathbf{h}$ for small $|s|$. Show that

$$\frac{\partial\, (\mathbf{g}^{(s)})^{ij}}{\partial s}\Big|_{s=0} = -\,\eta_{ik}\,\mathbf{h}_{k\ell}\,\eta_{\ell j}.$$

For a fixed $s \in (a, b)$, let $\Gamma^{(s)}{}^k_{ij}$ denote the Christoffel symbols corresponding to the Levi-Civita connection on M induced by the metric $\mathbf{g}^{(s)}$. Show that

$$\frac{\partial \Gamma^{(s)}{}^k_{ij}}{\partial s}\Big|_{s=0} := \frac{\eta^{k\ell}}{2}(\partial_{x^i}\mathbf{h}_{j\ell} + \partial_{x^j}\mathbf{h}_{i\ell} - \partial_{x^\ell}\mathbf{h}_{ij}). \qquad (\star)$$

Suppose that the cosmological constant is $\Lambda = 0$, the energy-momentum tensor field is $\mathbf{T} = 0$, and that each of the metrics $\mathbf{g}^{(s)}$, for $s \in (a, b)$, satisfies the vacuum field equation $\mathbf{Ric}^{(s)} = 0$ (where the superscript '(s)' indicates the dependence of the Ricci curvature tensor field on the parameter s). Differentiating this with respect to s and using the above, one obtains, after setting $s=0$, that

$$\frac{\partial \mathbf{Ric}^{(s)}}{\partial s}\Big|_{s=0} = 0.$$

Using the expression for the Ricci curvature tensor in terms of the Christoffel symbols, and $(\star)$ above, one obtains an equation for $\mathbf{h}$, which is referred to as the *linearised field equation in vacuum:*

$$\eta^{k\ell}(\partial_{x^k}\partial_{x^i}\mathbf{h}_{j\ell} + \partial_{x^k}\partial_{x^j}\mathbf{h}_{i\ell} - \partial_{x^i}\partial_{x^j}\mathbf{h}_{k\ell} - \partial_{x^k}\partial_{x^\ell}\mathbf{h}_{ij}) = 0.$$

Exercise 14.7. (Gauge transformations.) Consider $M=\mathbb{R}^4$ as a smooth manifold with the standard smooth structure. Let $\mathbf{g}$ be a Lorentzian metric. Let $V \in T_0^1 M$ be a complete vector field with a flow $\{\psi_s : s \in \mathbb{R}\}$. Define $\mathbf{g}^{(s)} := \psi_s^* \mathbf{g}$ for $s \in \mathbb{R}$. Show that if

$$\mathbf{h}^V := \frac{\partial \mathbf{g}_{ij}^{(s)}}{\partial s}\Big|_{s=0} dx^i \otimes dx^j,$$

then $\mathbf{h}^V = \mathcal{L}_V \mathbf{g}$. Now suppose that $(M, \mathbf{g})$ is Ricci-flat, that is, it satisfies the vacuum field equation with the cosmological constant $\Lambda = 0$. Then it can be shown that $(M, \mathbf{g}^{(s)})$ is also Ricci-flat. Thus $\mathcal{L}_V \mathbf{g}$ satisfies the linearised field equation. We conclude that a solution $\mathbf{h}$ to the linearised field equation is not unique: $\mathbf{h} + \mathcal{L}_V \mathbf{g}$ also satisfies the linearised field equation for arbitrary complete $V \in T_0^1 M$. Such transformations of $\mathbf{h}$ are called *gauge transformations.*

Remark 14.3. (Gravitational waves[15].) By an appropriate choice of gauge, one gets a wave equation for the perturbation $\mathbf{h}$ of $\mathbf{g}^{(0)}$, namely $\Box\, \mathbf{h}_{ij} = 0$, where the wave operator $\Box := \eta^{ij}\partial_{x^i}\partial_{x^j}$. For time-independent perturbations one gets the Laplace equation for $\mathbf{h}$, which is consistent with the Poisson equation for $\rho=0$. The perturbation is thought of as a 'gravitational wave' around a nominal metric $\mathbf{g}=\mathbf{g}^{(0)}$. The exact gravitational wave in Exercise 14.3 (p.306) is actually a solution of the linearised field equation. *

Exercise 14.8. (+ and $\times$ polarised monochromatic gravitational waves.) The linearised field equation in vacuum (Exercise 14.6, p.313) can be used to describe gravitational waves received by an observer far away from the source of the gravitational waves. Just as the polarised monochromatic light waves considered in Remark 12.1 on p.286, we introduce polarised monochromatic gravitational waves in this exercise, and consider their effect on 'dust'. Consider $M=\mathbb{R}^4$ with the standard smooth structure, and the metric $\mathbf{g} = \mathbf{g}^{(0)} + \mathbf{h}$, where $\mathbf{g}_{ij}^{(0)} = \eta_{ij}$, $\mathbf{h}=\alpha\mathbf{h}^{(+)} + \beta\mathbf{h}^{(-)}$, $\alpha, \beta \in \mathbb{R}$ satisfy $\alpha^2 + \beta^2 = 1$,

$$\begin{aligned}\mathbf{h}^{(+)} &= h\cos(kz-\omega t)(dx\otimes dx - dy\otimes dy),\\ \mathbf{h}^{(-)} &= h\cos(kz-\omega t)(dx\otimes dy + dy\otimes dx),\end{aligned}$$

$k, \omega > 0$, $h > 0$. Then $(M, \mathbf{g})$ is called a *monochromatic gravitational plane wave.* If $\beta=0$ (or $\alpha=0$), then $(M, \mathbf{g})$ is said to be + (respectively, $\times$) *polarised.* Assume h is small, and use the approximation $h^2 \approx 0$. It turns out that although the spacetime $(M, \mathbf{g})$ is not flat, it satisfies the linearised field equation in vacuum (and hence is approximately Ricci flat). For a fixed $\mathbf{p} = (x_0, y_0, 0) \in \mathbb{R}^3$, define $\gamma_{\mathbf{p}}(t) = (t, \mathbf{p})$, $t \in \mathbb{R}$. Check that $\gamma_{\mathbf{p}}$ is a geodesic in $(M, \mathbf{g})$. Assume we are

[15] See e.g. [Misner, Thorne and Wheeler (2017), Part VIII] and [Natário (2021), §6.4] for a detailed exposition of gravitational waves.

the observer with the worldline $\gamma_{\mathbf{0}}$. Let $R > 0$ be fixed. Suppose we are surrounded by particles whose worldlines are $\gamma_{\mathbf{p}(\phi)}$, where $\mathbf{p}(\phi) = (R\cos\phi, R\sin\phi, 0)$ for all $\phi \in [0, 2\pi)$. Thus in $(M, \mathbf{g}^{(0)})$, at any event $q = \gamma_{\mathbf{0}}(t) \in M$, we perceive the particles as lying in a ring of radius R around us in our perceived space $S(t) := v_{\gamma_{\mathbf{0}},q}^{\perp}$ at q. Suppose that a gravitational wave, say with a $+$ polarisation, arrives at some t_0. For $t \geqslant t_0$, at $q = \gamma_{\mathbf{0}}(t)$, we perceive the 'spatial distance' $d(\phi)$ to the particle at $\gamma_{\mathbf{p}(\phi)}(t)$ to be approximately the length $\sqrt{\mathbf{g}(q)(w,w)}$ of the 'separation' $w := R(\cos\phi)\partial_{x,q} + R(\sin\phi)\partial_{y,q} \in S(t)$, that is, $d(\phi) := \sqrt{(1 + h\cos(\omega t))(R\cos\phi)^2 + (1 - h\cos(\omega t))(R\sin\phi)^2}$. If $\widetilde{x}(t) := d(\phi)\cdot\cos\phi$ and $\widetilde{y}(t) := d(\phi)\cdot\sin\phi$, then show that

$$\Big(\frac{\widetilde{x}(t)}{R\sqrt{1 + h\cos(\omega t)}}\Big)^2 + \Big(\frac{\widetilde{y}(t)}{R\sqrt{1 - h\cos(\omega t)}}\Big)^2 \approx 1.$$

Consequently, we see the ring 'wobble' in time, giving a sequence of elliptical snapshots, as illustrated below. The major and minor axes of the ellipses lie along the original x and y axes, justifying the label $+$ for this polarisation.

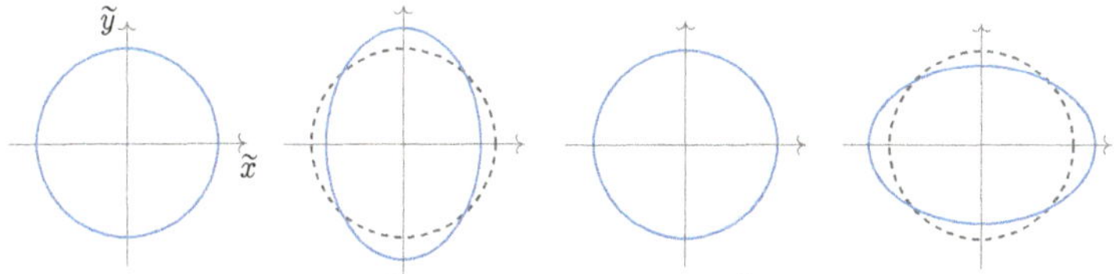

A similar effect can be observed with the $\times$ polarised wave, but the major and minor axes of the ellipses lie along the $x = y$ and $x = -y$ lines. In 2015, LIGO (Laser Interferometer Gravitational wave Observatory) detected a gravitational wave deduced to arise from the merger of $29M_\odot$ and $36M_\odot$ black holes, 1.2 billion light years from the Earth.

Remark 14.4. Any system that has a time-varying 'quadrupole moment'[16], i.e., time-variation in the mass distributions which is not spherically symmetric, can act as a source of gravitational radiation. Analogous to the energy flux of an electromagnetic wave[17], an expression for the gravitational energy flux can be derived (see, e.g., [Wald (1984), §4.4b]).

Exercise 14.9. Consider a straight rod of mass m and length ℓ rotating with an angular speed ω about its centre. The rate of gravitational energy loss mentioned in the previous remark turns out to be[18] (after reinstating the gravitational constant $G_{\rm N}$ and the speed of light c) $\dfrac{dE}{dt} = \dfrac{8G_{\rm N}m^2\ell^4\omega^6}{5c^5}$.

If $\ell = 1\,\mathrm{m}$, $m = 1\,\mathrm{kg}$, $\omega = 10^3\,\mathrm{s}^{-1}$, then determine the power loss in Watts.

[16]If $U \subset \mathbb{R}^3$ is the spatial region containing the source, then $I_{ij} = \int_U x^i x^j \rho(t, \mathbf{x})\, dV(\mathbf{x})$ is sometimes referred to as the *quadrupole moment* of the mass distribution.

[17]The energy flux of an electromagnetic wave per unit area per unit time is given by the magnitude of the Poynting vector $\vec{\mathbf{E}} \times \vec{\mathbf{B}}$ (see e.g. [Feynman (1964), §27-3]).

[18]The ω is expected to decrease over time, but the above expression is the one obtained in the linearised theory.

Exercise 14.10. In the Newtonian viewpoint, gravity is an *attractive* force. The aim of this exercise is to show a corresponding fact in the geometry-based view of gravity: For the perfect fluid matter model, the 'average tidal force' along any 'geodesic reference frame' X is negative, so that fluid particles are seen to be clumping inwards by the observers along the integral curves of X.

A *geodesic reference frame* in an open subset U of a spacetime M is a vector field $X \in T^1_0 U$ whose integral curves are timelike, future-pointing, geodesics, that are parametrised by proper time (so $\mathbf{g}(X, X) = -1$).

The *average tidal force* of X is the function $f_X : U \to \mathbb{R}$ given by

$$f_X(p) = \tfrac{1}{3}\operatorname{trace} F_{X_p} \text{ for all } p \in U,$$

where F_{X_p} is the tidal force $F_{X_p} : X_p^\perp \to X_p^\perp$ (see Definition 9.9, p.200).

(1) Let $S^2 := \{\mathbf{x} \in \mathbb{R}^3 : \|\mathbf{x}\| = 1\}$. Let $\{e_1, e_2, e_3\}$ an orthonormal basis for $X_p^\perp$. For $\mathbf{y} = (y^1, y^2, y^3) \in S^2$, set $v_{\mathbf{y}} := y^i e_i \in X_p^\perp \subset T_p M$. Then $\mathbf{g}(p)(v_{\mathbf{y}}, v_{\mathbf{y}}) = 1$. Let dA be the usual area measure on the sphere $S^2 \subset \mathbb{R}^3$. So $\int_{\mathbf{y} \in S^2} 1\, dA = 4\pi$.

Show that $f_X(p) = \dfrac{\int_{\mathbf{y} \in S^2} \mathbf{g}(p)(F_{X_p} v_{\mathbf{y}}, v_{\mathbf{y}})\, dA}{\int_{\mathbf{y} \in S^2} 1\, dA}$.

(This justifies the name for f_X, since we take an average over all directions $\mathbf{y} \in S^2$, of the relative acceleration/tidal force component in the direction of the instantaneous spatial separation $v_{\mathbf{y}}$.)

(2) Suppose the cosmological constant is 0. From Exercise 9.20 (p.202), and from the fact that $\mathbf{g}(X, X) = -1$, it follows that $f_X = -\frac{1}{3}\mathbf{Ric}(X, X)$. Using this, show that $f_X = -\frac{8\pi}{3}(\frac{\varrho}{2} + \frac{3}{2}\mathrm{p}) < 0$ pointwise. *Hint:* Use $(\mathbf{g}(X, V))^2 \geqslant 1$, which can be seen by the reversed Cauchy-Schwarz inequality.

14.2 Some symmetries of spacetime

In this section, we present a spacetime which can be thought of as being '(spatially) spherically symmetric' and 'stationary'. We begin with the following definitions.

Definition 14.1. (Symmetry of a spacetime; stationary metric.)
Let $(M, \mathbf{g})$ be a spacetime. A *symmetry of* $(M, \mathbf{g})$ is a Killing vector field $V \in T^1_0 M$ for $\mathbf{g}$, that is, $\mathcal{L}_V \mathbf{g} = 0$. The spacetime metric $\mathbf{g}$ is called *stationary* if there exists a symmetry $V \in T^1_0 M$ of $(M, \mathbf{g})$ which is everywhere timelike, that is, $\mathbf{g}(p)(V_p, V_p) < 0$ for all $p \in M$.

A spherically symmetric stationary spacetime metric

Let us consider the smooth manifold $M = \mathbb{R} \times \Omega$, where

$$\Omega = \{(x, y, z) \in \mathbb{R}^3 \mid r_{\min} < r := \sqrt{x^2 + y^2 + z^2}\},$$

where $r_{\min} \geqslant 0$. Then M is diffeomorphically[19] identified with the smooth manifold $\mathbb{R} \times (r_{\min}, \infty) \times S^2$. We use the global chart (M, id). We will also

[19]Under the diffeomorphism $M \ni (t, (x, y, z)) \mapsto (t, r, \frac{(x,y,z)}{r})$, where $r = \sqrt{x^2 + y^2 + z^2}$.

use the notation (x^0, x^1, x^2, x^3) or (t, x, y, z) for the component functions of this (identity) chart map. The smooth manifold M is given the metric

$$\mathbf{g} = -f(r)\, dt \otimes dt + (k(r)-1)\, dr \otimes dr + dx \otimes dx + dy \otimes dy + dz \otimes dz,$$

where $f(r), k(r) > 0$ for all $r > r_{\min}$. If we use the usual spherical polar coordinates in Ω, then the metric $dx \otimes dx + dy \otimes dy + dz \otimes dz$ on Ω can be expressed (see Exercise 5.14, p.83) in the spherical polar coordinates (r, θ, ϕ) as $dr \otimes dr + r^2 d\theta \otimes d\theta + r^2 (\sin\theta)^2 d\phi \otimes d\phi$. Consequently, the metric $\mathbf{g}$ on M in the (t, r, θ, ϕ)-coordinates is given by

$$\mathbf{g} = -f(r)\, dt \otimes dt + k(r)\, dr \otimes dr + r^2 d\theta \otimes d\theta + r^2 (\sin\theta)^2 d\phi \otimes d\phi.$$

It is clear that $(M, \mathbf{g})$ is a Lorentzian manifold since $f(r), k(r) > 0$ for all $r > r_{\min}$. The time-orientation on M is $V = \partial_t \in T_0^1 M$.

Lemma 14.1. *For* $(M, \mathbf{g})$ *as above,* $\mathcal{L}_{\partial_t} \mathbf{g} = 0$, *that is, the spacetime metric is stationary.*

Proof. We use the chart (M, id). Then none of the components of $\mathbf{g}$ have t-dependence. Indeed, $dr = (1/r)(x\, dx + y\, dy + z\, dz)$, so that

$$dr \otimes dr = r^{-2}(x\, dx + y\, dy + z\, dz) \otimes (x\, dx + y\, dy + z\, dz).$$

Inserting this in the expression for $\mathbf{g}$, and collecting like terms $dx^i \otimes dx^j$, we see that the coefficients of these do not have any t-dependence. Using $\partial_t(\mathbf{g}_{ij}) = 0$ and $[\partial_{x^i}, \partial_{x^j}] = 0$, we obtain

$$\begin{aligned}(\mathcal{L}_{\partial_t} \mathbf{g})(\partial_{x^i}, \partial_{x^j}) &= \partial_t(\mathbf{g}(\partial_{x^i}, \partial_{x^j})) - \mathbf{g}([\partial_t, \partial_{x^i}], \partial_{x^j}) - \mathbf{g}(\partial_{x^i}, [\partial_t, \partial_{x^i}]) \\ &= \partial_t(\mathbf{g}_{ij}) - 0 - 0 = 0. \end{aligned}$$

□

We recall from Exercise 3.24 (p.55) the three vector fields L_x, L_y, L_z, which can be thought of the 'generators of rotations', since their flow maps were found to be rotations about the x-, y-, z-axis, respectively. Thus in light of Theorem 6.3 (p.131), the vanishing of the Lie derivatives of $\mathbf{g}$ with respect to these three vector fields means that $(M, \mathbf{g})$ is 'spherically symmetric'.

Lemma 14.2. *If* $(M, \mathbf{g})$ *is as above, and the vector fields* $L_x, L_y, L_z \in T_0^1 M$ *are given by*

$$\begin{aligned} L_x &= y\, \partial_z - z\, \partial_y \\ L_y &= z\, \partial_x - x\, \partial_z \\ L_z &= x\, \partial_y - y\, \partial_x, \end{aligned}$$

then $\mathcal{L}_{L_x} \mathbf{g} = \mathcal{L}_{L_y} \mathbf{g} = \mathcal{L}_{L_z} \mathbf{g} = 0$.

Proof. Let $X, Y \in T_0^1 M$, and decompose $X = X^i \partial_{x^i}$, $Y = Y^j \partial_{x^j}$, where $X^i, Y^j \in C^\infty(M)$. Below we use (t, x, y, z) and (x^0, x^1, x^2, x^3) interchangeably.

For any smooth function $\varphi \in C^\infty(M)$,

$$\begin{aligned}
&[L_x, X]\varphi \\
&= y\,\partial_z(X^i\partial_{x^i}\varphi) - z\,\partial_y(X^i\partial_{x^i}\varphi) - X^i\partial_{x^i}(y\,\partial_z\varphi) + X^i\partial_{x^i}(z\,\partial_y\varphi) \\
&= y(\partial_z X^i)\partial_{x^i}\varphi - z(\partial_y X^i)(\partial_{x^i}\varphi) - X^i(\partial_{x^i}y)(\partial_z\varphi) + X^i(\partial_{x^i}z)(\partial_y\varphi) \\
&\quad + y\,X^i\partial_z\partial_{x^i}\varphi - z\,X^i\partial_y\partial_{x^i}\varphi - X^i y\,\partial_{x^i}\partial_z\varphi + X^i z\,\partial_{x^i}\partial_y\varphi \\
&= y(\partial_z X^i)(\partial_{x^i}\varphi) - X^i(\partial_{x^i}y)(\partial_z\varphi) - z(\partial_y X^i)(\partial_{x^i}\varphi) + X^i(\partial_{x^i}z)(\partial_y\varphi) \\
&= \big(y(\partial_z X^i)\partial_{x^i} - (Xy)\partial_z - z(\partial_y X^i)\partial_{x^i} + (Xz)\partial_y\big)\varphi \\
&= \big((L_x X^i)\partial_{x^i} - X^y\partial_z + X^z\partial_y\big)\varphi.
\end{aligned}$$

Thus $[L_x, X] = (L_x X^i)\partial_{x^i} - X^y\partial_z + X^z\partial_y$. Also, we have

$$\sum_{i=1}^{3} X^i L_x x^i = \sum_{i=1}^{3} X^i(y\,\partial_z - z\,\partial_y)x^i = y\,X^z - z\,X^y.$$

We have

$$(x\,dx + y\,dy + z\,dz)((L_x X^i)\partial_{x^i} - X^y\partial_z + X^z\partial_y) = -zX^y + yX^z + \sum_{i=1}^{3} x^i L_x X^i,$$

$$(x\,dx + y\,dy + z\,dz)Y = xY^x + yY^y + zY^z = \sum_{i=1}^{3} x^i Y^i.$$

Using $dr \otimes dr = \frac{1}{r^2}(x\,dx + y\,dy + z\,dz) \otimes (x\,dx + y\,dy + z\,dz)$, we obtain

$$\begin{aligned}
\mathbf{g}([L_x,X], Y) &= -f(r)Y^0 L_x X^0 + \frac{k(r)-1}{r^2}\Big(\sum_{i=1}^{3} x^i L_x X^i + yX^z - X^y z\Big)\sum_{j=1}^{3} Y^j x^j \\
&\quad + \sum_{i=1}^{3} Y^i L_x X^i + X^z Y^y - X^y Y^z \\
&= -f(r)Y^0 L_x X^0 + \frac{k(r)-1}{r^2}\sum_{i=1}^{3}(x^i L_x X^i + X^i L_x x^i)\sum_{j=1}^{3} Y^j x^j \\
&\quad + \sum_{i=1}^{3} Y^i L_x X^i + X^z Y^y - X^y Y^z.
\end{aligned}$$

Also,

$$\begin{aligned}
\mathbf{g}(X, [L_x, Y]) &= -f(r)X^0 L_x Y^0 + \frac{k(r)-1}{r^2}\sum_{i=1}^{3} X^i x^i \sum_{j=1}^{3}(x^j L_x Y^j + Y^j L_x x^j) \\
&\quad + \sum_{i=1}^{3} X^i L_x Y^i + Y^z X^y - Y^y X^z, \\
L_x(\mathbf{g}(X,Y)) &= L_x\Big(-f(r)X^0 Y^0 + \frac{k(r)-1}{r^2}\sum_{i=1}^{3} X^i x^i \sum_{j=1}^{3} Y^j x^j + \sum_{i=1}^{3} X^i Y^i\Big).
\end{aligned}$$

We claim $L_x(f(r)) = 0$ and $L_x\big(\frac{k(r)-1}{r^2}\big) = 0$: For any smooth function φ,

$$\begin{aligned}
L_x(\varphi(r)) &= y\,\partial_z(\varphi(\sqrt{x^2+y^2+z^2})) - z\,\partial_y(\varphi(\sqrt{x^2+y^2+z^2})) \\
&= y\varphi'(r)\frac{2z}{2r} - z\varphi'(r)\frac{2y}{2r} = 0.
\end{aligned}$$

So we get $(\mathcal{L}_{L_x}\mathbf{g})(X,Y) = L_x(\mathbf{g}(X,Y)) - \mathbf{g}([L_x,X],Y) - \mathbf{g}(X,[L_x,Y]) = 0$, by using the above. Hence $\mathcal{L}_{L_x}\mathbf{g} = 0$. Cycling through x, y, z, also $\mathcal{L}_{L_y}\mathbf{g} = 0$ and $\mathcal{L}_{L_z}\mathbf{g} = 0$. □

14.3 Schwarzschild spacetime

The field equations were published in 1915, but besides the Minkowski spacetime solution, no other exact solutions were known then. In 1916, Schwarzschild derived another solution, by making the simple symmetry assumptions of 'time-independence' and 'spherical symmetry', and that the spacetime metric looks like the Minkowski metric far away from the source. The resulting Schwarzschild spacetime is a simple model of a spacetime containing a single star. It is a good model for the exterior region in the proximity of a star or a planet. Moreover, if the matter source is sufficiently dense, then it also gives a model of the simplest possible 'black hole'.

Consider the smooth manifold $M=\mathbb{R}\times(r_{\min},\infty)\times S^2$ as in the previous section, where $r_{\min}\geqslant 0$. We have seen a metric $\mathbf{g}$ that is stationary and spherically symmetric, and which, in the (t,r,θ,ϕ)-coordinates, is given by

$$\mathbf{g} = -f(r)\,dt\otimes dt + k(r)\,dr\otimes dr + r^2 d\theta\otimes d\theta + r^2(\sin\theta)^2 d\phi\otimes d\phi. \quad (14.3)$$

In this section, we will use the field equation to solve for the functions f,k. We label the coordinate functions (t,r,θ,ϕ) as (x^0,x^1,x^2,x^3). First, we note that the inverse of the diagonal matrix $[\mathbf{g}_{ij}]$ is again a diagonal matrix with the diagonal entries

$$\mathbf{g}^{00}=-\frac{1}{f(r)},\quad \mathbf{g}^{11}=\frac{1}{k(r)},\quad \mathbf{g}^{22}=\frac{1}{r^2},\quad \mathbf{g}^{33}=\frac{1}{r^2(\sin\theta)^2}.$$

We can now compute the Christoffel coefficients. We suppress writing the argument r for the functions f,k below. For example,

$$\Gamma^0_{01}=\Gamma^0_{10}=\frac{\mathbf{g}^{tt}}{2}(\partial_t\mathbf{g}_{tr}+\partial_r\mathbf{g}_{tt}-\partial_t\mathbf{g}_{tr}) = -\frac{1}{2f}(\partial_r(-f)) = \frac{f'}{2f}.$$

Similarly, the other nonzero Γ-symbols can be found to be

$$\Gamma^1_{00}=\frac{f'}{2k},\qquad \Gamma^1_{11}=\frac{k'}{2k},\qquad \Gamma^1_{22}=-\frac{r}{k},\qquad \Gamma^1_{33}=-\frac{r(\sin\theta)^2}{k},$$

$$\Gamma^2_{12}=\Gamma^2_{21}=\frac{1}{r},\quad \Gamma^2_{33}=-(\sin\theta)\cos\theta,\quad \Gamma^3_{13}=\Gamma^3_{31}=\frac{1}{r},\quad \Gamma^3_{23}=\Gamma^3_{32}=\cot\theta.$$

Knowing the Christoffel coefficients, we can compute the components of the Ricci curvature tensor field. For example,

$$\begin{aligned}
\mathbf{Ric}_{00} &= \partial_{x^k}\Gamma^k_{00}-\partial_t\Gamma^k_{0k}+\Gamma^i_{00}\Gamma^k_{ik}-\Gamma^i_{0k}\Gamma^k_{i0}\\
&= \partial_r\Gamma^1_{00}-0+\Gamma^1_{00}(\Gamma^0_{10}+\Gamma^1_{11}+\Gamma^2_{12}+\Gamma^3_{13})-\Gamma^1_{00}\Gamma^0_{10}-\Gamma^0_{01}\Gamma^1_{00}\\
&= \partial_r\Big(\frac{f'}{2k}\Big)+\frac{f'}{2k}\Big(\frac{f'}{2f}+\frac{k'}{2k}+\frac{1}{r}+\frac{1}{r}\Big)-\frac{f'}{2k}\frac{f'}{2f}-\frac{f'}{2f}\frac{f'}{2k}\\
&= \frac{f''}{2k}-\frac{f'k'}{4k^2}-\frac{(f')^2}{4kf}+\frac{f'}{kr},
\end{aligned}$$

where $\cdot' := \dfrac{d}{dr}$.

In a similar manner, we can compute the other nonzero components of **Ric**:

$$\begin{aligned}
\mathbf{Ric}_{11} &= -\frac{f''}{2f}+\frac{(f')^2}{4f^2}+\frac{k'f'}{4kf}+\frac{k'}{kr},\\
\mathbf{Ric}_{22} &= -\frac{1}{k}+1-\frac{rf'}{2kf}+\frac{rk'}{2k^2},\\
\mathbf{Ric}_{33} &= (\sin\theta)^2\,\mathbf{Ric}_{22}.
\end{aligned}$$

In the exterior of the star, $\mathbf{T}=0$. We also assume that the cosmological constant is $\Lambda=0$ or in any case negligible. Then the field equation given in Exercise 14.1 (p.306) reduces simply to $\mathbf{Ric}=0$. In particular, $\mathbf{Ric}_{ii}=0$ for $i=0,1$, and so also $\frac{k}{f}\mathbf{Ric}_{00}+\mathbf{Ric}_{11}=0$, which yields

$$\frac{f''}{2f}-\frac{f'k'}{4kf}-\frac{(f')^2}{4f^2}+\frac{f'}{rf}-\frac{f''}{2f}+\frac{(f')^2}{4f^2}+\frac{f'k'}{4kf}+\frac{k'}{kr}=0,\ \text{ i.e., }\ \frac{f'}{f}+\frac{k'}{k}=0.$$

Hence $f'k+k'f=0$, that is, $(fk)'=0$. Thus fk is constant, say C. But as $r\to\infty$, we expect that the metric expression (14.3) for $\mathbf{g}$ approaches the Minkowski spacetime metric (in spherical coordinates), so that

$$\lim_{r\to\infty} f(r)=1=\lim_{r\to\infty} k(r).$$

Passing the limit $r\to\infty$ in the relation $f(r)k(r)=C$, we get $C=1$. Thus

$$k(r)=\frac{1}{f(r)}.$$

Substituting for k in $\mathbf{Ric}_{22}=0$ gives

$$-f+1-\frac{rf'}{2}+\frac{rf^2}{2}\Big(\frac{1}{f}\Big)'=0.$$

So $rf'+f=1$, i.e., $(rf)'=1$. Thus by integrating, we obtain $rf(r)=r+A$, where $A\in\mathbb{R}$ is an integration constant. This yields $f(r)=1+\frac{A}{r}$. So far, we have not specified $r_{\min}$ ($\geqslant 0$). We want $f(r)>0$ on $(r_{\min},\infty)$, with $r_{\min}\geqslant 0$ the smallest possible such number. We have the following cases:

1° $A\geqslant 0$. Then $r_{\min}:=0$.

2° $A<0$. Then $r_{\min}:=-A$.

We will soon argue on physical grounds that $A=:-2m<0$, where $m>0$ is thought of as the mass of the star. For now, given an $A\in\mathbb{R}$, we just take $r_{\min}$ as specified above. Then,

$$f(r)=1+\frac{A}{r}\quad\text{and}\quad k(r)=\Big(1+\frac{A}{r}\Big)^{-1}.$$

We can plug these expressions back and check that indeed $\mathbf{Ric}_{ii}=0$ for $i=0,1,2,3$, so that the field equations do hold (the other Ricci curvature field components $\mathbf{Ric}_{ij}$, $i\neq j$, were already known to be zero for arbitrary f,k). Thus we obtain the *Schwarzschild metric*

$$\mathbf{g}=-\Big(1+\frac{A}{r}\Big)dt\otimes dt+\Big(1+\frac{A}{r}\Big)^{-1}dr\otimes dr+r^2d\theta\otimes d\theta+r^2(\sin\theta)^2d\phi\otimes d\phi$$

for the spacetime region $\mathbb{R}\times(r_{\min},\infty)\times S^2$ (thought of as the exterior of a star), under the assumptions that the metric is stationary[20], spherically symmetric, and it approaches the Minkowski spacetime metric as $r\to\infty$.

We now give an argument to justify that $A=:-2m<0$, and that then $r_{\min}=-A=2m$, where m can be interpreted to be the mass of the star. Recall that in §14.1, we had found that in the Newtonian limit, if we write the metric $\mathbf{g}$ on $\mathbb{R}\times\mathbb{R}^3$ as $\mathbf{g}_{ij}=\eta_{ij}+\mathbf{h}_{ij}$, then $\mathbf{h}_{00}=-2\Phi$, where Φ is the gravitational potential. But at a distance r from the source of mass m (assumed here to be a homogeneous ball of a finite radius $R<r$), the Newtonian gravitational potential is (see Exercise 13.1, p.291)

$$\Phi=-\frac{m}{r}.$$

So for large r, writing the Schwarzschild metric as $\mathbf{g}_{ij}=\eta_{ij}+\mathbf{h}_{ij}$, we have

$$\mathbf{g}_{00}=-1+\mathbf{h}_{00}\approx-1-2\Phi=-1-2\frac{-m}{r}=-\big(1-\frac{2m}{r}\big).$$

Thus if we identify A with $-2m$, i.e., $m:=-A/2$, then this m corresponds to the mass of the star in the classical viewpoint, and in particular $m>0$. So $A=:-2m>0$, and $r_{\min}=2m$, where m has the interpretation of being the mass of the star.

Alternatively, we could look at the geodesic equation to justify this correspondence. We do this in the following exercise.

Exercise 14.11. (Radial geodesic in Schwarzschild spacetime.)
Let $(M,\mathbf{g})$ be the Schwarzschild spacetime. Let $\theta_0\in(0,\pi)$ and $\phi_0\in(0,2\pi)$ be fixed, and $\mathbf{p}\subset S^2$ be the point with the spherical polar coordinates (θ_0,ϕ_0).
Let $\gamma:I\to M$, $I\ni\tau\mapsto(t(\tau),r(\tau),\mathbf{p})$, be the worldline of a free-falling particle, parametrised by the arclength/proper time. We have that $\theta\equiv\theta_0$ and $\phi\equiv\phi_0$ for all $\tau\in I$ satisfy the θ and ϕ components of the geodesic equation because $\theta''+2\Gamma^{\theta}_{r\theta}\theta'r'+\Gamma^{\theta}_{\phi\phi}\phi'\phi'=0$, and also $\phi''+2\Gamma^{\phi}_{\theta\phi}\theta'\phi'+2\Gamma^{\phi}_{r\phi}r'\phi'=0$. Show that if $\cdot':=\frac{d}{d\tau}$, then:

$$-\big(1-\frac{2m}{r}\big)(t')^2+\big(1-\frac{2m}{r}\big)^{-1}(r')^2=-1,\quad t''=-\frac{2m}{r^2}\big(1-\frac{2m}{r}\big)^{-1}t'r',\quad r''=-\frac{m}{r^2}.$$

Let γ pass through $p\in M$. Consider a normalised instantaneous observer

$$v=\sqrt{\big(1-\frac{2m}{r}\big)^{-1}}\;\partial_{t,p}\in T_pM.$$

Show that v perceives the relative speed of the particle γ to be

$$u=\frac{r'}{t'}\sqrt{\big(1-\frac{2m}{r}\big)^{-1}}\;\partial_{r,p}.$$

Determine the magnitude $|u|:=\sqrt{\mathbf{g}(p)(u,u)}$ of u as perceived by v.

[20]Birkhoff's theorem says that the assumptions of stationarity and that the metric approaches the Minkowski metric away from the source, are superfluous here. For a precise statement and a proof, we refer the reader to [Kriele (2001), Theorem 7.2.1].

Now suppose that $r \gg 2m$, and $|u| \ll 1$ (that is, low speed as reckoned by a 'stationary observer' far away from the source). Conclude that $|r'| \ll t'$, and $t' \approx 1$. (In physics parlance, $t' \approx 1$ is expressed as 'coordinate time is proper time'.) Thus if the map h denotes the inverse of the map $\tau \mapsto t(\tau)$, then $\gamma \circ h$ is the map $t \mapsto (t, r(h(t)), \theta_0, \phi_0)$. Show that

$$\frac{d^2}{dt^2}(r \circ h) \approx -\frac{m}{r^2}.$$

From the Newtonian viewpoint, in the spacetime $\mathbb{R} \times \mathbb{R}^3$, a radially freely falling particle in the gravitational field of a mass m at the origin experiences an acceleration which matches with the above. So the constant m in the Schwarzschild spacetime can be thought of as being approximately the mass of the star.

Exercise 14.12. (Radial escape velocity in Schwarzschild spacetime.) In continuation of the previous exercise, determine the *escape velocity* of the radial geodesic, defined as the infimum of the relative speed $|u|$ with respect to the normalised instantaneous observer v such that $r'(\tau) > 0$ for all $\tau > 0$. Is the escape velocity the same as that obtained from classical considerations? (See Exercise 15.2, p.338.) *Hint:* First show that $E := f(r)t'$ is constant along the radial geodesic.

We note that the Schwarzschild metric can be used only for $r > 2m$, since if $r = 2m$, the coefficient of $dt \otimes dt$ becomes zero, and also the coefficient of $dr \otimes dr$ is not defined. For a star like our Sun, we have $2m \approx 2M_\odot \approx 3\,\text{km}$, and so when we are outside the body of the Sun, we do have $r > 2m$. For the Earth, $2m \approx 2M_\oplus \approx 9\,\text{mm}$, which is again much smaller than its own radius. For a body of mass m, we define its *Schwarzschild radius* R_S to be $2m$. A body with a radius smaller than its own Schwarzschild radius is thought of as a 'black hole', and will be discussed in the next chapter.

Exercise 14.13. Find the minimum density needed for a body with radius equal to that of the Earth ($R_\oplus = 6400$km), so that its Schwarzschild radius is at least as large as $R_\oplus$. Compare it with the density of water, $\rho_\text{water} = 1000\,\text{kg/m}^3$.

Exercise 14.14. (Reissner-Nordström spacetime.)
Recall the Reissner-Nordström spacetime $(M, \mathbf{g})$ from Exercise 12.15 (p.287), and the $(0,2)$-tensor field $\mathbf{T} \in T_2^0 M$ given in Exercise 13.7 (p.303). Show that the field equation is satisfied, assuming that the cosmological constant $\Lambda = 0$.

14.4 Perihelion precession

We had seen two effects within the spacetime-geometry viewpoint of gravity, which could not be explained classically, and which have been tested experimentally/by observation:

- the gravitational red-shift (Example 5.16, p.96), and
- the deflection of light (Section 8.4, p.163).

We now learn about a third observational verification, which is on the trajectories of planets in the solar system.

In particular, the effect is most pronounced for Mercury, and the amount of 'precession of its orbit' predicted matches very well with the observed value. We begin by discussing orbits of planets within the classical framework in the idealised case when we have a single planet around a star.

Newtonian description of planetary motion

Let us first consider planetary motion as described in Newtonian gravity. We will learn that the trajectory of a planet like Mercury around the Sun is in an ellipse, with the Sun at one of the foci.

Let the Sun be fixed at the origin of $\mathbb{R}^3$. It generates a gravitational field in which the test matter, namely Mercury, moves according to the law of gravitation. Here we assume that (since the Sun is much more massive than Mercury) the common center of mass of the Sun and Mercury is essentially located at the center of the Sun, that is, the origin.

If $\mathbf{x}(t) = (x(t), y(y), z(t)) \in \mathbb{R}^3$ is the position of the planet at time $t \in \mathbb{R}$, then the equation of motion is

$$\ddot{\mathbf{x}} = -\frac{m}{\|\mathbf{x}\|^3}\mathbf{x}.$$

Here m is the mass of the Sun, and $\dot{} = \frac{d}{dt}$. If $\mathbf{v} = \dot{\mathbf{x}}$ is the velocity, then

$$\frac{d}{dt}(\mathbf{x} \times \mathbf{v}) = \dot{\mathbf{x}} \times \mathbf{v} + \mathbf{x} \times \dot{\mathbf{v}} = \mathbf{v} \times \mathbf{v} + \mathbf{x} \times \ddot{\mathbf{x}} = \mathbf{0} - \frac{m}{\|\mathbf{x}\|^3}(\mathbf{x} \times \mathbf{x}) = \mathbf{0}.$$

Thus the 'angular momentum' vector $\mathbf{L} = \mathbf{x} \times \mathbf{v}$ is a constant. We assume henceforth that $\mathbf{L} \neq \mathbf{0}$. Then the planet is confined[21] to move in the plane which contains the 'initial' position vector $\mathbf{x}(0)$ of the planet, and the initial velocity $\mathbf{v}(0)$ of the planet. See the following picture.

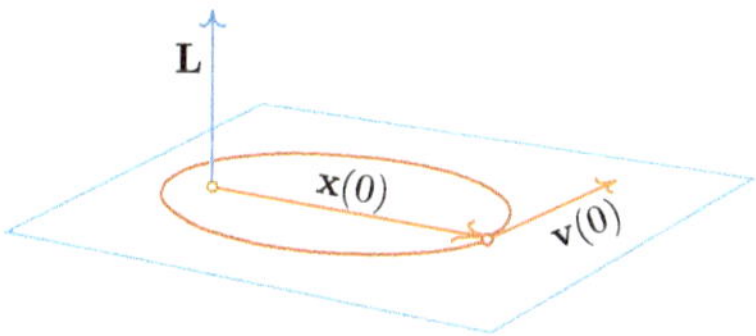

By a suitable rotation of coordinates in $\mathbb{R}^3$, we can assume that the plane of planetary motion is the xy-plane, and we will use planar polar coordinates (r, ϕ) in this plane. Then $\mathbf{x} = (r\cos\phi,\ r\sin\phi,\ 0)$, and so

$$\mathbf{v} = (\dot{r}\cos\phi - r\dot{\phi}\sin\phi,\ \dot{r}\sin\phi + r\dot{\phi}\cos\phi,\ 0).$$

We know the angular momentum vector $\mathbf{L} = \mathbf{x} \times \mathbf{v}$ is constant. We have

$$\begin{aligned}\mathbf{L} &= \mathbf{x} \times \mathbf{v} = (r\cos\phi,\ r\sin\phi,\ 0) \times (\dot{r}\cos\phi - r\dot{\phi}\sin\phi,\ \dot{r}\sin\phi + r\dot{\phi}\cos\phi,\ 0) \\ &= (0,\ 0,\ r^2\dot{\phi}).\end{aligned}$$

[21]Indeed, $\mathbf{x} \cdot \mathbf{L} = \mathbf{x} \cdot (\mathbf{x} \times \mathbf{v}) = 0$, showing that $\mathbf{x}(t)$ lies in the plane perpendicular to $\mathbf{L}$.

So $L:=r^2\dot{\phi}$ is a constant[22]. Differentiating $r^2=\langle \mathbf{x},\mathbf{x}\rangle$ with respect to time, we get $2r\dot{r}=2\langle \mathbf{x},\dot{\mathbf{x}}\rangle=2\langle \mathbf{x},\mathbf{v}\rangle$, so that $r\dot{r}=\langle \mathbf{x},\mathbf{v}\rangle$. Differentiating again,

$$r\ddot{r}+\dot{r}^2=\langle \mathbf{v},\mathbf{v}\rangle+\langle \mathbf{x},\ddot{\mathbf{x}}\rangle=\|\mathbf{v}\|^2+\Big\langle \mathbf{x},-\frac{m}{\|\mathbf{x}\|^3}\mathbf{x}\Big\rangle=\|\mathbf{v}\|^2-\frac{m}{\|\mathbf{x}\|}. \tag{14.4}$$

But $\|\mathbf{x}\|=r$ and $\|\mathbf{v}\|^2=(\dot{r}\cos\phi-r\dot{\phi}\sin\phi)^2+(\dot{r}\sin\phi+r\dot{\phi}\cos\phi)^2=\dot{r}^2+r^2\dot{\phi}^2$. Hence (14.4) becomes $r\ddot{r}+\dot{r}^2=\dot{r}^2+r^2\dot{\phi}^2-\frac{m}{r}$, that is,

$$r\ddot{r}=r^2\dot{\phi}^2-\frac{m}{r}=\frac{(r^2\dot{\phi})^2}{r^2}-\frac{m}{r}=\frac{L^2}{r^2}-\frac{m}{r}. \tag{14.5}$$

We want r as a function of ϕ in order to see what the trajectory looks like in the xy-plane. So let us suppose that (locally) $t\mapsto\phi(t)$ is a diffeomorphism with an inverse h, and set $u(\phi)=\frac{1}{(r\circ h)(\phi)}$. (It turns out it is more convenient to work with the reciprocal of r.) Then

$$\frac{du}{d\phi}=-\frac{1}{(r(h(\cdot)))^2}\frac{dr}{dt}(h(\cdot))\frac{dh}{d\phi}=-\frac{1}{(r(h(\cdot)))^2}\dot{r}(h(\cdot))\frac{1}{\dot{\phi}(h(\cdot))}=-\frac{\dot{r}(h(\cdot))}{L}.$$

Similarly

$$\begin{aligned}\frac{d^2u}{d\phi^2}&=-\frac{1}{L}\frac{d(\dot{r}\circ h)}{d\phi}=-\frac{1}{L}\ddot{r}(h(\cdot))\frac{dh}{d\phi}=-\frac{1}{L}\ddot{r}(h(\cdot))\frac{1}{\dot{\phi}(h(\cdot))}\\&=-\frac{1}{L}\ddot{r}(h(\cdot))\frac{(r(h(\cdot)))^2}{(r(h(\cdot)))^2\dot{\phi}(h(\cdot))}=-\frac{(r(h(\cdot)))^2\ddot{r}(h(\cdot))}{L^2}=-\frac{\ddot{r}(h(\cdot))}{u^2L^2}.\end{aligned}$$

Substituting this in (14.5), we obtain

$$\frac{d^2u}{d\phi^2}+u=\frac{m}{L^2}. \tag{14.6}$$

The general solution to the homogeneous equation

$$\frac{d^2u}{d\phi^2}+u=0$$

is given by $u_{\text{hom}}=A\cos(\phi-\phi_0)$, where $A\geqslant 0$ and $\phi_0\in[0,2\pi)$. A particular solution to (14.6) is the constant function $u_{\text{par}}=\frac{m}{L^2}$. Thus the general solution to (14.6) is given by

$$u(\phi)=A\cos(\phi-\phi_0)+\frac{m}{L^2}.$$

We are interested in what this trajectory looks like. We may assume that the xy-coordinate plane has been rotated about the z axis if necessary to have $\phi_0=0$. Then the above gives with $t=h(\phi)$ that

$$r(t)=\frac{1}{u(\phi)}=\frac{1}{\frac{m}{L^2}+A\cos\phi}=\frac{\frac{L^2}{m}}{1+\frac{AL^2}{m}\cos\phi}=\frac{k}{1+e\cos\phi}, \tag{14.7}$$

where $k:=\frac{L^2}{m}>0$ and $e:=\frac{AL^2}{m}\geqslant 0$.

[22] This one of Kepler's laws of planetary motion: A line joining a planet and the Sun sweeps out equal areas during equal intervals of time.

Now we will show that if $0<e<1$, then $\phi \mapsto \frac{1}{u(\phi)}$ describes an ellipse.

Consider an ellipse in the xy-plane with foci at $(0,0)$ and $(-2c,0)$ where $c>0$. The centre C of the ellipse is at $(-c,0)$. If the lengths of the major and minor axes are $2a$, respectively $2b$ $(<2a)$, then the ellipse is given by

$$\frac{(x+c)^2}{a^2}+\frac{y^2}{b^2}=1. \tag{14.8}$$

Recall also that the ellipse is the locus of a movable point P such that the sum S of the distances of P to the two foci is a constant. By taking the point P along the major axis, we obtain that $S=c+a+(a-c)=2a$. Next taking the movable point P along the minor axis, and using Pythagoras theorem in the right angled triangle formed by P, C and the origin, we obtain $b^2+c^2=(S/2)^2=a^2$. Thus the *eccentricity* e of the ellipse is

$$e:=\frac{\text{distance between foci}}{\text{major axis length}}=\frac{2c}{2a}=\frac{c}{a}=\frac{\sqrt{a^2-b^2}}{a}=\sqrt{1-\frac{b^2}{a^2}}\in(0,1).$$

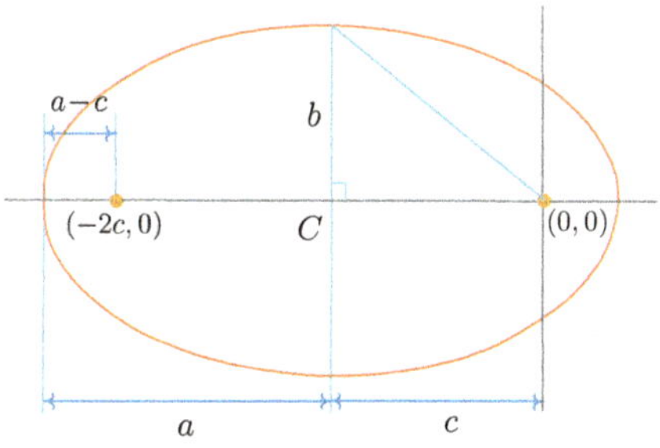

Rewriting (14.7) as $r+er\cos\phi=k$, we have $r+ex=k$, i.e., $r=k-ex$. Squaring, $x^2+y^2=r^2=e^2x^2-2ekx+k^2$, i.e., $(1-e^2)x^2+2ekx+y^2=k^2$. Completing the square in the x terms,

$$(1-e^2)\Big(x+\frac{ek}{1-e^2}\Big)^2+y^2=k^2+\frac{e^2k^2}{1-e^2}=\frac{k^2}{1-e^2},$$

that is,

$$\frac{\Big(x+\frac{ek}{1-e^2}\Big)^2}{\Big(\frac{k}{1-e^2}\Big)^2}+\frac{y^2}{\Big(\frac{k}{\sqrt{1-e^2}}\Big)^2}=1,$$

which matches (14.8) with $a=\frac{k}{1-e^2}>b=\frac{k}{\sqrt{1-e^2}}>0$, and $c=\frac{ek}{1-e^2}=ae$. So

$$\phi\mapsto\frac{1}{u(\phi)}=\frac{k}{1+e\cos\phi}$$

describes an ellipse, with $(0,0)$ as one of its foci. Consequently, the trajectory of the planet is an ellipse with the Sun its focus.

Planetary motion as a spacetime geodesic

We consider the exterior region of the Sun as Schwarzschild spacetime $(M, \mathbf{g})$, where $\mathbf{g}$ is the Schwarzschild metric, and $2m \approx 2M_\odot = 3\text{km}$. The planet Mercury is considered as a 'test particle' which does not influence the geometry (so its energy-momentum contribution in the field equation is negligible). Thus its trajectory γ, parametrised by proper time, is a geodesic $I \ni \tau \mapsto (t(\tau), r(\tau), \mathbf{p}(\tau))$, where $\mathbf{p}(\tau) \in S^2$ has the spherical polar coordinates $(\theta(\tau), \phi(\tau))$. So γ satisfies the geodesic equation. In particular, for the θ-component of γ, with $\cdot' := \frac{d}{d\tau}$, we have

$$0 = \theta'' + 2\Gamma^{\theta}_{r\theta} r'\theta' + \Gamma^{\theta}_{\phi\phi}\phi'\phi' = \theta'' + \frac{2}{r}r'\theta' - (\sin\theta)(\cos\theta)\,\phi'\phi'.$$

Suppose $0 \in I$, and consider the initial conditions $\theta(0) = \frac{\pi}{2}$ and $\frac{d\theta}{d\tau}(0) = 0$. Intuitively, this means that the motion begins in the equatorial plane with the 'initial velocity' also lying in the equatorial plane. Since $\cos\frac{\pi}{2} = 0$, the above equation shows that $\theta \equiv \frac{\pi}{2}$ is *the* solution to the above equation (thanks to uniqueness). So the planet moves in the plane, just as we had deduced via the Newtonian viewpoint earlier in this section.

The ϕ-component of the geodesic equation gives, when $\theta \equiv \frac{\pi}{2}$, that

$$0 = \phi'' + 2\Gamma^{\phi}_{\theta\phi}\theta'\phi' + 2\Gamma^{\phi}_{r\phi}r'\phi' = \phi'' + 0 + 2\frac{1}{r}r'\phi',$$

i.e., $\phi'' = -\frac{2}{r}r'\phi'$. Hence $(r^2\phi')' = 2rr'\phi' + r^2\phi'' = 2rr'\phi' + r^2(-\frac{2}{r}r'\phi') = 0$. Thus $L := r^2\phi'$ is a constant[23]. The r-component of the geodesic equation gives (with $\theta \equiv \frac{\pi}{2}$):

$$\begin{aligned}
0 &= r'' + \Gamma^{r}_{tt}t'^2 + \Gamma^{r}_{rr}r'^2 + \Gamma^{r}_{\theta\theta}\theta'^2 + \Gamma^{r}_{\phi\phi}\phi'^2 \\
&= r'' + \frac{m}{r^2}\Big(1 - \frac{2m}{r}\Big)t'^2 - \frac{m}{r^2}\Big(1 - \frac{2m}{r}\Big)^{-1}r'^2 + 0 - r\Big(1 - \frac{2m}{r}\Big)\Big(\sin\frac{\pi}{2}\Big)^2\phi'^2 \\
&= r'' + \frac{m}{r^2}f(r)t'^2 - \frac{m}{r^2 f(r)}r'^2 - rf(r)\phi'^2
\end{aligned}$$

where $f(r) := 1 - \frac{2m}{r}$. Thus

$$r'' = -\frac{m}{r^2}\Big(f(r)t'^2 - \frac{1}{f(r)}r'^2\Big) + rf(r)\phi'^2. \qquad (\star)$$

We have $v_{\gamma,\gamma(\tau)} = t'\partial_{t,\gamma(\tau)} + r'\partial_{r,\gamma(\tau)} + 0 + \phi'\partial_{\phi,\gamma(\tau)}$. As γ is parametrised by proper time, we have $\mathbf{g}(\gamma(\tau))(v_{\gamma,\gamma(\tau)}, v_{\gamma,\gamma(\tau)}) = -1$, that is,

$$\begin{aligned}
-1 &= -f(r)t'^2 + \frac{1}{f(r)}r'^2 + 0 + r^2\Big(\sin\frac{\pi}{2}\Big)^2\phi'^2 \\
&= -f(r)t'^2 + \frac{1}{f(r)}r'^2 + r^2\phi'^2. \qquad (\star\star)
\end{aligned}$$

[23] So, as in the Newtonian picture, we get Kepler's second law of planetary motion: For the planet, the line segment joining it to the Sun sweeps equal areas in equal times.

Thus ($\star$) and ($\star\star$) together give

$$r'' = -\frac{m}{r^2}(1+r^2\phi'^2) + rf(r)\phi'^2 = -\frac{m}{r^2} + \Big(-m + r\big(1-\frac{2m}{r}\big)\Big)\phi'^2$$
$$= -\frac{m}{r^2} + (r-3m)\phi'^2. \tag{14.9}$$

Again we will work locally in an interval J, where the map $J \ni \tau \mapsto \phi(\tau)$ admits an inverse, and this inverse will be denoted by h. Set $u(\phi) = \frac{1}{(r\circ h)(\phi)}$. Then analogous to what we had done earlier, we have

$$\frac{du}{d\phi} = -\frac{1}{(r(h(\cdot)))^2}\frac{dr}{d\tau}(h(\cdot))\frac{dh}{d\phi} = -\frac{1}{(r(h(\cdot)))^2}r'(h(\cdot))\frac{1}{\phi'(h(\cdot))} = -\frac{r'(h(\cdot))}{L}.$$

Similarly

$$\frac{d^2u}{d\phi^2} = -\frac{1}{L}\frac{d(r'\circ h)}{d\phi} = -\frac{1}{L}r''(h(\cdot))\frac{dh}{d\phi} = -\frac{1}{L}r''(h(\cdot))\frac{1}{\phi'(h(\cdot))}$$
$$= -\frac{1}{L}r''(h(\cdot))\frac{(r(h(\cdot)))^2}{(r(h(\phi)))^2\phi'(h(\cdot))} = -\frac{(r(h(\cdot)))^2 r''(h(\cdot))}{L^2} = -\frac{r''(h(\cdot))}{u^2L^2}.$$

Substituting this in (14.9), and using $\phi' = \frac{r^2\phi'}{r^2} = \frac{L}{r^2} = L(u\circ h^{-1})^2$, we get

$$-L^2u^2\frac{d^2u}{d\phi^2} = -mu^2 + \Big(\frac{1}{u} - 3m\Big)L^2u^4, \quad \text{i.e.,} \quad \frac{d^2u}{d\phi^2} + u = \frac{m}{L^2} + 3mu^2. \tag{14.10}$$

Comparing with (14.6), we see that there is now an extra term $3mu^2$. As opposed to (14.6), which we could solve explicitly, now because of the extra term, we can no longer solve the equation in an elementary manner. But since $2m \approx 3$ km, and for planetary orbits r is in the order of millions[24] of kilometers, it follows that $3mu \ll 1$. As $3mu^2 = 3muu \ll u$, in equation (14.10), the term $3mu^2$ is quite small as compared to the u on the left-hand side. Also, assuming that the orbital speed $r\phi'$ is much less than the speed of light, we have $(mu^2)/\frac{m}{L^2} = \frac{L^2}{r^2} = (r\phi')^2 \ll 1$. So we proceed perturbatively[25] as in §8.4. With $\epsilon := 3m$, we seek an approximation of the form $u_0 + \epsilon u_1$ to the solution u to

$$\frac{d^2u}{d\phi^2} + u = \frac{m}{L^2} + \epsilon u^2, \tag{14.11}$$

where $u_0 = \frac{m}{L^2}(1 + e\cos\phi)$ is a solution to

$$\frac{d^2u_0}{d\phi^2} + u_0 = \frac{m}{L^2}, \quad \text{with the initial conditions} \quad \begin{cases} u_0(0) = \frac{m}{L^2}(1+e) \\ \frac{du_0}{d\phi}(0) = 0. \end{cases}$$

---

[24]For example, the closest distance of Mercury to the Sun is $\sim$ 46 million kilometers.

[25]Set $w = ku$, for some $k \gg 3m$, e.g. $k = R_\odot$. Then w satisfies $w'' + w = \frac{km}{L^2} + \varepsilon w^2$, where $\varepsilon := \frac{3m}{k} \ll 1$. Writing $w = w_0 + \varepsilon w_1 + \varepsilon^2 w_2 + \cdots$, and using the initial conditions $w_0(0) = \frac{km}{L^2}(1+e)$ and $w_0'(0) = 0$, an approximate solution w is given by $w_0 + \varepsilon w_1$ with w_1 satisfying $w_1'' + w_1 = w_0^2$. Scaling back by $\frac{1}{k}$, we get an approximation of u of the form $u_0 + \epsilon u_1$, $\epsilon := 3m$, for a suitable u_0, u_1 as given in the subsequent analysis here.

Substituting the series expansion for u into the differential equation (14.11), and by comparing terms in ϵ, we get that

$$\frac{d^2u_1}{d\phi^2}+u_1=u_0^2=\frac{m^2}{L^4}(1+e\cos\phi)^2=\frac{m^2}{L^4}\Big(1+\frac{e^2}{2}+2e\cos\phi+\frac{e^2}{2}\cos(2\phi)\Big).$$

Set $u_{1,\mathrm{par}}=a+\phi(b_1\cos\phi+b_2\sin\phi)+c\cos(2\phi)$ in order to find a particular solution $u_{1,\mathrm{par}}$. Plugging this into the equation, and using the independence of $1, \sin\phi, \cos\phi, \cos(2\phi)$, we can determine a, b_1, b_2, c, and obtain

$$u_{1,\mathrm{par}} = \frac{m^2}{L^4}\Big(1+\frac{e^2}{2}+e\phi\sin\phi-\frac{e^2}{6}\cos(2\phi)\Big).$$

Then $\frac{du_{1,\mathrm{par}}}{d\phi}(0)=0$. So if $\widetilde{u}(\phi):=u_0(\phi)+\epsilon u_{1,\mathrm{par}}(\phi)$, then, as[26], $\frac{m^2}{L^2}\ll 1$,

$$\widetilde{u}(0)=\frac{m}{L^2}(1+e)+\frac{3m^3}{L^4}\Big(1+\frac{e^2}{3}\Big)\approx\frac{m}{L^2}(1+e)=u_0(0),\quad \frac{d\widetilde{u}}{d\phi}(0)=0.$$

So we may take $\widetilde{u}$ as an approximation of the solution u to (14.10) with

$$u(0)=\frac{m}{L^2}(1+e)\ \text{ and }\ \frac{du}{d\phi}(0)=0.$$

The *perihelion* of a planet's trajectory is the point along the trajectory that is closest to the Sun. Thus r is minimised, and this corresponds to u being maximised. In the case of the ellipse described by $\phi\mapsto 1/u_0(\phi)$, where

$$u_0(\phi) = \frac{m}{L^2}(1+e\cos\phi),$$

the perihelion occurs at $\phi=0, \pm 2\pi, \pm 4\pi, \cdots$. To determine the local maximum of $\widetilde{u}$, we seek solutions to $\frac{d\widetilde{u}}{d\phi}(\phi)=0$. We have

$$\frac{d\widetilde{u}}{d\phi}(\phi) = -\frac{me}{L^2}\sin\phi+\frac{3m^3e}{L^4}\Big(\sin\phi+\phi\cos\phi+\frac{e}{3}\sin(2\phi)\Big).$$

We have $\frac{d\widetilde{u}}{d\phi}(0)=0$, so there is a perihelion at $\phi=0$. The next perihelion is not at 2π (as opposed to the case of the ellipse), since

$$\frac{d\widetilde{u}}{d\phi}(2\pi) = \frac{3m^3e}{L^4}2\pi \neq 0.$$

But if we suppose that the perihelion is instead at $2\pi+\delta$, with a $|\delta|\ll 1$, then this δ satisfies approximately that

$$0\approx -\frac{me}{L^2}\delta+\frac{3m^3e}{L^4}\Big(\delta+(2\pi+\delta)+\frac{e}{3}2\delta\Big).$$

Since $\frac{m^2}{L^2}\ll 1$, we obtain $\delta\approx\frac{6m^2\pi}{L^2}$.

[26] As $u_0=\frac{m}{L^2}(1+e\cos\phi)$, and since the eccentricity of Mercury is ~ 0.2, we have that $\frac{m^2}{L^2}\lesssim\frac{5}{4}m\frac{m}{L^2}(1+e\cos\phi)=\frac{5}{4}mu_0\sim\frac{m}{r}\sim\frac{3}{40\times 10^6}\sim 10^{-7}\ll 1$.

Thus the Newtonian elliptical planetary orbit is now prevented from closing in on itself, and the resulting trajectory resembles a rosette pattern, as shown (much exaggerated) below, with an advancement of the perihelion in each traversal of a 'petal'. This is called the perihelion precession[27].

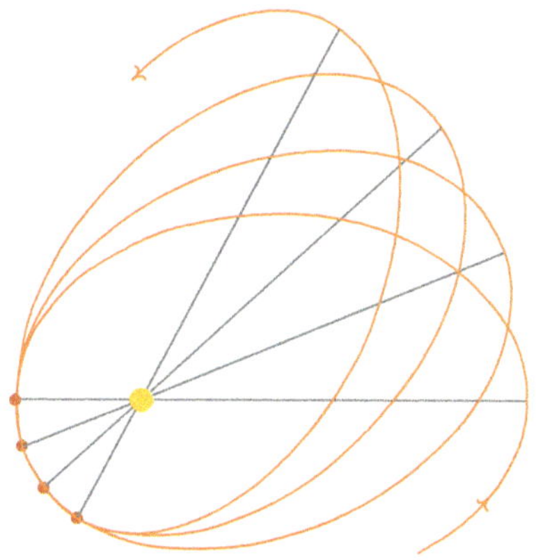

Remark 14.5. In fact, in the Newtonian viewpoint as well, the orbit of Mercury precesses owing to the gravitational influence of the other planets; see e.g. [Landau and Lifshitz (1976), Problem 3, p.40]. Indeed, in our Newtonian description, we looked a test particle moving in the field created by a single body (the Sun), while in reality we have an n-body (with $n > 1$) problem in our solar system, because of the presence of the other planets which *do* affect the orbit of Mercury. This amounts to about[28] 532 arcseconds[29] per century. However, the classical viewpoint could not account for a discrepancy of about $43''$ with the observed value of about $575''$. Before the spacetime geodesic viewpoint, it was proposed that perhaps there is another planet, called 'Vulcan', whose orbit lies between that of Mercury and the Sun, which explains the discrepancy from the classical value of precession. However, Vulcan does not exist. On the other hand, the calculation from the spacetime geodesic viewpoint predicts a value of $43.03''$ per century, in excellent agreement with the observed value. This early result (1916) due to Einstein is considered as the first experimental verification of spacetime-geometric description of gravitation. $\ast$

[27]The terminology is borrowed from the mechanics of a spinning top in Earth's gravity, where the axis of rotation itself rotates, spanning a cone, and one says the axis 'precesses'.

[28]See for example [Misner, Thorne and Wheeler (2017), p.1113].

[29]An arcsecond (denoted henceforth by $\cdot''$) is $\frac{1}{3600}$ of a degree, i.e., $3600'' = 1^\circ$.

Chapter 15

Black holes

In Section 14.3, we saw that the smooth manifold $M = \mathbb{R} \times (r_{\min}, \infty) \times S^2$, where $r_{\min} = 2m > 0$, is a spacetime[1] with the Schwarzschild metric $\mathbf{g}$. In a spherical coordinate chart described there, $\mathbf{g}$ is given by

$$\mathbf{g} = -\Big(1 - \frac{2m}{r}\Big) dt \otimes dt + \Big(1 - \frac{2m}{r}\Big)^{-1} dr \otimes dr + r^2 d\theta \otimes d\theta + r^2 (\sin\theta)^2 d\phi \otimes d\phi.$$

We had remarked that for a typical star, $2m$ is much smaller than its radius, and so the above describes the exterior region of the star, where we assumed that the cosmological constant is $\Lambda = 0$, and that there is vacuum, that is, the energy-momentum tensor field is $\mathbf{T} = 0$.

But now let us consider the situation, where all the mass is located 'at $r = 0$'. We do not think of this necessarily as a point, but rather simply something not in our manifold (where in the manifold we have $r > 0$). Let $M_{\text{BH}} := \mathbb{R} \times J \times S^2$, where $J = (0, 2m) \cup (2m, \infty) \subset \mathbb{R}$. Then M_{BH} is a Lorentzian manifold with the same metric $\mathbf{g}$ given above, and again the field equations are satisfied in M_{BH} with $\Lambda = 0$ and with the vacuum energy-momentum tensor $\mathbf{T} = 0$. While ∂_t is timelike in the region $r > 2m$, we have ∂_r is timelike in the region $r < 2m$. Thus we cannot have a particle trajectory with $r =$ constant in the region $r < 2m$. The r-coordinate of the particle in this region must necessarily increase or decrease, depending on what time-orientation is chosen in this part.

By looking at the behaviour of lightlike geodesics, we will see that light-rays emanating from the region $0 < r < 2m$ cannot travel to the exterior region, while null geodesics are allowed to fall in. Thus no light emerges from such an object, and it is legitimate to call it a *black hole*. We will only study Schwarzschild black holes, roughly collapsed objects characterised by a unique parameter, namely their mass m.

[1]A time-orientation is given by ∂_t, and an orientation is induced by the orientation $[\text{vol}_{\mathbf{g}}]$ on $\mathbb{R} \times \mathbb{R}^3$, where $\text{vol}_{\mathbf{g}} = \sqrt{|\det[\mathbf{g}_{ij}]|}\, dx \wedge dy \wedge dz \wedge dt$.

15.1 The spacetime $(M_{\rm BH}, \mathbf{g})$

In this section, we will show that the spacetime $(M_{\rm BH}, \mathbf{g})$ is 'singular' with a singularity at $r=0$. But first, we will introduce the spacetime $(M_{\rm BH}, \mathbf{g})$ in more precise terms than the description given in the introduction.

Definition 15.1. (The spacetime $(M_{\rm BH}, \mathbf{g})$.)
Let $M_{\rm BH} = \mathbb{R} \times J \times S^2$, where $J = (0, 2m) \cup (2m, \infty) \subset \mathbb{R}$. The open set $W := \{(t, x, y, z) \in \mathbb{R}^4 \mid 0 < r := \sqrt{x^2+y^2+z^2} \neq 2m\}$ of $\mathbb{R}^4$ is given the smooth structure induced from the standard smooth structure on $\mathbb{R}^4$. The smooth manifold $M_{\rm BH}$ is identified with W, via the diffeomorphism $W \ni (t, x, y, z) \mapsto (t, r, \frac{(x,y,z)}{r}) \in M_{\rm BH}$. We endow $M_{\rm BH}$ with the Lorentzian metric $\mathbf{g}$ given in the global chart $(W, (t, x, y, z))$ by

$$\mathbf{g} = -f(r)\, dt \otimes dt + \Big(\frac{1}{f(r)} - 1\Big) dr \otimes dr + dx \otimes dx + dy \otimes dy + dz \otimes dz,$$

where $f(r) = 1 - \frac{2m}{r}$. Let (U, φ) be the spherical coordinate chart for S^2 from Example 1.8 (p.7). In the chart $(\mathbb{R} \times J \times U, (t, r, \mathbf{p}) \mapsto (t, r, \varphi(\mathbf{p})))$,

$$\mathbf{g} = -\Big(1 - \frac{2m}{r}\Big) dt \otimes dt + \Big(1 - \frac{2m}{r}\Big)^{-1} dr \otimes dr + r^2 d\theta \otimes d\theta + r^2 (\sin\theta)^2 d\phi \otimes d\phi.$$

An orientation is induced on $M_{\rm BH}$ by the restriction to $M_{\rm BH}$ of the top-form field $\mathrm{vol}_{\mathbf{g}} = \sqrt{|\det[\mathbf{g}_{ij}]|}\, dx \wedge dy \wedge dz \wedge dt$ on $\mathbb{R} \times \mathbb{R}^3$. A time-orientation is given by ∂_t if $r > 2m$, and by $-\partial_r$ if $0 < r < 2m$.

In the above, we have chosen 'the direction of decreasing r' in the region $0 < r < 2m$, as being future-pointing. This will allow us to talk about in-falling lightlike geodesics.

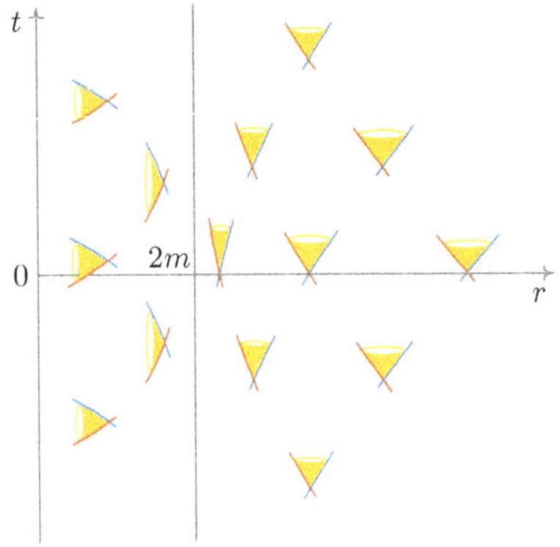

The expression for the metric $\mathbf{g}$ seems to break down at $r = 2m$. But we shall see that this is a mere artefact of the chosen coordinates. To explain what we mean by this, let us first consider the following example.

Example 15.1. Consider the Minkowski plane $\mathbb{R}^2$ with the standard smooth structure, and the metric $\mathbf{g} = -\,dt \otimes dt + dx \otimes dx$ in the global admissible chart $(\mathbb{R}^2, (t,x) \mapsto (t,x))$. Then $(\mathbb{R}^2, \mathbf{g})$ is a Lorentzian manifold. Consider the admissible[2] chart $(\mathbb{R}^2 \backslash \{(0,x) : x \in \mathbb{R}\},\ (t,x) \mapsto (\tau(t,x) := t^3, x))$.

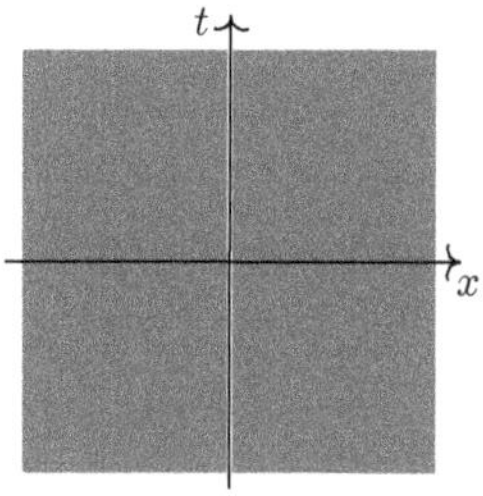

In this new chart, the metric is given by $\mathbf{g} = -\frac{1}{9\tau^{4/3}} d\tau \otimes d\tau + dx \otimes dx$. The coefficient of $d\tau \otimes d\tau$ above, namely $\mathbb{R}^2 \backslash \{(0,x) : x \in \mathbb{R}\} \ni (t,x) = p \mapsto -\frac{1}{9t^4}$, does not stay bounded as $p = (t,x) \to (0,0)$. ◇

How do we detect if there is a genuine 'singularity' at a point or if it is the case that awkward coordinates have been chosen? To do this, we first need to agree upon what we consider as a singularity for a given manifold M. Intuitively, we could demand from a non-singular manifold that geodesics don't stop or run into an 'obstacle' (a 'hole' or a 'missing point'), but have their affine parameter living on the whole of $\mathbb{R}$. This motivates the following definition.

Definition 15.2. (Singular spacetime.)
A spacetime$(M, \mathbf{g})$ is called *singular* if it is not geodesically complete.

Example 15.2. The Minkowski plane $(M, \mathbf{g})$ from Example 15.1 above is geodesically complete, since each geodesic is a straight line

$$\gamma(t) = (x_0, y_0) + t\,\mathbf{v} \quad (t \in \mathbb{R}),$$

for some point $(x_0, y_0) \in \mathbb{R}^2$ and some vector $\mathbf{v} \in \mathbb{R}^2$. This follows from the fact that the Levi-Civita connection in this case is just the flat connection on $\mathbb{R}^2$, and we had determined the geodesics for the flat connection in Example 8.1 (p.152).

Also, from Example 8.4 (p.161), the 4-dimensional Minkowski spacetime is geodesically complete. ◇

[2]As opposed to Example 1.6 (p.5), where we also had the chart map $t \mapsto t^3$, we took the chart domain as the whole manifold, while we now remove the problematic points (i.e. the t-axis), so that there is no obstacle for the chart transition maps to be smooth.

One way[3] to detect if the spacetime is singular is to find out if the curvature goes to infinity along a geodesic as the affine parameter approaches a finite value. In the example below, we will see in this manner that the Schwarzschild Lorentzian manifold $(M_{\rm BH}, \mathbf{g})$ is singular, by showing that a radial geodesic in the interior region $(r < 2m)$ hits $r = 0$ in finite proper time, while the Kretschmann scalar goes to ∞.

Example 15.3. In the spacetime $M_{\rm BH}$, let $\gamma: I \to M_{\rm BH}$ be a radial timelike geodesic parametrised by proper time, with constant $\theta \equiv \theta_0$ and $\phi \equiv \phi_0$. Then we have $\phi' = \theta' = 0$. The connection coefficients are given by the expressions in Example 6.8 (p.121) even in the interior region. As in Exercise 14.11 (p.321), the geodesic equations for the ϕ and θ components are satisfied by $\theta \equiv \theta_0$ and $\phi \equiv \phi_0$. The geodesic equations for the r and t components are given by

$$r'' = -\frac{m}{r^2}\Big(f(r)t'^2 - \frac{1}{f(r)}r'^2\Big), \tag{15.1}$$

$$t'' = -\frac{2m}{r^2 f(r)}t'r' \tag{15.2}$$

where $f(r) := 1 - \frac{2m}{r}$. Since $\mathbf{g}(\gamma(\tau))(v_{\gamma,\gamma(\tau)}, v_{\gamma,\gamma(\tau)}) = -1$, we also have

$$-1 = -f(r)t'^2 + \frac{1}{f(r)}r'^2. \tag{15.3}$$

We will now construct a solution to the above equations. Let $r_0 \in (0, 2m)$, and let r satisfy the initial value problem

$$r' = -\sqrt{\frac{2m}{r}}, \qquad r(0) = r_0 \in (0, 2m).$$

So $r(\tau) = \big(r_0^{3/2} - \frac{3\sqrt{2m}}{2}\tau\big)^{2/3}$, and $r'^2 = \frac{2m}{r}$. Let $t' = \frac{1}{f(r)}$, and $t(0) = t_0$, i.e.,

$$t(\tau) = t_0 + \int_0^\tau \Big(1 - \frac{2m}{\big(r_0^{3/2} - \frac{3\sqrt{2m}}{2}s\big)^{2/3}}\Big)^{-1} ds, \qquad \text{for } 0 \leqslant \tau < \frac{2r_0^{3/2}}{3\sqrt{2m}}.$$

We check that (15.1), (15.2), (15.3) are satisfied. Indeed, (15.3) holds, as

$$-f(r)t'^2 + \frac{1}{f(r)}r'^2 = -f(r)\frac{1}{(f(r))^2} + \frac{1}{f(r)}\frac{2m}{r} = -\frac{1}{f(r)}\Big(1 - \frac{2m}{r}\Big) = -1.$$

[3]However, this is not a foolproof way. The curvature tensor field may behave benignly even for geodesically incomplete semi-Riemannian manifolds. This can happen if a point is missing and cannot be added, or has been viciously removed. For example, consider the double cone $C_* = \{(x, y, z) \in \mathbb{R}^3 : x^2 + y^2 = z^2\} \setminus \{(0,0,0)\} \subset \mathbb{R}^3$ as a smooth manifold. Equipping C_* with the metric induced from the Euclidean inner product on $\mathbb{R}^3$, the curvature tensor field of C_* can be shown to be everywhere zero. But if $p \in C_*$, then the line $L(t) = tp$, $t \in (0, \infty)$ can be shown to be a maximal geodesic, and leaves C_* at $t = 0$.

Moreover, (15.2) holds, since

$$-\frac{2m}{r^2 f(r)} t' r' = -\frac{2m}{r^2 f(r)} \frac{1}{f(r)} r' = -\frac{2m}{r^2} r' \frac{1}{(f(r))^2}$$
$$= \Big(\frac{1}{f(r)}\Big)' = (t')' = t''.$$

Finally differentiating $r'^2 = \frac{2m}{r}$ gives

$$2r'r'' = -\frac{2m}{r^2} r',$$

and so using (15.3) (already proved)

$$r'' = -\frac{m}{r^2} = -\frac{m}{r^2}(1) = -\frac{m}{r^2}\Big(f(r)t'^2 - \frac{1}{f(r)} r'^2\Big).$$

Thus γ, with the above constructed r, t, is a geodesic. We note that $r=0$ is reached in the finite proper time

$$\tau = \frac{2r_0^{3/2}}{3\sqrt{2m}}.$$

On the other hand, it turns out[4] that the Kretschmann invariant/scalar (Exercise 9.13, p.190) is

$$\mathbf{K} = \frac{48m^2}{r^6}, \tag{15.4}$$

and so

$$\lim_{r \searrow 0} \mathbf{K} = +\infty.$$

Hence $(M_{\mathrm{BH}}, \mathbf{g})$ is singular. ◇

Thus '$r=0$' is a genuine singularity for the Lorentzian manifold $(M_{\mathrm{BH}}, \mathbf{g})$. What about $r = 2m$? From the above expression for the Kretschmann scalar, we suspect that the problem at $r = 2m$ for the expression for $\mathbf{g}$ might very well be just a coordinate artefact. In order to see what happens when $r=2m$, we study null geodesics in $(M_{\mathrm{BH}}, \mathbf{g})$. We refer to the 'surface' $\mathbb{R} \times \{2m\} \times S^2 \subset \mathbb{R} \times (0, \infty) \times S^2$ as the *event horizon*. The Schwarzschild radius is $r_{\mathrm{S}} := 2m$ (i.e., $\frac{G_{\mathrm{N}}}{c^2} 2m$, after G_{N} and c have been reinstated).

[4]See e.g. [Grave and Mueller (2010), (2.2.5), p.18] or [O'Neill (1983), Exercise 8, p.399]. We will not carry out this computation here. We remark that the scalar curvature is not an indicator of the presence of the singularity: $\mathbf{S} \equiv 0$. This is because the Ricci curvature tensor field $\mathbf{Ric} \equiv 0$ for the Schwarzschild spacetime. We had seen this in the exterior region ($r > 2m$) while deriving the Schwarzschild metric (p.320). But since the connection coefficients in the interior region ($0 < r < 2m$) are given by the same expressions as those in the exterior region, the Ricci curvature tensor field is zero everywhere.

15.2 Null geodesics and the event horizon

Proposition 15.1. *Let $\theta_0 \in (0, \pi)$ and $\phi_0 \in (0, 2\pi)$ be fixed, determining a radial direction in S^2, and $\mathbf{p} \in S^2$ have spherical polar coordinates (θ_0, ϕ_0). Then the corresponding radial lightlike geodesics in $(M_{\text{BH}}, \mathbf{g})$ are affine reparametrisations of $\gamma : I \to M_{\text{BH}}$, $\gamma(r) = (t(r), r, \mathbf{p})$ for all $r \in I$, where the interval $I \subset J$, and I and t are given in the table below:*

Region	I	$t(r)$	$t(r)$
Interior	$(0, 2m)$	$r + 2m \log(2m - r) + c$	$-r - 2m \log(2m - r) + c$
Exterior	$(2m, \infty)$	$-r - 2m \log(r - 2m) + c$	$r + 2m \log(r - 2m) + c$

The constant $c \in \mathbb{R}$ is arbitrary in the above table.

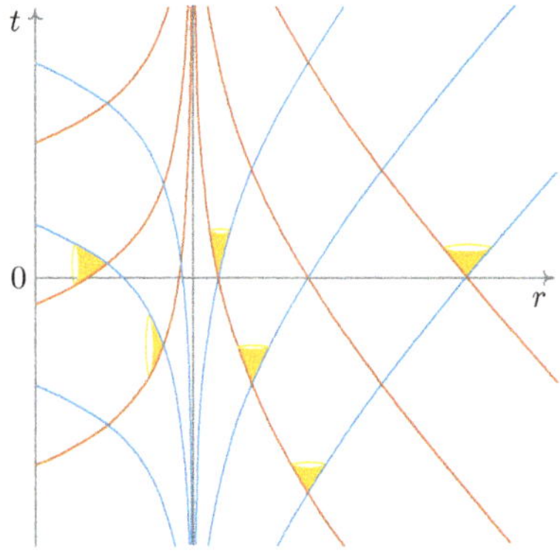

Proof. First consider the radial lightlike geodesic $\widetilde{\gamma} : \widetilde{I} \to M_{\text{BH}}$ described by an affine parameter $\tau \in \widetilde{I}$: $\widetilde{\gamma}(\tau) = (\widetilde{t}(\tau), r(\tau), \mathbf{p})$ for all $\tau \in \widetilde{I}$. With $\cdot' = \frac{d}{d\tau}$, its ϕ- and θ-components satisfy $\phi' = 0$ and $\theta' = 0$. The geodesic equations for the ϕ- and θ-components are satisfied by $\theta \equiv \theta_0$ and $\phi \equiv \phi_0$. The t-component of the geodesic equation for $\widetilde{\gamma}$ gives

$$\widetilde{t}'' = -\frac{2m}{r^2 f(r)} \widetilde{t}' r'$$

where $f(r) := 1 - \frac{2m}{r}$. As γ is lightlike, $\mathbf{g}(\gamma(\tau))(v_{\gamma,\gamma(\tau)}, v_{\gamma,\gamma(\tau)}) = 0$, and so

$$-f(r)\widetilde{t}'^2 + \frac{1}{f(r)} r'^2 = 0. \qquad (15.5)$$

We first note that $E := f(r)\widetilde{t}'$ is a constant, since

$$(f(r)\widetilde{t}')' = (f(r))'\widetilde{t}' + f(r)\widetilde{t}'' = \frac{2m}{r^2} r'\widetilde{t}' - f(r)\frac{2m}{r^2 f(r)} \widetilde{t}' r' = 0. \qquad (15.6)$$

Now using (15.5), $-f(r)\dfrac{E^2}{(f(r))^2} + \dfrac{1}{f(r)} r'^2 = 0$. Consequently, $r'^2 = E^2$.

So $r' = a \in \{-E, E\}$. Thus $r(\tau) = a\tau + b$, $\tau \in \widetilde{I}$, which is an affine map. Set $I = a\widetilde{I} + b$, and define $\tau : I \to \widetilde{I}$ by $\tau(r) = \frac{r-b}{a}$, $r \in I$. Theorem 8.2 (p.160) implies $\gamma = \widetilde{\gamma} \circ \tau : I \to M_{\rm BH}$ is a geodesic. As $\gamma(r) = (\widetilde{t}(\frac{r-b}{a}), r, \theta_0, \phi_0)$ $(r \in I)$,

$$\frac{d}{dr}\widetilde{t}\Big(\frac{r-b}{a}\Big) = \widetilde{t}'\Big(\frac{r-b}{a}\Big)\frac{1}{a} = \frac{E}{f(r)}\frac{1}{a} = \pm\frac{1}{f(r)} = \pm\Big(1 + \frac{2m}{r-2m}\Big).$$

Integrating, $t(r) := \widetilde{t}(\dfrac{r-b}{a}) = \pm(r + 2m\log|r-2m|) + c$ for some $c \in \mathbb{R}$. □

Let us now see how things appear to an observer outside the event horizon.

Definition 15.3. (Schwarzschild observer.)
Let $r_0 > 2m$ and $\mathbf{p} \in S^2$. The worldline $\gamma_{\rm S} : \mathbb{R} \to M_{\rm BH}$ given by

$$\gamma_{\rm S}(\tau) = \Big(\tau\sqrt{\big(1-\tfrac{2m}{r_0}\big)^{-1}},\ r_0,\ \mathbf{p}\Big), \quad \tau \in \mathbb{R},$$

is called a *Schwarzschild observer.*

For a Schwarzschild observer $\gamma_{\rm S}$, we have $v_{\gamma_{\rm S},\gamma_{\rm S}(\tau)} = \sqrt{\big(1-\frac{2m}{r_0}\big)^{-1}}\,\partial_{t,\gamma_{\rm S}(\tau)}$.

So $\mathbf{g}(\gamma_{\rm S}(\tau))(v_{\gamma_{\rm S},\gamma_{\rm S}(\tau)}, v_{\gamma_{\rm S},\gamma_{\rm S}(\tau)}) = -1$ for all $\tau \in \mathbb{R}$. Thus proper time for $\gamma_{\rm S}$ between two events $p = \gamma_{\rm S}(\tau_p)$ and $q = \gamma_{\rm S}(\tau_q)$ along $\gamma_{\rm S}$ is given by

$$\Big|\int_{\tau_p}^{\tau_q} -1\, d\tau\Big| = |\tau_p - \tau_q| = \sqrt{1-\frac{2m}{r_0}}\,|t(p) - t(q)|.$$

Now consider a radial infalling lightlike geodesic λ as shown below.

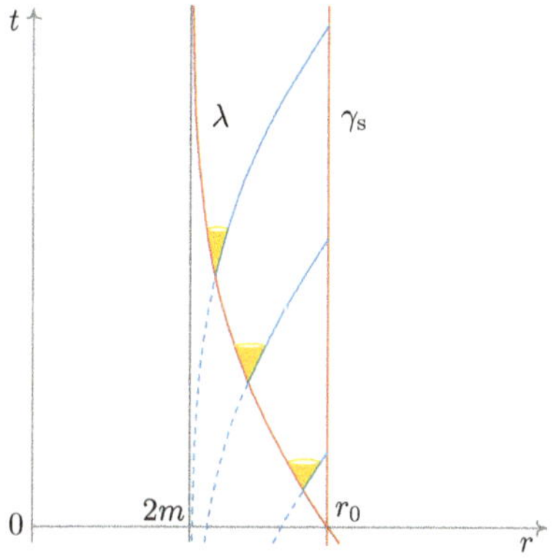

The radial light geodesics emanating from the events along λ will reach the worldline of $\gamma_{\rm S}$ at later and later times, so that the Schwarzschild observer will think that the photon never reaches the event horizon. On the other hand, if we consider the photon parametrised[5] by the affine parameter r, then starting from the parameter value $r_0 > 2m$, the photon reaches the event horizon at the finite affine parameter value $r_0 - 2m$.

[5]This fixes its energy/frequency as measured e.g. by the Schwarzschild observer at the common event $\lambda(r_0)$.

Exercise 15.1. A similar phenomenon as above is valid with a radial *timelike* geodesic. For example, we show this by constructing a geodesic similar to the one in Example 15.3. Let r satisfy

$$\frac{dr}{d\tau} = -\sqrt{\frac{2m}{r}}, \qquad r(0) = r_0 > 2m. \qquad (\star)$$

So $r(\tau) = \big(r_0^{\frac{3}{2}} - \frac{3\sqrt{2m}}{2}\tau\big)^{\frac{2}{3}}$. For $T = \frac{2}{3\sqrt{2m}}(r_0^{\frac{3}{2}} - (2m)^{\frac{3}{2}}) < \infty$, $r(T) = 2m$. Let t satisfy

$$\frac{dt}{d\tau} = \frac{1}{f(r)} = \Big(1 - \frac{2m}{\big(r_0^{3/2} - \frac{3\sqrt{2m}}{2}\tau\big)^{2/3}}\Big)^{-1}, \qquad t(0) = 0. \qquad (\star\star)$$

Let $\mathbf{p} \in S^2$. The same steps as in Example 15.3 show that $\gamma(\tau) := (t(\tau), r(\tau), \mathbf{p})$, $\tau \in (0, T)$ is a timelike geodesic (particle) starting at the event $(0, r_0, \mathbf{p})$ along the Schwarzschild observer's worldline, and reaching the event horizon in the particle's proper time $T < \infty$. Let $\tau : (2m, r_0) \to (0, T)$ be the inverse of the map r, that is, $r \circ \tau = \mathrm{id}_{(2m, r_0)}$. Using $(\star)$ and $(\star\star)$, and setting $\widetilde{t} = t \circ \tau$,

$$\frac{d\widetilde{t}}{dr}(r) = t'(\tau(r))\frac{1}{r'(\tau(r))} = \frac{\frac{1}{f(r)}}{-\sqrt{\frac{2m}{r}}} = -\frac{1}{(1 - \frac{2m}{r})}\frac{1}{\sqrt{\frac{2m}{r}}}.$$

Find $\widetilde{t}$ as a function of r. In the rt-plane, plot its graph for $2m = 1$, and $r_0 = 2$.

No light rays emanating from the region inside the event horizon, reach the Schwarzschild observer γ_S, and also do not reach the events along any timelike curve living in the exterior of the event horizon. Thus a black hole appears black. Of course the starry light from elsewhere, e.g. 'behind' the black hole, can travel to the observers outside the event horizon, as shown.

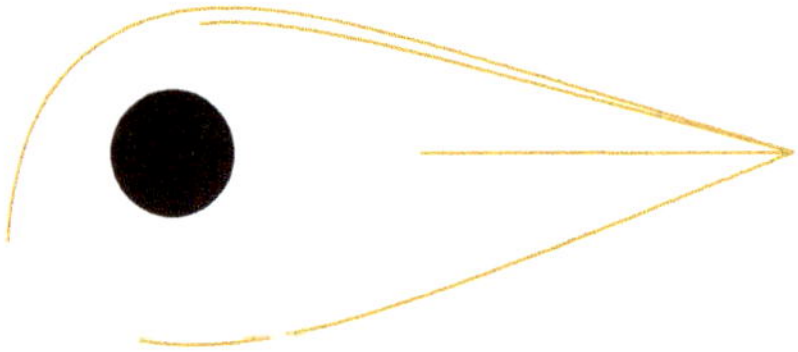

Exercise 15.2. (Schwarzschild radius pre-Schwarzschild.)
The aim of this exercise is to point out that the Schwarzschild radius already appeared in Laplace's work in 1798. Classically, the gravitational potential energy due to a star at distance a r from its centre, is minus the amount of work done to bring an object from separation r to far away ('at infinity, $r = \infty$'). By the Newton gravitation law, the force experienced by a mass $\widetilde{m}$ at a separation r is $F(r) = G_N M m / r^2$ where m is the star's mass. The work done to move a test mass $\widetilde{m}$ at r radially through a small distance dr is (force)$\cdot$(displacement) $= F(r)\,dr$. So the gravitational potential energy $V(r)$ at r is given by

$$V(r) := -\int_r^\infty \frac{G_N m \widetilde{m}}{r^2}\,dr = -\frac{G_N m \widetilde{m}}{r}.$$

The *escape velocity* $v(r)$ at a separation r from the star's centre is defined so that it imparts enough kinetic energy to the object in order to overcome the gravitation potential energy $V(r)$. Show that

$$v(r) = \sqrt{\frac{2G_\text{N}m}{r}}.$$

Assuming that m is the mass of a star, for what radius does the escape velocity equal c, the speed of light? This radius r_s is the Schwarzschild radius. Show that if the density of the star is ρ, then the condition that the star does not release light is that its radius R satisfies

$$R > \sqrt{\frac{3}{8\pi}\frac{c^2}{G_\text{N}\rho}}.$$

Laplace concluded that the largest objects in the universe are invisible.

15.3 Kruskal extension of M_BH

We have seen that while the Kretschmann scalar goes to infinity in finite proper time as $r \to 0$ for a radially infalling particle starting from the region inside the event horizon, there is no such problem at $r = 2m$ for a radially infalling particle starting from outside the event horizon: it reaches the event horizon in finite proper time (Exercise 15.1) but the Kretschmann scalar remains finite. One may wonder if the trajectory of the radially infalling particle starting from the outside can be patched with the one from the inside, so that although it looks discontinuous in the coordinates chosen hitherto, in the patched-up spacetime, with better coordinates, it appears as a single smooth curve. So we ask:

> Is there an extension of M_BH to a larger spacetime $\widetilde{M}_\text{BH}$, in which a copy of M_BH can be diffeomorphically embedded, so that there is no singularity corresponding to the event horizon in $\widetilde{M}_\text{BH}$?

Yes, the Kruskal[6] spacetime. We give an abridged description here[7].

Transformations in the outer region

There are four steps, in each of which we maintain the θ, ϕ coordinates, and change the original pair (r,t) to get a final pair (u,v) through the sequence $(r,t) \to (\widetilde{r},t) \to (U,V) \to (\widetilde{u},\widetilde{v}) \to (u,v)$. We specify these below:

- In the first step, we keep the Schwarzschild coordinate time t, but replace r by $\widetilde{r}$, given by $\widetilde{r} = r + 2m\log(r-2m)$. This has the effect of mapping the half-plane $\{(r,t) : r > 2m\}$ to the entire plane $\mathbb{R}^2$, in such a way that the radial incoming and outgoing lightlike geodesics become 45° straight lines, as shown in the picture on page 341: Indeed, a point (r,t) on such a curve in the (r,t)-plane is mapped to the point $(\widetilde{r},t)$, where $t = \pm(r+2m\log(r-2m))+c = \pm\widetilde{r}+c$.

[6] After Martin Kruskal (1925–2006), an American mathematician and physicist.
[7] A detailed description can be found for example in [O'Neill (1983), Chapter 13].

- In the second step, the transformation is $(\widetilde{r}, t) \mapsto (t - \widetilde{r}, t + \widetilde{r}) =: (U, V)$, which is a 45° anticlockwise rotation, followed by a reflection in the vertical axis, followed by a dilation by $\sqrt{2}$.
- In the third step, we compress the entire planar region to the second quadrant via the map $(U, V) \mapsto (-e^{\frac{-U}{4m}}, e^{\frac{V}{4m}}) =: (\widetilde{u}, \widetilde{v})$.
- In the final step, we use again the composition of a 45° anticlockwise rotation, a reflection in the vertical axis, and a dilation, now by $1/\sqrt{2}$: $(\widetilde{u}, \widetilde{v}) \mapsto \frac{1}{2}(\widetilde{v} - \widetilde{u}, \widetilde{v} + \widetilde{u}) =: (u, v)$.

Altogether, incoming and outgoing light rays run along the 45° half-lines $v + u = c$, respectively $v = u - c$ in the (u, v) plane. We have brought the original $r = 2m$ event horizon in the (r, t)-plane (along which lied the edges of light cones in the (r, t)-plane) along a 45° ray L, where $u = v$, in the first quadrant of the (u, v)-plane. Soon, we will transform the interior region $0 < r < 2m$ in a similar manner, and glue the so-obtained new region in the (u, v)-plane along this ray L.

We now calculate the metric in the new coordinates (v, u, θ, ϕ). First, we find the overall transformation $(t, r) \mapsto (v, u)$:

$$\begin{aligned}(u, v) &= \frac{1}{2}(e^{\frac{V}{4m}} + e^{\frac{-U}{4m}}, e^{\frac{V}{4m}} - e^{\frac{-U}{4m}}) = \frac{1}{2}(e^{\frac{t+\widetilde{r}}{4m}} + e^{\frac{-t+\widetilde{r}}{4m}}, e^{\frac{t+\widetilde{r}}{4m}} - e^{\frac{-t+\widetilde{r}}{4m}}) \\ &= \frac{e^{\frac{\widetilde{r}}{4m}}}{2}(e^{\frac{t}{4m}} + e^{-\frac{t}{4m}}, e^{\frac{t}{4m}} - e^{-\frac{t}{4m}}) \\ &= e^{\frac{r}{4m}} e^{\frac{1}{2}\log(r-2m)}\left(\cosh\frac{t}{4m}, \sinh\frac{t}{4m}\right) \\ &= e^{\frac{r}{4m}}\sqrt{r-2m}\left(\cosh\frac{t}{4m}, \sinh\frac{t}{4m}\right). \end{aligned} \tag{15.7}$$

Thus $v^2 - u^2 = -e^{\frac{r}{2m}}(r - 2m) = (2m - r)e^{\frac{r}{2m}}$. With $f(r) := 1 - \frac{2m}{r}$,

$$v^2 - u^2 = -r e^{\frac{r}{2m}} f(r). \tag{15.8}$$

An expression for t in terms of u, v is obtained by dividing v by u:

$$t(u, v) = 4m \tanh^{-1}\frac{v}{u}.$$

Obtaining r in terms of u, v is not possible in such a straightforward manner, as it involves solving a transcendental equation. We will need the following result.

Lemma 15.1. *The map $h : (-1, \infty) \to (-\frac{1}{e}, \infty)$, given by $h(x) = xe^x$, is bijective, smooth, and has a smooth inverse.*

Proof. For $x > -1$, $h'(x) = e^x + xe^x = (1+x)e^x > 0$. So h is strictly increasing (and thus injective) on $(-1, \infty)$. As $xe^x|_{x=-1} = -\frac{1}{e}$, and $xe^x \to \infty$ when $x \to \infty$, we have $h((-1,\infty)) = (-\frac{1}{e}, \infty)$, and h is onto $(-\frac{1}{e}, \infty)$. Thus h is bijective. The map h is C^∞, and h' is positive on $(-1,\infty)$. By the inverse function theorem, the inverse map[8] $h^{-1} : (-\frac{1}{e}, \infty) \to (-1, \infty)$ is smooth. □

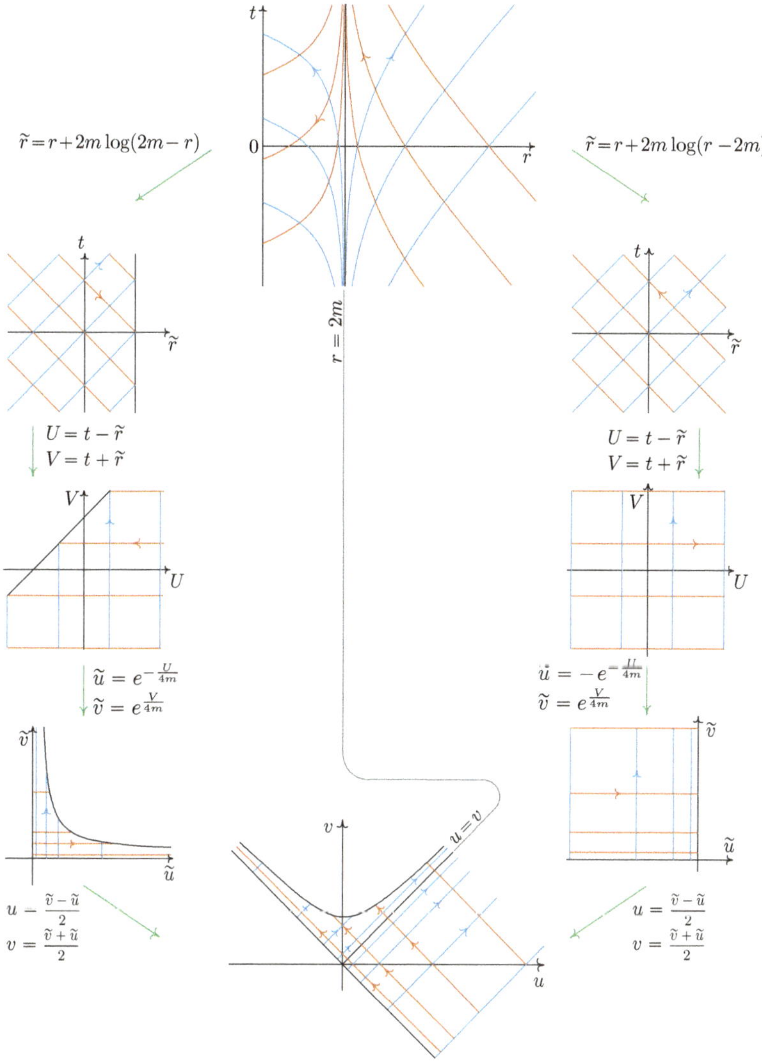

[8]We remark that for $y \geqslant 0$, the unique $x \geqslant 0$ such that $xe^x = y$ is sometimes denoted by $W_0(y)$, and the so-obtained function W_0 is called the *Lambert W function*; see e.g. [Corless, Gonnet, Hare, Jeffrey and Knuth (1996)].

With h as in Lemma 15.1, it can be seen from (15.8) that

$$r(u,v) = 2m\big(1 + h^{-1}(\tfrac{u^2-v^2}{2me})\big). \tag{15.9}$$

Using (15.7), we have

$$\begin{aligned} \frac{\partial u}{\partial t} &= \frac{v}{4m}, & \frac{\partial v}{\partial t} &= \frac{u}{4m}, \\ \frac{\partial u}{\partial r} &= \frac{u}{4m}\frac{1}{f(r)}, & \frac{\partial v}{\partial r} &= \frac{v}{4m}\frac{1}{f(r)}. \end{aligned}$$

Using these, we obtain

$$\begin{aligned} \partial_t &= \frac{v}{4m}\partial_u + \frac{u}{4m}\partial_v \\ \partial_r &= \frac{u}{4m}\frac{1}{f(r)}\partial_u + \frac{v}{4m}\frac{1}{f(r)}\partial_v. \end{aligned}$$

From these two equations, using $v^2 - u^2 = -re^{\frac{r}{2m}}f(r)$, we solve for ∂_u, ∂_v:

$$\begin{aligned} \partial_u &= \frac{4m}{re^{\frac{r}{2m}}}\big(u\,\partial_r - \frac{v}{f(r)}\partial_t\big) \\ \partial_v &= \frac{4m}{re^{\frac{r}{2m}}}\big(\frac{u}{f(r)}\partial_t - v\,\partial_r\big). \end{aligned}$$

Hence the components of $\mathbf{g}$ in the new coordinates are given by

$$\begin{aligned} \mathbf{g}(\partial_u,\partial_u) &= \big(\frac{4m}{re^{\frac{r}{2m}}}\big)^2\mathbf{g}\big(u\,\partial_r - \frac{v}{f(r)}\partial_t,\ u\,\partial_r - \frac{v}{f(r)}\partial_t\big) \\ &= \frac{16\,m^2}{r^2e^{\frac{r}{m}}}\big(u^2\frac{1}{f(r)} + \frac{v^2}{(f(r))^2}(-f(r))\big) \\ &= \frac{16\,m^2}{r^2e^{\frac{r}{m}}}\frac{1}{f(r)}(u^2 - v^2) \\ &= \frac{16\,m^2}{r^2e^{\frac{r}{m}}}\frac{1}{f(r)}re^{\frac{r}{2m}}f(r) \\ &= \frac{16m^2}{re^{\frac{r}{2m}}}. \end{aligned}$$

Similarly, $\mathbf{g}(\partial_v,\partial_v) = -\dfrac{16\,m^2}{re^{\frac{r}{2m}}}$ and

$$\begin{aligned} \mathbf{g}(\partial_u,\partial_v) = \mathbf{g}(\partial_v,\partial_u) &= \frac{16\,m^2}{r^2e^{\frac{r}{m}}}\mathbf{g}\big(\frac{u}{f(r)}\partial_t - v\,\partial_r,\ u\,\partial_r - \frac{v}{f(r)}\partial_t\big) \\ &= \frac{16\,m^2}{r^2e^{\frac{r}{m}}}\big(0 - \frac{uv}{(f(r))^2}(-f(r)) - \frac{uv}{f(r)} + 0\big) = 0. \end{aligned}$$

Thus,

$$\mathbf{g} = \frac{16\,m^2}{re^{\frac{r}{2m}}}(-\,dv\otimes dv + du\otimes du) + r^2d\theta\otimes d\theta + r^2(\sin\theta)^2d\phi\otimes d\phi, \tag{15.10}$$

where $r = r(u,v)$ is given by (15.9).

Transformations in the inner region

Again there are four steps, in each of which we maintain the θ, ϕ coordinates, and change the original pair (r,t) to get a final pair (u,v) through the sequence $(r,t) \to (\widetilde{r},t) \to (U,V) \to (\widetilde{u},\widetilde{v}) \to (u,v)$ as follows:

- In the first step, as before, we keep the Schwarzschild coordinate time t, and set $\widetilde{r} = r + 2m\log(2m-r)$. This sends the strip $\{(r,t) : 0 < r < 2m\}$ to the left half-plane $\{(\widetilde{r},t) : \widetilde{r} < 2m\log(2m)\}$ in the $(\widetilde{r},t)$-plane.
- In the second step, we use the same transformation as used in the second step for the exterior region, namely, $(\widetilde{r},t) \mapsto (t-\widetilde{r}, t+\widetilde{r}) =: (U,V)$, which has the effect of mapping the previously obtained left half-plane to a 'lower-triangular' right half-plane in the (U,V)-plane.
- In the third step, the map $(U,V) \mapsto (e^{\frac{-U}{4m}}, e^{\frac{V}{4m}}) =: (\widetilde{u},\widetilde{v})$ compresses the triangular region obtained above onto the region in the first quadrant of the $(\widetilde{u},\widetilde{v})$ plane lying below the branch of the hyperbola $\widetilde{u}\widetilde{v} = 2m$.
- In the final step, we use again a $45°$ anticlockwise rotation, followed by a reflection in the vertical axis, followed by a dilation by $1/\sqrt{2}$, that is, $(\widetilde{u},\widetilde{v}) \mapsto \frac{1}{2}(\widetilde{v}-\widetilde{u}, \widetilde{v}+\widetilde{u}) =: (u,v)$.

Altogether, analogous to the calculation done to obtain (15.7), we obtain

$$(u,v) = e^{\frac{r}{4m}}\sqrt{2m-r}\Big(\sinh\frac{t}{4m},\ \cosh\frac{t}{4m}\Big).$$

Indeed,

$$\begin{aligned}(u,v) &= \frac{1}{2}(e^{\frac{V}{4m}} - e^{\frac{-U}{4m}},\ e^{\frac{V}{4m}} + e^{\frac{-U}{4m}}) = \frac{1}{2}(e^{\frac{t+\widetilde{r}}{4m}} - e^{\frac{-t+\widetilde{r}}{4m}},\ e^{\frac{t+\widetilde{r}}{4m}} + e^{\frac{-t+\widetilde{r}}{4m}})\\ &= e^{\frac{\widetilde{r}}{4m}}\big(\sinh\frac{t}{4m},\ \cosh\frac{t}{4m}\big) = e^{\frac{r}{4m}}e^{\frac{1}{2}\log(2m-r)}\big(\sinh\frac{t}{4m},\ \cosh\frac{t}{4m}\big)\\ &= e^{\frac{r}{4m}}\sqrt{2m-r}\big(\sinh\frac{t}{4m},\ \cosh\frac{t}{4m}\big).\end{aligned}$$

Thus again, we have $v^2 - u^2 = (2m-r)e^{\frac{r}{2m}} = -re^{\frac{r}{2m}}f(r)$, and so $r(u,v)$ is given by (15.9). The equations after (15.9), all the way to the metric expression in (15.10) remain the same.

Consequently, we take $\widetilde{M}_{\text{BH}}$ as the smooth manifold $\Omega \times S^2$, where $\Omega = \{(u,v) \in \mathbb{R}^2 : v > -u \text{ and } v^2 - u^2 < 2m\}$, with the metric given by

$$\mathbf{g} = \frac{16m^2}{re^{\frac{r}{2m}}}(-dv \otimes dv + du \otimes du) + r^2 d\theta \otimes d\theta + r^2(\sin\theta)^2 d\phi \otimes d\phi,$$

where $r = r(u,v)$ is given by

$$r(u,v) = 2m\big(1 + h^{-1}(\frac{u^2-v^2}{2me})\big),$$

and the function $h : (-1,\infty) \to (-\frac{1}{e},\infty)$ is given by $h(x) = xe^x$.

The interior region of $M_{\rm BH}$, where $0 < r < 2m$, is mapped to the region $\Omega_{\rm in} \times S^2$, where

$$\Omega_{\rm in} := \{(u, v) \in \Omega : v > u\}.$$

The exterior region of $M_{\rm BH}$, where $r > 2m$, is mapped to $\Omega_{\rm ex} \times S^2$, with

$$\Omega_{\rm ex} := \{(u, v) \in \Omega : u > v\}.$$

As each of the maps in the four steps used for the interior/exterior region were diffeomorphisms, we see that $(\Omega_{\rm in} \cup \Omega_{\rm ex}) \times S^2$ with the above metric $\mathbf{g}$ can be considered to be a diffeormorphic 'copy' of $(M_{\rm BH}, \mathbf{g})$ lying inside $\widetilde{M}_{\rm BH}$. But now we have also glued the original interior and exterior regions, by mapping the event horizon $r = 2m$ along $u = v$. The above expression for $\mathbf{g}$ for the extended spacetime $\widetilde{M}_{\rm BH}$ has no singularity at $u = v$. Also, the function h^{-1} is C^∞ at 0 (Lemma 15.1), and so $\mathbf{g}$ is a metric for $\widetilde{M}_{\rm BH}$. By continuity, the metric $\mathbf{g}$ solves the field equation also on the event horizon.

In the Kruskal spacetime, radially incoming free falling photons in the exterior region are straight lines at right angles to the $u = v$ line (corresponding to the event horizon at $r = 2m$) in the (u, v)-plane, which simply continue through the $u = v$ line, and move in the interior region also as a straight line, eventually reaching the singularity at $r = 0$ (corresponding to $v^2 - u^2 = 2m$ in the (u, v)-plane). Also, it is now clear that particles starting inside the black hole region cannot escape to the exterior region.

Remark 15.1. (Maximal extension of $M_{\rm BH}$.[9])
The spacetime $\widetilde{M}_{\rm BH}$ is not a 'maximal' extension of $M_{\rm BH}$. There is a larger smooth manifold in which $\widetilde{M}_{\rm BH}$ can be isometrically embedded. Formally, this is obtained by gluing a time-reversed copy of $\widetilde{M}_{\rm BH}$ to the original $\widetilde{M}_{\rm BH}$. This produces a new singularity (called a *white hole*, since light can only come out of it), and a corresponding new exterior region. The Schwarzschild spacetime is used to model the exterior of a non-rotating star, which eventually may collapse to form a black hole. Inside the star, the spacetime metric can't be modelled by the Schwarzschild spacetime, since the energy-momentum tensor field is not zero there. During the star's lifetime, the r-component $r_\star(t)$ of the worldline γ of a particle on the surface of the collapsing star changes with t, and eventually goes to 0 if a black hole is formed. This worldline γ lies in $\widetilde{M}_{\rm BH}$, and so does the region $r > r_\star(t)$. Thus the part of the maximal extension of $\widetilde{M}_{\rm BH}$, in the complement of the region where $r > r_\star(t)$, can be discarded based on these physical grounds. $\ast$

[9]See [Natário (2021), §2.3] and [Wald (1984), §6.6] for a detailed exposition.

15.4 Miscellanea

In this last section of this chapter on black holes, we make brief remarks about some topics which will not be discussed here.

Existence and formation of black holes

Based on models of stellar evolution, black holes are theoretically expected to be formed at the end of the life history of a (massive enough) star when it has burnt up all of its nuclear fuel, and there is no outward pressure to stop it from 'collapsing' inwards under gravity. We give an outline below[10].

Stars are formed from giant clouds of hydrogen gas, which shrink due to gravity, causing the heated core to ignite, undergoing a thermonuclear burn (nuclear fusion, converting lighter elements, here hydrogen, into heavier elements, here helium) if the mass of the initial cloud is large enough. The radiation pressure from the nuclear reaction prevents the cloud from collapsing completely, and a star is thus born. The initial nuclear reaction is that of conversion of hydrogen to helium (called the *p-p cycle*). Thus we may define a *star* as a cosmic gaseous cloud held in equilibrium by the balance of the outward (radiation and gas) pressure from the thermonuclear reaction in the core trying to distend the star, and the inward gravitational attraction trying to crush it.

When the initial *p-p* cycle ends, the gas pressure starts dwindling, and gravity, gaining the upper hand, starts shrinking the star, which heats the outer hydrogen shell surrounding the burnt-up 'core'. Due to this heating, the shell now gets ignited into a thermonuclear burning, resulting in an expansion of the outer mantle. Meanwhile the core contracts, while the outer shell expands to a huge size, making it look like a (typically red) giant star. The core, under contraction, then begins to heat up so that a new thermonuclear cycle starts, that of conversion of helium into heavier elements. This process continues until the elements feeding into the nuclear fusion reaction are exhausted. Then the burning ceases, and the core begins to shrink due to gravity. The collapse is prevented when the next cycle of thermonuclear reactions gets triggered. Such an evolution of the star, where the core elements get heavier and heavier, continues until the core becomes iron. Then fusion is no longer possible due to an energy barrier, that is, while the earlier fusion reactions (e.g. hydrogen to helium, helium to carbon, etc.) all release energy[11] during the fusion process, when iron starts

[10] For a detailed exposition, see, e.g., [Misner, Thorne and Wheeler (2017), Chapter 32].
[11] That is, they are 'exothermic' reactions.

to fuse into heavier elements, it absorbs energy[12], cooling down the core. Thus the core succumbs to the gravitational pressure and starts collapsing. The star is then destroyed, and if the mass of the star is big enough, it becomes a supernova, throwing off much of its mass into the universe (hence we are all made of star dust). If sufficient matter is not shed in such an explosion, then what is left can become a white dwarf or neutron star or a black hole. In the case of a white dwarf or a neutron star, the collapse of a dead star gets arrested by an outward pressure called the 'degeneracy pressure', which is due to a quantum mechanical effect, and which can exist even at absolute zero temperature. We elaborate on this below.

Matter inside a white dwarf can be thought of as an electron gas and a sea of nuclei. As electrons are fermions, an electron gas is described by Fermi-Dirac quantum statistical mechanics (rather than the Boltzmann classical statistical mechanics applicable to ordinary gases), and the quantum mechanical pressure (called *degeneracy pressure*) that arises is due to the *Pauli exclusion principle* (preventing two or more identical fermions from being in the same quantum state). In the context of dying stars, this was first considered by Ralph Fowler[13], but he did not take relativistic effects into account, which ought not to be ignored owing to the high speeds of the electrons. The fundamental result incorporating relativity was obtained by Subrahmanyan Chandrasekhar[14] in the early 1930s. He proved that if the mass of the star is larger than $1.4M_{\odot}$ (now called the *Chandrasekhar limit*), then the degeneracy pressure is not enough to prevent gravitational collapse.

What happens if the remnant core has a mass greater than the Chandrasekhar limit? Then there is the formation of a neutron star. Roughly, the mechanism for the formation of a neutron star is as follows. Under collapse, the protons and electrons come close and a stage is reached when they start forming neutrons. Neutrons are unstable, meaning that they are radioactive, and decay into protons and electrons. There is a maximum energy $E_{\max}$ that can be possessed by the electrons produced by neutron decay. They start occupying the available energy levels, but again each quantum level can accommodate at most one electron. When all the levels up to $E_{\max}$ are occupied, the electrons produced by the neutron decay cannot go anywhere, implying that there can be no further neutron decay. At this stage, most of the core comprises only neutrons. Neutrons,

---

[12]That is, they are 'endothermic' reactions.
[13]British physicist (1889–1944).
[14]Indian-American physicist (1910–1995).

being fermions, exert degeneracy pressure, which could exactly match the inward gravitational pressure, arresting the gravitational collapse. In this case, a *neutron star* is formed. Analogous to the Earth's magnetic field, the neutron star produces a magnetic field, that is much stronger[15] than the Earth's magnetic field. Electrons escaping from the crust of the neutron star spiral in the magnetic field, emitting electromagnetic radiation. Due to the rotation of the star, the resulting radio waves received here on Earth are in the form of periodic pulses, and so neutron stars are also called *pulsars*. Analogous to the Chandrasekhar limit for white dwarf stars, there is also a mass limit for neutron stars.

We also briefly describe the reason behind the phenomenon of the explosion of a star (*supernova*). When a star starts rapidly collapsing due to gravity (e.g. when the radiation pressure stops due to the cessation of nucleosynthesis, or when the degeneracy pressure is not sufficient), the interior gets compressed suddenly and a lot of heat is generated. As a result of the collapse, the core gets converted into neutrons, and the violent heating during this core formation produces a shock wave that tears off the outer layers of the star, hurtling them into space in a supernova explosion. The luminosity of a supernova can be comparable to that of the entire galaxy, and can last for several months. The core of a supernova can be a neutron star, and if the original mass of the dying star is greater than $5-10M_{\odot}$, then it ends up as a black hole, since then the contraction cannot be arrested either at the white dwarf or the neutron star phase. Based on our study of lightlike geodesics in Schwarzschild spacetime, we know that we will just observe the star getting dimmer and dimmer, and will never see it become a black hole, while for the collapsing stellar matter, the collapse just takes finite proper time.

A spacetime diagram (with one dimension suppressed) of a star's collapse to a black hole is shown in the picture on page 348.

The existence of black holes is by now widely believed, based on multiple lines of evidence (X-ray observations of the surrounding matter, astrometric measurement of the motions of nearby stars, detection of gravitational waves from black hole mergers, and even direct interferometric radio images by the Event Horizon Telescope). Astronomical observations also suggest that there exists a 'supermassive' black hole at the centre of each galaxy.

[15]For example, they can be a billion or a trillion times stronger.

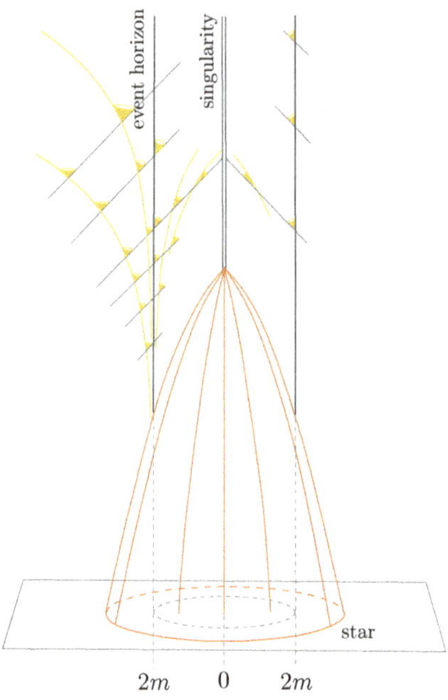

Rotating black holes

The Schwarzschild solution to the field equation was found already in 1916, and as we have seen in this chapter, it also describes a stationary black hole. However, in 1963, Roy Kerr[16] found a generalisation, which can be thought of as a 'rotating' black hole, described by two parameters, m and a (a representing the angular momentum per unit mass). Thus when $a=0$, the metric for the Kerr spacetime[17] reduces to that of the Schwarzschild spacetime. Kerr black holes are considered more realistic as most stars rotate, and should eventually end up as rotating black holes.

Black hole thermodynamics

A 1975 paper by Stephen Hawking showed, based on quantum field theoretic considerations, that black holes have a temperature[18], now called the *Hawking temperature*, given by

$$T_{\rm H} = \frac{\hbar c^3}{8\pi G_{\rm N} k_{\rm B} m} = 6.18 \times 10^{-1} \frac{M_\odot}{m} \text{Kelvin}. \qquad (\star)$$

Here, $k_{\rm B}$ is the Boltzmann constant, $\hbar = h/(2\pi)$, h being the Planck con-

[16]New Zealand mathematician, born in 1934.
[17]See [O'Neill (2014)] for a geometrical treatment of Kerr black holes.
[18]See [Wald (1984), Chapter 14].

stant, m is the mass of the black hole, and $M_{\odot}$ is the mass of our Sun.

Exercise 15.3. When an object falls into a black hole, information is lost. Lack of information is associated with entropy in statistical physics. Entropy is related to temperature. Let us use these ideas to give a 'back of the envelope' heuristic derivation of the Hawking temperature formula.

It was shown by Hawking in 1971 that if two black holes collide forming a new black hole, then the area of the event horizon surface is bigger than the sum of that of the original black holes. This result is reminiscent of the second law of thermodynamics about entropy increasing in any dynamical process. This led J. Bekenstein[19] to guess that the entropy of a black hole is an increasing function of the area. But as entropy is additive, and since in the Hawking inequality mentioned above, the right-hand side takes the *sum* of the areas, the simplest guess is to take $S \propto A$. As entropy is k_{B} times a dimensionless quantity, we divide the area A by the square of an appropriate 'universal' length.

The elementary particles, with kinematics governed by the constants c and $\hbar$, do not have a length scale, because these two constants cannot be combined to form a quantity with the dimension of length. But using mass, we do obtain length, for example the *Compton length*

$$\ell_{\text{C}} := \frac{\hbar}{mc}.$$

Similarly G_{N} and c also cannot be combined to give length. Adding mass again rescues the situation, since $G_{\text{N}}m/c^2$ has dimensions of length. Taking the geometric mean of the lengths $\hbar/(mc)$ and $G_{\text{N}}m/c^2$, the pesky m cancels, and we obtain the universal length scale, the so-called *Planck length*, given in terms of the universal physical constants by

$$\ell_{\text{P}} = \sqrt{\frac{\hbar G_{\text{N}}}{c^3}}.$$

Using this, we have $S = \alpha k_{\text{B}} A/\ell_{\text{P}}^2$, where α is a numerical proportionality constant, which was shown in Hawking's 1975 work to be equal to $\frac{1}{4}$. As we are only giving a heuristic derivation, let us use this numerical constant, and see if we recover the rest of the terms in the formula ($\star$) on p.348 for T_{H}.

A particle of mass Δm has 'rest energy' $\Delta U = c^2 \Delta m$. If this enters the event horizon, then, using $T\Delta S = \Delta U$, we obtain $T\Delta S = c^2\Delta m$. Using $A = 4\pi R_{\text{S}}^2$, where $R_{\text{S}} := 2G_{\text{N}}m/c^2$ is the Schwarzschild radius, derive ($\star$) on p.348.

Remark 15.2. (Emergent gravity paradigm.) We have seen two instances in spacetime physics, where the existence of a 'horizon' creates a thermodynamics temperature: the Unruh effect in Exercise 5.27 (p.100) and the black hole temperature (above). In the classical viewpoint, the field equations of spacetime are fundamental, and the thermodynamics of spacetime is a consequence. In the 'emergent gravity paradigm', roughly speaking, one tries to go the other way, obtaining[20] the field equations as an effective macroscopic description of a more fundamental microscopic theory ('atoms of spacetime'/associated degrees of freedom). $\ast$

[19] 1947–2015, a Mexican-born Israeli-American theoretical physicist.
[20] See [Padmanabhan (2016)].

Exercise 15.4. (Rotating black holes.) The Kerr spacetime models a rotating black hole. In addition to the parameter $m > 0$, now there is an extra parameter a corresponding to its angular momentum $J := ma$ (perceived to be possessed by the black hole by a far away observer against the starry sky). Let $0 < a < m$ ('slowly rotating' case). Define

$$r_+ := m + \sqrt{m^2 - a^2} > r_- := m - \sqrt{m^2 - a^2} > 0.$$

Let $J = (0, r_-) \cup (r_-, r_+) \cup (r_+, \infty) \subset \mathbb{R}$, and $M := \mathbb{R} \times J \times S^2$, the product of the smooth manifolds $\mathbb{R}$, J, S^2. The smooth manifold M is equipped with a Lorentzian metric $\mathbf{g}$, and we describe it below in the chart[21] $(\mathbb{R} \times J \times U, \mathrm{id}_{\mathbb{R}} \times \mathrm{id}_J \times \varphi)$, where (U, φ) is the spherical coordinate chart for S^2 from Example 1.8 (p.7). Let $\rho := \sqrt{r^2 + a^2(\cos\theta)^2}$ and $\Delta := r^2 - 2mr + a^2 = (r - r_-)(r - r_+)$. Then

$$\begin{aligned}\mathbf{g} &= -\big(1 - \frac{2mr}{\rho^2}\big) dt \otimes dt - \frac{2mra(\sin\theta)^2}{\rho^2}(dt \otimes d\phi + d\phi \otimes dt) + \frac{\rho^2}{\Delta} dr \otimes dr \\ &\quad + \rho^2 d\theta \otimes d\theta + \big(r^2 + a^2 + \frac{2mra^2(\sin\theta)^2}{\rho^2}\big)(\sin\theta)^2 d\phi \otimes d\phi.\end{aligned}$$

In the limit when $a \to 0$, the metric reduces to Schwarzschild spacetime metric. It can be shown[22] that the Kerr metric is Lorentzian and a solution to the field equation in vacuum. Write $J = J_1 \cup J_2 \cup J_3$, where $J_1 = (r_+, \infty)$, $J_2 = (r_-, r_+)$, $J_3 = (0, r_-)$. Then M is the union of the three open sets $M_i := \mathbb{R} \times J_i \times S^2$, $i = 1, 2, 3$. Analogous to the Kruskal extension we obtained for the Schwarzschild black hole, it turns out that there is a connected Lorentzian manifold $\widetilde{M}$ such that each M_i is isometric to an open subset of $\widetilde{M}$. We use the notation $\partial_t, \partial_r, \partial_\theta, \partial_\phi$ analogous to what was done in Schwarzschild spacetime on page 87.

(1) Show that $V := (r^2 + a^2)\partial_t + a\partial_\phi$ is timelike at each point of M_1 and M_3.

(2) Show that ∂_r is timelike at each point of M_2.

(3) Show that ∂_t is timelike at each point for which $2mr < \rho^2$.
In particular, if for a point p, we have $r \in (r_+, r_0)$, where
$$r_0 := m + \sqrt{m^2 - a^2(\cos\theta)^2} > m + \sqrt{m^2 - a^2} = r_+,$$
then show that $\partial_{t,p}$ is not timelike, but is spacelike.

(4) Show that ∂_t is a Killing vector field.
By Exercise 8.2 (p.153), it follows that for any timelike geodesic $\gamma : I \to M$, the quantity $E := -\mathbf{g}(\gamma(\cdot))(v_{\gamma,\gamma(\cdot)}, \partial_{t,\gamma(\cdot)})$ is constant on I.

---

[21]These are called the *Boyer-Lindquist coordinates.* The coordinate ϕ can be thought of as the standard spherical coordinate in 'space' surrounding the blackhole. If $m = 0$, but $a \neq 0$, then the Kerr metric reduces to the Minkowski metric in 'spheroidal/oblate-polar coordinates'. This motivates (see e.g. [Visser (2009)]) relating the coordinates r and θ with the standard Cartesian coordinates x, y by $x = \sqrt{r^2 + a^2}(\sin\theta)\cos\phi$ and $y = \sqrt{r^2 + a^2}(\sin\theta)\sin\phi$. The coordinate r is still a kind of radial coordinate, but $r = 0$ does not identify a unique point. The metric is singular when $\rho = 0$ and when $\Delta = 0$. The first of these is a genuine physical singularity in the sense that the Kretschmann scalar tends to infinity when ρ tends to 0, while the second is just a coordinate artefact. With the aforementioned relation between the Cartesian coordinates and the Boyer-Lindquist coordinates r and θ, the singularity $\rho = 0$, i.e., $r = 0$ and $\theta = \pi/2$, corresponds to the circle $z = 0$, $x^2 + y^2 = a^2$ in the equatorial plane. It is thus called a 'ring singularity'.

[22]See e.g., [O'Neill (2014), p.61, p.99].

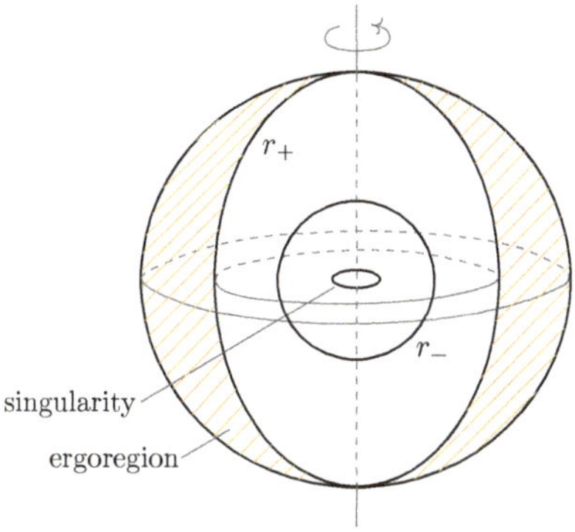

Remark 15.3. (Ergoregion and the Penrose process.) In the previous exercise, the region $\mathcal{E}$ of M consisting of all points for which the r coordinate lies between r_+ and r_0 is called the *ergoregion*:

$$\mathcal{E} := \{p \in M : r_+ < r(p) < r_0\}.$$

The Penrose process[23] is a theoretical mechanism which allows the extraction of energy from a black hole, thereby reducing its angular speed (and hence also the size of its ergoregion). Having done the groundwork calculations in the above exercise, we now give a description of the process.

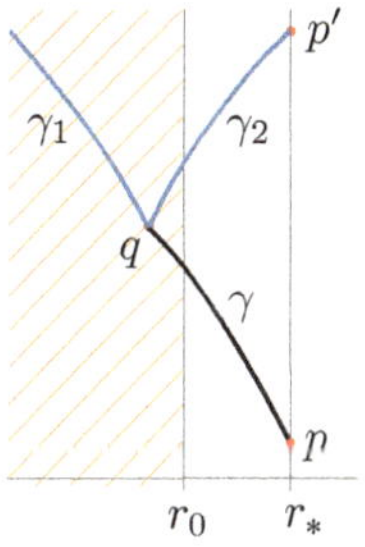

In the region where $r > r_0$, ∂_t is timelike at each point, since it is outside the ergoregion. Suppose we are along an integral curve of ∂_t with a fixed radial coordinate $r = r_* > r_0$. Also, we assume that the curves $\gamma, \gamma_1, \gamma_2$ below are parametrised by their arclength. At the event $p \in M$, we send a particle (γ, m) along a timelike geodesic γ towards the black hole. As ∂_t is a Killing vector field, and since γ is a geodesic, $Q := -\mathbf{g}(p)(v_{\gamma,\gamma(p)}, \partial_{t,p}) > 0$ is constant along γ. The energy perceived by us at p of the particle is

$$E = \frac{mQ}{\sqrt{-\mathbf{g}(p)(\partial_{t,p}, \partial_{t,p})}}.$$

Suppose that we have arranged (with carefully 'timed explosives' within) that the particle splits into two particles, (γ_1, m_1) and (γ_2, m_2), when it is at an event q belonging to the ergoregion $\mathcal{E}$. Moreover, it is arranged that the quantity $Q_1 := -\mathbf{g}(q)(v_{\gamma_1,\gamma_1(q)}, \partial_{t,q}) < 0$. This is possible because $\partial_{t,q}$ is spacelike in $\mathcal{E}$.

[23]See [Penrose (1969)].

Suppose that γ_2 is a geodesic returning the second fragment to us, at the event p'. Since the r coordinate at p and p' is the same, namely r_*, and also the θ coordinate is the same (suppose, for e.g., we stay in the equatorial plane with $\theta = \frac{\pi}{2}$), we have

$$\mathbf{g}(p)(\partial_{t,p}, \partial_{t,p}) = -(1 - \frac{2mr_*}{\rho^2}) = \mathbf{g}(p')(\partial_{t,p'}, \partial_{t,p'}).$$

The quantity $Q_2 := -\mathbf{g}(q)(v_{\gamma_1,\gamma_1(q)}, \partial_{t,q})$ is constant along γ_2 because ∂_t is a Killing vector field. The energy perceived by us at p' of the returning fragment is

$$E_2 = \frac{m_2 Q_2}{\sqrt{-\mathbf{g}(p')(\partial_{t,p'}, \partial_{t,p'})}} = \frac{m_2 Q_2}{\sqrt{-\mathbf{g}(p)(\partial_{t,p}, \partial_{t,p})}}.$$

By the law of conservation of energy-momentum, we have

$$mv^\flat_{\gamma,\gamma(q)} = m_1 v^\flat_{\gamma_1,\gamma_1(q)} + m_2 v^\flat_{\gamma_2,\gamma_2(q)}.$$

Acting on $\partial_{t,q}$ yields $E\sqrt{-\mathbf{g}(p)(\partial_{t,p}, \partial_{t,p})} = m_1 Q_1 + E_2\sqrt{-\mathbf{g}(p)(\partial_{t,p}, \partial_{t,p})}$. Thus we have extracted[24] the energy

$$E_2 - E = -\frac{m_1 Q_1}{\sqrt{-\mathbf{g}(p)(\partial_{t,p}, \partial_{t,p})}} > 0$$

from the black hole. *

Exercise 15.5. (Frame dragging.) Consider the spacetime region around a slowly rotating star, modelled by the smooth manifold $M = \mathbb{R} \times (2m, \infty) \times S^2$, as with Schwarzschild spacetime, but with the metric given by

$$[\mathbf{g}_{ij}] = \begin{bmatrix} -(1-\frac{2m}{r}) & 0 & 0 & -\frac{2J}{r}(\sin\theta)^2 \\ 0 & 1+\frac{2m}{r} & 0 & 0 \\ 0 & 0 & (1+\frac{2m}{r})r^2 & 0 \\ -\frac{2J}{r}(\sin\theta)^2 & 0 & 0 & (1+\frac{2m}{r})r^2(\sin\theta)^2 \end{bmatrix}$$

in the chart $\mathbb{R} \times (2m, \infty) \times U$, with respect to the basis vectors $\partial_t, \partial_r, \partial_\theta, \partial_\phi$, in this order, and where (U, φ) is the chart for S^2 from Example 1.8 (p.7). Here $m, J > 0$ are parameters corresponding to the mass and the angular momentum, respectively, of the star. This metric may be viewed as an approximation to the Kerr metric. Let I be an open interval in $\mathbb{R}$ containing 0. Let $\gamma : I \to M$ be a geodesic parametrised by proper time, and let $(t(\tau), r(\tau), \theta(\tau), \phi(\tau))$ be the coordinates of $\gamma(\tau)$, $\tau \in I$, in the chart used above. Let $\theta(0) = \frac{\pi}{2}$ and $\theta'(0) = 0$, where $' = \frac{d}{d\tau}$. Then it can be shown that $\theta \equiv 0$ for $\tau > 0$, that is, γ stays in the equatorial plane.

(1) Show that $E := -\mathbf{g}(\gamma(\tau))(\partial_{t,\gamma(\tau)}, v_{\gamma,\gamma(\tau)})$ and $L := \mathbf{g}(\gamma(\tau))(\partial_{\phi,\gamma(\tau)}, v_{\gamma,\gamma(\tau)})$ are constant along γ.

(2) Let $t'(0), \phi'(0)$ be such that $E > 0$ and $L = 0$. Show that for all $\tau > 0$, $\frac{d\phi}{d\tau} > 0$.

[24]It can be shown (see, e.g., [Wald (1984), §2.4]) that the fragment falling in with the worldline γ_1 has the effect of reducing the angular momentum $J = ma$ of the black hole. So the energy extraction is possible up until $J = 0$, at which stage the ergosphere disappears, ceasing the Penrose process.

(3) Let $\phi'(0)=0$ and $r(0)>2m$. Express E and L in terms of $r(0)$ and $t'(0)>0$.

Show that $\dfrac{d\phi}{d\tau}(\tau)=\dfrac{2J\left(\frac{1}{r(\tau)}-\frac{1}{r(0)}\right)}{\left(1-\left(\frac{2m}{r(\tau)}\right)^2\right)(r(\tau))^2+\left(\frac{2J}{r(\tau)}\right)^2}$.

Thus a particle that starts moving radially outward ($r'(0)>0$), is deflected against the rotation direction of the source. Also, a particle that starts moving radially inward ($r'(0)<0$) is deflected in the same direction as that in which the source rotates.

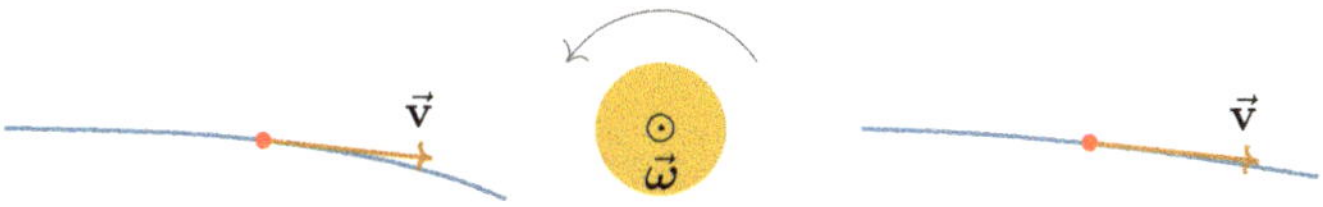

If we wrap the fingers of our right hand in the direction of rotation of the source of the gravitational field, so that the thumb points in the direction of the angular velocity vector $\vec{\omega}$ (perpendicular to the page, towards the reader), then it is as if the particle trajectory experiences a deflection 'force' in direction given by $\vec{\mathbf{v}}\times\vec{\omega}$, where $\vec{\mathbf{v}}$ is the velocity of the particle.

Exercise 15.6. (Equilibrium equation for a star.) Let the density of the star be ρ (assumed constant within the body of the star). Let S_r denote a spherical surface of radius r and centre at the centre of the star. Let $M(r)$ denote the mass inside S_r, and $P(r)$ denote the outward (radiation and gas) pressure on S_r. Show that

$$\frac{1}{r^2}\frac{d}{dr}\left(\frac{r^2}{\rho}\frac{dP}{dr}\right)=-4\pi G_{\mathrm{N}}\rho.$$

Remark 15.4. (Chandrasekhar's white dwarf equation.) To adapt the equation from Exercise 15.6 to the case of a relativistic Fermi gas, Chandrasekhar introduced the variable $x=p_{\mathrm{F}}/(m_{\mathrm{e}}c)$, where m_{e} is the mass of the electron, and p_{F} is the 'Fermi-momentum' (so that x is the Fermi-momentum expressed in units of $m_{\mathrm{e}}c$). While in a Boltzmann gas all particles have zero momentum at absolute zero, in contrast, in a Fermi gas, the particles can have a nonzero momentum even at absolute zero, up to a maximum value p_{F}, called the *Fermi-momentum*. Chandrasekhar derived that $P=Af(x)$, $\rho=Bx^3$, where A,B are suitable constants and $f(x)=x(2x^2-3)\sqrt{x^2+1}-3\sinh^{-1}x$. When inserted into the equation from Exercise 15.6, one can derive

$$\frac{1}{\eta^2}\frac{d}{d\eta}\left(\eta^2\frac{d\varphi}{d\eta}\right)=-\left(\varphi^2-\frac{1}{y_0^2}\right),$$

in terms of the new variables η (a dimensionless radial variable), φ (corresponding to the density of the white dwarf), and the constant y_0 (related to the density of the white dwarf at the centre). The high density limit corresponds to $y_0\to\infty$, and Chandrasekhar derived that the mass is still finite, which he denoted by M_3. As the density goes to infinity, this implies the radius ought to go to 0. If the star is more massive than $M_3\sim1.4M_\odot$, then even the degeneracy pressure in the white dwarf core is not enough to arrest the collapse, and the white dwarf will go on collapsing until a neutron star is formed or explode in a supernova due to the rapid collapse. $\ast$

Chapter 16

Cosmology

In this chapter, we will see that in the spacetime-geometry based theory of gravitation, the field equation is able to give information about the universe as a whole. To achieve this, we ignore the local details (planets, stars, galaxies, etc.), and instead adopt a global view by making reasonable assumptions about the energy-momentum tensor field $\mathbf{T}$ in the 'large scale'. This is akin to how we think of a fluid when we model its flow, ignoring the individual molecules or particles in the fluid. In smoothed-and-averaged form, we will assume that the energy-momentum tensor field is that of a perfect fluid. The picture below is meant to indicate the scales of distances in the visible universe.

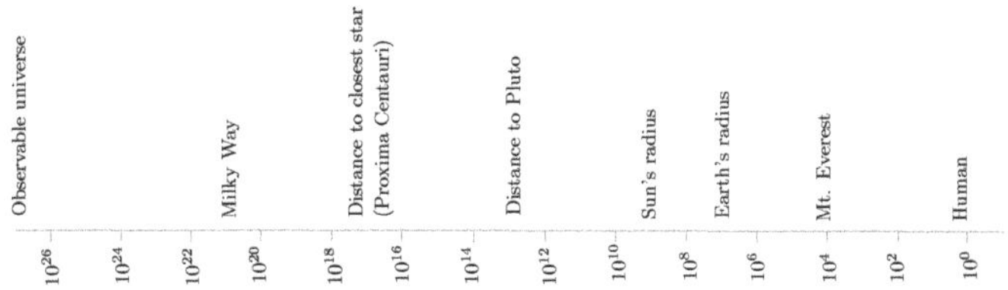

So far we know only a small portion of the universe, since we just have access to the light signals that have reached us (roughly the events in our past light 'cone'). So in light of this lack of observational knowledge, it is inevitable that in our cosmological model we will have to make certain assumptions. (Then after deriving a model using these assumptions, one makes predictions or computations, which can be compared with certain observations to examine the validity of the model.) The astronomical observational data and the radiation we receive from the cosmos is essentially 'spatially isotropic', that is, it is the same in all spatial directions, and no particular direction is favoured. Thus in our universe $(M, \mathbf{g})$, if we are at the point $p \in M$, and we are the normalised instantaneous observer $v \in T_pM$, then our perceived space $v^{\perp}$ appears the same in all directions. We will use this as a guiding principle in the development of a cosmological model, the FLRW spacetime. (The FLRW spacetime so far had 'flat' spatial sections, but now we will consider more general FLRW spacetimes.)

We assume that the spacetime M is $I \times S$, where $I \subset \mathbb{R}$ is an open interval in $\mathbb{R}$, and $(S, \boldsymbol{\sigma})$ is a 3-dimensional connected Riemannian manifold. If $(U, \mathbf{x})$ is an admissible chart for S, then the metric $\mathbf{g}$ on M in the chart $(I \times U, (t, p) \mapsto (t, \mathbf{x}(p)))$ is

$$\mathbf{g} = -dt \otimes dt + (a(t))^2 \sum_{1 \leqslant i,j \leqslant 3} \boldsymbol{\sigma}_{ij} dx^i \otimes dx^j.$$

As ∂_t is timelike, if we evolve along the integral curve γ of ∂_t given by $\gamma(t) = (t, \mathbf{p})$, $t \in I$, for some fixed $\mathbf{p} \in S$, then $(v_{\gamma,\gamma(t)})^\perp$ can be identified with $T_\mathbf{p}S$. The isotropy assumption implies in particular, that there is no preferred direction in $T_\mathbf{p}S$. We will see that this says something about the curvature of S, by using the notion of sectional curvatures.

16.1 Sectional curvatures

Definition 16.1. (Sectional curvature.)
Let $(S, \boldsymbol{\sigma})$ be a Riemannian manifold of dimension at least 2. Let $p \in S$. Let $\mathrm{Gr}(2, T_pS)$ be[1] the collection of all 2-dimensional subspaces V of T_pS. The *sectional curvature of S at p* is the map $K_p : \mathrm{Gr}(2, T_pS) \to \mathbb{R}$, where

$$K_p(V) = \frac{\boldsymbol{\sigma}(p)(R(p)(v,w)w, v)}{\boldsymbol{\sigma}(p)(v,v)\,\boldsymbol{\sigma}(p)(w,w) - (\boldsymbol{\sigma}(p)(v,w))^2},$$

where $V = \mathrm{span}\{v, w\} \in \mathrm{Gr}(2, T_pS)$.

Here $R(X,Y)Z = \nabla_X\nabla_Y Z - \nabla_Y\nabla_X Z - \nabla_{[X,Y]}Z$ is the curvature operator (Exercise 6.20, p.122 and Definition 9.1, p.181). The linear independence of v, w shows that the denominator is positive, as shown below. We have

$$z := v - \frac{\boldsymbol{\sigma}(p)(v,w)}{\boldsymbol{\sigma}(p)(w,w)} w \neq 0$$

thanks to the linear independence of v, w, and so

$$\begin{aligned} 0 < \boldsymbol{\sigma}(p)(z,z) &= \boldsymbol{\sigma}(p)(v,v) + \frac{(\boldsymbol{\sigma}(p)(v,w))^2}{\boldsymbol{\sigma}(p)(w,w)} - 2\frac{(\boldsymbol{\sigma}(p)(v,w))^2}{\boldsymbol{\sigma}(p)(w,w)} \\ &= \boldsymbol{\sigma}(p)(v,v) - \frac{(\boldsymbol{\sigma}(v,w))^2}{\boldsymbol{\sigma}(p)(w,w)}. \end{aligned}$$

We also need to check that K_p is well-defined, i.e., that it does not depend on the spanning set $\{v, w\}$ chosen. We first note that

$$\begin{aligned} \boldsymbol{\sigma}(p)(R(p)(v,w)w, v) = v^\flat(R(p)(v,w)w) &= \mathbf{R}(p)(v^\flat, v, w, w) \\ &= \widetilde{\mathbf{R}}(p)(v, w, w, v). \end{aligned}$$

Now let $v', w' \in T_pS =: V$ be such that $\mathrm{span}\{v', w'\} = V = \mathrm{span}\{v, w\}$. Then $\{v, w\}$ as well as $\{v', w'\}$ form a basis for the 2-dimensional vector space V, and so there are numbers $a, b, c, d \in \mathbb{R}$ such that $v' = av + bw$, $w' = cv + dw$.

[1]We use this notation, as it is customary to call the set consisting of all k-dimensional subspaces of a vector space V as the *Grassmannian*, and this set is denoted by $\mathrm{Gr}(k, V)$.

Also, the change of basis matrix

$$A = \begin{bmatrix} a & b \\ c & d \end{bmatrix}$$

is invertible, so that $\det A = ab - cd \neq 0$. Using the anti-symmetry properties of the covariant Riemann curvature tensor field $\widetilde{\mathbf{R}}$ in the first two and last two slots, we obtain

$$\begin{aligned} &\boldsymbol{\sigma}(p)(R(p)(v', w')w', v') \\ &= \widetilde{\mathbf{R}}(p)(v', w', w', v') = \widetilde{\mathbf{R}}(p)(av + bw, cv + dw, cv + dw, av + bw) \\ &= (ad - bc)\,\widetilde{\mathbf{R}}(p)(av+bw, cv+dw, w, v) \\ &= (ad - bc)^2\,\widetilde{\mathbf{R}}(p)(v, w, w, v) = (\det A)^2\boldsymbol{\sigma}(p)(R(p)(v, w)w, v). \end{aligned}$$

The denominator in the definition of $K_p(V)$ also picks up the same factor:

$$\begin{aligned} &\boldsymbol{\sigma}(p)(v', v')\boldsymbol{\sigma}(p)(w', w') - (\boldsymbol{\sigma}(p)(v', w'))^2 = \det\begin{bmatrix} \boldsymbol{\sigma}(p)(v', v') & \boldsymbol{\sigma}(p)(v', w') \\ \boldsymbol{\sigma}(p)(w', v') & \boldsymbol{\sigma}(p)(w', w') \end{bmatrix} \\ &= \det\begin{bmatrix} a\boldsymbol{\sigma}(p)(v, v') + b\boldsymbol{\sigma}(p)(w, v') & a\boldsymbol{\sigma}(p)(v, w') + b\boldsymbol{\sigma}(p)(w, w') \\ c\boldsymbol{\sigma}(p)(v, v') + d\boldsymbol{\sigma}(p)(w, v') & c\boldsymbol{\sigma}(p)(v, w') + d\boldsymbol{\sigma}(p)(w, w') \end{bmatrix} \\ &= \det\left(\begin{bmatrix} a & b \\ c & d \end{bmatrix}\begin{bmatrix} \boldsymbol{\sigma}(p)(v, v') & \boldsymbol{\sigma}(p)(v, w') \\ \boldsymbol{\sigma}(p)(w, v') & \boldsymbol{\sigma}(p)(w, w') \end{bmatrix}\right) \\ &= (\det A)\det\begin{bmatrix} \boldsymbol{\sigma}(p)(v, v') & \boldsymbol{\sigma}(p)(v, w') \\ \boldsymbol{\sigma}(p)(w, v') & \boldsymbol{\sigma}(p)(w, w') \end{bmatrix} \\ &= (\det A)\det\begin{bmatrix} a\boldsymbol{\sigma}(p)(v, v) + b\boldsymbol{\sigma}(p)(v, w) & c\boldsymbol{\sigma}(p)(v, v) + d\boldsymbol{\sigma}(p)(v, w) \\ a\boldsymbol{\sigma}(p)(w, v) + b\boldsymbol{\sigma}(p)(w, w) & c\boldsymbol{\sigma}(p)(w, v) + d\boldsymbol{\sigma}(p)(w, w) \end{bmatrix} \\ &= (\det A)\det\left(\begin{bmatrix} \boldsymbol{\sigma}(p)(v, v) & \boldsymbol{\sigma}(p)(v, w) \\ \boldsymbol{\sigma}(p)(w, v) & \boldsymbol{\sigma}(p)(w, w) \end{bmatrix}\begin{bmatrix} a & b \\ c & d \end{bmatrix}^{\mathrm{t}}\right) \\ &= (\det A)(\det(A^{\mathrm{t}}))\det\begin{bmatrix} \boldsymbol{\sigma}(p)(v, v) & \boldsymbol{\sigma}(p)(v, w) \\ \boldsymbol{\sigma}(p)(w, v) & \boldsymbol{\sigma}(p)(w, w) \end{bmatrix} \\ &= (\det A)^2\big(\boldsymbol{\sigma}(p)(v, v)\boldsymbol{\sigma}(w, w) - (\boldsymbol{\sigma}(p)(v, w))^2\big). \end{aligned}$$

We now show that if at a point p, all the sectional curvatures match, and have the common value $k(p)$, then $k(p)$ determines the curvature $\mathbf{R}(p)$ at p. Later in the spacetime context of $M = I \times S$, the isotropy assumption will imply a constant sectional curvature for S at any point $p \in S$ because otherwise there would be a plane in T_pS where the sectional curvature is the largest, producing a distinguished direction.

Proposition 16.1. *Let $(S, \boldsymbol{\sigma})$ be a Riemannian manifold with the metric $\boldsymbol{\sigma}$. Let $p \in S$ be such that at p, $K_p : \mathrm{Gr}(2, T_pS) \to \mathbb{R}$ is a constant function having the value $k(p) \in \mathbb{R}$. Then for all $x, y, z, w \in T_pS$*

$$\widetilde{\mathbf{R}}(p)(x, y, z, w) = k(p)\big(\boldsymbol{\sigma}(p)(x, w)\boldsymbol{\sigma}(p)(y, z) - \boldsymbol{\sigma}(p)(x, z)\boldsymbol{\sigma}(p)(y, w)\big).$$

Proof. The bracketed expression on the right-hand side defines a $(0,4)$-tensor at p, which we call $\mathbf{D}(p) \in T_4^0 S(p)$: For $x, y, z, w \in T_p S$,

$$\mathbf{D}(p)(x,y,z,w) := \boldsymbol{\sigma}(p)(x,w)\boldsymbol{\sigma}(p)(y,z) - \boldsymbol{\sigma}(p)(x,z)\boldsymbol{\sigma}(p)(y,w).$$

$\mathbf{D}(p)$ enjoys the same symmetry properties of $\widetilde{\mathbf{R}}(p)$:

$$\begin{aligned}
\mathbf{D}(p)(y,x,z,w) &= \boldsymbol{\sigma}(p)(y,w)\boldsymbol{\sigma}(p)(x,z) - \boldsymbol{\sigma}(p)(y,z)\boldsymbol{\sigma}(p)(x,w)\\
&= -\mathbf{D}(p)(x,y,z,w),\\
\mathbf{D}(p)(x,y,w,z) &= \boldsymbol{\sigma}(p)(x,z)\boldsymbol{\sigma}(p)(y,w) - \boldsymbol{\sigma}(p)(x,w)\boldsymbol{\sigma}(p)(y,z)\\
&= -\mathbf{D}(p)(x,y,z,w),\\
\mathbf{D}(p)(z,w,x,y) &= \boldsymbol{\sigma}(p)(z,y)\boldsymbol{\sigma}(p)(w,x) - \boldsymbol{\sigma}(p)(z,x)\boldsymbol{\sigma}(p)(w,y)\\
&= \mathbf{D}(p)(x,y,z,w).
\end{aligned}$$

$\mathbf{D}(p)$ also satisfies the first Bianchi identity (Exercise 9.4, p.183):

$$\begin{aligned}
&\mathbf{D}(p)(x,y,z,w) + \mathbf{D}(p)(y,z,x,w) + \mathbf{D}(p)(z,x,y,w)\\
= \;&\boldsymbol{\sigma}(p)(x,w)\boldsymbol{\sigma}(p)(y,z) - \boldsymbol{\sigma}(p)(x,z)\boldsymbol{\sigma}(p)(y,w)\\
&+\boldsymbol{\sigma}(p)(y,w)\boldsymbol{\sigma}(p)(z,x) - \boldsymbol{\sigma}(p)(y,x)\boldsymbol{\sigma}(p)(z,w)\\
&+\boldsymbol{\sigma}(p)(z,w)\boldsymbol{\sigma}(p)(x,y) - \boldsymbol{\sigma}(p)(z,y)\boldsymbol{\sigma}(p)(x,w)\\
= \;&0.
\end{aligned}$$

So the $(0,4)$-tensor $T(p) := \widetilde{\mathbf{R}}(p) - k(p)\mathbf{D}(p)$ also has the above properties. Let $v, w \in T_p S$ be any linearly independent vectors, and set $V = \operatorname{span}\{v, w\}$. As the sectional curvature at p has the constant value $k(p)$,

$$\begin{aligned}
T(p)(v,w,w,v) &= \widetilde{\mathbf{R}}(p)(v,w,w,v) - k(p)\mathbf{D}(p)(v,w,w,v)\\
&= (K_p(V) - k(p))\,\mathbf{D}(p)(v,w,w,v) = 0.
\end{aligned}$$

Also, for linearly dependent v, w we have $T(p)(v,w,w,v) = 0$, thanks to the skew-symmetry of $T(p)$ in the first two slots. So $T(p)(v,w,w,v) = 0$ for *all* $v, w \in T_p S$. We want to show $T(p) = 0$, that is, $T(p)(x,y,z,w) = 0$ for all $x, y, z, w \in T_p S$. First, we note that

$$\begin{aligned}
0 &= T(p)(x+z,y,y,x+z) = 0 + T(p)(x,y,y,z) + T(p)(z,y,y,x) + 0\\
&= -T(p)(x,y,z,y) - T(p)(x,y,z,y) = -2T(p)(x,y,z,y).
\end{aligned}$$

So $T(p)(x,y,z,y) = 0$ for all $x, y, z \in T_p S$. Hence

$$\begin{aligned}
0 &= T(p)(x,y+w,z,y+w) = 0 + T(p)(x,y,z,w) + T(p)(x,w,z,y) + 0\\
&= T(p)(x,y,z,w) - T(p)(y,z,x,w),
\end{aligned}$$

so that $T(p)(x,y,z,w) = T(p)(y,z,x,w)$. Cycling through the (arbitrary) x, y, z, we obtain $T(p)(x,y,z,w) = T(p)(y,z,x,w) = T(p)(z,x,y,w)$. The first Bianchi identity for $T(p)$ then yields $3T(p)(x,y,z,w) = 0$, completing the proof. □

We have seen above that a constant sectional curvature at a point implies that the point evaluation of the curvature tensor field is determined by the value of the sectional curvature at that point. But now suppose that at *each* point of S, we have a constant sectional curvature at the point. (This is a 'homogeneity' assumption in our spacetime scenario, namely that at *each* spacetime point $(t, \mathbf{p}) \in I \times S$, we assume we have isotropy for instantaneous observers there.) Then the above result tells us that the curvature tensor field of S ought to be governed essentially by a *function* $p \mapsto k(p) : S \to \mathbb{R}$. A remarkable 'local implies global' phenomenon now happens: the function k is in fact forced to be a constant! This is called Schur's theorem[2].

Theorem 16.1. (Schur).
Let $(S, \boldsymbol{\sigma})$ be a connected Riemannian manifold of dimension $n \geqslant 3$. Suppose that at each $p \in S$, $K_p : \mathrm{Gr}(2, T_pS) \to \mathbb{R}$ is a constant function, i.e., there exists a function $k : S \to \mathbb{R}$ such that for all $p \in S$, $K_p(V) = k(p)$ for all $V \in \mathrm{Gr}(2, T_pS)$. Then the function k is constant. Moreover,

$$\mathbf{Ric} = (n-1)\, k\, \boldsymbol{\sigma}.$$

Proof. We first note that k is smooth. From Proposition 16.1, $\widetilde{\mathbf{R}} = k\,\mathbf{D}$, where $\mathbf{D}$ is the $(0,4)$-tensor field given as follows: for all $X, Y, Z, W \in T_0^1 S$,

$$\mathbf{D}(X, Y, Z, W) := \boldsymbol{\sigma}(X, W)\,\boldsymbol{\sigma}(Y, Z) - \boldsymbol{\sigma}(X, Z)\,\boldsymbol{\sigma}(Y, W).$$

If $\mathrm{span}\{v, w\} = V \in \mathrm{Gr}(2, T_pS)$, then we have

$$k(p) = K_p(V) = \frac{\widetilde{\mathbf{R}}(p)(v, w, w, v)}{\mathbf{D}(p)(v, w, w, v)}. \tag{16.1}$$

Let $(U, \mathbf{x})$ be an admissible chart. For $p \in U$, and $i \neq j$, we have $v := \partial_{x^i, p}$, $w := \partial_{x^j, p}$ are linearly independent. Consequently, $\mathbf{D}(p)(v, w, w, v) > 0$, and $U \ni p \mapsto \mathbf{D}(p)(v, w, w, v)$ is smooth. It follows from (16.1) that $k \in C^\infty(U)$. As this happens with each admissible chart, $k \in C^\infty(S)$.

By Proposition 16.1, $\widetilde{\mathbf{R}} = k\,\mathbf{D}$, and so in an admissible chart

$$\begin{aligned} \mathbf{Ric}_{ij} &= \mathbf{R}^{\ell}_{\ell ij} = \widetilde{\mathbf{R}}_{\ell ijr}\boldsymbol{\sigma}^{\ell r} = k\,\mathbf{D}_{\ell ijr}\boldsymbol{\sigma}^{\ell r} = k\,(\boldsymbol{\sigma}_{\ell r}\boldsymbol{\sigma}_{ij} - \boldsymbol{\sigma}_{\ell j}\boldsymbol{\sigma}_{ir})\,\boldsymbol{\sigma}^{\ell r} \\ &= k\,(\delta^{\ell}_{\ell}\boldsymbol{\sigma}_{ij} - \delta^{r}_{j}\boldsymbol{\sigma}_{ir}) = k\,(n\boldsymbol{\sigma}_{ij} - \boldsymbol{\sigma}_{ij}) = k\,(n-1)\,\boldsymbol{\sigma}_{ij}. \end{aligned}$$

Hence

$$\mathbf{Ric} = k\,(n-1)\,\boldsymbol{\sigma}. \tag{16.2}$$

Taking the trace, we obtain[3] $\mathbf{S} = k\,(n-1)\,n$. So

$$d\mathbf{S} = (n-1)\,n\,dk. \tag{16.3}$$

[2]After Axel Schur (1891–1930), a German mathematician.

[3]That trace $\boldsymbol{\sigma} = n$ similarly as in the solution on p.424 to Exercise 5.30, p.103.

Taking the divergence in (16.2), we obtain

$$\frac{1}{2}d\mathbf{S} = (n-1)\,\mathrm{div}\,(k\boldsymbol{\sigma}) \overset{(\star)}{=} (n-1)\,dk. \tag{16.4}$$

To check ($\star$), we work locally in an admissible chart $(U, \mathbf{x})$: for all $W \in T_0^1 S$,

$$\begin{aligned}
&(\mathrm{div}\,(k\boldsymbol{\sigma}))(W)\\
&= \mathbf{C}(T_1^0 S \times T_0^1 S \ni (\Omega, V) \mapsto (\nabla_V(k\boldsymbol{\sigma}))(\Omega^\sharp, W))\\
&= \mathbf{C}(T_1^0 S \times T_0^1 S \ni (\Omega, V) \mapsto (Vk)\boldsymbol{\sigma}(\Omega^\sharp, W) + k\,(\nabla_V\boldsymbol{\sigma})(\Omega^\sharp, W))\\
&= \mathbf{C}(T_1^0 S \times T_0^1 S \ni (\Omega, V) \mapsto (Vk)\boldsymbol{\sigma}(\Omega^\sharp, W) + k0)\\
&= (\partial_{x^i}k)\boldsymbol{\sigma}((dx^i)^\sharp, W) = (\partial_{x^i}k)\boldsymbol{\sigma}(\boldsymbol{\sigma}^{ij}\partial_{x^j}, W^\ell\partial_{x^\ell}) = (\partial_{x^i}k)\,W^\ell\boldsymbol{\sigma}^{ij}\boldsymbol{\sigma}_{j\ell}\\
&= (\partial_{x^i}k)\,W^\ell\delta_\ell^i = (\partial_{x^i}k)\,W^i = (W^i\partial_{x^i})k = Wk = (dk)W.
\end{aligned}$$

From (16.3) and (16.4), $2(n-1)\,dk = (n-1)\,n\,dk$, that is, $(n-1)(n-2)\,dk = 0$. As $n \geqslant 3$, this implies that $dk = 0$. Hence in any admissible chart $(U, \mathbf{x})$, we have $\partial_{x^i}k = 0$ for $1 \leqslant i \leqslant n$, implying the local constancy of k. Since S is connected, it follows that k is constant on S. □

Exercise 16.1. (Einstein manifolds.)
A Riemannian manifold $(S, \boldsymbol{\sigma})$ is called an *Einstein manifold* if there exists a $\lambda \in \mathbb{R}$ such that $\mathbf{Ric} = \lambda\,\boldsymbol{\sigma}$. Below, we suppose that $(S, \boldsymbol{\sigma})$ is a connected Riemannian manifold.

(1) Let $\dim S \geqslant 3$, and let $f \in C^\infty(S)$ be such that $\mathbf{Ric} = f\boldsymbol{\sigma}$.
Show that $(S, \boldsymbol{\sigma})$ is an Einstein manifold.

(2) Let $\dim S = 3$. Show that $(S, \boldsymbol{\sigma})$ is an Einstein manifold if and only if at each $p \in S$, $K_p : \mathrm{Gr}\,(2, T_pS) \to \mathbb{R}$ is a constant function.

Some constant sectional curvature manifolds in $\mathbb{R}^3$

Consider the smooth manifold S to be $\mathbb{R}^3$ with its standard smooth structure. We will determine a metric $\boldsymbol{\sigma}$ that gives constant sectional curvatures at each point, and hence a globally constant value, say k, for the pointwise sectional curvature. We use a chart $(U, \mathbf{x})$, where the chart map $\mathbf{x}$ comprises the usual spherical polar coordinates (r, θ, ϕ) on the set

$$U = \mathbb{R}^3 \backslash \{(x, y, z) \in \mathbb{R}^3 : x \geqslant 0 \text{ and } y = 0\}.$$

Suppose that $\boldsymbol{\sigma}$ is given in the chart $(U, \mathbf{x})$ by

$$\boldsymbol{\sigma} = b(r)\,dr \otimes dr + r^2 d\theta \otimes d\theta + r^2(\sin\theta)^2 d\phi \otimes d\phi, \tag{16.5}$$

where we want to determine[4] $b(r)$. A calculation analogous to the one done in Section 14.3 (p.319) (formally replace $k(r)$ there by $b(r)$, and set

[4]We use 'b' rather than 'a', since, in our discussion of the FLRW spacetime so far, the expression for the 'spatial part of the metric' of the metric $\mathbf{g}$ of the full spacetime M, already has a function $a(t)$, and we want to avoid confusion with that function.

$f(r) = 0$) gives the nonzero Ricci curvature tensor field components:

$$\mathbf{Ric}_{rr} = \frac{b'(r)}{r\,b(r)}, \quad \mathbf{Ric}_{\theta\theta} = -\frac{1}{b(r)} + 1 + \frac{r\,b'(r)}{2(b(r))^2}, \quad \mathbf{Ric}_{\phi\phi} = (\sin\theta)^2\,\mathbf{Ric}_{\theta\theta}.$$

We have $\mathbf{Ric} = k(n-1)\boldsymbol{\sigma} = 2k\boldsymbol{\sigma}$. In particular, $\mathbf{Ric}_{rr} = 2k\,\boldsymbol{\sigma}_{rr}$, i.e.,

$$\frac{b'(r)}{r\,b(r)} = 2k\,b(r). \tag{16.6}$$

Also, $\mathbf{Ric}_{\theta\theta} = 2k\,\boldsymbol{\sigma}_{\theta\theta}$ gives

$$-\frac{1}{b(r)} + 1 + \frac{r\,b'(r)}{2(b(r))^2} = 2k\,r^2. \tag{16.7}$$

By (16.6),

$$b'(r) = 2k\,r(b(r))^2.$$

Using this in (16.7), and solving for $b(r)$,

$$b(r) = \frac{1}{1-k\,r^2}.$$

If $k > 0$, then we restrict the r coordinate to $r \in (0, 1/\sqrt{k})$ and consider only an open ball $B = \{\mathbf{x} \in \mathbb{R}^3 : \|\mathbf{x}\| < 1/\sqrt{k}\}$ in $\mathbb{R}^3$.

Thus we have seen that if we have a constant sectional curvature k everywhere for a metric $\boldsymbol{\sigma}$ of the form (16.5) in the chart $(U, \mathbf{x})$, then

$$b(r) = \frac{1}{1-k\,r^2},$$

and so the metric is

$$\boldsymbol{\sigma} = \frac{1}{1-k\,r^2}dr \otimes dr + r^2 d\theta \otimes d\theta + r^2(\sin\theta)^2 d\phi \otimes d\phi. \tag{16.8}$$

Vice versa, it can be checked that the sectional curvature is equal to k everywhere for the metric above. One way to do this is to compute the components of the covariant curvature tensor field $\widetilde{\mathbf{R}}$, and using them, show that we have $\widetilde{\mathbf{R}} = k\,\mathbf{D}$. This implies that for each $p \in S$, K_p is a constant function assuming the value[5] k. We will not carry out this computation.

Exercise 16.2. Let $\boldsymbol{\sigma}$ be the metric as described above in (16.8). Let $\theta_0 \in (0, \pi)$ and $\phi_0 \in (0, 2\pi)$ be fixed. Let $I = (0, \infty)$ if $k \leqslant 0$, and $I = (0, 1/\sqrt{k})$ if $k > 0$. Consider the radial curve $\gamma : I \to \mathbb{R}^3$ given by $\mathbf{x}(\gamma(r)) = (r, \theta_0, \phi_0)$, $r \in I$. Determine the arclength of γ as a function of $r \in I$. (This will show that if $k \neq 0$, then we should not interpret r as the radial distance to the origin.)

[5]These metrics are those of the 3-sphere of radius $\sqrt{k}$ if $k > 0$, or Euclidean space when $k = 0$, or the 'hyperbolic space of radius $\sqrt{-k}$' if $k < 0$. The *hyperbolic space of radius* $a > 0$ is the open half-space $\{(x^1, x^2, x^3) \in \mathbb{R}^3 : x^3 > 0\}$ equipped with the Riemannian metric given by $\mathbf{g}_{ij} = a^2\delta_{ij}/(x^3)^2$; see e.g. [Godinho and Natário (2014), Example 4.2] or [Lee (2018), pp.62 – 67]. We have seen the 2-dimensional version in Exercises 5.24 (p.93), 8.6 (p.158) and 9.11 (p.188).

16.2 FLRW spacetime

As mentioned earlier, the isotropy of the spatial section for each instantaneous observer motivates the spacetime model $M = I \times S$, where S has constant sectional curvature at each point. So we consider the following.

Definition 16.2. (FLRW spacetime.)
Let $a : I \to (0, \infty)$ be a smooth function, where $I \subset \mathbb{R}$ is an open interval. Let $k \in \mathbb{R}$, and

$$S := \mathbb{R}^3 \text{ if } k \leqslant 0,$$
$$S := \{(x,y,z) \in \mathbb{R}^3 : x^2+y^2+z^2 < \tfrac{1}{k}\} \text{ if } k > 0,$$

with the standard smooth structure from $\mathbb{R}^3$. Let M be the smooth manifold $M = I \times S$. We equip M with the Lorentzian metric $\mathbf{g}$, given as follows: In the chart $(I \times U, \mathrm{id}_I \times \mathbf{x})$, where $(U, \mathbf{x})$ is the spherical coordinate chart[6],

$$\mathbf{g} = -dt \otimes dt + (a(t))^2 \Big(\frac{1}{1-kr^2} dr \otimes dr + r^2 d\theta \otimes d\theta + r^2 (\sin\theta)^2 d\phi \otimes d\phi\Big).$$

M is given the orientation $[\Omega]$, where $\Omega = dx \wedge dy \wedge dz \wedge dt \in \mathbf{\Omega}^4 M$, where (t,x,y,z) are the components of the chart map $\mathrm{id}_{I\times S}$ in the global chart $(M, \mathrm{id}_{I \times S})$. Also, M is given the time-orientation $V := \partial_t \in T_0^1 M$. Then $(M, \mathbf{g})$ is called the *FLRW spacetime*.

Analogous to the computation done in Example 6.7 (p.121), where k was 0, we can determine connection coefficients in the chart $I \times U$, with spherical coordinates used in U. The nonzero ones are listed below:

$$\Gamma^t_{rr} = \frac{a\dot{a}}{1-kr^2} \quad \Gamma^t_{\theta\theta} = a\dot{a}r^2 \quad \Gamma^t_{\phi\phi} = a\dot{a}r^2(\sin\theta)^2$$

$$\Gamma^r_{tr} = \Gamma^r_{rt} = \frac{\dot{a}}{a} \quad \Gamma^r_{rr} = \frac{kr}{1-kr^2} \quad \Gamma^r_{\theta\theta} = -r(1-kr^2) \quad \Gamma^r_{\phi\phi} = -r(1-kr^2)(\sin\theta)^2$$

$$\Gamma^\theta_{t\theta} = \Gamma^\theta_{\theta t} = \frac{\dot{a}}{a} \quad \Gamma^\theta_{r\theta} = \Gamma^\theta_{\theta r} = \frac{1}{r} \quad \Gamma^\theta_{\phi\phi} = -(\sin\theta)\cos\theta$$

$$\Gamma^\phi_{t\phi} = \Gamma^\phi_{\phi t} = \frac{\dot{a}}{a} \quad \Gamma^\phi_{r\phi} = \Gamma^\phi_{\phi r} = \frac{1}{r} \quad \Gamma^\phi_{\theta\phi} = \Gamma^\phi_{\phi\theta} = \cot\theta.$$

Here $\dot{} = \frac{d}{dt}$. Using the above, we can calculate the components of the Ricci curvature tensor field, and the nonzero components are given as follows:

$$\mathbf{Ric}_{tt} = -3\frac{\ddot{a}}{a}, \qquad \mathbf{Ric}_{rr} = \frac{a\ddot{a} + 2\dot{a}^2 + 2k}{1-kr^2},$$

$$\mathbf{Ric}_{\theta\theta} = (a\ddot{a} + 2\dot{a}^2 + 2k)r^2, \quad \mathbf{Ric}_{\phi\phi} = (\sin\theta)^2 \mathbf{Ric}_{\theta\theta}.$$

Thus the scalar curvature is given by $\mathbf{S} = 6\,\dfrac{a\ddot{a} + \dot{a}^2 + k}{a^2}$.

[6] $U := \mathbb{R}^3 \backslash \{(x,y,z) \in \mathbb{R}^3 : x \geqslant 0 \text{ and } y = 0\}$ and $\mathbf{x} := (r, \theta, \phi)$, where r, θ, ϕ are the usual spherical polar coordinates.

So we have now collected all the information for writing the 'geometry' side of the field equation.

Example 16.1. (Galaxy geodesics.)
In FLRW spacetime, let $\gamma : I \to M$ be given by $\gamma(t) = (t, \mathbf{x}^{-1}(r_0, \theta_0, \phi_0))$ for $t \in I$, where $(U, \mathbf{x})$ is the spherical coordinate chart from Definition 16.2. We have $v_{\gamma,\gamma(t)} = \partial_{t,\gamma(t)}$. So $v_{\gamma,\gamma(t)}$ is timelike and future-pointing everywhere. Also, γ is an integral curve of ∂_t. Moreover,

$$(\nabla_{V_\gamma} V_\gamma)(t) = (\nabla_{\partial_t}\partial_t)_{\gamma(t)} = \Gamma^i_{tt}(\gamma(t))\,\partial_{x^i,\gamma(t)} = 0,$$

and so γ is a geodesic. The curve γ is the worldline of a freely-falling galaxy in the FLRW spacetime. ◇

Exercise 16.3. (A geodesic in a spatial slice.)
Consider smooth manifold $S \subset \mathbb{R}^3$ as in Definition 16.2, with the Riemannian metric $\boldsymbol{\sigma}$ defined as follows: Let $t \in I$, and let

$$\boldsymbol{\sigma} = (a(t))^2\Big(\frac{1}{1-kr^2}dr \otimes dr + r^2 d\theta \otimes d\theta + r^2(\sin\theta)^2 d\phi \otimes d\phi\Big),$$

in the spherical coordinate chart $(U, \mathbf{x})$. It can be shown that the nonzero connection coefficients of the Levi-Civita connection are

$$\begin{aligned}
&\Gamma^r_{rr} = \frac{kr}{1-kr^2} && \Gamma^r_{\theta\theta} = -r(1-kr^2) && \Gamma^r_{\phi\phi} = -r(1-kr^2)(\sin\theta)^2\\
&\Gamma^\theta_{r\theta} = \Gamma^\theta_{\theta r} = \frac{1}{r} && \Gamma^\theta_{\phi\phi} = -(\sin\theta)\cos\theta &&\\
&\Gamma^\phi_{r\phi} = \Gamma^\phi_{\phi r} = \frac{1}{r} && \Gamma^\phi_{\theta\phi} = \Gamma^\phi_{\phi\theta} = \cot\theta. &&
\end{aligned}$$

The aim in this exercise is show that a certain curve μ, joining the origin in S to the point $\mathbf{p}$ in S with spherical coordinates (r_0, θ_0, ϕ_0), is a geodesic. We will calculate the 'length' of μ, which will then serve as a notion of distance between the origin and the point $\mathbf{p}$. Let $s_0 > 0$, and let r be the solution to

$$\begin{aligned}
\frac{dr}{ds}(s) &= \frac{\sqrt{1-k(r(s))^2}}{a(t)} \quad (0 \leqslant s \leqslant s_0),\\
r(0) &= 0.
\end{aligned}$$

Let $r_0 := r(s_0)$. Define $\mu : [0, s_0] \to S$ by $\mu(s) = \mathbf{x}^{-1}(r(s), \theta_0, \phi_0)$ for $0 \leqslant s \leqslant s_0$. Determine $v_{\mu,\mu(s)}$, and $\boldsymbol{\sigma}(\mu(s))(v_{\mu,\mu(s)}, v_{\mu,\mu(s)})$. Show that the length $d(t)$ of μ,

$$d(t) := \int_0^{s_0} \sqrt{\boldsymbol{\sigma}(\mu(s))(v_{\mu,\mu(s)}, v_{\mu,\mu(s)})}\, ds = s_0 = a(t)\int_0^{r_0} \frac{1}{\sqrt{1-kr^2}}\, dr.$$

Show that μ is a geodesic in $(S, \boldsymbol{\sigma})$.

Remark 16.1. A *space slice* in the FLRW spacetime $(M, \mathbf{g})$ is a level set of $t : M \to I$. Identifying a space slice $\{t\} \times S$ with S, it is reasonable to define the distance between $(t, \mathbf{0}), (t, \mathbf{p}) \in \{t\} \times S \subset M$ as $d(t)$. Indeed, if the distance $d(t)$ between $\mathbf{0}$ and $\mathbf{p}$ is defined as the length of the shortest curve joining $\mathbf{0}$ to $\mathbf{p}$, then $d(t)$ is the length of the geodesic joining these points by an analogue of Theorem 8.5 (p.175) in the Riemannian manifold context; see, e.g. [Lee (2018), Chapter 6 and Theorem 6.4]. ∗

Exercise 16.4. (Hubble's law.)
With the notation from Definition 16.2, let $\mathbf{p} \in S$ be such that $\mathbf{x}(\mathbf{p}) = (r_0, \theta_0, \phi_0)$, where $(U, \mathbf{x})$ is the spherical coordinate chart. The *distance* between two galaxies γ_0, γ_1 in a space slice $\{t\} \times S$, where $\gamma_0(s) = (s, \mathbf{0})$ and $\gamma_1(s) = (s, \mathbf{p})$, $s \in I$, is defined by

$$d(t) = a(t) \int_0^{r_0} \frac{1}{\sqrt{1-kr^2}}\, dr.$$

Define the *Hubble constant*[7] by $H(t) = \dfrac{\dot{a}(t)}{a(t)}$, $t \in I$. Show that $\dot{d}(t) = H(t)\, d(t)$.

Remark 16.2. This is *Hubble's law*[8]: The 'recessional speed' of a galaxy (i.e., $\dot{d}(t)$) is directly proportional to its distance ($d(t)$). $\ast$

16.3 Field equations

We assume that on the matter side, we have a perfect fluid (ρ, p, V), where $V = \partial_t$ in the chart $I \times U$ described in the previous section. Here p and ρ are assumed to be functions of t only. This is again due to our homogeneity assumption that in each 'spatial slice', one point is not more special than the other. Then the energy-momentum tensor field $\mathbf{T} = (\rho + \mathrm{p}) V^\flat \otimes V^\flat + \mathrm{p}\mathbf{g}$ has the nonzero components given by

$$\mathbf{T}_{tt} = \rho(t), \quad \mathbf{T}_{rr} = \frac{\mathrm{p}(t)(a(t))^2}{1-kr^2}, \quad \mathbf{T}_{\theta\theta} = \mathrm{p}(t)(a(t))^2 r^2, \quad \mathbf{T}_{\phi\phi} = \mathbf{T}_{\theta\theta}(\sin\theta)^2.$$

We also assume that the cosmological constant Λ is absorbed into the matter side as described in Remark 13.2 (p.300). Then the field equation becomes

$$\mathbf{Ric} - \frac{\mathbf{S}}{2}\mathbf{g} = 8\pi\mathbf{T}.$$

The tt-component gives $-3\dfrac{\ddot{a}}{a} - 3\dfrac{a\ddot{a} + \dot{a}^2 + k}{a^2}(-1) = 8\pi\rho$. Rearranging,[9]

$$\frac{\dot{a}^2}{a^2} + \frac{k}{a^2} = \frac{8\pi}{3}\rho. \tag{16.9}$$

The rr component of the field equation gives

$$\frac{a\ddot{a} + 2\dot{a}^2 + 2k}{1-kr^2} - 3\frac{a\ddot{a} + \dot{a}^2 + k}{a^2}\frac{a^2}{1-kr^2} = 8\pi\frac{\mathrm{p}a^2}{1-kr^2},$$

which reduces to the equation[10]

$$2\frac{\ddot{a}}{a} + \frac{\dot{a}^2}{a^2} + \frac{k}{a^2} = -8\pi\mathrm{p}. \tag{16.10}$$

---

[7]Not really a constant, since it depends on t! The constancy is within the spatial slice.

[8]After Edwin Hubble (1889–1953), an American astronomer. This was also discovered by Georges Lemaître (1894–1966), a Belgian physicist, two years prior to Hubble. The law shows that the spatial slices are 'expanding'. Historically, this was an important milestone, since it ended the then-prevalent static view of the universe.

[9]This is the *Friedmann equation*, named after the Russian mathematical physicist A. Friedmann (1888–1925). It was also derived independently by G. Lemaître.

[10]This is sometimes called the *acceleration equation* because of the presence of the double derivative with respect to the t-coordinate.

The $\theta\theta$, $\phi\phi$ components of the field equation also give the same equation as above, while the mixed components just reduce to $0 = 0$.

The pair of equations (16.9) and (16.10) have three unknowns: a, p, ρ. So we need one more equation if we hope to determine each of these functions. The extra equation is often an equation of state, giving the pressure as a function of the density. In a linear model $\mathrm{p} = w\rho$, where w is a constant, and the table below gives the interpretation of the type of matter[11]:

w	matter type
$\frac{1}{3}$	radiation
0	dust
-1	cosmological constant

Exercise 16.5. Show, using (16.9) and (16.10), that $\dot{\rho} + 3(\mathrm{p}+\rho)\dfrac{\dot{a}}{a} = 0$.
Hint: Differentiate (16.9) with respect to t, and substitute for $\ddot{a}$ in (16.10).
Assuming $\mathrm{p} = w\rho$, show $\rho = Ca^{-3(1+w)}$, where C is a constant. *Hint:* $\frac{d}{dt}\log a = \frac{\dot{a}}{a}$.
Thus conclude that for
- radiation, $\rho \propto a^{-4}$,
- dust, $\rho \propto a^{-3}$,
- cosmological constant, ρ is a constant.

Exercise 16.6. (de Sitter spacetime[12].)
Let the energy-momentum tensor field of FLRW spacetime be the 'matter' due to the cosmological constant, i.e., $\mathrm{p} = -\rho = -\frac{\Lambda}{8\pi}$.
Let $\Lambda > 0$, $k = 1$, $\ell = \sqrt{3/\Lambda}$, $I = \mathbb{R}$ and $a(t) = \ell\cosh(t/\ell)$ for $t \in I$. Show that the field equations are satisfied. This spacetime is called the *de Sitter spacetime*[13].
Now let $\Lambda < 0$, $k = -1$, $\ell = \sqrt{-3/\Lambda}$, $I = (0, \ell\pi)$, and $a(t) = \ell\sin(t/\ell)$ for $t \in I$. Show that the field equations are satisfied[14].

[11]For a *static* FLRW universe (i.e., $\dot{a} \equiv 0$), the Friedmann equations imply that we must have $\frac{k}{a^2} = -8\pi\mathrm{p}$ and $\frac{k}{a^2} = \frac{8\pi}{3}\rho$. But these have no solution for a perfect fluid with $\rho > 0$ and $\mathrm{p} \geqslant 0$. This had prompted Einstein to introduce the cosmological constant Λ as a remedy, as the prevalent view then was that the universe is static. This changed with Hubble's discovery (around 1930) that $\dot{a} > 0$ (see Exercise 16.4, p.364), and Λ was again set to 0. Later at the end of the 20th century, this view was revised again, since it was discovered that the expansion of the universe is accelerating, and so in the ΛCDM model (see Exercise 16.7, p.369), it is assumed that $\Lambda > 0$.

[12]After Willem de Sitter (1872–1934), a Dutch mathematician and astronomer.

[13]A famous open problem in cosmology is the 1977 'cosmic no hair conjecture' due to Gibbons and Hawking, roughly speaking saying that all solutions to the Einstein equations with a positive cosmological constant 'eventually' look alike, in the sense that they resemble the de Sitter spacetime. The 'no-hair' is picturesque language suggesting 'no distinguishing features'.

[14]This spacetime corresponds to an open region of a bigger spacetime, namely the 'anti-de Sitter spacetime'. We refer the reader to [Natário (2021), §2.4.3] for details.

16.4 Simple case solutions and the big bang

In this section, we will assume matter to be dust, so that $w = 0$, i.e., the fluid is pressureless. In this simple scenario, we can solve the Friedmann equations explicitly. In each of the cases of constant spatial sectional curvatures $k<0$, $k=0$, $k>0$, we will see that the solution $a : I \to (0,\infty)$ living on a 'maximal' interval $I = (t_{\min}, t_{\max})$ tends to 0 as $t \to t_{\min}$. For simplicity, by translating the t-coordinate, we make this happen at $t_{\min} = 0$. We will also see that then the scalar curvature $\mathbf{S}$ tends to ∞ as $t \searrow 0$, showing that there is a 'singularity' at this end point '$t = 0$', which is referred to as the 'big bang'. We first show the following.

Lemma 16.1. *$C := \rho a^3$ is a constant.*

Proof. As $\mathrm{p} = w\rho = 0\rho = 0$, using the equation from Exercise 16.5 (p. 365),

$$\frac{d}{dt}(\rho a^3) = \dot{\rho} a^3 + 3\rho a^2 \dot{a} = a^3(\dot{\rho} + 3(0+\rho)\frac{\dot{a}}{a}) = a^3(\dot{\rho} + 3(\mathrm{p}+\rho)\frac{\dot{a}}{a}) = a^3 0 = 0. \quad \square$$

Below, we will assume that the universe is not empty, so that $\rho a^3 = C > 0$.

The case $k = 0$.

Equation (16.9) becomes $\frac{\dot{a}^2}{a^2} = \frac{8\pi}{3}\rho = \frac{8\pi}{3}\frac{C}{a^3}$, and so $a\dot{a}^2 = \frac{8\pi}{3}C$. Thus[15]

$$\frac{d}{dt}a^{\frac{3}{2}} = \frac{3}{2}\sqrt{a}\,\dot{a} = \frac{3}{2}\sqrt{\frac{8\pi}{3}C} =: \alpha. \tag{16.11}$$

Solving this on $I = (0,\infty)$ such that[16] $\lim_{t\to 0} a(t) = 0$, we get, with $\beta := \alpha^{\frac{2}{3}}$,

$$a(t) = \beta t^{\frac{2}{3}}.$$

The scalar curvature is $\mathbf{S} = 6\,\frac{a\ddot{a} + \dot{a}^2 + 0}{a^2} = \frac{4}{3t^2}$. Hence $\mathbf{S} \to \infty$ as $t \searrow 0$. After a rescaling of the t variable, the metric is a constant multiple of the one considered in §9.2 (p.192).

The case $k > 0$.

We will see that the solution a starts from 0, grows to a maximum value, and reduces back to 0 at a value $t_{\max} < \infty$. Rather than solving (16.9) and (16.10), we will just check that a function given below solves them.

Let $t_{\max} = \frac{8\pi^2}{3}\frac{C}{k\sqrt{k}}$, and $h : (0, 2\pi) \to (0, t_{\max}) := I$ be given by

$$h(\tau) = \frac{4\pi}{3}\frac{C}{k\sqrt{k}}(\tau - \sin\tau), \quad \tau \in (0, 2\pi).$$

---

[15] We assume that pointwise $\dot{a} > 0$.

[16] It is clear that every solution a to $\frac{d}{dt}a^{\frac{3}{2}} = \alpha$ becomes 0 at some finite $t_0 \in \mathbb{R}$, and we choose this $t_0 = 0$ by shifting the t-coordinate.

Then $h'(\tau)=\frac{4\pi}{3}\frac{C}{k\sqrt{k}}(1-\cos\tau)>0$, and so h is a diffeomorphism. Moreover, $h(0,2\pi)=I$. Set

$$a(t) = \frac{4\pi}{3}\frac{C}{k}(1-\cos(h^{-1}t)), \quad t \in I.$$

Then it follows that

$$\dot{a}(t)=\sqrt{k}\frac{\sin(h^{-1}t)}{1-\cos(h^{-1}t)} \quad \text{and}$$

$$\ddot{a}(t)=-\frac{k^2}{\frac{4\pi}{3}C(1-\cos(h^{-1}t))^2}.$$

Using these expressions, (16.9) holds after substituting $\rho = \dfrac{C}{a^3}$: Indeed,

$$\begin{aligned}
&\frac{\dot{a}^2}{a^2}+\frac{k}{a^2}\\
&= \frac{k(\sin(h^{-1}t))^2}{(1-\cos(h^{-1}t))^2}\frac{k^2}{(\frac{4\pi}{3}C)^2(1-\cos(h^{-1}t))^2}+k\frac{k^2}{(\frac{4\pi}{3}C)^2(1-\cos(h^{-1}t))^2}\\
&= \frac{2k^3}{(\frac{4\pi}{3}C)^2(1-\cos(h^{-1}t))^3} = \frac{8\pi}{3}\frac{C}{a^3} = \frac{8\pi}{3}\rho.
\end{aligned}$$

Also, (16.10) is satisfied: Using the above expression for $\dfrac{\dot{a}^2}{a^2}+\dfrac{k}{a^2}$, we have

$$\begin{aligned}
&2\frac{\ddot{a}}{a}+\frac{\dot{a}^2}{a^2}+\frac{k}{a^2}\\
&=2\frac{(-k^2)}{\frac{4\pi}{3}C(1-\cos(h^{-1}t))^2}\cdot\frac{k}{\frac{4\pi}{3}C(1-\cos(h^{-1}t))}+\frac{2k^3}{(\frac{4\pi}{3}C)^2(1-\cos(h^{-1}t))^3}\\
&=0=-8\pi\mathrm{p}.
\end{aligned}$$

The graph of $t\mapsto a(t)$ is a 'scaled version' of a cycloid. To see this, we note that if a circle of radius R rolls on the x-axis, then the coordinates of the point P on the circle, starting from the origin, are given by

$$(R\tau - R\sin\tau,\ R - R\cos\tau),$$

where τ is the angle that OP makes with the vertical, as shown in the following picture.

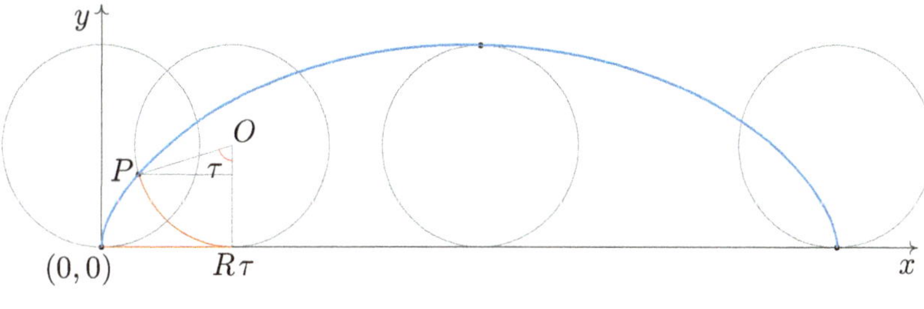

In our case, define $R=\frac{4\pi}{3}\frac{C}{k}$. Also, for $t\in(0,t_{\max})$, if $\tau\in(0,2\pi)$ is such that $h(\tau)=t$, then

$$(t,a(t))=(h(\tau),a(h(\tau)))=\Big(\frac{1}{\sqrt{k}}(R\tau-R\sin\tau),\ R-R\cos\tau\Big),$$

which, after a dilation of the x-axis, is a point on the cycloid. As $t\to 0$, we have $\tau:=h^{-1}t\to 0$, and $a(t)\to 0$. Also, we compute

$$\mathbf{S}=6\,\frac{a\,\ddot{a}+\dot{a}^2+k}{a^2}=\frac{6k^3}{(\frac{4\pi}{3}C)^2}\frac{1}{(1-\cos(h^{-1}t))^3}\overset{t\to 0}{\longrightarrow}\infty.$$

Thus again we have a singularity at $t=0$.

The case $k<0$.

This is similar to what we did above, except that we use hyperbolic functions instead of sin / cos. Define $h:(0,\infty)\to(0,\infty)=:I$ by

$$h(\tau)=\frac{4\pi}{3}\frac{C}{(-k)\sqrt{-k}}((\sinh\tau)-\tau),\quad \tau>0.$$

So $h'(\tau)=\frac{4\pi}{3}\frac{C}{(-k)\sqrt{-k}}((\cosh\tau)-1)>0$, and h is a diffeomorphism. Define

$$a(t)=\frac{4\pi}{3}\frac{C}{(-k)}(\cosh(h^{-1}t)-1),\quad t\in I.$$

Then it follows that

$$\dot{a}(t)=\sqrt{-k}\,\frac{\sinh(h^{-1}t)}{\cosh(h^{-1}t)-1}\quad\text{and}\quad \ddot{a}(t)=-\frac{k^2}{\frac{4\pi}{3}C(\cosh(h^{-1}t)-1)^2}.$$

Using these expressions, one verifies in the same manner as in the $k>0$ case, that (16.9) holds after substituting $\rho=\frac{C}{a^3}$:

$$\frac{\dot{a}^2}{a^2}+\frac{k}{a^2}=\frac{2(-k)^3}{(\frac{4\pi}{3}C)^2(\cosh(h^{-1}t)-1)^3}=\frac{8\pi}{3}\frac{C}{a^3}=\frac{8\pi}{3}\rho.$$

Also, (16.10) is satisfied: Using the above expression for $\frac{\dot{a}^2}{a^2}+\frac{k}{a^2}$, we have

$$2\frac{\ddot{a}}{a}+\frac{\dot{a}^2}{a^2}+\frac{k}{a^2}=0=-8\pi\mathrm{p}.$$

We have a singularity at $t=0$, since

$$\mathbf{S}=6\,\frac{a\,\ddot{a}+\dot{a}^2+k}{a^2}=\frac{6(-k)^3}{(\frac{4\pi}{3}C)^2}\frac{1}{(\cosh(h^{-1}t)-1)^3}\overset{t\to 0}{\longrightarrow}\infty.$$

In the following picture, qualitative plots of a versus t are shown in each of the three cases ($k<0$, $k=0$, $k>0$).

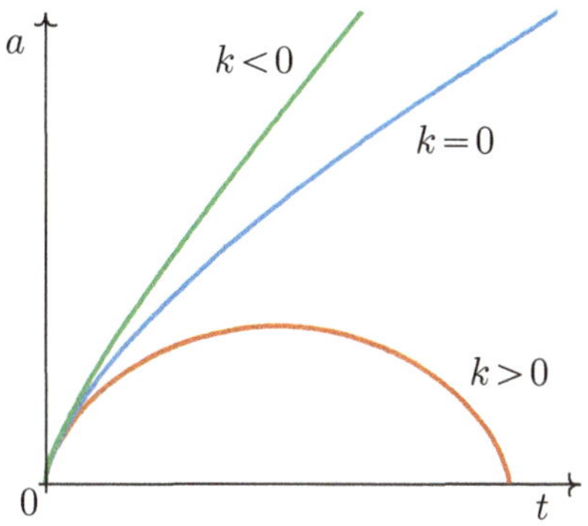

Exercise 16.7. (ΛCDM model.)
One of currently adopted models of the universe is the FLRW spacetime model with $k=0$ and $\Lambda>0$. We assume the presence of dust[17] (perfect fluid with $\mathrm{p}=0$). Recall the Friedmann equations from Example 13.1 (p.302):

$$\dot{a}^2=\Big(\frac{8\pi}{3}\rho+\frac{\Lambda}{3}\Big)a^2, \quad \text{and} \quad \ddot{a}=\Big(\frac{\Lambda}{3}-\frac{4\pi}{3}(\rho+3\mathrm{p})\Big)a.$$

We assume that ρ is a function of t alone. Also, we suppose that $\rho, a, \dot{a}$ have limits as $t\to 0$, and consider their continuous extensions defined on $[0,\infty)$. From Exercise 13.6 (p.302), it follows that ρa^3 is a constant (since $\mathrm{p}=0$). Then

$$M:=\frac{8\pi}{3}\rho a^3$$

is a constant, assumed positive. Set $H=\frac{\dot{a}}{a}$, $A:=a(0)$, and $L=\sqrt{\frac{3}{\Lambda}}$. Show that

$$H^2=\frac{M}{a^3}+\frac{1}{L^2}, \quad \text{and} \quad \dot{H}=\frac{3}{2}\Big(\frac{1}{L^2}-H^2\Big).$$

Since we have[18] $H(0)>\frac{1}{L}$, we can locally define the function $q(t)=\dfrac{1}{\frac{1}{L}-H}$.

Use the differential equation for H to show that q satisfies the differential equation

$$\dot{q}=\frac{3}{L}q-\frac{3}{2}.$$

Solving for q, show that $H(t)=\dfrac{1}{L}\dfrac{(\frac{1}{L}+H(0))e^{\frac{3}{L}t}-(\frac{1}{L}-H(0))}{(\frac{1}{L}+H(0))e^{\frac{3}{L}t}+(\frac{1}{L}-H(0))}$.

Prove that $a(t)=\dfrac{A}{2^{2/3}}\Big(\big(1+\sqrt{1+\frac{ML^2}{A^3}}\big)e^{\frac{3}{2L}t}+\big(1-\sqrt{1+\frac{ML^2}{A^3}}\big)e^{-\frac{3}{2L}t}\Big)^{\frac{2}{3}}$.

[17]The 'CDM' part of the abbreviation, standing for 'cold dark matter'.
[18]Indeed, $M>0$ gives $a(0)>0$ too, which yields $H(0)^2=\frac{M}{A^3}+\frac{1}{L^2}>\frac{1}{L^2}$.

Exercise 16.8. (Particle horizon.)
Consider the FLRW spacetime with $k=0$. Suppose that we are on the integral curve of ∂_t given by $\gamma_0(t)=(t,\mathbf{0})$, $t \in I$. Let $\theta_0 \in (0,\pi)$ and $\phi_0 \in (0,2\pi)$ be fixed. Let $\lambda : [\alpha,\beta] \to M$ be a radial null geodesic, given by $\lambda(s)=(t(s),\mathbf{p}(s))$, $s \in [\alpha,\beta]$, where $\mathbf{p}(s) \in \mathbb{R}^3$ has spherical coordinates $(r(s),\theta_0,\phi_0)$. Moreover, suppose that pointwise $t'>0$ and $r'<0$, and $0<t(\alpha)=t_0<T=t(\beta)$, $r(\alpha)=R$, $r(\beta)=0$ (so that the light emitted at t_0 and from a point with the radial coordinate R, reaches us at time T). Prove that

$$R = \int_{t_0}^{T} \frac{1}{a(t)}\,dt < \int_{0}^{T} \frac{1}{a(t)}\,dt =: R_{\text{H}}(T).$$

$R_{\text{H}}(T)$ is called the *particle horizon at time* T. The above inequality means that, for us to receive a radially emitted light signal from the past, the radial coordinate of the emitter at the moment of emission can be at most $R_{\text{H}}(T)$. In other words, we can receive no signals at time T from those galaxies whose radial coordinate in the past was always bigger than $R_{\text{H}}(T)$.

Exercise 16.9. Consider FLRW spacetime $(M,\mathbf{g})$. Let $J \subset \mathbb{R}$ be an open interval, and let $\gamma : J \to M$ be a lightlike signal. Let $\theta_0 \in (0,\pi)$ and $\phi_0 \in (0,2\pi)$ be fixed, and $\gamma(s)=\big(t(s),\mathbf{x}^{-1}(r(s),\theta_0,\phi_0)\big)$ for all $s \in J$, where $(U,\mathbf{x})$ is the spherical coordinate chart from Definition 16.2 (p.362).

(1) Show that $\dfrac{t'(s)}{r'(s)} = \dfrac{a(t(s))}{\sqrt{1-k(r(s))^2}}$ assuming that $t'(s), r'(s) > 0$ for all $s \in J$.

(2) In §9.2 (p.192), we had seen that if γ emanates at event $A=(t(s_A),\mathbf{0})=\gamma(s_A)$ from a distant galaxy, and reaches us at the event $B=(t(s_B),\mathbf{p})=\gamma(s_B)$ belonging to our worldline, then its red-shift z is given by

$$\mathrm{z}+1 = \frac{a(t(s_B))}{a(t(s_A))} = \frac{E_A}{E_B},$$

where E_A (respectively, E_B) is the energy/frequency of the light signal perceived by the distant galaxy at A (respectively by us at B). Define $R = a(t(s_B))r(s_B)$. Express R in terms of $(t(s_B)$ and) z for the following FLRW models:

(a) $k=-\beta^2$ and $a(t)=\beta t$, for $\beta>0$.
(b) $k=0$ and $a(t)=e^{Ht}$, $H>0$.
(c) $k=0$ and $a(t)=\beta t^p$, $\beta>0$, $p \neq 1$.

Data points $(\mathrm{z}_i, R(\mathrm{z}_i))$, $i=1,\cdots,n$, are obtained from the observations of brightness of galaxies of known luminosity, and the appropriate a in the FLRW model is chosen as the one for which the resulting R best fits the observational data.

Solutions

Notation and Terminology

Solution to Exercise 0.1

(1) $5\frac{\text{km}}{\text{hr}} = 5\frac{10^3}{60\times 60}\frac{\text{m}}{\text{s}}$. Thus in units where $c = 3\times 10^8\frac{\text{m}}{\text{s}}$ is 1, the speed $5\frac{\text{km}}{\text{hr}}$ corresponds to $5\frac{10^3}{60\times 60}\frac{1}{3\times 10^8} = 4.6\times 10^{-9} = 4.6\times 10^{-9}\text{ m}^0$.

(2) $24\,\text{hr} = 3600\times 24\,\text{s} \equiv 3600\times 24\times 3\times 10^8\,\text{m} = 2.6\times 10^{13}\text{m}$.

(3) In units where $G_{\text{N}} = 1$, we have $6.67\times 10^{-11}\frac{\text{Nm}^2}{\text{kg}^2} = 1$, where N stands for the unit of force, Newton, and $1\,\text{N} = 1\,\text{kg}\frac{\text{m}}{\text{s}^2}$. Thus $1\,\text{kg} \equiv 6.67\times 10^{-11}\frac{\text{m}^3}{\text{s}^2}$. Hence, $M_\oplus = 6\times 10^{24}\text{kg} = 6\times 10^{24}\times \frac{6.67\times 10^{-11}}{(3\times 10^8)^2}\frac{\text{m}^3}{\text{m}^2} = 4.44\times 10^{-3}\text{m}$.

(4) Recall that $1\,\text{Joule} = 1\,\text{Nm}$, and so $h = 6.63\times 10^{-34}\,\text{kg}\frac{\text{m}^2}{\text{s}}$ corresponds to $6.63\times 10^{-34}\times 6.67\times 10^{-11}\times \frac{1}{(3\times 10^8)^3}\text{m}^2 = 1.64\times 10^{-69}\,\text{m}^2$.

(5) We have $10^3\frac{\text{kg}}{\text{m}^3} = \frac{10^3\times 6.67\times 10^{-11}}{(3\times 10^8)^2}\text{m}^{-2} = 7.4\times 10^{-25}\text{m}^{-2}$.

(6) We have $9.8\frac{\text{m}}{\text{s}^2}$ corresponds to $\frac{9.8}{(3\times 10^8)^2}\text{m}^{-1} = 1.1\times 10^{-16}\text{m}^{-1}$.

Solution to Exercise 0.2

(1) We have 3.7×10^{-5} in units where $c = 1$ corresponds to the speed of $3.7\times 10^{-5}\times 3\times 10^8\frac{\text{m}}{\text{s}} = 11200\frac{\text{m}}{\text{s}} = 11.2\frac{\text{km}}{\text{s}}$.

(2) We have that $5.4\times 10^{24}\text{m}$ corresponds to $\frac{5.4\times 10^{24}}{3\times 10^8}\text{s}$.
In billions of years, this is $\frac{5.4\times 10^{24}}{3\times 10^8\times 3600\times 24\times 365\times 10^9} = 13.7$.

(3) We have $1\,\text{kg} \equiv \frac{6.67\times 10^{-11}}{(3\times 10^8)^2}\text{m}$.
So the mass of the Sun in kg is $1500\times \frac{(3\times 10^8)^2}{6.67\times 10^{-11}} = 2\times 10^{30}\text{kg}$.

(4) $1\text{W} = 1\frac{\text{J}}{\text{s}} = 1\frac{\text{Nm}}{\text{s}} = 1\frac{\text{kgm}^2}{\text{s}^3}$, which corresponds to $\frac{6.67\times 10^{-11}}{(3\times 10^8)^5}$.
Thus the power of the LED bulb is $\frac{2.74\times 10^{-52}\times (3\times 10^8)^5}{6.67\times 10^{-11}}\text{W} = 10\,\text{W}$.

(5) $1\text{Nm}^{-2} = 1\frac{\text{kg}}{\text{ms}^2}$ corresponds to $\frac{6.67\times 10^{-11}}{(3\times 10^8)^2}\text{m}^{-1}$.
So the atmospheric pressure is $8.34\times 10^{-40}\times \frac{(3\times 10^8)^4}{6.67\times 10^{-11}}\frac{\text{N}}{\text{m}^2} = 1.01\times 10^5\frac{\text{N}}{\text{m}^2}$.

(6) The acceleration $3\times 10^{-20}\text{m}^{-1}$ in units where $c = 1$, corresponds to $3\times 10^{-20}\times (3\times 10^8)^2\frac{\text{m}}{\text{s}^2} = 0.0027\frac{\text{m}}{\text{s}^2}$.

Solution to Exercise 0.3

The magnitude of the electrostatic force between a proton and an electron at a distance d metres apart is given by

$$F_e = \frac{1}{4\pi\epsilon_0}\frac{e^2}{d^2} = 9 \times 10^9 \times \frac{(1.6 \times 10^{19})^2}{d^2}\mathrm{N}.$$

The magnitude of the gravitational force between a proton and an electron at a distance d metres apart is given by

$$F_g = G_{\mathrm{N}}\frac{m_p m_e}{d^2} = 6.67 \times 10^{-11} \times \frac{1.6 \times 10^{-27} \times 9.1 \times 10^{-31}}{d^2}\mathrm{N}.$$

So their ratio is

$$\frac{F_e}{F_g} = \frac{9 \times 10^9 \times (1.6 \times 10^{19})^2}{6.67 \times 10^{-11} \times 1.6 \times 10^{-27} \times 9.1 \times 10^{-31}} = 2.4 \times 10^{39}.$$

The order of the dimensionless number $\dfrac{F_e}{F_g}$ is 10^{40}.

If T is the age of the universe, c is the speed of light, and r_e is the classical electron radius, then the dimensionless number $\dfrac{cT}{r_e}$ has the value

$$\frac{3 \times 10^8 \times 13.7 \times 10^9 \times 365 \times 24 \times 3600}{3.7 \times 10^{-16}} = 3.5 \times 10^{41},$$

which is of order 10^{42}.

Chapter 1

Solution to Exercise 1.1

By the definition of $\mathbf{v}_{pq}$ and $\mathbf{v}_{qr}$, we have $q = p + \mathbf{v}_{pq}$, $r = q + \mathbf{v}_{qr}$. Thus $r = q + \mathbf{v}_{qr} = (p + \mathbf{v}_{pq}) + \mathbf{v}_{qr} = p + (\mathbf{v}_{pq} + \mathbf{v}_{qr})$, where the last equality follows from (A1). By (A3), $\mathbf{v}_{pq} + \mathbf{v}_{qr} = \mathbf{v}_{pr}$.

Solution to Exercise 1.2

Let $(u, v) \in \mathbb{R}^2$. We wish to find the point $p = (x, y, z) \in S^2 \setminus \{\mathbf{n}\}$ such that $\varphi_{\mathbf{n}}((x, y, z)) = (u, v)$, i.e., $\left(\frac{x}{1-z}, \frac{y}{1-z}\right) = (u, v)$. For such a triple (x, y, z),

$$u^2 + v^2 = \frac{x^2}{(1-z)^2} + \frac{y^2}{(1-z)^2} = \frac{x^2+y^2}{(1-z)^2} = \frac{1-z^2}{(1-z)^2} = \frac{1+z}{1-z},$$

and so $z = \frac{u^2+v^2-1}{u^2+v^2+1}$. Also,

$$x = u(1-z) = u\left(1 - \frac{u^2+v^2-1}{u^2+v^2+1}\right) = \frac{2u}{u^2+v^2+1}$$

and $y = v(1-z) = \frac{2v}{u^2+v^2+1}$. So we define

$$p = (x, y, z) = \left(\frac{2u}{u^2+v^2+1}, \frac{2v}{u^2+v^2+1}, \frac{u^2+v^2-1}{u^2+v^2+1}\right).$$

Then $x^2+y^2+z^2 = \frac{4u^2+4v^2+u^4+v^4+1+2u^2v^2-2u^2-2v^2}{(u^2+v^2+1)^2} = \frac{(u^2+v^2+1)^2}{(u^2+v^2+1)^2} = 1.$

So $p \in S^2$. Also $p \neq \mathbf{n}$, since otherwise

$$\left(\frac{2u}{u^2+v^2+1}, \frac{2v}{u^2+v^2+1}, \frac{u^2+v^2-1}{u^2+v^2+1}\right) = (0, 0, 1),$$

and comparing the first two entries, we get $u = 0$ and $v = 0$, implying $\frac{u^2+v^2-1}{u^2+v^2+1} = -1 \neq 1$, a contradiction. Hence $p \in S^2 \setminus \{\mathbf{n}\}$. Finally,

$$\varphi_{\mathbf{n}}(p) = \frac{1}{1 - \frac{u^2+v^2-1}{u^2+v^2+1}}\left(\frac{2u}{u^2+v^2+1}, \frac{2v}{u^2+v^2+1}\right) = (u, v).$$

Hence $(\varphi_{\mathbf{n}})^{-1}(u, v) = \left(\frac{2u}{u^2+v^2+1}, \frac{2v}{u^2+v^2+1}, \frac{u^2+v^2-1}{u^2+v^2+1}\right)$.

Solution to Exercise 1.3

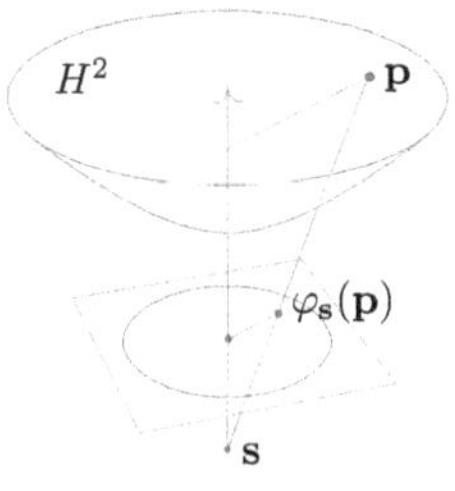

If $\mathbf{p} \neq (0, 0, 1)$ and $(u, v) := \varphi_{\mathbf{s}}(\mathbf{p})$, then using the similarity of the two triangles shown in the picture above, we get $\frac{1+t}{1} = \frac{x}{u} = \frac{y}{v}$. If $\mathbf{p} = (0, 0, 1)$, then it is clear that $\varphi_{\mathbf{s}}(\mathbf{p}) = (0,0) = \frac{1}{1+0}(0,0)$. Consider the open unit disc in $\mathbb{R}^2$, $\mathbb{D} := \{(u,v) \in \mathbb{R}^2 : u^2+v^2 < 1\}$. We will show that $\varphi_{\mathbf{s}}(H^2) = \mathbb{D}$. For $(u,v) \in \mathbb{D}$, let $\mathbf{p} := (x,y,t) := (\frac{2u}{1-u^2-v^2}, \frac{2v}{1-u^2-v^2}, \frac{1+u^2+v^2}{1-u^2-v^2})$. Then $t > 0$, and it is easy to check that $x^2+y^2-t^2 = -1$ (so that $\mathbf{p} \in H^2$) and $\varphi_{\mathbf{s}}(\mathbf{p}) = (u,v)$ (so that $\varphi_{\mathbf{s}}$ is onto). Next we show the injectivity of $\varphi_{\mathbf{s}}$. Let $\varphi_{\mathbf{s}}(x,y,t) = \varphi_{\mathbf{s}}(x',y',t')$. Then $\frac{x}{1+t} = \frac{x'}{1+t'}$ and $\frac{y}{1+t} = \frac{y'}{1+t'}$, giving $-1+t^2 = x^2+y^2 = (\frac{1+t}{1+t'})^2(x'^2+y'^2) = (\frac{1+t}{1+t'})^2(-1+t'^2)$. Simplifying, we get $t = t'$. Thus $x = \frac{1+t}{1+t'}x' = x'$, and similarly, $y = y'$.

Solution to Exercise 1.4

Let $\varphi(x) = x^3$, $x \in \mathbb{R}$. As $\varphi'(x) = 3x^2 > 0$ for $x \neq 0$, φ is strictly increasing on $(0,\infty)$ and on $(-\infty,0)$. It follows that φ is strictly increasing on $\mathbb{R}$, and hence is injective. We have $\varphi((0,\infty)) = (0,\infty)$, $\varphi((-\infty,0)) = (-\infty,0)$, and $\varphi(0) = 0$, so that $\varphi(\mathbb{R}) = \mathbb{R}$. Also, $\varphi(\mathbb{R}) = \mathbb{R}$ is open. Thus $(\mathbb{R},\varphi)$ is a chart on $\mathbb{R}$.

Solution to Exercise 1.5

Let $(x,y,z) \in U_{x+}$. Then $x > 0$, and as $x^2+y^2 = 1$, we get $|y| < 1$. So we have $\varphi_{x+}(U_{x+}) \subset (-1,1)\times\mathbb{R}$. We will show $\varphi_{x+} : U_{x+} \to (-1,1)\times\mathbb{R}$ is a bijection.

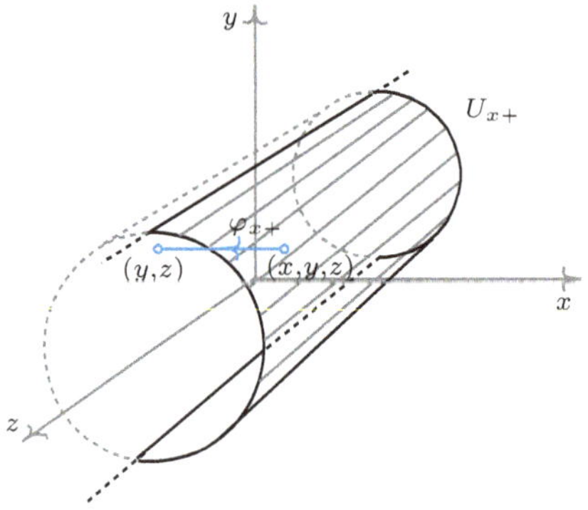

If $\varphi_{x+}(x,y,z) = \varphi_{x+}(\widetilde{x},\widetilde{y},\widetilde{z})$, then $(y,z) = (\widetilde{y},\widetilde{z})$, and so $y = \widetilde{y}$, $z = \widetilde{z}$. As $(x,y,z), (\widetilde{x},\widetilde{y},\widetilde{z}) \in C$, $x^2+y^2 = 1$ and $\widetilde{x}^2+\widetilde{y}^2 = 1$. But since $x, \widetilde{x} > 0$, we obtain $\widetilde{x} = \sqrt{1-\widetilde{y}^2} = \sqrt{1-y^2} = x$ as well. So φ_{x+} is injective.

Now let $(y,z) \in (-1,1)\times\mathbb{R}$. Define $x := \sqrt{1-y^2} > 0$. Then $x^2+y^2 = 1$ and $x > 0$, so that $(x,y,z) \in U_{x+}$. Also $\varphi_{x+}(x,y,z) = (y,z)$. Thus φ_{x+} is also surjective onto $(-1,1)\times\mathbb{R}$. Finally, $\varphi_{x+}(U_{x+}) = (-1,1)\times\mathbb{R}$, which is open in $\mathbb{R}^2$.

Solution to Exercise 1.6

(A1) is trivially satisfied. As $\varphi_{\mathbf{s}}(H^2) = \mathbb{D}$ and $p(H^2) = \mathbb{R}^2$ are open, (A2) holds. For $(u, v) \in \mathbb{D}$, we have $(p \circ \varphi_{\mathbf{s}}^{-1})(u, v) = (\frac{2u}{1-u^2-v^2}, \frac{2v}{1-u^2-v^2})$, and so $p \circ \varphi_{\mathbf{s}}^{-1} \in C^\infty(\mathbb{D})$. For $(x, y) \in \mathbb{R}^2$, we have

$$(\varphi_{\mathbf{s}} \circ p^{-1})(x, y) = \varphi_{\mathbf{s}}(x, y, \sqrt{1+x^2+y^2}) = \frac{1}{1+\sqrt{1+x^2+y^2}}(x, y).$$

It can be argued that $\varphi_{\mathbf{s}} \circ p^{-1} \in C^\infty(\mathbb{R}^2)$ as follows. The polynomial map f_1, $\mathbb{R}^2 \ni (x, y) \mapsto 1 + x^2 + y^2$ is C^∞ on $\mathbb{R}^2$, and its range is contained in $[1, \infty) \subset (-1, \infty)$. The function f_2, $(-1, \infty) \ni t \mapsto \sqrt{1+t}$ is C^∞ on $(-1, \infty)$ and its range is contained in $(0, \infty)$. Finally, the map f_3 given by $(0, \infty) \ni s \mapsto (1 + s)^{-1}$ is C^∞ on $(0, \infty)$. Thus the composition of these three maps, $f := f_3 \circ f_2 \circ f_1 : \mathbb{R}^2 \to \mathbb{R}$, is well-defined and C^∞ on $\mathbb{R}^2$. Then the pointwise product of this f with x or with y is C^∞ as well. It follows that $\varphi_{\mathbf{s}} \circ p^{-1} \in C^\infty(\mathbb{R}^2)$ because $(\varphi_{\mathbf{s}} \circ p^{-1})(x, y) = (xf(x, y), yf(x, y))$ for all $(x, y) \in \mathbb{R}^2$. So (A3) holds.

Solution to Exercise 1.7

$$\begin{aligned}
\text{Let } U_{x+} &= \{(x,y,z) \in C : x > 0\}, \ \varphi_{x+} : U_{x+} \to \mathbb{R}^2, \ \varphi_{x+}(x,y,z) = (y,z),\\
U_{x-} &= \{(x,y,z) \in C : x < 0\}, \ \varphi_{x-} : U_{x-} \to \mathbb{R}^2, \ \varphi_{x-}(x,y,z) = (y,z),\\
U_{y+} &= \{(x,y,z) \in C : y > 0\}, \ \varphi_{y+} : U_{y+} \to \mathbb{R}^2, \ \varphi_{y+}(x,y,z) = (x,z),\\
U_{y-} &= \{(x,y,z) \in C : y < 0\}, \ \varphi_{y-} : U_{y-} \to \mathbb{R}^2, \ \varphi_{y-}(x,y,z) = (x,z).
\end{aligned}$$

Then $U_{x+} \cup U_{x-} \cup U_{y+} \cup U_{y-} = C$: For all $(x, y, z) \in C$, $x^2 + y^2 = 1$, and so that it cannot be that both x and y are 0, i.e., either $x \neq 0$ (in which case either $x > 0$ or $x < 0$) or $y \neq 0$ (and then either $y > 0$ or $y < 0$). We have $U_{x+} \cap U_{x-} = \varnothing$ and $U_{y+} \cap U_{y-} = \varnothing$. So we have the eight transition maps

$$\begin{aligned}
&\varphi_{y+} \circ \varphi_{x+}^{-1} : \varphi_{x+}(U_{x+} \cap U_{y+}) \to \varphi_{y+}(U_{x+} \cap U_{y+}), \ (y, z) \mapsto (\sqrt{1-y^2}, z),\\
&\varphi_{y-} \circ \varphi_{x+}^{-1} : \varphi_{x+}(U_{x+} \cap U_{y-}) \to \varphi_{y-}(U_{x+} \cap U_{y-}), \ (y, z) \mapsto (\sqrt{1-y^2}, z),\\
&\varphi_{y+} \circ \varphi_{x-}^{-1} : \varphi_{x-}(U_{x-} \cap U_{y+}) \to \varphi_{y+}(U_{x-} \cap U_{y+}), \ (y, z) \mapsto (-\sqrt{1-y^2}, z),\\
&\varphi_{y-} \circ \varphi_{x-}^{-1} : \varphi_{x-}(U_{x-} \cap U_{y-}) \to \varphi_{y-}(U_{x-} \cap U_{y-}), \ (y, z) \mapsto (-\sqrt{1-y^2}, z),\\
&\varphi_{x+} \circ \varphi_{y+}^{-1} : \varphi_{y+}(U_{y+} \cap U_{x+}) \to \varphi_{x+}(U_{y+} \cap U_{x+}), \ (x, z) \mapsto (\sqrt{1-x^2}, z),\\
&\varphi_{x-} \circ \varphi_{y+}^{-1} : \varphi_{y+}(U_{y+} \cap U_{x-}) \to \varphi_{x-}(U_{y+} \cap U_{x-}), \ (x, z) \mapsto (\sqrt{1-x^2}, z),\\
&\varphi_{x+} \circ \varphi_{y-}^{-1} : \varphi_{y-}(U_{y-} \cap U_{x+}) \to \varphi_{x+}(U_{y-} \cap U_{x+}), \ (x, z) \mapsto (-\sqrt{1-x^2}, z),\\
&\varphi_{x-} \circ \varphi_{y-}^{-1} : \varphi_{y-}(U_{y-} \cap U_{x-}) \to \varphi_{x-}(U_{y-} \cap U_{x-}), \ (x, z) \mapsto (-\sqrt{1-x^2}, z),
\end{aligned}$$

and each of these is a C^∞ map. For example, we give the details below for the map $\varphi_{y+} \circ \varphi_{x-}^{-1}$. Firstly, $U_{x-} \cap U_{y+} = \{(x, y, z) \in C : x < 0, \ y > 0\}$. Thus $\varphi_{x-}(U_{x-} \cap U_{y+}) = \{(y, z) \in \mathbb{R}^2 : 1 > y > 0\}$. So for $(y, z) \in \varphi_{x-}(U_{x-} \cap U_{y+})$, $\varphi_{x-}^{-1}(y, z) = (-\sqrt{1-y^2}, y, z)$, and $(\varphi_{y+} \circ \varphi_{x-}^{-1})(y, z) = (-\sqrt{1-y^2}, z)$. Hence $\varphi_{y+} \circ \varphi_{x-}^{-1}$ is C^∞ on $\varphi_{x-}(U_{x-} \cap U_{y+}) = \{(y, z) \in \mathbb{R}^2 : 1 > y > 0\}$.

Solution to Exercise 1.8

$\cdot$ is associative: For $(\mathbf{b}_i, A_i) \in \mathbb{R}^m \times \mathrm{GL}_m(\mathbb{R})$, $i = 1, 2, 3$, we have

$$\begin{aligned}(\mathbf{b}_3, A_3)\cdot((\mathbf{b}_2, A_2)\cdot(\mathbf{b}_1, A_1)) &= (\mathbf{b}_3, A_3)\cdot(\mathbf{b}_2 + A_2\mathbf{b}_1, A_2A_1)\\ &= (\mathbf{b}_3 + A_3(\mathbf{b}_2 + A_2\mathbf{b}_1), A_3(A_2A_1))\\ &= ((\mathbf{b}_3 + A_3\mathbf{b}_2) + (A_3A_2)\mathbf{b}_1, (A_3A_2)A_1)\\ &= (\mathbf{b}_3 + A_3\mathbf{b}_2, A_3A_2)\cdot(\mathbf{b}_1, A_1)\\ &= ((\mathbf{b}_3, A_3)\cdot(\mathbf{b}_2, A_2))\cdot(\mathbf{b}_1, A_1).\end{aligned}$$

Let $(\mathbf{b}, A) \in \mathbb{R}^m \times \mathrm{GL}_m(\mathbb{R})$. With the zero vector $\mathbf{0} \in \mathbb{R}^m$ and the identity matrix $I \in \mathrm{GL}_m(\mathbb{R})$, we have $(\mathbf{0}, I)\cdot(\mathbf{b}, A) = (\mathbf{0} + I\,\mathbf{b}, IA) = (\mathbf{b}, A)$, and also $(\mathbf{b}, A)\cdot(\mathbf{0}, I) = (\mathbf{b} + A\,\mathbf{0}, AI) = (\mathbf{b}, A)$. So $(\mathbf{0}, I)$ serves as the identity.

Finally, for $(\mathbf{b}, A) \in \mathbb{R}^m \times \mathrm{GL}_m(\mathbb{R})$, $(-A^{-1}\mathbf{b}, A^{-1}) \in \mathbb{R}^m \times \mathrm{GL}_m(\mathbb{R})$. Moreover, $(\mathbf{b}, A)\cdot(-A^{-1}\mathbf{b}, A^{-1}) = (\mathbf{b} + A(-A^{-1}\mathbf{b}), AA^{-1}) = (\mathbf{0}, I)$, and $(-A^{-1}\mathbf{b}, A^{-1})\cdot(\mathbf{b}, A) = (-A^{-1}\mathbf{b} + A^{-1}\mathbf{b}, A^{-1}A) = (\mathbf{0}, I)$. So every element in $\mathbb{R}^m \times \mathrm{GL}_m(\mathbb{R})$ has an inverse with respect to $\cdot$.

Solution to Exercise 1.9

Denote the compatibility relation by $\sim$. Reflexivity ($\mathcal{A} \cup \mathcal{A} = \mathcal{A}$ is an atlas on M for every atlas $\mathcal{A}$ on M) and symmetry ($\mathcal{A}_1 \cup \mathcal{A}_2 = \mathcal{A}_2 \cup \mathcal{A}_1$ is an atlas on M for atlases $\mathcal{A}_1, \mathcal{A}_2$ on M such that $\mathcal{A}_1 \sim \mathcal{A}_2$) are clear. We now show transitivity. Let $\mathcal{A}_1, \mathcal{A}_2, \mathcal{A}_3$ be atlases on M such that $\mathcal{A}_1 \sim \mathcal{A}_2$ and $\mathcal{A}_2 \sim \mathcal{A}_3$. Thus $\mathcal{A}_1 \cup \mathcal{A}_2$ and $\mathcal{A}_2 \cup \mathcal{A}_3$ are atlases on M. As the charts in $\mathcal{A}_1$ alone cover M, certainly the charts in $\mathcal{A}_1 \cup \mathcal{A}_3$ cover M, and so (A1) is true. Also, clearly (A2) holds. So to show $\mathcal{A}_1 \cup \mathcal{A}_3$ is an atlas, we only need to show (A3), i.e., the transition maps are C^∞. Clearly, if both charts are from $\mathcal{A}_1$, or if both are from $\mathcal{A}_3$, then there is nothing to prove. So let $(U, \varphi) \in \mathcal{A}_1$, $(V, \psi) \in \mathcal{A}_3$, and $U \cap V \neq \varnothing$. Let $p \in U \cap V$. Then there exists a chart $(W, \sigma) \in \mathcal{A}_2$ such that $p \in W$. As $p \in U \cap W$, and since $\mathcal{A}_1 \cup \mathcal{A}_2$ is an atlas, the map $\sigma \circ \varphi^{-1} : \varphi(U \cap W) \to \sigma(U \cap W)$ is C^∞. Similarly, as $p \in W \cap V$, and as $\mathcal{A}_2 \cup \mathcal{A}_3$ is an atlas, the map $\psi \circ \sigma^{-1} : \sigma(W \cap V) \to \psi(W \cap V)$ is C^∞. Now the map $\psi \circ \varphi^{-1} : \varphi(U \cap V \cap W) \to \psi(U \cap V \cap W)$ is C^∞ because $\psi \circ \varphi^{-1} = (\psi \circ \sigma^{-1}) \circ (\sigma \circ \varphi^{-1})$ is the composition of two C^∞ maps. Hence $\psi \circ \varphi^{-1}$ is C^∞ in a neighbourhood of $\varphi(p) \in \varphi(U \cap V)$. As $p \in U \cap V$ was arbitrary, $\psi \circ \varphi^{-1}$ is C^∞ on $\varphi(U \cap V)$. Similarly, $\varphi \circ \psi^{-1} : \psi(U \cap V) \to \varphi(U \cap V)$ is C^∞. Hence all the transition maps in $\mathcal{A}_1 \cup \mathcal{A}_3$ are smooth, and so $\mathcal{A}_1 \cup \mathcal{A}_3$ forms an atlas. Thus $\mathcal{A}_1 \sim \mathcal{A}_3$.

Solution to Exercise 1.10

The maps $\varphi_+, \varphi_-, \psi_+, \psi_-$ are clearly injective, and moreover we have $\varphi_+(U_+) = \varphi_-(U_-) = \psi_+(V_+) = \psi_-(V_-) = (-1, 1)$ is open in $\mathbb{R}$.

Also $U_+ \cup U_- \cup V_+ \cup V_- = S$, and

$$\begin{array}{lll} \varphi_+(U_+ \cap U_-) = \varnothing, & \varphi_-(U_+ \cap U_-) = \varnothing, & \varphi_+(U_+ \cap V_+) = (0,1), \\ \psi_+(U_+ \cap V_+) = (0,1), & \varphi_+(U_+ \cap V_-) = (-1,0), & \psi_-(U_+ \cap V_-) = (0,1), \\ \varphi_-(U_- \cap V_+) = (0,1), & \psi_+(U_- \cap V_+) = (-1,0), & \varphi_-(U_- \cap V_-) = (-1,0), \\ \psi_-(U_- \cap V_-) = (-1,0), & \psi_+(V_+ \cap V_-) = \varnothing, & \psi_-(V_+ \cap V_-) = \varnothing \end{array}$$

are all open. As the transition maps given below are all smooth, $(S, [\mathcal{A}])$ is a smooth manifold.

$\psi_+ \circ (\varphi_+)^{-1} : \varphi_+(U_+ \cap V_+) \to \psi_+(U_+ \cap V_+)$ is $(0,1) \ni \beta \mapsto 1 - \beta \in (0,1)$,
$\varphi_+ \circ (\psi_+)^{-1} : \psi_+(U_+ \cap V_+) \to \varphi_+(U_+ \cap V_+)$ is $(0,1) \ni \alpha \mapsto 1 - \alpha \in (0,1)$,
$\psi_- \circ (\varphi_+)^{-1} : \varphi_+(U_+ \cap V_-) \to \psi_-(U_+ \cap V_-)$ is $(-1,0) \ni \beta \mapsto 1 + \beta \in (0,1)$,
$\varphi_+ \circ (\psi_-)^{-1} : \psi_-(U_+ \cap V_-) \to \varphi_+(U_+ \cap V_-)$ is $(0,1) \ni \alpha \mapsto \alpha - 1 \in (-1,0)$,
$\psi_+ \circ (\varphi_-)^{-1} : \varphi_-(U_- \cap V_+) \to \psi_+(U_- \cap V_+)$ is $(0,1) \ni \beta \mapsto \beta - 1 \in (-1,0)$,
$\varphi_- \circ (\psi_+)^{-1} : \psi_+(U_- \cap V_+) \to \varphi_-(U_- \cap V_+)$ is $(-1,0) \ni \alpha \mapsto 1 + \alpha \in (0,1)$,
$\psi_- \circ (\varphi_-)^{-1} : \varphi_-(U_- \cap V_-) \to \psi_-(U_- \cap V_-)$ is $(-1,0) \ni \beta \mapsto -1 - \beta \in (-1,0)$,
$\varphi_- \circ (\psi_-)^{-1} : \psi_-(U_- \cap V_-) \to \varphi_-(U_- \cap V_-)$ is $(-1,0) \ni \alpha \mapsto -1 - \alpha \in (-1,0)$.

Solution to Exercise 1.11

Firstly:

- $U_i \times V_j \ni (u,v) \overset{\varphi_i \times \psi_j}{\mapsto} (\varphi_i u, \psi_j v) \in \mathbb{R}^{m+n}$ is injective.
  Since if $(\varphi_i u, \psi_j v) = (\varphi_i \widetilde{u}, \psi_j \widetilde{v})$, then we have $\varphi_i u = \varphi_i \widetilde{u}$, giving $u = \widetilde{u}$, and also $\psi_j v = \psi_j \widetilde{v}$, which implies $v = \widetilde{v}$.
- $(\varphi_i \times \psi_j)(U_i \times V_j)$ is open.
  $(\varphi_i \times \psi_j)(U_i \times V_j) = \{(\varphi_i u, \psi_j v),\ u \in U_i,\ v \in V_j\} = \varphi_i(U_i) \times \psi_j(V_j)$ is open in $\mathbb{R}^{m+n}$, since $\varphi_i(U_i) \subset \mathbb{R}^m$ and $\psi_j(V_j) \subset \mathbb{R}^n$ are open.

Thus $(U_i \times V_j, \varphi_i \times \psi_j)$, $i \in I$, $j \in J$, are charts for $M \times N$.

If $(p,q) \in M \times N$, then there exist a $U_i \in \mathcal{A}_M$ such that $p \subset M$ and there exists a $V_j \in \mathcal{A}_N$ such that $q \in V_j$. So we have $(p,q) \in U_i \times V_j$. Thus $\{U_i \times V_j,\ i \in I,\ j \in J\}$ covers $M \times N$.

If $i, i' \in I$, $j, j' \in J$, then

$$(\varphi_i \times \psi_j)((U_i \times V_j) \cap (U_{i'} \times V_{j'})) = (\varphi_i(U_i \cap U_{i'})) \times (\psi_j(V_j \times V_{j'}))$$

is open in $\mathbb{R}^{m+n}$, because $\varphi_i(U_i \cap U_{i'})$ is open in $\mathbb{R}^m$, and $\psi_j(V_j \times V_{j'})$ is open in $\mathbb{R}^n$.

Now suppose that $(U_{i_1} \times V_{j_i}) \cap (U_{i_2} \times V_{j_2}) \neq \varnothing$. Clearly we have that if $(p,q) \in (U_{i_1} \times V_{j_i}) \cap (U_{i_2} \times V_{j_2})$, then $p \in U_{i_1} \cap U_{i_2}$ and $q \in V_{j_i} \cap V_{j_2}$. The chart transition maps

$$\begin{array}{l} \varphi_{i_2} \circ (\varphi_{i_1})^{-1} : \varphi_{i_1}(U_{i_1} \cap U_{i_2}) \to \varphi_{i_2}(U_{i_1} \cap U_{i_2}), \\ \varphi_{i_1} \circ (\varphi_{i_2})^{-1} : \varphi_{i_2}(U_{i_1} \cap U_{i_2}) \to \varphi_{i_1}(U_{i_1} \cap U_{i_2}), \\ \psi_{j_2} \circ (\psi_{j_1})^{-1} : \psi_{j_1}(V_{j_1} \cap V_{j_2}) \to \psi_{j_2}(V_{j_1} \cap V_{j_2}), \\ \psi_{j_1} \circ (\psi_{j_2})^{-1} : \psi_{j_2}(V_{j_1} \cap V_{j_2}) \to \psi_{j_1}(V_{j_1} \cap V_{j_2}), \end{array}$$

are C^∞.

So it follows that the maps

$$\begin{array}{c}(\varphi_{i_1}\times\psi_{j_1})((U_{i_1}\times V_{j_1})\cap(U_{j_2}\times V_{j_2}))\ni(\alpha,\beta)\\ \Big\downarrow{\scriptstyle(\varphi_{i_2}\times\psi_{j_2})\circ(\varphi_{i_1}\times\psi_{j_1})^{-1}}\\ (\varphi_{i_2}\times\psi_{j_2})((U_{i_1}\times V_{j_1})\cap(U_{j_2}\times V_{j_2}))\ni((\varphi_{i_2}\circ\varphi_{i_1}^{-1})(\alpha),(\psi_{j_2}\circ\psi_{j_1}^{-1})(\beta))\end{array}$$

and

$$\begin{array}{c}(\varphi_{i_2}\times\psi_{j_2})((U_{i_1}\times V_{j_1})\cap(U_{j_2}\times V_{j_2}))\ni(\alpha,\beta)\\ \Big\downarrow{\scriptstyle(\varphi_{i_1}\times\psi_{j_1})\circ(\varphi_{i_2}\times\psi_{j_2})^{-1}}\\ (\varphi_{i_1}\times\psi_{j_1})((U_{i_1}\times V_{j_1})\cap(U_{j_2}\times V_{j_2}))\ni((\varphi_{i_1}\circ\varphi_{i_2}^{-1})(\alpha),(\psi_{j_1}\circ\psi_{j_2}^{-1})(\beta))\end{array}$$

are C^∞ maps.

Solution to Exercise 1.12

Let $\mathcal{O}$ denote the topology induced on $\mathbb{R}^m$ by $[\mathcal{A}]$, where $\mathcal{A}$ is the atlas $\mathcal{A} := \{(\mathbb{R}^m, \mathrm{id})\}$, and let $\mathcal{O}_{\mathrm{E}}$ denote the topology on $\mathbb{R}^m$ induced by the Euclidean metric.

Let $U \subset \mathbb{R}^m$ be such that $U \in \mathcal{O}$. Then $\mathrm{id}(U \cap \mathbb{R}^m) = U$ is open in $(\mathbb{R}^m, \mathcal{O}_{\mathrm{E}})$. So $\mathcal{O} \subset \mathcal{O}_{\mathrm{E}}$.

Let $U \in \mathcal{O}_{\mathrm{E}}$. Then U is open in $(\mathbb{R}^m, \mathcal{O}_{\mathrm{E}})$, and so $\mathrm{id}(U \cap \mathbb{R}^m) = U$ is open in $(\mathbb{R}^m, \mathcal{O}_{\mathrm{E}})$. Thus $U \in \mathcal{O}$. Hence $\mathcal{O}_{\mathrm{E}} \subset \mathcal{O}$.

Solution to Exercise 1.13

Suppose there exists an atlas $\mathcal{A}$ on C with the desired property. Then $\mathbf{0} := (0,0,0) \in U$ for some chart $(U, \varphi) \in \mathcal{A}$. Let $D \subset \varphi(U) \subset \mathbb{R}^2$ be an open disc of radius $\epsilon > 0$ with center $\varphi(\mathbf{0})$.

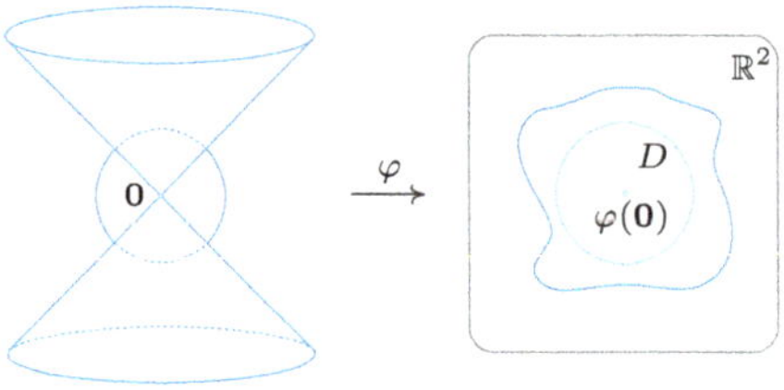

Then $\varphi^{-1}D$ is open in C in the subspace topology. So $\varphi^{-1}D = C \cap V$, where V is an open subset of $\mathbb{R}^3$. But as $\mathbf{0} \in \varphi^{-1}D = C \cap V \subset V$, there exists a ball B in $\mathbb{R}^3$ with center $\mathbf{0}$ and radius $\delta > 0$ such that $B \subset V$. Then $\{\mathbf{0}\} \subsetneq C \cap B \subset C \cap V = \varphi^{-1}D$. If $\pi : \mathbb{R}^3 \to \mathbb{R}^2$ is the projection map $\mathbb{R}^3 \ni (x,y,z) \mapsto (x,y) \in \mathbb{R}^2$, then take any nonzero $(x,y) \in \pi(C \cap B)$, and define p, q by

$$\begin{aligned}\mathbf{0} \neq p &:= (x, y, \sqrt{x^2+y^2}) \;\in C \cap B \subset \varphi^{-1}D,\\ \mathbf{0} \neq q &:= (x, y, -\sqrt{x^2+y^2}) \in C \cap B \subset \varphi^{-1}D.\end{aligned}$$

Now $\varphi(p), \varphi(q) \in D \backslash \{\varphi(\mathbf{0})\}$, and suppose that $\gamma : [0,1] \to D\backslash\{\varphi(\mathbf{0})\}$ is any continuous path joining $\varphi(p)$ to $\varphi(q)$ while avoiding $\varphi(\mathbf{0})$. (If $\varphi(\mathbf{0})$ does not lie on the line segment joining $\varphi(p)$ and $\varphi(q)$, then their convex combination will do. Otherwise, we first move along a concentric circular arc starting from $\varphi(p)$, and then follow a straight line segment to $\varphi(q)$.) Consider the continuous map $f = z \circ \varphi^{-1} \circ \gamma : [0,1] \to \mathbb{R}$. Then f is continuous, and

$$f(0) = (z \circ \varphi^{-1} \circ \gamma)(0) = z \circ \varphi^{-1}(\varphi(p)) = z(p) > 0,$$

while

$$f(1) = (z \circ \varphi^{-1} \circ \gamma)(1) = z \circ \varphi^{-1}(\varphi(q)) = z(q) < 0.$$

Hence, by the intermediate value theorem, there exists a $c \in [0,1]$ such that $f(c) = 0$, that is, $z(\varphi^{-1}(\gamma(c))) = 0$, and so we have $\varphi^{-1}(\gamma(c)) = \mathbf{0}$. Thus $\gamma(c) = \varphi(\mathbf{0})$, a contradiction.

Solution to Exercise 1.14

Let $\mathcal{A} = \{(V_\alpha, \psi_\alpha) : \alpha \in A\}$. Since U is open in M, $\psi_\alpha(U \cap V_\alpha)$ is open in $\mathbb{R}^m$. Also, $\psi_\alpha|_{U\cap V_\alpha}$ is injective. So $(U \cap V_\alpha, \psi_\alpha|_{U\cap V_\alpha})$ is a chart on U. These charts cover U because $\bigcup\limits_{\alpha\in A} (U \cap V_\alpha) = U \cap (\bigcup\limits_{\alpha\in A} V_\alpha) = U \cap M = U$.

Let $\alpha, \beta \in A$. Since the sets U, V_β are open in M, so is $U \cap V_\beta$. But then $\psi_\alpha((U \cap V_\beta) \cap V_\alpha)$ is open in $\mathbb{R}^m$, that is, $\psi_\alpha((U \cap V_\alpha) \cap (U \cap V_\beta))$ is open in $\mathbb{R}^m$. Also, the maps $\psi_\beta \circ \psi_\alpha^{-1} : \psi_\alpha((U \cap V_\alpha) \cap (U \cap V_\beta)) \to \mathbb{R}^m$ and $\psi_\alpha \circ \psi_\beta^{-1} : \psi_\beta((U \cap V_\alpha) \cap (U \cap V_\beta)) \to \mathbb{R}^m$, being the restrictions to open subsets of the two C^∞ maps $\psi_\beta \circ \psi_\alpha^{-1} : \psi_\alpha(V_\alpha \cap V_\beta) \to \mathbb{R}^m$ and $\psi_\alpha \circ \psi_\beta^{-1} : \psi_\beta(V_\alpha \cap V_\beta) \to \mathbb{R}^m$, are C^∞ too. So $[\mathcal{A}_U]$ is an atlas for U.

If (W, σ) is admissible for M, then for all $\alpha \in A$, $(V_\alpha \cap W, \psi_\alpha)$, $(V_\alpha \cap W, \sigma)$ are admissible for M. So $\psi_\alpha(U \cap (V_\alpha \cap W)) = \psi_\alpha((U \cap V_\alpha) \cap (U \cap W))$ and $\sigma(U \cap (V_\alpha \cap W)) = \sigma((U \cap V_\alpha) \cap (U \cap W))$ are open. Moreover, we have $\psi_\alpha \circ \sigma^{-1} : \sigma((U \cap V_\alpha) \cap (U \cap W)) \to \mathbb{R}^m$, $\sigma \circ \psi_\alpha^{-1} : \psi_\alpha((U \cap V_\alpha) \cap (U \cap W)) \to \mathbb{R}^m$, being the restrictions of the smooth maps $\psi_\alpha \circ \sigma^{-1} : \sigma(V_\alpha \cap W) \to \mathbb{R}^m$ and $\sigma \circ \psi_\alpha^{-1} : \psi_\alpha(V_\alpha \cap W) \to \mathbb{R}^m$ to open subsets, are C^∞ too.

Solution to Exercise 1.15

To show the continuity of f, we will show that each $p \in M$ has an open neighbourhood U such that $f|_U$ is continuous. As $f : M \to N$ is smooth, given $p \in M$, there exists an admissible chart (U, φ) for M such that $p \in U$, an admissible chart (V, ψ) for N such that $f(U) \subset V$, and also the map $\psi \circ f \circ \varphi^{-1} : \varphi(U) \to \psi(V)$ is smooth, and in particular continuous. Since $\varphi : U \to \varphi(U)$ and $\psi : V \to \psi(V)$ are homeomorphisms, it follows that the map $\psi^{-1} \circ (\psi \circ f \circ \varphi^{-1}) \circ \varphi : U \to V$ is continuous, that is, $f|_U : U \to V$ is continuous.

Solution to Exercise 1.16

Let $m \in M_1$, and

- (U_i, φ_i) be admissible for M_i, $i = 1, 2$, such that we have $m \in U_1$, $f_{12}(U_1) \subset U_2$, and $\varphi_2 \circ f_{12} \circ \varphi_1^{-1} \in C^\infty(\varphi_1(U_1))$,
- (V_i, ψ_i) be admissible for M_i, $i = 2, 3$, such that we have $f_{12}m \in V_2$, $f_{23}(V_2) \subset V_3$, and $\psi_3 \circ f_{23} \circ \psi_2^{-1} \in C^\infty(\psi_2(V_2))$.

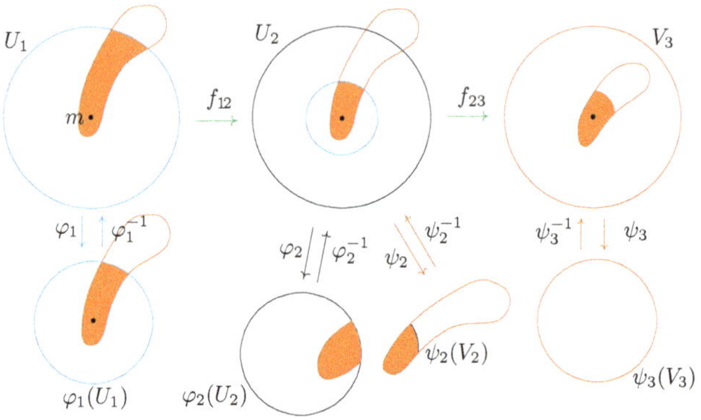

We know $U_2 \cap V_2$ is open in M_2. As f_{12} is continuous, $f_{12}^{-1}(U_2 \cap V_2)$ is open in M_1. We consider the new charts

- $(U_1 \cap f_{12}^{-1}(U_2 \cap V_2), \varphi_1|_{U_1 \cap f_{12}^{-1}(U_2 \cap V_2)})$ that contains m,
- (V_3, ψ_3), which is admissible for M_3, and furthermore we have $(f_{23} \circ f_{12})(U_1 \cap f_{12}^{-1}(U_2 \cap V_2)) \subset f_{23}(U_2 \cap V_2) \subset f_{23}V_2 \subset V_3$.

Finally, $\psi_3 \circ (f_{23} \circ f_{12}) \circ (\varphi_1|_{U_1 \cap f_{12}^{-1}(U_2 \cap V_2)})^{-1} \in C^\infty(\varphi_1(U_1 \cap f_{12}^{-1}(U_2 \cap V_2)))$.

Indeed, $\psi_3 \circ f_{23} \circ f_{12} \circ (\varphi_1|_{U_1 \cap f_{12}^{-1}(U_2 \cap V_2)})^{-1}$

$$= \psi_3 \circ f_{23} \circ \psi_2^{-1} \circ \psi_2 \circ \varphi_2^{-1} \circ \varphi_2 \circ f_{12} \circ (\varphi_1|_{U_1 \cap f_{12}^{-1}(U_2 \cap V_2)})^{-1}$$

is smooth, because it is the composition of the following three smooth maps:

$$\psi_3 \circ f_{23} \circ \psi_2^{-1} \in C^\infty(\psi_2(V_2)), \quad \psi_2 \circ \varphi_2^{-1} \in C^\infty(\varphi_2(U_2 \cap V_2)), \text{ and}$$
$$\varphi_2 \circ f_{12} \circ (\varphi_1|_{U_1 \cap f_{12}^{-1}(U_2 \cap V_2)})^{-1} \in C^\infty(\varphi_1(U_1 \cap f_{12}^{-1}(U_2 \cap V_2))).$$

Solution to Exercise 1.17

Suppose that the point $(p, q) \in M \times N$. Then there exist admissible charts $(U, \varphi), (V, \psi)$ for M, N, respectively such that $p \in U$, $q \in V$. Consider

- $(U \times V, \varphi \times \psi)$, which is admissible for $M \times N$, and we have $(p, q) \in U \times V$,
- (U, φ), which is admissible for M, $\pi_M(U \times V) = U$, and moreover for $(\alpha, \beta) \in (\varphi \times \psi)(U \times V) = (\varphi(U)) \times (\psi(V))$, we have $(\varphi \circ \pi_M \circ (\varphi \times \psi)^{-1})(\alpha, \beta) = (\varphi \circ \pi_M)(\varphi^{-1}(\alpha), \psi^{-1}(\beta)) = \varphi(\varphi^{-1}(\alpha)) = \alpha$, and so $\varphi \circ \pi_M \circ (\varphi \times \psi)^{-1} \in C^\infty((\varphi \times \psi)(U \times V))$.

As $(p,q) \in M \times N$ was arbitrary, π_M is smooth.

Smoothness of i_q: Let $p \in M$, and $(U, \mathbf{x})$ be an admissible chart for M containing p. Let $(V, \mathbf{y})$ be an admissible chart for N containing q. Then $(U \times V, \mathbf{x} \times \mathbf{y})$ is an admissible chart for $M \times N$ containing (p,q). Let the dimensions of M, N be m, n, respectively.

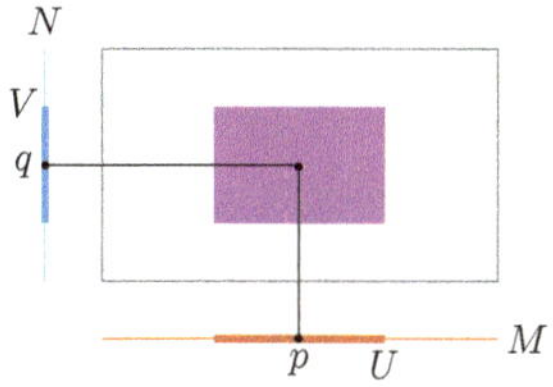

Then $i_q(U) = \{(\widetilde{p}, q) : \widetilde{p} \in U\} \subset U \times V$. Also, $(\mathbf{x} \times \mathbf{y}) \circ i_q \circ \mathbf{x}^{-1} : \mathbf{x}(U) \to \mathbb{R}^{m+n}$ is given by

$$\begin{aligned}((\mathbf{x} \times \mathbf{y}) \circ i_q \circ \mathbf{x}^{-1})(\xi) &= (\mathbf{x} \times \mathbf{y})(i_q(\mathbf{x}^{-1}(\xi))) = (\mathbf{x} \times \mathbf{y})(\mathbf{x}^{-1}(\xi), q) \\ &= (\mathbf{x}(\mathbf{x}^{-1}(\xi)), \mathbf{y}(q)) = (\xi, \mathbf{y}(q))\end{aligned}$$

for $\xi \in \mathbf{x}(U)$. The map $\mathbf{x}(U) \ni \xi \mapsto (\xi, \mathbf{y}(q)) \in \mathbb{R}^{m+n}$ is C^∞. As $p \in M$ was arbitrary, i_q is smooth.

Solution to Exercise 1.18

The 'if' part is clear by taking $U = M$. We now show the 'only if' part. Suppose $p \in U$. Let $(\widetilde{U}, \widetilde{\varphi})$ be an admissible chart for M such that $p \in \widetilde{U}$, and (V, ψ) be an admissible chart for N such that $f(\widetilde{U}) \subset V$ and the map $\psi \circ f \circ \widetilde{\varphi}^{-1} : \widetilde{\varphi}(\widetilde{U}) \to \mathbb{R}^n$ is C^∞. Firstly, $(\widetilde{U} \cap U, \widetilde{\varphi})$ is an admissible chart for U. Also, $f(\widetilde{U} \cap U) \subset f(\widetilde{U}) \subset V$, and the map $\psi \circ f \circ \widetilde{\varphi}^{-1} : \widetilde{\varphi}(\widetilde{U} \cap U) \to \mathbb{R}^n$ is C^∞ on $\widetilde{\varphi}(\widetilde{U} \cap U)$, being the restriction of $\psi \circ f \circ \widetilde{\varphi}^{-1} : \widetilde{\varphi}(\widetilde{U}) \to \mathbb{R}^n$ to the open set $\widetilde{\varphi}(\widetilde{U} \cap U)$.

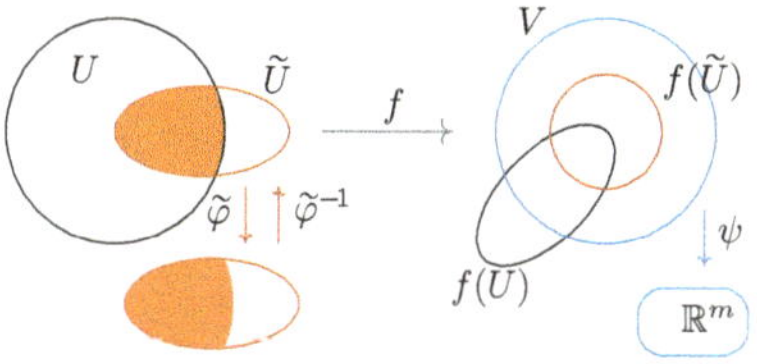

Solution to Exercise 1.19

Let $t \in \mathbb{R}$. Let us take

- $(U, \varphi) = (\mathbb{R}, \mathrm{id})$, an admissible chart for $\mathbb{R}$, with $t \in \mathbb{R} = U$.
- Let $\{\mathbf{e}_1, \cdots, \mathbf{e}_m\}$ be a basis for V, and let $\psi : M \to \mathbb{R}^m$ be the map $q \mapsto \psi(q) := (y^1, \cdots, y^m)$, where $q = p + \mathbf{v}_{pq}$ and $\mathbf{v}_{pq} = y^i \mathbf{e}_i$. Then we had seen that (M, ψ) is an admissible chart for M. Let $\mathbf{v} = v^i \mathbf{e}_i$.

We have $\gamma_{\mathbf{v}}(U) = \gamma_{\mathbf{v}}(\mathbb{R}) \subset M$, and for $s \in \varphi(U) = \mathbb{R}$,

$$\begin{aligned}(\psi \circ \gamma_{\mathbf{v}} \circ \varphi^{-1})(s) &= (\psi \circ \gamma_{\mathbf{v}})(s) = \psi(p + s\mathbf{v}) = \psi(p + sv^i\mathbf{e}_i) \\ &= (sv^1, \cdots, sv^m).\end{aligned}$$

Clearly, the map $\mathbb{R} \ni s \mapsto (sv^1, \cdots, sv^m) \in \mathbb{R}^m$ is C^∞.

As the choice of t was arbitrary, it follows that $\gamma_{\mathbf{v}}$ is smooth.

Solution to Exercise 1.20

Let $f : U \to V$ be C^∞, and $p \in U$. Then

- p belongs to U, and $(U, \varphi := \mathrm{id}_U)$ is an admissible chart for $M = U$,
- $(V, \psi := \mathrm{id}_V)$ is an admissible chart for $N = V$ such that $f(U) \subset V$,
- $\psi \circ f \circ \varphi^{-1} = \mathrm{id}_V \circ f \circ (\mathrm{id}_U)^{-1} = f : (\varphi(U) =)U \to \mathbb{R}^n$ is C^∞.

As $p \in U$ was arbitrary, it follows that f is smooth.

Suppose $f : U \to V$ is smooth. Let $p \in U$. Then there exist an admissible chart $(\widetilde{U}, \widetilde{\varphi})$ for U such that $p \in \widetilde{U}$, an admissible chart $(\widetilde{V}, \widetilde{\psi})$ for V such that $f(\widetilde{U}) \subset \widetilde{V}$, and $\widetilde{\psi} \circ f \circ \widetilde{\varphi}^{-1} : \widetilde{\varphi}(\widetilde{U}) \to \mathbb{R}^n$ is C^∞. As $(\widetilde{U}, \widetilde{\varphi})$ is admissible for U, it follows that the map $\widetilde{\varphi} = \widetilde{\varphi} \circ \mathrm{id}_U^{-1} : (\mathrm{id}_U(U \cap \widetilde{U}) =)\,\widetilde{U} \to \widetilde{\varphi}(\widetilde{U})$ is C^∞. Similarly, $\widetilde{\psi}^{-1} = \mathrm{id}_V \circ \widetilde{\psi}^{-1} : (\widetilde{\psi}(\widetilde{V} \cap V) =)\,\widetilde{\psi}(\widetilde{V}) \to \widetilde{V}$ is C^∞. Thus $f = \widetilde{\psi}^{-1} \circ (\widetilde{\psi} \circ f \circ \widetilde{\varphi}^{-1}) \circ \widetilde{\varphi} : \widetilde{U} \to \mathbb{R}^n$ is C^∞. As $p \in U$ was arbitrary, we conclude that f is C^∞.

Solution to Exercise 1.21

Suppose f, g are given by $f = (x \mapsto x^{\frac{1}{3}}) : M = (\mathbb{R}, [\mathcal{A}_1]) \to (\mathbb{R}, [\mathcal{A}_2]) = N$, $g = (y \mapsto y^3) : N = (\mathbb{R}, [\mathcal{A}_2]) \to (\mathbb{R}, [\mathcal{A}_1]) = M$. Then we have

$$(f \circ g)(y) = f(y^3) = (y^3)^{\frac{1}{3}} = y \text{ for all } y \in N, \text{ and}$$
$$(g \circ f)(x) = g(x^{\frac{1}{3}}) = (x^{\frac{1}{3}})^3 = x \text{ for all } x \in M.$$

Hence f is a bijection from M to N, with the inverse $f^{-1} = g : N \to M$.

$$\begin{array}{ccc} M = \mathbb{R} & \underset{g\, =\, (y \mapsto y^3)}{\overset{f\, =\, (x \mapsto x^{\frac{1}{3}})}{\rightleftarrows}} & \mathbb{R} = N \\ {\scriptstyle \varphi_1 = (x \mapsto x)}\big\downarrow & & \big\downarrow{\scriptstyle \varphi_2 = (y \mapsto y^3)} \\ \mathbb{R} & & \mathbb{R} \end{array}$$

f is smooth: Let $x \in M$. With

- $(U = \mathbb{R}, \varphi_1 = \mathrm{id})$, an admissible chart for M with $x \in U$;
- $(V = \mathbb{R}, \varphi_2 = (y \mapsto y^3))$, an admissible chart for N, $f(U) = f(\mathbb{R}) = \mathbb{R} = V$.
  For $\alpha \in \varphi_1(U) = \mathbb{R}$, $(\varphi_2 \circ f \circ \varphi_1^{-1})(\alpha) = (\varphi_2 \circ f)(\alpha) = \varphi_2(\alpha^{\frac{1}{3}}) = (\alpha^{\frac{1}{3}})^3 = \alpha$.
  So $\varphi_2 \circ f \circ \varphi_1^{-1} = \mathrm{id}_{\mathbb{R}} : \varphi_1(U) = \mathbb{R} \to \mathbb{R}$, which is C^∞.

As x was arbitrary, f is smooth.

g is smooth: Let $y \in N$. With

- $(U = \mathbb{R}, \varphi_2 = (y \mapsto y^3))$, an admissible chart for N with $y \in U = \mathbb{R}$;
- $(V = \mathbb{R}, \varphi_1 = \mathrm{id})$, an admissible chart for M, $g(U) = g(\mathbb{R}) = \mathbb{R} = V$.
  For $\beta \in \varphi_2(U) = \mathbb{R}$, $(\varphi_1 \circ g \circ \varphi_2^{-1})(\beta) = (\mathrm{id} \circ g)(\beta^{\frac{1}{3}}) = g(\beta^{\frac{1}{3}}) = (\beta^{\frac{1}{3}})^3 = \mathrm{id}(\beta)$.
  So $\psi \circ g \circ \varphi^{-1} = \mathrm{id} : \varphi(U) = \mathbb{R} \to \mathbb{R}$, which is C^∞.

As y was arbitrary, g is smooth. So $(\mathbb{R}, [\mathcal{A}_1])$ is diffeomorphic to $(\mathbb{R}, [\mathcal{A}_2])$.

Solution to Exercise 1.22

Let (V, ψ) be an admissible chart for N such that $f(U) \cap V \neq \varnothing$. As (U, φ) is an admissible chart for M, U is open in M. So $f(U) = (f^{-1})^{-1}(U)$ is open in N (because $f : M \to N$ being a diffeomorphism implies that $f^{-1} : N \to M$ is a smooth map, and in particular, continuous). Also, since (V, ψ) is an admissible chart for N, V is open in N. Thus $f(U) \cap V$ is an open set in N. As $\psi : V \to \psi(V) \subset \mathbb{R}^m$ is a homeomorphism, it follows that $\psi(f(U) \cap V)$ is open in $\mathbb{R}^m$. We have that $(\varphi \circ f^{-1})(f(U) \cap V)$ is open in $\mathbb{R}^m$, because $f^{-1}(f(U) \cap V)$ is an open subset of U, and $\varphi : U \to \mathbb{R}^m$ is a homeomorphism. Finally, since restrictions of smooth maps to open sets are smooth, and since $\psi : V \to \psi(V)$, $f : M \to N$, and $\varphi : U \to \varphi(U)$ are diffeomorphisms, $\psi \circ (\varphi \circ f^{-1})^{-1} : (\varphi \circ f^{-1})(f(U) \cap V) \to \psi(f(U) \cap V)$ and $\varphi \circ f^{-1} \circ \psi^{-1} : \psi(f(U) \cap V) \to (\varphi \circ f^{-1})(f(U) \cap V)$ are smooth, and so C^∞.

Solution to Exercise 1.23

For $f, g \in \mathrm{Diff}(M)$, we have that $f \circ g$ is a smooth map and a bijection. Also, $(f \circ g)^{-1} = g^{-1} \circ f^{-1}$, which is smooth since g^{-1} and f^{-1} are smooth. Thus $f \circ g \in \mathrm{Diff}(M)$. Composition of maps is associative. The identity map $\mathrm{id}_M : M \to M$ is a diffeomorphism, and it serves as the identity element for $(\mathrm{Diff}(M), \circ)$ because for all $f \in \mathrm{Diff}(M)$, we have $f \circ \mathrm{id}_M = f = \mathrm{id}_M \circ f$. If $f \in \mathrm{Diff}(M)$, then f^{-1} is also a smooth bijection, with $(f^{-1})^{-1}$ being f, which is smooth again. Thus $f^{-1} \in \mathrm{Diff}(M)$, and $f^{-1} \circ f = \mathrm{id}_M = f \circ f^{-1}$. Consequently, $(\mathrm{Diff}(M), \circ)$ forms a group.

Solution to Exercise 1.24

Let $p \in G$. The inclusion map $i_p : G \to G \times G$, $i_p(q) = (p, q)$ for $q \in G$, is smooth by Exercise 1.17. Thus the composition $\cdot \circ i_p$ of this smooth map with the multiplication map is smooth too. But this is the map L_p, since $(\cdot \circ i_p)(q) = \cdot(p, q) = p \cdot q = L_p q$ for all $q \in G$. As $p \in G$ was arbitrary, $L_{p^{-1}}$ is smooth too. But $L_p \circ L_{p^{-1}} = \mathrm{id}_G$, as (with e denoting the identity element in G) $(L_p \circ L_{p^{-1}})(q) = L_p(p^{-1}q) = p(p^{-1}q) = (pp^{-1})q = eq = q = \mathrm{id}_G q$. Thus $L_{p^{-1}} \circ L_p = L_{p^{-1}} \circ L_{(p^{-1})^{-1}} = \mathrm{id}_G$. So L_p is a bijection with the inverse $(L_p)^{-1} = L_{p^{-1}}$. As L_p and $L_{p^{-1}}$ are smooth, L_p is a diffeomorphism.

Solution to Exercise 1.25

We begin by showing that $(U \cap N, \pi \circ \varphi|_{U\cap N})$ is a chart. We have that $(\pi \circ \varphi)(U \cap N) = \widetilde{U}$ is open in $\mathbb{R}^n$. Also, if for $p, p' \in U \cap N$, we have $(\pi\circ\varphi)(p) = (\pi\circ\varphi)(p')$, then $\varphi(p) = ((\pi\circ\varphi)(p), \mathbf{0}) = ((\pi\circ\varphi)(p'), \mathbf{0}) = \varphi(p')$, and so $p = p'$ by the injectivity of φ. Thus $\pi\circ\varphi|_{U\cap N}$ is injective on $U \cap N$.

For each $p \in N$, there is an allowed chart (U, φ) for N that contains p. So $p \in U \cap N \in \mathcal{A}_N$. Thus the chart domains of charts from $\mathcal{A}_N$ cover N.

Let $(U, \varphi), (V, \psi)$ be allowed charts for N. Then $\varphi(U) = \widetilde{U} \times \{\mathbf{0}\}$ and $\psi(V) = \widetilde{V} \times \{\mathbf{0}\}$, where $\widetilde{U}, \widetilde{V}$ are open subsets of $\mathbb{R}^n$. As φ is injective, $\varphi((U \cap N) \cap (V \cap N)) = \varphi(U \cap V) \cap (\widetilde{U} \times \{0\})$. As $\varphi(U \cap V)$ is open in $\mathbb{R}^m$, and $\widetilde{U}$ is open in $\mathbb{R}^n$, $\pi(\varphi(U \cap V) \cap (\widetilde{U} \times \{0\}))$ is open in $\mathbb{R}^n$. Also, $(\pi\circ\psi)\circ(\pi\circ\varphi)^{-1} : (\pi\circ\varphi)((U\cap N)\cap(V\cap N)) \to \mathbb{R}^n$ is $\boldsymbol{\alpha} \mapsto (\pi\circ\psi\circ\varphi^{-1})(\boldsymbol{\alpha}, \mathbf{0})$, which is C^∞, as $\mathbb{R}^n \ni \boldsymbol{\alpha} \mapsto (\boldsymbol{\alpha}, \mathbf{0}) \in \mathbb{R}^m$, π, and $\psi \circ \varphi^{-1}$ are all C^∞.

Let $p \in N$, and (U, φ) be an allowed chart for N with $p \in U$. Then $(U \cap N, \pi \circ \varphi|_{U\cap N}) \in \mathcal{A}_N$, and its image under $i : N \to M$ is $U \cap N \subset U$. For $\boldsymbol{\alpha} \in (\pi \circ \varphi)(U \cap N)$, $(\varphi \circ i \circ (\pi \circ \varphi)^{-1})(\boldsymbol{\alpha}) = (\varphi \circ i)\varphi^{-1}(\boldsymbol{\alpha}, \mathbf{0}) = (\boldsymbol{\alpha}, \mathbf{0})$, which is C^∞. Thus $i : N \to M$ is smooth.

Solution to Exercise 1.26

Let the dimensions of M_1, M_2, N_1, N_2 be m_1, m_2, n_1, n_2, respectively.

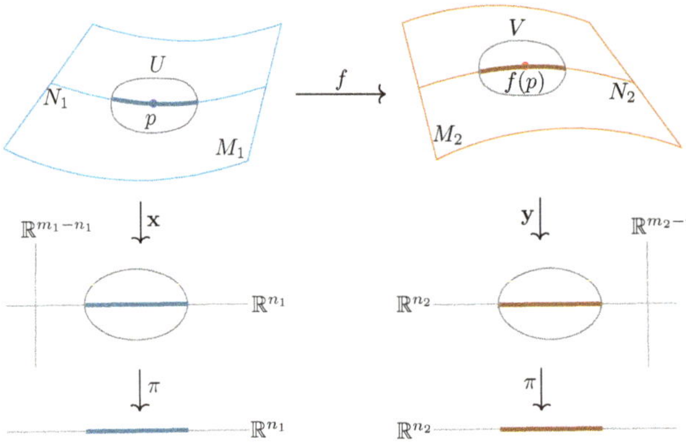

Let $p \in N_1 \subset M_1$. As $f : M_1 \to M_2$ is smooth, there exist admissible charts $(\widetilde{U}, \widetilde{\mathbf{x}})$ and $(\widetilde{V}, \widetilde{\mathbf{y}})$ such that $p \in \widetilde{U}$, $f(\widetilde{U}) \subset \widetilde{V}$, and $\widetilde{\mathbf{y}} \circ f \circ \widetilde{x}^{-1} : \widetilde{\mathbf{x}}(U) \to \mathbb{R}^{m_2}$ is C^∞. Let $(\tilde{\tilde{U}}, \mathbf{x})$ be an allowed chart for N_1 with $p \in \tilde{\tilde{U}}$. For $f(p) \in N_2$, let $(\tilde{\tilde{V}}, \mathbf{y})$ be an allowed chart for N_2 with $f(p) \in \tilde{\tilde{V}}$. As $f : M_1 \to M_2$ is smooth, and in particular continuous, $f^{-1}(\tilde{\tilde{V}})$ is open in M_1. Moreover $p \in f^{-1}(\tilde{\tilde{V}})$. Thus $(U := \widetilde{U} \cap \tilde{\tilde{U}} \cap f^{-1}(\tilde{\tilde{V}}), \mathbf{x}|_U)$ is also an allowed chart for N_1

containing p. Also, $(V := \widetilde{V} \cap \widetilde{\widetilde{V}}, \mathbf{y}|_V)$ is an allowed chart for N_2 containing $f(p)$, and $f(U) \subset f(\widetilde{U} \cap \widetilde{\widetilde{U}}) \cap f(f^{-1}(\widetilde{\widetilde{V}})) \subset f(\widetilde{U}) \cap \widetilde{\widetilde{V}} \subset \widetilde{V} \cap \widetilde{\widetilde{V}} = V$. The map sending $(\mathbf{a}, \mathbf{b}) \in \mathbf{x}(U) \subset \mathbb{R}^{n_1} \times \mathbb{R}^{m_1 - n_1}$ to

$$\begin{aligned}(\mathbf{y} \circ f \circ \mathbf{x}^{-1})(\mathbf{a}, \mathbf{b}) &= ((\mathbf{y} \circ \widetilde{\mathbf{y}}^{-1}) \circ (\widetilde{\mathbf{y}} \circ f \circ \widetilde{\mathbf{x}}^{-1}) \circ (\widetilde{\mathbf{x}} \circ \mathbf{x}^{-1}))(\mathbf{a}, \mathbf{b}) \\ &=: (\varphi(\mathbf{a}, \mathbf{b}), \psi(\mathbf{a}, \mathbf{b}))\end{aligned}$$

is C^∞ (being the composition of the C^∞ map $\widetilde{\mathbf{y}} \circ f \circ \widetilde{\mathbf{x}}^{-1}$ with the smooth chart transition maps $\widetilde{\mathbf{x}} \circ \mathbf{x}^{-1}$ and $\mathbf{y} \circ \widetilde{\mathbf{y}}^{-1}$), and so its 'component' functions φ, ψ are C^∞ too. For any $\mathbf{a} \in (\pi \circ \mathbf{x})(U \cap N_1)$, $(\pi \circ \mathbf{x})|_{U \cap N_1}^{-1}(\mathbf{a}) = \mathbf{x}^{-1}(\mathbf{a}, \mathbf{0})$. Here $\pi : \mathbb{R}^{m_1} \to \mathbb{R}^{n_1}$ is the projection map onto the first n_1 components. We denote the projection $\mathbb{R}^{m_2} \to \mathbb{R}^{n_2}$ onto the first n_2 components also by π. We have that $(U \cap N_1, \pi \circ \mathbf{x}|_{U \cap N_1})$ is an admissible chart for N_1 containing p, and $(V \cap N_2, \pi \circ \mathbf{y}|_{V \cap N_2})$ is an admissible chart for N_2 such that $f|_{N_1}(U \cap N_1) \subset f(U) \cap f(N_1) \subset V \cap N_2$. Finally,

$$\begin{aligned}((\pi \circ \mathbf{y})|_{V \cap N_2} \circ f \circ ((\pi \circ \mathbf{x})|_{U \cap N_1}^{-1}))(\mathbf{a}) &= ((\pi \circ \mathbf{y})|_{V \cap N_2} \circ f)(\mathbf{x}^{-1}(\mathbf{a}, \mathbf{0})) \\ &= (\pi \circ (\mathbf{y} \circ f \circ \mathbf{x}^{-1}))(\mathbf{a}, \mathbf{0}) \\ &= \pi(\varphi(\mathbf{a}, \mathbf{0}), \psi(\mathbf{a}, \mathbf{0})) = \varphi(\mathbf{a}, \mathbf{0}),\end{aligned}$$

and so we conclude that the map $f|_{N_1} : N_1 \to N_2$ is smooth.

Chapter 2

Solution to Exercise 2.1

If $v, w \in T_pM$, then for smooth $f, g \in C^\infty(M)$ and $c \in \mathbb{R}$, we have

$$\begin{aligned}(v+w)(f+cg) &= v(f+cg)+w(f+cg) = vf+cvg+wf+cwg\\ &= (v+w)f + c(v+w)g, \text{ and}\\ (v+w)(f\cdot g) &= v(f\cdot g)+w(f\cdot g)\\ &= f(p)vg+g(p)vf+f(p)wg+g(p)wf\\ &= f(p)(vg+wg)+g(p)(vf+wf)\\ &= f(p)\,(v+w)g+g(p)\,(v+w)f,\end{aligned}$$

and so $v+w \in T_pM$.

If $v \in T_pM$, $\alpha \in \mathbb{R}$, then for $f, g \in C^\infty(M)$, and $c \in \mathbb{R}$, we have

$$\begin{aligned}(\alpha\cdot v)(f+cg) &= \alpha v(f+cg) = \alpha(vf+cvg) = \alpha vf + \alpha cvg\\ &= (\alpha\cdot v)f + c(\alpha\cdot v)g, \text{ and}\\ (\alpha\cdot v)(f\cdot g) &= \alpha v(f\cdot g) = \alpha(f(p)vg+g(p)vf) = \alpha f(p)vg+\alpha g(p)vf\\ &= f(p)\,(\alpha\cdot v)g + g(p)\,(\alpha\cdot v)f,\end{aligned}$$

and so $\alpha\cdot v \in T_pM$.

Solution to Exercise 2.2

Let $f(p)=c$ for all $p \in M$. Let $\mathbf{1} \in C^\infty(M)$ be given by $\mathbf{1}(p)=1$ for $p \in M$. Then $v(\mathbf{1}) = v(\mathbf{1}\cdot\mathbf{1}) = \mathbf{1}(p)\,v(\mathbf{1}) + \mathbf{1}(p)\,v(\mathbf{1}) = 1\,v(\mathbf{1}) + 1\,v(\mathbf{1}) = 2v(\mathbf{1})$, and so $v(\mathbf{1})=0$. Thus $v(f) = v(c\mathbf{1}) = cv(\mathbf{1}) = c0 = 0$.

Solution to Exercise 2.3

For $f \in C^\infty(\mathbb{R})$, we have $\gamma_1(0)=-1$, $\gamma_2(0)=-1$, and

$$v_{\gamma_1,-1}f = \frac{d(f\circ\gamma_1)}{dt}(0) = \frac{d(f(t-1))}{dt}(0) = f'(t-1)\,1|_{t=0} = f'(-1),$$

$$v_{\gamma_2,-1}f = \frac{d(f\circ\gamma_2)}{dt}(0) = \frac{d(f(t+3t^2-1))}{dt}(0) = f'(t+3t^2-1)\,(1+6t)|_{t=0} = f'(-1).$$

Thus for all $f \in C^\infty(M)$, we have $v_{\gamma_1,-1}f = v_{\gamma_2,-1}f$. Hence $v_{\gamma_1,-1} = v_{\gamma_2,-1}$.

Solution to Exercise 2.4

For $f \in C^\infty(M)$, and $s \in J$, we have

$$\begin{aligned}v_{\gamma\circ h,(\gamma\circ h)(s)}f &= \frac{d(f\circ(\gamma\circ h))}{dt}(s) = \frac{d((f\circ\gamma)\circ h)}{dt}(s) = \frac{d(f\circ\gamma)}{dt}(h(s))\,\dot{h}(s)\\ &= \dot{h}(s)\,v_{\gamma,\gamma(h(s))}f.\end{aligned}$$

Hence $v_{\gamma\circ h,(\gamma\circ h)(s)} = \dot{h}(s)\,v_{\gamma,\gamma(h(s))}$.

Solution to Exercise 2.5

(1) Let $L := \lim_{t\to 0} g'(t)$. If $t \neq 0$, then applying the mean value theorem for g on the interval with endpoints 0 and t, we obtain

$$\frac{g(t)-g(0)}{t-0} = g'(\theta_t t)$$

for some $\theta_t \in (0,1)$. Given $\epsilon > 0$, let $\delta > 0$ be such that if $0 < |\tau - 0| < \delta$, $|g'(\tau) - L| < \epsilon$. Thus for $0 < |t-0| < \delta$, we have $0 < |\theta t - 0| < \theta\delta$ for all $\theta \in (0,1)$, and so

$$\Big|\frac{g(t)-g(0)}{t-0} - L\Big| = |g'(\theta_t t) - L| < \epsilon.$$

So $g'(0) = L = \lim_{t\to 0} g'(t)$. Thus g is differentiable at 0 and moreover g' is continuous at 0. Hence g is continuously differentiable on $\mathbb{R}$.

(2) $g^{(n-1)} : \mathbb{R} \to \mathbb{R}$ is continuous on $\mathbb{R}$ and continuously differentiable on $\mathbb{R}_*$. Moreover, $\lim_{t\to 0}(g^{(n-1)})'(t)$ exists. By (1), $g^{(n-1)}$ is continuously differentiable on $\mathbb{R}$. So g is n times continuously differentiable on $\mathbb{R}$.

(3) We have for $t > 0$ that $f'(t) = \frac{1}{t^2}e^{-\frac{1}{t}} = R_1(t)e^{-\frac{1}{t}}$, where $R_1(t) = \frac{1}{t^2}$. Suppose $f^{(n)} = \frac{p_n}{q_n}e^{-\frac{1}{t}}$ for some n, where p_n, q_n are polynomials. Then

$$\begin{aligned} f^{(n+1)} &= \frac{(p_n' e^{-\frac{1}{t}} + \frac{1}{t^2}p_n e^{-\frac{1}{t}})q_n - q_n' p_n e^{-\frac{1}{t}}}{q_n^2} \\ &= \frac{(p_n' + \frac{p_n}{t^2})q_n - q_n' p_n}{q_n^2} e^{-\frac{1}{t}} =: R_{n+1}e^{-\frac{1}{t}}, \end{aligned}$$

for some rational function R_{n+1}. By induction, $f^{(n)} = R_n f$ for all n.

For a fixed n, if $R_n = \frac{p_n}{q_n}$, and $q_n = t^k Q(t)$ with $k \in \{0,1,2,3,\cdots\}$, and Q a polynomial such that $Q(0) \neq 0$, then we have

$$\lim_{t\searrow 0} f^{(n)} = \lim_{t\searrow 0} R_n f = \lim_{t\searrow 0} \frac{p_n(t)}{q_n(t)}e^{-\frac{1}{t}} = \lim_{t\searrow 0} \frac{p_n(t)}{Q(t)}\frac{e^{-\frac{1}{t}}}{t^k} = \frac{p_n(0)}{Q(0)}0 = 0.$$

(To see that $\lim_{t\searrow 0} \frac{e^{-1/t}}{t^k} = 0$, we note that for $t > 0$,

$$e^{1/t} = 1 + \frac{1}{1!}\frac{1}{t} + \cdots + \frac{1}{(k+1)!}\frac{1}{t^{k+1}} + \cdots > \frac{1}{(k+1)!}\frac{1}{t^{k+1}},$$

and so $0 < \frac{e^{-\frac{1}{t}}}{t^k} < t(k+1)! \overset{t\searrow 0}{\to} 0$. Hence $\lim_{t\searrow 0}\frac{e^{-\frac{1}{t}}}{t^k} = 0$.)

Thus for all n, $\lim_{t\searrow 0} f^{(n)}(t) = 0 = \lim_{t\nearrow 0} f^{(n)}(t)$, and so $\lim_{t\to 0} f^{(n)}(t)$ exists. As $f \in C^\infty(\mathbb{R}_*)$, it follows from (1) and (2) that $f \in C^\infty(\mathbb{R})$. (Note that f is continuous on $\mathbb{R}$: $\lim_{t\nearrow 0} f(t) = 0$, and $\lim_{t\searrow 0} f(t) = \lim_{t\searrow 0} e^{-\frac{1}{t}} = 0$. Thus $\lim_{t\to 0} f(t) = 0 = f(0)$.)

Solution to Exercise 2.6

With f as in Exercise 2.5, we know that $f \in C^\infty(\mathbb{R})$. Then $f(-\cdot) \in C^\infty(\mathbb{R})$, and $tf(-\cdot), tf(\cdot) \in C^\infty(\mathbb{R})$ too. We take $\gamma(t) = (tf(-t), tf(t))$, $t \in \mathbb{R}$. Then γ is smooth. We note that for $t > 0$, $f'(t) = \frac{e^{-1/t}}{t^2} > 0$, and so f is strictly increasing for $t > 0$. Thus tf is strictly increasing for $t > 0$. Also, while f approaches 1 as $t \to \infty$, $tf \to \infty$ as $t \to \infty$, and so the range of tf is $(0, \infty)$ for $t \in (0, \infty)$.

We see that $t \mapsto tf(-t)$ is strictly increasing for $t < 0$. Indeed, for $t_1 < t_2 < 0$, we have that $-t_1 > -t_2 > 0$, and so $f(-t_1) > f(-t_2)$, giving $(-t_1)f(-t_1) > (-t_2)f(-t_2)$, i.e., $t_1 f(-t_1) < t_2 f(-t_2)$.

Moreover, the range of $tf(-\cdot)$ on $(-\infty, 0)$ is $(-\infty, 0)$. Finally, we have $\gamma(0) = (0f(-0), 0f(0)) = (0, 0)$.

Solution to Exercise 2.7

The partial derivatives of the chart transition map $\psi \circ (\mathrm{id})^{-1}$ are

$$\frac{\partial r}{\partial x} = \frac{x}{\sqrt{x^2+y^2}}, \qquad \frac{\partial r}{\partial y} = \frac{y}{\sqrt{x^2+y^2}},$$
$$\frac{\partial \theta}{\partial x} = \frac{-y}{x^2+y^2}, \qquad \frac{\partial \theta}{\partial y} = \frac{x}{x^2+y^2}.$$

If $\partial_{r,p} = v^1 \partial_{x,p} + v^2 \partial_{y,p}$, then

$$\begin{bmatrix} 1 \\ 0 \end{bmatrix} = \begin{bmatrix} \frac{x}{\sqrt{x^2+y^2}} & \frac{y}{\sqrt{x^2+y^2}} \\ \frac{-y}{x^2+y^2} & \frac{x}{x^2+y^2} \end{bmatrix} \begin{bmatrix} v^1 \\ v^2 \end{bmatrix}, \text{ giving } \begin{bmatrix} v^1 \\ v^2 \end{bmatrix} = \begin{bmatrix} \frac{x}{\sqrt{x^2+y^2}} \\ \frac{y}{\sqrt{x^2+y^2}} \end{bmatrix}.$$

So $\partial_{r,p} = \frac{x}{\sqrt{x^2+y^2}} \partial_{x,p} + \frac{y}{\sqrt{x^2+y^2}} \partial_{y,p}$.

Also if, $\partial_{\theta,p} = w^1 \partial_{x,p} + w^2 \partial_{y,p}$, then

$$\begin{bmatrix} 0 \\ 1 \end{bmatrix} = \begin{bmatrix} \frac{x}{\sqrt{x^2+y^2}} & \frac{y}{\sqrt{x^2+y^2}} \\ \frac{-y}{x^2+y^2} & \frac{x}{x^2+y^2} \end{bmatrix} \begin{bmatrix} w^1 \\ w^2 \end{bmatrix}, \text{ giving } \begin{bmatrix} w^1 \\ w^2 \end{bmatrix} = \begin{bmatrix} -y \\ x \end{bmatrix},$$

and so $\partial_{\theta,p} = -y\,\partial_{x,p} + x\,\partial_{y,p}$.

Solution to Exercise 2.8

$v \oplus w \in T_{(p,q)}(M \times N)$: For all $f, g \in C^\infty(M \times N)$ and all $c \in \mathbb{R}$, we have

$$\begin{aligned}(v \oplus w)(f + cg) &= v((f+cg)(\cdot, q)) + w((f+cg)(p, \cdot)) \\ &= v(f(\cdot, q) + cg(\cdot, q)) + w(f(p, \cdot) + cg(p, \cdot)) \\ &= v(f(\cdot, q)) + cv(g(\cdot, q)) + w(f(p, \cdot)) + cw(g(p, \cdot)) \\ &= (v \oplus w)(f) + c(v \oplus w)(g).\end{aligned}$$

For all $f, g \in C^\infty(M \times N)$, we also have

$$\begin{aligned}(v \oplus w)(fg) &= v((fg)(\cdot, q)) + w((fg)(p, \cdot)) \\ &= v(f(\cdot, q)g(\cdot, q)) + w(f(p, \cdot)g(p, \cdot)) \\ &= f(p, q)v(g(\cdot, q)) + g(p, q)v(f(\cdot, q)) \\ &\quad + f(p, q)w(g(p, \cdot)) + g(p, q)w(f(p, \cdot)) \\ &= f(p, q)(v \oplus w)g + g(p, q)(v \oplus w)f.\end{aligned}$$

Linearity of $(v, w) \overset{T}{\mapsto} v \oplus w$:
For $v_1, v_2 \in T_pM$, $w_1, w_2 \in T_qN$, $c \in \mathbb{R}$, and $f \in C^\infty(M \times N)$,

$$\begin{aligned}((v_1 \oplus w_1) + c(v_2 \oplus w_2))f &= (v_1 \oplus w_1)f + c(v_2 \oplus w_2)f \\ &= v_1 f(\cdot, q) + w_1 f(p, \cdot) + c v_2 f(\cdot, q) + c w_2 f(p, \cdot) \\ &= (v_1 + c v_2)(f(\cdot, q)) + (w_1 + c w_2)(f(p, \cdot)) \\ &= ((v_1 + c v_2) \oplus (w_1 + c w_2))f.\end{aligned}$$

So $(v_1 \oplus w_1) + c(v_2 \oplus w_2) = (v_1 + c v_2) \oplus (w_1 + c w_2)$, that is,

$$T(v_1, w_1) + cT(v_2, w_2) = T(v_1 + c v_2, w_1 + c w_2) = T((v_1, w_1) + c(v_2, w_2)).$$

Injectivity: Let $(v, w) \in T_pM \times T_qN$ be such that $T(v, w) = 0$, that is, $v \oplus w = 0 \in T_{(p,q)}(M \times N)$. Then $(v \oplus w)f = 0$, i.e., $v(f(\cdot, q)) + w(f(p, \cdot)) = 0$ for all $f \in C^\infty(M \times N)$. If $g \in C^\infty(M)$, then $f := g \circ \pi_M \in C^\infty(M \times N)$, and so

$$\begin{aligned}0 &= v(f(\cdot, q)) + w(f(p, \cdot)) = v((\circ \pi_M)(\cdot, q)) + w((g \circ \pi_M)(p, \cdot)) \\ &= v(g) + w(g(p)) = v(g) + 0 = v(g).\end{aligned}$$

Hence for all $g \in C^\infty(M)$, $vg = 0$. Thus $v = 0 \in T_pM$. Similarly, for all $h \in C^\infty(N)$, $wh = 0$ (by considering $f := h \circ \pi_N$). So $w = 0 \in T_qN$. Hence $(v, w) = (0, 0) \in T_pM \times T_qN$. So T is injective.

Surjectivity: We have

$$m + n = \dim(T_pM \times T_qN) = \dim(\ker T) + \dim(\operatorname{ran} T) = 0 + \dim(\operatorname{ran} T)$$

using the rank-nullity theorem. Thus

$$\dim \operatorname{ran} T = m + n = \dim(M \times N) = \dim T_{(p,q)}(M \times N).$$

As $\operatorname{ran} T \subset T_{(p,q)}(M \times N)$, $\operatorname{ran} T = T_{(p,q)}(M \times N)$, i.e., T is surjective.

Solution to Exercise 2.9

Recall that the inclusion map $i : N \hookrightarrow M$ is smooth. Let $v \in T_pN$. Define $Iv : C^\infty(M) \to \mathbb{R}$ by $(Iv)(f) = v(f \circ i)$ for all $f \in C^\infty(M)$. Then it is easy to check that Iv is linear and obeys the Leibniz rule, so that $Iv \in T_pM$. Also the map $v \mapsto Iv : T_pN \to T_pM$ is linear. Indeed, for all $f \in C^\infty(M)$, $v, w \in T_pN$, $c \in \mathbb{R}$, we have

$$\begin{aligned}(I(v + cw))(f) &= (v + cw)(f \circ i) = v(f \circ i) + cw(f \circ i) \\ &= (Iv)(f) + c(Iw)(f) = ((Iv) + c(Iv))(f).\end{aligned}$$

Finally, we show that I is injective. Suppose that $v \in T_pN$ is such that $Iv = 0$. Let p belong to the admissible chart (U, φ) of M, which is allowed for N. Then, as for each $j \in \{1, \cdots, n\}$, the component map $\varphi^j \in C^\infty(M)$, we have $0 = 0(\varphi^j) = (Iv)(\varphi^j) = v(\varphi^j \circ i) = v(\varphi^j) = v((\pi \circ \varphi)^j)$. Thus $v = v((\pi \circ \varphi)^j)\partial_{(\pi\circ\varphi)^j,p} = 0$.

Solution to Exercise 2.10

Let $v, w \in T_pM$, $c \in \mathbb{R}$, and $g \in C^\infty(N)$. Then

$$\begin{aligned}((df_p)(v+cw))(g) &= (v+cw)(g \circ f) = v(g \circ f) + cw(g \circ f)\\ &= (df_p(v))(g) + c(df_p(w))(g) = (df_p(v) + cdf_p(w))(g).\end{aligned}$$

As $g \in C^\infty(N)$ was arbitrary, $df_p(v+cw) = df_p(v) + cdf_p(w)$.

Solution to Exercise 2.11

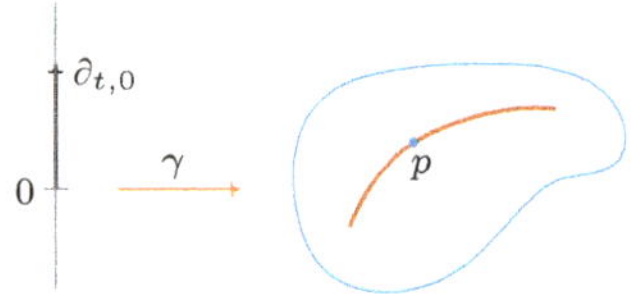

For $f \in C^\infty(M)$,

$$\begin{aligned}(d\gamma_0(\partial_{t,0}))(f) &= \partial_{t,0}(f \circ \gamma) = \frac{d(f \circ \gamma \circ \mathrm{id}_{\mathbb{R}}^{-1})}{du}(\mathrm{id}_{\mathbb{R}}(0)) = \frac{d(f \circ \gamma)}{du}(0)\\ &= v_{\gamma,\gamma(0)}f = v_{\gamma,p}f.\end{aligned}$$

As $f \in C^\infty(M)$ was arbitrary, $d\gamma_0(\partial_{t,0}) = v_{\gamma,p}$.

Solution to Exercise 2.12

The Jacobian matrix of $\mathrm{id}_{\mathbb{R}^2} \circ \psi_t \circ \mathrm{id}_{\mathbb{R}^2}^{-1} = \psi_t$ is given by

$$\begin{bmatrix} \dfrac{\partial(x-yt)}{\partial x}(x,y) & \dfrac{\partial(x-yt)}{\partial y}(x,y) \\ \dfrac{\partial y}{\partial x}(x,y) & \dfrac{\partial y}{\partial y}(x,y) \end{bmatrix} = \begin{bmatrix} 1 & -t \\ 0 & 1 \end{bmatrix}.$$

Thus $(d\psi_t)_p(\partial_{y,p}) = a\partial_{x,\psi_t(p)} + b\partial_{y,\psi_t(p)}$, where

$$\begin{bmatrix} a \\ b \end{bmatrix} = \begin{bmatrix} 1 & -t \\ 0 & 1 \end{bmatrix}\begin{bmatrix} 0 \\ 1 \end{bmatrix} = \begin{bmatrix} -t \\ 1 \end{bmatrix}$$

by Theorem 2.2. Hence $(d\psi_t)_p(\partial_{y,p}) = -t\partial_{x,\psi_t(p)} + \partial_{y,\psi_t(p)}$.

Solution to Exercise 2.13

Let $v \in T_pM$. Then for all $\varphi \in C^\infty(M_3)$,

$$\begin{aligned}(d(g \circ f)_p(v))(\varphi) &= v(\varphi \circ (g \circ f)) = v((\varphi \circ g) \circ f) = ((df_p)(v))(\varphi \circ g)\\ &= (dg_{f(p)}(df_p(v)))\varphi.\end{aligned}$$

So $d(g\circ f)_p(v) = (dg_{f(p)} \circ df_p)(v)$ for all $v \in T_pM$. Thus $d(g\circ f)_p = dg_{f(p)} \circ df_p$.

Solution to Exercise 2.14

Considering the composition $f^{-1} \circ f = \mathrm{id}_M : M \to M$, we have by Exercise 2.13 that $(df^{-1})_{f(p)} \circ df_p = d(\mathrm{id}_M)_p$.

Claim. $d(\mathrm{id}_M)_p = \mathrm{id}_{T_pM}$.

Proof. Let $v \in T_pM$. For all $\varphi \in C^\infty(M)$,

$$(d(\mathrm{id}_M)_p(v))(\varphi) = v(\varphi \circ \mathrm{id}_M) = v(\varphi) = (\mathrm{id}_{T_pM}(v))(\varphi).$$

Thus $d(\mathrm{id}_M)_p(v) = \mathrm{id}_{T_pM}(v)$ for all $v \in T_pM$. Hence $d(\mathrm{id}_M)_p = \mathrm{id}_{T_pM}$. □

Consequently, $(df^{-1})_{f(p)} \circ df_p = d(\mathrm{id}_M)_p = \mathrm{id}_{T_pM}$. Similarly, considering $f \circ f^{-1} = \mathrm{id}_N : N \to N$, we also conclude that $df_p \circ (df^{-1})_{f(p)} = \mathrm{id}_{T_{f(p)}N}$. So it follows that $df_p : T_pM \to T_{f(p)}N$ is an isomorphism. In particular, $\dim M = \dim T_pM = \dim T_{f(p)}N = \dim N$.

Solution to Exercise 2.15

Let $p \in f^{-1}\{q\}$. As the linear transformation $f'(p) : \mathbb{R}^m \to \mathbb{R}^n$ is surjective, it follows from the rank-nullity theorem that $\dim(\ker f'(p)) = m - n$. Let $\{e_1, \cdots, e_{m-n}\}$ be an orthonormal basis for $\ker f'(p)$, and extend it to an orthonormal basis $\{e_1, \cdots, e_{m-n}, f_1, \cdots, f_n\}$ for $\mathbb{R}^m$. (This is possible by starting with a basis for $\ker f'(p)$, extending it to a basis for $\mathbb{R}^m$, and using Gram-Schmidt orthogonalisation.) Let $\pi_p : \mathbb{R}^m \to \mathbb{R}^{m-n}$ be the 'orthogonal projection' onto $\ker f'(p)$, that is, $\pi_p(x) = (\langle x, e_1\rangle, \cdots, \langle x, e_{m-n}\rangle)$ for all $x \in \mathbb{R}^m$. Define $g : \mathbb{R}^m \to \mathbb{R}^n \times \mathbb{R}^{m-n}$ by $g(x) = (f(x), \pi_p(x))$, $x \in \mathbb{R}^m$. Then $g'(p) : \mathbb{R}^m \to \mathbb{R}^m$ is the linear map $(g'(p))h = (f'(p)h, \pi_p h)$, $h \in \mathbb{R}^m$. We claim that $g'(p)$ is invertible. It is enough to show that $g'(p)$ is injective. If $g'(p)h = 0$ for some $h \in \mathbb{R}^m$, then $f'(p)h = 0$ and $\pi_p h = 0$, and so we get $h \in \ker f'(p) \cap (\ker f'(p))^\perp = \{0\}$. Now the C^∞ inverse function theorem implies that there exist open sets $U \subset \mathbb{R}^m$ and $V \subset \mathbb{R}^n$ such that $p \in U$, $g(p) \in V$, and the C^∞ map $\varphi := g|_U : U \to V$ is bijective, with a C^∞ inverse $(g|_U)^{-1} : V \to U$. Then (U, φ) is an admissible chart for $\mathbb{R}^m$ (with the standard smooth structure). Moreover,

$$\varphi(U \cap M) = (\{q\} \times \mathbb{R}^{m-n}) \cap V = \{q\} \times \widetilde{U},$$

where $\widetilde{U} := \{y \in \mathbb{R}^{m-n} : (q, y) \in V\}$. Then $\widetilde{U}$ is open in $\mathbb{R}^{m-n}$. (Indeed, if $y_0 \in \widetilde{U}$, then $(q, y_0) \in V$. As V is open, there exists an $r > 0$ such that $\{z \in \mathbb{R}^m : \|z - (q, y_0)\|_2 < r\} \subset V$. Then $B := \{y \in \mathbb{R}^{m-n} : \|y - y_0\| < r\} \subset \widetilde{U}$ because for each $y \in B$, we have $\|(q, y) - (q, y_0)\|_2 = \|y - y_0\|_2 < r$, showing that $(q, y) \in V$, and hence $y \in \widetilde{U}$.) Let $\widetilde{\varphi}(x) := \varphi(x) - (q, 0)$, $x \in U$. Then $(U, \widetilde{\varphi})$ is admissible for $\mathbb{R}^m$, and $\widetilde{\varphi}(U \cap M) = \{0\} \times \widetilde{U} \subset \{0\} \times \mathbb{R}^{m-n}$. Hence (U, φ) is an allowed chart for M (see Exercise 1.25, p.15). Thus M is a submanifold of $\mathbb{R}^m$ of dimension $m - n$.

Solution to Exercise 2.16

The map $\mathbb{R}^{n\times n} \ni [a_{ij}] \mapsto \det[a_{ij}] \in \mathbb{R}$ is C^∞, as it is a polynomial function of the arguments a_{ij}. Let E_{ij} denote the matrix with the entry in the ith row and j^{th} column equal to 1, and all other entries zero. Then it is clear that $\det(I+tE_{ij}) = 1$ whenever $i \neq j$, and $\det(I+tE_{ii}) = 1+t$. Hence for the map $[a_{ij}] \mapsto \det[a_{ij}]$, the partial derivatives at I are given by

$$\frac{\partial \det}{\partial a_{ij}}(I) = \begin{cases} 0 & \text{if } i \neq j, \\ 1 & \text{if } i = j. \end{cases}$$

Consequently, $(\det' I)(h) = h_{11} + \cdots + h_{nn} = \operatorname{trace}(h)$ for all $h \in \mathbb{R}^{n\times n}$.

Let $A \in \mathrm{SL}_n(\mathbb{R})$. We claim $(\det' A)(h) = \operatorname{trace}(A^{-1}h)$ for all $h \in \mathbb{R}^{n\times n}$. Let $\|\!\cdot\!\|$ be the operator norm on $\mathbb{R}^{n\times n}$ (induced by equipping $\mathbb{R}^n$ with the Euclidean norm $\|\cdot\|$), i.e., $\|h\| = \sup\{\|hx\| : x \in \mathbb{R}^n,\ \|x\| \leqslant 1\}$. Let $\epsilon > 0$. Then $\epsilon' := \epsilon/\|A^{-1}\| > 0$. As $\det' I = \operatorname{trace}$, there exists a $\delta > 0$ such that for all $h \in \mathbb{R}^{n\times n}$ satisfying $\|h\| < \delta$, $|\det(I+h) - 1 - \operatorname{trace}(h)| < \epsilon'\|h\|$. For $h \in \mathbb{R}^{n\times n}$ satisfying $0 < \|h\| < \delta\|A^{-1}\|^{-1}$, we have $\|A^{-1}h\| \leqslant \|A^{-1}\|\,\|h\| < \delta$, and so

$$\begin{aligned} |\det(A+h) - 1 - \operatorname{trace}(A^{-1}h)| &= |(\det A)\det(I+A^{-1}h) - 1 - \operatorname{trace}(A^{-1}h)| \\ &= |\det(I+A^{-1}h) - 1 - \operatorname{trace}(A^{-1}h)| \\ &< \epsilon'\|A^{-1}h\| < \epsilon'\|A^{-1}\|\,\|h\| = \epsilon\|h\|, \end{aligned}$$

proving the claim.

For $A \in \mathrm{SL}_n(\mathbb{R})$, we claim that $\det' A = \operatorname{trace}(A^{-1}\cdot)$ is a surjective map. As $\det' A : \mathbb{R}^{n\times n} \to \mathbb{R}$ is a linear map, it suffices to show that the range contains a nonzero number. In fact, $(\det' A)(A) = \operatorname{trace} I = n \neq 0$. Since $\mathrm{SL}_n(\mathbb{R}) = \det^{-1}\{1\}$, it follows by Exercise 2.15 that $\mathrm{SL}_n(\mathbb{R})$ is a submanifold of dimension $n^2 - 1$.

Solution to Exercise 2.17

We have

$$\begin{aligned} \omega'_i &= \omega \partial_{x'^i,p} = (\omega_j (dx^j)_p)\partial_{x'^i,p} = \omega_j (dx^j)_p \partial_{x'^i,p} \\ &= \omega_j \partial_{x'^i,p} x^j = \omega_j \frac{\partial (x^j \circ \mathbf{x}'^{-1})}{\partial u^i}(\mathbf{x}'(p)) = \omega_j \frac{\partial x^j}{\partial x'^i}(\boldsymbol{\xi}'). \end{aligned}$$

Solution to Exercise 2.18

With $\mathbf{x}'(p) = (x,y,z)$ for $p = (x,y,z) \in U' = \mathbb{R}^3$, and $\boldsymbol{\xi} = \mathbf{x}(p)$,

$$\left[\frac{\partial x'^i}{\partial x^j}(\boldsymbol{\xi})\right] = \left[\frac{\partial (x'^i \circ \mathbf{x}^{-1})}{\partial u^j}(\mathbf{x}(p))\right] = \begin{bmatrix} (\cos\phi)\sin\theta & r(\cos\phi)\cos\theta & -r(\sin\phi)\sin\theta \\ (\sin\phi)\sin\theta & r(\sin\phi)\cos\theta & r(\cos\phi)\sin\theta \\ \cos\theta & -r\sin\theta & 0 \end{bmatrix}.$$

So multiplying from the left by the row vectors $(1,0,0)$, $(0,1,0)$, $(0,0,1)$, respectively, we can read off the claimed relations.

Solution to Exercise 2.19

We have the entry in the ith row and k^{th} column of AB is given by

$$\begin{aligned}
A^i_j B^j_k &= \frac{\partial(x'^i \circ \mathbf{x}^{-1})}{\partial u^j}(\mathbf{x}(p))\, \frac{\partial(x^j \circ \mathbf{x}'^{-1})}{\partial u^k}(\mathbf{x}'(p)) \\
&= \frac{\partial(x'^i \circ \mathbf{x}^{-1})}{\partial u^j}((\mathbf{x} \circ \mathbf{x}'^{-1})(\mathbf{x}'(p))\, \frac{\partial(x^j \circ \mathbf{x}'^{-1})}{\partial u^k}(\mathbf{x}'(p)) \\
&= \left.\frac{\partial[i^{\text{th}} \text{ component of } \mathbf{u} \mapsto (\mathbf{x}' \circ \mathbf{x}^{-1})((\mathbf{x} \circ \mathbf{x}'^{-1})(\mathbf{u}))]}{\partial u^k}\right|_{\mathbf{u}=\mathbf{x}'(p)} \\
&= \left.\frac{\partial u^i}{\partial u^k}\right|_{\mathbf{u}=\mathbf{x}'(p)} = \delta^i_k.
\end{aligned}$$

Hence $AB = I_m$. As A is an $m \times m$ matrix, it follows that also $BA = I_m$.

Solution to Exercise 2.20

For any $v \in T_pM$,

$$\begin{aligned}
d(f \cdot g)_p(v) &= v(f \cdot g) = f(p)vg + g(p)vf = f(p)dg_pv + g(p)df_pv \\
&= (f(p)dg_p + g(p)df_p)(v).
\end{aligned}$$

Hence $d(f \cdot g)_p = f(p)dg_p + g(p)df_p$.

Solution to Exercise 2.21

Consider the chart $(\mathbb{R}, u \mapsto u)$ on $\mathbb{R}$. By Theorem 2.2, the matrix for the linear transformation $df_t : T_t\mathbb{R} \mapsto T_{f(t)}\mathbb{R}^2$ with respect to the bases $\{\partial_{u,t}\}$ and $\{\partial_{x,f(t)}, \partial_{y,f(t)}\}$ is given by the Jacobian matrix of $t \mapsto \begin{bmatrix}\cos t & \sin t\end{bmatrix}^{\mathrm{t}}$ at t, which is $\begin{bmatrix}-\sin t & \cos t\end{bmatrix}^{\mathrm{t}}$. So $df_t\partial_{u,t} = -(\sin t)\partial_{x,f(t)} + (\cos t)\partial_{y,f(t)}$. Thus

$$\begin{aligned}
&((df_t)^*\omega)(\partial_{u,t}) \\
&= \omega(df_t(\partial_{u,t})) = \omega(-(\sin t)\partial_{x,f(t)} + (\cos t)\partial_{y,f(t)}) \\
&= (-(\sin t)(dx)_{f(t)} + (\cos t)(dy)_{f(t)})(-(\sin t)\partial_{x,f(t)} + (\cos t)\partial_{y,f(t)}) \\
&= (-\sin t)(-\sin t)(dx)_{f(t)}\partial_{x,f(t)} + (\cos t)(\cos t)(dy)_{f(t)}\partial_{y,f(t)} \\
&\quad -(\sin t)(\cos t)(dx)_{f(t)}\partial_{y,f(t)} + (\cos t)(-(\sin t))(dy)_{f(t)}\partial_{x,f(t)} \\
&= (-\sin t)(-\sin t)\partial_{x,f(t)}x + (\cos t)(\cos t)\partial_{y,f(t)}y \\
&\quad -(\sin t)(\cos t)\partial_{y,f(t)}x + (\cos t)(-(\sin t))\partial_{x,f(t)}y \\
&= (-\sin t)(-\sin t)1 + (\cos t)(\cos t)1 - (\sin t)(\cos t)0 + (\cos t)(-(\sin t))0 \\
&= (-\sin t)(-\sin t) + (\cos t)(\cos t) = 1 = \partial_{u,t}u = (du)_t(\partial_{u,t}),
\end{aligned}$$

and so $(df_t)^*\omega = (du)_t$.

Chapter 3

Solution to Exercise 3.1

Let $(p, v) \in TM$. Let (U, φ) be any admissible chart for M such that $p \in U$. Let $(\mathbf{U}, \boldsymbol{\varphi})$ be the lift of (U, φ). Clearly, $\pi(\mathbf{U}) = U$. If $\varphi = (x^1, \cdots, x^m)$, $(\boldsymbol{\alpha}, \boldsymbol{\beta}) \in \boldsymbol{\varphi}(\mathbf{U})$, and $\boldsymbol{\beta} = (\beta^1, \cdots, \beta^m)$, then

$$(\varphi \circ \pi \circ \boldsymbol{\varphi}^{-1})(\boldsymbol{\alpha}, \boldsymbol{\beta}) = (\varphi \circ \pi)(\varphi^{-1}\boldsymbol{\alpha}, \beta^i \partial_{x^i, \varphi^{-1}\boldsymbol{\alpha}}) = \varphi(\varphi^{-1}\boldsymbol{\alpha}) = \boldsymbol{\alpha}.$$

So $\varphi \circ \pi \circ \boldsymbol{\varphi}^{-1} : \boldsymbol{\varphi}(\mathbf{U}) \to \mathbb{R}^m$ is the map $(\boldsymbol{\alpha}, \boldsymbol{\beta}) \mapsto \boldsymbol{\alpha}$, which is C^∞. Hence $\pi : TM \to M$ is smooth.

Solution to Exercise 3.2

Let $(p, v) \in TM$. Since f is smooth, there are admissible charts (U, φ), (V, ψ), for M, N, respectively, containing $p, f(p)$, respectively, such that $f(U) \subset V$, and $\psi \circ f \circ \varphi^{-1} \in C^\infty(\varphi(U))$. Let $(\mathbf{U}, \boldsymbol{\varphi})$, $(\mathbf{V}, \boldsymbol{\psi})$ be lifts of $(U, \varphi), (V, \psi)$, respectively. Then the charts $(\mathbf{U}, \boldsymbol{\varphi})$, $(\mathbf{V}, \boldsymbol{\psi})$ are admissible for TM, respectively TN, and for each $(q, w) \in \mathbf{U}$, we have

$$\mathbf{df}(q, w) = (f(q), (df)_q w) \in \{f(q)\} \times T_{f(q)}N \subset \mathbf{V}.$$

Let M, N have dimensions m, n, respectively, and denote the component functions of φ, ψ by $(x^1, \cdots, x^m)$, $(y^1, \cdots, y^n)$, respectively. For $\boldsymbol{\alpha} \in \varphi(U)$, the components of the vector $(df)_{\varphi^{-1}\boldsymbol{\alpha}} \partial_{x^i, \varphi^{-1}\boldsymbol{\alpha}}$ with respect to the basis $\partial_{y^j, f(\varphi^{-1}\boldsymbol{\alpha})}$, $1 \leqslant j \leqslant n$, are given by

$$((df)_{\varphi^{-1}\boldsymbol{\alpha}} \partial_{x^i, \varphi^{-1}\boldsymbol{\alpha}})(y^j) = \partial_{x^i, \varphi^{-1}\boldsymbol{\alpha}}(y^j \circ f) \qquad (1 \leqslant j \leqslant n).$$

Using the linearity of $df_{\varphi^{-1}\boldsymbol{\alpha}}$, we get for $(\boldsymbol{\alpha}, \boldsymbol{\beta}) \in \boldsymbol{\varphi}(\mathbf{U})$ that

$$\begin{aligned}
&(\boldsymbol{\psi} \circ \mathbf{df} \circ \boldsymbol{\varphi}^{-1})(\boldsymbol{\alpha}, \boldsymbol{\beta}) \\
&= (\boldsymbol{\psi} \circ \mathbf{df})(\varphi^{-1}\boldsymbol{\alpha}, \beta^i \partial_{x^i, \varphi^{-1}\boldsymbol{\alpha}}) = \boldsymbol{\psi}\big(f(\varphi^{-1}\boldsymbol{\alpha}), \beta^i \partial_{x^i, \varphi^{-1}\boldsymbol{\alpha}}(y^j \circ f) \partial_{y^j, f(\varphi^{-1}\boldsymbol{\alpha})}\big) \\
&= \Big((\psi \circ f \circ \varphi^{-1})\boldsymbol{\alpha}, \beta^i \frac{\partial(\psi \circ f \circ \varphi^{-1})^1}{\partial u^i}(\boldsymbol{\alpha}), \cdots, \beta^i \frac{\partial(\psi \circ f \circ \varphi^{-1})^n}{\partial u^i}(\boldsymbol{\alpha})\Big),
\end{aligned}$$

and so the map $\boldsymbol{\varphi}(\mathbf{U}) \ni (\boldsymbol{\alpha}, \boldsymbol{\beta}) \mapsto (\boldsymbol{\psi} \circ \mathbf{df} \circ \boldsymbol{\varphi}^{-1})(\boldsymbol{\alpha}, \boldsymbol{\beta})$ is C^∞. Thus $\mathbf{df} : TM \to TN$ is smooth.

Solution to Exercise 3.3

For all $p \in M$,

$$\begin{aligned}
(V(fg))(p) &= V_p(fg) = f(p)V_p g + g(p)V_p f \\
&= f(p)(Vg)(p) + g(p)(Vf)(p) \\
&= (f \cdot (Vg))(p) + (g \cdot (Vf))(p) \\
&= (f \cdot (Vg) + g \cdot (Vf))(p).
\end{aligned}$$

Since $p \in M$ was arbitrary, $V(fg) = f \cdot (Vg) + g \cdot (Vf)$.

Also, $V : C^\infty(M) \to C^\infty(M)$ is $\mathbb{R}$-linear, as $V(f+c\,g) = Vf + c\,Vg$ for $f, g \in C^\infty(M)$ and $c \in \mathbb{R}$: Indeed for all $p \in M$, we have

$$\begin{aligned}(V(f+cg))(p) &= V_p(f+cg) = V_p(f) + cV_p(g)\\ &= (Vf)(p) + c(Vg)(p)\\ &= (Vf+cVg)(p).\end{aligned}$$

Solution to Exercise 3.4

We will first show that f is locally constant. As M is connected, it will then follow that f is also globally constant. Let $p \in M$, $(U, \mathbf{x})$ be an admissible chart containing p, and $\chi \in C^\infty(M)$ be a bump function such that $\chi \equiv 1$ in a neighbourhood $\widetilde{U} \subset U$ of p, and $\chi \equiv 0$ outside a neighbourhood W, such that $\overline{W} \subset U$. Then for $1 \leqslant i \leqslant m$, $V := \chi \partial_{x^i} \in T_0^1 M$, and $\chi \partial_{x^i} f = Vf = 0$. So for all $q \in \widetilde{U}$,

$$0 = (\chi \partial_{x^i} f)(q) = \chi(q) \partial_{x^i, q} f = 1 \frac{\partial (f \circ \mathbf{x}^{-1})}{\partial u^i}(\mathbf{x}(q)),\ 1 \leqslant i \leqslant m.$$

Thus $f \circ \mathbf{x}^{-1}$ is constant on $\mathbf{x}(\widetilde{U})$, and so f is constant in $\widetilde{U}$.

Solution to Exercise 3.5

We have

$$\begin{aligned}(fV)_p &= (x^2 y)|_{(1,1)} V_p = 1 V_p = V_p\\ &= (xy)|_{(1,1)} \partial_{x,p} + x^2|_{(1,1)} \partial_{y,p}\\ &= 1 \partial_{x,p} + 1 \partial_{y,p} = \partial_{x,p} + \partial_{y,p}.\end{aligned}$$

Also, $(Vf)(p) = V_p f = (\partial_{x,p} + \partial_{y,p})(x^2 y) = (2xy + x^2)|_{(1,1)} = 2 + 1 = 3$. Finally, we have

$$\begin{aligned}(df)_p(V_p) &= (df)_p(\partial_{x,p} + \partial_{y,p})\\ &= ((2xy)|_{(1,1)} + (x^2)|_{(1,1)}) \partial_{u, f(p)}\\ &= (2+1)\partial_{u,1} = 3\,\partial_{u,1}.\end{aligned}$$

Solution to Exercise 3.6

We will use Theorem 3.1 (p.40), and check that for all $g \in C^\infty(N)$, we have $(f_* V) g \in C^\infty(N)$. For $q \in N$, we have

$$\begin{aligned}((f_* V)g)(q) &= (f_* V)_q g = ((df)_{f^{-1}q} V_{f^{-1}q})(g)\\ &= V_{f^{-1}q}(g \circ f) = (V(g \circ f))(f^{-1}q)\\ &= ((V(g \circ f)) \circ f^{-1})(q).\end{aligned}$$

So $(f_* V)g = (V(g \circ f)) \circ f^{-1}$, which is smooth, since firstly, f is a diffeomorphism, and secondly, as $V \in T_0^1 M$ and $g \circ f \in C^\infty(M)$, by Theorem 3.1, $(V(g \circ f)) \in C^\infty(M)$.

Solution to Exercise 3.7

Suppose that $V \in T_0^1 G$ is left-invariant. Then for all $q \in G$, we have

$$((L_p)_* V)_q = ((L_p)_* V)_{L_p(p^{-1}q)} = (dL_p)_{p^{-1}q} V_{p^{-1}q} = V_{p(p^{-1}q)} = V_q,$$

and so $(L_p)_* V = V$.

On the other hand, if $(L_p)_* V = V$, then for all $q \in G$, we have, by evaluating both sides at the point $L_p q \in G$, that

$$(dL_p)_q V_q = ((L_p)_* V)_{L_p q} = V_{L_p q} = V_{p \cdot q}.$$

$V = (G \ni p \mapsto (dL_p)_e v)$ is left-invariant: First, note that $L_p \circ L_q = L_{p \cdot q}$. Indeed, for all $r \in G$, $(L_p \circ L_q) r = L_p(q \cdot r) = p \cdot (q \cdot r) = (p \cdot q) \cdot r = L_{p \cdot q} r$. By Exercise 2.13 (p.31), $(d(L_p \circ L_q))_e = (dL_p)_q (dL_q)_e$. V is left-invariant as

$$(dL_p)_q V_q = (dL_p)_q (dL_q)_e v = (d(L_p \circ L_q))_e v = (dL_{p \cdot q})_e v = V_{p \cdot q}.$$

$\mathbf{I}$ is surjective: Let $V \in \mathfrak{g}$. Then $V_e \in T_e G$, and $(\mathbf{I} V_e)_p = (dL_p)_e V_e = V_{p \cdot e}$ (as V is left-invariant) for all $p \in G$. So $(\mathbf{I} V_e)_p = V_{p \cdot e} = V_p$ for all $p \in G$, that is, $\mathbf{I} V_e = V$.

$\mathbf{I}$ is injective: Let $v \in T_e G$ and $\mathbf{I} v = 0 \in T_0^1 G$. In particular, we have that $(\mathbf{I} v)_e = 0 \in T_e G$, and so $0 = (\mathbf{I} v)_e = (dL_e)_e v = (d(\mathrm{id}_G))_e v = \mathrm{id}_{T_e G} v = v$. (Here we used $L_e q = e \cdot q = q = \mathrm{id}_G q$ for all $q \in G$.)

Solution to Exercise 3.8

(1) We claim that for any matrix $A \in \mathbb{C}^{n \times n}$, $\det(e^A) = e^{\operatorname{trace} A}$. First, we show by induction on n that there exists an invertible matrix P such that $P^{-1} A P = U$, where U is an upper triangular matrix. Indeed, A has at least one eigenvector v with eigenvalue λ, so that when we extend it to a basis $\{v, v_2, \cdots, v_n\}$, then with P defined to be the matrix with the columns $v, v_2, \cdots, v_n$, we get $AP = \begin{bmatrix} \lambda v & Av_2 & \cdots & Av_n \end{bmatrix}$, and so

$$P^{-1} A P = P^{-1} \begin{bmatrix} v & Av_2 & \cdots & Av_n \end{bmatrix} \begin{bmatrix} \lambda & \\ & I_{n-1} \end{bmatrix} = \begin{bmatrix} 1 & * \\ \mathbf{0} & * \end{bmatrix} \begin{bmatrix} \lambda & \\ & I_{n-1} \end{bmatrix} = \begin{bmatrix} \lambda & * \\ \mathbf{0} & A_{n-1} \end{bmatrix}$$

and we repeat the process with the matrix $A_{n-1} \in \mathbb{R}^{(n-1) \times (n-1)}$ in the bottom right corner of the right-hand side. Next, it is clear that the eigenvalues of A and U are identical, but since U is upper triangular, the eigenvalues of U are its diagonal entries, say $\lambda_1, \cdots, \lambda_n$. Thanks to the upper triangular structure of U, we have that e^U is upper triangular too, with diagonal entries $e^{\lambda_1}, \cdots, e^{\lambda_n}$. Thus

$$\det e^A = \det(P e^U P^{-1}) = \det(e^U) = e^{\lambda_1} \cdots e^{\lambda_n} = e^{\lambda_1 + \cdots + \lambda_n} = e^{\operatorname{trace} A}.$$

Next, if $\alpha \in \mathfrak{sl}_n(\mathbb{R})$, then viewing it as a complex matrix, and applying the result above, we get $\det(e^\alpha) = e^{\operatorname{trace} \alpha} = e^0 = 1$, and as e^α has real entries, $e^\alpha \in \mathrm{SL}_n(\mathbb{R})$.

(2) Recall from the solution on p.392 to Exercise 2.16, that $\det' I = \text{trace}$. As $(\det \circ \gamma)(t) = 1$ for all $t \in (-\epsilon, \epsilon)$, we have that the derivative of the constant function $\det \circ \gamma$ at 0 is 0, that is,

$$0 = \frac{d(\det \circ \gamma)}{dt}(0) = (\det'(\gamma(0)))(\dot{\gamma}(0)) = (\det'(I))(\dot{\gamma}(0)) = \text{trace}(\dot{\gamma}(0)).$$

(3) Let $v \in T_{I_n}(\text{SL}_n(\mathbb{R}))$. Then there exists an $\epsilon > 0$ and a smooth curve $\gamma : (-\epsilon, \epsilon) \to \text{SL}_n(\mathbb{R})$ such that $\gamma(0) = I_n$ and $v_{\gamma, I_n} = v$. As the inclusion map $\iota : \text{SL}_n(\mathbb{R}) \to \mathbb{R}^{n \times n}$ is smooth, $\iota \circ \gamma : (-\epsilon, \epsilon) \to \mathbb{R}^{n \times n}$ is a smooth curve, and moreover, $(\iota \circ \gamma)(0) = I_n$ and $\det((\iota \circ \gamma)(t)) = 1$ for all $t \in (-\epsilon, \epsilon)$. Thus $[v_{\iota \circ \gamma, I_n}] \in \mathfrak{sl}_n(\mathbb{R})$. But $v_{\iota \circ \gamma, I_n} = (d\iota)_{I_n}(v_{\gamma, I_n})$, because for all $f \in C^\infty(\mathbb{R}^{n \times n})$, we have

$$((d\iota)_{I_n} v)(f) = ((d\iota)_{I_n} v_{\gamma, I_n})(f) = v_{\gamma, I_n}(f \circ \iota) = \frac{d(f \circ \iota \circ \gamma)}{dt}(0) = v_{\iota \circ \gamma, I_n}(f).$$

We have $\dim T_{I_n}(\text{SL}_n(\mathbb{R})) = \dim \text{SL}_n(\mathbb{R}) = n^2 - 1$ by Exercise 2.16 (p.33). As $\mathfrak{sl}_n(\mathbb{R})$ is the kernel of the nonzero linear map $\text{trace} : \mathbb{R}^{n \times n} \to \mathbb{R}$, it follows by the rank-nullity theorem that $\dim \mathfrak{sl}_n(\mathbb{R}) = n^2 - 1$. The map $T_{I_n}(\text{SL}_n(\mathbb{R})) \ni v \mapsto [(d\iota)_{I_n}(v)] \in \mathfrak{sl}_n(\mathbb{R})$ is clearly linear. To show that it is an isomorphism, it is suffices to show that it is surjective. To this end, let $\alpha \in \mathfrak{sl}_n(\mathbb{R})$, and consider the curve $\gamma : \mathbb{R} \to \text{SL}_n(\mathbb{R})$ given by $\gamma(t) = e^{t\alpha}$, $t \in \mathbb{R}$. Then $\gamma(0) = I_n$ and

$$[(d\iota)_{I_n}(v_{\gamma, I_n})] = [v_{\iota \circ \gamma, I_n}] = \frac{d(e^{t\alpha})}{dt}(0) = \alpha.$$

Solution to Exercise 3.9

For all $p \in M$, we have

$$\begin{aligned}(((f_* V)\varphi) \circ f)(p) &= ((f_* V)\varphi)(f(p)) = (f_* V)_{f(p)}\varphi \\ &= ((df)_p V_p)\varphi = V_p(\varphi \circ f) = (V(\varphi \circ f))(p),\end{aligned}$$

and so $((f_* V)\varphi) \circ f = V(\varphi \circ f)$.

Solution to Exercise 3.10

For $p \in \mathbb{R}^2$, using Thm. 2.2 (p.30), we determine the matrix of the linear map $(d\psi_{-t})_p$ with respect to the ordered bases $(\partial_{x,p}, \partial_{y,p})$ and $(\partial_{x,\psi_{-t}p}, \partial_{y,\psi_{-t}p})$:

$$[(d\psi_{-t})_p] = \begin{bmatrix} \cos t & -\sin t \\ \sin t & \cos t \end{bmatrix}.$$

Then, as $\partial_{y,p}$ is represented by the column vector $\mathbf{e}_2 \in \mathbb{R}^2$ with respect to the ordered basis $(\partial_{x,p}, \partial_{y,p})$, the representation of $(d\psi_{-t})_p \partial_{y,p}$ with respect to the basis $(\partial_{x,\psi_{-t}p}, \partial_{y,\psi_{-t}p})$ is

$$\begin{bmatrix} \cos t & -\sin t \\ \sin t & \cos t \end{bmatrix} \begin{bmatrix} 0 \\ 1 \end{bmatrix} = \begin{bmatrix} -\sin t \\ \cos t \end{bmatrix}.$$

So
$$((\psi_{-t})_*V)_{\psi_{-t}p} = (d\psi_{-t})_p V_p = (-\sin t)\,\partial_{x,\psi_{-t}p} + (\cos t)\,\partial_{y,\psi_{-t}p}$$
$$= ((-\sin t)\,\partial_x + (\cos t)\,\partial_y)_{\psi_{-t}p}.$$
Hence, as $\psi_{-t} : \mathbb{R}^2 \to \mathbb{R}^2$ is surjective, $(\psi_{-t})_*V = (-\sin t)\,\partial_x + (\cos t)\,\partial_y$.

Solution to Exercise 3.11

We will use Exercise 3.9 (p.43). For all $\varphi \in C^\infty(N)$, and $q \in N$,
$$\begin{aligned}((f_*[X,Y])\varphi)(q) &= ((f_*[X,Y])\varphi)(f(f^{-1}q)) = (f_*[X,Y])_{f(f^{-1}q)}\varphi \\ &= ((df)_{f^{-1}q}[X,Y]_{f^{-1}q})(\varphi) = [X,Y]_{f^{-1}q}(\varphi \circ f) \\ &= X_{f^{-1}q}(Y(\varphi\circ f)) - Y_{f^{-1}q}(X(\varphi \circ f)) \\ &= X_{f^{-1}q}(((f_*Y)\varphi)\circ f) - Y_{f^{-1}q}(((f_*X)\varphi)\circ f) \\ &= ((df)_{f^{-1}q}X_{f^{-1}q})((f_*Y)\varphi) - ((df)_{f^{-1}q}Y_{f^{-1}q})((f_*X)\varphi) \\ &= (f_*X)_q((f_*Y)\varphi) - (f_*Y)_q((f_*X)\varphi) \\ &= ((f_*X)((f_*Y)\varphi) - (f_*Y)((f_*X)\varphi))(q) \\ &= ([f_*X, f_*Y]\varphi)(q),\end{aligned}$$
and so $f_*[X,Y] = [f_*X, f_*Y]$.

Solution to Exercise 3.12

For $p \in G$, $(L_p)_*[V,W] = [(L_p)_*V, (L_p)_*W] = [V,W]$. So $[V,W] \in \mathfrak{g}$.

Solution to Exercise 3.13

For all $g \in C^\infty(M)$,
$$\begin{aligned}[fX,Y]g &= fX(Yg) - Y(f(Xg)) = fX(Yg) - (Yf)(Xg) - fY(Xg) \\ &= f(X(Yg) - Y(Xg)) - ((Yf)X)g = f[X,Y]g - ((Yf)X)g \\ &= (f[X,Y] - (Yf)X)g.\end{aligned}$$
Thus $[fX,Y] = f[X,Y] - (Yf)X$.

In particular, for $M = \mathbb{R}^2$, $X = \partial_x$, $Y = \partial_y$, $f = 1 + y$,
$$\begin{aligned}[(1+y)\partial_x, \partial_y] &= [fX,Y] = f[X,Y] - (Yf)X \\ &= (1+y)[\partial_x, \partial_y] - (\partial_y(1+y))\partial_x \\ &= (1+y)0 - 1\partial_x = -\partial_x.\end{aligned}$$

Solution to Exercise 3.14

We have
$$\begin{aligned}[V,W]x^j &= V(Wx^j) - W(Vx^j) = V(W^i\partial_{x^i}x^j) - W(V^i\partial_{x^i}x^j) \\ &= V(W^i\delta_i^j) - W(V^i\delta_i^j) = V(W^j) - W(V^j) \\ &= V^i\partial_{x^i}W^j - W^i\partial_{x^i}V^j.\end{aligned}$$
Thus $[V,W] = ([V,W]x^j)\partial_{x^j} = (V^i\partial_{x^i}W^j - W^i\partial_{x^i}V^j)\partial_{x^j}$.

Solution to Exercise 3.15

For $p \in \mathrm{GL}_n(\mathbb{R})$ and $f \in C^\infty(\mathrm{GL}_n(\mathbb{R}))$, $V_p f = ((dL_p)_{I_n}(v))(f) = v(f(p\,\cdot))$, and so $V_p(x^{ij}) = ((dL_p)_{I_n}(v))(x^{ij}) = v(p^{ik}x^{kj}) = p^{ik}v^{kj}$. Thus

$$\begin{aligned}[V,W]_{I_n}x^{ij} &= V_{I_n}(Wx^{ij}) - W_{I_n}(Vx^{ij}) = v(x^{ik}w^{kj}) - w(x^{ik}v^{kj}) \\ &= v^{ik}w^{kj} - w^{ik}v^{kj}.\end{aligned}$$

Thus $[V,W]_{I_n} = (v^{ik}w^{kj} - w^{ik}v^{kj})\partial_{x^{ij}} = \alpha^{ij}\partial_{x^{ij}}$, where $[\alpha^{ij}] = [v,w]$.

Solution to Exercise 3.16

Suppose $(p,\omega) \in TM^*$. Then there exists an admissible chart (U,φ) for M containing p. Let the chart $(\mathbf{U},\boldsymbol{\varphi})$ for TM^* be the lift of (U,φ). Clearly, $\pi(\mathbf{U}) = U$. Let $x^1,\cdots,x^m$ be the components of φ. For $(\boldsymbol{\alpha},\boldsymbol{\beta}) \in \boldsymbol{\varphi}(\mathbf{U})$, with $\boldsymbol{\beta} = (\beta_1,\cdots,\beta_m)$, we have

$$(\varphi\circ\pi\circ\boldsymbol{\varphi}^{-1})(\boldsymbol{\alpha},\boldsymbol{\beta}) = (\varphi\circ\pi)(\varphi^{-1}\boldsymbol{\alpha},\beta_i(dx^i)_{\varphi^{-1}\boldsymbol{\alpha}}) = \varphi(\varphi^{-1}\boldsymbol{\alpha}) = \boldsymbol{\alpha},$$

and so $\varphi\circ\pi\circ\boldsymbol{\varphi}^{-1} : \boldsymbol{\varphi}(\mathbf{U}) \to \mathbb{R}^m$ is given by $(\boldsymbol{\alpha},\boldsymbol{\beta}) \mapsto \boldsymbol{\alpha}$, which is clearly C^∞. Hence $\pi : TM^* \to M$ is smooth.

Solution to Exercise 3.17

For all $p \in M$, we have

$$\begin{aligned}(\Omega(V+fW))(p) &= \Omega_p(V+fW)_p = \Omega_p(V_p+f(p)W_p) \\ &= \Omega_pV_p+f(p)\Omega_pW_p = (\Omega V)(p)+f(p)(\Omega W)(p) \\ &= (\Omega V+f(\Omega W))(p).\end{aligned}$$

As $p \in M$ was arbitrary, it follows that $\Omega(V+fW) = \Omega V+f(\Omega W)$.

Solution to Exercise 3.18

For $V \in T_0^1M$ and $p \in M$, $((df)V)(p) = (df)_pV_p = V_pf = (Vf)(p)$.
As $V \in T_0^1M$, the map $p \mapsto (Vf)(p)$ is smooth by Theorem 3.1 (p.40).
So $(df)V \in C^\infty(M)$. By Theorem 3.3 (p.48), we conclude that $df \in T_1^0M$.

Solution to Exercise 3.19

Let $\Omega = x\,dy \in T_1^0\mathbb{R}^2$. Then we have $\Omega\,\partial_x = 0$, $\Omega\,\partial_y = x$, and

$$\partial_x(\Omega\,\partial_y) = \partial_x x = 1 \neq 0 = \partial_y 0 = \partial_y(\Omega\,\partial_x).$$

So Ω cannot be df for any $f \in C^\infty(M)$.

Solution to Exercise 3.20

For all $p \in M$, we have

$$\begin{aligned}(f^*(g\cdot\Omega))_p &= (df_p)^*(g\cdot\Omega)_{f(p)} = (df_p)^*(g(f(p))\,\Omega_{f(p)}) \\ &= g(f(p))\cdot(df_p)^*\Omega_{f(p)} = (g\circ f)(p)\cdot(f^*\Omega)_p \\ &= ((g\circ f)\cdot(f^*\Omega))_p.\end{aligned}$$

So $f^*(g\cdot\Omega) = (g\circ f)\cdot(f^*\Omega)$.

Solution to Exercise 3.21

For all $p \in M$ and all $v \in T_pM$, we have

$$\begin{aligned}(f^*(dg))_p v &= ((df_p)^*(dg)_{f(p)})v = (dg)_{f(p)}((df_p)v)\\ &= ((df_p)v)(g) = v(g \circ f)\\ &= (d(g\circ f))_p v.\end{aligned}$$

Thus $f^*(dg) = d(g \circ f)$.

Solution to Exercise 3.22

The ordinary differential equation describing the integral curve is

$$\frac{du}{dt} = u^2,$$

with the initial condition $u(0) = p = \gamma(0) > 0$. Since

$$\frac{du}{dt} = u^2 \geqslant 0,$$

u is increasing. As $u(0) > 0$, we get $\gamma(t) = u(t) \geqslant u(0) > 0$ for $t \geqslant 0$. Thus the above gives

$$\int_p^{\gamma(t)} \frac{d}{du}\Big(\frac{1}{u}\Big)\, du = -\int_0^t dt.$$

So $\frac{1}{\gamma(t)} - \frac{1}{p} = -t$. Thus $\gamma(t) = \frac{p}{1-pt}$ for $t < \frac{1}{p}$. (As $t \nearrow \frac{1}{p}$, $\gamma(t) \nearrow +\infty$.)

Solution to Exercise 3.23

Suppose that $\gamma_p : \mathbb{R} \to \mathbb{R}^2$ is the integral curve of V with $\gamma_p(0) = p$. Writing $\gamma_p(t) = (x(t), y(t))$, $t \in \mathbb{R}$, we obtain the initial value problem

$$\begin{aligned}\dot{x}(t) &= 1 + y(t), & x(0) &= x_0,\\ \dot{y}(t) &= 0, & y(0) &= y_0.\end{aligned}$$

Integrating both sides of the second equation from 0 to t, $y(t) - y_0 = 0$, i.e., $y(t) = y_0$, $t \in \mathbb{R}$. Inserting this in the first equation, and integrating from 0 to t yields $x(t) - x_0 = (1 + y_0)t$, so that $x(t) = x_0 + (1 + y_0)t$. Thus $\psi_t(p) = \gamma_p(t) = (x_0 + (1 + y_0)t, y_0)$, $t \in \mathbb{R}$. The flow of V consists of the flow maps $\psi_t : \mathbb{R}^2 \to \mathbb{R}^2$ given by $\psi_t(x, y) = (x + (1 + y)t, y)$ for all $(x, y) \in \mathbb{R}^2$ and $t \in \mathbb{R}$.

We have $\psi_{-t}(x, y) = (x - (1 + y)t, y)$, $t \in \mathbb{R}$. The matrix representation of the linear transformation $(d\psi_{-t})_p : T_p\mathbb{R}^2 \to T_{\psi_{-t}p}\mathbb{R}^2$ with respect to the ordered basis $(\partial_{x,p}, \partial_{y,p})$ and $(\partial_{x,f(p)}, \partial_{y,f(p)})$ is

$$\begin{bmatrix} 1 & -t \\ 0 & 1 \end{bmatrix}.$$

Hence the action of $(d\psi_{-t})_p$ on $\partial_{y,p}$ is the multiplication of the above matrix with the vector $\mathbf{e}_2 \in \mathbb{R}^2$. Consequently, $(d\psi_{-t})_p(\partial_{y,p}) = -t\,\partial_{x,\psi_{-t}p} + \partial_{y,\psi_{-t}p}$.

Solution to Exercise 3.24

For all $f \in C^\infty(\mathbb{R}^2)$, we have

$$
\begin{aligned}
&[L_x, L_y]f \\
&= (y\,\partial_z - z\,\partial_y)(z\,\partial_x f - x\,\partial_z f) - (z\,\partial_x - x\,\partial_z)(y\,\partial_z f - z\,\partial_y f) \\
&= y(\partial_z z)(\partial_x f) + \underline{yz\,\partial_z\partial_x f} - \underset{\sim\sim\sim}{xy\,\partial_z^2 f} - 0 - \underset{------}{z^2\partial_y\partial_x f} - 0 + \underset{\cdots\cdots}{zx\,\partial_y\partial_z f} + 0 \\
&\quad -\underline{zy\,\partial_x\partial_z f} - 0 + \underset{------}{z^2\partial_x\partial_y f} + 0 + \underset{\sim\sim\sim}{xy\,\partial_z^2 f} + 0 - x(\partial_z z)(\partial_y f) - \underset{\cdots\cdots}{xz\,\partial_z\partial_y f} \\
&= y\,\partial_x f - x\,\partial_y f = (y\,\partial_x - x\,\partial_y)f = -L_z f.
\end{aligned}
$$

Thus $[L_x, L_y] = -L_z$. Using the cycling sequence $x \to y \to z \to x$ in the formula for L_x, we see that we obtain $L_x \to L_y \to L_z \to L_x$, and so from the above, $[L_y, L_z] = -L_x$ and $[L_z, L_x] = -L_y$. Clearly, $[L_a, L_a] = 0$ for $a = x, y, z$, and $[L_a, L_b] = -[L_b, L_a]$ for all $a, b \in \{x, y, z\}$.

Let $\gamma_p : \mathbb{R} \to \mathbb{R}^3$ be the integral curve of L_x such that $\gamma_p(0) = p = (x_0, y_0, z_0)$. Then with $\gamma_p(t) = (x(t), y(t), z(t))$, $t \in \mathbb{R}$, we obtain the initial value problem

$$
\begin{aligned}
\dot{x}(t) &= 0, & x(0) &= x_0, \\
\dot{y}(t) &= -z(t), & y(0) &= y_0, \\
\dot{z}(t) &= y(t), & z(0) &= z_0.
\end{aligned}
$$

Integrating both sides of the first equation from 0 to t yields $x(t) - x_0 = 0$, that is, $x(t) = x_0$ for all $t \in \mathbb{R}$. From the second and third differential equations, we obtain $\ddot{y} = -\dot{z} = -y$, that is, $\ddot{y} + y = 0$. This has the general solution $y(t) = A\cos t + B\sin t$ for some constants A, B. Since $y(0) = y_0$, we have $A = y_0$. Also, $z(t) = -\dot{y}(t) = -(-A\sin t + B\cos t) = A\sin t - B\cos t$. We have $z_0 = z(0) = -B$. So $\gamma_p(t) = (x_0, y_0\cos t - z_0\sin t, y_0\sin t + z_0\cos t)$, $t \in \mathbb{R}$. Thus the flow of L_x is given by $\{\xi_t : \mathbb{R}^3 \to \mathbb{R}^3, t \in \mathbb{R}\}$, where

$$
\xi_t \begin{bmatrix} x \\ y \\ z \end{bmatrix} = \begin{bmatrix} 1 & & \\ & \cos t & -\sin t \\ & \sin t & \cos t \end{bmatrix} \begin{bmatrix} x \\ y \\ z \end{bmatrix}, \quad \text{for all } \begin{bmatrix} x \\ y \\ z \end{bmatrix} \in \mathbb{R}^3.
$$

We recognize this as a rotation about the x-axis through an angle t, in a sense determined by the right-hand rule: If the thumb points in the positive direction of the x-axis, then the other fingers show the direction of rotation (i.e., looking at the yz-plane from the $x > 0$ half-space in $\mathbb{R}^3$, we will record the rotation as being anticlockwise). By cyclically permuting the x, y, z, we obtain the flows $\{\eta_t : \mathbb{R}^3 \to \mathbb{R}^3, t \in \mathbb{R}\}$, $\{\zeta_t : \mathbb{R}^3 \to \mathbb{R}^3, t \in \mathbb{R}\}$, of L_y, L_z, respectively, where

$$
\eta_t \begin{bmatrix} x \\ y \\ z \end{bmatrix} = \begin{bmatrix} \cos t & & \sin t \\ & 1 & \\ -\sin t & & \cos t \end{bmatrix} \begin{bmatrix} x \\ y \\ z \end{bmatrix}, \quad \zeta_t \begin{bmatrix} x \\ y \\ z \end{bmatrix} = \begin{bmatrix} \cos t & -\sin t & \\ \sin t & \cos t & \\ & & 1 \end{bmatrix} \begin{bmatrix} x \\ y \\ z \end{bmatrix}.
$$

Chapter 4

Solution to Exercise 4.1

Suppose for some $p \in M$, $V_p \neq 0$. Let $(U, \mathbf{x})$ be an admissible chart such that $p \in U$. Then $V_p = v^i \partial_{x^i,p}$, and as $V_p \neq 0$, not all the v^i are zero. Let $i_* \in \{1, \cdots, m\}$ be such that $v^{i_*} \neq 0$. Let χ be a bump function such that $\chi \equiv 1$ in a neighbourhood of p, $\chi \equiv 0$ outside U_0, and $\overline{U_0} \subset U$. Then $f := x^{i_*}\chi \in C^\infty(M)$, and $df \in T_1^0 M$. We have with $\Omega := df$ that $T_V \Omega = \Omega V = df(V)$, and so

$$\begin{aligned}(T_V\Omega)(p) &= (df(V))(p) = (df)_p V_p \\ &= V_p f = V_p(\chi x^{i_*}) \\ &= \chi(p)\, V_p x^{i_*} + x^{i_*}(p)\, V_p\chi = 1\, V_p x^{i_*} + x^{i_*}(p)\cdot 0 \\ &= V_p x^{i_*} = v^i \partial_{x^i,p} x^{i_*} \\ &= v^{i_*} \neq 0.\end{aligned}$$

Thus the function $T_V\Omega$ is not identically zero, a contradiction to $T_V = 0$.

Solution to Exercise 4.2

For $V_1, V_2 \in T_0^1 M$, we have

$$T(V_1+V_2) = (V_1+V_2)(Wg) = V_1(Wg) + V_2(Wg) = T(V_1) + T(V_2).$$

For $V \in T_0^1 M$ and $f \in C^\infty(M)$, we have

$$T(fV) = (fV)(Wg) = f\cdot(V(Wg)) = f\cdot T(V).$$

So T is $C^\infty(M)$-linear. Hence T is a $(0,1)$-tensor field.

Solution to Exercise 4.3

For $V_1, V_2, W \in T_0^1 M$ and $f \in C^\infty(M)$,

$$\begin{aligned}(\Omega\otimes\Theta)(V_1+fV_2, W) &= (\Omega(V_1+fV_2))(\Theta W) = (\Omega(V_1)+f\Omega(V_2))(\Theta W) \\ &= (\Omega(V_1))(\Theta W) + f\Omega(V_2)(\Theta W) \\ &= (\Omega\otimes\Theta)(V_1, W) + f(\Omega\otimes\Theta)(V_2, W).\end{aligned}$$

For $V, W_1, W_2 \in T_0^1 M$ and $f \in C^\infty(M)$,

$$\begin{aligned}(\Omega\otimes\Theta)(V, W_1+fW_2) &= (\Omega V)\Theta(W_1+fW_2) = (\Omega V)(\Theta W_1 + f\Theta W_2) \\ &= (\Omega V)\Theta(W_1) + f\Omega(V)\Theta(W_2) \\ &= (\Omega\otimes\Theta)(V, W_1) + f(\Omega\otimes\Theta)(V, W_2).\end{aligned}$$

So $\Omega\otimes\Theta$ is a $(0,2)$-tensor field.

We have $\Omega\wedge\Theta = \Omega\otimes\Theta - \Theta\otimes\Omega = -(\Theta\otimes\Omega - \Omega\otimes\Theta) = -\Theta\wedge\Omega$.

For $V \in T_0^1 M$,

$$\begin{aligned}(\Omega\wedge\Theta)(V,V) &= (\Omega\otimes\Theta)(V,V) - (\Theta\otimes\Omega)(V,V) \\ &= (\Omega V)(\Theta V) - (\Theta V)(\Omega V) = 0.\end{aligned}$$

Solution to Exercise 4.4

For $V, W \in T_0^1 M$,

$$\begin{aligned}(f(\Omega\otimes\Theta))(V,W) &= f\cdot((\Omega\otimes\Theta)(V,W))\\ &\overset{(*)}{=} f\cdot(\Omega V)\cdot\Theta W\\ &= ((f\Omega)V)\cdot\Theta W\\ &= ((f\Omega)\otimes\Theta)(V,W)\\ &\overset{(*)}{=} (\Omega V)\cdot((f\Theta)W)\\ &= (\Omega\otimes(f\Theta))(V,W),\end{aligned}$$

and so $f(\Omega\otimes\Theta) = (f\Omega)\otimes\Theta = \Omega\otimes f\Theta$.

Also for $V, W \in T_0^1 M$,

$$\begin{aligned}((\Omega_1+\Omega_2)\otimes\Theta)(V,W) &= ((\Omega_1+\Omega_2)V)\cdot\Theta W\\ &= (\Omega_1 V+\Omega_2 V)\cdot\Theta W\\ &= (\Omega_1 V)\Theta W + (\Omega_2 V)\Theta W\\ &= (\Omega_1\otimes\Theta)(V,W) + (\Omega_2\otimes\Theta)(V,W)\\ &= (\Omega_1\otimes\Theta+\Omega_2\otimes\Theta)(V,W), \quad \text{and}\\ (\Theta\otimes(\Omega_1+\Omega_2))(V,W) &= (\Theta V)\cdot((\Omega_1+\Omega_2)W)\\ &= (\Theta V)\cdot(\Omega_1 W+\Omega_2 W)\\ &= (\Theta V)\Omega_1 W + (\Theta V)\Omega_2 W\\ &= (\Theta\otimes\Omega_1)(V,W) + (\Theta\otimes\Omega_2)(V,W)\\ &= (\Theta\otimes\Omega_1+\Theta\otimes\Omega_2)(V,W),\end{aligned}$$

and so

$$\begin{aligned}&(\Omega_1+\Omega_2)\otimes\Theta = \Omega_1\otimes\Theta+\Omega_2\otimes\Theta \text{ and}\\ &\Theta\otimes(\Omega_1+\Omega_2) = \Theta\otimes\Omega_1+\Theta\otimes\Omega_2.\end{aligned}$$

Solution to Exercise 4.5

For $\boldsymbol{\omega}, \boldsymbol{\omega}' \in (\mathbb{R}^3)^*$, $\mathbf{v}, \mathbf{v}', \mathbf{w}, \mathbf{w}' \in \mathbb{R}^3$, and $c \in \mathbb{R}$, we have

$$\begin{aligned}\mathcal{T}(\boldsymbol{\omega}+c\boldsymbol{\omega}',\mathbf{v},\mathbf{w}) &= (\boldsymbol{\omega}+c\boldsymbol{\omega}')(\mathbf{v}\times\mathbf{w}) = \boldsymbol{\omega}(\mathbf{v}\times\mathbf{w}) + c\boldsymbol{\omega}'(\mathbf{v}\times\mathbf{w})\\ &= \mathcal{T}(\boldsymbol{\omega},\mathbf{v},\mathbf{w}) + c\mathcal{T}(\boldsymbol{\omega}',\mathbf{v},\mathbf{w}),\\ \mathcal{T}(\boldsymbol{\omega},\mathbf{v}+c\mathbf{v}',\mathbf{w}) &= \boldsymbol{\omega}((\mathbf{v}+c\mathbf{v}')\times\mathbf{w})\\ &= \boldsymbol{\omega}(\mathbf{v}\times\mathbf{w}+c\mathbf{v}'\times\mathbf{w})\\ &= \boldsymbol{\omega}(\mathbf{v}\times\mathbf{w}) + c\boldsymbol{\omega}(\mathbf{v}'\times\mathbf{w})\\ &= \mathcal{T}(\boldsymbol{\omega},\mathbf{v},\mathbf{w}) + c\mathcal{T}(\boldsymbol{\omega},\mathbf{v}',\mathbf{w}), \quad \text{and}\\ \mathcal{T}(\boldsymbol{\omega},\mathbf{v},\mathbf{w}+c\mathbf{w}') &= \boldsymbol{\omega}(\mathbf{v}\times(\mathbf{w}+c\mathbf{w}')) = \boldsymbol{\omega}(\mathbf{v}\times\mathbf{w}+c\mathbf{v}\times\mathbf{w}')\\ &= \boldsymbol{\omega}(\mathbf{v}\times\mathbf{w}) + c\boldsymbol{\omega}(\mathbf{v}\times\mathbf{w}')\\ &= \mathcal{T}(\boldsymbol{\omega},\mathbf{v},\mathbf{w}) + c\mathcal{T}(\boldsymbol{\omega},\mathbf{v},\mathbf{w}').\end{aligned}$$

Solution to Exercise 4.6

Let τ, τ', τ'' be (r,s)-, (r',s')-, (r'',s'')-tensors, respectively, on V. For all $\omega^1, \cdots, \omega^{r+r'+r''} \in V^*$ and all $v_1, \cdots, v_{s+s'+s''} \in V$, we have

$$\begin{aligned}
&(\tau \otimes (\tau' \otimes \tau''))(\omega^1, \cdots, \omega^{r+r'+r''}, v_1, \cdots, v_{s+s'+s''}) \\
&= \tau(\omega^1, \cdots, \omega^r, v_1, \cdots, v_s)(\tau' \otimes \tau'')(\omega^{r+1}, \cdots, \omega^{r+r'+r''}, v_{s+1}, \cdots, v_{s+s'+s''}) \\
&= \tau(\omega^1, \cdots, \omega^r, v_1, \cdots, v_s) \cdot (\tau'(\omega^{r+1}, \cdots, \omega^{r+r'}, v_{s+1}, \cdots, v_{s+s'}) \cdot \\
&\quad \tau''(\omega^{r+r'+1}, \cdots, \omega^{r+r'+r''}, v_{s+s'+1}, \cdots, v_{s+s'+s''})) \\
&= (\tau(\omega^1, \cdots, \omega^r, v_1, \cdots, v_s) \cdot \tau'(\omega^{r+1}, \cdots, \omega^{r+r'}, v_{s+1}, \cdots, v_{s+s'})) \cdot \\
&\quad \tau''(\omega^{r+r'+1}, \cdots, \omega^{r+r'+r''}, v_{s+s'+1}, \cdots, v_{s+s'+s''}) \\
&= (\tau \otimes \tau')(\omega^1, \cdots, \omega^{r+r'}, v_1, \cdots, v_{s+s'}) \cdot \\
&\quad \tau''(\omega^{r+r'+1}, \cdots, \omega^{r+r'+r''}, v_{s+s'+1}, \cdots, v_{s+s'+s''}) \\
&= ((\tau \otimes \tau') \otimes \tau'')(\omega^1, \cdots, \omega^{r+r'+r''}, v_1, \cdots, v_{s+s'+s''}).
\end{aligned}$$

Thus $\tau \otimes (\tau' \otimes \tau'') = (\tau \otimes \tau') \otimes \tau''$.

Solution to Exercise 4.7

First, we show the spanning property. Let τ be an (r,s)-tensor on V. For any $v \in V$, we have $v = \epsilon^i(v)e_i$, and for $\omega \in V^*$, $\omega = \omega(e_i)\epsilon^i$. Thus for $\omega^1, \cdots, \omega^r \in V^*$ and $v_1, \cdots, v_s \in V$, we have

$$\begin{aligned}
&\tau(\omega^1, \cdots, \omega^r, v_1, \cdots, v_s) \\
&= \tau(\omega^1(e_{i_1})\epsilon^{i_1}, \cdots, \omega^r(e_{i_r})\epsilon^{i_r}, \epsilon^{j_1}(v_1)e_{j_1}, \cdots, \epsilon^{j_s}(v_s)e_{j_s}) \\
&= \tau(\epsilon^{i_1}, \cdots, \epsilon^{i_r}, e_{j_1}, \cdots, e_{j_s})\omega^1(e_{i_1}) \cdots \omega^r(e_{i_r})\epsilon^{j_1}(v_1) \cdots \epsilon^{j_s}(v_s) \\
&= \tau(\epsilon^{i_1}, \cdots, \epsilon^{i_r}, e_{j_1}, \cdots, e_{j_s}) \cdot \\
&\quad (e_{i_1} \otimes \cdots \otimes e_{i_r} \otimes \epsilon^{j_1} \cdots \otimes \epsilon^{j_s})(\omega^1, \cdots, \omega^r, v_1, \cdots, v_s).
\end{aligned}$$

So $\tau = \tau(\epsilon^{i_1}, \cdots, \epsilon^{i_r}, e_{j_1}, \cdots, e_{j_s}) \cdot e_{i_1} \otimes \cdots \otimes e_{i_r} \otimes \epsilon^{j_1} \otimes \cdots \otimes \epsilon^{j_s}$.

Next, we show the independence. Let $c^{i_1 \cdots i_r}_{j_1 \cdots j_s} e_{i_1} \otimes \cdots \otimes c_{i_r} \otimes c^{j_1} \cdots \otimes \epsilon^{j_s} = 0$. Operating on $(\epsilon^{k_1}, \cdots, \epsilon^{k_r}, e_{\ell_1}, \cdots, e_{\ell_s}) \in (V^*)^r \times V^s$, we get

$$\begin{aligned}
0 &= c^{i_1 \cdots i_r}_{j_1 \cdots j_s} e_{i_1} \otimes \cdots \otimes e_{i_r} \otimes \epsilon^{j_1} \cdots \otimes \epsilon^{j_s}(\epsilon^{k_1}, \cdots, \epsilon^{k_r}, e_{\ell_1}, \cdots, e_{\ell_s}) \\
&= c^{i_1 \cdots i_r}_{j_1 \cdots j_s} e_{i_1}(\epsilon^{k_1}) \cdots e_{i_r}(\epsilon^{k_r})\epsilon^{j_1}(e_{\ell_1}) \cdots \epsilon^{j_s}(e_{\ell_s}) \\
&= c^{i_1 \cdots i_r}_{j_1 \cdots j_s} \epsilon^{k_1}(e_{i_1}) \cdots \epsilon^{k_r}(e_{i_r})\delta^{j_1}_{\ell_1} \cdots \delta^{j_s}_{\ell_s} \\
&= c^{i_1 \cdots i_r}_{\ell_1 \cdots \ell_s} \delta^{k_1}_{i_1} \cdots \delta^{k_r}_{i_r} \\
&= c^{k_1 \cdots k_r}_{\ell_1 \cdots \ell_s}.
\end{aligned}$$

A $1 \leqslant k_1, \cdots, k_r, \ell_1, \cdots, \ell_s \leqslant m$ were arbitrary, all the coefficients $c^{i_1 \cdots i_r}_{j_1 \cdots j_s}$ are zero, showing the linear independence of B. Finally, it is clear that the number of elements in B is m^{r+s}, since it is the number of ways of choosing the $r+s$ (not necessarily all distinct) numbers $i_1, \cdots, i_r, j_1, \cdots, j_s$ from the set $\{1, \cdots, m\}$. Thus $\dim T^r_s V = m^{r+s}$.

Solution to Exercise 4.8

Suppose that $p \in U$. Let χ be a bump function such that $\chi \equiv 1$ in a neighbourhood U_1 of p, $\chi \equiv 0$ outside U_0, and $\overline{U_0} \subset U$. Then for all $q \in U_1$,

$$\begin{aligned} T(q)^{i_1\cdots i_r}_{j_1\cdots j_s} &= T(q)((\chi dx^{i_1})_q, \cdots, (\chi dx^{i_r})_q, \chi\partial_{x^{j_1},q}, \cdots, \chi\partial_{x^{j_s},q}) \\ &= (T(\chi dx^{i_1}, \cdots, \chi dx^{i_r}, \chi\partial_{x^{j_1}}, \cdots, \chi\partial_{x^{j_s}}))(q). \end{aligned}$$

As the map

$$q \overset{g}{\mapsto} (T(\chi dx^{i_1}, \cdots, \chi dx^{i_r}, \chi\partial_{x^{j_1}}, \cdots, \chi\partial_{x^{j_s}}))(q)$$

belongs to $C^\infty(M)$, and since U_1 is open, $g|_{U_1}$ is smooth. Since smoothness is a local property, the map

$$U \ni p \mapsto T(p)^{i_1\cdots i_r}_{j_1\cdots j_s}$$

belongs to $C^\infty(U)$.

Solution to Exercise 4.9

We have

$$\begin{aligned} T(p)^i_j &= T(p)((dx^i)_p, \partial_{x^j,p}) \\ &= (T(\chi dx^i, \chi\partial_{x^j}))(p), \end{aligned}$$

where χ is a bump function, identically 1 in a neighbourhood of p, and 0 outside U_0, such that $\overline{U_0} \subset U$. Then

$$\begin{aligned} T(p)^i_j &= (T(\chi dx^i, \chi\partial_{x^j}))(p) \\ &= (\chi dx^i)_p(\chi\partial_{x^j})_p \\ &= (dx^i)_p(\chi\partial_{x^j})_p \\ &= \chi(p)\,\partial_{x^j,p}x^i \\ &= 1\,\delta^i_j. \end{aligned}$$

So $T^i_j \equiv \delta^i_j$.

Solution to Exercise 4.10

For $p \in U \cap U'$, we have

$$\begin{aligned} T'^{i_1\cdots i_r}_{j_1\cdots j_s}(p) &= T(p)((dx'^{i_1})_p, \cdots, (dx'^{i_r})_p, \partial_{x'^{j_1},p}, \cdots, \partial_{x'^{j_s},p}) \\ &= T(p)\Big(\frac{\partial x'^{i_1}}{\partial x^{i_1'}}(\mathbf{x}(p))(dx^{i_1'})_p, \cdots, \frac{\partial x'^{i_r}}{\partial x^{i_r'}}(\mathbf{x}(p))(dx^{i_r'})_p, \\ &\qquad \frac{\partial x^{j_1'}}{\partial x'^{j_1}}(\mathbf{x}'(p))\partial_{x^{j_1'},p}, \cdots, \frac{\partial x^{j_s'}}{\partial x'^{j_s}}(\mathbf{x}'(p))\partial_{x^{j_s'},p}\Big) \\ &= \frac{\partial x'^{i_1}}{\partial x^{i_1'}}(\mathbf{x}(p)) \cdots \frac{\partial x'^{i_r}}{\partial x^{i_r'}}(\mathbf{x}(p)) \frac{\partial x^{j_1'}}{\partial x'^{j_1}}(\mathbf{x}'(p)) \cdots \frac{\partial x^{j_s'}}{\partial x'^{j_s}}(\mathbf{x}'(p)) \cdot T^{i_1'\cdots i_r'}_{j_1'\cdots j_s'}(p). \end{aligned}$$

Solution to Exercise 4.11

We will first show well-definedness.

We have

$$\begin{aligned}
& T^{i_1\cdots i_r;U'}_{j_1\cdots j_s}(\mathbf{x}'(p))\cdot \\
& (\Omega^1)_p(\partial_{x'^{i_1},p})\cdots(\Omega^r)_p(\partial_{x'^{i_r},p})\cdot \\
& (dx'^{j_1})((V_1)_p)\cdots(dx'^{j_s})((V_s)_p) \\
= & \frac{\partial x'^{i_1}}{\partial x^{i'_1}}(\mathbf{x}(p))\cdots\frac{\partial x'^{i_r}}{\partial x^{i'_r}}(\mathbf{x}(p))\frac{\partial x^{j'_1}}{\partial x'^{j_1}}(\mathbf{x}'(p))\cdots\frac{\partial x^{j'_s}}{\partial x'^{j_s}}(\mathbf{x}'(p))T^{i'_1\cdots i'_r;U}_{j'_1\cdots j'_s}(\mathbf{x}(p))\cdot \\
& (\Omega^1)_p\Big(\frac{\partial x^{i''_1}}{\partial x'^{i_1}}(\mathbf{x}'(p))\partial_{x^{i''_1},p}\Big)\cdots(\Omega^r)_p\Big(\frac{\partial x^{i''_r}}{\partial x'^{i_r}}(\mathbf{x}'(p))\partial_{x^{i''_r},p}\Big)\cdot \\
& \frac{\partial x'^{j_1}}{\partial x^{j''_1}}(\mathbf{x}(p))(dx^{j''_1})_p((V_1)_p)\cdots\frac{\partial x'^{j_s}}{\partial x^{j''_s}}(\mathbf{x}(p))(dx^{j''_s})_p((V_s)_p) \\
= & \frac{\partial x'^{i_1}}{\partial x^{i'_1}}(\mathbf{x}(p))\frac{\partial x^{i''_1}}{\partial x'^{i_1}}(\mathbf{x}'(p))\cdots\frac{\partial x'^{i_r}}{\partial x^{i'_r}}(\mathbf{x}(p))\frac{\partial x^{i''_r}}{\partial x'^{i_r}}(\mathbf{x}'(p))\cdot \\
& \frac{\partial x^{j'_1}}{\partial x'^{j_1}}(\mathbf{x}'(p))\frac{\partial x'^{j_1}}{\partial x^{j''_1}}(\mathbf{x}(p))\cdots\frac{\partial x^{j'_s}}{\partial x'^{j_s}}(\mathbf{x}'(p))\frac{\partial x'^{j_s}}{\partial x^{j''_s}}(\mathbf{x}(p))\cdot \\
& T^{i'_1\cdots i'_r;U}_{j'_1\cdots j'_s}(\mathbf{x}(p))\cdot(\Omega^1)_p(\partial_{x^{i''_1},p})\cdots(\Omega^r)_p(\partial_{x^{i''_r},p})\cdot \\
& (dx^{j''_1})_p((V_1)_p)\cdots(dx^{j''_s})_p((V_s)_p) \\
= & \delta^{i''_1}_{i'_1}\cdots\delta^{i''_r}_{i'_r}\delta^{j'_1}_{j''_1}\cdots\delta^{j'_s}_{j''_s} \\
& T^{i'_1\cdots i'_r;U}_{j'_1\cdots j'_s}(\mathbf{x}(p))\cdot(\Omega^1)_p\partial_{x^{i''_1},p}\cdots(\Omega^r)_p\partial_{x^{i''_r},p}\cdot \\
& (dx^{j''_1})_p(V_1)_p\cdots(dx^{j''_s})_p(V_s)_p \\
= & T^{i'_1\cdots i'_r;U}_{j'_1\cdots j'_s}(\mathbf{x}(p))\cdot \\
& (\Omega^1)_p(\partial_{x^{i'_1},p})\cdots(\Omega^r)_p(\partial_{x^{i'_r},p})\cdot \\
& (dx^{j'_1})_p((V_1)_p)\cdots(dx^{j'_s})_p((V_s)_p).
\end{aligned}$$

So T is well-defined.

Since smoothness is a local property (Exercise 1.18, p.14), and by the definition of $T(\Omega^1,\cdots,\Omega^r,V_1,\cdots,V_s)$ in a chart, it follows that the map $M\ni p\mapsto T(\Omega^1,\cdots,\Omega^r,V_1,\cdots,V_s)$ is smooth. Hence T is a map from $(T^0_1M)^r\times(T^1_0M)^s$ to $C^\infty(M)$. The multilinearity of T in each component follows from the observations that

$$\begin{aligned}
(\Omega+f\widetilde{\Omega})_p\partial_{x^{i_k},p} &= \Omega_p\partial_{x^{i_k},p}+f(p)\cdot\widetilde{\Omega}_p\partial_{x^{i_k},p}, \\
(dx^{j_\ell})_p(V+f\widetilde{V})_p &= (V+f\widetilde{V})_px^{j_\ell}=V_px^{j_\ell}+f(p)\cdot\widetilde{V}_px^{j_\ell} \\
&= (dx^{j_\ell})_pV_p+f(p)\cdot(dx^{j_\ell})_p\widetilde{V}_p
\end{aligned}$$

for all $\Omega,\widetilde{\Omega}\in T^0_1M$, $V,\widetilde{V}\in T^1_0M$, $f\in C^\infty(M)$ and $p\in U$. So $T\in T^r_sM$.

Solution to Exercise 4.12

Suppose that $\omega^1,\cdots,\omega^{r+r'}\in(T_pM)^*$, and $v_1,\cdots,v_{s+s'}\in T_pM$.
Let $\Omega^1,\cdots,\Omega^{r+r'}\in T^0_1M$ be such that $(\Omega^1)_p=\omega^1,\cdots,(\Omega^{r+r'})_p=\omega^{r+r'}$, and $V_1,\cdots,V_{s+s'}\in T^1_0M$ be such that $(V_1)_p=v_1,\cdots,(V_{s+s'})_p=v_{s+s'}$.

Then

$$\begin{aligned}
&(T\otimes T')(p)(\omega^1,\cdots,\omega^{r+r'},v_1,\cdots,v_{s+s'})\\
&=((T\otimes T')(\Omega^1,\cdots,\Omega^{r+r'},V_1,\cdots,V_{s+s'}))(p)\\
&=(T(\Omega^1,\cdots,\Omega^r,V_1,\cdots,V_s)\cdot T'(\Omega^{r+1},\cdots,\Omega^{r+r'},V_{s+1},\cdots,V_{s+s'}))(p)\\
&=(T(\Omega^1,\cdots,\Omega^r,V_1,\cdots,V_s))(p)\cdot\\
&\quad(T'(\Omega^{r+1},\cdots,\Omega^{r+r'},V_{s+1},\cdots,V_{s+s'}))(p)\\
&=T(p)((\Omega^1)_p,\cdots,(\Omega^r)_p,(V_1)_p,\cdots,(V_s)_p)\cdot\\
&\quad T'(p)((\Omega^{r+1})_p,\cdots,(\Omega^{r+r'})_p,(V_{s+1})_p,\cdots,(V_{s+s'})_p)\\
&=T(p)(\omega^1,\cdots,\omega^r,v_1,\cdots,v_s)\cdot T'(p)(\omega^{r+1},\cdots,\omega^{r+r'},v_{s+1},\cdots,v_{s+s'})\\
&=(T(p)\otimes T'(p))(\omega^1,\cdots,\omega^{r+r'},v_1,\cdots,v_{s+s'}).
\end{aligned}$$

Thus $(T\otimes T')(p)=T(p)\otimes T'(p)$.

Solution to Exercise 4.13

We have $(dx\otimes dy)(\partial_x,\partial_y)=dx(\partial_x)\cdot dy(\partial_y)=(\partial_x x)(\partial_y y)=1\cdot 1=1$, while $(dy\otimes dx)(\partial_x,\partial_y)=dy(\partial_x)\cdot dx(\partial_y)=(\partial_x y)(\partial_y x)=0\cdot 0=0$.

Solution to Exercise 4.14

$$\begin{aligned}
(T\otimes S)^{i_1\cdots i_{r+r'}}_{j_1\cdots j_{s+s'}}(p)&=(T\otimes S)(p)((dx^{i_1})_p,\cdots,(dx^{i_{r+r'}})_p,\partial_{x^{j_1},p},\cdots,\partial_{x^{j_{s+s'}},p})\\
&=((T\otimes S)(\chi\, dx^{i_1},\cdots,\chi\, dx^{i_{r+r'}},\chi\,\partial_{x^{j_1}},\cdots,\chi\,\partial_{x^{j_{s+s'}}}))(p),
\end{aligned}$$

where $p\in U$, and $\chi\in C^\infty(M)$ is identically 1 in a neighbourhood of p, and identically 0 outside U_0 with $\overline{U_0}\subset U$. Thus

$$\begin{aligned}
(T\otimes S)^{i_1\cdots i_{r+r'}}_{j_1\cdots j_{s+s'}}(p)&=((T\otimes S)(\chi\, dx^{i_1},\cdots,\chi\, dx^{i_{r+r'}},\chi\,\partial_{x^{j_1}},\cdots,\chi\,\partial_{x^{j_{s+s'}}}))(p)\\
&=(T(\chi\, dx^{i_1},\cdots,\chi\, dx^{i_r},\chi\,\partial_{x^{j_1}},\cdots,\chi\,\partial_{x^{j_s}}))(p)\cdot\\
&\quad(S(\chi\, dx^{i_{r+1}},\cdots,\chi\, dx^{i_{r+r'}},\chi\,\partial_{x^{j_{s+1}}},\cdots,\chi\,\partial_{x^{j_{s+s'}}}))(p)\\
&=T(p)((dx^{i_1})_p,\cdots,(dx^{i_r})_p,\partial_{x^{j_1},p},\cdots,\partial_{x^{j_s},p})\cdot\\
&\quad S(p)((dx^{i_{r+1}})_p,\cdots,(dx^{i_{r+r'}})_p,\partial_{x^{j_{s+1}},p},\cdots,\partial_{x^{j_{s+s'}},p})\\
&=T^{i_1\cdots i_r}_{j_1\cdots j_s}(p)\cdot S^{i_{r+1}\cdots i_{r+r'}}_{j_{s+1}\cdots j_{s+s'}}(p).
\end{aligned}$$

Thus $(T\otimes S)^{i_1\cdots i_{r+r'}}_{j_1\cdots j_{s+s'}}=T^{i_1\cdots i_r}_{j_1\cdots j_s}S^{i_{r+1}\cdots i_{r+r'}}_{j_{s+1}\cdots j_{s+s'}}$.

Solution to Exercise 4.15

For $p\in U$, $(\mathbf{C}^1_2 T)_{jk}(p)=(\mathbf{C}^1_2 T)(p)(\partial_{x^j,p},\partial_{x^k,p})=((\mathbf{C}^1_2 T)(\chi\,\partial_{x^j},\chi\,\partial_{x^k}))(p)$, where χ is a bump function, identically 1 in a neighbourhood of p, and identically 0 outside U_0 with $\overline{U_0}\subset U$. We have

$$\begin{aligned}
((\mathbf{C}^1_2 T)(\chi\,\partial_{x^j},\chi\,\partial_{x^k}))(p)&=\mathbf{C}(T(\cdot,\chi\,\partial_{x^j},\cdot,\chi\,\partial_{x^k}))(p)\\
&=T(\chi\, dx^i,\chi\,\partial_{x^j},\chi\,\partial_{x^i},\chi\,\partial_{x^k})(p)\\
&=T(p)((dx^i)_p,\partial_{x^j,p},\partial_{x^i,p},\partial_{x^k,p})=T^i_{jik}(p).
\end{aligned}$$

Thus $(\mathbf{C}^1_2 T)_{jk}=T^i_{jik}$.

Solution to Exercise 4.16

$\mathbf{C}T = \operatorname{trace}\tau$, because for all $p \in \mathbb{R}^m$, we have

$$\begin{aligned}(\mathbf{C}T)(p) &= T(p)((dx^i)_p, \partial_{x^i,p}) = (T(dx^i, \partial_{x^i}))(p)\\ &= (dx^i)_{\tau(p)}(\tau_j^k \delta_i^j \partial_{x^k,\tau(p)})\\ &= \delta_k^i \tau_j^k \delta_i^j = \tau_i^i\\ &= \operatorname{trace}\tau.\end{aligned}$$

Solution to Exercise 4.17

In general, there are 18 different contraction operations possible: We have

- $\mathbf{C}_1^1$, $\mathbf{C}_2^1$, $\mathbf{C}_3^1$, $\mathbf{C}_1^2$, $\mathbf{C}_2^2$, $\mathbf{C}_3^2$, which give a $(1,2)$-tensor field, and
- $\mathbf{C}_1^1\mathbf{C}_1^1$, $\mathbf{C}_1^1\mathbf{C}_2^1$, $\mathbf{C}_1^1\mathbf{C}_3^1$, $\mathbf{C}_1^1\mathbf{C}_1^2$, $\mathbf{C}_1^1\mathbf{C}_2^2$, $\mathbf{C}_1^1\mathbf{C}_3^2$, $\mathbf{C}_2^1\mathbf{C}_1^1$, $\mathbf{C}_2^1\mathbf{C}_2^1$, $\mathbf{C}_2^1\mathbf{C}_3^1$, $\mathbf{C}_2^1\mathbf{C}_1^2$, $\mathbf{C}_2^1\mathbf{C}_2^2$, $\mathbf{C}_2^1\mathbf{C}_3^2$, which give a $(0,1)$-tensor field.

But some of these contraction operations in the second row produce the *same* $(0,1)$-tensor field. To see which ones are identical, we compute components: If T_{abc}^{ij} are the components of T in an admissible chart $(U, \mathbf{x})$, then the contraction operations in the second row above result in tensor fields having the following chart-induced components, respectively:

$$\begin{array}{cccccc} \underline{T_{ijc}^{ij}}, & \underset{\sim\sim}{T_{jic}^{ij}}, & \underset{--}{T_{jbi}^{ij}}, & \underset{\sim\sim}{T_{jic}^{ij}}, & \underline{T_{ijc}^{ij}}, & \underline{\underline{T_{ibj}^{ij}}}, \\ \underline{\underline{T_{ibj}^{ij}}}, & \underset{\cdots}{T_{aij}^{ij}}, & T_{aji}^{ij}, & \underset{--}{T_{jbi}^{ij}}, & T_{aji}^{ij}, & \underset{\cdots}{T_{aij}^{ij}}. \end{array}$$

So altogether, we have $6+6 = 12$ distinct tensor fields produced by the contraction operations.

Solution to Exercise 4.18

First, we show that $\mathbf{C}: T_1^1 M \to C^\infty(M)$ is $C^\infty(M)$-linear. Let $(U, \mathbf{x})$ be an admissible chart. Let $p \in U$, and let $\chi \in C^\infty(M)$ be a bump function which is identically 1 in a neighbourhood of p and identically 0 outside U_0 with $\overline{U_0} \subset U$. We have

$$\begin{aligned}(\mathbf{C}(T + f\,S))(p) &= ((T + f\,S)(p))((dx^i)_p, \partial_{x^i,p})\\ &= ((T + f\,S)(\chi\, dx^i, \chi\, \partial_{x^i}))(p)\\ &= (T(\chi\, dx^i, \chi\, \partial_{x^i}) + (f\,S)(\chi\, dx^i, \chi\, \partial_{x^i}))(p)\\ &= T(\chi\, dx^i, \chi\, \partial_{x^i})(p) + f(p)S(\chi\, dx^i, \chi\, \partial_{x^i})(p)\\ &= T(p)((dx^i)_p, \partial_{x^i,p}) + f(p)S(p)((dx^i)_p, \partial_{x^i,p})\\ &= (\mathbf{C}T)(p) + f(p)(\mathbf{C}S)(p)\\ &= (\mathbf{C}T)(p) + (f\cdot(\mathbf{C}S))(p)\\ &= ((\mathbf{C}T) + f\cdot(\mathbf{C}S))(p).\end{aligned}$$

Thus $\mathbf{C}(T + f\,S) = \mathbf{C}T + f\cdot\mathbf{C}S$.

For fixed $\Omega^1, \cdots, \Omega^{r-1} \in T_1^0 M$ and $V_1, \cdots, V_{s-1} \in T_0^1 M$, we have

$$\begin{aligned}
&(\mathbf{C}_j^i(T+fS))(\Omega^1, \cdots, \Omega^{r-1}, V_1, \cdots, V_{s-1}) \\
&= \mathbf{C}((T+fS)(\Omega^1, \cdots, \Omega^{i-1}, \bullet, \Omega^i, \cdots, \Omega^{r-1}, V_1, \cdots, V_{j-1}, \bullet, V_j, \cdots, V_{s-1})) \\
&= \mathbf{C}(T(\Omega^1, \cdots, \Omega^{i-1}, \bullet, \Omega^i, \cdots, \Omega^{r-1}, V_1, \cdots, V_{j-1}, \bullet, V_j, \cdots, V_{s-1}) \\
&\quad + f\cdot S(\Omega^1, \cdots, \Omega^{i-1}, \bullet, \Omega^i, \cdots, \Omega^{r-1}, V_1, \cdots, V_{j-1}, \bullet, V_j, \cdots, V_{s-1})) \\
&= \mathbf{C}(T(\Omega^1, \cdots, \Omega^{i-1}, \bullet, \Omega^i, \cdots, \Omega^{r-1}, V_1, \cdots, V_{j-1}, \bullet, V_j, \cdots, V_{s-1})) \\
&\quad + f\cdot\mathbf{C}(S(\Omega^1, \cdots, \Omega^{i-1}, \bullet, \Omega^i, \cdots, \Omega^{r-1}, V_1, \cdots, V_{j-1}, \bullet, V_j, \cdots, V_{s-1})) \\
&= (\mathbf{C}_j^i T)(\Omega^1, \cdots, \Omega^{r-1}, V_1, \cdots, V_{s-1}) + f\cdot(\mathbf{C}_j^i S)(\Omega^1, \cdots, \Omega^{r-1}, V_1, \cdots, V_{s-1}) \\
&= ((\mathbf{C}_j^i T) + f\cdot(\mathbf{C}_j^i S))(\Omega^1, \cdots, \Omega^{r-1}, V_1, \cdots, V_{s-1}).
\end{aligned}$$

Consequently, $\mathbf{C}_j^i(T + fS) = (\mathbf{C}_j^i T) + f\cdot(\mathbf{C}_j^i S)$.

Solution to Exercise 4.19

Consider the chart $(M, (r,\theta) \mapsto (r,\theta))$. Then for $q = (r,\theta) \in M$,

$$\begin{aligned}
((f^*T)(\partial_r, \partial_r))(q) &= \frac{1}{r^2(\cos\theta)^2}\big((\partial_{r,q} r)(\partial_{r,q} r)(\partial_{r,q} r\sin\theta)(\partial_{r,q} r\sin\theta) \\
&\quad + (\partial_{r,q} r)(\partial_{r,q}\theta)(\partial_{r,q} r\sin\theta)(\partial_{\theta,q} r\sin\theta) \\
&\quad + \partial_{r,q}(\theta)\partial_{r,q}(r)\partial_{\theta,q}(r\sin\theta)\partial_{r,q}(r\sin\theta) \\
&\quad + \partial_{r,q}(\theta)\partial_{r,q}(\theta)\partial_{\theta,q}(r\sin\theta)\partial_{\theta,q}(r\sin\theta)\big) \\
&= \frac{1}{r^2(\cos\theta)^2}(\sin\theta)^2 = \frac{1}{r^2}(\tan\theta)^2.
\end{aligned}$$

Similarly,

$$\begin{aligned}
((f^*T)(\partial_r, \partial_\theta))(q) &= \frac{1}{r^2(\cos\theta)^2}\big(0 + (\partial_{r,q} r)(\partial_{\theta,q}\theta)(\partial_{r,q} r\sin\theta)(\partial_{\theta,q} r\sin\theta) + 0 + 0\big) \\
&= \frac{1}{r^2(\cos\theta)^2}(\sin\theta) r\cos\theta = \frac{1}{r}\tan\theta, \\
((f^*T)(\partial_\theta, \partial_r))(q) &= \frac{1}{r^2(\cos\theta)^2}\big(0 + 0 + (\partial_{\theta,q}\theta)(\partial_{r,q} r)(\partial_{\theta,q} r\sin\theta)(\partial_{r,q} r\sin\theta) + 0\big) \\
&= \frac{1}{r^2(\cos\theta)^2}(r\cos\theta)\sin\theta = \frac{1}{r}\tan\theta, \\
((f^*T)(\partial_\theta, \partial_\theta))(q) &= \frac{1}{r^2(\cos\theta)^2}\big(0 + 0 + 0 + (\partial_{\theta,q}\theta)(\partial_{\theta,q}\theta)(\partial_{\theta,q} r\sin\theta)(\partial_{\theta,q} r\sin\theta)\big) \\
&= \frac{1}{r^2(\cos\theta)^2}(r\cos\theta)(r\cos\theta) = 1.
\end{aligned}$$

Thus

$$f^*T = \frac{1}{r^2}(\tan\theta)^2 dr \otimes dr + \frac{1}{r}(\tan\theta)(dr \otimes d\theta + d\theta \otimes dr) + d\theta \otimes d\theta$$

in the chart $(M, (r,\theta) \mapsto (r,\theta))$.

Chapter 5

Solution to Exercise 5.1

By the linearity of g in the first slot, and by the symmetry of g, g is linear in the second slot too. Thus for each $v \in V$, $v^\flat : V \to \mathbb{R}$ is a linear map, that is, $v^\flat \in V^*$.

The map $\cdot^\flat : V \to V^*$ is linear: For all $v, w \in V$, $c \in \mathbb{R}$,

$$(v + cw)^\flat = g(v + cw, \cdot) = g(v, \cdot) + cg(w, \cdot) = v^\flat + cw^\flat.$$

Moreover, the map $\cdot^\flat : V \to V^*$ is injective. Indeed, if $v \in V$ is such that $v^\flat = 0 \in V^*$, then for all $w \in V$, $v^\flat w = 0$, that is, $g(v, w) = 0$, and as g is nondegenerate, $v = 0$.

Finally, we show that $\cdot^\flat : V \to V^*$ is an isomorphism. Since V, V^* have the same dimension, it follows that the injective map $v \mapsto v^\flat$ is also surjective, and hence it is an isomorphism.

Solution to Exercise 5.2

We have $\mathbf{w} = (x, t) \in \mathbf{v}^\perp$ if and only if $g(\mathbf{w}, \mathbf{v}) = 0$, that is, if and only if $ax - bt = 0$. Thus $\mathbf{v}^\perp = \{(x, t) \in \mathbb{R}^2 : ax - bt = 0\}$, which represents a straight line in the (x, t)-plane. Since the points $(0, 0)$ and (b, a) belong to $\mathbf{v}^\perp$, it follows that $\mathbf{v}^\perp$ is the line passing through the origin and the point (b, a) obtained by reflecting the point (a, b) in the $x = t$ line.

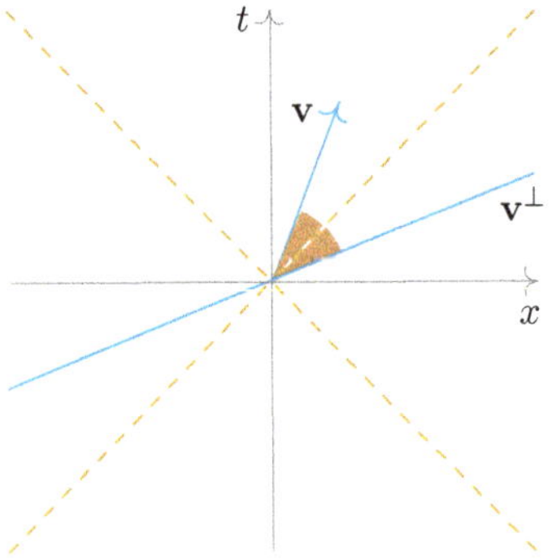

Solution to Exercise 5.3

We have by definition that $U^\perp = \{w \in V : g(w, u) = 0 \text{ for all } u \in U\}$, and $(U^\perp)^\perp = \{v \in V : g(v, w) = 0 \text{ for all } w \in U^\perp\}$. Clearly, if we take a $u \in U$, then for all $w \in U^\perp$, $g(w, u) = 0$, and so $u \in (U^\perp)^\perp$. Thus $U \subset (U^\perp)^\perp$.

Solution to Exercise 5.4

Let $\{e_0, e_1, \cdots, e_{m-1}\}$ be an orthonormal basis for V with $g(e_0, e_0) = -1$, and $g(e_i, e_i) = 1$ for $1 \leqslant i \leqslant m-1$. Decompose

$$u = u^0 e_0 + \sum_{i=1}^{m-1} u^i e_i, \quad v = v^0 e_0 + \sum_{i=1}^{m-1} v^i e_i.$$

As u, v are lightlike, we have

$$0 = g(u,u) = -(u^0)^2 + \|\mathbf{u}\|^2, \quad \text{and} \quad 0 = g(v,v) = -(v^0)^2 + \|\mathbf{v}\|^2,$$

where $\mathbf{u} = (u^1, \cdots, u^{m-1}) \in \mathbb{R}^{m-1}$, $\mathbf{v} = (v^1, \cdots, v^{m-1}) \in \mathbb{R}^{m-1}$. If u, v are orthogonal, then $-u^0 v^0 + \langle \mathbf{u}, \mathbf{v} \rangle = 0$. So $|\langle \mathbf{u}, \mathbf{v} \rangle| = |u^0 v^0| = |u^0||v^0| = \|\mathbf{u}\|\|\mathbf{v}\|$, that is, the Cauchy-Schwarz inequality for the Euclidean inner product on $\mathbb{R}^{m-1}$ happens to be an equality. Thus $\mathbf{u}, \mathbf{v}$ are linearly dependent. (Note that as u, v are lightlike, we have $u \neq 0$ and $v \neq 0$. So $\mathbf{u} \neq \mathbf{0}$, since otherwise $(u^0)^2 = \|\mathbf{u}\|^2 = 0$, and then $u = 0$, a contradiction. Similarly, $\mathbf{v} \neq \mathbf{0}$.)

Because $\mathbf{u}, \mathbf{v}$ are linearly dependent, and each is nonzero, there exists a $t \in \mathbb{R}\backslash\{0\}$ such that $\mathbf{u} = t\mathbf{v}$. We have

$$\begin{aligned} u^0 &= \frac{u^0 v^0}{v^0} \quad (\text{note that } (v^0)^2 = \|\mathbf{v}\|^2 > 0) \\ &= \frac{\langle \mathbf{u}, \mathbf{v} \rangle}{v^0} = \frac{t\langle \mathbf{v}, \mathbf{v} \rangle}{v^0} = \frac{t(v^0)^2}{v^0} = t v^0. \end{aligned}$$

Thus $u = u^0 e_0 + \sum_{i=1}^{m-1} u^i e_i = t v^0 e_0 + \sum_{i=1}^{m-1} t v^i e_i = t v$.

Conversely, if $u = tv$ with $t \in \mathbb{R}\backslash\{0\}$, then $g(u,v) = g(tv, v) = t g(v,v) = t \cdot 0 = 0$, where we have used $g(v,v) = 0$ (since v is lightlike).

Solution to Exercise 5.5

If $v \in T$, then $g(v,v) < 0$ and so $v \sim v$. So $\sim$ is reflexive.

If $v, w \in T$, and $v \sim w$, then $g(w,v) = g(v,w) < 0$, and so $w \sim v$. Thus $\sim$ is symmetric.

Finally, suppose that $v, w, x \in T$ and $v \sim w$, $w \sim x$. Then $g(v,w) < 0$ and $g(w,x) < 0$. So with $k := \frac{g(v,w)}{g(x,w)} \left(= \frac{-}{-}\right) > 0$, we have $k g(x,w) = g(v,w)$, that is, $g(v - kx, w) = 0$. As w is timelike, and $v - kx$ is orthogonal to w, by Lemma 5.1, either $v - kx = 0$ or $v - kx$ is spacelike.

1° $v - kx = 0$, i.e., $v = kx$. Then $g(v,x) = \frac{g(v,kx)}{k} = \frac{g(v,v)}{k} \left(= \frac{-}{+}\right) < 0$. So $v \sim x$.

2° $v - kx$ is spacelike, i.e., $0 < g(v-kx, v-kx) = g(v,v) - 2k g(v,x) + k^2 g(x,x)$. Hence we have $g(v,x) < \frac{1}{2k} g(v,v) + \frac{k}{2} g(x,x) < 0 + 0 = 0$, and so $v \sim x$.

So $\sim$ is transitive.

Consequently, $\sim$ is an equivalence relation on T.

Solution to Exercise 5.6

By the reversed Cauchy-Schwarz inequality,

$$(g(u,v))^2 \geqslant g(u,u)g(v,v) = (-g(u,u))(-g(v,v)) = (\tau(u))^2(\tau(v))^2.$$

So $|g(u,v)| \geqslant \tau(u)\tau(v)$. As $g(u,v) < 0$, this gives $-g(u,v) \geqslant \tau(u)\tau(v)$. Thus

$$\begin{aligned} -g(u+v,u+v) &= -g(u,u) - g(v,v) - 2g(u,v) \\ &\geqslant (\tau(u))^2 + (\tau(v))^2 + 2\tau(u)\tau(v) = (\tau(u)+\tau(v))^2 > 0. \end{aligned}$$

Thus $g(u+v,u+v) < 0$, that is, $u+v$ is timelike, and moreover, we obtain $(\tau(u+v))^2 \geqslant (\tau(u)+\tau(v))^2$, giving $\tau(u+v) \geqslant \tau(u)+\tau(v)$ (as $\tau(u+v)$, $\tau(u)$, $\tau(v)$ are all positive).

Solution to Exercise 5.7

(1)⇒(2): Let $\Lambda \in O(1,m-1)$. For all $u,v \in \mathbb{R}^m$, $\eta(\Lambda u,\Lambda v) = \eta(u,v)$, that is, $(\Lambda u)^{\mathrm{t}}[\eta]\Lambda v = u^{\mathrm{t}}[\eta]v$, that is, $u^{\mathrm{t}}\Lambda^{\mathrm{t}}[\eta]\Lambda v = u^{\mathrm{t}}[\eta]v$. For $1 \leqslant k \leqslant m$, let $e_k \in \mathbb{R}^m$ be the column vector with all entries equal to zero, except for the kth entry, which is equal to one. Taking successively $v = e_k$, $1 \leqslant k \leqslant m$, we obtain $u^{\mathrm{t}}\Lambda^{\mathrm{t}}[\eta]\Lambda = u^{\mathrm{t}}[\eta]$. Taking successively $u = e_k$, $1 \leqslant k \leqslant m$, we obtain $\Lambda^{\mathrm{t}}[\eta]\Lambda = [\eta]$.

(2)⇒(3): The (i,j)-entry of $\Lambda^{\mathrm{t}}[\eta]\Lambda$ is $(\Lambda^{\mathrm{t}})^i_k\eta_{k\ell}\Lambda^\ell_j = \Lambda^k_i\eta_{k\ell}\Lambda^\ell_j$. Thus we obtain $\eta_{ij} = \Lambda^k_i\Lambda^\ell_j\eta_{k\ell}$.

(3)⇒(1): $\Lambda \in O(1,m-1)$, because for all $u,v \in \mathbb{R}^m$, we have

$$\begin{aligned} \eta(\Lambda u,\Lambda v) &= (\Lambda u)^{\mathrm{t}}[\eta]\Lambda v = u^{\mathrm{t}}\Lambda^{\mathrm{t}}[\eta]\Lambda v = u^i\Lambda^k_i\eta_{k\ell}\Lambda^\ell_j v^j \\ &= u^i\Lambda^k_i\Lambda^\ell_j\eta_{k\ell}v^j = u^i\eta_{ij}v^j = \eta(u,v). \end{aligned}$$

Solution to Exercise 5.8

If $\Lambda \in O(1,m-1)$, then $\Lambda^{\mathrm{t}}[\eta]\Lambda = [\eta]$. Thus taking determinants,

$$\begin{aligned} -1 = \det[\eta] &= \det(\Lambda^{\mathrm{t}}[\eta]\Lambda) = (\det(\Lambda^{\mathrm{t}}))(\det[\eta])\det\Lambda = (\det\Lambda)(-1)\det\Lambda \\ &= -(\det\Lambda)^2. \end{aligned}$$

So $\det\Lambda \in \{-1,1\}$. As $\det\Lambda \neq 0$, Λ is invertible. So $O(1,m-1) \subset \mathrm{GL}_m(\mathbb{R})$.

Solution to Exercise 5.9

$I_m \in O(1,m-1)$ because $\eta(\Lambda u,\Lambda v) = \eta(Iu,Iv) = \eta(u,v)$ for all $u,v \in \mathbb{R}^m$. If $\Lambda_1,\Lambda_2 \in O(1,m-1)$, then $\Lambda_1\Lambda_2 \in O(1,m-1)$ since for all $u,v \in \mathbb{R}^m$,

$$\eta((\Lambda_1\Lambda_2)u,(\Lambda_1\Lambda_2)v) = \eta(\Lambda_1(\Lambda_2 u),\Lambda_1(\Lambda_2 v)) = \eta(\Lambda_2 u,\Lambda_2 v) = \eta(u,v).$$

If $\Lambda \in O(1,m-1) \subset \mathrm{GL}_m(\mathbb{R})$, then $\Lambda^{-1} \in O(1,m-1)$: For $u,v \in \mathbb{R}^m$, with $u' := \Lambda^{-1}u$, $v' := \Lambda^{-1}v$,

$$\begin{aligned} \eta(\Lambda^{-1}u,\Lambda^{-1}v) &= \eta(u',v') = \eta(\Lambda u',\Lambda v') \quad (\text{as } \Lambda \in O(1,m-1)) \\ &= \eta(\Lambda\Lambda^{-1}u,\Lambda\Lambda^{-1}v) = \eta(Iu,Iv) = \eta(u,v). \end{aligned}$$

Thus $O(1,m-1)$ is a subgroup of $\mathrm{GL}_m(\mathbb{R})$.

Solution to Exercise 5.10

For each of the given matrices Λ, we will check that $\Lambda^{\mathrm{t}}[\eta]\Lambda = [\eta]$, to conclude that $\Lambda \in O(1,3)$.

We have, using $(\cosh\phi)^2 - (\sinh\phi)^2 = 1$, that

$$\begin{bmatrix} \cosh\phi & \sinh\phi & & \\ \sinh\phi & \cosh\phi & & \\ & & 1 & \\ & & & 1 \end{bmatrix} \begin{bmatrix} -1 & & & \\ & 1 & & \\ & & 1 & \\ & & & 1 \end{bmatrix} \begin{bmatrix} \cosh\phi & \sinh\phi & & \\ \sinh\phi & \cosh\phi & & \\ & & 1 & \\ & & & 1 \end{bmatrix} = \begin{bmatrix} -1 & & & \\ & 1 & & \\ & & 1 & \\ & & & 1 \end{bmatrix}.$$

We have, using $(\cos\theta)^2 + (\sin\theta)^2 = 1$, that

$$\begin{bmatrix} 1 & & & \\ & \cos\theta & \sin\theta & \\ & -\sin\theta & \cos\theta & \\ & & & 1 \end{bmatrix} \begin{bmatrix} -1 & & & \\ & 1 & & \\ & & 1 & \\ & & & 1 \end{bmatrix} \begin{bmatrix} 1 & & & \\ & \cos\theta & -\sin\theta & \\ & \sin\theta & \cos\theta & \\ & & & 1 \end{bmatrix} = \begin{bmatrix} -1 & & & \\ & 1 & & \\ & & 1 & \\ & & & 1 \end{bmatrix}.$$

We have $\begin{bmatrix} -1 & \\ & I_3 \end{bmatrix} \begin{bmatrix} -1 & \\ & I_3 \end{bmatrix} \begin{bmatrix} -1 & \\ & I_3 \end{bmatrix} = \begin{bmatrix} -1 & \\ & I_3 \end{bmatrix}$.

Finally, $\begin{bmatrix} 1 & & \\ & -1 & \\ & & I_2 \end{bmatrix} \begin{bmatrix} -1 & & \\ & 1 & \\ & & I_2 \end{bmatrix} \begin{bmatrix} 1 & & \\ & -1 & \\ & & I_2 \end{bmatrix} = \begin{bmatrix} -1 & & \\ & 1 & \\ & & I_2 \end{bmatrix}$.

Solution to Exercise 5.11

Let $\mathrm{Sym}_m = \{A \in \mathbb{R}^{m\times m} : A^{\mathrm{t}} = A\}$ be the subspace of $\mathbb{R}^{m\times m}$ of all symmetric matrices. Then $\dim \mathrm{Sym}_m = m(m+1)/2$, and so Sym_m can be identified with $\mathbb{R}^{m(m+1)/2}$ after choosing a basis for Sym_m. We will use the result from Exercise 2.15 (p.32). Consider the C^∞ map $f : \mathbb{R}^{m\times m} \to \mathrm{Sym}_m$ defined by $f(\Lambda) = \Lambda^{\mathrm{t}}[\eta]\Lambda$ for all $\Lambda \in \mathbb{R}^{m\times m}$. Then $O(1, m-1) = f^{-1}\{[\eta]\}$. Also, $f'(\Lambda)h = \Lambda^{\mathrm{t}}[\eta]h + h^{\mathrm{t}}[\eta]\Lambda$ for all $\Lambda, h \in \mathbb{R}^{m\times m}$. We will now show that for all $\Lambda \in O(1, m-1)$, the map $f'(\Lambda) : \mathbb{R}^{m\times m} \to \mathrm{Sym}_m$ is surjective. Let $A \in \mathrm{Sym}_m$. Define $h = \frac{1}{2}\Lambda^{\mathrm{t}}[\eta]A \in \mathbb{R}^{m\times m}$. Then

$$\begin{aligned} f'(\Lambda)h &= \Lambda^{\mathrm{t}}[\eta](\tfrac{1}{2}\Lambda[\eta]A) + (\tfrac{1}{2}\Lambda[\eta]A)^{\mathrm{t}}[\eta]\Lambda \\ &= \tfrac{1}{2}\Lambda^{\mathrm{t}}[\eta]\Lambda[\eta]A + \tfrac{1}{2}A^{\mathrm{t}}[\eta]^{\mathrm{t}}\Lambda^{\mathrm{t}}[\eta]\Lambda \\ &= \tfrac{1}{2}[\eta][\eta]A + \tfrac{1}{2}A[\eta][\eta] = A. \end{aligned}$$

By Exercise 2.15, it follows that $O(1, m-1) = f^{-1}\{[\eta]\}$ is a submanifold of $\mathbb{R}^{m\times m} \simeq \mathbb{R}^{m^2}$ of dimension $m^2 - \frac{m(m+1)}{2} = \frac{m(m-1)}{2}$.

Solution to Exercise 5.12

For all $p \in M$, $(\mathbf{g}(V,W))(p) = (\mathbf{g}(\widetilde{V},W))(p)$, i.e., $\mathbf{g}(p)(V_p, W_p) = \mathbf{g}(p)(\widetilde{V}_p, W_p)$. Let $w \in T_pM$. Suppose that the vector field $W \in T_0^1M$ is such that $W_p = w$. (Such a vector field W exists, by Lemma 3.1, p.41.) Then the equality

$\mathbf{g}(p)(V_p, W_p) = \mathbf{g}(p)(\widetilde{V}_p, W_p)$ above gives $\mathbf{g}(p)(V_p, w) = \mathbf{g}(p)(\widetilde{V}_p, w)$, that is, $\mathbf{g}(p)(V_p - \widetilde{V}_p, w) = 0$. Since $w \in T_pM$ was arbitrary, $\mathbf{g}(p)(V_p - \widetilde{V}_p, w) = 0$ for all $w \in T_pM$. As $\mathbf{g}(p)$ is nondegenerate, it follows that $V_p - \widetilde{V}_p = 0$, that is, $V_p = \widetilde{V}_p$. Since $p \in M$ was arbitrary, we conclude that $V = \widetilde{V}$.

Solution to Exercise 5.13

The identity map id_M is clearly an isometry. Let $f, h \in \mathrm{Iso}(M)$. We have $f \circ h^{-1} \in \mathrm{Iso}(M)$, because for $V, W \in T_0^1M$ and $p \in M$,

$$\begin{aligned}
&(((f \circ h^{-1})^*\mathbf{g})(V, W))(p)\\
&= \mathbf{g}((f \circ h^{-1})p)\big((d(f \circ h^{-1}))_pV_p, (d(f \circ h^{-1}))_pW_p\big)\\
&= \mathbf{g}(f(h^{-1}p))\big((df)_{h^{-1}p}((dh^{-1})_pV_p), (df)_{h^{-1}p}((dh^{-1})_pW_p)\big)\\
&= \mathbf{g}(h^{-1}p)\big((dh^{-1})_pV_p, (dh^{-1})_pW_p\big) \quad (\text{since } f^*\mathbf{g} = \mathbf{g})\\
&= \mathbf{g}(hh^{-1}p)\big((dh)_{h^{-1}p}(dh^{-1})_pV_p, (dh)_{h^{-1}p}(dh^{-1})_pW_p\big) \quad (\text{as } h^*\mathbf{g} = \mathbf{g})\\
&= \mathbf{g}(p)\big((d(h \circ h^{-1}))_pV_p, (d(h \circ h^{-1}))_pW_p\big)\\
&= \mathbf{g}(p)\big((d(\mathrm{id}_M))_pV_p, (d(\mathrm{id}_M))_pW_p\big)\\
&= \mathbf{g}(p)\big(\mathrm{id}_{T_pM}V_p, \mathrm{id}_{T_pM}W_p\big) = \mathbf{g}(p)\big(V_p, W_p\big)\\
&= (\mathbf{g}(V, W))(p).
\end{aligned}$$

Solution to Exercise 5.14

Using the properties of the tensor product $\otimes$, and Exercise 2.18 (p.35),

$$\begin{aligned}
dx \otimes dx &= (\cos\phi)^2(\sin\theta)^2 dr \otimes dr + r^2(\cos\phi)^2(\cos\theta)^2 d\theta \otimes d\theta\\
&\quad + r^2(\sin\phi)^2(\sin\theta)^2 d\phi \otimes d\phi\\
&\quad + r(\cos\phi)^2(\sin\theta)(\cos\theta)(dr \otimes d\theta + d\theta \otimes dr)\\
&\quad - r(\cos\phi)(\sin\phi)(\sin\theta)^2(dr \otimes d\phi + d\phi \otimes dr)\\
&\quad - r^2(\cos\phi)(\sin\phi)(\cos\theta)(\sin\theta)(d\theta \otimes d\phi + d\phi \otimes d\theta)\\
dy \otimes dy &= (\sin\phi)^2(\sin\theta)^2 dr \otimes dr + r^2(\sin\phi)^2(\cos\theta)^2 d\theta \otimes d\theta\\
&\quad + r^2(\cos\phi)^2(\sin\theta)^2 d\phi \otimes d\phi\\
&\quad + r(\sin\phi)^2(\sin\theta)(\cos\theta)(dr \otimes d\theta + d\theta \otimes dr)\\
&\quad + r(\sin\phi)(\cos\phi)(\sin\theta)^2(dr \otimes d\phi + d\phi \otimes dr)\\
&\quad + r^2(\cos\phi)(\sin\phi)(\cos\theta)(\sin\theta)(d\theta \otimes d\phi + d\phi \otimes d\theta)\\
dz \otimes dz &= (\cos\theta)^2 dr \otimes dr + r^2(\sin\theta)^2 d\theta \otimes d\theta\\
&\quad - r(\cos\theta)(\sin\theta)(dr \otimes d\theta + d\theta \otimes dr).
\end{aligned}$$

Adding these, we obtain

$$dx \otimes dx + dy \otimes dy + dz \otimes dz = dr \otimes dr + r^2 d\theta \otimes d\theta + r^2(\sin\theta)^2 d\phi \otimes d\phi.$$

To show that ψ_t is an isometry, we must show that $(\psi_t)^*\mathbf{g} = \mathbf{g}$. For any $p = (x, y, z) \in \mathbb{R}^3$, and any vector field $V = V^x\partial_x + V^y\partial_y + V^z\partial_z \in T_0^1\mathbb{R}^3$, using the result from Theorem 2.2 (p.30), we have

$$\begin{aligned}(d\psi_t)_p V_p &= (d\psi_t)_p (V^x(p)\partial_{x,p} + V^y(p)\partial_{y,p} + V^z(p)\partial_{z,p}) \\ &= V^x(p)\partial_{x,\psi_t p} + ((\cos t)V^y(p) - (\sin t)V^z(p))\partial_{y,\psi_t p} \\ &\quad + ((\sin t)V^y(p) + (\cos t)V^z(p))\partial_{z,\psi_t p}.\end{aligned}$$

Hence, for $V, W \in T_0^1\mathbb{R}^3$, we have

$$\begin{aligned}&(((\psi_t)^*\mathbf{g})(V,W))(p) \\ &= \mathbf{g}(\psi_t p)((d\psi_t)_p V_p, (d\psi_t)_p W_p) \\ &= ((dx)_{\psi_t p} \otimes (dx)_{\psi_t p} + (dy)_{\psi_t p} \otimes (dy)_{\psi_t p} + (dz)_{\psi_t p} \otimes (dz)_{\psi_t p}) \\ &\quad ((d\psi_t)_p V_p, (d\psi_t)_p W_p) \\ &= V^x(p)W^x(p) \\ &\quad + ((\cos t)V^y(p) - (\sin t)V^z(p))((\cos t)W^y(p) - (\sin t)W^z(p)) \\ &\quad + ((\sin t)V^y(p) + (\cos t)V^z(p))((\sin t)W^y(p) + (\cos t)W^z(p)) \\ &= V^x(p)W^x(p) + ((\cos t)^2 + (\sin t)^2)V^y(p)W^y(p) \\ &\quad + ((-\sin t)^2 + (\cos t)^2)V^z(p)W^z(p) \\ &= \mathbf{g}(p)(V_p, W_p) \\ &= (\mathbf{g}(V,W))(p).\end{aligned}$$

Consequently, $(\psi_t)^*\mathbf{g} = \mathbf{g}$, i.e., ψ_t is an isometry.

Solution to Exercise 5.15

With $p = (u_0, v_0, u_0 v_0)$, we have

$$\begin{aligned}\mathbf{g}(p)_{uu} &= (\partial_{u,p}x)^2 + (\partial_{u,p}y)^2 + (\partial_{u,p}z)^2 \\ &= \Big(\frac{\partial u}{\partial u}(p)\Big)^2 + \Big(\frac{\partial v}{\partial u}(p)\Big)^2 + \Big(\frac{\partial (uv)}{\partial u}(p)\Big)^2 \\ &= 1^2 + 0^2 + v_0^2 = 1 + v_0^2, \\ \mathbf{g}(p)_{vv} &= (\partial_{v,p}x)^2 + (\partial_{v,p}y)^2 + (\partial_{v,p}z)^2 \\ &= \Big(\frac{\partial u}{\partial v}(p)\Big)^2 + \Big(\frac{\partial v}{\partial v}(p)\Big)^2 + \Big(\frac{\partial (uv)}{\partial v}(p)\Big)^2 \\ &= 0^2 + 1^2 + u_0^2 = 1 + u_0^2, \\ \mathbf{g}(p)_{uv} &= \mathbf{g}(p)_{vu} = (\partial_{u,p}x)(\partial_{v,p}x) + (\partial_{u,p}y)(\partial_{v,p}y) + (\partial_{u,p}z)(\partial_{v,p}z) \\ &= \Big(\frac{\partial u}{\partial u}(p)\Big)\Big(\frac{\partial u}{\partial v}(p)\Big) + \Big(\frac{\partial v}{\partial u}(p)\Big)\Big(\frac{\partial v}{\partial v}(p)\Big) + \Big(\frac{\partial (uv)}{\partial u}(p)\Big)\Big(\frac{\partial (uv)}{\partial v}(p)\Big) \\ &= 1\cdot 0 + 0\cdot 1 + v_0 \cdot u_0 = u_0 v_0.\end{aligned}$$

Thus

$$\begin{bmatrix}\mathbf{g}(p)_{uu} & \mathbf{g}(p)_{uv} \\ \mathbf{g}(p)_{vu} & \mathbf{g}(p)_{vv}\end{bmatrix} = \begin{bmatrix}1+v_0^2 & u_0v_0 \\ u_0v_0 & 1+u_0^2\end{bmatrix}.$$

If $\mathbf{v} = (a,b) \in \mathbb{R}^2\backslash\{(0,0)\}$, then $\mathbf{v}^{\mathrm{t}}[\mathbf{g}_{ij}(p)]\mathbf{v} = a^2 + b^2 + (av_0 + bu_0)^2 > 0$, and so the metric $\mathbf{g}$ is Riemannian.

Solution to Exercise 5.16

Let $p = \varphi_{\mathbf{s}}^{-1}(u,v) = \big(\frac{2u}{1-u^2-v^2}, \frac{2v}{1-u^2-v^2}, \frac{1+u^2+v^2}{1-u^2-v^2}\big)$.

Then we have

$$\begin{aligned}
\mathbf{g}(p)_{uu} &= (\partial_{u,p}x)^2 + (\partial_{u,p}y)^2 - (\partial_{u,p}z)^2 \\
&= \Big(\frac{2(1+u^2-v^2)}{(1-u^2-v^2)^2}\Big)^2 + \Big(\frac{4uv}{(1-u^2-v^2)^2}\Big)^2 - \Big(\frac{4u}{(1-u^2-v^2)^2}\Big)^2 = \frac{4}{(1-u^2-v^2)^2}, \\
\mathbf{g}(p)_{vv} &= (\partial_{v,p}x)^2 + (\partial_{v,p}y)^2 - (\partial_{v,p}z)^2 \\
&= \Big(\frac{4uv}{(1-u^2-v^2)^2}\Big)^2 + \Big(\frac{2(1-u^2+v^2)}{(1-u^2-v^2)^2}\Big)^2 - \Big(\frac{4v}{(1-u^2-v^2)^2}\Big)^2 = \frac{4}{(1-u^2-v^2)^2}, \\
\mathbf{g}(p)_{uv} &= \mathbf{g}(p)_{vu} = (\partial_{u,p}x)(\partial_{v,p}x) + (\partial_{u,p}y)(\partial_{v,p}y) - (\partial_{u,p}z)(\partial_{v,p}z) \\
&= \frac{2(1+u^2-v^2)\cdot 4uv}{(1-u^2-v^2)^4} + \frac{4uv\cdot 2(1-u^2+v^2)}{(1-u^2-v^2)^4} - \frac{4u\cdot 4v}{(1-u^2-v^2)^4} = 0.
\end{aligned}$$

Thus

$$\begin{bmatrix} \mathbf{g}(p)_{uu} & \mathbf{g}(p)_{uv} \\ \mathbf{g}(p)_{vu} & \mathbf{g}(p)_{vv} \end{bmatrix} = \frac{4}{(1-u^2-v^2)^2}\begin{bmatrix} 1 & 0 \\ 0 & 1 \end{bmatrix},$$

which is positive definite. Hence $(H^2, \mathbf{g})$ is a Riemannian manifold.

Solution to Exercise 5.17

Tensorial property of $\mathbf{g}$: Let $f \in C^\infty(M_1 \times M_2)$, $X, X', X'' \in T_0^1(M_1 \times M_2)$ and $(p_1, p_2) \in M_1 \times M_2$. Below, we will write $X_{(p_1,p_2)} = V_{p_1} \oplus W_{p_2}$, where $V_{p_1} \in T_{p_1}M_1$, $W_{p_2} \in T_{p_2}M_2$, etc. We have

$$\begin{aligned}
&(\mathbf{g}(X + fX', X''))(p_1,p_2) \\
&= \mathbf{g}(p_1,p_2)((X+fX')_{(p_1,p_2)}, X''_{(p_1,p_2)}) \\
&= \mathbf{g}(p_1,p_2)(X_{(p_1,p_2)} + f(p_1,p_2)X'_{(p_1,p_2)}, X''_{(p_1,p_2)}) \\
&= \mathbf{g}(p_1,p_2)((V_{p_1} \oplus W_{p_2}) + f(p_1,p_2)(V'_{p_1} \oplus W'_{p_2}), V''_{p_1} \oplus W''_{p_2}) \\
&= \mathbf{g}(p_1,p_2)((V_{p_1} + f(p_1,p_2)V'_{p_1}) \oplus (W_{p_2} + f(p_1,p_2)W'_{p_2}), V''_{p_1} \oplus W''_{p_2}) \\
&= \mathbf{g}_1(p_1)(V_{p_1} + f(p_1,p_2)V'_{p_1}, V''_{p_1}) + \mathbf{g}_2(p_2)(W_{p_2} + f(p_1,p_2)W'_{p_2}, W''_{p_2}) \\
&= \mathbf{g}_1(p_1)(V_{p_1}, V''_{p_1}) + \mathbf{g}_2(p_2)(W_{p_2}, W''_{p_2}) \\
&\quad + f(p_1,p_2)\,(\mathbf{g}_1(p_1)(V'_{p_1}, V''_{p_1}) + \mathbf{g}_2(p_2)(W'_{p_2}, W''_{p_2})) \\
&= \mathbf{g}(p_1,p_2)(V_{p_1} \oplus W_{p_2}, V''_{p_1} \oplus W''_{p_2}) \\
&\quad + f(p_1,p_2)\,\mathbf{g}(p_1,p_2)(V'_{p_1} \oplus W'_{p_2}, V''_{p_1} \oplus W''_{p_2}) \\
&= \mathbf{g}(p_1,p_2)(X_{(p_1,p_2)}, X''_{(p_1,p_2)}) + f(p_1,p_2)\,\mathbf{g}(p_1,p_2)(X'_{(p_1,p_2)}, X''_{(p_1,p_2)}) \\
&= \mathbf{g}(X, X'')(p_1,p_2) + f(p_1,p_2)\,\mathbf{g}(X', X'')(p_1,p_2) \\
&= (\mathbf{g}(X, X'') + f\cdot\mathbf{g}(X', X''))(p_1,p_2).
\end{aligned}$$

The $C^\infty(M_1 \times M_2)$-linearity in the second slot of $\mathbf{g}$ is shown similarly. So $\mathbf{g} \in T_2^0(M_1 \times M_2)$.

$\mathbf{g}(p_1,p_2)$ is a scalar product on $T_{(p_1,p_2)}(M_1\times M_2)$, $(p_1,p_2)\in M_1\times M_2$: Bilinearity is a consequence of the fact that $\mathbf{g}$ is a $(0,2)$-tensor field, established above. We show symmetry:

$$\begin{aligned}\mathbf{g}(p_1,p_2)(v_1\oplus w_1,v_2\oplus w_2) &= \mathbf{g}_1(p_1)(v_1,v_2)+\mathbf{g}_2(p_2)(w_1,w_2)\\ &= \mathbf{g}_1(p_1)(v_2,v_1)+\mathbf{g}_2(p_2)(w_2,w_1)\\ &= \mathbf{g}(p_1,p_2)(v_2\oplus w_2,v_1\oplus w_1).\end{aligned}$$

Nondegeneracy: Suppose that for all $v_2\oplus w_2\in T_{(p_1,p_2)}(M_1\times M_2)$,

$$\mathbf{g}(p_1,p_2)(v_1\oplus w_1,v_2\oplus w_2)=0.$$

Then in particular, for all $v_2\in T_{p_1}M_1$, $v_2\oplus 0\in T_{(p_1,p_2)}(M_1\times M_2)$, and so

$$\begin{aligned}0 &= \mathbf{g}(p_1,p_2)(v_1\oplus w_1,v_2\oplus 0)=\mathbf{g}_1(p_1)(v_1,v_2)+\mathbf{g}_2(p_2)(w_1,0)\\ &= \mathbf{g}_1(p_1)(v_1,v_2)+0=\mathbf{g}_1(p_1)(v_1,v_2).\end{aligned}$$

Thus by the nondegeneracy of $\mathbf{g}_1(p_1)$, $v_1=0$. For all $w_2\in T_{p_2}M_2$, we have $0\oplus w_2\in T_{(p_1,p_2)}(M_1\times M_2)$, and so

$$\begin{aligned}0 &= \mathbf{g}(p_1,p_2)(v_1\oplus w_1,0\oplus w_2)=\mathbf{g}_1(p_1)(v_1,0)+\mathbf{g}_2(p_2)(w_1,w_2)\\ &= \mathbf{g}_2(p_2)(w_1,w_2).\end{aligned}$$

So we conclude that $w_1=0$, as $\mathbf{g}_2(p_2)$ is nondegenerate. Consequently, we have $v_1\oplus w_1=0\oplus 0\in T_{(p_1,p_2)}(M_1\times M_2)$.

Finally, we show that $\iota(\mathbf{g})=1$. Let $(p_1,p_2)\in M_1\times M_2$. Let $\{e_1,\cdots,e_{m_1}\}$ be an orthonormal basis for $T_{p_1}M_1$ such that $\mathbf{g}_1(p_1)(e_1,e_1)=-1$ and $\mathbf{g}_1(p_1)(e_2,e_2)=\cdots=\mathbf{g}_1(p_1)(e_{m_1},e_{m_1})=1$. Similarly, let $\{f_1,\cdots,f_{m_2}\}$ be an orthonormal basis for $T_{p_2}M_2$. Then $\{e_1\oplus 0,\cdots,e_{m_1}\oplus 0,0\oplus f_1,\cdots,0\oplus f_{m_2}\}$ spans $T_{(p_1,p_2)}(M_1\times M_2)$, and

$$\begin{aligned}&\mathbf{g}(p_1,p_2)(e_1\oplus 0,e_1\oplus 0)=\mathbf{g}_1(p_1)(e_1,e_1)+0=-1,\\ &\mathbf{g}(p_1,p_2)(e_2\oplus 0,e_2\oplus 0)=\cdots=\mathbf{g}(p_1,p_2)(e_{m_1}\oplus 0,e_{m_1}\oplus 0)=1,\\ &\mathbf{g}(p_1,p_2)(0\oplus f_1,0\oplus f_1)=\cdots=\mathbf{g}(p_1,p_2)(0\oplus f_{m_2},0\oplus f_{m_2})=1,\\ &\mathbf{g}(p_1,p_2)(e_i\oplus 0,e_j\oplus 0)=\mathbf{g}_1(p_1)(e_i,e_j)=0 \text{ for } i\neq j,\\ &\mathbf{g}(p_1,p_2)(e_i\oplus 0,0\oplus f_j)=0+0=0 \text{ for all } i,j,\\ &\mathbf{g}(p_1,p_2)(0\oplus f_i,0\oplus f_j)=\mathbf{g}_2(p_2)(f_i,f_j)=0 \text{ for } i\neq j.\end{aligned}$$

By Theorem 5.4 (p.78), $\mathfrak{i}_{\mathbf{g}(p_1,p_2)}=1$. But as (p_1,p_2) was arbitrary, $\iota(\mathbf{g})=1$.

Solution to Exercise 5.18

Let $p\in U$. Choose an orthonormal basis $B=\{e_1,\cdots,e_m\}$ for T_pM with respect to the scalar product $\mathbf{g}(p)$: $\mathbf{g}(p)(e_i,e_j)=\epsilon_i\delta_{ij}$, where

$$\epsilon_i:=\begin{cases}-1 & \text{if } i=1,\\ 1 & \text{if } i\in\{2,\cdots,m\}.\end{cases}$$

Let α_i^k be such that $\partial_{x^i,p}=\alpha_i^k e_k$, $1\leqslant i,k\leqslant m$. Then $A=[\alpha_i^j]$ is invertible.

As $\mathbf{g}_{ij}(p) = \mathbf{g}(p)(\partial_{x^i,p}, \partial_{x^j,p})$, we have

$$\mathbf{g}_{ij}(p) = \mathbf{g}(p)(\alpha_i^k e_k, \alpha_j^\ell e_\ell) = \alpha_i^k \alpha_j^\ell \epsilon_k \delta_{k\ell} = \alpha_i^k \epsilon_k \delta_{k\ell} \alpha_j^\ell = (A^{\mathrm{t}} E A)_{ij},$$

where E is the diagonal matrix with the diagonal entries $\epsilon_1, \cdots, \epsilon_m$. So $G(p) = A^{\mathrm{t}} E A$, and $\det G(p) = (\det(A^{\mathrm{t}}))(-1)(1)^{m-1} \det A = -(\det A)^2 < 0$. (As A is invertible, $\det A \neq 0$.)

Solution to Exercise 5.19

We have $\mathbf{g}_{ij} = \mathbf{g}(\partial_{x^i}, \partial_{x^j}) \in C^\infty(U)$. The matrix $G(p)$ is symmetric for all $p \in U$ because $\mathbf{g}_{ij} = \mathbf{g}(\partial_{x^i}, \partial_{x^j}) = \mathbf{g}(\partial_{x^j}, \partial_{x^i}) = \mathbf{g}_{ji}$.

Let $v \in \mathbb{R}^m$ be such that $G(p)v = 0$. Let $w^j \partial_{x^j,p} \in T_pM$. Then with $w^{\mathrm{t}} := \begin{bmatrix} w^1 \cdots w^m \end{bmatrix}$, $w^{\mathrm{t}} G(p) v = w^{\mathrm{t}} 0 = 0$, i.e., $v^i w^j \mathbf{g}_{ji}(p) = 0$, and so $\mathbf{g}(p)(w^j \partial_{x^j,p}, v^i \partial_{x^i,p}) = 0$ for all $w^j \partial_{x^i,p} \in T_pM$. As $\mathbf{g}(p)$ is nondegenerate, $v^i \partial_{x^i,p} = 0$, i.e., $v = 0$. Thus $G(p)$ is invertible, and $\det(G(p)) \neq 0$, $p \in U$.

As G has $C^\infty(U)$ entries, and since $C^\infty(U)$ forms an algebra with pointwise operations, we conclude that $\det G \in C^\infty(U)$. Also for all $p \in U$, $\det(G(p)) \neq 0$. Thus $\frac{1}{\det G} \in C^\infty(U)$. Let $\operatorname{adj} G(p)$ be the matrix with entry $(-1)^{i+j} \det \widetilde{G}(p)_{ij}$ in the ith row and j^{th} column, where $\widetilde{G}(p)_{ij}$ is the matrix obtained from $G(p)$ by deleting its ith column and j^{th} row. Then (by Cramer's rule) $(G(p))^{-1} = \frac{1}{\det G(p)} \operatorname{adj} G(p)$. As $p \mapsto [\operatorname{adj} G(p)]_{ij}$ belongs to $C^\infty(U)$, also $p \mapsto [(G(p))^{-1}]_{ij}$ is in $C^\infty(U)$.

Solution to Exercise 5.20

Recall (Exercise 2.19, p.35) that if

$$A = [A_j^i] = \Big[\frac{\partial(x'^i \circ \mathbf{x}^{-1})}{\partial u^j}(\mathbf{x}(\cdot))\Big], \quad \text{and} \quad B = [B_j^i] = \Big[\frac{\partial(x^i \circ \mathbf{x}'^{-1})}{\partial u^j}(\mathbf{x}'(\cdot))\Big],$$

then $AB = BA = I_m$. Let $\mathbf{g}_{ij}, \mathbf{g}'_{ij}$ denote the components of $\mathbf{g}$ in the charts $(U, \mathbf{x}), (U', \mathbf{x}')$, respectively. Since $\mathbf{g}$ is a $(0,2)$-tensor field, by Exercise 4.10 (p.66), we get

$$\mathbf{g}'_{ij} = \frac{\partial x^k}{\partial x'^i} \frac{\partial x^\ell}{\partial x'^j} \mathbf{g}_{k\ell} = \frac{\partial x^k}{\partial x'^i} \mathbf{g}_{k\ell} \frac{\partial x^\ell}{\partial x'^j} = (B^{\mathrm{t}})_k^i \mathbf{g}_{k\ell} B_j^\ell.$$

Thus if $G = [\mathbf{g}_{ij}]$ and $G' = [\mathbf{g}'_{ij}]$, then $G' = B^{\mathrm{t}} G B$. This gives

$$(G')^{-1} = B^{-1} G^{-1} (B^{\mathrm{t}})^{-1} = B^{-1} G^{-1} (B^{-1})^{\mathrm{t}} = A G^{-1} A^{\mathrm{t}}.$$

Hence $\mathbf{g}'^{ij} = [(G')^{-1}]^{ij} = A_k^i (G^{-1})^{k\ell} (A^{\mathrm{t}})_j^\ell = \dfrac{\partial x'^i}{\partial x^k} \mathbf{g}^{k\ell} \dfrac{\partial x'^j}{\partial x^\ell}$.

Solution to Exercise 5.21

Let $p \in \mathrm{GL}_n(\mathbb{R})$ and $v, w \in \mathbb{R}^{n \times n}$. As $\operatorname{trace}(\alpha\beta) = \operatorname{trace}(\beta\alpha)$ for any two matrices $\alpha, \beta \in \mathbb{R}^{n \times n}$, we have

$$\begin{aligned} \mathbf{g}(p)(v, w) &= \operatorname{trace}(p^{-1} v p^{-1} w) = \operatorname{trace}(w p^{-1} v p^{-1}) = \operatorname{trace}(p^{-1} w p^{-1} v) \\ &= \mathbf{g}(p)(w, v). \end{aligned}$$

Next, if $v, \widetilde{v}, w \in \mathbb{R}^{n\times n}$, and $c \in \mathbb{R}$, then

$$\begin{aligned}\mathbf{g}(p)(v+c\widetilde{v}, w) &= \text{trace}(p^{-1}(v+c\widetilde{v})p^{-1}w) = \text{trace}(p^{-1}vp^{-1}w + cp^{-1}\widetilde{v}p^{-1}w)\\ &= \text{trace}(p^{-1}vp^{-1}w) + c\,\text{trace}(p^{-1}\widetilde{v}p^{-1}w)\\ &= \mathbf{g}(p)(v,w) + c\,\mathbf{g}(p)(\widetilde{v}, w).\end{aligned}$$

Finally, we show nondegeneracy. Suppose $v \in \mathbb{R}^{n\times n}$ is such that for all $w \in \mathbb{R}^{n\times n}$, we have $\mathbf{g}(p)(v,w) = 0$. Taking $w := pv^{\text{t}}p$, we obtain

$$\begin{aligned}0 = \mathbf{g}(p)(v,w) &= \text{trace}(p^{-1}vp^{-1}pv^{\text{t}}p) = \text{trace}(p^{-1}vv^{\text{t}}p)\\ &= \text{trace}(pp^{-1}vv^{\text{t}}) = \text{trace}(vv^{\text{t}}).\end{aligned}$$

But $\text{trace}(vv^{\text{t}})$ is the sum of the squares of the entries of v, and so we conclude $v = 0$, as required.

Next, we show that the index of $\mathbf{g}(p)$ is $n(n-1)/2$. Let S be the subspace of $\mathbb{R}^{n\times n}$ consisting of all symmetric matrices, and A be the subspace of all skew-symmetric matrices. Then $pS := \{p\sigma : \sigma \in S\}$ is a subspace of $\mathbb{R}^{n\times n}$ with the same dimension as S, namely $n(n+1)/2$. Also, $pA := \{p\alpha : \alpha \in A\}$ is a subspace of $\mathbb{R}^{n\times n}$ with the same dimension as A, namely $n(n-1)/2$. Let $v \in \mathbb{R}^{n\times n}$. Then $\sigma := \frac{1}{2}(p^{-1}v + (p^{-1}v)^{\text{t}}) \in S$ and $\alpha := \frac{1}{2}(p^{-1}v - (p^{-1}v)^{\text{t}}) \in A$, and $p^{-1}v = \sigma + \alpha$, i.e., $v = p\sigma + p\alpha \in pS + pA$. Thus pS and pA span $\mathbb{R}^{n\times n}$. If $\sigma \in S$, then $\mathbf{g}(p)(p\sigma, p\sigma) = \text{trace}(p^{-1}p\sigma p^{-1}p\sigma) = \text{trace}(\sigma\sigma^{\text{t}}) \geqslant 0$. If $\alpha \in A$, then $\mathbf{g}(p)(p\alpha, p\alpha) = \text{trace}(p^{-1}p\alpha p^{-1}p\alpha) = \text{trace}(\alpha(-\alpha^{\text{t}})) \leqslant 0$. Finally, for $\sigma \in S$ and $\alpha \in A$,

$$\begin{aligned}\mathbf{g}(p)(p\sigma, p\alpha) &= \text{trace}(p^{-1}p\sigma p^{-1}p\alpha) = \text{trace}(\sigma\alpha) = \text{trace}(\sigma^{\text{t}}(-\alpha^{\text{t}}))\\ &= -\text{trace}((\alpha\sigma)^{\text{t}}) = -\text{trace}(\alpha\sigma) = -\text{trace}(p^{-1}p\alpha p^{-1}p\sigma)\\ &= -\mathbf{g}(p)(p\alpha, p\sigma) = -\mathbf{g}(p)(p\sigma, p\alpha),\end{aligned}$$

and so $\mathbf{g}(p)(p\sigma, p\alpha) = 0$. By Theorem 5.4 (p.78), $\mathfrak{i}_g = \dim(pA) = n(n-1)/2$.

Solution to Exercise 5.22

Let v be any timelike vector in T_pM. Set $k := \sqrt{-\mathbf{g}(p)(v,v)} > 0$, and $e_0 := k^{-1}v$. Then $\mathbf{g}(p)(e_0, e_0) = -1$. Also, using Theorems 5.2 (p.76) and 5.1 (p.75), it follows that $\mathbf{g}|_{v^\perp \times v^\perp}$ is positive definite , and $v^\perp$ is 3-dimensional. Let $\{e_1, e_2, e_3\}$ be a basis for $v^\perp$. Then $\{e_0, e_1, e_2, e_3\}$ is an orthonormal basis for T_pM with the Minkowski scalar product $\mathbf{g}(p)$.

(1) False. The spacelike vectors e_0+2e_1, $-2e_1$ have the timelike sum e_0.

(2) False. The lightlike vectors $e_0 + e_1$, $e_0 - e_1$ have the timelike sum $2e_0$.

(3) True. Let v be the given timelike vector. Define $e_0 = k^{-1}v$, and construct $\{e_1, e_2, e_3\}$ as done at the outset of the solution. Then the vectors $n_+ := (k/2)(e_0 + e_1)$ and $n_- := (k/2)(e_0 - e_1)$ are lightlike, and their sum is $n_+ + n_- = ke_0 = v$.

(4) False. The vectors $2e_0 + e_1$ and $-2e_0$ are both timelike, but their sum e_1 is spacelike.

(5) True. Suppose that u, v are two future-pointing timelike vectors. Then $\mathbf{g}(p)(u, V_p) < 0$ and $\mathbf{g}(p)(v, V_p) < 0$. By Ex. 5.5 (p.81), $\mathbf{g}(p)(u, v) < 0$. Thus $\mathbf{g}(p)(u+v, u+v) = \mathbf{g}(p)(u, u) + 2\mathbf{g}(p)(u, v) + \mathbf{g}(p)(v, v) < 0$, since each summand is negative. So $u + v$ is timelike. Moreover, $u + v$ is future-pointing, because $\mathbf{g}(p)(u+v, V_p) = \mathbf{g}(p)(u, V_p) + \mathbf{g}(p)(v, V_p) < 0$.

Solution to Exercise 5.23

We have $v_{\gamma,\gamma(t)} = -d\,\partial_{x^1,\gamma(t)} + \frac{d}{u}\partial_{t,\gamma(t)}$. Since for all $t \in \mathbb{R}$,

$$\begin{aligned}\mathbf{g}(\gamma(t))(v_{\gamma,\gamma(t)}, \partial_{t,\gamma(t)}) &= -\frac{d}{u} < 0, \quad \text{and} \\ \mathbf{g}(\gamma(t))(v_{\gamma,\gamma(t)}, v_{\gamma,\gamma(t)}) &= -\frac{d^2}{u^2} + d^2 = -d^2\Big(\frac{1}{u^2} - 1\Big) < 0,\end{aligned}$$

So γ is an observer. We have $\gamma(0) = p + d\mathbf{e}_1 = P$, $\gamma(1) = p + \frac{d}{u}\mathbf{e}_0 = Q$, and

$$\begin{aligned}\tau_\gamma(P, Q) &= \int_0^1 \sqrt{-\mathbf{g}(\gamma(t))(v_{\gamma,\gamma(t)}, v_{\gamma,\gamma(t)})}\, dt = \sqrt{d^2\Big(\frac{1}{u^2} - 1\Big)} \\ &< \sqrt{d^2\Big(1 + \frac{T^2}{d^2} - 1\Big)} = T.\end{aligned}$$

(With $T = 660\,\text{m}$, $d = 10 \times 10^3\,\text{m}$, we have $1/\sqrt{1 + \frac{T^2}{d^2}} \approx 0.99783$.

So if $u = 0.998$, then $1 > u > 1/\sqrt{1 + \frac{T^2}{d^2}}$. Thus $\tau_\gamma(P, Q) < T$.

In fact, $\tau_\gamma(P, Q) = \sqrt{d^2\big(\frac{1}{u^2} - 1\big)} \approx 633.4\,\text{m} = 2.11135 \times 10^{-6}\,\text{sec} < 2.2 \times 10^{-6}\,\text{sec}$.

So the muon survives the atmospheric transit.)

Solution to Exercise 5.24

(1) We have $v_{\lambda,\lambda(y)} = 0\partial_{x,\lambda(y)} + 1\partial_{y,\lambda(y)} = \partial_{y,\lambda(y)}$, and so

$$L(\lambda) = \int_y^{y_0} \sqrt{\mathbf{g}(\lambda(y))(v_{\lambda,\lambda(y)}, v_{\lambda,\lambda(y)})}\, dy = \int_y^{y_0} \sqrt{\frac{1}{y^2}}\, dy = \int_y^{y_0} \frac{1}{y}\, dy = \log\frac{y_0}{y}.$$

We have $L(\lambda) \to \infty$ as $y \searrow 0$.

(2) We have $v_{\sigma,\sigma(\theta)} = r(-\sin\theta)\partial_{x,\sigma(\theta)} + r(\cos\theta)\partial_{y,\sigma(\theta)}$, and so

$$\begin{aligned}L(\sigma) &= \int_\theta^{\theta_0} \sqrt{\mathbf{g}(\sigma(\theta))(v_{\sigma,\sigma(\theta)}, v_{\sigma,\sigma(\theta)})}\, d\theta = \int_\theta^{\theta_0} \sqrt{\frac{1}{(\sin\theta)^2}((-\sin\theta)^2 + (\cos\theta)^2)}\, d\theta \\ &= \int_\theta^{\theta_0} \frac{1}{\sin\theta}\, d\theta = \frac{1}{2}\log\Big(\frac{\tan(\theta_0/2)}{\tan(\theta/2)}\Big)^2.\end{aligned}$$

We have $L(\sigma) \to \infty$ as $\theta \searrow 0$.

Solution to Exercise 5.25

We have $v_{\gamma,\gamma(s)} = \mathrm{t}'(s)\,\partial_{t,\gamma(s)} + \phi'(s)\,\partial_{\phi,\gamma(s)}$. As γ is an observer, $v_{\gamma,\gamma(s)}$ is future-pointing, so that $\mathrm{t}'(s) > 0$ for all $s \in \mathbb{R}$. (Here we use the time-orientation on $(M, \mathbf{g})$ given in Example 5.14, p.90.) We have

$$\begin{aligned}\mathbf{g}(\gamma(s))(v_{\gamma,\gamma(s)}, v_{\gamma,\gamma(s)}) &= -\big(1-\frac{2m}{r_0}\big)(\mathrm{t}'(s))^2 + r_0^2\big(\sin\frac{\pi}{2}\big)^2(\phi'(s))^2\\ &= -\big(1-\frac{2m}{r_0}\big)(\mathrm{t}'(s))^2 + r_0^2\frac{m}{r_0^3}(\mathrm{t}'(s))^2\\ &= -\big(1-\frac{3m}{r_0}\big)(\mathrm{t}'(s))^2.\end{aligned}$$

Hence the proper time between p, q experienced by γ is

$$\begin{aligned}T_{\text{orbit}} &= \int_0^{s_0}\sqrt{-\mathbf{g}(\gamma(s))(v_{\gamma,\gamma(s)}, v_{\gamma,\gamma(s)})}\,ds = \int_0^{s_0}\sqrt{\big(1-\frac{3m}{r_0}\big)(\mathrm{t}'(s))^2}\,ds\\ &= \sqrt{\big(1-\frac{3m}{r_0}\big)}\int_0^{s_0}\mathrm{t}'(s)\,ds = \sqrt{\big(1-\frac{3m}{r_0}\big)}\,(\mathrm{t}(s_0)-\mathrm{t}(0))\\ &= \sqrt{\big(1-\frac{3m}{r_0}\big)}\,(t(\gamma(s_0))-t(\gamma(0))) = \sqrt{\big(1-\frac{3m}{r_0}\big)}\,(t(q)-t(p)).\end{aligned}$$

In the above, the map $t: M \to \mathbb{R}$ is the t-coordinate map, that is, the map $M = \mathbb{R}\times(2m,\infty)\times S^2 \ni (t, r, \mathbf{p}) \mapsto t$.

Next, we determine the proper time experienced by $\widetilde{\gamma}$ between the events p, q on its worldline. We have $v_{\widetilde{\gamma},\widetilde{\gamma}(s)} = \partial_{t,\widetilde{\gamma}(s)}$. We note that

$$\begin{aligned}\widetilde{\gamma}(t(p)) &= (t(p), r_0, \varphi^{-1}((\tfrac{\pi}{2}, \phi_0))) = p,\\ \widetilde{\gamma}(t(q)) &= (t(q), r_0, \varphi^{-1}((\tfrac{\pi}{2}, \phi_0))) = q.\end{aligned}$$

Thus

$$\begin{aligned}T_{\text{stationary}} &= \int_{t(p)}^{t(q)}\sqrt{-\mathbf{g}(\widetilde{\gamma}(s))(v_{\widetilde{\gamma},\widetilde{\gamma}(s)}, v_{\widetilde{\gamma},\widetilde{\gamma}(s)})}\,ds = \int_{t(p)}^{t(q)}\sqrt{\big(1-\frac{2m}{r_0}\big)(1)^2}\,ds\\ &= \sqrt{\big(1-\frac{2m}{r_0}\big)}\,(t(q)-t(p)).\end{aligned}$$

Consequently,

$$T_{\text{orbit}} = \sqrt{\big(1-\frac{3m}{r_0}\big)}\,(t(q)-t(p)) < \sqrt{\big(1-\frac{2m}{r_0}\big)}\,(t(q)-t(p)) = T_{\text{stationary}}.$$

Solution to Exercise 5.26

We have, using the geodesic equation involving t'' that

$$\Big(\big(1-\frac{2m}{r}\big)t'\Big)' = \frac{2m}{r^2}r't' + \big(1-\frac{2m}{r}\big)t'' = \frac{2m}{r^2}t'r' - \frac{2m}{r^2}t'r' = 0.$$

So $(1-\frac{2m}{r})t'$ is constant, say E. As the time-orientation is given by ∂_t, and since $v_{\gamma_*,\gamma_*(\tau)}$ is future-pointing, we have $t' > 0$ everywhere. So $E > 0$.

Moreover, multiplying $-(1-\frac{2m}{r})(t')^2+(1-\frac{2m}{r})^{-1}(r')^2=-1$ by $1-\frac{2m}{r}$, and using $(1-\frac{2m}{r})t' = E$ then gives $-E^2+(r')^2 = -(1-\frac{2m}{r})$. So we obtain $(r')^2 = E^2-1+\frac{2m}{r}$. If $r(\tau_1) = r_1$ for $\tau_1 \in I$, then the r-maximisation assumption gives $r'(\tau_1) = 0$, so that $0^2 = E^2-1+\frac{2m}{r_1}$. Thus $E = \sqrt{1-\frac{2m}{r_1}}$. So $\left(\frac{dr}{d\tau}\right)^2 = E^2-1+\frac{2m}{r} = \frac{2m}{r}-\frac{2m}{r_1}$, and while r increases from $r=r_0$ to r_1,

$$\frac{dr}{d\tau} = \sqrt{\frac{2m}{r}-\frac{2m}{r_1}},$$

and on the return trip from $r = r_1$ to $r = r_0$, we have a minus sign in front of the square root. Thus the total proper time to make the round trip is

$$\Delta\tau = 2\int_{r_0}^{r_1} \frac{1}{\sqrt{\frac{2m}{r}-\frac{2m}{r_1}}}dr.$$

Let $\tau_0, \tau_2 \in I$ be such that $r(\tau_0)=r_0=r(\tau_2)$. If s denotes the inverse of the (increasing) function $r:(\tau_0,\tau_1)\to(r_0,r_1)$, then

$$\left.\frac{d(t\circ s)}{du}\right|_{r=r(\tau)} = t'(\tau)\frac{1}{r'(\tau)} = \frac{\sqrt{1-\frac{2m}{r_1}}}{1-\frac{2m}{r}}\frac{1}{\sqrt{\frac{2m}{r}-\frac{2m}{r_1}}}.$$

So the coordinate time lapse to go from $r = r_0$ to $r = r_1$ is given by

$$(\Delta t)_1 = \int_{r_0}^{r_1} \frac{\sqrt{1-\frac{2m}{r_1}}}{(1-\frac{2m}{r})\sqrt{\frac{2m}{r}-\frac{2m}{r_1}}}dr.$$

A similar calculation using the (decreasing) function $r:(\tau_1,\tau_2)\to(r_0,r_1)$ shows that the coordinate time lapse to go from $r=r_1$ to $r=r_0$ is equal to $(\Delta t)_1$. Hence the coordinate time lapse for the entire round trip is $\Delta t = 2(\Delta t)_1$, as wanted.

Finally, to show the inequality, we first note that

$$\begin{aligned}
\Delta t &= 2\int_{r_0}^{r_1} \frac{\sqrt{1-\frac{2m}{r_1}}}{(1-\frac{2m}{r})\sqrt{\frac{2m}{r}-\frac{2m}{r_1}}}dr \\
&> 2\int_{r_0}^{r_1} \frac{\sqrt{1-\frac{2m}{r_1}}}{(1-\frac{2m}{r_1})}\frac{\sqrt{rr_1}}{\sqrt{2m}\sqrt{r_1-r}}dr \\
&> 2\int_{r_0}^{r_1} \frac{1}{\sqrt{1-\frac{2m}{r_1}}}\frac{r_0}{\sqrt{2m}\sqrt{r_1-r}}dr = \frac{2r_0}{\sqrt{1-\frac{2m}{r_1}}\sqrt{2m}}2\sqrt{r_1-r_0} \\
&= \frac{4r_0}{\sqrt{1-\frac{2m}{r_1}}}\frac{\sqrt{r_1r_0}}{2m}\sqrt{\frac{2m}{r_0}-\frac{2m}{r_1}} \geqslant \frac{2r_0}{\sqrt{1-\frac{2m}{r_1}}}\frac{r_0}{m}\sqrt{\frac{2m}{r_0}-\frac{2m}{r_1}}. \qquad (\star)
\end{aligned}$$

Also, we compute

$$\frac{d(\Delta\tau)}{dr_0} = -2\frac{1}{\sqrt{\frac{2m}{r_0}-\frac{2m}{r_1}}},$$

$$\frac{d(\sqrt{1-\frac{2m}{r_0}}\Delta t)}{dr_0} = \frac{1}{2\sqrt{1-\frac{2m}{r_0}}}\frac{2m}{r_0^2}\Delta t + \sqrt{1-\frac{2m}{r_0}}\,\frac{(-2\sqrt{1-\frac{2m}{r_1}})}{(1-\frac{2m}{r_0})\sqrt{\frac{2m}{r_0}-\frac{2m}{r_1}}}.$$

Using ($\star$) and the above expressions, we obtain (after some algebra)

$$\frac{d(\sqrt{1-\frac{2m}{r_0}}\Delta t)}{dr_0} - \frac{d(\Delta\tau)}{dr_0} > \frac{2}{\sqrt{1-\frac{2m}{r_1}}\sqrt{\frac{2m}{r_0}-\frac{2m}{r_1}}}\Big(\sqrt{1-\frac{2m}{r_1}}-\sqrt{1-\frac{2m}{r_0}}\Big) > 0.$$

Thus, the function

$$r_0 \mapsto \sqrt{1-\frac{2m}{r_0}}\Delta t - \Delta\tau,$$

is increasing. But the value when r_0 equals r_1 is 0, and so the value at $r_0 < r_1$ must be negative, which gives the desired inequality.

Solution to Exercise 5.27

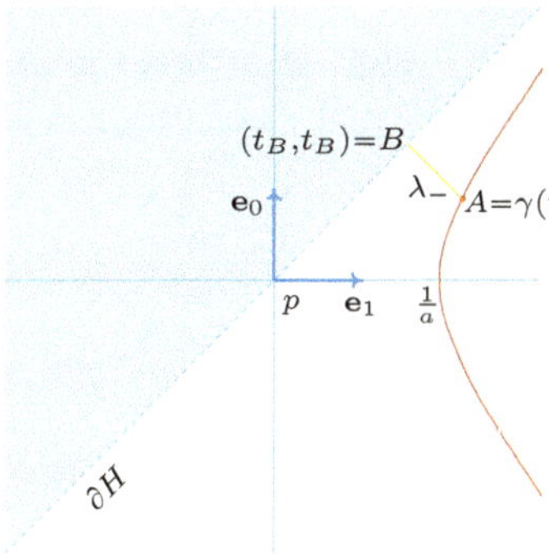

The slope of the segment AB is -1, giving $\dfrac{t_B - \frac{1}{a}\sinh(t_0 a)}{t_B - \frac{1}{a}\cosh(t_0 a)} = -1$, that is,

$$t_B = \frac{1}{2a}(\cosh(t_0 a) + \sinh(t_0 a)) = \frac{e^{t_0 a}}{2a}.$$

So the time Δt it takes for the photon λ_- to reach the horizon in the $(U, \mathbf{x})$-coordinates is

$$\Delta t = t_B - \frac{1}{a}\sinh(t_0 a) = \frac{e^{t_0 a}}{2a} - \frac{e^{t_0 a} - e^{-t_0 a}}{2a} = \frac{e^{-t_0 a}}{2a}.$$

Hence

$$E = \frac{\hbar}{2\Delta t} = \frac{\hbar}{2\frac{e^{-t_0 a}}{2a}} = \frac{\hbar a}{e^{-t_0 a}}.$$

We have[19] $v_{\lambda_+,\lambda_+(t_0)} = E(\mathbf{e}_0 + \mathbf{e}_1)$ and $v_{\gamma,\gamma(t)} = \cosh(at_0)\mathbf{e}_0 + \sinh(at_0)\mathbf{e}_1$. So

$$\begin{aligned} E_{\text{observed}} &= -\frac{\eta\big(E(\mathbf{e}_0+\mathbf{e}_1), \cosh(at_0)\mathbf{e}_0 + \sinh(at_0)\mathbf{e}_1\big)}{\sqrt{-\eta\big(\cosh(at_0)\mathbf{e}_0 + \sinh(at_0)\mathbf{e}_1, \cosh(at_0)\mathbf{e}_0 + \sinh(at_0)\mathbf{e}_1\big)}} \\ &= -\frac{(-\cosh(at_0) + \sinh(at_0))\,\frac{\hbar a}{e^{-t_0 a}}}{\sqrt{-\big(-(\cosh(at_0))^2 + (\sinh(at_0))^2\big)}} = \hbar a. \end{aligned}$$

Solution to Exercise 5.28

We have $\mathbf{g}(\operatorname{grad} f, V) = (\operatorname{grad} f)^\flat V = ((df)^\sharp)^\flat V = df(V) = Vf$.

For the given global chart on $\mathbb{R}^m$, $\mathbf{g}_{ij} = \delta_{ij}$, and so $\mathbf{g}^{ij} = \delta^{ij}$. We have $(df)_i = df(\partial_{x^i}) = \partial_{x^i} f$. Thus $(\operatorname{grad} f)^i = \delta^{ji}\partial_{x^j} f = \partial_{x^i} f$.

Solution to Exercise 5.29

As $(v_i)^\flat \in V^*$, and $\{\omega^1, \cdots, \omega^m\}$ forms a basis for V^*, there exist $c_{ij} \in \mathbb{R}$ such that $(v_i)^\flat = c_{ij}\omega^j$. Acting on v_k yields

$$g(v_i, v_k) = (v_i)^\flat v_k = c_{ij}\omega^j v_k = c_{ij}\delta^j_k = c_{ik}.$$

So $(v_i)^\flat = c_{ij}\omega^j = g(v_i, v_j)\omega^j$.

Formula for $\omega^\sharp$ when $\{v_1, \cdots, v_m\}$ is orthonormal: Writing $\omega = c_i\omega^i$, and operating on v_k, we get $c_k = c_i\delta^i_k = c_i\omega^i(v_k) = \omega v_k$. So (dropping the summation convention)

$$\omega = \sum_{i=1}^m c_i\omega^i = \sum_{i=1}^m (\omega v_i)\omega^i = \sum_{i=1}^m (\omega v_i)(\iota(v_i)\cdot(v_i)^\flat),$$

giving $\omega^\sharp = \sum_{i=1}^m (\omega v_i) g(v_i, v_i)((v_i)^\flat)^\sharp = \sum_{i=1}^m (\omega v_i) g(v_i, v_i) v_i$.

Solution to Exercise 5.30

In any admissible chart $(U, \mathbf{x})$, we have

$$\begin{aligned} \operatorname{trace}\mathbf{g} &= \mathbf{C}(\mathbf{g}^\sharp) = \mathbf{g}^\sharp(dx^i, \partial_{x^i}) = \mathbf{g}((dx^i)^\sharp, \partial_{x^i}) \\ &= \mathbf{g}(\mathbf{g}^{kj}\delta^i_k\partial_{x^j}, \partial_{x^i}) = \mathbf{g}(\mathbf{g}^{ij}\partial_{x^j}, \partial_{x^i}) \\ &= \mathbf{g}^{ij}\mathbf{g}_{ji} = \delta^i_i = 4. \end{aligned}$$

[19] More precisely, with $p := \lambda_+(t_0)$, we ought to write $v_{\lambda_+,p} = \partial_{x^0,p} + \partial_{x^1,p} = I(\mathbf{e}_0 + \mathbf{e}_1)$, where $I : V \to T_pM$ is the isomorphism described in Example 2.2 (p.27).

Chapter 6

Solution to Exercise 6.1

Let $V = V^i\partial_{x^i}$, $W = W^j\partial_{x^j}$, $X = X^k\partial_{x^k}$, $c \in \mathbb{R}$, $f \in C^\infty(\mathbb{R}^m)$. Then:

(C1)
$$\begin{aligned}\overline{\nabla}_V(W+cX) &= V(W^i+cX^i)\partial_{x^i}\\ &= (VW^i+cVX^i)\partial_{x^i}\\ &= VW^i\partial_{x^i}+cVX^i\partial_{x^i}\\ &= \overline{\nabla}_V W+c\overline{\nabla}_V X.\end{aligned}$$

(C2)
$$\begin{aligned}\overline{\nabla}_V(fW) &= V(fW^i)\partial_{x^i} = ((Vf)W^i + fV(W^i))\partial_{x^i}\\ &= (Vf)W^i\partial_{x^i}+f(VW^i)\partial_{x^i}\\ &= (Vf)W+f\overline{\nabla}_V W.\end{aligned}$$

(C3)
$$\begin{aligned}\overline{\nabla}_{fV+X}W &= ((fV+X)W^i)\partial_{x^i}\\ &= (fVW^i+XW^i)\partial_{x^i}\\ &= f(VW^i)\partial_{x^i}+(XW^i)\partial_{x^i}\\ &= f\overline{\nabla}_V W+\overline{\nabla}_X W.\end{aligned}$$

(C4) $\overline{\nabla}_V W - \overline{\nabla}_W V = (VW^i)\partial_{x^i} - (WV^i)\partial_{x^i}$. Also,
$$\begin{aligned}[V,W]f &= V^i\partial_{x^i}(W^j\partial_{x^j}f) - W^j\partial_{x^j}(V^i\partial_{x^i}f)\\ &= V^i(\partial_{x^i}W^j)(\partial_{x^j}f) + V^iW^j\partial_{x^i}\partial_{x^j}f\\ &\quad -W^j(\partial_{x^j}V^i)(\partial_{x^i}f) - W^jV^i\partial_{x^j}\partial_{x^i}f\\ &= ((V(W^j) - W(V^j))\partial_{x^j})f.\end{aligned}$$
Thus $[V,W] = (V(W^j) - W(V^j))\partial_{x^j} = \overline{\nabla}_V W - \overline{\nabla}_W V$.

(C5) Finally, we have
$$\begin{aligned}&\langle\overline{\nabla}_V W, X\rangle + \langle W, \overline{\nabla}_V X\rangle\\ &= \langle V(W^i)\partial_{x^i}, X^j\partial_{x^j}\rangle + \langle W^i\partial_{x^i}, V(X^j)\partial_{x^j}\rangle\\ &= V(W^i)X^j\langle\partial_{x^i}, \partial_{x^j}\rangle + W^iV(X^j)\langle\partial_{x^i}, \partial_{x^j}\rangle\\ &= (V(W^i)X^j + W^iV(X^j))\langle\partial_{x^i}, \partial_{x^j}\rangle\\ &= (V(W^iX^j))\langle\partial_{x^i}, \partial_{x^j}\rangle\\ &= (V(W^iX^j))\langle\partial_{x^i}, \partial_{x^j}\rangle + 0\\ &= (V(W^iX^j))\langle\partial_{x^i}, \partial_{x^j}\rangle + W^iX^jV(\delta_{ij})\\ &= (V(W^iX^j))\langle\partial_{x^i}, \partial_{x^j}\rangle + W^iX^jV(\langle\partial_{x^i}, \partial_{x^j}\rangle)\\ &= V(W^iX^j\langle\partial_{x^i}, \partial_{x^j}\rangle)\\ &= V\langle W^i\partial_{x^i}, X^j\partial_{x^j}\rangle\\ &= V\langle W, X\rangle.\end{aligned}$$
(In the above, we used $V(\delta_{ij}) = 0$, since δ_{ij} is a constant.)

Solution to Exercise 6.2

We have

$$\begin{aligned}\overline{\nabla}_V W &= ((yz\,\partial_x + zx\,\partial_y + xy\,\partial_z)(xy^2+z))\partial_x \\ &\quad +((yz\,\partial_x + zx\,\partial_y + xy\,\partial_z)(y^2-x))\partial_y \\ &\quad +((yz\,\partial_x + zx\,\partial_y + xy\,\partial_z)(x+z^3))\partial_z \\ &= (yz(y^2) + zx(x2y) + xy(1))\partial_x \\ &\quad +(yz(-1) + zx(2y) + xy(0))\partial_y \\ &\quad +(yz(1) + zx(0) + xy(3z^2))\partial_z \\ &= (zy^3 + 2x^2yz + xy)\partial_x + (2xyz - yz)\partial_y + (yz + 3xyz^2)\partial_z.\end{aligned}$$

So $\overline{\nabla}_{V_p} W = (\overline{\nabla}_V W)(p) = 4\,\partial_{x,p} + \partial_{y,p} + 4\,\partial_{z,p}$.

Solution to Exercise 6.3

Define $\widetilde{\nabla}_V W = V^j W^i \Gamma^k_{ij}\partial_{x^k}$. Then $\nabla_V = \widetilde{\nabla}_V + \overline{\nabla}_V$.
We know that $\overline{\nabla}$ satisfies (C1), (C2), (C3). We have:

(1) $\widetilde{\nabla}_V(W+cX) = V^j(W^i+cX^i)\Gamma^k_{ij}\partial_{x^k} = V^jW^i\Gamma^k_{ij}\partial_{x^k} + cV^jX^i\Gamma^k_{ij}\partial_{x^k}$
$= \widetilde{\nabla}_V W + c\widetilde{\nabla}_V X.$

(2) $\widetilde{\nabla}_V(fW) = V^j(fW)^i\Gamma^k_{ij}\partial_{x^k} = fV^jW^i\Gamma^k_{ij}\partial_{x^k} = f\widetilde{\nabla}_V W.$

(3) $\widetilde{\nabla}_{fV+X} W = (fV+X)^jW^i\Gamma^k_{ij}\partial_{x^k} = fV^jW^i\Gamma^k_{ij}\partial_{x^k} + X^jW^i\Gamma^k_{ij}\partial_{x^k}$
$= f\widetilde{\nabla}_V W + \widetilde{\nabla}_X W.$

Hence:

(C1) $\nabla_V(W+cX) = \widetilde{\nabla}_V(W+cX) + \overline{\nabla}_V(W+cX)$
$\overset{(1)}{=} \widetilde{\nabla}_V W + c\widetilde{\nabla}_V X + \overline{\nabla}_V W + c\overline{\nabla}_V X = \nabla_V W + c\nabla_V X.$

(C2) $\nabla_V(fW) = \widetilde{\nabla}_V(fW) + \overline{\nabla}_V(fW)$
$\overset{(2)}{=} f\widetilde{\nabla}_V W + (Vf)W + f\overline{\nabla}_V W = f\nabla_V W + (Vf)W.$

(C3) $\nabla_{fV+X} W = \widetilde{\nabla}_{fV+X} W + \overline{\nabla}_{fV+X} W$
$\overset{(3)}{=} f\widetilde{\nabla}_V W + \widetilde{\nabla}_X W + f\overline{\nabla}_V W + \overline{\nabla}_X W = f\nabla_V W + \nabla_X W.$

So ∇ is a connection.

Solution to Exercise 6.4

First, we note that if $\varphi, f \in C^\infty(M)$ and $V, W \in T^1_0 M$, then

$$\begin{aligned}[fV,W]\varphi &= fV(W\varphi) - W(fV\varphi) = fV(W\varphi) - (Wf)(V\varphi) - fW(V\varphi) \\ &= f(V(W\varphi) - W(V\varphi)) - (Wf)(V\varphi) \\ &= f([V,W]\varphi) - (Wf)(V\varphi) = (f[V,W] - (Wf)V)\varphi.\end{aligned}$$

Thus $[fV,W] = f[V,W] - (Wf)V$.

Now for $f \in C^\infty(M)$, $V, W, X \in T_0^1 M$,

$$\begin{aligned} B(fV+X,W) &= \nabla_{fV+X}W - \nabla_W(fV+X) - [fV+X,W] \\ &= f\nabla_V W + \nabla_X W - (Wf)V - f\nabla_W V - \nabla_W X \\ &\quad -f[V,W] + (Wf)V - [X,W] \\ &= f(\nabla_V W - \nabla_W V - [V,W]) + \nabla_X W - \nabla_W X - [X,W] \\ &= fB(V,W) + B(X,W). \end{aligned}$$

Also,

$$B(V,W) = \nabla_V W - \nabla_W V - [V,W] = -(\nabla_W V - \nabla_V W - [W,V]) = -B(W,V).$$

Thus B is $C^\infty(M)$-bilinear. Finally, T is $C^\infty(M)$-linear in the first slot:

$$\begin{aligned} T(f\Omega+\Theta,V,W) &= (f\Omega+\Theta)(B(V,W)) = f\Omega(B(V,W)) + \Theta(B(V,W)) \\ &= fT(\Omega,V,W) + T(\Theta,V,W), \end{aligned}$$

and also in the latter two slots, e.g.

$$\begin{aligned} T(\Omega,V,fW+X) &= \Omega(B(V,fW+X)) = \Omega(-B(fW+X,V)) \\ &= \Omega(-fB(W,V)-B(X,V)) = \Omega(fB(V,W)+B(V,X)) \\ &= f\Omega(B(V,W)) + \Omega(B(V,X)) = fT(\Omega,V,W) + T(\Omega,V,X). \end{aligned}$$

So T is a $(1,2)$-tensor field.

Solution to Exercise 6.5

For $V, W, X \in T_0^1 M$, $f \in C^\infty(M)$, $c \in \mathbb{R}$, and with $\nabla^{\mathrm{c}} := (1-\rho)\nabla + \rho\widetilde{\nabla}$:

(C1)
$$\begin{aligned} \nabla^{\mathrm{c}}{}_V(W+cX) &= (1-\rho)\nabla_V(W+cX) + \rho\widetilde{\nabla}_V(W+cX) \\ &= (1-\rho)(\nabla_V W + c\nabla_V X) + \rho(\widetilde{\nabla}_V W + c\widetilde{\nabla}_V X) \\ &= (1-\rho)\nabla_V W + \rho\widetilde{\nabla}_V W + c((1-\rho)\nabla_V X + \rho\widetilde{\nabla}_V X) \\ &= \nabla^{\mathrm{c}}{}_V W + c\nabla^{\mathrm{c}}{}_V X. \end{aligned}$$

(C2)
$$\begin{aligned} \nabla^{\mathrm{c}}{}_V(fW) &= (1-\rho)\nabla_V(fW) + \rho\widetilde{\nabla}_V(fW) \\ &= (1-\rho)((Vf)W + f\nabla_V W) + \rho((Vf)W + f\widetilde{\nabla}_V W) \\ &= (1-\rho+\rho)(Vf)W + f((1-\rho)\nabla_V W + \rho\widetilde{\nabla}_V W) \\ &= (Vf)W + f\nabla^{\mathrm{c}}{}_V W. \end{aligned}$$

(C3)
$$\begin{aligned} \nabla^{\mathrm{c}}{}_{fV+X}W &= (1-\rho)\nabla_{fV+X}W + \rho\widetilde{\nabla}_{fV+X}W \\ &= (1-\rho)(f\nabla_V W + \nabla_X W) + \rho(f\widetilde{\nabla}_V W + \widetilde{\nabla}_X W) \\ &= f((1-\rho)\nabla_V W + \rho\widetilde{\nabla}_V W) + (1-\rho)\nabla_X W + \rho\widetilde{\nabla}_X W \\ &= f\nabla^{\mathrm{c}}{}_V W + \nabla^{\mathrm{c}}{}_X W. \end{aligned}$$

Solution to Exercise 6.6

With $f := x$, $V := \partial_x$, $W := \partial_x$, $2\overline{\nabla}_V(f\cdot W) = 2\overline{\nabla}_{\partial_x}(x\partial_x) = 2(\partial_x x)\partial_x = 2\partial_x$, while $(Vf)W + f\,2\overline{\nabla}_V W = (\partial_x x)\partial_x + x\,2(\partial_x 1)\partial_x = \partial_x + 0 = \partial_x$. So we have $2\overline{\nabla}_V(f\cdot W) = 2\,\partial_x \neq \partial_x = (Vf)W + f\,2\overline{\nabla}_V W$.

Solution to Exercise 6.7

For $f \in C^\infty(M)$, $V, W, X \in T_0^1 M$, $\Omega, \Theta \in T_1^0 M$, we have:

$$\begin{aligned} T(\Omega + f\Theta, V, W) &= (\Omega + f\Theta)(\nabla_V W - \widetilde{\nabla}_V W) \\ &= \Omega(\nabla_V W - \widetilde{\nabla}_V W) + f\Theta(\nabla_V W - \widetilde{\nabla}_V W) \\ &= T(\Omega, V, W) + fT(\Theta, V, W), \\ T(\Omega, fV + X, W) &= \Omega(\nabla_{fV+X} W - \widetilde{\nabla}_{fV+X} W) \\ &= \Omega(f\nabla_V W + \nabla_X W - f\widetilde{\nabla}_V W - \widetilde{\nabla}_X W) \\ &= f\Omega(\nabla_V W - \widetilde{\nabla}_V W) + \Omega(\nabla_X W - \widetilde{\nabla}_X W) \\ &= fT(\Omega, V, W) + T(\Omega, X, W). \end{aligned}$$

Finally,

$$\begin{aligned} &T(\Omega, V, fW + X) \\ &= \Omega(\nabla_V(fW + X) - \widetilde{\nabla}_V(fW + X)) \\ &= \Omega((Vf)W + f\nabla_V W + \nabla_V X - (Vf)W - f\widetilde{\nabla}_V W - \widetilde{\nabla}_V X) \\ &= \Omega(f(\nabla_V W - \widetilde{\nabla}_V W) + \nabla_V X - \widetilde{\nabla}_V X) \\ &= f\Omega(\nabla_V W - \widetilde{\nabla}_V W) + \Omega(\nabla_V X - \widetilde{\nabla}_V X) = fT(\Omega, V, W) + T(\Omega, V, X). \end{aligned}$$

So T is a $(1, 2)$-tensor field.

Solution to Exercise 6.8

Firstly, for $V, W \in T_0^1 M$, and $f, \varphi \in C^\infty(M)$, we have

$$\begin{aligned} [V, fW]\varphi &= V(fW\varphi) - fW(V\varphi) = (Vf)(W\varphi) + fV(W\varphi) - fW(V\varphi) \\ &= (Vf)(W\varphi) + f[V, W]\varphi. \end{aligned}$$

So $\mathcal{L}_V(fW) = (Vf)W + f\mathcal{L}_V W$, that is, (C2) holds. Moreover, we have for $V, W, X \in T_0^1 M$ and $c \in \mathbb{R}$, that

$$\mathcal{L}_V(W + cX) = [V, W + cX] = [V, W] + c[V, X] = \mathcal{L}_V W + c\mathcal{L}_V X,$$

and so (C1) holds.

With $f = x$, $W = \partial_x = V$ and $X = 0$, we have

$$\mathcal{L}_{fV+X} W = \mathcal{L}_{fV} W = \mathcal{L}_{x\partial_x}\partial_x = [x\partial_x, \partial_x].$$

For $\varphi \in C^\infty(\mathbb{R}^2)$,

$$\begin{aligned} [x\partial_x, \partial_x]\varphi &= x\partial_x(\partial_x\varphi) - \partial_x(x\partial_x\varphi) \\ &= x\partial_x(\partial_x\varphi) - (\partial_x x)(\partial_x\varphi) - x\partial_x\partial_x\varphi = -1\,\partial_x\varphi. \end{aligned}$$

So $[x\partial_x, \partial_x] = -\partial_x$. Thus $\mathcal{L}_{fV+X} W = -\partial_x$. But

$$\begin{aligned} \mathcal{L}_{fV+X} W = -\partial_x \neq 0 &= x0 + 0 = x[\partial_x, \partial_x] + [0, \partial_x] \\ &= x\mathcal{L}_{\partial_x}\partial_x + \mathcal{L}_0\partial_x = f\mathcal{L}_V W + \mathcal{L}_X W. \end{aligned}$$

Hence (C3) fails to hold.

Solution to Exercise 6.9

We have $\partial_u = (\partial_u x)\partial_x + (\partial_u y)\partial_y + (\partial_u z)\partial_z = 1\partial_x + 0\partial_y + 0\partial_z = \partial_x$, which corresponds to the vector $(1,0,0) \in \mathbb{R}^3$. Similarly, $\partial_v = 0\partial_x + 1\partial_y + 0\partial_z = \partial_y$, corresponding to $(0,1,0) \in \mathbb{R}^3$. So $\overline{\nabla}_{\partial_u}\partial_u = \overline{\nabla}_{\partial_x}\partial_x = 0$, $\overline{\nabla}_{\partial_u}\partial_v = \overline{\nabla}_{\partial_x}\partial_y = 0$, $\overline{\nabla}_{\partial_v}\partial_u = \overline{\nabla}_{\partial_y}\partial_x = 0$, and $\overline{\nabla}_{\partial_v}\partial_v = \overline{\nabla}_{\partial_y}\partial_y = 0$. For $V = V^u\partial_u + V^v\partial_v$ and $W = W^u\partial_u + W^v\partial_v$, we have $\overline{\nabla}_V W = V(W^u)\partial_u + V(W^v)\partial_v$, which corresponds to $(V(W^u), V(W^v), 0) \in \mathbb{R}^3$. At each point in M, we have

$$\widehat{\mathbf{n}} = \frac{\frac{\partial \mathbf{r}}{\partial u} \times \frac{\partial \mathbf{r}}{\partial v}}{\left\|\frac{\partial \mathbf{r}}{\partial u} \times \frac{\partial \mathbf{r}}{\partial v}\right\|} = \frac{(0,0,1)}{\|(0,0,1)\|} = (0,0,1).$$

So $\overline{\nabla}_V W \equiv (V(W^u), V(W^v), 0)$ is orthogonal to $\widehat{\mathbf{n}} = (0,0,1)$, and hence $\nabla_V W = \overline{\nabla}_V W - \langle \overline{\nabla}_V W, \widehat{\mathbf{n}}\rangle\, \widehat{\mathbf{n}} = \overline{\nabla}_V W = V(W^u)\partial_u + V(W^v)\partial_v$. This is a connection of the form given in Exercise 6.3 (p.108) on $\mathbb{R}^2$, where the functions $\Gamma^u_{uu} = \Gamma^u_{uv} = \Gamma^u_{vu} = \Gamma^u_{vv} = 0$ and $\Gamma^v_{uu} = \Gamma^v_{uv} = \Gamma^v_{vu} = \Gamma^v_{vv} = 0$. Thus the connection ∇ is just the flat connection $\overline{\nabla}$ on $\mathbb{R}^2$.

Solution to Exercise 6.10

We have by Lemma 6.1 (p.112) that $(\nabla_{\widetilde{V}}\widetilde{W})_p = (\nabla_{\widetilde{V}} W)_p$. We want to show that $(\nabla_{\widetilde{V}} W)_p = (\nabla_V W)_p$. We have

$$(\nabla_V W)_p - (\nabla_{\widetilde{V}} W)_p = (\nabla_V W - \nabla_{\widetilde{V}} W)_p \overset{(\mathrm{C3})}{=} (\nabla_{V-\widetilde{V}} W)_p.$$

Let $(U, \mathbf{x})$ be a chart such that $p \in U$. Set $V - \widetilde{V} = X$. Writing $X = X^i\partial_{x^i}$ in U, we have $X^i(p)\,\partial_{x^i,p} = X_p = V_p - \widetilde{V}_p = 0$. So $X^i(p) = 0$, $1 \leqslant i \leqslant m$. By Proposition 6.1 (p.112),

$$\begin{aligned}(\nabla_X W)_p &= ((\nabla_X W)|_U)_p = (\nabla^U_{X|_U} W|_U)_p = (\nabla^U_{X^i\partial_{x^i}} W|_U)_p \\ &= (X^i \nabla^U_{\partial_{x^i}} W|_U)_p = X^i(p)(\nabla^U_{\partial_{x^i}} W|_U)_p = 0\cdot(\nabla^U_{\partial_{x^i}} W|_U)_p = 0.\end{aligned}$$

Thus we have $(\nabla_V W)_p - (\nabla_{\widetilde{V}} W)_p = (\nabla_{V-\widetilde{V}} W)_p = (\nabla_X W)_p = 0$, that is, $(\nabla_V W)_p = (\nabla_{\widetilde{V}} W)_p$. Consequently, $(\nabla_{\widetilde{V}}\widetilde{W})_p = (\nabla_{\widetilde{V}} W)_p = (\nabla_V W)_p$.

Solution to Exercise 6.11

We have that in $(U, \mathbf{x})$,

$$\begin{aligned}\operatorname{div} V &= dx^i(\nabla_{\partial_{x^i}} V) = dx^i(\nabla_{\partial_{x^i}}(V^j\partial_{x^j})) = dx^i(\partial_{x^i}V^j\partial_{x^j} + V^j\nabla_{\partial_{x^i}}\partial_{x^j}) \\ &= dx^i(\partial_{x^i}V^j\partial_{x^j} + V^j\Gamma^k_{ji}\partial_{x^k}) = \partial_{x^i}V^k\delta^i_k + V^j\Gamma^k_{ji}\delta^i_k = \partial_{x^i}V^i + V^j\Gamma^i_{ji}.\end{aligned}$$

Solution to Exercise 6.12

Recall from Lemma 3.3 (p.45) that $[\partial_{x^i}, \partial_{x^j}] = 0$. From the torsion-freeness,

$$\nabla_{\partial_{x^i}}\partial_{x^j} - \nabla_{\partial_{x^j}}\partial_{x^i} = [\partial_{x^i}, \partial_{x^j}] = 0.$$

So $\Gamma^k_{ij} = dx^k(\nabla_{\partial_{x^j}}\partial_{x^i}) = dx^k(\nabla_{\partial_{x^i}}\partial_{x^j}) = \Gamma^k_{ji}$.

Solution to Exercise 6.13

Uniqueness: Let $\nabla_V \partial_{x^i} = (\Omega_i^j V)\partial_{x^j}$. Acting by dx^k, we obtain

$$\begin{aligned}(\Omega_i^j V)\delta_j^k &= dx^k((\Omega_i^j V)\partial_{x^j}) \\ &= dx^k(\nabla_V \partial_{x^i}),\end{aligned}$$

that is,

$$\begin{aligned}(\Omega_i^k V) &= dx^k(\nabla_V \partial_{x^i}) \\ &= dx^k(\nabla_{V^j\partial_{x^j}} \partial_{x^i}) \\ &= V^j \Gamma_{ij}^\ell \delta_\ell^k \\ &= V^j \Gamma_{ij}^k \\ &= \Gamma_{ij}^k V^\ell \delta_\ell^j \\ &= \Gamma_{ij}^k V^\ell dx^j(\partial_{x^\ell}) \\ &= \Gamma_{ij}^k dx^j(V^\ell \partial_{x^\ell}) \\ &= \Gamma_{ij}^k dx^j(V).\end{aligned}$$

Thus $\Omega_i^k = \Gamma_{ij}^k dx^j$.
Existence: Define $\Omega_i^k \in T_1^0 U$ by $\Omega_i^k = \Gamma_{ij}^k dx^j$. If $V \in T_0^1 M$, then $V|_U \in T_0^1 U$, and we can decompose $V|_U = V^\ell \partial_{x^\ell}$, giving

$$\begin{aligned}(\Omega_i^j V)\partial_{x^j} &= \Gamma_{ir}^j dx^r(V^\ell \partial_{x^\ell})\partial_{x^j} \\ &= \Gamma_{ir}^j V^\ell \delta_\ell^r \partial_{x^j} \\ &= \Gamma_{ir}^j V^r \partial_{x^j} \\ &= V^r \nabla_{\partial_{x^r}} \partial_{x^i} \\ &= \nabla_{V^r \partial_{x^r}} \partial_{x^i} \\ &= \nabla_V \partial_{x^i}.\end{aligned}$$

Recovering Γ-symbols from the connection 1-form matrix $[\Omega_j^i]$: We have

$$\begin{aligned}\Omega_i^k \partial_{x^j} &= \Gamma_{i\ell}^k dx^\ell \partial_{x^j} \\ &= \Gamma_{i\ell}^k \delta_j^\ell \\ &= \Gamma_{ij}^k.\end{aligned}$$

Solution to Exercise 6.14

If $W \in T_0^1 M$, then $\mathbf{g}((df)^\sharp, W) = ((df)^\sharp)^\flat W = (df)W = Wf$. For all $V \in T_0^1 M$,

$$\begin{aligned}\mathbf{g}(\nabla_{(df)^\sharp}((df)^\sharp), V) &= (df)^\sharp(\mathbf{g}((df)^\sharp, V)) - \mathbf{g}((df)^\sharp, \nabla_{(df)^\sharp} V) \\ &= (df)^\sharp(Vf) - \mathbf{g}((df)^\sharp, \nabla_V((df)^\sharp)) - \mathbf{g}((df)^\sharp, [(df)^\sharp, V]) \\ &= (df)^\sharp(Vf) - \tfrac{1}{2} V(\mathbf{g}((df)^\sharp, (df)^\sharp)) - [(df)^\sharp, V]f \\ &= (df)^\sharp(Vf) - \tfrac{1}{2}0 - (df)^\sharp(Vf) - V((df)^\sharp f) \\ &= -V((df)^\sharp f) \\ &= -V(\mathbf{g}((df)^\sharp, (df)^\sharp)) = 0.\end{aligned}$$

Solution to Exercise 6.15

For all $Z \in T_0^1 M$, and $p \in M$, we have

$$\begin{aligned}
&2\mathbf{h}(f(p))((f_*(\nabla^{\mathbf{g}}_X Y))_{f(p)}, (f_*Z)_{f(p)}) \\
&= 2\mathbf{h}(f(p))(df_p(\nabla^{\mathbf{g}}_X Y)_p, df_p Z_p) \\
&= 2(f^*\mathbf{h})(p)((\nabla^{\mathbf{g}}_X Y)_p, Z_p) \\
&= 2\mathbf{g}(p)((\nabla^{\mathbf{g}}_X Y)_p, Z_p)) \\
&= X_p\mathbf{g}(Y,Z) + Y_p\mathbf{g}(Z,X) - Z_p\mathbf{g}(X,Y) \\
&\quad -\mathbf{g}(p)([X,Z]_p, Y_p) - \mathbf{g}(p)([Y,X]_p, Z_p) + \mathbf{g}(p)([Z,Y]_p, X_p).
\end{aligned}$$

Also,

$$\begin{aligned}
&2\mathbf{h}(f(p))(\nabla^{\mathbf{h}}_{f_*X}(f_*Y)_{f(p)}, (f_*Z)_{f(p)}) \\
&= (f_*X)_{f(p)}\mathbf{h}(f_*Y, f_*Z) + (f_*Y)_{f(p)}\mathbf{h}(f_*Z, f_*X) \\
&\quad -(f_*Z)_{f(p)}\mathbf{h}(f_*X, f_*Y) \\
&\quad -\mathbf{h}(f(p))([f_*X, f_*Z]_{f(p)}, (f_*Y)_{f(p)}) - \mathbf{h}(f(p))([f_*Y, f_*X]_{f(p)}, (f_*Z)_{f(p)}) \\
&\quad +\mathbf{h}(f(p))([f_*Z, f_*Y]_{f(p)}, (f_*X)_{f(p)}) \\
&= (df_p(X_p))(\mathbf{h}(f_*Y, f_*Z)) + (df_p(Y_p))(\mathbf{h}(f_*Z, f_*X)) \\
&\quad -(df_p(Z_p))(\mathbf{h}(f_*X, f_*Y)) \\
&\quad -\mathbf{h}(f(p))((f_*[X,Z])_{f(p)}, (f_*Y)_{f(p)}) - \mathbf{h}(f(p))((f_*[Y,X])_{f(p)}, (f_*Z)_{f(p)}) \\
&\quad +\mathbf{h}(f(p))((f_*[Z,Y])_{f(p)}, (f_*X)_{f(p)}) \\
&= X_p(\mathbf{h}(f(\cdot))((f_*Y)_{f(\cdot)}, (f_*Z)_{f(\cdot)}) + Y_p(\mathbf{h}(f(\cdot))((f_*Z)_{f(\cdot)}, (f_*X)_{f(\cdot)}) \\
&\quad -Z_p(\mathbf{h}(f(\cdot))((f_*X)_{f(\cdot)}, (f_*Y)_{f(\cdot)}) \\
&\quad -\mathbf{g}([X,Z]_p, Y_p) - \mathbf{g}(p)([Y,X]_p, Z_p) \\
&\quad +\mathbf{g}(p)([Z,Y]_p, X_p) \\
&= X_p(\mathbf{g}(Y,Z)) + Y_p(\mathbf{g}(Z,X)) - Z_p(\mathbf{g}(X,Y)) \\
&\quad -\mathbf{g}([X,Z]_p, Y_p) - \mathbf{g}(p)([Y,X]_p, Z_p) + \mathbf{g}(p)([Z,Y]_p, X_p).
\end{aligned}$$

Thus we obtain that for all $Z \in T_0^1 M$, and all $p \in M$, we have

$$\mathbf{h}(f(p))((f_*(\nabla^{\mathbf{g}}_X Y))_{f(p)}, (f_*Z)_{f(p)}) = \mathbf{h}(f(p))(\nabla^{\mathbf{h}}_{f_*X}(f_*Y)_{f(p)}, (f_*Z)_{f(p)}). \quad (\star)$$

But as f is a diffeomorphism, for every vector field $\widetilde{Z} \in T_0^1 N$, there exists a vector field $Z \in T_0^1 M$ such that $\widetilde{Z} = f_*Z$. Indeed, setting $Z := (f^{-1})_*\widetilde{Z}$, we have that for all $p \in M$,

$$\begin{aligned}
(f_*Z)_{f(p)} &= (df)_p Z_p = (df)_p((d(f^{-1}))_{f(p)}(\widetilde{Z}_{f(p)})) \\
&= (d(f \circ f^{-1}))_{f(p)}(\widetilde{Z}_{f(p)}) = (d(\mathrm{id}_N))_{f(p)}(\widetilde{Z}_{f(p)}) \\
&= \mathrm{id}_{T_{f(p)}N}(\widetilde{Z}_{f(p)}) = \widetilde{Z}_{f(p)},
\end{aligned}$$

showing that $\widetilde{Z} = f_*Z$. So from $(\star)$, it follows that $f_*(\nabla^{\mathbf{g}}_X Y) = \nabla^{\mathbf{h}}_{f_*X}(f_*Y)$, as wanted.

Solution to Exercise 6.16

If $X, Y \in \mathfrak{g}$, then for all $p \in M$,

$$\begin{aligned}\mathbf{g}(p)(X_p, Y_p) &= \mathbf{g}(pe)(X_{pe}, Y_{pe}) = \mathbf{g}(L_p(e))((dL_p)_e X_e, (dL_p)_e Y_e) \\ &= (L_p^* \mathbf{g})(e)(X_e, Y_e) = \mathbf{g}(e)(X_e, Y_e).\end{aligned}$$

So for $X, Y \in \mathfrak{g}$, $\mathbf{g}(X, Y)$ is a constant function. Thus for $X, Y, Z \in \mathfrak{g}$, and all $p \in G$, $(X(\mathbf{g}(Y, Z)))(p) = X_p(\mathbf{g}(Y, Z)) = X_p(\text{constant function}) = 0$. By the Koszul formula, and using the above observation, we have

$$\begin{aligned}2\mathbf{g}(\nabla_X Y, Z) &= (X\mathbf{g}(Y, Z) + Y\mathbf{g}(Z, X) - Z\mathbf{g}(X, Y) \\ &\qquad - \mathbf{g}([X, Z], Y) - \mathbf{g}([Y, X], Z) + \mathbf{g}([Z, Y], X)) \\ &= (0 + 0 - 0 + \mathbf{g}([Z, X], Y) + \mathbf{g}([X, Y], Z) + \mathbf{g}([Z, Y], X)) \\ &= \mathbf{g}([X, Y], Z) + \mathbf{g}([Z, X], Y) + \mathbf{g}(X, [Z, Y]),\end{aligned}$$

for all $X, Y, Z \in \mathfrak{g}$, as wanted.

Now suppose that $\mathbf{g}([Z, X], Y) = -\mathbf{g}(X, [Z, Y])$ for all $X, Y, Z \in \mathfrak{g}$. Then $2\mathbf{g}(\nabla_X Y, Z) = \mathbf{g}([X, Y], Z) + \cancel{\mathbf{g}([Z, X], Y)} + \cancel{\mathbf{g}(X, [Z, Y])} = \mathbf{g}([X, Y], Z)$. Let $p \in G$, and $v \in T_p G$. From Exercise 3.7, $Z_q := (dL_q)_e((dL_{p^{-1}})_p v)$, $q \in G$, defines an element $Z \in \mathfrak{g}$. Moreover, $Z_p = v$. We obtain from the above that $2\mathbf{g}(p)((\nabla_X Y)_p, v) = \mathbf{g}(p)([X, Y]_p, v)$. As v was arbitrary, we get $2(\nabla_X Y)_p = [X, Y]_p$. Since $p \in G$ was arbitrary, $2\nabla_X Y = [X, Y] = \mathcal{L}_X Y$.

Solution to Exercise 6.17

We have

$$\begin{aligned}&-\widetilde{\mathbf{g}}([X, Z], Y) - \widetilde{\mathbf{g}}([Y, X], Z) + \widetilde{\mathbf{g}}([Z, Y], X) \\ &= f(-\mathbf{g}([X, Z], Y) - \mathbf{g}([Y, X], Z) + \mathbf{g}([Z, Y], X)).\end{aligned}$$

Also $X\widetilde{\mathbf{g}}(Y, Z) = X(f\mathbf{g}(Y, Z)) = (Xf)\mathbf{g}(Y, Z) + fX\mathbf{g}(Y, Z)$. Thus

$$\begin{aligned}&X\widetilde{\mathbf{g}}(Y, Z) + Y\widetilde{\mathbf{g}}(Z, X) - Z\widetilde{\mathbf{g}}(X, Z) \\ &= (Xf)\mathbf{g}(Y, Z) + (Yf)\mathbf{g}(Z, X) - (Zf)\mathbf{g}(X, Y) \\ &\quad + f(X\mathbf{g}(Y, Z) + Y\mathbf{g}(Z, X) - Z\mathbf{g}(X, Y)).\end{aligned}$$

Using the Koszul formula, we obtain

$$\begin{aligned}2f\mathbf{g}(\widetilde{\nabla}_X Y, Z) &= 2\widetilde{\mathbf{g}}(\widetilde{\nabla}_X Y, Z) \\ &= X\widetilde{\mathbf{g}}(Y, Z) + Y\widetilde{\mathbf{g}}(Z, X) - Z\widetilde{\mathbf{g}}(X, Z) \\ &\quad - \widetilde{\mathbf{g}}([X, Z], Y) - \widetilde{\mathbf{g}}([Y, X], Z) + \widetilde{\mathbf{g}}([Z, Y], X) \\ &= (Xf)\mathbf{g}(Y, Z) + (Yf)\mathbf{g}(Z, X) - (Zf)\mathbf{g}(X, Y) \\ &\quad + f(X\mathbf{g}(Y, Z) + Y\mathbf{g}(Z, X) - Z\mathbf{g}(X, Y) \\ &\quad - \mathbf{g}([X, Z], Y) - \mathbf{g}([Y, X], Z) + \mathbf{g}([Z, Y], X)) \\ &= (Xf)\mathbf{g}(Y, Z) + (Yf)\mathbf{g}(Z, X) - (Zf)\mathbf{g}(X, Y) \\ &\quad + 2f\mathbf{g}(\nabla_X Y, Z).\end{aligned}$$

Thus

$$\begin{aligned}2f\mathbf{g}(\widetilde{\nabla}_X Y-\nabla_X Y, Z) &= (Xf)\mathbf{g}(Y,Z)+(Yf)\mathbf{g}(Z,X)-(Zf)\mathbf{g}(X,Y)\\ &= (Xf)\mathbf{g}(Y,Z)+(Yf)\mathbf{g}(X,Z)-\mathbf{g}(X,Y)\mathbf{g}(\operatorname{grad} f,\, Z)\\ &= \mathbf{g}((Xf)Y+(Yf)X-\mathbf{g}(X,Y)\operatorname{grad} f,\, Z).\end{aligned}$$

As $Z \in T_0^1 M$ was arbitrary, the result follows.

Solution to Exercise 6.18

$\det G = \sum\limits_{\pi\in S_m} (\operatorname{sign}\pi)\,\mathbf{g}_{1\pi(1)}\cdots\mathbf{g}_{m\pi(m)}$, and so by the product rule

$$\begin{aligned}\partial_{x^k}\det G &= \sum_{\pi\in S_m}\operatorname{sign}\pi\sum_{i=1}^m \mathbf{g}_{1\pi(1)}\cdots\mathbf{g}_{(i-1)\pi(i-1)}(\partial_{x^k}\mathbf{g}_{i\pi(i)})\mathbf{g}_{(i+1)\pi(i+1)}\cdots\mathbf{g}_{m\pi(m)}\\ &= \sum_{i=1}^m\sum_{\pi\in S_m}(\operatorname{sign}\pi)\mathbf{g}_{1\pi(1)}\cdots\mathbf{g}_{(i-1)\pi(i-1)}(\partial_{x^k}\mathbf{g}_{i\pi(i)})\mathbf{g}_{(i+1)\pi(i+1)}\cdots\mathbf{g}_{m\pi(m)}\\ &= \sum_{i=1}^m\det\begin{bmatrix}\mathbf{g}_{11} & \cdots & \mathbf{g}_{1m}\\ \vdots & & \vdots\\ \mathbf{g}_{(i-1)1} & \cdots & \mathbf{g}_{(i-1)m}\\ \partial_{x^k}\mathbf{g}_{i1} & \cdots & \partial_{x^k}\mathbf{g}_{im}\\ \mathbf{g}_{(i+1)1} & \cdots & \mathbf{g}_{(i+1)m}\\ \vdots & & \vdots\\ \mathbf{g}_{m1} & \cdots & \mathbf{g}_{mm}\end{bmatrix} = \sum_{i=1}^m\sum_{j=1}^m(-1)^{i+j}(\partial_{x^k}\mathbf{g}_{ij})\det G^{ij},\end{aligned}$$

where the summation convention is not used, and G^{ij} is the (i,j)-minor of G, i.e., the matrix obtained from G by deleting its ith row and j^{th} column.

As G is invertible, we have $G^{-1} = \frac{1}{\det G}[\operatorname{adj} G]$, where adj G is the adjugate of G (the matrix whose entry in the ith row and j^{th} column is $(-1)^{i+j}\det G^{ji}$. Since G is a symmetric matrix, we have $G^{ji}=(G^{ij})^{\mathrm{t}}$, and so $\det G^{ij}=\det((G^{ji})^{\mathrm{t}})=\det G^{ji}$. Hence $(G^{-1})^{ij}=\frac{(-1)^{i+j}}{\det G}\det G^{ij}$.

So (not using the summation convention), we have

$$\partial_{x^k}\det G=\sum_{i=1}^m\sum_{j=1}^m(-1)^{i+j}(\partial_{x^k}\mathbf{g}_{ij})(\det G)(G^{-1})^{ij}(-1)^{i+j}=\det G\sum_{i=1}^m\sum_{j=1}^m\mathbf{g}^{ij}\partial_{x^k}\mathbf{g}_{ij}.$$

Recall that $\Gamma_{ij}^k = \frac{\mathbf{g}^{k\ell}}{2}(\partial_{x^i}\mathbf{g}_{j\ell}+\partial_{x^j}\mathbf{g}_{i\ell}-\partial_{x^\ell}\mathbf{g}_{ij})$. Thus

$$\begin{aligned}\sum_{j=1}^m\Gamma_{ij}^j &= \sum_{j=1}^m\sum_{\ell=1}^m\frac{\mathbf{g}^{j\ell}}{2}(\partial_{x^i}\mathbf{g}_{j\ell}+\partial_{x^j}\mathbf{g}_{i\ell}-\partial_{x^\ell}\mathbf{g}_{ij})\\ &= \sum_{j=1}^m\sum_{\ell=1}^m\frac{\mathbf{g}^{j\ell}}{2}\partial_{x^i}\mathbf{g}_{j\ell}+\sum_{j=1}^m\sum_{\ell=1}^m\frac{\mathbf{g}^{j\ell}}{2}(\partial_{x^j}\mathbf{g}_{i\ell}-\partial_{x^\ell}\mathbf{g}_{ij})=\sum_{j=1}^m\sum_{\ell=1}^m\frac{\mathbf{g}^{j\ell}}{2}\partial_{x^i}\mathbf{g}_{j\ell}+0,\end{aligned}$$

where we used $\mathbf{g}^{j\ell}=\mathbf{g}^{\ell j}$, and exchanged the dummy indices j,ℓ in order to get 0 in the line above. Indeed, if $S:=\sum\limits_{j=1}^m\sum\limits_{\ell=1}^m\frac{\mathbf{g}^{j\ell}}{2}(\partial_{x^j}\mathbf{g}_{i\ell}-\partial_{x^\ell}\mathbf{g}_{ij})$, then with the $j\leftrightarrow\ell$ exchange, we get $S=0$ because

$$S=\sum_{\ell=1}^m\sum_{j=1}^m\frac{\mathbf{g}^{\ell j}}{2}(\partial_{x^\ell}\mathbf{g}_{ij}-\partial_{x^j}\mathbf{g}_{i\ell})=\sum_{j=1}^m\sum_{\ell=1}^m\frac{\mathbf{g}^{j\ell}}{2}(-1)(\partial_{x^j}\mathbf{g}_{i\ell}-\partial_{x^\ell}\mathbf{g}_{ij})=-S.$$

On the other hand, also

$$\partial_{x^i}(\log\sqrt{-\det G}) = \partial_{x^i}\frac{\log(-\det G)}{2} = \frac{1}{-2\det G}(-\det G)\sum_{j=1}^{m}\sum_{\ell=1}^{m}\mathbf{g}^{j\ell}\partial_{x^i}\mathbf{g}_{j\ell}$$
$$= \sum_{j=1}^{m}\sum_{\ell=1}^{m}\frac{\mathbf{g}^{j\ell}}{2}\partial_{x^i}\mathbf{g}_{j\ell}.$$

Here we used, for $h \in C^\infty(U)$ and a real-valued f smooth on an open set containing $h(U)$, that $\partial_{x^i}(f\circ h) = (f'\circ h)\,\partial_{x^i}h$: for all $p \in U$,

$$(\partial_{x^i}(f\circ h))(p) = \frac{\partial(f\circ h\circ \mathbf{x}^{-1})}{\partial u^i}(\mathbf{x}(p)) = f'(h(p))\,\partial_{x^i,p}h = ((f'\circ h)\,\partial_{x^i}h)(p).$$

Solution to Exercise 6.19

We recall that $G = \begin{bmatrix}\mathbf{g}_{uu} & \mathbf{g}_{uv}\\ \mathbf{g}_{vu} & \mathbf{g}_{vv}\end{bmatrix} = \begin{bmatrix}1+v^2 & uv\\ uv & 1+u^2\end{bmatrix}$. Thus

$$G^{-1} = \begin{bmatrix}\mathbf{g}^{uu} & \mathbf{g}^{uv}\\ \mathbf{g}^{vu} & \mathbf{g}^{vv}\end{bmatrix} = \frac{1}{(1+v^2)(1+u^2)-u^2v^2}\begin{bmatrix}1+u^2 & -uv\\ -uv & 1+v^2\end{bmatrix}.$$

Hence

$$\Gamma^u_{uu} = \frac{\mathbf{g}^{uv}}{2}(\partial_u\mathbf{g}_{uv}+\partial_u\mathbf{g}_{uv}-\partial_v\mathbf{g}_{uu}) + \frac{\mathbf{g}^{uu}}{2}(\partial_u\mathbf{g}_{uu}+\partial_u\mathbf{g}_{uu}-\partial_u\mathbf{g}_{uu})$$
$$= \frac{-uv}{2(1+u^2+v^2)}(v+v-2v) + \frac{1+u^2}{2(1+u^2+v^2)}0 = 0,$$
$$\Gamma^u_{uv} = \Gamma^u_{vu} = \frac{\mathbf{g}^{uu}}{2}(\partial_u\mathbf{g}_{uv}+\partial_v\mathbf{g}_{uu}-\partial_u\mathbf{g}_{uv}) + \frac{\mathbf{g}^{uv}}{2}(\partial_u\mathbf{g}_{vv}+\partial_v\mathbf{g}_{uv}-\partial_v\mathbf{g}_{uv})$$
$$= \frac{1+u^2}{2(1+u^2+v^2)}(2v) + \frac{-uv}{2(1+u^2+v^2)}(2u) = \frac{v}{1+u^2+v^2}.$$

By the $u \leftrightarrow v$ symmetry, $\Gamma^v_{uv} = \Gamma^v_{vu} = \frac{u}{1+u^2+v^2}$, and $\Gamma^v_{vv} = 0$. Finally,

$$\Gamma^u_{vv} = \frac{\mathbf{g}^{uu}}{2}(\partial_v\mathbf{g}_{uv}+\partial_v\mathbf{g}_{uv}-\partial_u\mathbf{g}_{vv}) + \frac{\mathbf{g}^{uv}}{2}(\partial_v\mathbf{g}_{vv}+\partial_v\mathbf{g}_{vv}-\partial_v\mathbf{g}_{vv})$$
$$= \frac{1+u^2}{2(1+u^2+v^2)}(u+u-2u) + \frac{-uv}{2(1+u^2+v^2)}0 = 0,$$

and so again by the $u \leftrightarrow v$ symmetry, $\Gamma^v_{uu} = 0$.

Solution to Exercise 6.20

The additivity is clear. We simply show the $C^\infty(M)$-homogeneity. We have for $X, Y, Z \in T^1_0M$ and $f \in C^\infty(M)$ that

$$\begin{aligned}R(fX,Y)Z &= \nabla_{fX}\nabla_Y Z - \nabla_Y\nabla_{fX}Z - \nabla_{[fX,Y]}Z\\ &= f\nabla_X\nabla_Y Z - \nabla_Y f\nabla_X Z - \nabla_{f[X,Y]-(Yf)X}Z\\ &= f\nabla_X\nabla_Y Z - (Yf)\nabla_X Z - f\nabla_Y\nabla_X Z - f\nabla_{[X,Y]}Z + (Yf)\nabla_X Z\\ &= f(\nabla_X\nabla_Y Z - \nabla_Y\nabla_X Z - \nabla_{[X,Y]}Z) = fR(X,Y)Z.\end{aligned}$$

Since we have the symmetry property $R(X,Y)Z = -R(Y,X)Z$, it follows that $R(X,fY)Z = -R(fY,X)Z = -fR(Y,X)Z = fR(X,Y)Z$. Finally,

$$\begin{aligned}
&R(X,Y)fZ \\
&= \nabla_X\nabla_Y(fZ) - \nabla_Y\nabla_X(fZ) - \nabla_{[X,Y]}(fZ) \\
&= \nabla_X((Yf)Z+f\nabla_Y Z)-\nabla_Y((Xf)Z+f\nabla_X Z)-([X,Y]f)Z-f\nabla_{[X,Y]}Z \\
&= X(Yf)Z+(Yf)\nabla_X Z+(Xf)\nabla_Y Z+f\nabla_X\nabla_Y Z-Y(Xf)Z-(Xf)\nabla_Y Z \\
&\quad -(Yf)\nabla_X Z - f\nabla_Y\nabla_X Z - X(Yf)Z + Y(Xf)Z - f\nabla_{[X,Y]}Z \\
&= f(\nabla_X\nabla_Y Z - \nabla_Y\nabla_X Z - \nabla_{[X,Y]}Z) = fR(X,Y)Z.
\end{aligned}$$

To show the well-definition of $R(p)$, we proceed as in the proof of Theorem 4.1 (p.63). Let $(U,\mathbf{x})$ be an admissible chart containing p. Let χ be a bump function that is identically 1 in a neighbourhood of p, and identically zero outside an open set U_0 whose closure $\overline{U_0}$ is contained in U. Let $X,\widetilde{X},Y,\widetilde{Y},Z,\widetilde{Z}$ be vector fields such that $X_p = \widetilde{X}_p = x$, $Y_p = \widetilde{Y}_p = y$, and $Z_p = \widetilde{Z}_p = z$. We decompose $X = X^i\partial_{x^i}$, etc. Then we note that $X^i(p) = \widetilde{X}^i(p)$, etc. We have

$$\begin{aligned}
(R(X,Y)Z)_p &= 1(R(X,Y)Z)_p = (\chi(p))^6(R(X,Y)Z)_p = (\chi^6 R(X,Y)Z)_p \\
&= (R(\chi^2 X,\chi^2 Y)(\chi^2 Z))_p = (R(\chi^2 X^i\partial_{x^i},\chi^2 Y^j\partial_{x^j})(\chi^2 Z^k\partial_{x^k}))_p \\
&= (\chi X^i)(p)(\chi Y^j)(p)(\chi Z^k)(p)(R(\chi\partial_{x^i},\chi\partial_{x^j})\chi\partial_{x^k})_p \\
&= 1^3 X^i(p)Y^i(p)Z^i(p)(R(\chi\partial_{x^i},\chi\partial_{x^j})\chi\partial_{x^k})_p \\
&= 1^3 \widetilde{X}^i(p)\widetilde{Y}^i(p)\widetilde{Z}^i(p)(R(\chi\partial_{x^i},\chi\partial_{x^j})\chi\partial_{x^k})_p \\
&= \cdots \text{ (retrace the steps) } \cdots = (R(\widetilde{X},\widetilde{Y})\widetilde{Z})_p.
\end{aligned}$$

In any affine chart $(U,\mathbf{x})$, $\Gamma_{ij}^k = 0$ for all i,j,k. Also using $[\partial_{x^i},\partial_{x^j}] = 0$,

$$\begin{aligned}
dx^\ell(R(\partial_{x^i},\partial_{x^j})\partial_{x^k}) &= dx^\ell(\nabla_{\partial_{x^i}}\nabla_{\partial_{x^j}}\partial_{x^k} - \nabla_{\partial_{x^j}}\nabla_{\partial_{x^i}}\partial_{x^k} - \nabla_{[\partial_{x^i},\partial_{x^j}]}\partial_{x^k}) \\
&= dx^\ell(\nabla_{\partial_{x^i}}(\Gamma_{kj}^r\partial_{x^r}) - \nabla_{\partial_{x^j}}(\Gamma_{ki}^s\partial_{x^s}) - 0) = dx^\ell(0) = 0.
\end{aligned}$$

Thus $\mathbf{R}_{ijk}^\ell = 0$ in the chart $(U,\mathbf{x})$, for $1 \leqslant i,j,k,\ell \leqslant m$. As these charts cover M, $\mathbf{R} = \mathbf{0} \in T_3^1 M$.

We have, in the case of the sphere with the chart (U,φ), that:

$$\begin{aligned}
R(\partial_\theta,\partial_\phi)\partial_\theta &= \nabla_{\partial_\theta}\nabla_{\partial_\phi}\partial_\theta - \nabla_{\partial_\phi}\nabla_{\partial_\theta}\partial_\theta - \nabla_{[\partial_\theta,\partial_\phi]}\partial_\theta \\
&= \nabla_{\partial_\theta}(\Gamma_{\theta\phi}^\theta\partial_\theta + \Gamma_{\theta\phi}^\phi\partial_\phi) - \nabla_{\partial_\phi}(\Gamma_{\theta\theta}^\theta\partial_\theta + \Gamma_{\theta\theta}^\phi\partial_\phi) \\
&= \nabla_{\partial_\theta}((\cot\theta)\partial_\phi) - 0 = (\partial_\theta\cot\theta)\partial_\phi + (\cot\theta)\nabla_{\partial_\theta}\partial_\phi \\
&= -(\sin\theta)^{-2}\partial_\phi + (\cot\theta)(\Gamma_{\phi\theta}^\theta\partial_\theta + \Gamma_{\phi\theta}^\phi\partial_\phi) \\
&= -(\sin\theta)^{-2}\partial_\phi + (\cot\theta)^2\partial_\phi = -\partial_\phi.
\end{aligned}$$

Hence $\mathbf{R}(d\phi,\partial_\theta,\partial_\phi,\partial_\theta) = d\phi(-\partial_\phi) = -\partial_\phi\phi = -1$. (So $\mathbf{R}_{\theta\phi\theta}^\phi = -1 \neq 0$.)

Solution to Exercise 6.21

We have for all $W \in T_0^1 M$,

$$\begin{aligned}(\nabla_V(f\Omega))W &= V(f\Omega W) - f\Omega(\nabla_V W)\\ &= (Vf)\Omega W + fV(\Omega W) - f\Omega(\nabla_V W) = ((Vf)\Omega + f(\nabla_V \Omega))W.\end{aligned}$$

Thus $\nabla_V(f\Omega) = (Vf)\Omega + f(\nabla_V\Omega)$. Also,

$$\begin{aligned}(\nabla_V(\Omega+\Theta))W &= V((\Omega+\Theta)W) - (\Omega+\Theta)(\nabla_V W)\\ &= V(\Omega W + \Theta W) - \Omega(\nabla_V W) - \Theta(\nabla_V W)\\ &= V(\Omega W) + V(\Theta W) - \Omega(\nabla_V W) - \Theta(\nabla_V W)\\ &= (\nabla_V\Omega)W + (\nabla_V\Theta)W = (\nabla_V\Omega + \nabla_V\Theta)W.\end{aligned}$$

Thus $\nabla_V(\Omega+\Theta) = \nabla_V\Omega + \nabla_V\Theta$.
So $\nabla_V(f\Omega+\Theta) = \nabla_V(f\Omega) + \nabla_V\Theta = (Vf)\Omega + f\nabla_V\Omega + \nabla_V\Theta$.
We have for $X \in T_0^1 M$

$$\begin{aligned}(\nabla_{fV+W}\Omega)(X) &= (fV+W)(\Omega X) - \Omega(\nabla_{fV+W}X)\\ &= fV(\Omega X) + W(\Omega X) - \Omega(f\nabla_V X + \nabla_W X)\\ &= fV(\Omega X) + W(\Omega X) - f\Omega(\nabla_V X) - \Omega(\nabla_W X)\\ &= f(V(\Omega X) - \Omega(\nabla_V X)) + W(\Omega X) - \Omega(\nabla_W X)\\ &= f(\nabla_V\Omega)(X) + (\nabla_W\Omega)(X) = (f\nabla_V\Omega + \nabla_W\Omega)(X).\end{aligned}$$

Consequently, $\nabla_{fV+W}\Omega = f\nabla_V\Omega + \nabla_W\Omega$.

Solution to Exercise 6.22

For all $X \in T_0^1 M$, we have, using the metric compatibility, that

$$\begin{aligned}(\nabla_V(W^\flat))X &= V(W^\flat X) - W^\flat(\nabla_V X) = V(\mathbf{g}(W,X)) - \mathbf{g}(W,\nabla_V X)\\ &= V(\mathbf{g}(W,X)) + \mathbf{g}(\nabla_V W, X) - V(\mathbf{g}(W,X))\\ &= \mathbf{g}(\nabla_V W, X) = (\nabla_V W)^\flat X.\end{aligned}$$

Solution to Exercise 6.23

Using Definition 6.4 (p.125), we have

$$\begin{aligned}&(\nabla_V\mathbf{R})(\Omega,X,Y,Z)\\ &= V(\mathbf{R}(\Omega,X,Y,Z)) - \mathbf{R}(\nabla_V\Omega,X,Y,Z) - \mathbf{R}(\Omega,\nabla_V X,Y,Z)\\ &\quad -\mathbf{R}(\Omega,X,\nabla_V Y,Z) - \mathbf{R}(\Omega,X,Y,\nabla_V Z)\\ &= V(-\mathbf{R}(\Omega,Y,X,Z)) + \mathbf{R}(\nabla_V\Omega,Y,X,Z) + \underline{\mathbf{R}(\Omega,Y,\nabla_V X,Z)}\\ &\quad +\underline{\mathbf{R}(\Omega,\nabla_V Y,X,Z)} + \mathbf{R}(\Omega,Y,X,\nabla_V Z)\\ &= -(V(\mathbf{R}(\Omega,Y,X,Z)) - \mathbf{R}(\nabla_V\Omega,Y,X,Z) - \underline{\mathbf{R}(\Omega,\nabla_V Y,X,Z)}\\ &\quad -\underline{\mathbf{R}(\Omega,Y,\nabla_V X,Z)} - \mathbf{R}(\Omega,Y,X,\nabla_V Z))\\ &= -(\nabla_V\mathbf{R})(\Omega,Y,X,Z).\end{aligned}$$

Solution to Exercise 6.24

1° $r \geqslant 2$ or $s \geqslant 1$.

For all $\Omega^1, \cdots, \Omega^r \in T_1^0 M$ and $W_1, \cdots, W_s \in T_0^1 M$,

$$
\begin{aligned}
&(\nabla_V(fT))(\Omega^1, \cdots, \Omega^r, W_1, \cdots, W_s) \\
&= V(fT(\Omega^1, \cdots, \Omega^r, W_1, \cdots, W_s)) \\
&\quad - \sum_{i=1}^{r} fT(\Omega^1, \cdots, \nabla_V \Omega^i, \cdots, \Omega^r, W_1, \cdots, W_s) \\
&\quad - \sum_{j=1}^{s} fT(\Omega^1, \cdots, \Omega^r, W_1, \cdots, \nabla_V W_j, \cdots, W_s) \\
&= (Vf)T(\Omega^1, \cdots, \Omega^r, W_1, \cdots, W_s) \\
&\quad + fV(T(\Omega^1, \cdots, \Omega^r, W_1, \cdots, W_s)) \\
&\quad - f \sum_{i=1}^{r} T(\Omega^1, \cdots, \nabla_V \Omega^i, \cdots, \Omega^r, W_1, \cdots, W_s) \\
&\quad - f \sum_{j=1}^{s} T(\Omega^1, \cdots, \Omega^r, W_1, \cdots, \nabla_V W_j, \cdots, W_s) \\
&= (Vf)T(\Omega^1, \cdots, \Omega^r, W_1, \cdots, W_s) \\
&\quad + f(\nabla_V T)(\Omega^1, \cdots, \Omega^r, W_1, \cdots, W_s) \\
&= ((Vf)T + f(\nabla_V T))(\Omega^1, \cdots, \Omega^r, W_1, \cdots, W_s).
\end{aligned}
$$

Thus $\nabla_V(fT) = (Vf)T + f(\nabla_V T)$.

2° $r = s = 0$.

Let $T =: g$. Then

$$
\begin{aligned}
\nabla_V(fg) = V(fg) &= (Vf)g + f(Vg) \\
&= (Vf)g + f\nabla_V g.
\end{aligned}
$$

3° $r = 1$, $s = 0$.

Let $T =: W$.

Then $\nabla_V(fW) = (Vf)W + f\nabla_V W$ is precisely (C2).

Solution to Exercise 6.25

1° $r = s = 0$.

Let $T =: g$. Then

$$
\begin{aligned}
\nabla_{fV+W} g &= (fV + W)g \\
&= fVg + Wg \\
&= f\nabla_V g + \nabla_W g.
\end{aligned}
$$

2° $r = 1$, $s = 0$.

Let $T =: X$.

Then $\nabla_{fV+W} X = f\nabla_V X + \nabla_W X$ is precisely (C3).

3° $r \geqslant 2$ or $s \geqslant 1$. For all $\Omega^1, \cdots, \Omega^r \in T_1^0 M$ and $X_1, \cdots, X_s \in T_0^1 M$,

$$\begin{aligned}
&(\nabla_{fV+W}T)(\Omega^1, \cdots, \Omega^r, X_1, \cdots, X_s)\\
&= (fV+W)(T(\Omega^1, \cdots, \Omega^r, X_1, \cdots, X_s))\\
&\quad -\sum_{i=1}^{r} T(\Omega^1, \cdots, \nabla_{fV+W}\Omega^i, \cdots, \Omega^r, X_1, \cdots, X_s)\\
&\quad -\sum_{j=1}^{s} T(\Omega^1, \cdots, \Omega^r, X_1, \cdots, \nabla_{fV+W}X_j, \cdots, X_s)\\
&= fV(T(\Omega^1, \cdots, \Omega^r, X_1, \cdots, X_s)) + W(T(\Omega^1, \cdots, \Omega^r, X_1, \cdots, X_s))\\
&\quad -\sum_{i=1}^{r} T(\Omega^1, \cdots, f\nabla_V\Omega^i + \nabla_W\Omega^i, \cdots, \Omega^r, X_1, \cdots, X_s)\\
&\quad -\sum_{j=1}^{s} T(\Omega^1, \cdots, \Omega^r, X_1, \cdots, f\nabla_V X_j + \nabla_W X_j, \cdots, X_s)\\
&= fV(T(\Omega^1, \cdots, \Omega^r, X_1, \cdots, X_s)) + W(T(\Omega^1, \cdots, \Omega^r, X_1, \cdots, X_s))\\
&\quad -f\sum_{i=1}^{r} T(\Omega^1, \cdots, \nabla_V\Omega^i, \cdots, \Omega^r, X_1, \cdots, X_s)\\
&\quad -\sum_{i=1}^{r} T(\Omega^1, \cdots, \nabla_W\Omega^i, \cdots, \Omega^r, X_1, \cdots, X_s)\\
&\quad -f\sum_{j=1}^{s} T(\Omega^1, \cdots, \Omega^r, X_1, \cdots, \nabla_V X_j, \cdots, X_s)\\
&\quad -\sum_{j=1}^{s} T(\Omega^1, \cdots, \Omega^r, X_1, \cdots, \nabla_W X_j, \cdots, X_s)\\
&= f(\nabla_V T)(\Omega^1, \cdots, \Omega^r, X_1, \cdots, X_s) + (\nabla_W T)(\Omega^1, \cdots, \Omega^r, X_1, \cdots, X_s)\\
&= \big(f(\nabla_V T) + (\nabla_W T)\big)(\Omega^1, \cdots, \Omega^r, X_1, \cdots, X_s).
\end{aligned}$$

Thus $\nabla_{fV+W}T = f\nabla_V T + \nabla_W T$.

Solution to Exercise 6.26

As $dx^i = \delta^i_r dx^r$, by Example 6.9 (p.126),

$$\nabla_V dx^i = (V\delta^i_r - \delta^i_\ell V^s \Gamma^\ell_{rs})dx^r.$$

So

$$\begin{aligned}
(\nabla_V T)^i_{jk} &= (\nabla_V T)(dx^i, \partial_{x^j}, \partial_{x^k})\\
&= V(T(dx^i, \partial_{x^j}, \partial_{x^k})) - T(\nabla_V dx^i, \partial_{x^j}, \partial_{x^k})\\
&\quad -T(dx^i, \nabla_V \partial_{x^j}, \partial_{x^k}) - T(dx^i, \partial_{x^j}, \nabla_V \partial_{x^k})\\
&= VT^i_{jk} - T((V\delta^i_r - \delta^i_\ell V^s\Gamma^\ell_{rs})dx^r, \partial_{x^j}, \partial_{x^k})\\
&\quad -T(dx^i, V^s\Gamma^r_{js}\partial_{x^r}, \partial_{x^k}) - T(dx^i, \partial_{x^j}, V^s\Gamma^r_{ks}\partial_{x^r})\\
&= VT^i_{jk} + V^s\Gamma^i_{rs}T^r_{jk} - V^s\Gamma^r_{js}T^i_{rk} - V^s\Gamma^r_{ks}T^i_{jr}\\
&= VT^i_{jk} + \Gamma^i_{rs}T^r_{jk}V^s - \Gamma^r_{js}T^i_{rk}V^s - \Gamma^r_{ks}T^i_{jr}V^s.
\end{aligned}$$

Solution to Exercise 6.27

For $\Omega, \Theta \in T_1^0 M$ and $X \in T_0^1 M$,

$$\begin{aligned}
&(\nabla_V(W \otimes T))(\Omega, \Theta, X)\\
&= V(W(\Omega)T(\Theta, X)) - (W \otimes T)(\nabla_V \Omega, \Theta, X)\\
&\quad -(W \otimes T)(\Omega, \nabla_V \Theta, X) - (W \otimes T)(\Omega, \Theta, \nabla_V X)\\
&= V(W\Omega)T(\Theta, X) + (W\Omega)V(T(\Theta, X)) - W(\nabla_V \Omega)T(\Theta, X)\\
&\quad -(W\Omega)T(\nabla_V \Theta, X) - (W\Omega)T(\Theta, \nabla_V X)\\
&= \big(V(W\Omega) - W(\nabla_V \Omega)\big)T(\Theta, X)\\
&\quad +W(\Omega)\big(V(T(\Theta, X)) - T(\nabla_V \Theta, X) - T(\Theta, \nabla_V X)\big)\\
&= \big(V(\Omega W) - (\nabla_V \Omega)W\big)T(\Theta, X) + W(\Omega)(\nabla_V T)(\Theta, X)\\
&= \big(V(\Omega W) - V(\Omega W) + \Omega(\nabla_V W)\big)T(\Theta, X) + (W \otimes (\nabla_V T))(\Omega, \Theta, X)\\
&= (\nabla_V W)(\Omega)T(\Theta, X) + (W \otimes (\nabla_V T))(\Omega, \Theta, X)\\
&= ((\nabla_V W) \otimes T)(\Omega, \Theta, X) + (W \otimes (\nabla_V T))(\Omega, \Theta, X)\\
&= \big((\nabla_V W) \otimes T + W \otimes (\nabla_V T)\big)(\Omega, \Theta, X).
\end{aligned}$$

Solution to Exercise 6.28

We have

$$\nabla_V(\mathbf{g} \otimes T) = (\nabla_V \mathbf{g}) \otimes T + \mathbf{g} \otimes \nabla_V T = 0 \otimes T + \mathbf{g} \otimes \nabla_V T = \mathbf{g} \otimes \nabla_V T$$
$$\nabla_V(T \otimes \mathbf{g}) = (\nabla_V T) \otimes \mathbf{g} + T \otimes \nabla_V \mathbf{g} = (\nabla_V T) \otimes \mathbf{g} + T \otimes 0 = (\nabla_V T) \otimes \mathbf{g}.$$

Solution to Exercise 6.29

(1) For $\Omega \in T_1^0 M$, $V \in T_0^1 M$,

$$(T^\flat)^\sharp(\Omega, V) = T^\flat(\Omega^\sharp, V) = T((\Omega^\sharp)^\flat, V) = T(\Omega, V).$$

So $(T^\flat)^\sharp = T$.

(2) For $V, W \in T_0^1 M$, $(T^\sharp)^\flat(V, W) = T^\sharp(V^\flat, W) = T((V^\flat)^\sharp, W) = T(V, W)$.
So $(T^\sharp)^\flat = T$.

(3) Let $(U, \mathbf{x})$ be an admissible chart for M.
Then $(\partial_{x^i})^\flat = (\partial_{x^i})^\flat(\partial_{x^k})dx^k = \mathbf{g}(\partial_{x^i}, \partial_{x^k})dx^k = \mathbf{g}_{ik}dx^k$, and so

$$\begin{aligned}
(T^\flat)_{ij} &= T^\flat(\partial_{x^i}, \partial_{x^j}) = T((\partial_{x^i})^\flat, \partial_{x^j}) = T(\mathbf{g}_{ik}dx^k, \partial_{x^j})\\
&= \mathbf{g}_{ik}T(dx^k, \partial_{x^j}) = \mathbf{g}_{ik}T^k_j.
\end{aligned} \tag{17.1}$$

For $\Omega \in T_1^0 M$, $V, X, W \in T_0^1 M$, we have

$$(\mathbf{g} \otimes T)(\Omega, V, X, W) = \mathbf{g}(V, X)T(\Omega, W).$$

Hence $(\mathbf{C}_2^1(\mathbf{g} \otimes T))(V, W) = \mathbf{g}(V, \partial_{x^k})T(dx^k, W)$. Thus

$$(\mathbf{C}_2^1(\mathbf{g} \otimes T))_{ij} = \mathbf{g}(\partial_{x^i}, \partial_{x^k})T(dx^k, \partial_{x^j}) = \mathbf{g}_{ik}T^k_j. \tag{17.2}$$

From (17.1) and (17.2), it follows that $T^\flat = \mathbf{C}_2^1(\mathbf{g} \otimes T)$.

(4) Let $T \in T^1_1 M$. Then

$$\begin{aligned}\nabla_V(T^\flat) &= \nabla_V(\mathbf{C}^1_2(\mathbf{g}\otimes T)) = \mathbf{C}^1_2(\nabla_V(\mathbf{g}\otimes T))\\ &= \mathbf{C}^1_2(\mathbf{g}\otimes \nabla_V T) = (\nabla_V T)^\flat.\end{aligned}$$

(5) Let $T \in T^0_2 M$. Then $\nabla_V T \in T^0_2 M$ and

$$\nabla_V T \overset{(2)}{=} \nabla_V((T^\sharp)^\flat) \overset{(4)}{=} (\nabla_V(T^\sharp))^\flat. \qquad (\star)$$

We have $\nabla_V(T^\sharp) \in T^1_1 M$, and $(\nabla_V T)^\sharp \overset{(\star)}{=} ((\nabla_V(T^\sharp))^\flat)^\sharp \overset{(1)}{=} \nabla_V(T^\sharp)$.

Solution to Exercise 6.30

For all $W \in T^1_0 M$, we have

$$\begin{aligned}(\mathcal{L}_V(f\Omega+\Theta))(W) &= V((f\Omega+\Theta)(W)) - (f\Omega+\Theta)[V,W]\\ &= V(f\Omega W + \Theta W) - f\Omega[V,W] - \Theta[V,W]\\ &= (Vf)(\Omega W) + fV(\Omega W) + V(\Theta W) - f\Omega[V,W] - \Theta[V,W]\\ &= (Vf)(\Omega W) + f(\mathcal{L}_V\Omega)(W) + (\mathcal{L}_V\Theta)(W)\\ &= ((Vf)\Omega + f\mathcal{L}_V\Omega + \mathcal{L}_V\Theta)(W).\end{aligned}$$

Solution to Exercise 6.31

The first claim follows immediately from $\mathcal{L}_V\mathbf{g} = 0$, since this implies that for all $X, Y \in T^1_0 M$,

$$\begin{aligned}0 &= (\mathcal{L}_V\mathbf{g})(X,Y) = V(\mathbf{g}(X,Y)) - \mathbf{g}(\mathcal{L}_V X, Y) - \mathbf{g}(X, \mathcal{L}_V Y)\\ &= V(\mathbf{g}(X,Y)) - \mathbf{g}([V,X],Y) - \mathbf{g}(X,[V,Y]).\end{aligned}$$

For the second claim, we start with the above, and use the torsion-freeness (C4) and metric compatibility (C5) of ∇:

$$\begin{aligned}0 &= V(\mathbf{g}(X,Y)) - \mathbf{g}([V,X],Y) - \mathbf{g}(X,[V,Y])\\ &\overset{(C5)}{=} \mathbf{g}(\nabla_V X, Y) + \mathbf{g}(X, \nabla_V Y) - \mathbf{g}([V,X],Y) - \mathbf{g}(X,[V,Y])\\ &= \mathbf{g}(\nabla_V X - [V,X], Y) + \mathbf{g}(X, \nabla_V Y - [V,Y])\\ &\overset{(C4)}{=} \mathbf{g}(\nabla_X V, Y) + \mathbf{g}(X, \nabla_Y V),\end{aligned}$$

proving the second claim.

Solution to Exercise 6.32

We use the global chart $(\mathbb{R}^3, \mathrm{id})$, and label the coordinate maps x^1, x^2, x^3 by x, y, z. We have

$$\begin{aligned}(\mathcal{L}_{L_x}\mathbf{g})_{ij} &= (\mathcal{L}_{L_x}\mathbf{g})(\partial_{x^i}, \partial_{x^j})\\ &= L_x(\mathbf{g}(\partial_{x^i}, \partial_{x^j})) - \mathbf{g}([L_x, \partial_{x^i}], \partial_{x^j}) - \mathbf{g}(\partial_{x^i}, [L_x, \partial_{x^j}])\\ &= L_x(\delta_{ij}) - \mathbf{g}([y\partial_z - z\partial_y, \partial_{x^i}], \partial_{x^j}) - \mathbf{g}(\partial_{x^i}, [y\partial_z - z\partial_y, \partial_{x^j}])\\ &= 0 - \mathbf{g}([y\partial_z - z\partial_y, \partial_{x^i}], \partial_{x^j}) - \mathbf{g}(\partial_{x^i}, [y\partial_z - z\partial_y, \partial_{x^j}]).\end{aligned}$$

Using Exercise 3.13 (p.45), we have

$$\begin{aligned}[y\partial_z - z\partial_y, \partial_{x^i}] &= [y\partial_z, \partial_{x^i}] - [z\partial_y, \partial_{x^i}] \\ &= y[\partial_z, \partial_{x^i}] - (\partial_{x^i} y)\partial_z - z[\partial_y, \partial_{x^i}] + (\partial_{x^i} z)\partial_y \\ &= y0 - \delta_{i2}\partial_z - z0 + \delta_{i3}\partial_y = \delta_{i3}\partial_y - \delta_{i2}\partial_z.\end{aligned}$$

So $\mathcal{L}_{L_x}\mathbf{g} = 0$, since for $1 \leqslant i, j \leqslant 3$, we have

$$\begin{aligned}(\mathcal{L}_{L_x}\mathbf{g})_{ij} &= -\mathbf{g}(\delta_{i3}\partial_y - \delta_{i2}\partial_z, \partial_{x^j}) - \mathbf{g}(\partial_{x^i}, \delta_{j3}\partial_y - \delta_{j2}\partial_z) \\ &= -\cancel{\delta_{i3}\delta_{j2}} + \cancel{\delta_{i2}\delta_{j3}} - \cancel{\delta_{i2}\delta_{j3}} + \cancel{\delta_{i3}\delta_{j2}} = 0.\end{aligned}$$

That $\mathcal{L}_{L_y}\mathbf{g} = 0$ and $\mathcal{L}_{L_z}\mathbf{g} = 0$ can be shown analogously, or by employing the symmetry in the expression for $\mathbf{g} = dx \otimes dx + dy \otimes dy + dz \otimes dz$ and cycling through $x \to y \to z \to x$ to go from L_x to L_y to L_z.

Solution to Exercise 6.33

For any vector field X, we have

$$\begin{aligned}[\partial_{x^k}, V]^i &= [\partial_{x^k}, V]x^i \\ &= \partial_{x^k}(V(x^i)) - V(\partial_{x^k}(x^i)) \\ &= \partial_{x^k}(V^i) - V(\delta_i^k) \\ &= \partial_{x^k}(V^i) - 0 = \partial_{x^k}(V^i).\end{aligned}$$

Next, for vector fields V, W, we have

$$\begin{aligned}(\mathcal{L}_{\partial_{x^k}}\mathbf{g})(V, W) &= \partial_{x^k}(\mathbf{g}(V, W)) - \mathbf{g}([\partial_{x^k}, V], W) - \mathbf{g}(V, [\partial_{x^k}, W]) \\ &= \partial_{x^k}(g_{ij}V^iW^j) - \mathbf{g}((\partial_{x^k}V^i)\partial_{x^i}, W) - \mathbf{g}(V, (\partial_{x^k}W^j)\partial_{x^j}) \\ &= (\partial_{x^k}g_{ij})V^iW^j + \cancel{g_{ij}(\partial_{x^k}V^i)W^j} + \cancel{g_{ij}V^i(\partial_{x^k}W^j)} \\ &\quad -\cancel{g_{ij}(\partial_{x^k}V^i)W^j} - \cancel{g_{ij}V^i(\partial_{x^k}W^j)} \\ &= (\partial_{x^k}g_{ij})V^iW^j = 0.\end{aligned}$$

Solution to Exercise 6.34

For $V, W \in T_0^1 M$, we have

$$\begin{aligned}(\mathcal{L}_X\mathcal{L}_Y\mathbf{g})(V, W) &= X(\mathcal{L}_Y\mathbf{g})(V, W) - (\mathcal{L}_Y\mathbf{g})([X, V], W) - (\mathcal{L}_Y\mathbf{g})(V, [X, W]) \\ &= X(Y\mathbf{g}(V, W) - \mathbf{g}([Y, V], W) - \mathbf{g}(V, [Y, W])) \\ &\quad -Y\mathbf{g}([X, V], W) + \mathbf{g}([Y, [X, V]], W) + \mathbf{g}([X, V], [Y, W]) \\ &\quad -Y\mathbf{g}(V, [X, W]) + \mathbf{g}([Y, V], [X, W]) + \mathbf{g}(V, [Y, [X, W]]).\end{aligned}$$

Exchanging X and Y, we also get

$$\begin{aligned}(\mathcal{L}_Y\mathcal{L}_X\mathbf{g})(V, W) &= Y(X\mathbf{g}(V, W) - \mathbf{g}([X, V], W) - \mathbf{g}(V, [X, W])) \\ &\quad -X\mathbf{g}([Y, V], W) + \mathbf{g}([X, [Y, V]], W) + \mathbf{g}([Y, V], [X, W]) \\ &\quad -X\mathbf{g}(V, [Y, W]) + \mathbf{g}([X, V], [Y, W]) + \mathbf{g}(V, [X, [Y, W]]).\end{aligned}$$

We shall use the following, which are obtained from the Jacobi identity:

$$\begin{aligned}[Y, [X, V]] + [X, [V, Y]] &= [V, [X, Y]] \\ [Y, [X, W]] + [X, [W, Y]] &= [W, [X, Y]].\end{aligned}$$

Subtracting the above expressions for $\mathcal{L}_X\mathcal{L}_Y\mathbf{g}$ and $\mathcal{L}_Y\mathcal{L}_X\mathbf{g}$, we get

$$\begin{aligned}&(\mathcal{L}_X\mathcal{L}_Y\mathbf{g} - \mathcal{L}_Y\mathcal{L}_X\mathbf{g})(V, W)\\ &= [X, Y]\mathbf{g}(V, W) + \mathbf{g}([V, [X, Y]], W) + \mathbf{g}(V, [W, [X, Y]])\\ &= [X, Y]\mathbf{g}(V, W) - \mathbf{g}(\mathcal{L}_{[X,Y]}V, W) - \mathbf{g}(V, \mathcal{L}_{[X,Y]}W) = \mathcal{L}_{[X,Y]}\mathbf{g}.\end{aligned}$$

If X and Y are Killing vector fields, then $\mathcal{L}_X\mathbf{g} = \mathbf{0} = \mathcal{L}_Y\mathbf{g}$, and so

$$\mathcal{L}_{[X,Y]}\mathbf{g} = \mathcal{L}_X\mathcal{L}_Y\mathbf{g} - \mathcal{L}_Y\mathcal{L}_X\mathbf{g} = \mathcal{L}_X\mathbf{0} - \mathcal{L}_Y\mathbf{0} = \mathbf{0}.$$

So $[X, Y]$ is a Killing vector field.

Chapter 7

Solution to Exercise 7.1

Let $Z(t) = Z^i(t)\,\partial_{x^i,\gamma(t)}$, $t \in I$. Then

$$W(s) = Z^i(h(s))\,\partial_{x^i,\gamma(h(s))} \quad (s \in J).$$

For an $s \in J$, let $(U, \mathbf{x})$ be an admissible chart containing $\gamma(h(s))$. We have

$$(\nabla_{V_{\gamma\circ h}} W)(s) = \big(\dot{W}^k(s) + \Gamma^k_{ij}(\gamma(h(s)))\,(\overset{\bullet}{\overbrace{\gamma \circ h}})^j(s) W^i(s)\big)\partial_{x^k,(\gamma\circ h)(s)}. \quad (17.3)$$

But

$$\begin{aligned}\dot{W}^k(s) &= \frac{d(Z^k(h(\cdot)))}{ds}(s) \\ &= \frac{dZ^k}{ds}(h(s))\,\dot{h}(s) \\ &= \dot{h}(s)\,\dot{Z}^k(h(s)).\end{aligned} \quad (17.4)$$

Also, by Exercise 2.4 (p.23),

$$\begin{aligned}v_{\gamma\circ h,(\gamma\circ h)(s)} &= \dot{h}(s)\,v_{\gamma,\gamma(h(s))} \\ &= \dot{h}(s)\,\dot{\gamma}^i(h(s))\,\partial_{x^i,\gamma(h(s))} \\ &= \dot{h}(s)\,\dot{\gamma}^i(h(s))\,\partial_{x^i,(\gamma\circ h)(s)},\end{aligned}$$

so that

$$(\overset{\bullet}{\overbrace{\gamma \circ h}})^j(s) = \dot{h}(s)\,\dot{\gamma}^j(h(s)). \quad (17.5)$$

Substituting (17.4) and (17.5) in (17.3), we obtain

$$\begin{aligned}&(\nabla_{V_{\gamma\circ h}} W)(s) \\ &= \big(\dot{h}(s)\,\dot{Z}^k(h(s)) + \Gamma^k_{ij}(\gamma(h(s)))\,\dot{h}(s)\,\dot{\gamma}^j(h(s))\,Z^i(h(s))\big)\,\partial_{x^k,(\gamma\circ h)(s)} \\ &= \dot{h}(s)\big(\dot{Z}^k(h(s)) + \Gamma^k_{ij}(\gamma(h(s)))\,\dot{\gamma}^j(h(s))\,Z^i(h(s))\big)\,\partial_{x^k,\gamma(h(s))} \\ &= \dot{h}(s)\,(\nabla_{V_\gamma} Z)(h(s)).\end{aligned}$$

Hence for all $s \in J$, $(\nabla_{V_{\gamma\circ h}} W)(s) = \dot{h}(s)\,(\nabla_{V_\gamma} Z)(h(s))$.

Solution to Exercise 7.2

Define the vector field $X \in T^1_0\gamma$ by $X(t) = W_{\gamma(t)} = \widetilde{W}_{\gamma(t)}$ for $t \in I$. We have

$$(\nabla_{V_\gamma} X)(0) = (\nabla_V W)_{\gamma(0)} = (\nabla_V W)_p. \quad (17.6)$$

Also,

$$(\nabla_{V_\gamma} X)(0) = (\nabla_{\widetilde{V}} \widetilde{W})_{\gamma(0)} = (\nabla_{\widetilde{V}} \widetilde{W})_p. \quad (17.7)$$

From (17.6) and (17.7), it follows that $(\nabla_V W)_p = (\nabla_{\widetilde{V}} \widetilde{W})_p$.

Solution to Exercise 7.3

$Y(t) = \partial_{\phi,\mu(t)}$, and so $Y^\theta \equiv 0$, $Y^\phi \equiv 1$, giving $\dot{Y}^\theta \equiv 0 \equiv \dot{Y}^\phi$. We had seen in Example 7.3 that $\dot{\mu}^\theta \equiv 0$ and $\dot{\mu}^\phi \equiv 1$. Thus

$$\begin{aligned}
(\nabla_{V_\mu} Y)(t) &= \big(\dot{Y}^\theta(t) + \Gamma^\theta_{ij}(\mu(t))\,\dot{\mu}^j(t)\,Y^i(t)\big)\,\partial_{\theta,\mu(t)} \\
&\quad + \big(\dot{Y}^\phi(t) + \Gamma^\phi_{ij}(\mu(t))\,\dot{\mu}^j(t)\,Y^i(t)\big)\,\partial_{\phi,\mu(t)} \\
&= \big(\Gamma^\theta_{\phi\theta}(\mu(t))\,\dot{\mu}^\theta(t)\,Y^\phi(t) + \Gamma^\theta_{\phi\phi}(\mu(t))\,\dot{\mu}^\phi(t)\,Y^\phi(t)\big)\,\partial_{\theta,\mu(t)} \\
&\quad + \big(\Gamma^\phi_{\phi\theta}(\mu(t))\,\dot{\mu}^\theta(t)\,Y^\phi(t) + \Gamma^\phi_{\phi\phi}(\mu(t))\,\dot{\mu}^\phi(t)\,Y^\phi(t)\big)\,\partial_{\phi,\mu(t)} \\
&= \big(0\,\dot{\mu}^\theta(t)\,Y^\phi(t) - (\sin\theta)(\cos\theta)|_{\theta=\frac{\pi}{2}}\,\dot{\mu}^\phi(t)Y^\phi(t)\big)\,\partial_{\theta,\mu(t)} \\
&\quad + \big((\cot\theta)|_{\theta=\frac{\pi}{2}}\,\dot{\mu}^\theta(t)Y^\phi(t) + 0\,\dot{\mu}^\phi(t)Y^\phi(t)\big)\,\partial_{\phi,\mu(t)} \\
&= \big(0 - (1)(0)\cdot 1\cdot 1\big)\,\partial_{\theta,\mu(t)} + \big(0\cdot 0\cdot 1 + 0\big)\,\partial_{\phi,\mu(t)} = 0 + 0 = 0.
\end{aligned}$$

So Y is parallel along μ.

Finally, for all $t \in (0, 2\pi)$, we have

$$(\nabla_{V_\mu}(\alpha X + \beta Y))(t) = \alpha(\nabla_{V_\mu} X)(t) + \beta(\nabla_{V_\mu} Y)(t) = 0 + 0 = 0.$$

Hence $\alpha X + \beta Y$ is parallel along μ.

Chapter 8

Solution to Exercise 8.1

Let $\gamma : I \to U$ be a geodesic, and set $(\theta(t), \phi(t)) = \varphi(\gamma(t))$ for all $t \in I$. Then $\gamma(t)= ((\sin\theta(t))(\cos\phi(t)), (\sin\theta(t))(\sin\phi(t)), \cos\theta(t))$ for all $t \in I$. The geodesic equation gives:

$$\ddot{\theta}(t) + \Gamma^{\theta}_{\theta\theta}\dot{\theta}(t)\dot{\theta}(t) + 2\Gamma^{\theta}_{\theta\phi}\dot{\theta}(t)\dot{\phi}(t) + \Gamma^{\theta}_{\phi\phi}\dot{\phi}(t)\dot{\phi}(t) = 0$$
$$\ddot{\phi}(t) + \Gamma^{\phi}_{\theta\theta}\dot{\theta}(t)\dot{\theta}(t) + 2\Gamma^{\phi}_{\theta\phi}\dot{\theta}(t)\dot{\phi}(t) + \Gamma^{\phi}_{\phi\phi}\dot{\phi}(t)\dot{\phi}(t) = 0$$

that is,

$$\ddot{\theta}(t) - (\sin\theta(t))(\cos\theta(t))(\dot{\phi}(t))^2 = 0$$
$$\ddot{\phi}(t) + 2(\cot\theta(t))\dot{\theta}(t)\dot{\phi}(t) = 0.$$

We had seen that in (U, φ), the component matrix for $\mathbf{g}$ is given by

$$\begin{bmatrix} \mathbf{g}_{\theta\theta} & \mathbf{g}_{\theta\phi} \\ \mathbf{g}_{\phi\theta} & \mathbf{g}_{\phi\phi} \end{bmatrix} = \begin{bmatrix} 1 & 0 \\ 0 & (\sin\theta)^2 \end{bmatrix},$$

and so for $v_{\gamma,\gamma(t)} = \dot{\theta}\,\partial_{\theta,\gamma(t)} + \dot{\phi}\,\partial_{\phi,\gamma(t)}$, we have

$$1 = \mathbf{g}(\gamma(t))(v_{\gamma,\gamma(t)}, v_{\gamma,\gamma(t)}) = \dot{\theta}^2 + (\sin\theta)^2\dot{\phi}^2. \qquad (\star)$$

The second geodesic equation is $\ddot{\phi} = -2(\cot\theta)\dot{\theta}\dot{\phi}$, that is,

$$\frac{d}{dt}\log\dot{\phi} = \frac{\ddot{\phi}}{\dot{\phi}} = -2\frac{\cos\theta}{\sin\theta}\dot{\theta} = -2\frac{d}{dt}\log(\sin\theta).$$

Integrating from $t=0$ to t, we get $\log\dfrac{\dot{\phi}(t)}{\dot{\phi}(0)} = -2\log\dfrac{\sin\theta(t)}{\sin\theta_0}$, that is,

$$\dot{\phi}(t) = \frac{\dot{\phi}(0)(\sin\theta_0)^2}{(\sin\theta(t))^2} = \frac{c}{(\sin\theta(t))^2}, \qquad (\star\star)$$

where $c := \dot{\phi}(0)(\sin\theta_0)^2$. Substituting $\dot{\phi}$ from $(\star\star)$ in $(\star)$ gives

$$\dot{\theta}^2 + (\sin\theta)^2\frac{c^2}{(\sin\theta)^4} = 1.$$

So

$$\dot{\theta} = \sqrt{1 - \frac{c^2}{(\sin\theta)^2}}, \qquad (*)$$

where we consider the portion of the geodesic traversed in the 'increasing θ direction', so that $\dot{\theta} > 0$. Thus we can invert the function θ. Let $h = \theta^{-1}$. We can then view ϕ as a function of the 'θ-variable' by considering the composition $\phi \circ h$. Dividing $(\star\star)$ and $(*)$ yields by the chain rule

$$\frac{d(\phi\circ h)}{d\theta}(\theta) = \frac{\dot{\phi}(h(\theta))}{\dot{\theta}(h(\theta))} = \frac{c}{(\sin\theta)^2}\frac{1}{\sqrt{1 - c^2/(\sin\theta)^2}} = \frac{c(\operatorname{cosec}\theta)^2}{\sqrt{1 - c^2 - c^2(\cot\theta)^2}}.$$

Hence, using the substitution $t = \frac{c}{\sqrt{1-c^2}} \cot\theta$, we obtain

$$\int_{\theta_0}^{\theta} \frac{d(\phi \circ h)}{d\theta}(\theta)\, d\theta = \int_{\theta_0}^{\theta} \frac{c(\operatorname{cosec}\theta)^2}{\sqrt{1-c^2}\sqrt{1-\frac{c^2}{1-c^2}(\cot\theta)^2}}\, d\theta = \int_{\frac{c}{\sqrt{1-c^2}}\cot\theta_0}^{\frac{c}{\sqrt{1-c^2}}\cot\theta} \frac{-1}{\sqrt{1-t^2}}\, dt.$$

With a slight abuse of notation, we denote $(\phi \circ h)(\theta)$ simply by ϕ. Then

$$\phi - \phi_0 = \sin^{-1}\Big(\frac{c}{\sqrt{1-c^2}} \cot\theta_0\Big) - \sin^{-1}\Big(\frac{c}{\sqrt{1-c^2}} \cot\theta\Big),$$

i.e., $\sin^{-1}\big(\frac{c}{\sqrt{1-c^2}} \cot\theta\big) = \alpha - \phi$, where $\alpha := \phi_0 + \sin^{-1}\big(\frac{c}{\sqrt{1-c^2}} \cot\theta_0\big)$. So

$$\frac{c}{\sqrt{1-c^2}} \frac{\cos\theta}{\sin\theta} = \sin(\alpha - \phi) = (\sin\alpha)(\cos\phi) - (\cos\alpha)\sin\phi.$$

We rewrite this as

$$(\sin\theta)(\cos\phi)(\sin\alpha) - (\sin\theta)(\sin\phi)(\cos\alpha) - \frac{c}{\sqrt{1-c^2}}\cos\theta = 0.$$

Thus, with $\mathbf{n} := \big(\sin\alpha, -\cos\alpha, -\frac{c}{\sqrt{1-c^2}}\big) \in \mathbb{R}^3$, we have

$$\begin{aligned}\langle \gamma(t), \mathbf{n}\rangle_{\mathbb{R}^3} &= \Big\langle \big((\sin\theta)\cos\phi, (\sin\theta)\sin\phi, \cos\theta\big), \big(\sin\alpha, -\cos\alpha, -\frac{c}{\sqrt{1-c^2}}\big)\Big\rangle \\ &= (\sin\theta)(\cos\phi)(\sin\alpha) - (\sin\theta)(\sin\phi)(\cos\alpha) - \frac{c}{\sqrt{1-c^2}}\cos\theta = 0.\end{aligned}$$

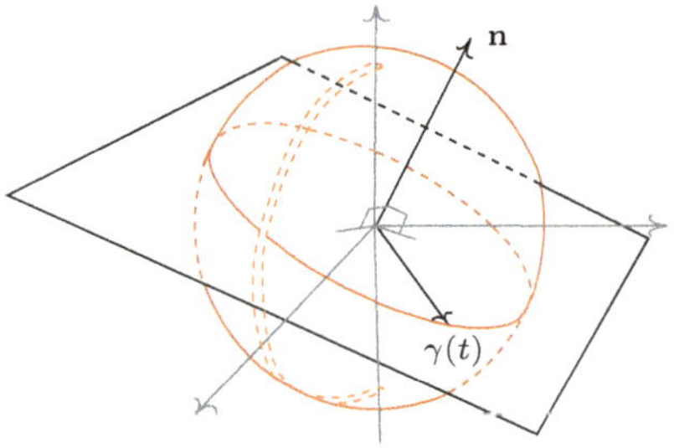

Solution to Exercise 8.2

By Proposition 7.2 (p.142), and using $\nabla_{V_\gamma} V_\gamma \equiv 0$ (as γ is a geodesic),

$$\begin{aligned}\frac{d\mathbf{g}(\gamma(\cdot))(v_{\gamma,\gamma(\cdot)}, X_{\gamma(\cdot)})}{dt} &= \mathbf{g}(\gamma(\cdot))(\nabla_{V_\gamma} V_\gamma, X_{\gamma(\cdot)}) + \mathbf{g}(\gamma(\cdot))(v_{\gamma,\gamma(\cdot)}, (\nabla_{V_\gamma} X_{\gamma(\cdot)})(\cdot)) \\ &= \mathbf{g}(\gamma(\cdot))(0, X_{\gamma(\cdot)}) + \mathbf{g}(\gamma(\cdot))(v_{\gamma,\gamma(\cdot)}, (\nabla_{V_\gamma} X_{\gamma(\cdot)})(\cdot)) \\ &= \mathbf{g}(\gamma(\cdot))(v_{\gamma,\gamma(\cdot)}, (\nabla_{V_\gamma} X_{\gamma(\cdot)})(\cdot).\end{aligned}$$

By Remark 7.1 (p.140), for each $\gamma(t)$, $t \in I$, there exists an open set U and a $V \in T_0^1 U$ which extends V_γ in U, i.e., $V_{\gamma(t)} = V_\gamma(t)$ for all $t \in I$. By the Killing equation (Exercise 6.31, p.129), $0 = \mathbf{g}(\nabla_V X, V) + \mathbf{g}(V, \nabla_V X) = 2\mathbf{g}(V, \nabla_V X)$. Thus $\mathbf{g}(\gamma(t))(v_{\gamma,\gamma(t)}, (\nabla_{V_\gamma} X_{\gamma(\cdot)})(t)) = (\mathbf{g}(V, \nabla_V X))(\gamma(t)) = 0$.

Solution to Exercise 8.3

Let $\gamma : I \to M$ be a geodesic. For $t_0 \in I$, let (U, ϕ) be an admissible chart for M containing $\gamma(t_0)$, and let $\widetilde{V} \in T_0^1 U$ be such that $\widetilde{V}_{\gamma(t)} = v_{\gamma,\gamma(t)}$ for all $t \in \gamma^{-1}U$. Then $f_*\widetilde{V} \in T_0^1(f(U))$ and

$$(f_*\widetilde{V})_{(f\circ\gamma)(t)} = df_{\gamma(t)}\widetilde{V}_\gamma(t) = df_{\gamma(t)}(v_{\gamma,\gamma(t)}) = v_{f\circ\gamma,(f\circ\gamma)(t)}.$$

Let $\nabla^{\mathbf{g}}, \nabla^{\mathbf{h}}$ denote the Levi-Civita connections in $(M, \mathbf{g}), (N, \mathbf{h})$. So

$$\begin{aligned}(\nabla^{\mathbf{h}}_{V_{f\circ\gamma}} V_{f\circ\gamma})(t) &= (\nabla^{\mathbf{h}}_{f_*\widetilde{V}} f_*\widetilde{V})_{(f\circ\gamma)(t)} = (f_*(\nabla^{\mathbf{g}}_{\widetilde{V}}\widetilde{V}))_{(f\circ\gamma)(t)} \\ &= df_{\gamma(t)}((\nabla^{\mathbf{g}}_{\widetilde{V}}\widetilde{V})_{\gamma(t)}) = df_{\gamma(t)}((\nabla^{\mathbf{g}}_{V_\gamma} V_\gamma)(t)) = df_{\gamma(t)}(0) = 0.\end{aligned}$$

Solution to Exercise 8.4

Let $t_0 \in I$, and define the point p along the geodesic by $p := \gamma(t_0)$. Proposition 8.2 (p.153) implies that the map $I \ni t \mapsto \mathbf{g}(\gamma(t))(v_{\gamma,\gamma(t)}, v_{\gamma,\gamma(t)})$ is a constant map. Thus for all $t \in I$, $\mathbf{g}(\gamma(t))(v_{\gamma,\gamma(t)}, v_{\gamma,\gamma(t)}) = \mathbf{g}(p)(v_{\gamma,p}, v_{\gamma,p})$. So depending on whether $\mathbf{g}(p)(v_{\gamma,p}, v_{\gamma,p}) > 0$, $= 0$, or < 0, it follows that for all $t \in I$, $\mathbf{g}(\gamma(t))(v_{\gamma,\gamma(t)}, v_{\gamma,\gamma(t)}) > 0$, $= 0$, or < 0, respectively, and hence γ is spacelike, lightlike, or timelike, respectively.

Solution to Exercise 8.5

We take a chart as in Example 5.10 (p.87) assuming without loss of generality that $\mathbf{p} \in U (\subset S^2)$. We have $v_{\gamma,\gamma(t)} = \partial_{t,\gamma(t)}$. For the components (t, r, θ, ϕ) of γ, with $\dot{} = \frac{d}{dt}$, we have $\dot{t} \equiv 1$, $\dot{r} \equiv 0$, $\dot{\theta} \equiv 0$, $\dot{\phi} \equiv 0$. Using the connection coefficients from Example 6.8 (p.121), the r-component of the geodesic equation is

$$\begin{aligned}0 &= \ddot{r} + \Gamma^r_{tt}(\dot{t})^2 + \Gamma^r_{rr}(\dot{r})^2 + \Gamma^r_{\theta\theta}(\dot{\theta})^2 + \Gamma^r_{\phi\phi}(\dot{\phi})^2 \\ &= 0 + \Gamma^r_{tt}(1)^2 + \Gamma^r_{rr}(0)^2 + \Gamma^r_{\theta\theta}(0)^2 + \Gamma^r_{\phi\phi}(0)^2 = \Gamma^r_{tt} = \frac{m}{r^2}\Big(1 - \frac{2m}{r}\Big),\end{aligned}$$

so that $r = 2m$, a contradiction.

Solution to Exercise 8.6

Let $\lambda^x := 0$, $\lambda^y := e^t$ be the chart-induced components of λ. Then

$$\ddot{\lambda^x} + \Gamma^x_{xy}\dot{\lambda^x}\dot{\lambda^y} + \Gamma^x_{yx}\dot{\lambda^y}\dot{\lambda^x} = 0 - 2\frac{1}{e^t}0e^t = 0, \quad \text{and}$$

$$\ddot{\lambda^y} + \Gamma^y_{yy}(\dot{\lambda^y})^2 + \Gamma^y_{xx}(\dot{\lambda^x})^2 = e^t - \frac{1}{e^t}(e^t)^2 + \frac{1}{e^t}(0)^2 = 0.$$

Hence λ satisfies the geodesic equation, and so λ is a geodesic.

The chart-induced components of σ are $\sigma^x = \tanh t$, $\sigma^y = (\cosh t)^{-1}$. Then

$$\begin{aligned}\dot{\sigma^x} &= \frac{1}{(\cosh t)^2}, & \ddot{\sigma^x} &= -\frac{2\sinh t}{(\cosh t)^3}, \\ \dot{\sigma^y} &= -\frac{\sinh t}{(\cosh t)^2}, & \ddot{\sigma^y} &= \frac{2(\sinh t)^2 - (\cosh t)^2}{(\cosh t)^3} = \frac{(\sinh t)^2 - 1}{(\cosh t)^3}.\end{aligned}$$

So $\ddot{\sigma^x} + \Gamma^x_{xy}\dot{\sigma^x}\dot{\sigma^y} + \Gamma^x_{yx}\dot{\sigma^y}\dot{\sigma^x} = -\frac{2\sinh t}{(\cosh t)^3} - 2(\cosh t)\frac{1}{(\cosh t)^2}(-\frac{\sinh t}{(\cosh t)^2}) = 0$, and

$\ddot{\sigma^y}+\Gamma^y_{yy}(\dot{\sigma^y})^2+\Gamma^y_{xx}(\dot{\sigma^x})^2 = \frac{(\sinh t)^2-1}{(\cosh t)^3} - (\cosh t)(-\frac{\sinh t}{(\cosh t)^2})^2 + (\cosh t)\frac{1}{(\cosh t)^4} = 0.$

Thus σ satisfies the geodesic equation, and so σ is a geodesic.

The picture below shows that if p is a point outside λ, then there are infinitely many geodesics that pass through p which do not intersect λ.

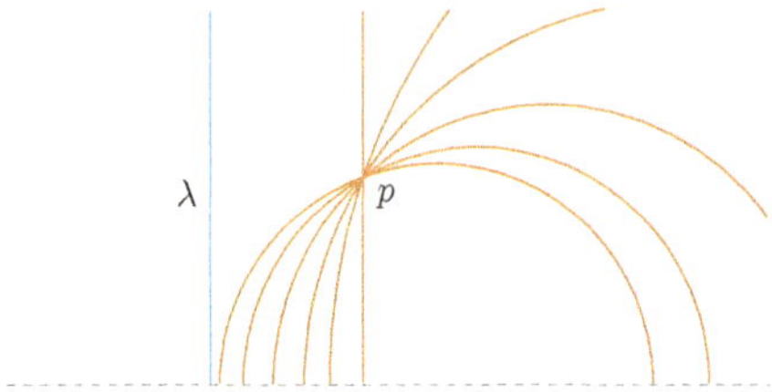

Solution to Exercise 8.7

Define $\widetilde{\gamma} : \mathbb{R} \to M$ by $\widetilde{\gamma}(t) = p$ for all $t \in \mathbb{R}$. For any $f \in C^\infty(M)$,

$$v_{\widetilde{\gamma},\widetilde{\gamma}(t)} f = \frac{d(f \circ \widetilde{\gamma})}{dt}(t) = \frac{d(t \mapsto f(p))}{dt}(t) = 0.$$

Thus $V_{\widetilde{\gamma}} = 0$ for all $t \in \mathbb{R}$. Hence

$$(\nabla_{V_{\widetilde{\gamma}}} V_{\widetilde{\gamma}})(t) = (\nabla_{V_{\widetilde{\gamma}}} 0)(t) = (\nabla_{V_{\widetilde{\gamma}}} (0 \cdot 0))(t) = 0(\nabla_{V_{\widetilde{\gamma}}} 0)(t) = 0.$$

(In the term $0 \cdot 0$ above, the first 0 is the scalar $0 \in \mathbb{R}$.) So $\widetilde{\gamma}$ is a geodesic. Moreover, $\widetilde{\gamma}(0) = p$ and $v_{\widetilde{\gamma},p} = 0$. Thus $\widetilde{\gamma}$ is a maximal geodesic passing through p with initial velocity 0. By Cor. 8.1 (p.156), this geodesic is unique. Thm. 8.1 (p.155) implies $\widetilde{\gamma}|_I = \gamma$, i.e., $\gamma(t) = \widetilde{\gamma}(t) = p$ for all $t \in I$.

Solution to Exercise 8.8

Let $\gamma : I \to \mathbb{R}^2$ be the maximal geodesic passing through $p = (a, b) \in \mathbb{R}^2$ in the direction of $v = (\alpha, \beta) \in \mathbb{R}^2$. Then with $\gamma(t) = (x(t), y(t))$, the geodesic equation becomes

$$\left\{ \begin{array}{r} \ddot{x} + 2\dot{x}\dot{y} = 0 \\ \ddot{y} = 0 \end{array} \right\}, \text{ with the initial conditions } \left\{ \begin{array}{l} x(0) = a \\ \dot{x}(0) = \alpha \end{array} \right\}, \left\{ \begin{array}{l} y(0) = b \\ \dot{y}(0) = \beta \end{array} \right\}.$$

The y-equation gives $y(t) = b + \beta t$, $t \in \mathbb{R}$. Then the x-equation is $\ddot{x} + 2\dot{x}\beta = 0$. Multiplying by $e^{2\beta t}$ gives $\frac{d}{dt}(e^{2\beta t}\dot{x}) = e^{2\beta t}\ddot{x} + 2\beta e^{2\beta t}\dot{x} = e^{2\beta t}(\ddot{x} + 2\beta\dot{x}) = 0$. Integrating from 0 to t, we get $e^{2\beta t}\dot{x} - 1\alpha = 0$, and so $\dot{x} = \alpha e^{-2\beta t}$. Again integrating from 0 to t yields

$$x(t) = a + \alpha \int_0^t e^{-2\beta\tau}\, d\tau = \begin{cases} a + \alpha t & \text{if } \beta = 0, \\ a + \alpha \frac{1 - e^{-2\beta t}}{2\beta} & \text{if } \beta \neq 0. \end{cases}$$

It can be checked that the y and x obtained above solve the geodesic equation and the initial conditions.

Let $p=(a,b)$ and $q=(c,d)$. We consider the two possible cases:

1° $b=d$. Take e.g. $\beta=0$, and $\alpha=1$. We solve for $T\in\mathbb{R}$ so that $x(T)=c$:
$c=x(T)=a+\alpha T=a+1T$ yields $T=c-a$.
Then $y(T)=b+\beta T=b+0T=b=d$.

2° $b\neq d$. Take for example $\beta=1$ and $T=d-b$.
Then $y(T)=b+\beta T=b+1\,(d-b)=d$. We solve for α so that $x(T)=c$:
$c=x(T)=a+\alpha\frac{1-e^{-2\beta T}}{2\beta}=a+\alpha\frac{1-e^{2(d-b)}}{2}$ gives $\alpha=\frac{2(c-a)}{1-e^{-2(d-b)}}$.

So given any $p,q\in\mathbb{R}^2$, there is a geodesic which passes through p and q. The plots of a few geodesics are shown below.

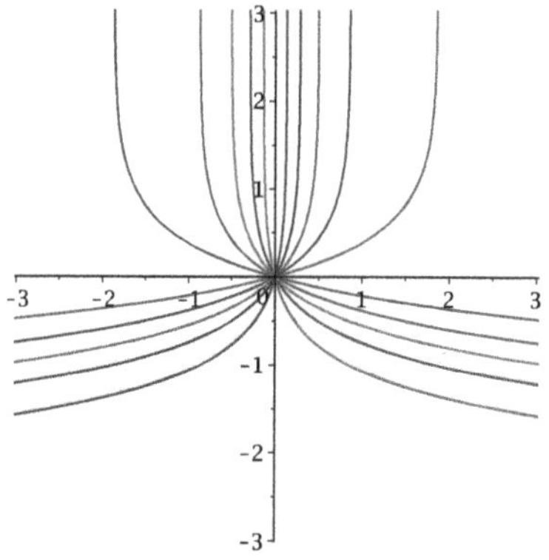

Solution to Exercise 8.9

Recall from Example 6.3 (p.116) that the only nonzero connection coefficients in the chart $(V,\mathbf{y})$ are $\Gamma^r_{\theta\theta}=-r$ and $\Gamma^\theta_{r\theta}=\Gamma^\theta_{\theta r}=\frac{1}{r}$. If $\gamma:I\to\mathbb{R}^2$ denotes a geodesic, where $I\subset\mathbb{R}$ is an open interval, and $(r(t),\theta(t)):=\mathbf{y}(\gamma(t))$, then the geodesic equation for γ gives $0=\ddot{r}+\Gamma^r_{\theta\theta}\dot{\theta}\dot{\theta}=\ddot{r}-r\dot{\theta}^2$, and $0=\ddot{\theta}+\Gamma^\theta_{r\theta}\dot{r}\dot{\theta}+\Gamma^\theta_{\theta r}\dot{\theta}\dot{r}=\ddot{\theta}+2\frac{1}{r}\dot{r}\dot{\theta}$, as wanted.

If $r(t):=\sqrt{y_0^2+t^2}$, then $\dot{r}(t)=\frac{t}{\sqrt{y_0^2+t^2}}$, and $\ddot{r}(t)=\frac{y_0^2}{(y_0^2+t^2)\sqrt{y_0^2+t^2}}$.

If $\theta(t):=\cos^{-1}\frac{t}{\sqrt{y_0^2+t^2}}$, then $\dot{\theta}(t)=-\frac{y_0}{y_0^2+t^2}$, and $\ddot{\theta}(t)=\frac{2y_0t}{(y_0^2+t^2)^2}$.

Thus $\ddot{r}-r\dot{\theta}^2=\frac{y_0^2}{(y_0^2+t^2)\sqrt{y_0^2+t^2}}-\sqrt{y_0^2+t^2}(-\frac{y_0}{y_0^2+t^2})^2=0$.

Also, $\ddot{\theta}+2\frac{1}{r}\dot{r}\dot{\theta}=\frac{2y_0t}{(y_0^2+t^2)^2}+2\frac{1}{\sqrt{y_0^2+t^2}}\frac{t}{\sqrt{y_0^2+t^2}}(-\frac{y_0}{y_0^2+t^2})=0$.

For all $t\in\mathbb{R}$, we have

$$\begin{aligned}\gamma(t)&=\mathbf{y}(r(t),\theta(t))=\big(r(t)\cos(\theta(t)),r(t)\sin(\theta(t))\big)\\&=\Big(r(t)\frac{t}{r(t)},r(t)\sqrt{1-\frac{t^2}{(r(t))^2}}\Big)=(t,y_0).\end{aligned}$$

Solution to Exercise 8.10

With $J = (a-c, b-c)$ and $h(s) = s+c$, $s \in J$, we have $h(J) = (a,b)$. For $s \in J$, $h(s) \in (a,b)$, and as γ is a geodesic, $(\nabla_{V_\gamma} V_\gamma)(h(s)) = 0$, giving

$$\begin{aligned}(\nabla_{V_{\gamma\circ h}} V_{\gamma\circ h})(s) &= \ddot{h}(s)\, V_\gamma(h(s)) + (\dot{h}(s))^2 (\nabla_{V_\gamma} V_\gamma)(h(s)) \\ &= 0\cdot V_\gamma(h(s)) + (\dot{h}(s))^2 \cdot 0 = 0 + 0 = 0.\end{aligned}$$

So $\gamma \circ h : J \to M$ is a geodesic. Also, $(\gamma \circ h)(0) = \gamma(c) = p$, and

$$v_{\gamma\circ h, \gamma\circ h(0)} = \dot{h}(0)\, v_{\gamma,\gamma(h(0))} = 1\, v_{\gamma,\gamma(c)} = 0.$$

By Exercise 8.7, $(\gamma \circ h)(s) = p$ for all $s \in J$, i.e., $\gamma(s+c) = p$ for all $s \in J$. Given any $t \in (a,b)$, $t - c \in J$ (as $a - c < t - c < b - c$) and we have that $\gamma(t) = \gamma(h(t-c)) = (\gamma \circ h)(t-c) = p$. So γ is constant.

Solution to Exercise 8.11

We have

$$\begin{aligned}\ddot{\gamma}^0(t) &= \frac{d^2 t}{dt^2} = 0 \\ \ddot{\gamma}^1(t) &= \frac{d^2}{dt^2}(v(\cos\alpha)t) = 0 \\ \ddot{\gamma}^2(t) &= \frac{d^2}{dt^2}\big(v(\sin\alpha)t - \tfrac{g}{2}t^2\big) = -g \\ \ddot{\gamma}^3(t) &= \frac{d^2 0}{dt^2} = 0.\end{aligned}$$

Thus $(\nabla_{V_\gamma} V_\gamma)(t) = -g\, \partial_{x^2,\gamma(t)}$. Recalling the isomorphism from Example 2.2 (p.27), the acceleration can be identified with $-g\mathbf{f}_2$ everywhere along the curve. So γ is not a geodesic in Minkowski spacetime.

Solution to Exercise 8.12

Using $r' \equiv 0$ and $\theta' \equiv 0$, and the connection coefficients from Example 6.8 (p.121), we obtain from the geodesic equation that the t- and ϕ-components of γ satisfy

$$\begin{aligned}0 &= t'' + 2\Gamma^t_{rt} t' r' = t'' + 2\Gamma^t_{rt} t' 0 = t'' \\ 0 &= \phi'' + 2\Gamma^\phi_{r\phi} r' \phi' + 2\Gamma^\phi_{\theta\phi} \theta' \phi' = \phi'' + 2\Gamma^\phi_{r\phi} 0 \phi' + 2\Gamma^\phi_{\theta\phi} 0 \phi' = \phi''.\end{aligned}$$

This gives the first two equations. For the third equation, we look at the r-component in the geodesic equation:

$$\begin{aligned}0 &= r'' + \Gamma^r_{tt} t'^2 + \Gamma^r_{rr} r'^2 + \Gamma^r_{\theta\theta}\theta'^2 + \Gamma^r_{\phi\phi}\phi'^2 \\ &= \Gamma^r_{tt} t'^2 + \Gamma^r_{\phi\phi}\phi'^2 \\ &= \frac{m}{r^2}\Big(1 - \frac{2m}{r}\Big)t'^2 - r\Big(1 - \frac{2m}{r}\Big)\Big(\sin\frac{\pi}{2}\Big)^2 \phi'^2 \\ &= \frac{m}{r^2}\Big(1 - \frac{2m}{r}\Big)t'^2 - r\Big(1 - \frac{2m}{r}\Big)\phi'^2.\end{aligned}$$

A rearrangement gives the third equation,

$$\phi'^2 = \frac{m}{r^3}t'^2. \qquad (\star)$$

As we have that $v_{\gamma,\gamma(\tau)} = t'(\tau)\,\partial_{t,\gamma(\tau)} + \phi'(\tau)\,\partial_{\phi,\gamma(\tau)}$, and since γ is timelike and parametrised by proper time, $\mathbf{g}(\gamma(\tau))(v_{\gamma,\gamma(\tau)}, v_{\gamma,\gamma(\tau)}) = -1$, giving

$$-1 = -t'^2\big(1 - \frac{2m}{r}\big) + r^2\big(\sin\frac{\pi}{2}\big)^2\phi'^2 = -t'^2\big(1 - \frac{2m}{r}\big) + r^2\phi'^2.$$

Using $(\star)$ to replace ϕ'^2 in the above yields, after a rearrangement,

$$\big(1 - \frac{3m}{r}\big)t'^2 = 1.$$

Let $P_2 = \gamma(0), \widetilde{P}_2 = \gamma(T_{\text{orbit}}) \in M$ be the events along γ corresponding to the completion of one circular orbit. Then

$$t(\widetilde{P}_2) - t(P_2) = \int_0^{T_{\text{orbit}}} t'(\tau)\,d\tau = \frac{T_{\text{orbit}}}{\sqrt{1-\frac{3m}{r}}}.$$

Let γ_1 be the observer at rest on the surface of the Earth, that is, $\gamma_1(t) = (t, R_\oplus, \varphi^{-1}(\frac{\pi}{2}, \phi_0))$, where (U, φ) is the spherical coordinate chart (Example 5.7, p.85), and $\phi_0 = \phi(P_2) = \phi(\widetilde{P}_2)$. Let $P_1, \widetilde{P}_1$ be the events at which radial light signals from P_2, $\widetilde{P}_2$, respectively, reach the observer γ_1. By the computation done in Example 5.16 (p.96) on the gravitational redshift, it follows that the proper time elapsed for the observer γ_1 on Earth between the events $P_1, \widetilde{P}_1$ is

$$T_{\text{Earth}} = \sqrt{1-\frac{2m}{R_\oplus}}(t(\widetilde{P}_1)-t(P_1)) = \sqrt{1-\frac{2m}{R_\oplus}}(t(\widetilde{P}_2)-t(P_2)) = \sqrt{1-\frac{2m}{R_\oplus}}\,\frac{T_{\text{orbit}}}{\sqrt{1-\frac{3m}{r}}}.$$

Using $m \ll r$ and $m \ll R_\oplus$, we obtain

$$\frac{T_{\text{orbit}}}{T_{\text{Earth}}} = \frac{\sqrt{1-\frac{3m}{r}}}{\sqrt{1-\frac{2m}{R_\oplus}}} \approx \big(1 - \frac{1}{2}\frac{3m}{r}\big)\big(1 - \big(-\frac{1}{2}\big)\frac{2m}{R_\oplus}\big) \approx 1 - \frac{3}{2}\frac{m}{r} + \frac{m}{R_\oplus}.$$

With the given values, $\dfrac{T_{\text{orbit}} - T_{\text{Earth}}}{T_{\text{Earth}}} = \dfrac{T_{\text{orbit}}}{T_{\text{Earth}}} - 1 \approx m\big(\dfrac{1}{R_\oplus} - \dfrac{3}{2r}\big) \approx 4.531\times 10^{-10}$.

Solution to Exercise 8.13

We have $\mathbf{g}(\gamma(\cdot))(V_\gamma(\cdot), V_\gamma(\cdot)) = 0$ (γ lightlike), and $\nabla_{V_\gamma}V_\gamma = 0$ (γ geodesic). Using the result from Exercise 6.17 (p.119), it follows that for all $t \in I$,

$$\begin{aligned}(\widetilde{\nabla}_{V_\gamma}V_\gamma)(t) &= (\nabla_{V_\gamma}V_\gamma)(t) + \frac{2(V_\gamma(t)f)V_\gamma(t) - \mathbf{g}(\gamma(t))(V_\gamma(t), V_\gamma(t))(\text{grad } f)(\gamma(t))}{2f(\gamma(t))}\\ &= 0 + \frac{2(V_\gamma(t)f)V_\gamma(t) - 0(\text{grad } f)(\gamma(t))}{2f(\gamma(t))}\\ &= \frac{(V_\gamma(t)f)V_\gamma(t)}{f(\gamma(t))}.\end{aligned}$$

Thus γ may not be a geodesic with respect to $\widetilde{\nabla}$. We show below that it is possible to find a reparametrisation $h : J \to I$ such that $\gamma \circ h$ is a lightlike with respect to $\widetilde{\mathbf{g}}$ and also a geodesic with respect to $\widetilde{\nabla}$. Irrespective of the specific form of h, we will have $V_{\gamma\circ h} = \dot{h}V_\gamma(h(\cdot))$, and so for all $s \in J$,

$$\begin{aligned}
&\widetilde{\mathbf{g}}((\gamma \circ h)(s))(V_{\gamma\circ h}(s), V_{\gamma\circ h}(s)) \\
&= (f\mathbf{g})((\gamma \circ h)(s))(\dot{h}(s)V_\gamma(h(s)), \dot{h}(s)V_\gamma(h(s))) \\
&= f(\gamma(h(s)))(\dot{h}(s))^2\mathbf{g}(\gamma(h(s)))(V_\gamma(h(s)), V_\gamma(h(s))) \\
&= f(\gamma(h(s)))(\dot{h}(s))^2 0 = 0.
\end{aligned}$$

Thus $\gamma \circ h$ is lightlike. We will now choose a specific h so that $\gamma \circ h$ is also a geodesic. By Lemma 8.1 (p.159), we have

$$\begin{aligned}
(\widetilde{\nabla}_{V_{\gamma\circ h}} V_{\gamma\circ h})(s) &= \ddot{h}(s)V_\gamma(h(s)) + (\dot{h}(s))^2 \frac{(V_\gamma(h(s))f)V_\gamma(h(s))}{f(\gamma(h(s)))} \\
&= \Big(\ddot{h}(s) + (\dot{h}(s))^2 \frac{V_\gamma(h(s))f}{f(\gamma(h(s)))}\Big)V_\gamma(h(s)).
\end{aligned}$$

In order to obtain geodesy with respect to $\widetilde{\nabla}$, we select h so that

$$\ddot{h}(s) + (\dot{h}(s))^2 \frac{V_\gamma(h(s))f}{f(\gamma(h(s)))} = 0. \qquad (\star)$$

This is a second order ordinary differential equation, with the initial conditions $h(0) = 0$ (so that $(\gamma \circ h)(0) = p$) and $\dot{h}(0) = 1$ (as $v_{\gamma\circ h,p} = v$). It can be rewritten as a first order system for $(x, y) := (h, \dot{h})$ with the initial condition $(x(0), y(0)) := (0, 1)$. To obtain the existence of a local smooth solution to this initial value problem, we proceed as follows. We will show below, see $(*)$, that the dependence on x and y of the function that describes the first order differential equation system is C^∞. In particular, it is Lipschitz[20] in the vicinity of $(0, 1)$ guaranteeing a C^1-solution $(x, y) = (h, \dot{h})$. In particular, h is C^2, and so $(\star)$ implies h is in fact C^3, and this process can be continued to show that h is C^∞.

$(*)$: To see that the dependence on x and y of the function that describes the first order differential equation system is C^∞, first we note that as $f \circ \gamma$ is smooth and pointwise positive, $x \mapsto (f(\gamma(x))^{-1}$ is C^∞ in a neighbourhood of $x = 0$. Also $y \mapsto y^2$ is C^∞ in a neighbourhood of $y = 1$. Finally, we have $V_\gamma(h(s))f = v_{\gamma,\gamma(h(s))}f = \frac{d(f\circ\gamma)}{dt}(h(s))$, and so, as the map $f \circ \gamma \in C^\infty(I)$, the map $x \mapsto \frac{d(f\circ\gamma)}{dt}(x)$ is C^∞ in a neighbourhood of $x = 0$.

Thus there exists a neighbourhood J of 0 such that $h : J \to I$ is a smooth map which satisfies the differential equation $(\star)$ above, with the initial conditions $h(0) = 0$ and $\dot{h}(0) = 1$.

[20]See, e.g., [Apostol (1969), Theorem 7.19, p.229].

Solution to Exercise 8.14

The curve P is a geodesic because it satisfies the geodesic equation. Indeed,

$$\ddot{P} - \dot{P}P^{-1}\dot{P} = A^2e^{tA} - Ae^{tA}e^{-tA}Ae^{tA} = A^2e^{tA} - A^2I_2e^{tA} = 0.$$

Thus the exponential map $\exp_{I_2} : T_{I_2}(\mathrm{GL}_2(\mathbb{R})) \to \mathrm{GL}_2(\mathbb{R})$ is given by $\exp_{I_2}(A) = P(1) = e^{1A} = e^A$ for all $A \in \mathbb{R}^{2\times 2} \simeq T_{I_2}(\mathrm{GL}_2(\mathbb{R}))$.

Solution to Exercise 8.15

For $f \in C^\infty(M)$, we have (with the summation convention suspended)

$$\begin{aligned}\partial_{\widetilde{x}^i,p}f &= \frac{\partial(f\circ\widetilde{\mathbf{x}}^{-1})}{\partial u^i}(\widetilde{\mathbf{x}}(p)) = \frac{\partial(f\circ\mathbf{x}^{-1}\circ D^{-1})}{\partial u^i}(D(\mathbf{x}(p)))\\ &= \sum_{j=1}^{m}\frac{\partial(f\circ\mathbf{x}^{-1})}{\partial u^j}(D^{-1}D\mathbf{x}(p))\frac{1}{k_j}\delta_i^j = \frac{1}{k_i}\partial_{x^i,p}f = e_i(f).\end{aligned}$$

Thus $\partial_{\widetilde{x}^i,p} = e_i$ for all $i \in \{1,\cdots,m\}$, as wanted.

Noting that $\Gamma_{ij}^k(p) = 0$, $1 \leqslant i,j,k \leqslant m$, it follows from Proposition 6.2 (p.115) that $\widetilde{\Gamma}_{ij}^k(p) = (\partial_{x^\ell,p}\widetilde{x}^k)\partial_{\widetilde{x}^j,p}(\partial_{\widetilde{x}^i}x^\ell)$. But

$$\partial_{\widetilde{x}^i}x^\ell = \frac{\partial(\mathbf{x}\circ\widetilde{\mathbf{x}}^{-1})^\ell}{\partial u^i}(\widetilde{\mathbf{x}}(\cdot)) = \begin{cases}1/k_i & \text{if } \ell = i,\\ 0 & \text{if } \ell \neq i.\end{cases}$$

So $\partial_{\widetilde{x}^j,p}(\partial_{\widetilde{x}^i}x^\ell) = 0$. Thus $\widetilde{\Gamma}_{ij}^k(p) = 0$, $1 \leqslant i,j,k \leqslant m$.

Solution to Exercise 8.16

Writing $\boldsymbol{\gamma}(t) = (\gamma^1(t),\cdots,\gamma^m(t)) = \mathbf{x}\circ\gamma$ and $g_{ij} = \mathbf{g}_{ij}\circ\mathbf{x}^{-1}$, the Euler-Lagrange equation is given by

$$\frac{\partial g_{ij}}{\partial u^k}(\boldsymbol{\gamma}(t))\dot{\gamma}^i(t)\dot{\gamma}^j(t) - \frac{d}{dt}\big(g_{kj}(\boldsymbol{\gamma}(t))\dot{\gamma}^j(t) + g_{ik}(\boldsymbol{\gamma}(t))\dot{\gamma}^i(t)\big) = 0, \quad i = 1,\cdots,m.$$

Expanding the derivative of the term in the brackets $\big(\cdots\big)$ above, we get for all $k = 1,\cdots,m$, that (suppressing the argument t):

$$\frac{\partial g_{ij}}{\partial u^k}(\boldsymbol{\gamma})\dot{\gamma}^i\dot{\gamma}^j - \frac{\partial g_{kj}}{\partial u^\ell}\dot{\gamma}^\ell\dot{\gamma}^j - g_{kj}(\boldsymbol{\gamma})\ddot{\gamma}^j - \frac{\partial g_{ik}}{\partial u^\ell}\dot{\gamma}^\ell\dot{\gamma}^i - g_{ik}(\boldsymbol{\gamma})\ddot{\gamma}^i = 0.$$

Using the symmetries $g_{ij} = g_{ji}$ and by renaming the dummy summation indices appropriately, we can rewrite the above as follows

$$\begin{aligned}0 &= 2g_{ik}\ddot{\gamma}^i + \Big(\frac{\partial g_{kj}}{\partial u^i}(\boldsymbol{\gamma}) + \frac{\partial g_{ik}}{\partial u^j}(\boldsymbol{\gamma}) - \frac{\partial g_{ij}}{\partial u^k}(\boldsymbol{\gamma})\Big)\dot{\gamma}^i\dot{\gamma}^j\\ &= 2\mathbf{g}_{ik}(\gamma)\ddot{\gamma}^i + ((\partial_{x^i}\mathbf{g}_{kj})(\gamma) + (\partial_{x^j}\mathbf{g}_{ik})(\gamma) - (\partial_{x^k}\mathbf{g}_{ij})(\gamma))\dot{\gamma}^i\dot{\gamma}^j.\end{aligned}$$

Multiplying by $\mathbf{g}^{k\ell}(\gamma)/2$ (and summing over k), we get

$$\begin{aligned}0 &= \mathbf{g}^{k\ell}\mathbf{g}_{ik}(\gamma)\ddot{\gamma}^i + \frac{\mathbf{g}^{k\ell}}{2}\big((\partial_{x^i}\mathbf{g}_{kj})(\gamma) + (\partial_{x^j}\mathbf{g}_{ik})(\gamma) - (\partial_{x^k}\mathbf{g}_{ij})(\gamma)\big)\dot{\gamma}^i\dot{\gamma}^j\\ &= \ddot{\gamma}^\ell + \Gamma_{ij}^\ell(\gamma)\dot{\gamma}^i\dot{\gamma}^j\end{aligned}$$

i.e., we obtain the geodesic equation $\ddot{\gamma}^\ell(t) + \Gamma_{ij}^\ell(\gamma(t))\dot{\gamma}^i(t)\dot{\gamma}^j(t) = 0$ for all $t \in (a,b)$ and all $\ell \in \{1,\cdots,m\}$.

Solution to Exercise 8.17

Let E_{ij} denote the matrix all of whose entries are zeroes except the entry in the ith row and j^{th} column, which is equal to 1. Let $\mathcal{L} : \mathrm{GL}_2(\mathbb{R}) \times \mathbb{R}^{2\times 2} \to \mathbb{R}$ be given by $\mathcal{L}(P, Q) = \mathrm{trace}(P^{-1}QP^{-1}Q)$ for all $P \in \mathrm{GL}_2(\mathbb{R})$ and $Q \in \mathbb{R}^{2\times 2}$. Then using[21] $(I + hP^{-1}E_{ij})^{-1} = I - hP^{-1}E_{ij} + h^2(P^{-1}E_{ij})^2 - + \cdots$ for all real h with small enough $|h|$, we have

$$\begin{aligned}\frac{\partial \mathcal{L}}{\partial p_{ij}} &= \lim_{h\to 0} \mathrm{trace}\Big(\frac{(P+hE_{ij})^{-1}Q(P+hE_{ij})^{-1}Q - P^{-1}QP^{-1}Q}{h}\Big)\\ &= \lim_{h\to 0} \mathrm{trace}\Big(\frac{(I+hP^{-1}E_{ij})^{-1}P^{-1}Q(I+hP^{-1}E_{ij})^{-1}P^{-1}Q - P^{-1}QP^{-1}Q}{h}\Big)\\ &= \mathrm{trace}(-P^{-1}E_{ij}P^{-1}QP^{-1}Q - P^{-1}QP^{-1}E_{ij}P^{-1}Q)\\ &= -2\,\mathrm{trace}(P^{-1}QP^{-1}QP^{-1}E_{ij}).\end{aligned}$$

In particular, $\frac{\partial \mathcal{L}}{\partial p_{ij}}(P, \dot{P}) = -2\,\mathrm{trace}(P^{-1}\dot{P}P^{-1}\dot{P}P^{-1}E_{ij})$. Next, we have

$$\begin{aligned}\frac{\partial \mathcal{L}}{\partial q_{ij}} &= \lim_{h\to 0} \mathrm{trace}\Big(\frac{P^{-1}(Q+hE_{ij})P^{-1}(Q+hE_{ij}) - P^{-1}QP^{-1}Q}{h}\Big)\\ &= \mathrm{trace}(P^{-1}E_{ij}P^{-1}Q + P^{-1}QP^{-1}E_{ij}) = 2\,\mathrm{trace}(P^{-1}QP^{-1}E_{ij}).\end{aligned}$$

In particular,

$$\frac{\partial \mathcal{L}}{\partial \dot{p}_{ij}}(P, \dot{P}) = 2\,\mathrm{trace}(P^{-1}\dot{P}P^{-1}E_{ij}).$$

As $P(t)(P(t))^{-1} = I$, we get upon differentiating with respect to t, that

$$\dot{P}P^{-1} + P\frac{d}{dt}(P^{-1}) = 0,$$

and so $\frac{d}{dt}(P^{-1}) = -P^{-1}\dot{P}P^{-1}$. Thus we obtain

$$\begin{aligned}&\frac{d}{dt}\frac{\partial \mathcal{L}}{\partial \dot{p}_{ij}}(P, \dot{P}) = \frac{d}{dt}(2\,\mathrm{trace}(P^{-1}\dot{P}P^{-1}E_{ij}))\\ &= 2\,\mathrm{trace}(-P^{-1}\dot{P}P^{-1}\dot{P}P^{-1}E_{ij} + P^{-1}\ddot{P}P^{-1}E_{ij} - P^{-1}\dot{P}P^{-1}\dot{P}P^{-1}E_{ij}).\end{aligned}$$

So the Euler-Lagrange equation is

$$\begin{aligned}0 &= \frac{\partial \mathcal{L}}{\partial p_{ij}}(P, \dot{P}) - \frac{d}{dt}\frac{\partial \mathcal{L}}{\partial \dot{p}_{ij}}(P, \dot{P})\\ &= -2\,\mathrm{trace}(P^{-1}\dot{P}P^{-1}\dot{P}P^{-1}E_{ij})\\ &\quad -2\,\mathrm{trace}(-2\,P^{-1}\dot{P}P^{-1}\dot{P}P^{-1}E_{ij} + P^{-1}\ddot{P}P^{-1}E_{ij})\\ &= 2\,\mathrm{trace}(P^{-1}\dot{P}P^{-1}\dot{P}P^{-1}E_{ij} - P^{-1}\ddot{P}P^{-1}E_{ij}).\end{aligned}$$

But for any matrix Q, we have that $(QE_{ij})_{\ell k} = q_{\ell m}\delta_i^m \delta_j^k = q_{\ell i}\delta_j^k$, and so

[21]By the Neumann series theorem, see, e.g., [Sasane (2017), Theorem 2.9].

$\text{trace}(QE_{ij}) = q_{\ell i}\delta_j^\ell = q_{ji}$. So if for all i, j, $\text{trace}(QE_{ij}) = 0$, then $Q = 0$. Thus from the above, we get

$$0 = P^{-1}\dot{P}P^{-1}\dot{P}P^{-1} - P^{-1}\ddot{P}P^{-1} = P^{-1}(\dot{P}P^{-1}\dot{P} - \ddot{P})P^{-1},$$

which is equivalent to $\ddot{P} - \dot{P}P^{-1}\dot{P} = 0$, as wanted.

Chapter 9

Solution to Exercise 9.1

Recall Jacobi's identity: $[X,[Y,Z]]+[Y,[Z,X]]+[Z,[X,Y]]=0$. We have

$$\begin{aligned} \mathcal{L}_X\mathcal{L}_Y Z-\mathcal{L}_Y\mathcal{L}_X Z &= [X,[Y,Z]]-[Y,[X,Z]] \\ &= [X,[Y,Z]]+[Y,[Z,X]] \\ &= -[Z,[X,Y]] \quad \text{(Jacobi identity)} \\ &= [[X,Y],Z] = \mathcal{L}_{[X,Y]}Z. \end{aligned}$$

Solution to Exercise 9.2

From Exercise 6.16 (p.119), we know that $2\nabla_X Y = \mathcal{L}_X Y$ for all $X, Y \in \mathfrak{g}$. So for all $X, Y, Z \in \mathfrak{g}$, we have

$$\begin{aligned} R(X,Y)Z &= \nabla_X(\nabla_Y Z)-\nabla_Y(\nabla_X Z)-\nabla_{[X,Y]}Z \\ &= \tfrac{1}{2}[X,\tfrac{1}{2}[Y,Z]]-\tfrac{1}{2}[Y,\tfrac{1}{2}[X,Z]]-\tfrac{1}{2}[[X,Y],Z] \\ &= \tfrac{1}{4}[X,[Y,Z]]+\tfrac{1}{4}[Y,[Z,X]]-\tfrac{1}{2}[[X,Y],Z] \\ &= \tfrac{1}{4}[[X,Y],Z]-\tfrac{1}{2}[[X,Y],Z] \quad \text{(Jacobi identity)} \\ &= -\tfrac{1}{4}[[X,Y],Z]. \end{aligned}$$

Solution to Exercise 9.3

We have that $dx^r(R(\partial_{x^i},\partial_{x^j})\partial_{x^k}) = \mathbf{R}(dx^r,\partial_{x^i},\partial_{x^j},\partial_{x^k}) = \mathbf{R}^r_{ijk}$. We also have that $dx^r(\mathbf{R}^\ell_{ijk}\partial_{x^\ell}) = \mathbf{R}^\ell_{ijk}\delta^r_\ell = \mathbf{R}^r_{ijk}$. Consequently, the vector fields $R(\partial_{x^i},\partial_{x^j})\partial_{x^k} \in T^1_0 U$ and $\mathbf{R}^\ell_{ijk}\partial_{x^\ell} \in T^1_0 U$ coincide (as their components, $dx^r(R(\partial_{x^i},\partial_{x^j})\partial_{x^k})$, respectively $dx^r(\mathbf{R}^\ell_{ijk}\partial_{x^\ell})$, $1 \leqslant r \leqslant m$, coincide).

Solution to Exercise 9.4

We have

$$\begin{aligned} &\widetilde{\mathbf{R}}(X,Y,Z,W)+\widetilde{\mathbf{R}}(Y,Z,X,W)+\widetilde{\mathbf{R}}(Z,X,Y,W) \\ &= \mathbf{g}(W,R(X,Y)Z)+\mathbf{g}(W,R(Y,Z)X)+\mathbf{g}(W,R(Z,X)Y) \\ &= \mathbf{g}(W,R(X,Y)Z+R(Y,Z)X+R(Z,X)Y) = \mathbf{g}(W,0) = 0. \end{aligned}$$

Solution to Exercise 9.5

We have $\mathbf{R}(\Omega,X,Y,Z) = \mathbf{R}((\Omega^\sharp)^\flat,X,Y,Z) = \widetilde{\mathbf{R}}(X,Y,Z,\Omega^\sharp)$. Thus

$$\begin{aligned} \mathbf{R}^\ell_{ijk} &= \mathbf{R}(dx^\ell,\partial_{x^i},\partial_{x^j},\partial_{x^k}) = \widetilde{\mathbf{R}}(\partial_{x^i},\partial_{x^j},\partial_{x^k},(dx^\ell)^\sharp) \\ &= \widetilde{\mathbf{R}}(\partial_{x^i},\partial_{x^j},\partial_{x^k},\mathbf{g}^{\ell r}\partial_{x^r}) = \mathbf{g}^{\ell r}\widetilde{\mathbf{R}}(\partial_{x^i},\partial_{x^j},\partial_{x^k},\partial_{x^r}) = \mathbf{g}^{\ell r}\widetilde{\mathbf{R}}_{ijkr}. \end{aligned}$$

Alternatively, as $\widetilde{\mathbf{R}}_{ijkr} = \mathbf{g}_{rs}\mathbf{R}^s_{ijk}$, $\mathbf{g}^{\ell r}\widetilde{\mathbf{R}}_{ijkr} = \mathbf{g}^{\ell r}\mathbf{g}_{rs}\mathbf{R}^s_{ijk} = \delta^\ell_s\mathbf{R}^s_{ijk} = \mathbf{R}^\ell_{ijk}$.

Solution to Exercise 9.6

$(\mathbf{R}\otimes\mathbf{g})(\Omega, X, Y, Z, V, W) = \mathbf{R}(\Omega, X, Y, Z)\mathbf{g}(V, W)$, and so in any admissible chart $(U, \mathbf{x})$, $(\mathbf{C}^1_5(\mathbf{R}\otimes\mathbf{g}))(X, Y, Z, V) = \mathbf{R}(dx^r, X, Y, Z)\mathbf{g}(V, \partial_{x^r})$. So

$$\begin{aligned}(\mathbf{C}^1_5(\mathbf{R}\otimes\mathbf{g}))_{ijk\ell} &= (\mathbf{C}^1_5(\mathbf{R}\otimes\mathbf{g}))(\partial_{x^i}, \partial_{x^j}, \partial_{x^k}, \partial_{x^\ell}) \\ &= \mathbf{R}(dx^r, \partial_{x^i}, \partial_{x^j}, \partial_{x^k})\mathbf{g}(\partial_{x^\ell}, \partial_{x^r}) = \mathbf{R}^r_{ijk}\mathbf{g}_{\ell r} = \widetilde{\mathbf{R}}_{ijk\ell}.\end{aligned}$$

Hence $\mathbf{C}^1_5(\mathbf{R}\otimes\mathbf{g}) = \widetilde{\mathbf{R}}$. (As $\mathbf{g}$ is symmetric, we also have $\widetilde{\mathbf{R}} = \mathbf{C}^1_4(\mathbf{R}\otimes\mathbf{g})$.)

Solution to Exercise 9.7

We have

$$\begin{aligned}\mathbf{R}^k_{kij} &= \mathbf{g}^{k\ell}\widetilde{\mathbf{R}}_{kij\ell} = \mathbf{g}^{k\ell}\widetilde{\mathbf{R}}_{j\ell ki} = (-1)(-1)\mathbf{g}^{k\ell}\widetilde{\mathbf{R}}_{\ell jik} \\ &= \mathbf{g}^{k\ell}\widetilde{\mathbf{R}}_{\ell jik} = \mathbf{g}^{\ell k}\widetilde{\mathbf{R}}_{\ell jik} = \mathbf{R}^\ell_{\ell ji} = \mathbf{R}^k_{kji}.\end{aligned}$$

Solution to Exercise 9.8

Recall that $\Gamma^\theta_{\phi\phi} = -(\sin\theta)\cos\theta$, $\Gamma^\phi_{\theta\phi} = \cot\theta = \Gamma^\phi_{\phi\theta}$ are the only nonzero connection coefficients in the given chart. We have

$$\begin{aligned}\mathbf{R}^\theta_{\phi\theta\phi} &= \partial_\phi\Gamma^\theta_{\phi\theta} - \partial_\theta\Gamma^\theta_{\phi\phi} + \Gamma^\phi_{\phi\theta}\Gamma^\theta_{\phi\phi} - \Gamma^\theta_{\phi\phi}\Gamma^\theta_{\theta\theta} \\ &= \partial_\phi 0 - \partial_\theta(-(\sin\theta)\cos\theta) + (\cot\theta)(-(\sin\theta)\cos\theta) - \Gamma^\theta_{\phi\phi}0 \\ &= 0 + (\cos\theta)^2 - (\sin\theta)^2 - (\cos\theta)^2 - 0 = -(\sin\theta)^2.\end{aligned}$$

Thus $\widetilde{\mathbf{R}}_{\phi\theta\phi\theta} = \mathbf{g}_{\theta\theta}\mathbf{R}^\theta_{\phi\theta\phi} = 1\,(-(\sin\theta)^2) = -(\sin\theta)^2$. Then

$$\begin{array}{llll}
\widetilde{\mathbf{R}}_{\phi\phi\phi\phi} = 0 & \widetilde{\mathbf{R}}_{\phi\phi\phi\theta} = 0 & \widetilde{\mathbf{R}}_{\phi\phi\theta\phi} = 0 & \widetilde{\mathbf{R}}_{\phi\phi\theta\theta} = 0 \\
\widetilde{\mathbf{R}}_{\phi\theta\phi\phi} = 0 & \widetilde{\mathbf{R}}_{\phi\theta\phi\theta} = -(\sin\theta)^2 & \widetilde{\mathbf{R}}_{\phi\theta\theta\phi} = (\sin\theta)^2 & \widetilde{\mathbf{R}}_{\phi\theta\theta\theta} = 0 \\
\widetilde{\mathbf{R}}_{\theta\phi\phi\phi} = 0 & \widetilde{\mathbf{R}}_{\theta\phi\phi\theta} = (\sin\theta)^2 & \widetilde{\mathbf{R}}_{\theta\phi\theta\phi} = -(\sin\theta)^2 & \widetilde{\mathbf{R}}_{\theta\phi\theta\theta} = 0 \\
\widetilde{\mathbf{R}}_{\theta\theta\phi\phi} = 0 & \widetilde{\mathbf{R}}_{\theta\theta\phi\theta} = 0 & \widetilde{\mathbf{R}}_{\theta\theta\theta\phi} = 0 & \widetilde{\mathbf{R}}_{\theta\theta\theta\theta} = 0
\end{array}$$

are the 16 components of the covariant Riemann curvature tensor field in the given chart. Hence

$$\begin{array}{llll}
\mathbf{R}^\phi_{\phi\phi\phi} = 0 & \mathbf{R}^\phi_{\phi\phi\theta} = 0 & \mathbf{R}^\phi_{\phi\theta\phi} = 0 & \mathbf{R}^\phi_{\phi\theta\theta} = 1 \\
\mathbf{R}^\phi_{\theta\phi\phi} = 0 & \mathbf{R}^\phi_{\theta\phi\theta} = -1 & \mathbf{R}^\phi_{\theta\theta\phi} = 0 & \mathbf{R}^\phi_{\theta\theta\theta} = 0 \\
\mathbf{R}^\theta_{\phi\phi\phi} = 0 & \mathbf{R}^\theta_{\phi\phi\theta} = 0 & \mathbf{R}^\theta_{\phi\theta\phi} = -(\sin\theta)^2 & \mathbf{R}^\theta_{\phi\theta\theta} = 0 \\
\mathbf{R}^\theta_{\theta\phi\phi} = (\sin\theta)^2 & \mathbf{R}^\theta_{\theta\phi\theta} = 0 & \mathbf{R}^\theta_{\theta\theta\phi} = 0 & \mathbf{R}^\theta_{\theta\theta\theta} = 0.
\end{array}$$

Solution to Exercise 9.9

Let $(U, \mathbf{x})$ be an admissible chart for M. Then

$$\begin{aligned}\mathbf{Ric}_{ij} &= (\mathbf{C}^1_1\mathbf{R})_{ij} = \mathbf{R}(dx^k, \partial_{x^k}, \partial_{x^i}, \partial_{x^j}) = \mathbf{R}^k_{kij} = \mathbf{g}^{k\ell}\widetilde{\mathbf{R}}_{kij\ell} = -\mathbf{g}^{k\ell}\widetilde{\mathbf{R}}_{ikj\ell} \\ &= -\mathbf{g}^{k\ell}\mathbf{g}_{\ell r}\mathbf{R}^r_{ikj} = -\delta^k_r\mathbf{R}^r_{ikj} = -\mathbf{R}^k_{ikj} = -\mathbf{R}(dx^k, \partial_{x^i}, \partial_{x^k}, \partial_{x^j}) \\ &= -(\mathbf{C}^1_2\mathbf{R})(\partial_{x^i}, \partial_{x^j}) = -(\mathbf{C}^1_2\mathbf{R})_{ij}.\end{aligned}$$

Thus $\mathbf{Ric} = -\mathbf{C}^1_2\mathbf{R}$.

Solution to Exercise 9.10

The components of **Ric** in the chart (U, φ) are given by

$$\begin{aligned}
\mathbf{Ric}_{\theta\theta} &= \mathbf{R}^{\phi}_{\phi\theta\theta} + \mathbf{R}^{\theta}_{\theta\theta\theta} = 1 + 0 = 1, \\
\mathbf{Ric}_{\theta\phi} &= \mathbf{R}^{\phi}_{\phi\theta\phi} + \mathbf{R}^{\theta}_{\theta\theta\phi} = 0 + 0 = 0, \\
\mathbf{Ric}_{\phi\theta} &= \mathbf{Ric}_{\theta\phi} = 0, \\
\mathbf{Ric}_{\phi\phi} &= \mathbf{R}^{\phi}_{\phi\phi\phi} + \mathbf{R}^{\theta}_{\theta\phi\phi} = 0 + (\sin\theta)^2 = (\sin\theta)^2.
\end{aligned}$$

Thus the scalar curvature is given by

$$\begin{aligned}
\mathbf{S} &= \mathbf{Ric}_{\theta\theta}\mathbf{g}^{\theta\theta} + \mathbf{Ric}_{\theta\phi}\mathbf{g}^{\theta\phi} + \mathbf{Ric}_{\phi\phi}\mathbf{g}^{\phi\phi} + \mathbf{Ric}_{\phi\theta}\mathbf{g}^{\theta\phi} \\
&= \mathbf{Ric}_{\theta\theta}\cdot 1 + \mathbf{Ric}_{\theta\phi}\cdot 0 + \mathbf{Ric}_{\phi\phi}\cdot(\sin\theta)^{-2} + \mathbf{Ric}_{\phi\theta}\cdot 0 \\
&= 1\cdot 1 + 0 + (\sin\theta)^2\cdot(\sin\theta)^{-2} + 0 = 1 + 1 = 2.
\end{aligned}$$

Thus $\mathbf{S} \equiv 2$ on U. As every point in $S^2\backslash U$ is the limit of a sequence of points in U, it follows by the continuity of $\mathbf{S}$ that $\mathbf{S} \equiv 2$ on S^2.

Solution to Exercise 9.11

Using the first two symmetries from Proposition 9.3 (p.184), we conclude that the chart-induced components of $\widetilde{\mathbf{R}}$ satisfy $\widetilde{\mathbf{R}}_{jjk\ell} = 0$ and $\widetilde{\mathbf{R}}_{jk\ell\ell} = 0$. Thus the remaining components of $\widetilde{\mathbf{R}}$ are $\widetilde{\mathbf{R}}_{1212}, \widetilde{\mathbf{R}}_{1221}, \widetilde{\mathbf{R}}_{2112}, \widetilde{\mathbf{R}}_{2121}$. It is enough to determine $\widetilde{\mathbf{R}}_{1212}$, because, using the first two symmetries from Proposition 9.3 again, we get that $\widetilde{\mathbf{R}}_{1221} = -\widetilde{\mathbf{R}}_{1212}$, $\widetilde{\mathbf{R}}_{2112} = -\widetilde{\mathbf{R}}_{1212}$, and $\widetilde{\mathbf{R}}_{2121} = \widetilde{\mathbf{R}}_{1212}$. In fact, the same is true for any tensor field $\mathbf{T} \in T^0_4 M$ possessing the symmetries $\mathbf{T}(X,Y,Z,W) = -\mathbf{T}(Y,X,Z,W)$ and $\mathbf{T}(X,Y,Z,W) = -\mathbf{T}(X,Y,W,Z)$ for all $X,Y,Z,W \in T^1_0 M$. Clearly, $\mathbf{T}$ defined by $\mathbf{T}(X,Y,Z,W) = \mathbf{g}(X,W)\mathbf{g}(Y,Z) - \mathbf{g}(X,Z)\mathbf{g}(Y,W)$ for all vector fields $X,Y,Z,W \in T^1_0 M$ is a $(0,4)$-tensor field, and it has the symmetry properties above. If $\{e_1, e_2\}$ is an orthonormal basis for T_pM, then $\mathbf{T}(p)(e_1,e_2,e_1,e_2) = (\mathbf{g}(p)(e_1,e_2))^2 - \mathbf{g}(p)(e_1,e_1)\mathbf{g}(p)(e_2,e_2) = 0^2 - (\pm 1) \neq 0$. Define $k(p) := \widetilde{\mathbf{R}}(p)(e_1,e_2,e_1,e_2)/\mathbf{T}(p)(e_1,e_2,e_1,e_2)$. It follows that on T_pM, the $(0,4)$-tensor $\widetilde{\mathbf{R}}(p) = k(p)\mathbf{T}(p)$. By choosing an orthonormal basis in each T_pM, we get a global (not yet smooth) function $k : M \to \mathbb{R}$ such that $\widetilde{\mathbf{R}}(p) = k(p)\mathbf{T}(p)$ in each T_pM, for all $p \in M$. We have

$$\begin{aligned}
\mathbf{Ric}_{ij}(p) := \mathbf{g}^{k\ell}(p)\widetilde{\mathbf{R}}_{kij\ell}(p) &= \mathbf{g}^{k\ell}(p)\widetilde{\mathbf{R}}(p)(\partial_{x^k,p}, \partial_{x^i,p}, \partial_{x^j,p}, \partial_{x^\ell,p}) \\
&= \mathbf{g}^{k\ell}(p)k(p)\mathbf{T}(p)(\partial_{x^k,p}, \partial_{x^i,p}, \partial_{x^j,p}, \partial_{x^\ell,p}) \\
&= \mathbf{g}^{k\ell}(p)k(p)(\mathbf{g}_{k\ell}(p)\mathbf{g}_{ij}(p) - \mathbf{g}_{kj}(p)\mathbf{g}_{i\ell}(p)) \\
&= k(p)(\delta^k_k\mathbf{g}_{ij}(p) - \delta^\ell_j\mathbf{g}_{i\ell}(p)) = k(p)(2\mathbf{g}_{ij}(p) - \mathbf{g}_{ij}(p)) \\
&= k(p)\mathbf{g}_{ij}(p).
\end{aligned}$$

Finally, $\mathbf{S}(p) = \mathbf{g}^{ij}(p)\mathbf{Ric}_{ij}(p) = k(p)\mathbf{g}^{ij}(p)\mathbf{g}_{ij}(p) = k(p)\delta^i_i = 2k(p)$.
So $k = \frac{1}{2}\mathbf{S} \in C^\infty(M)$ (and now the smoothness of k is also established).

Using the result of Exercise 9.10, we immediately read-off that $k \equiv 1$ for the sphere $(S^2, \mathbf{g})$. For the Poincaré half-plane, we compute

$$\begin{aligned}\widetilde{\mathbf{R}}_{xyxy} &= \mathbf{g}_{yr}\mathbf{R}^r_{xyx} = \mathbf{g}_{yx}\mathbf{R}^x_{xyx} + \mathbf{g}_{yy}\mathbf{R}^y_{xyx} = 0 \cdot \mathbf{R}^x_{xyx} + \tfrac{1}{y^2}\mathbf{R}^y_{xyx} \\ &= \tfrac{1}{y^2}(\partial_x \Gamma^y_{xy} - \partial_y \Gamma^y_{xx} + \Gamma^r_{xy}\Gamma^y_{rx} - \Gamma^r_{xx}\Gamma^y_{ry}) \\ &= \tfrac{1}{y^2}(\partial_x 0 - \partial_y \tfrac{1}{y} + \Gamma^x_{xy}\Gamma^y_{xx} + \Gamma^y_{xy}\Gamma^y_{yx} - \Gamma^x_{xx}\Gamma^y_{xy} - \Gamma^y_{xx}\Gamma^y_{yy}) \\ &= \tfrac{1}{y^2}\big(\tfrac{1}{y^2} + (-\tfrac{1}{y})(\tfrac{1}{y}) + 0 \cdot \Gamma^y_{yx} - \Gamma^x_{xx} \cdot 0 - (\tfrac{1}{y})(-\tfrac{1}{y})\big) = \tfrac{1}{y^4}.\end{aligned}$$

So $\frac{1}{y^4} = \widetilde{\mathbf{R}}_{xyxy} = k\mathbf{T}_{xyxy} = k(0(0) - \frac{1}{y^2}\frac{1}{y^2})$, and $k \equiv -1$.

Solution to Exercise 9.12

We have

$$\begin{aligned}\mathbf{g}^{i\ell}\boldsymbol{C}_{ijk\ell} &= \mathbf{g}^{i\ell}\widetilde{\mathbf{R}}_{ijk\ell} - \tfrac{1}{m-2}\mathbf{g}^{i\ell}(\mathbf{g}_{i\ell}\mathbf{Ric}_{jk} + \mathbf{g}_{jk}\mathbf{Ric}_{i\ell} - \mathbf{g}_{ik}\mathbf{Ric}_{j\ell} - \mathbf{g}_{j\ell}\mathbf{Ric}_{ik}) \\ &\quad - \tfrac{1}{(m-1)(m-2)}\mathbf{g}^{i\ell}(\mathbf{g}_{ik}\mathbf{g}_{j\ell} - \mathbf{g}_{i\ell}\mathbf{g}_{jk})\mathbf{S} \\ &= \mathbf{Ric}_{jk} - \tfrac{1}{m-2}(m\mathbf{Ric}_{jk} + \mathbf{g}^{i\ell}\mathbf{g}_{jk}\mathbf{Ric}_{i\ell} - \mathbf{Ric}_{jk} - \mathbf{Ric}_{jk}) \\ &\quad - \tfrac{1}{(m-1)(m-2)}(\mathbf{g}_{jk} - m\mathbf{g}_{jk})\mathbf{S} \\ &= -\tfrac{1}{m-2}\mathbf{g}^{i\ell}\mathbf{g}_{jk}\mathbf{Ric}_{i\ell} + \tfrac{1}{m-2}\mathbf{g}_{jk}\mathbf{S} = -\tfrac{1}{m-2}\mathbf{g}_{jk}\mathbf{S} + \tfrac{1}{m-2}\mathbf{g}_{jk}\mathbf{S} = 0.\end{aligned}$$

Next, let $m = 3$. Let $p \in M$, and let $B = \{e_1, e_2, e_3\}$ be an orthogonal basis for T_pM with respect to $\mathbf{g}(p)$. Let $(U, \mathbf{x})$ be a normal chart starting with B. Then $\partial_{x^i,p} = e_i$ for all $i \in \{1,2,3\}$. We skip writing the dependence on p below to avoid clutter. Using the skew symmetry in the first two and last two arguments, we get $\boldsymbol{C}_{ii**} = 0 = \boldsymbol{C}_{**ii}$. By the orthogonality of e_1, e_2, e_3, we have $\mathbf{g}^{i\ell} \neq 0$ if and only if $i = \ell$. Using this and $\mathbf{g}^{i\ell}\boldsymbol{C}_{ijk\ell} = 0$,

$$\mathbf{g}^{11}\boldsymbol{C}_{1jk1} + \mathbf{g}^{22}\boldsymbol{C}_{2jk2} + \mathbf{g}^{33}\boldsymbol{C}_{3jk3} = 0. \qquad (\star)$$

Putting in $(\star)$ $jk = 11, 22, 33$ successively, and using (S1),(S2),(S3),

$$\begin{aligned} 0 \quad &+ \mathbf{g}^{22}\boldsymbol{C}_{2112} + \mathbf{g}^{33}\boldsymbol{C}_{3113} = 0 \\ \mathbf{g}^{11}\boldsymbol{C}_{2112} &+ 0 \qquad\quad + \mathbf{g}^{33}\boldsymbol{C}_{3223} = 0 \\ \mathbf{g}^{11}\boldsymbol{C}_{3113} &+ \mathbf{g}^{22}\boldsymbol{C}_{3223} + 0 \qquad\quad = 0.\end{aligned}$$

It follows that $\boldsymbol{C}_{1221} = \boldsymbol{C}_{2112} = \boldsymbol{C}_{1331} = \boldsymbol{C}_{3113} = \boldsymbol{C}_{2332} = \boldsymbol{C}_{3223} = 0$. Thus $\boldsymbol{C}_{ijji} = 0$. Putting in $(\star)$ $jk = 23$, we get $\boldsymbol{C}_{1231} = 0$. Cycling through the indices, we also obtain $\boldsymbol{C}_{ijki} = 0$ whenever $j \neq k$. Since we also know that $\boldsymbol{C}_{ijji} = 0$, we conclude that $\boldsymbol{C}_{i**i} = 0$. If $i, j, k, \ell \in \{1,2,3\}$, at least two have to be the same. The equal indices can be in the positions 12, 13, 14, 23, 24, 34, but in each case the component of $\boldsymbol{C}$ vanishes, because

$$\begin{aligned}&\boldsymbol{C}_{ii**} = 0, &&\boldsymbol{C}_{i*i\star} = -\boldsymbol{C}_{i**i} = 0, &&\boldsymbol{C}_{i\star\star i} = 0, \\ &\boldsymbol{C}_{*ii\star} = \boldsymbol{C}_{i*\star i} = 0, &&\boldsymbol{C}_{\star i*i} = -\boldsymbol{C}_{i\star*i} = 0, &&\boldsymbol{C}_{**ii} = 0.\end{aligned}$$

So all components of $\boldsymbol{C}$ at p vanish if $m = 3$. As $p \in M$ was arbitrary, $\boldsymbol{C} = \mathbf{0}$.

Solution to Exercise 9.13

In an admissible chart, we have $\widehat{\mathbf{R}}^{ijk\ell} = \mathbf{g}^{ia}\mathbf{g}^{jb}\mathbf{g}^{kc}\mathbf{g}^{\ell d}\widetilde{\mathbf{R}}_{abcd}$. Using the chart (U, φ) (Exercise 9.8), and noting that the component matrix for $\mathbf{g}$ is a diagonal matrix, the nonzero components of $\widehat{\mathbf{R}}$ are given by

$$\begin{aligned}
\widehat{\mathbf{R}}^{\phi\theta\phi\theta} &= \mathbf{g}^{\phi\phi}\mathbf{g}^{\theta\theta}\mathbf{g}^{\phi\phi}\mathbf{g}^{\theta\theta}\widetilde{\mathbf{R}}_{\phi\theta\phi\theta} = \frac{1}{(\sin\theta)^2}\frac{1}{1}\frac{1}{(\sin\theta)^2}\frac{1}{1}(-(\sin\theta)^2) = -\frac{1}{(\sin\theta)^2}\\
\widehat{\mathbf{R}}^{\phi\theta\theta\phi} &= \mathbf{g}^{\phi\phi}\mathbf{g}^{\theta\theta}\mathbf{g}^{\theta\theta}\mathbf{g}^{\phi\phi}\widetilde{\mathbf{R}}_{\phi\theta\theta\phi} = \frac{1}{(\sin\theta)^2}\frac{1}{1}\frac{1}{1}\frac{1}{(\sin\theta)^2}(\sin\theta)^2 = \frac{1}{(\sin\theta)^2}\\
\widehat{\mathbf{R}}^{\theta\phi\phi\theta} &= \mathbf{g}^{\theta\theta}\mathbf{g}^{\phi\phi}\mathbf{g}^{\phi\phi}\mathbf{g}^{\theta\theta}\widetilde{\mathbf{R}}_{\theta\phi\phi\theta} = \frac{1}{1}\frac{1}{(\sin\theta)^2}\frac{1}{(\sin\theta)^2}\frac{1}{1}(\sin\theta)^2 = \frac{1}{(\sin\theta)^2}\\
\widehat{\mathbf{R}}^{\theta\phi\theta\phi} &= \mathbf{g}^{\theta\theta}\mathbf{g}^{\phi\phi}\mathbf{g}^{\theta\theta}\mathbf{g}^{\phi\phi}\widetilde{\mathbf{R}}_{\theta\phi\theta\phi} = \frac{1}{1}\frac{1}{(\sin\theta)^2}\frac{1}{1}\frac{1}{(\sin\theta)^2}(-(\sin\theta)^2) = -\frac{1}{(\sin\theta)^2}.
\end{aligned}$$

Consequently,

$$\begin{aligned}
\mathbf{K} &= \widehat{\mathbf{R}}^{ijk\ell}\widetilde{\mathbf{R}}_{ijk\ell} = \widehat{\mathbf{R}}^{\phi\theta\phi\theta}\widetilde{\mathbf{R}}_{\phi\theta\phi\theta} + \widehat{\mathbf{R}}^{\phi\theta\theta\phi}\widetilde{\mathbf{R}}_{\phi\theta\theta\phi} + \widehat{\mathbf{R}}^{\theta\phi\phi\theta}\widetilde{\mathbf{R}}_{\theta\phi\phi\theta} + \widehat{\mathbf{R}}^{\theta\phi\theta\phi}\widetilde{\mathbf{R}}_{\theta\phi\theta\phi}\\
&= -\frac{1}{(\sin\theta)^2}(-(\sin\theta)^2) + \frac{1}{(\sin\theta)^2}(\sin\theta)^2 + \frac{1}{(\sin\theta)^2}(\sin\theta)^2 - \frac{1}{(\sin\theta)^2}(-(\sin\theta)^2)\\
&= 4.
\end{aligned}$$

Thus $\mathbf{K} \equiv 4$ on U. As every point in $S^2 \backslash U$ is the limit of a sequence of points in U, it follows by the continuity of $\mathbf{K}$ that $\mathbf{K} \equiv 4$ on S^2.

Solution to Exercise 9.14

We have for $V_1, \cdots, V_{s-1} \in T_0^1 M$ that

$$\begin{aligned}
&(\operatorname{div}(T + cS))(V_1, \cdots, V_{s-1})\\
&= \mathbf{C}((\Omega, V) \mapsto (\nabla_V (T + cS))(\Omega^\sharp, V_1, \cdots, V_{s-1}))\\
&= \mathbf{C}((\Omega, V) \mapsto ((\nabla_V T)(\Omega^\sharp, V_1, \cdots, V_{s-1}) + c(\nabla_V S)(\Omega^\sharp, V_1, \cdots, V_{s-1})))\\
&= \mathbf{C}((\Omega, V) \mapsto (\nabla_V T)(\Omega^\sharp, V_1, \cdots, V_{s-1}))\\
&\quad + c\,\mathbf{C}((\Omega, V) \mapsto (\nabla_V S)(\Omega^\sharp, V_1, \cdots, V_{s-1}))\\
&= (\operatorname{div} T)(V_1, \cdots, V_{s-1}) + c(\operatorname{div} S)(V_1, \cdots, V_{s-1})\\
&= ((\operatorname{div} T) + c\operatorname{div} S)(V_1, \cdots, V_{s-1}).
\end{aligned}$$

Consequently, $\operatorname{div}(T + cS) = (\operatorname{div} T) + c\operatorname{div} S$.

Solution to Exercise 9.15

In any admissible chart $(U, \mathbf{x})$, we have (using the metric compatibility of the Levi-Civita connection to obtain the equality in the second row):

$$\begin{aligned}
\operatorname{div}(V^\flat) &= (\nabla_{\partial_{x^i}}(V^\flat))((dx^i)^\sharp) = \partial_{x^i}(V^\flat(dx^i)^\sharp) - V^\flat(\nabla_{\partial_{x^i}}(dx^i)^\sharp)\\
&= \partial_{x^i}(\mathbf{g}(V, (dx^i)^\sharp)) - \mathbf{g}(V, \nabla_{\partial_{x^i}}(dx^i)^\sharp)\\
&= \cancel{\partial_{x^i}(\mathbf{g}(V, (dx^i)^\sharp))} + \mathbf{g}(\nabla_{\partial_{x^i}} V, (dx^i)^\sharp) - \cancel{\partial_{x^i}(\mathbf{g}(V, (dx^i)^\sharp))}\\
&= ((dx^i)^\sharp)^\flat(\nabla_{\partial_{x^i}} V) = dx^i(\nabla_{\partial_{x^i}} V)\\
&= \mathbf{C}((\Omega, X) \mapsto \Omega(\nabla_X V)) = \operatorname{div} V.
\end{aligned}$$

Solution to Exercise 9.16

We have $\frac{1}{5}=(\frac{T}{t})^{2/3}-1$. So the cosmological time t of emission is

$t=(\frac{5}{6})^{3/2}T\approx 0.76\cdot T=0.76\cdot(13.8 \text{ billion years})\approx 10.5$ billion years.

Hence the light was emitted $13.8-10.5=3.3$ billion years ago.

Solution to Exercise 9.17

Below, $(U,\mathbf{x})$ is any admissible chart containing the point $\Gamma(t,s)$.

(1) By the formula given in Theorem 7.1 (p.138), $\nabla_{J(\cdot,t)}V(\cdot,t)\in T_0^1\widetilde{\Gamma}_t$ is

$$(\nabla_{J(\cdot,t)}V(\cdot,t))(s)$$
$$=\Big(\frac{\partial^2(\mathbf{x}\circ\Gamma)^k}{\partial s\partial t}(s,t)+\Gamma_{ij}^k(\Gamma(s,t))\frac{\partial(\mathbf{x}\circ\Gamma)^i}{\partial t}(s,t)\frac{\partial^2(\mathbf{x}\circ\Gamma)^j}{\partial s^2}(s,t)\Big)\partial_{x^k,\Gamma(t,s)}.$$

So $(\nabla_{J(\cdot,t)}V(\cdot,t))(s)=W^k(t)\partial_{x^k,\Gamma_s(t)}$, where

$$W^k=\frac{\partial^2(\mathbf{x}\circ\Gamma)^k}{\partial s\partial t}(s,\cdot)+\Gamma_{ij}^k(\Gamma(s,\cdot))\frac{\partial(\mathbf{x}\circ\Gamma)^i}{\partial t}(s,\cdot)\frac{\partial^2(\mathbf{x}\circ\Gamma)^j}{\partial s^2}(s,\cdot)$$

is smooth on $\Gamma(s,\cdot)^{-1}U$. Hence $\big(I\ni t\mapsto(\nabla_{J(\cdot,t)}V(\cdot,t))(s)\big)\in T_0^1\Gamma_s$.

(2) Analogous to the proof of (1), we now have

$$(\nabla_{V(s,\cdot)}J(s,\cdot))(t)$$
$$=\Big(\frac{\partial^2(\mathbf{x}\circ\Gamma)^k}{\partial t\partial s}(s,t)+\Gamma_{ij}^k(\Gamma(s,t))\frac{\partial(\mathbf{x}\circ\Gamma)^i}{\partial s}(s,t)\frac{\partial^2(\mathbf{x}\circ\Gamma)^j}{\partial t^2}(s,t)\Big)\partial_{x^k,\Gamma(t,s)}.$$

so that $(\nabla_{V(\cdot,t)}J(\cdot,t))(s)=X^k(s)\partial_{x^k,\widetilde{\Gamma}_t(s)}$, where

$$X^k=\frac{\partial^2(\mathbf{x}\circ\Gamma)^k}{\partial s\partial t}(\cdot,t)+\Gamma_{ij}^k(\Gamma(\cdot,t))\frac{\partial(\mathbf{x}\circ\Gamma)^i}{\partial s}(\cdot,t)\frac{\partial^2(\mathbf{x}\circ\Gamma)^j}{\partial t^2}(\cdot,t)$$

is smooth on $\Gamma(\cdot,t)^{-1}U$. Thus $\big((-\epsilon,\epsilon)\ni s\mapsto(\nabla_{V(s,\cdot)}J(s,\cdot))(t)\big)\in T_0^1\widetilde{\Gamma}_t$.

Solution to Exercise 9.18

As Γ_s is a geodesic, $\nabla_{V(s,\cdot)}V(s,\cdot)=0$. Using Prop. 7.2 (p.142), we have

$$\frac{d}{dt}\big(\mathbf{g}(\Gamma_s(t))(J(s,t),V(s,t))\big)$$
$$=\mathbf{g}(\Gamma_s(t))\big((\nabla_{V(s,\cdot)}J(s,\cdot))(t),V(s,t)\big)+\mathbf{g}(\Gamma_s(t))\big(J(s,t),(\nabla_{V(s,\cdot)}V(s,\cdot))(t)\big)$$
$$=\mathbf{g}(\Gamma_s(t))\big((\nabla_{V(s,\cdot)}J(s,\cdot))(t),V(s,t)\big)+0.$$

We now use the Jacobi equation to get the equality $(\star)$ below. We have

$$\frac{d^2}{dt^2}\big(\mathbf{g}(\Gamma_s(t))(J(s,t),V(s,t))\big)=\frac{d}{dt}\big(\mathbf{g}(\Gamma_s(t))((\nabla_{V(s,\cdot)}J(s,\cdot))(t),V(s,t))\big)$$
$$=\mathbf{g}(\Gamma_s(t))\big((\nabla_{V(s,\cdot)}(\nabla_{V(s,\cdot)}J(s,\cdot)))(t),V(s,t)\big)+0$$
$$\overset{(\star)}{=}-(\mathbf{g}(R(\widetilde{J},\widetilde{V})\widetilde{V},\widetilde{V}))(\Gamma(s,t))=-(\widetilde{\mathbf{R}}(\widetilde{J},\widetilde{V},\widetilde{V},\widetilde{V}))(\Gamma(s,t))=0.$$

The last equality follows by the skew-symmetry of $\widetilde{\mathbf{R}}$ in its last two slots.

For vanishing initial conditions $\mathbf{g}(\Gamma_s(t_0))(J(s,t_0),V(s,t_0)) = 0$, and
$\frac{d(\mathbf{g}(\Gamma_s(\cdot))(J(s,\cdot),V(s,\cdot)))}{dt}(t_0) = \mathbf{g}(\Gamma_s(t_0))((\nabla_{V(s,\cdot)}J(s,\cdot))(t_0),V(s,t_0)) = 0$,
the ordinary differential equation

$$\frac{d^2}{dt^2}(\mathbf{g}(\Gamma_s(t))(J(s,t),V(s,t))) = 0 \quad (t \in I)$$

has the unique solution given by $\mathbf{g}(\Gamma_s(t))(J(s,t),V(s,t)) = 0$ for all $t \in I$.

Solution to Exercise 9.19

Let $x, y \in v^\perp$. Let $V, X, Y \in T_0^1 M$ be such that $V_p = v$, $X_p = x$ and $Y_p = y$.

$$\begin{aligned}\mathbf{g}(p)(F_p(x),y)\mathbf{g}(p)(v,v) &= \mathbf{g}(p)(R(p)(x,v)v,y) = (\mathbf{g}(R(X,V)V,Y))(p)\\ &= (Y^\flat(R(X,V)V))(p) = (\mathbf{R}(Y^\flat,X,V,V))(p)\\ &= (\widetilde{\mathbf{R}}(X,V,V,Y))(p) = (\widetilde{\mathbf{R}}(V,Y,X,V))(p)\\ &= (-\widetilde{\mathbf{R}}(Y,V,X,V))(p) = (-(-\widetilde{\mathbf{R}}(Y,V,V,X)))(p)\\ &= (\widetilde{\mathbf{R}}(Y,V,V,X))(p) = (\mathbf{R}(X^\flat,Y,V,V))(p)\\ &= (X^\flat(R(Y,V)V))(p) = (\mathbf{g}(X,R(Y,V)V))(p)\\ &= \mathbf{g}(p)(x,R(p)(y,v)v) = \mathbf{g}(p)(x,F_p(y))\mathbf{g}(p)(v,v).\end{aligned}$$

Thus $\mathbf{g}(p)(F_p(x),y) = \mathbf{g}(p)(x,F_p(y))$ for all $x, y \in v^\perp$, as wanted.

Solution to Exercise 9.20

If $V := \operatorname{span}\{v\} \subset T_pM$, then $\mathbf{g}(p)|_V$ is negative definite on V, i.e., $V \in \mathcal{N}_{\mathbf{g}(p)}$. By Thm. 5.2 (p.76), $\mathbf{g}(p)|_{V^\perp}$ is positive definite. Let $e_1, \cdots, e_{m-1}$ be an orthonormal basis for $V^\perp$. Then $(e_1)^\flat, \cdots, (e_{m-1})^\flat$ is the dual basis for $(V^\perp)^*$ since we have $(e_i)^\flat e_j = \mathbf{g}(p)(e_i,e_j) = \delta_{ij}$. Hence

$$\operatorname{trace} F_v = \frac{e_i^\flat(R(p)(e_i,v)v)}{\mathbf{g}(p)(v,v)} \overset{(*)}{=} \frac{(\mathbf{C}_1^1\mathbf{R})(p)(v,v)}{\mathbf{g}(p)(v,v)} = \frac{\mathbf{Ric}(p)(v,v)}{\mathbf{g}(p)(v,v)}.$$

Justification of $(*)$: Let $e_0 = kv$, where $k := 1/\sqrt{-\mathbf{g}(p)(v,v)}$. Then with respect to the scalar product $\mathbf{g}(p)$, $B := \{e_0, e_1, \cdots, e_{m-1}\}$ is an orthonormal basis for T_pM. By Corollary 8.3 (p.174), in the corresponding normal chart $(U, \mathbf{x})$, we have $\mathbf{g}_{ij}(p) = \eta_{ij}$, where the indices $i, j \in \{0, 1, \cdots, m-1\}$. Then

$$e_i^\flat(\partial_{x^j,p}) = \mathbf{g}(p)(e_i, \partial_{x^j,p}) = \mathbf{g}(p)(\partial_{x^i,p}, \partial_{x^j,p}) = \eta_{ij}.$$

In particular, $(e_i)^\flat = (dx^i)_p$ for $1 \leqslant i, j \leqslant m-1$. Recall that $e_i = \partial_{x^i,p}$ for all $i \in \{0, 1, \cdots, m-1\}$. We will also use the fact that

$$\begin{aligned}(e_0)^\flat(R(p)(e_0,v)v) &= \mathbf{R}(p)(e_0^\flat, e_0, v, v) = \mathbf{R}(p)(e_0^\flat, kv, v, v)\\ &= k\widetilde{\mathbf{R}}(p)(v,v,v,e_0) = k(0) = 0,\end{aligned}$$

by the skew-symmetry of $\widetilde{\mathbf{R}}$ in its first two slots.

Thus (suspending the summation convention)

$$\sum_{i=1}^{m-1}(e_i)^\flat(R(p)(e_i,v)v) = \sum_{i=0}^{m-1}(e_i)^\flat(R(p)(e_i,v)v) = \sum_{i=0}^{m-1}(dx^i)_p(R(p)(\partial_{x^i,p},v)v)$$
$$= \sum_{i=0}^{m-1}\mathbf{R}(p)((dx^i)_p,\partial_{x^i,p},v,v) = (\mathbf{C}_1^1\mathbf{R})(p)(v,v). \quad \square_{(*)}$$

(From Example 9.2, for the instantaneous observer $v = \partial_{t,p} \in T_pM$, where M is the Schwarzschild spacetime, we have

$$\operatorname{trace} F_v = \frac{2m}{r^3} - \frac{m}{r^3} - \frac{m}{r^3} = 0.$$

So for the timelike vectors $v = \partial_{t,p} \in T_pM$, we have that $\mathbf{Ric}(p)(v,v) = 0$. This holds for all p in the chart $\mathbb{R} \times (2m,\infty) \times U$. Consequently, $\mathbf{Ric}_{tt} = 0$. In fact, $\mathbf{Ric} = 0$ for the Schwarzschild spacetime, that is, it is 'Ricci-flat'.)

Solution to Exercise 9.21

We 'normalise' $\partial_{r,p}$ to get a unit length vector,

$$e_{r,p} := \frac{\partial_{r,p}}{\sqrt{\mathbf{g}(p)(\partial_{r,p},\partial_{r,p})}} = \frac{\partial_{r,p}}{\sqrt{\mathbf{g}_{rr}(p)}}.$$

If the astronaut's height is h, then the astronaut is represented by the vector $he_{r,p}$. The acceleration experienced by the astronaut at $r = 2m+\delta$, $\delta > 0$, is

$$F_v(he_{r,p}) = \frac{h}{\sqrt{\mathbf{g}_{rr}(p)}}F_v\partial_{r,p} = \Big(\frac{h}{\sqrt{\mathbf{g}_{rr}(p)}}\frac{2m}{r^3}\Big)\Big|_{r=2m+\delta}\partial_{r,p} = h\frac{2m}{(2m+\delta)^3}e_{r,p}.$$

The critical acceleration the astronaut can withstand is $h(100\,\mathrm{s}^{-2})$. Hence for survival,

$$h(100\,\mathrm{s}^{-2}) > h\frac{2m}{(2m+\delta)^3} \quad \text{for all } \delta > 0.$$

Thus $100\,\mathrm{s}^{-2} \geqslant \frac{2m}{(2m)^3} = \frac{1}{4m^2}$, i.e., $\frac{m}{\mathrm{M}_\odot} \geqslant \frac{1}{2}\frac{1}{10\,\mathrm{s}^{-1}}\frac{1}{5\times 10^{-6}\,\mathrm{s}} = \frac{1}{10^{-4}} = 10000.$

So the mass of the black hole should be at least ten thousand solar masses.

Solution to Exercise 9.22

We have $df = (\partial_x f)dx + (\partial_y f)dy + (\partial_z f)dz$, and so if $\Omega = df$, then

$$\Omega_x = \partial_x f$$
$$\Omega_y = \partial_y f$$
$$\Omega_z = \partial_z f.$$

By an application of the Schwarz theorem,

$$\partial_y\Omega_x = \partial_y\partial_x f = \partial_x\partial_y f = \partial_x\Omega_y,$$
$$\partial_z\Omega_y = \partial_z\partial_y f = \partial_y\partial_z f = \partial_y\Omega_z,$$
$$\partial_x\Omega_z = \partial_x\partial_z f = \partial_z\partial_x f = \partial_z\Omega_x.$$

This shows the 'only if' part.

Now suppose that

$$\begin{aligned}\partial_y\Omega_x &= \partial_x\Omega_y,\\ \partial_z\Omega_y &= \partial_y\Omega_z,\\ \partial_x\Omega_z &= \partial_z\Omega_x.\end{aligned}$$

These are exactly the consistency conditions for the system

$$(*)\begin{cases}\partial_x f = \Omega_x\\ \partial_y f = \Omega_y\\ \partial_z f = \Omega_z,\end{cases}$$

and so by the Frobenius theorem, there exists a smooth solution f satisfying $(*)$. But then

$$df = (\partial_x f)dx + (\partial_y f)dy + (\partial_z f)dz = \Omega_x dx + \Omega_y dy + \Omega_z dz = \Omega,$$

proving the ‘if’ part.

Chapter 10

Solution to Exercise 10.1

Let $w_i := v_{\pi(i)}$, $1 \leqslant i \leqslant k$. For all $v_1, \cdots, v_k \in V$, we have

$$\begin{aligned}(\pi(\sigma\tau))(v_1,\cdots,v_k) &= (\sigma\tau)(v_{\pi(1)},\cdots,v_{\pi(k)})\\ &= (\sigma\tau)(w_1,\cdots,w_k)\\ &= \tau(w_{\sigma(1)},\cdots,w_{\sigma(k)})\\ &= \tau(v_{\pi(\sigma(1))},\cdots,v_{\pi(\sigma(k))})\\ &= \tau(v_{(\pi\circ\sigma)(1)},\cdots,v_{(\pi\circ\sigma)(k)})\\ &= ((\pi\circ\sigma)\tau)(v_1,\cdots,v_k).\end{aligned}$$

Hence $\pi(\sigma\tau) = (\pi \circ \sigma)\tau$.

Solution to Exercise 10.2

We have

$$\begin{aligned}&(\operatorname{Alt}\tau)(x,y,z)\\ &= \frac{1}{3!}\big(\tau(x,y,z)-\tau(x,z,y)-\tau(y,x,z)+\tau(y,z,x)-\tau(z,y,x)+\tau(z,x,y)\big).\end{aligned}$$

Solution to Exercise 10.3

For all $v_1, \cdots, v_k$, we have

$$\begin{aligned}(\operatorname{Alt}\omega)(v_1,\cdots,v_k) &= \frac{1}{k!}\sum_{\pi\in S_k}(\operatorname{sign}\sigma)\,\omega(v_{\sigma(1)},\cdots,v_{\sigma(k)})\\ &= \frac{1}{k!}\sum_{\pi\in S_k}(\operatorname{sign}\sigma)\,(\operatorname{sign}\sigma)\cdot\omega(v_1,\cdots,v_k)\\ &= \frac{1}{k!}\sum_{\pi\in S_k}1\,\omega(v_1,\cdots,v_k)\end{aligned}$$

(since $\operatorname{sign}\sigma \in \{-1,1\}$, giving $(\operatorname{sign}\sigma)^2 = 1$). As S_k has $k!$ elements, we have for all $v_1, \cdots, v_k$,

$$(\operatorname{Alt}\omega)(v_1,\cdots,v_k) = \frac{1}{k!}k!\,\omega(v_1,\cdots,v_k) = \omega(v_1,\cdots,v_k).$$

Thus $\operatorname{Alt}\omega = \omega$.

Solution to Exercise 10.4

We have

$$\begin{aligned}(\epsilon^1\wedge\cdots\wedge\epsilon^n)(v_1,\cdots,v_n) &= n!\operatorname{Alt}(\epsilon^1\otimes\cdots\otimes\epsilon^n)(v_1,\cdots,v_n)\\ &= \sum_{\pi\in S_n}(\operatorname{sign}\pi)\,(\epsilon^1\otimes\cdots\otimes\epsilon^n)(v_{\pi(1)},\cdots,v_{\pi(n)})\\ &- \sum_{\pi\in S_n}(\operatorname{sign}\pi)\,\epsilon^1(v_{\pi(1)})\cdots\epsilon^n(v_{\pi(n)})\\ &= \det[\epsilon^i(v_j)].\end{aligned}$$

Solution to Exercise 10.5

For all $v_1, \cdots, v_k \in V$, we have

$$\begin{aligned}
&(\epsilon^{\pi(1)} \wedge \cdots \wedge \epsilon^{\pi(k)})(v_1, \cdots, v_k) \\
&= \det[\epsilon^{\pi(i)}(v_j)] = \sum_{\sigma \in S_k} (\operatorname{sign}\sigma)(\epsilon^{\pi(1)} v_{\sigma(1)}) \cdots (\epsilon^{\pi(k)} v_{\sigma(k)}) \\
&= \sum_{\tau \in S_k} (\operatorname{sign}(\tau \circ \pi))(\epsilon^{\pi(1)} v_{(\tau\circ\pi)(1)}) \cdots (\epsilon^{\pi(k)} v_{(\tau\circ\pi)(k)}) \\
&= (\operatorname{sign}\pi) \sum_{\tau \in S_k} (\operatorname{sign}\tau)(\epsilon^{\pi(1)} v_{\tau(\pi(1))}) \cdots (\epsilon^{\pi(k)} v_{\tau(\pi(k))}) \\
&= (\operatorname{sign}\pi) \sum_{\tau \in S_k} (\operatorname{sign}\tau)(\epsilon^{1} v_{\tau(1)}) \cdots (\epsilon^{k} v_{\tau(k)}) \\
&= (\operatorname{sign}\pi) \det[\epsilon^i v_j] = (\operatorname{sign}\pi)(\epsilon^1 \wedge \cdots \wedge \epsilon^k)(v_1, \cdots, v_k).
\end{aligned}$$

To get the equality in the third line above, we used the 'change of variable' $\sigma = \tau \circ \pi$: as σ runs over all elements of S_k, so does $\tau := \sigma \circ \pi^{-1}$. Also, for the equality in the second to last line, we note that the product $(\epsilon^{\pi(1)} v_{\tau(\pi(1))}) \cdots (\epsilon^{\pi(k)} v_{\tau(\pi(k))})$ contains each of the factors $\epsilon^i v_{\tau(i)}$, $i \in \{1, \cdots, k\}$ exactly once (because $\pi \in S_k$).

Alternatively, we can use the fact that every permutation in S_k can be written as a product of *adjacent transpositions*, i.e., permutations that swap only adjacent elements j, $j{+}1$, $1 \leqslant j < k$; see [Halmos (1987), Exercise 5, §27]. Using this, the claim follows from Proposition 10.2 (p.214).

Solution to Exercise 10.6

To show that $\mathcal{L}_V \Omega \in \mathbf{\Omega}^k M$, we require $\pi(\mathcal{L}_V \Omega) = (\operatorname{sign}\pi)\,(\mathcal{L}_V \Omega)$ for all $\pi \in S_k$. As we can write each permutation π as a product of transpositions, it is enough to show that for any transposition τ interchanging some distinct i, j, we have $\tau(\mathcal{L}_V \Omega) = -\mathcal{L}_V \Omega$. For all $V_1 \cdots, V_k \in T_0^1 M$,

$$V(\Omega(V_{\tau(1)}, \cdots, V_{\tau(k)})) = V(-\Omega(V_1, \cdots, V_k)) = -V(\Omega(V_1, \cdots, V_k)).$$

We have

$$\begin{aligned}
&\sum_{\ell \in \{1,\dots,k\}\setminus\{i,j\}} \Omega(V_{\tau(1)}, \cdots, V_{\tau(\ell-1)}, [V, V_{\tau(\ell)}], V_{\tau(\ell+1)}, \cdots, V_{\tau(k)}) \\
&= \sum_{\ell \in \{1,\dots,k\}\setminus\{i,j\}} (-1)\,\Omega(V_1, \cdots, V_{\ell-1}, [V, V_\ell], V_{\ell+1}, \cdots, V_k).
\end{aligned}$$

Also, we have

$$\begin{aligned}
&\sum_{\ell \in \{i,j\}} \Omega(V_{\tau(1)}, \cdots, V_{\tau(\ell-1)}, [V, V_{\tau(\ell)}], V_{\tau(\ell+1)}, \cdots, V_{\tau(k)}) \\
&= \Omega(V_1, \cdots, V_{i-1}, [V, V_j], V_{i+1}, \cdots, V_{j-1}, V_i, V_{j+1}, \cdots, V_k) \\
&\quad + \Omega(V_1, \cdots, V_{i-1}, V_j, V_{i+1}, \cdots, V_{j-1}, [V, V_i], V_{j+1}, \cdots, V_k) \\
&= -\Omega(V_1, \cdots, V_{i-1}, V_i, V_{i+1}, \cdots, V_{j-1}, [V, V_j], V_{j+1}, \cdots, V_k) \\
&\quad - \Omega(V_1, \cdots, V_{i-1}, [V, V_i], V_{i+1}, \cdots, V_{j-1}, V_j, V_{j+1}, \cdots, V_k) \\
&= \sum_{\ell \in \{i,j\}} (-1)\Omega(V_1, \cdots, V_{\ell-1}, [V, V_\ell], V_{\ell+1}, \cdots, V_k).
\end{aligned}$$

Thus

$$(\tau(\mathcal{L}_V\Omega))(V_1,\cdots,V_k) = (\mathcal{L}_V\Omega)(V_{\tau(1)},\cdots,V_{\tau(k)})$$

$$= V(\Omega(V_{\tau(1)},\cdots,V_{\tau(k)})) - \sum_{\ell=1}^{k} \Omega(V_{\tau(1)},\cdots,V_{\tau(\ell-1)},\mathcal{L}_V V_{\tau(\ell)},V_{\tau(\ell+1)},\cdots,V_{\tau(k)})$$

$$= (-1)\Big(V(\Omega(V_1,\cdots,V_k)) - \sum_{\ell=1}^{k} \Omega(V_1,\cdots,V_{\ell-1},\mathcal{L}_V V_\ell,V_{\ell+1},\cdots,V_k)\Big)$$

$$= -(\mathcal{L}_V\Omega)(V_1,\cdots,V_k).$$

Solution to Exercise 10.7

We have $\mathrm{Alt}\,(fT) = f\,(\mathrm{Alt}\,T)$, because for all $V_1,\cdots,V_k \in T_0^1M$,

$$\begin{aligned}(\mathrm{Alt}\,(fT))(V_1,\cdots,V_k) &= \frac{1}{k!}\sum_{\pi\in S_k}(\mathrm{sign}\,\pi)\,(fT)(V_{\pi(1)},\cdots,V_{\pi(k)})\\ &= \frac{1}{k!}\sum_{\pi\in S_k}(\mathrm{sign}\,\pi)f\,T(V_{\pi(1)},\cdots,V_{\pi(k)})\\ &= f\,((\mathrm{Alt}\,T)(V_1,\cdots,V_k))\\ &= (f\,\mathrm{Alt}\,T)(V_1,\cdots,V_k).\end{aligned}$$

Solution to Exercise 10.8

We give an inductive argument on the number of factors in the wedge product. For $n = 1$ there is nothing to prove since $\mathrm{Alt}\,\Omega^1 = \Omega^1$, and for $n = 2$, this is just the definition of the wedge product. Suppose that we have an $n > 2$, and that the result has been shown when the number of factors is $\leqslant n-1$. Then

$$\begin{aligned}&\Omega^1\wedge\Omega^2\wedge\cdots\wedge\Omega^n\\ &= \frac{(k_1+(k_2+\cdots+k_n))!}{k_1!\,(k_2+\cdots+k_n)!}\mathrm{Alt}\,(\Omega^1\otimes(\Omega^2\wedge\cdots\wedge\Omega^n))\\ &= \frac{(k_1+k_2+\cdots+k_n)!}{k_1!\,(k_2+\cdots+k_n)!}\frac{(k_2+\cdots+k_n))!}{k_2!\cdots k_n!}\mathrm{Alt}\,(\Omega^1\otimes\mathrm{Alt}\,(\Omega^2\otimes\cdots\otimes\Omega^n))\\ &= \frac{(k_1+k_2+\cdots+k_n)!}{k_1!k_2!\cdots k_n!}\mathrm{Alt}\,(\Omega^1\otimes(\Omega^2\otimes\cdots\otimes\Omega^n))\\ &= \frac{(k_1+\cdots+k_n)!}{k_1!\cdots k_n!}\mathrm{Alt}\,(\Omega^1\otimes\cdots\otimes\Omega^n).\end{aligned}$$

Solution to Exercise 10.9

As k is odd, k^2 is odd too, and so $(-1)^{k^2} = -1$. Proposition 10.2 (p.214) implies $\Omega\wedge\Omega = (-1)^{k^2}\Omega\wedge\Omega = -\Omega\wedge\Omega$, and so $2(\Omega\wedge\Omega) = 0$, i.e.,

$$\Omega\wedge\Omega = 0.$$

Solution to Exercise 10.10

We have

$$\begin{aligned}\Omega\wedge\Theta &= (x\,dx+y\,dy+z\,dz)\wedge(y\,dx+z\,dy+x\,dz)\\ &= (xy\,dx\wedge dx+xz\,dx\wedge dy+x^2\,dx\wedge dz\\ &\quad +y^2\,dy\wedge dx+yz\,dy\wedge dy+yx\,dy\wedge dz\\ &\quad +zy\,dz\wedge dx+z^2\,dz\wedge dy+zx\,dz\wedge dz)\\ &= (0+xz\,dx\wedge dy - x^2\,dz\wedge dx - y^2\,dx\wedge dy+0\\ &\quad +yx\,dy\wedge dz+zy\,dz\wedge dx - z^2\,dy\wedge dz+0)\\ &= (zx-y^2)\,dx\wedge dy+(xy-z^2)\,dy\wedge dz+(yz-x^2)\,dz\wedge dx.\end{aligned}$$

Solution to Exercise 10.11

We have

$$\begin{aligned}\Omega_{j_1\cdots j_k} &= \Omega(\partial_{x^{j_1}},\cdots,\partial_{x^{j_k}}) = f\,dx^{i_1}\wedge\cdots\wedge dx^{i_k}(\partial_{x^{j_1}},\cdots,\partial_{x^{j_k}})\\ &= f\,k!\,\mathrm{Alt}\,(dx^{i_1}\otimes\cdots\otimes dx^{i_k})(\partial_{x^{j_1}},\cdots,\partial_{x^{j_k}})\\ &= f\sum_{\pi\in S_k}(\operatorname{sign}\pi)\,(\pi(dx^{i_1}\otimes\cdots\otimes dx^{i_k}))(\partial_{x^{j_1}},\cdots,\partial_{x^{j_k}})\\ &= f\sum_{\pi\in S_k}(\operatorname{sign}\pi)\,(dx^{i_1}\otimes\cdots\otimes dx^{i_k})(\partial_{x^{j_{\pi(1)}}},\cdots,\partial_{x^{j_{\pi(k)}}})\\ &= f\sum_{\pi\in S_k}(\operatorname{sign}\pi)\,\delta^{i_1}_{j_{\pi(1)}}\cdots\delta^{i_k}_{j_{\pi(k)}}.\end{aligned}$$

Let the term $\delta^{i_1}_{j_{\pi(1)}}\cdots\delta^{i_k}_{j_{\pi(k)}}\neq 0$. Then $(i_1,\cdots,i_k)=(j_{\pi(1)},\cdots,j_{\pi(k)})$. Also, as $i_1<\cdots<i_k$, we have $j_{\pi(1)}<\cdots<j_{\pi(k)}$. Since $j_1<\cdots<j_k$, we conclude that π is the identity map (trivial permutation). Thus $I=J$. So

$$\Omega_{j_1\cdots j_k} = \begin{cases} f & \text{if } J=I,\\ 0 & \text{if } J\neq I.\end{cases}$$

Solution to Exercise 10.12

For all $p\in M$ and all $V_1,\cdots,V_{k+\ell}$, we have

$$\begin{aligned}&((f^*(\Omega\wedge\Theta))(V_1,\cdots,V_{k+\ell}))(p) = (\Omega\wedge\Theta)(f(p))(df_p(V_1)_p,\cdots,df_p(V_{k+\ell})_p)\\ &= \frac{1}{k!\,\ell!}\sum_{\pi\in S_{k+\ell}}(\operatorname{sign}\pi)\,(\Omega\otimes\Theta)(f(p))(df_p(V_{\pi(1)})_p,\cdots,df_p(V_{\pi(k+\ell)})_p)\\ &= \frac{1}{k!\,\ell!}\sum_{\pi\in S_{k+\ell}}(\operatorname{sign}\pi)\,\Omega(f(p))(df_p(V_{\pi(1)})_p,\cdots,df_p(V_{\pi(k)})_p)\cdot\\ &\qquad\qquad \Theta(f(p))(df_p(V_{\pi(k+1)})_p,\cdots,df_p(V_{\pi(k+\ell)})_p)\\ &= \frac{1}{k!\,\ell!}\sum_{\pi\in S_{k+\ell}}(\operatorname{sign}\pi)\,((f^*\Omega)(V_{\pi(1)},\cdots,V_{\pi(k)}))(p)\cdot\\ &\qquad\qquad ((f^*\Theta)(V_{\pi(k+1)},\cdots,V_{\pi(k+\ell)}))(p)\\ &= \frac{1}{k!\,\ell!}\sum_{\pi\in S_{k+\ell}}(\operatorname{sign}\pi)\,\big(((f^*\Omega)\otimes(f^*\Theta))(V_{\pi(1)},\cdots,V_{\pi(k+\ell)})\big)(p)\\ &= \big(((f^*\Omega)\wedge(f^*\Theta))(V_{\pi(1)},\cdots,V_{\pi(k+\ell)})\big)(p).\end{aligned}$$

Hence $f^*(\Omega\wedge\Theta) = (f^*\Omega)\wedge(f^*\Theta)$.

Solution to Exercise 10.13

We have, using Exercises 10.12 and 3.21 (p.51), that

$$\begin{aligned} f^*(dx \wedge dy \wedge dz) &= (f^*dx) \wedge (f^*dy) \wedge (f^*dz) = d(x \circ f) \wedge d(y \circ f) \wedge d(z \circ f) \\ &= d(r(\sin\theta)(\cos\phi)) \wedge d(r(\sin\theta)(\sin\phi)) \wedge d(r\cos\theta). \end{aligned}$$

Since the gradient of a function in a chart can be computed by letting the coordinate vector fields act on the function, we obtain

$$\begin{aligned} d(r(\sin\theta)(\cos\phi)) &= (\sin\theta)(\cos\phi)dr + r(\cos\theta)(\cos\phi)d\theta - r(\sin\theta)(\sin\phi)d\phi, \\ d(r(\sin\theta)(\sin\phi)) &= (\sin\theta)(\sin\phi)dr + r(\cos\theta)(\sin\phi)d\theta + r(\sin\theta)(\cos\phi)d\phi, \\ d(r\cos\theta) &= (\cos\theta)dr - r(\sin\theta)d\theta. \end{aligned}$$

So, using $dr \wedge dr = 0$, $dr \wedge d\theta = -d\theta \wedge dr$, etc., we get, after some algebraic manipulations, that

$$\begin{aligned} &d(r(\sin\theta)(\sin\phi)) \wedge d(r\cos\theta) \\ &= -r(\sin\phi)dr \wedge d\theta - r(\sin\theta)(\cos\theta)(\cos\phi)dr \wedge d\phi + r^2(\sin\theta)^2(\cos\phi)d\theta \wedge d\phi, \end{aligned}$$

and so

$$\begin{aligned} &d(r(\sin\theta)(\cos\phi)) \wedge d(r(\sin\theta)(\sin\phi)) \wedge d(r\cos\theta) \\ &= (r^2(\sin\theta)^3(\cos\phi)^2 + r^2(\sin\theta)(\cos\theta)^2(\cos\phi)^2 + r^2(\sin\theta)(\sin\phi)^2)dr \wedge d\theta \wedge d\phi \\ &= r^2(\sin\theta)\, dr \wedge d\theta \wedge d\phi. \end{aligned}$$

Thus $f^*(dx \wedge dy \wedge dz) = r^2(\sin\theta)\, dr \wedge d\theta \wedge d\phi$.

Solution to Exercise 10.14

Let $\pi \in S_k$ be such that $i_{\pi(1)} < \cdots < i_{\pi(k)}$. Then

$$\Omega = f\, dx^{i_1} \wedge \cdots \wedge dx^{i_k} = f(\operatorname{sign}\pi)\, dx^{i_{\pi(1)}} \wedge \cdots \wedge dx^{i_{\pi(k)}}.$$

By Exercise 10.11, if $1 \leqslant j_1 < \cdots < j_k \leqslant m$, then

$$\Omega_{j_1 \cdots j_k} = \begin{cases} f(\operatorname{sign}\pi) & \text{if } (j_1, \cdots, j_k) = (i_{\pi(1)}, \cdots, i_{\pi(k)}), \\ 0 & \text{if } (j_1, \cdots, j_k) \neq (i_{\pi(1)}, \cdots, i_{\pi(k)}). \end{cases}$$

So by the definition of the exterior derivative,

$$\begin{aligned} d\Omega &= \sum_{1 \leqslant j_1 < \cdots < j_k \leqslant m} (\partial_{x^i}\Omega_{j_1 \cdots j_k})\, dx^i \wedge dx^{j_1} \wedge \cdots \wedge dx^{j_k} \\ &= \partial_{x^i}(f(\operatorname{sign}\pi))\, dx^i \wedge dx^{i_{\pi(1)}} \wedge \cdots \wedge dx^{i_{\pi(k)}} \\ &= (\partial_{x^i} f)\, dx^i \wedge ((\operatorname{sign}\pi)\, dx^{i_{\pi(1)}} \wedge \cdots \wedge dx^{i_{\pi(k)}}) \\ &= (\partial_{x^i} f)\, dx^i \wedge dx^{i_1} \wedge \cdots \wedge dx^{i_k}. \end{aligned}$$

Solution to Exercise 10.15

We have

$$\begin{aligned} d\Omega = &\ (\partial_x x)\, dx \wedge dx + (\partial_y x)\, dy \wedge dx + (\partial_z x)\, dz \wedge dx \\ &+ (\partial_x y)\, dx \wedge dy + (\partial_y y)\, dy \wedge dy + (\partial_z y)\, dz \wedge dy \\ &+ (\partial_x z)\, dx \wedge dz + (\partial_y z)\, dy \wedge dz + (\partial_z z)\, dz \wedge dz, \end{aligned}$$

which is 0, since each of the summands on the left-hand side vanishes.

Similarly,

$$\begin{aligned} d\Theta =\ & (\partial_x y)\, dx\wedge dx + (\partial_y y)\, dy\wedge dx + (\partial_z y)\, dz\wedge dx \\ & + (\partial_x z)\, dx\wedge dy + (\partial_y z)\, dy\wedge dy + (\partial_z z)\, dz\wedge dy \\ & + (\partial_x x)\, dx\wedge dz + (\partial_y x)\, dy\wedge dz + (\partial_z x)\, dz\wedge dz \\ =\ & 1\, dy\wedge dx + 1\, dz\wedge dy + 1\, dx\wedge dz \\ =\ & -(dx\wedge dy + dy\wedge dz + dz\wedge dx). \end{aligned}$$

Solution to Exercise 10.16

In the global admissible chart $(\mathbb{R}^2, (x,y)\mapsto(x,y))$, write $\Omega=\Omega^x dx+\Omega^y dy$. So $d\Omega=\partial_y\Omega^x dy\wedge dx+\partial_x\Omega^y dx\wedge dy=(\partial_x\Omega^y-\partial_y\Omega^x)\, dx\wedge dy$. As Ω is closed,

$$\frac{\partial\Omega^y}{\partial x}=\frac{\partial\Omega^x}{\partial y}.$$

Set

$$f(x,y)=\int_0^x \Omega^x(\xi,y)\, d\xi+\int_0^y \Omega^y(0,\eta)\, d\eta.$$

Then $f\in C^\infty(\mathbb{R}^2)$, and $\partial_x f=\Omega^x+0=\Omega^x$. Also,

$$\begin{aligned} \partial_y f &= \int_0^x \frac{\partial\Omega^x}{\partial y}(\xi,y)\, d\xi + \Omega^y(0,y)=\int_0^x \frac{\partial\Omega^y}{\partial x}(\xi,y)\, d\xi + \Omega^y(0,y) \\ &=\Omega^y(x,y)-\Omega^y(0,y)+\Omega^y(0,y)=\Omega^y(x,y). \end{aligned}$$

Thus $\Omega=\Omega^x dx+\Omega^y dy=\partial_x f\, dx+\partial_y f\, dy=df$. So Ω is exact.

Solution to Exercise 10.17

The additivity is clear. Let $k\geqslant 2$. For all $\Omega\in\mathbf{\Omega}^k M$, $f\in C^\infty(M)$, and $W_1,\cdots,W_{k-1}\in T_0^1 M$, we have

$$\begin{aligned} (i_V(f\Omega))(W_1,\cdots,W_{k-1}) &= (f\Omega)(V,W_1,\cdots,W_{k-1}) \\ &= f\cdot\Omega(V,W_1,\cdots,W_{k-1}) \\ &= f\cdot(i_V\Omega)(W_1,\cdots,W_{k-1}) \\ &= (f(i_V\Omega))(W_1,\cdots,W_{k-1}). \end{aligned}$$

Thus $i_V(f\Omega)=f(i_V\Omega)$. For $k=1$, $i_V(f\Omega)=(f\Omega)(V)=f(\Omega V)=f(i_V\Omega)$.

Solution to Exercise 10.18

By Cartan's formula applied to $d\Omega$ in place of Ω,

$$\begin{aligned} \mathcal{L}_V(d\Omega) &= d(i_V(d\Omega))+i_V(d(d\Omega))=d(i_V(d\Omega))+i_V 0 \\ &= d(i_V(d\Omega))+0=d(i_V(d\Omega)). \end{aligned}$$

Applying d to both sides of Cartan's formula $\mathcal{L}_V\Omega=d(i_V\Omega)+i_V(d\Omega)$ yields

$$\begin{aligned} d(\mathcal{L}_V\Omega) &= d(d(i_V\Omega)+i_V(d\Omega))=d(d(i_V\Omega))+d(i_V(d\Omega)) \\ &= 0+d(i_V(d\Omega))=d(i_V(d\Omega)). \end{aligned}$$

Thus $\mathcal{L}_V(d\Omega)=d(i_V(d\Omega))=d(\mathcal{L}_V\Omega)$.

Chapter 11

Solution to Exercise 11.1

We have

$$\omega_B(e_1, \cdots, e_m) = (\epsilon^1 \wedge \cdots \wedge \epsilon^m)(e_1, \cdots, e_m) = \det[\epsilon^i(e_j)] = \det[\delta^i_j] = 1 > 0.$$

So $\omega_B \neq 0$, and $(e_1, \cdots, e_m)$ is positively oriented with respect to $[\omega_B]$. We note that

$$\begin{aligned}
\omega_B(e'_1, \cdots, e'_m) &= (\epsilon^1 \wedge \cdots \wedge \epsilon^m)(a_1^{i_1} e_{i_1}, \cdots, a_m^{i_m} e_{i_m}) \\
&= a_1^{i_1} \cdots a_m^{i_m} (\epsilon^1 \wedge \cdots \wedge \epsilon^m)(e_{i_1}, \cdots, e_{i_m}) \\
&= \sum_{\pi \in S_m} a_1^{\pi(1)} \cdots a_m^{\pi(m)} (\epsilon^1 \wedge \cdots \wedge \epsilon^m)(e_{\pi(1)}, \cdots, e_{\pi(m)}) \\
&= \sum_{\pi \in S_m} a_1^{\pi(1)} \cdots a_m^{\pi(m)} (\operatorname{sign} \pi)(\epsilon^1 \wedge \cdots \wedge \epsilon^m)(e_1, \cdots, e_m) \\
&= (\det[a^i_j]) \cdot 1 = \det[a^i_j].
\end{aligned}$$

We note that the third equality can be justified by noting that a summand will be nonzero only when $(i_1, \cdots, i_m)$ is a permutation of $(1, \cdots, m)$, thanks to the skew-symmetry of $\omega_B = \epsilon^1 \wedge \cdots \wedge \epsilon^m$.

(1)⇒(2): If B' is positively oriented with respect to ω_B, then we have that $\omega_B(e'_1, \cdots, e'_m) > 0$. From the above, $\det[a^i_j] = \omega_B(e'_1, \cdots, e'_m) > 0$.

(2)⇒(3): If $\det[a^i_j] > 0$, then $\omega_B(e'_1, \cdots, e'_m) > 0$. But $\omega_{B'}(e'_1, \cdots, e'_m) > 0$. The k relating these nonzero top forms, $\omega_{B'} = k\omega_B$, is easily seen to be > 0 by operating on the m-tuple $(e'_1, \cdots, e'_m)$. So $\omega_B \sim \omega'_B$.

(3)⇒(1): If $\omega_B \sim \omega'_B$, then we have $\omega_B = k\omega_{B'}$ for a $k > 0$. Hence we obtain that $\omega_B(e'_1, \cdots, e'_m) = k\omega_{B'}(e'_1, \cdots, e'_m) = k1 > 0$. Consequently, $(e'_1, \cdots, e'_m)$ is positively oriented with respect to $[\omega_B]$.

Solution to Exercise 11.2

The transition maps $\varphi_{\mathbf{s}} \circ \varphi_{\mathbf{n}}^{-1}, \varphi_{\mathbf{n}} \circ \varphi_{\mathbf{s}}^{-1} : \mathbb{R}^2 \setminus \{(0,0)\} \to \mathbb{R}^2 \setminus \{(0,0)\}$ are both given by $\mathbb{R}^2 \setminus \{(0,0)\} \ni (u,v) \mapsto \dfrac{(u,v)}{u^2+v^2}$, and so the Jacobian matrix is

$$J(u,v) = \begin{bmatrix} \dfrac{v^2-u^2}{(u^2+v^2)^2} & -\dfrac{2uv}{(u^2+v^2)^2} \\ -\dfrac{2uv}{(u^2+v^2)^2} & \dfrac{u^2-v^2}{(u^2+v^2)^2} \end{bmatrix}.$$

Thus $\det J(u,v) = -\dfrac{1}{(u^2+v^2)^2} < 0$. So $\mathcal{A}$ is not oriented.

In order to construct an oriented atlas $\widetilde{\mathcal{A}}$, we keep in $\widetilde{\mathcal{A}}$ the old chart $(U_{\mathbf{s}}, \varphi_{\mathbf{s}})$, and we replace the old chart $(U_{\mathbf{n}}, \varphi_{\mathbf{n}})$ by a new chart $(U_{\mathbf{n}}, \widetilde{\varphi}_{\mathbf{n}})$, where the new chart map $\widetilde{\varphi}_{\mathbf{n}} : U_{\mathbf{n}} \to \mathbb{R}^2$ is defined by $\widetilde{\varphi}_{\mathbf{n}} = R \circ \varphi_{\mathbf{n}}$, R being the reflection in the u-axis, i.e., $R(u,v) = (u,-v)$. The chart transition map $\widetilde{\varphi}_{\mathbf{n}} \circ \varphi_{\mathbf{s}}^{-1} = R \circ \varphi_{\mathbf{n}} \circ \varphi_{\mathbf{s}}^{-1}$ has the Jacobian matrix $R \cdot J(u,v)$, and its determinant is

$$(\det R)(\det J) = (-1) \cdot \frac{(-1)}{(u^2+v^2)^2} = \frac{1}{(u^2+v^2)^2} > 0.$$

Similarly, the chart transition map $\varphi_{\mathbf{s}} \circ (\widetilde{\varphi}_{\mathbf{n}})^{-1} = \varphi_{\mathbf{s}} \circ (R \circ \varphi_{\mathbf{n}})^{-1} = \varphi_{\mathbf{s}} \circ \varphi_{\mathbf{n}}^{-1} \circ R$ has the Jacobian matrix $J(u,v) \cdot R$, and so its determinant is again positive on $\mathbb{R}^2 \backslash \{(0,0)\}$. Thus $\widetilde{\mathcal{A}}$ is oriented.

To show that $\widetilde{\mathcal{A}} \cup \{(U,\varphi)\}$ is oriented, we note that $U \cap U_{\mathbf{n}} = U \cap U_{\mathbf{s}} = U$, and so it is enough to show that the determinant of the Jacobian of $\widetilde{\varphi}_{\mathbf{n}} \circ \varphi^{-1}$ is positive. (It will then follow that also the inverse $\varphi \circ \widetilde{\varphi}_{\mathbf{n}}^{-1}$, and the other transition maps $\varphi \circ \varphi_{\mathbf{s}}^{-1} = \varphi \circ \widetilde{\varphi}_{\mathbf{n}} \circ (\varphi_{\mathbf{s}} \circ \widetilde{\varphi}_{\mathbf{n}})^{-1}, \varphi_{\mathbf{s}} \circ \varphi^{-1}$, all have Jacobians having positive determinants.) Recalling $\varphi_{\mathbf{n}} \circ \varphi^{-1}$ from Example 1.8 (p.7), we can compute its Jacobian to be

$$\begin{bmatrix} \dfrac{-\cos\phi}{1-\cos\theta} & -\dfrac{(\sin\theta)(\sin\phi)}{1-\cos\theta} \\ \dfrac{-\sin\phi}{1-\cos\theta} & \dfrac{(\sin\theta)(\cos\phi)}{1-\cos\theta} \end{bmatrix},$$

having the determinant $\dfrac{-\sin\theta}{(1-\cos\theta)^2} < 0$. The determinant of the Jacobian of $\widetilde{\varphi}_{\mathbf{n}} \circ \varphi^{-1}$ is $(\det R) \cdot \dfrac{(-\sin\theta)}{(1-\cos\theta)^2} = \dfrac{\sin\theta}{(1-\cos\theta)^2} > 0$. So $\widetilde{\mathcal{A}} \cup \{(U,\varphi)\}$ is oriented.

Solution to Exercise 11.3

Clearly, $\bigcup_{n=0}^{\infty} \overline{V_{n+1}} \backslash V_n \subset M$. Let $p \in M$. As $M = \bigcup_{n=0}^{\infty} V_n$, there exists a first/smallest $m \in \mathbb{N}$ such that $p \in V_m$.

1° $m=1$. Then $p \in V_1 \subset \overline{V_1} = \overline{V_1} \backslash \varnothing = \overline{V_1} \backslash V_0 \subset \bigcup_{n=0}^{\infty} \overline{V_{n+1}} \backslash V_n$.

2° $m>1$. Then $p \notin V_{m-1}$ by the definition of m.

Thus $p \in V_m \backslash V_{m-1} \subset \overline{V_m} \backslash V_{m-1} \subset \bigcup_{n=0}^{\infty} \overline{V_{n+1}} \backslash V_n$.

Solution to Exercise 11.4

We know that $\mathcal{A} \cup \mathcal{A}'$ belongs to the smooth structure of M (as they are compatible). Suppose that $(U,\mathbf{x}) \in \mathcal{A}$ and $(V,\mathbf{y}) \in \mathcal{A}'$. We want to show that the determinant of the Jacobian matrix of the chart transition map is everywhere positive on $U \cap V$. In $U \cap V$, we have

$$\Omega^V dy^1 \wedge \cdots \wedge dy^m = \Omega = \Omega^U dx^1 \wedge \cdots \wedge dx^m = \Omega^U \det[\partial_{y^j} x^i]\, dy^1 \wedge \cdots \wedge dy^m.$$

So it follows that $\det[\partial_{y^j} x^i] = \Omega^V / \Omega^U = \frac{+}{+} > 0$. Then the inverse of this chart transition map also has an everywhere positive determinant for its Jacobian matrix.

Solution to Exercise 11.5

In $U_{\alpha(i)}$, we first decompose $\Omega = \Omega_k \, dx^1 \wedge \cdots \wedge \widehat{dx^k} \wedge \cdots \wedge dx^m$. So we have $\varphi_i \Omega = \varphi_i \Omega_k \, dx^1 \wedge \cdots \wedge \widehat{dx^k} \wedge \cdots \wedge dx^m$. Set $f_k := (-1)^{k-1} \varphi_i \Omega_k$, $1 \leqslant k \leqslant m$. We obtain the decomposition $\varphi_i \Omega = (-1)^{k-1} f_k \, dx^1 \wedge \cdots \wedge \widehat{dx^k} \wedge \cdots \wedge dx^m$. Thus

$$\begin{aligned} d(\varphi_i \Omega) &= d((-1)^{k-1} f_k \, dx^1 \wedge \cdots \wedge \widehat{dx^k} \wedge \cdots \wedge dx^m) \\ &= (-1)^{k-1} (\partial_{x^k} f_k) \, dx^k \wedge dx^1 \wedge \cdots \wedge \widehat{dx^k} \wedge \cdots \wedge dx^m \\ &= (\partial_{x^k} f_k) \, dx^1 \wedge \cdots \wedge dx^m. \end{aligned}$$

Since $g_k := f_k \circ \mathbf{x}_{\alpha(i)}^{-1}$ has a compact support contained in $\mathbf{x}_{\alpha(i)} U_{\alpha(i)}$, by extending it as being identically zero outside $\operatorname{supp} g_k$, we have

$$\begin{aligned} &\int_{\mathbf{x}_{\alpha(i)} U_{\alpha(i)}} (\partial_{x^k} f_k)(\mathbf{x}_{\alpha(i)}^{-1} \mathbf{u}) \, du^1 \cdots du^m = \int_{\mathbf{x}_{\alpha(i)} U_{\alpha(i)}} \frac{\partial (f_k \circ \mathbf{x}_{\alpha(i)}^{-1})}{\partial u^k} (\mathbf{u}) \, du^1 \cdots du^m \\ &= \int_{\mathbf{x}_{\alpha(i)} U_{\alpha(i)}} \frac{\partial g_k}{\partial u^k} (\mathbf{u}) \, du^1 \cdots du^m = \int_{\mathbb{R}^m} \frac{\partial g_k}{\partial u^k} (\mathbf{u}) \, du^1 \cdots du^m \\ &= \int_{\mathbb{R}^{m-1}} \int_{\mathbb{R}} \frac{\partial g_k}{\partial u^k} (\mathbf{u}) \, du^k du^1 \cdots \widehat{du^k} \cdots du^m \qquad \text{(Fubini)} \\ &= \int_{\mathbb{R}^{m-1}} g_k(\mathbf{u}) \Big|_{u^k = -\infty}^{u^k = \infty} du^1 \cdots \widehat{du^k} \cdots du^m = \int_{\mathbb{R}^{m-1}} 0 \, du^1 \cdots \widehat{du^k} \cdots du^m = 0. \end{aligned}$$

As Ω is compactly supported, and since $\{\operatorname{supp} \varphi_i : i \in I\}$ is locally finite, $\sum_i \varphi_i \Omega$ is a finite sum. Hence $\sum_i d(\varphi_i \Omega)$ is also a finite sum. We also note that $d(\varphi_i \Omega)$, just like $\varphi_i \Omega$, has compact support contained in $U_{\alpha(i)}$. Thus

$$\begin{aligned} \int_M d\Omega &= \int_M d(\textstyle\sum_i \varphi_i \Omega) = \displaystyle\int_M \sum_i d(\varphi_i \Omega) = \sum_i \int_M d(\varphi_i \Omega) \\ &\overset{(*)}{=} \sum_i \int_{U_{\alpha(i)}} d(\varphi_i \Omega) \\ &= \sum_i \int_{\mathbf{x}_{\alpha(i)} U_{\alpha(i)}} (\partial_{x^1} f_1 + \cdots + \partial_{x^m} f_m)(\mathbf{x}_{\alpha(i)}^{-1} \mathbf{u}) \, du^1 \cdots du^m \\ &= \sum_i \sum_{k=1}^m \int_{\mathbf{x}_{\alpha(i)} U_{\alpha(i)}} (\partial_{x^k} f_k)(\mathbf{x}_{\alpha(i)}^{-1} \mathbf{u}) \, du^1 \cdots du^m \\ &= \sum_i \sum_{k=1}^m 0 = 0. \end{aligned}$$

(In the fourth line of the above display, there is no implicit summation over k in the integrand: the explicit summation over k is outside the integral.) To see $(*)$, we note that the support of $\varphi_j d(\varphi_i \Omega) \subset (\operatorname{supp} \varphi_i) \cap (\operatorname{supp} \varphi_j)$ and that $\sum_j \varphi_j d(\varphi_i \Omega)$ is a finite sum, so that

$$\begin{aligned} \int_M d(\varphi_i \Omega) &= \sum_j \int_{U_{\alpha(j)}} \varphi_j d(\varphi_i \Omega) = \sum_j \int_{U_{\alpha(j)} \cap U_{\alpha(i)}} \varphi_j d(\varphi_i \Omega) \\ &= \sum_j \int_{U_{\alpha(i)}} \varphi_j d(\varphi_i \Omega) = \int_{U_{\alpha(i)}} \sum_j \varphi_j d(\varphi_i \Omega) = \int_{U_{\alpha(i)}} d(\varphi_i \Omega). \end{aligned}$$

Solution to Exercise 11.6

S^2 is compact. Let $\{\theta_{\mathbf{n}}, \theta_{\mathbf{s}}\}$ be partition of unity subordinate to $\{U_{\mathbf{n}}, U_{\mathbf{s}}\}$. Then

$$\begin{aligned}\mathrm{Vol}_{\mathbf{g}} S^2 &= \int_{U_{\mathbf{n}}} \theta_{\mathbf{n}} \mathrm{vol}_{\mathbf{g}} + \int_{U_{\mathbf{s}}} \theta_{\mathbf{s}} \mathrm{vol}_{\mathbf{g}} \\ &= \int_{\widetilde{\varphi}_{\mathbf{n}} U_{\mathbf{n}}} \theta_{\mathbf{n}}(\widetilde{\varphi}_{\mathbf{n}}^{-1}(u,v)) \sqrt{|\det[\mathbf{g}_{ij}^{U_{\mathbf{n}}} \widetilde{\varphi}_{\mathbf{n}}^{-1}(u,v)]|}\, du dv \\ &\quad + \int_{\varphi_{\mathbf{s}} U_{\mathbf{s}}} \theta_{\mathbf{s}}(\varphi_{\mathbf{s}}^{-1}(u,v)) \sqrt{|\det[\mathbf{g}_{ij}^{U_{\mathbf{s}}} \varphi_{\mathbf{s}}^{-1}(u,v)]|}\, du dv \\ &= \int_{\widetilde{\varphi}_{\mathbf{n}} U} \theta_{\mathbf{n}}(\widetilde{\varphi}_{\mathbf{n}}^{-1}(u,v)) \sqrt{|\det[\mathbf{g}_{ij}^{U_{\mathbf{n}}} \widetilde{\varphi}_{\mathbf{n}}^{-1}(u,v)]|}\, du dv \\ &\quad + \int_{\varphi_{\mathbf{s}} U} \theta_{\mathbf{s}}(\varphi_{\mathbf{s}}^{-1}(u,v)) \sqrt{|\det[\mathbf{g}_{ij}^{U_{\mathbf{s}}} \varphi_{\mathbf{s}}^{-1}(u,v)]|}\, du dv,\end{aligned}$$

since the contributions of the thin sets $\widetilde{\varphi}_{\mathbf{n}}(U_{\mathbf{n}} \backslash U)$ and $\varphi_{\mathbf{s}}(U_{\mathbf{s}} \backslash U)$ are 0. Here (U, φ) is the spherical coordinate chart from Example 1.8 (p.7).

We now use a change of variables given by $\widetilde{\Phi} = \widetilde{\varphi}_{\mathbf{n}} \circ \varphi^{-1}$ in the first integral, and $\Phi = \varphi_{\mathbf{s}} \circ \varphi^{-1}$ in the second integral. Using the transformation relations among $\mathbf{g}_{ij}^{U}$ and $\mathbf{g}_{ij}^{U_{\mathbf{n}}}$, respectively $\mathbf{g}_{ij}^{U_{\mathbf{s}}}$,

$$\begin{aligned}\mathrm{Vol}_{\mathbf{g}} S^2 &= \int_{\varphi U} \theta_{\mathbf{n}}(\varphi^{-1}(\theta,\phi)) \sqrt{|\det[\mathbf{g}_{ij}^{U} \varphi^{-1}(\theta,\phi)]|}\, d\theta d\phi \\ &\quad + \int_{\varphi U} \theta_{\mathbf{s}}(\varphi^{-1}(\theta,\phi)) \sqrt{|\det[\mathbf{g}_{ij}^{U} \varphi^{-1}(\theta,\phi)]|}\, d\theta d\phi \\ &= \int_{\varphi U} 1 \sqrt{|\det[\mathbf{g}_{ij}^{U} \varphi^{-1}(\theta,\phi)]|}\, d\theta d\phi = \int_0^{2\pi} \int_0^{\pi} \sqrt{(\sin\theta)^2}\, d\theta d\phi \\ &= \int_0^{2\pi} \int_0^{\pi} (\sin\theta)\, d\theta d\phi - \int_0^{2\pi} (-\cos\theta)\Big|_0^{\pi} d\phi = \int_0^{2\pi} -((-1)-1)\, d\phi = 4\pi,\end{aligned}$$

as expected.

Solution to Exercise 11.7

The chart $(\mathbf{U}, \mathbf{x}) = (\mathbb{R} \times (2m, \infty) \times U, p = (t, r, \mathbf{p}) \mapsto (r, \theta(\mathbf{p}), \varphi(\mathbf{p}), t))$ belongs to the atlas $\mathcal{A}_\star$. It is enough to show that Ω coincides with $\mathrm{vol}_{\mathbf{g}}$ on $(\mathbf{U}, \mathbf{x})$ (since then by continuity they coincide also on M). In $(\mathbf{U}, \mathbf{x})$, we have

$$[\mathbf{g}(\partial_{x^i}, \partial_{x^j})] = \begin{bmatrix} (1-\frac{2m}{r})^{-1} & & & \\ & r^2 & & \\ & & r^2(\sin\theta)^2 & \\ & & & -(1-\frac{2m}{r}) \end{bmatrix}.$$

So $\det[\mathbf{g}(\partial_{x^i}, \partial_{x^j})] = -r^4(\sin\theta)^2$, and $\sqrt{|\det[\mathbf{g}(\partial_{x^i}, \partial_{x^j})]|} = r^2 \sin\theta$. Thus in the chart $(\mathbf{U}, \mathbf{x})$, we have

$$\begin{aligned}\mathrm{vol}_{\mathbf{g}} &= r^2(\sin\theta) dr \wedge d\theta \wedge d\phi \wedge dt = r^2 dr \wedge ((\sin\theta) d\theta \wedge d\phi) \wedge dt \\ &= r^2 dr \wedge (\mathrm{vol}_{\mathbf{g}_S^2}|_U) \wedge dt.\end{aligned}$$

Solution to Exercise 11.8

For any vector $v \in V$, we have $v = v^i e_i$, and operating by ϵ^j, we have $v^j = \epsilon^j v$. Thus for $v \in V$, $(e_k)^\flat v = g(v, e_k) = g(\epsilon^j v e_j, e_k) = (\epsilon^j v)\, g(e_j, e_k) = \iota(k)\epsilon^k v$ (no sum over k), and so $(e_k)^\flat = \iota(k)\epsilon^k$. Hence

$$\begin{aligned}\mathrm{vol}_g &= (-1)^{\mathfrak{i}_g}(e_1)^\flat \wedge \cdots \wedge (e_m)^\flat = (-1)^{\mathfrak{i}_g}\iota(1)\cdots\iota(m)\,\epsilon^1 \wedge \cdots \wedge \epsilon^m \\ &= (-1)^{\mathfrak{i}_g}(-1)^{\mathfrak{i}_g}\epsilon^1 \wedge \cdots \wedge \epsilon^m = \epsilon^1 \wedge \cdots \wedge \epsilon^m.\end{aligned}$$

Solution to Exercise 11.9

Let $(e_1, \cdots, e_m)$ be a positively oriented ordered basis such that it is orthonormal with respect to g, and let $(\epsilon^1, \cdots, \epsilon^m)$ be the corresponding dual basis for V^*. Then $\omega^j = \omega^j_\ell \epsilon^\ell$ and $v_k = v^i_k e_i$ for some ω^i_j, v^i_j, $1 \leqslant i, j \leqslant m$. We have $\delta^j_k = \omega^j v_k = \omega^j_\ell v^i_k \epsilon^\ell e_i = \omega^j_\ell v^i_k \delta^\ell_i = \omega^j_i v^i_k$. Thus the matrices $[\omega^i_j]$ and $[v^i_j]$ are inverses of each other. Hence

$$\begin{aligned}&\sqrt{|\det[g(v_i, v_j)]|}\;\omega^1 \wedge \cdots \wedge \omega^m \\ &= \sqrt{|\det[g(v^r_i e_r, v^s_j e_s)]|}\;(\omega^1_{i_1}\epsilon^{i_1}) \wedge \cdots \wedge (\omega^m_{i_m}\epsilon^{i_m}) \\ &= \sqrt{|\det[v^r_i g(e_r, e_s) v^s_j]|}\;\omega^1_{i_1}\cdots\omega^m_{i_m}\,\epsilon^{i_1} \wedge \cdots \wedge \epsilon^{i_m} \\ &= \sqrt{|\det([v^r_i]^{\mathrm{t}}[g(e_r, e_s)][v^s_j])|}\;\sum_{\pi \in S_m}\omega^1_{\pi(1)}\cdots\omega^m_{\pi(m)}\,\epsilon^{\pi(1)} \wedge \cdots \wedge \epsilon^{\pi(m)} \\ &= \sqrt{(\det[v^r_i])^2|\det[g(e_r, e_s)]|}\;\sum_{\pi \in S_m}\omega^1_{\pi(1)}\cdots\omega^m_{\pi(m)}(\operatorname{sign}\pi)\epsilon^1 \wedge \cdots \wedge \epsilon^m \\ &= (\det[v^i_j])(\det[\omega^i_j])\,\mathrm{vol}_g = (\det I)\,\mathrm{vol}_g = (1)\,\mathrm{vol}_g = \mathrm{vol}_g.\end{aligned}$$

Note that in the above, we used $\det[v^i_j] > 0$, which follows from the fact that if $[\omega]$ is the orientation on V, then $\omega(e_1, \cdots, e_m) > 0$, and also

$$\begin{aligned}0 < \omega(v_1, \cdots, v_m) &= v^{i_1}_1 \cdots v^{i_m}_m \omega(e_{i_1}, \cdots, e_{i_m}) \\ &= \sum_{\pi \in S_m} v^{\pi(1)}_1 \cdots v^{\pi(m)}_m (\operatorname{sign}\pi)\,\omega(e_1, \cdots, e_m) \\ &= (\det[v^i_j])\,\omega(e_1, \cdots, e_m).\end{aligned}$$

Solution to Exercise 11.10

This is similar to the proof of Lemma 10.4 (p.230), and we will use Exercise 10.4 (p.214). For $v_1, \cdots, v_{k-1} \in V$, we have

$$\begin{aligned}(i_v(\omega^1 \wedge \cdots \wedge \omega^k))(v_1, \cdots, v_{k-1}) &= (\omega^1 \wedge \cdots \wedge \omega^k)(v, v_1, \cdots, v_{k-1}) \\ &= \det\begin{bmatrix}\omega^1 v & \omega^1 v_1 & \cdots & \omega^1 v_{k-1} \\ \vdots & \vdots & & \vdots \\ \omega^k v & \omega^k v_1 & \cdots & \omega^k v_{k-1}\end{bmatrix}.\end{aligned}$$

We shall expand the determinant down the first column.

Thus we have

$$
\begin{aligned}
&(i_v(\omega^1\wedge \cdots \wedge\omega^1))(v_1,\cdots ,v_{k-1})\\
&= \sum_{r=1}^{k}(-1)^{r-1}(\omega^r v)(\omega^1\wedge \cdots \wedge\widehat{\omega^r}\wedge \cdots \wedge\omega^k)(v_1,\cdots ,v_{k-1}).
\end{aligned}
$$

Consequently, $i_v(\omega^1 \wedge \cdots \wedge\omega^1)= \sum_{r=1}^{k}(-1)^{r-1}(\omega^r v)\omega^1 \wedge \cdots \wedge\widehat{\omega^r} \wedge \cdots \wedge\omega^k$.

Solution to Exercise 11.11

Let $\{e_1,\cdots ,e_m\}$ be a basis for V, and $\{\epsilon^1,\cdots ,\epsilon^m\}$ be the dual basis for V^*. Decompose $\omega=\omega_I\epsilon^I$, where $I=(i_1,\cdots ,i_k)$ and $\epsilon^I:=\epsilon^{i_1}\wedge \cdots \wedge\epsilon^{i_k}$, and similarly $\theta=\theta_J\epsilon^J$. Then

$$
\begin{aligned}
i_v(\omega \wedge \theta) &= \omega_I\theta_J\, i_v(\epsilon^{i_1} \wedge \cdots \wedge \epsilon^{i_k} \wedge \epsilon^{j_1} \wedge \cdots \wedge \epsilon^{j_\ell})\\
&= \omega_I\theta_J \sum_{r=1}^{k}(-1)^{r-1}\epsilon^{i_r}(v)\,\epsilon^{i_1} \wedge \cdots \wedge \widehat{\epsilon^{i_r}} \wedge \cdots \wedge \epsilon^{i_k} \wedge \epsilon^J\\
&\quad +\omega_I\theta_J \sum_{r=1}^{\ell}(-1)^{k+r-1}\epsilon^{j_r}(v)\,\epsilon^I \wedge \epsilon^{j_1} \wedge \cdots \wedge \widehat{\epsilon^{j_r}} \wedge \cdots \wedge \epsilon^{j_\ell}\\
&= (i_v\,\omega)\wedge\theta + (-1)^k\omega\wedge(i_v\,\theta).
\end{aligned}
$$

Solution to Exercise 11.12

This follows immediately from Lemma 11.6 by induction:

$$
\begin{aligned}
&(-1)^{k(m-1)}(\star\omega)\wedge v_1^\flat\wedge \cdots \wedge v_k^\flat\\
&= (-1)^{(k-1)(m-1)}(\star(i_{v_1}\omega))\wedge v_2^\flat\wedge \cdots \wedge v_k^\flat\\
&= (-1)^{(k-2)(m-1)}(\star(i_{v_2}(i_{v_1}\omega)))\wedge v_3^\flat\wedge \cdots \wedge v_k^\flat\\
&= \cdots\\
&= \star(i_{v_k}(\cdots(i_{v_1}\omega)\cdots)) = \star(\omega(v_1,\cdots ,v_k))\\
&= \omega(v_1,\cdots ,v_k)(\star 1) = \omega(v_1,\cdots ,v_k)((-1)^{\mathfrak{i}_g}\,\mathrm{vol}_g).
\end{aligned}
$$

Solution to Exercise 11.13

For $\Omega,\Theta\in\mathbf{\Omega}^kM$ and $f\in C^\infty(M)$, we have for all $p\in M$ that

$$
\begin{aligned}
(\star(\Omega+f\,\Theta))(p) &= \star((\Omega+f\,\Theta)(p)) = \star(\Omega(p)+f(p)(\Theta(p)))\\
&= \star(\Omega(p))+f(p)\star(\Theta(p)) = (\star\Omega)(p)+f(p)(\star\Theta)(p)\\
&= (\star\Omega)(p)+(f(\star\Theta))(p) = (\star\Omega+f(\star\Theta))(p).
\end{aligned}
$$

Thus $\star(\Omega+f\,\Theta)=\star\Omega+f(\star\Theta)$. So $\star:\mathbf{\Omega}^kM\to\mathbf{\Omega}^{m-k}M$ is $C^\infty(M)$-linear.

For $\Omega\in\mathbf{\Omega}^kM$, and for $p\in M$, we have

$$
\begin{aligned}
(\star\star\Omega)(p) &= \star((\star\Omega)(p)) = \star(\star(\Omega(p)))\\
&= (-1)^{k(m-1)+\mathfrak{i}_{\mathbf{g}(p)}}\Omega(p)\\
&= (-1)^{k(m-1)+\iota(\mathbf{g})}\Omega(p)\\
&= ((-1)^{k(m-1)+\iota(\mathbf{g})}\Omega)(p).
\end{aligned}
$$

Consequently, $\star\star\Omega = (-1)^{k(m-1)+\iota(\mathbf{g})}\Omega$.

Solution to Exercise 11.14

Let $(U, \mathbf{x})$ be an admissible chart for M containing p. Suppose that $(e_1, \cdots, e_m)$ is a positively oriented orthonormal basis for T_pM. Then $\mathrm{vol}_{\mathbf{g}}(p)(e_1, \cdots, e_m) > 0$ and also $\mathrm{vol}_{\mathbf{g}(p)}(e_1, \cdots, e_m) = 1$. It is enough to show that the action of $\mathrm{vol}_{\mathbf{g}}(p)$ and $\mathrm{vol}_{\mathbf{g}(p)}$ have the same absolute value on a basis. Let $\partial_{x^j,p} = \alpha_j^i e_i$. Then

$$\mathbf{g}_{ij}(p) = \mathbf{g}(p)(\partial_{x^i,p}, \partial_{x^j,p}) = \mathbf{g}(p)(\alpha_i^k e_k, \alpha_j^\ell e_\ell) = \alpha_i^k \alpha_j^\ell \mathbf{g}(p)(e_k, e_\ell).$$

As $\det[\mathbf{g}(p)(e_k, e_\ell)] = (-1)^{\mathfrak{i}_{\mathbf{g}(p)}}$, from above, $\det[\mathbf{g}_{ij}(p)] = (\det[\alpha_j^i])^2(-1)^{\mathfrak{i}_{\mathbf{g}(p)}}$. Hence

$$\begin{aligned}
|\det[\alpha_j^i]| &= \sqrt{|\det[\mathbf{g}_{ij}]|} = \sqrt{|\det[\mathbf{g}_{ij}(p)]|}\det[(dx^i)_p(\partial_{x^j,p})] \\
&= \sqrt{|\det[\mathbf{g}_{ij}(p)]|}((dx^1)_p \wedge \cdots \wedge (dx^m)_p)(\partial_{x^1,p}, \cdots, \partial_{x^m,p}) \\
&= \mathrm{vol}_{\mathbf{g}}(p)(\partial_{x^1,p}, \cdots, \partial_{x^m,p}).
\end{aligned}$$

On the other hand, we have

$$\begin{aligned}
&\mathrm{vol}_{\mathbf{g}(p)}(\partial_{x^1,p}, \cdots, \partial_{x^m,p}) \\
&= (-1)^{\mathfrak{i}_{\mathbf{g}(p)}}((e_1)^\flat \wedge \cdots \wedge (e_m)^\flat)(\alpha_1^{i_1} e_{i_1}, \cdots, \alpha_m^{i_m} e_{i_m}) \\
&= (-1)^{\mathfrak{i}_{\mathbf{g}(p)}} \alpha_1^{i_1} \cdots \alpha_m^{i_m} ((e_1)^\flat \wedge \cdots \wedge (e_m)^\flat)(e_{i_1}, \cdots, e_{i_m}) \\
&= (-1)^{\mathfrak{i}_{\mathbf{g}(p)}} \sum_{\pi \in S_m} \alpha_1^{\pi(1)} \cdots \alpha_m^{\pi(m)} ((e_1)^\flat \wedge \cdots \wedge (e_m)^\flat)(e_{\pi(1)}, \cdots, e_{\pi(m)}) \\
&= (-1)^{\mathfrak{i}_{\mathbf{g}(p)}} \sum_{\pi \in S_m} \alpha_1^{\pi(1)} \cdots \alpha_m^{\pi(m)} (\operatorname{sign} \pi)((e_1)^\flat \wedge \cdots \wedge (e_m)^\flat)(e_1, \cdots, e_m) \\
&= (-1)^{\mathfrak{i}_{\mathbf{g}(p)}} \sum_{\pi \in S_m} \alpha_1^{\pi(1)} \cdots \alpha_m^{\pi(m)} (\operatorname{sign} \pi) \det[(e_i)^\flat e_j] = \det[\alpha_j^i].
\end{aligned}$$

Solution to Exercise 11.15

Every $\Omega \in \mathbf{\Omega}^3 M$ can be decomposed as

$$\Omega = \omega_x \, dy \wedge dz \wedge dt + \omega_y \, dx \wedge dz \wedge dt + \omega_z \, dx \wedge dy \wedge dt + \omega_t \, dx \wedge dy \wedge dz$$

for some $\omega_x, \omega_y, \omega_z, \omega_t \in C^\infty(M)$. Then

$$\star \Omega = \omega_x(-dx) + \omega_y(dy) + \omega_z(-dz) + \omega_t(-dt),$$

and so

$$\begin{aligned}
\star \star \Omega &= -\omega_x \star (dx) + \omega_y \star (dy) - \omega_z \star (dz) - \omega_t \star (dt) \\
&= -\omega_x(-dy \wedge dz \wedge dt) + \omega_y(dx \wedge dz \wedge dt) \\
&\quad -\omega_z(-dx \wedge dy \wedge dt) - \omega_t(-dx \wedge dy \wedge dz) \\
&= \omega_x \, dy \wedge dz \wedge dt + \omega_y \, dx \wedge dz \wedge dt + \omega_z \, dx \wedge dy \wedge dt + \omega_t \, dx \wedge dy \wedge dz \\
&= \Omega.
\end{aligned}$$

With $k = 3$, $m = 4$, and $\iota(\mathbf{g}) = 1$, we have $(-1)^{k(m-1)+\iota(\mathbf{g})} = (-1)^{10} = 1$.

Chapter 12

Solution to Exercise 12.1

If $\lambda_n(s_n) = \gamma(t_n)$, then $p + \frac{n}{\nu}(\mathbf{e}_0 + (\tanh\theta)\mathbf{e}_1) + \big(s_n - \frac{n}{\nu}\big)(\mathbf{e}_0 - \mathbf{e}_1) = p + t_n\mathbf{e}_0$.
Comparing the coefficients of $\mathbf{e}_0$, we conclude that $s_n = t_n$.
Comparing the coefficients of $\mathbf{e}_1$ yields $\frac{n}{\nu}(\tanh\theta) = t_n - \frac{n}{\nu}$.
Thus $t_n = \frac{n}{\nu}(1 + \tanh\theta) = \frac{n}{\widetilde{\nu}}$, where $\widetilde{\nu} := \dfrac{\nu}{1+\tanh\theta} > 0$.
Consequently, $\frac{\widetilde{\nu}}{\nu} = (1+\tanh\theta)^{-1}$.

Solution to Exercise 12.2

Let s_+ be such that $\lambda_+(s_+) = \gamma(t_+)$, that is, $p + kw + s_+n_+ = p + t_+v$, and so $kw + s_+n_+ = t_+v$. Taking scalar product with v gives $s_+\mathbf{g}(p)(n_+, v) = -t_+$. Also, taking scalar product with w gives $k + s_+\mathbf{g}(p)(n_+, w) = 0$. Hence

$$t_+ = -s_+\mathbf{g}(p)(n_+, v) = k\frac{\mathbf{g}(p)(n_+, v)}{\mathbf{g}(p)(n_+, w)} = k\frac{-1}{1} = -k.$$

The proper time elapsed along γ between the events $\gamma(t_+)$ and $p = \gamma(0)$ is

$$\int_{t_+}^{0} \sqrt{-\mathbf{g}(\gamma(t))(v_{\gamma,\gamma(t)}, v_{\gamma,\gamma(t)})}\, dt = \int_{t_+}^{0} 1 dt = -t_+ = k.$$

Let s_- be such that $\lambda_-(s_-) = \gamma(t_-)$. Proceeding as above, we get

$$t_- = -s_-\mathbf{g}(p)(n_-, v) = k\frac{\mathbf{g}(p)(n_-, v)}{\mathbf{g}(p)(n_-, w)} = k\frac{-1}{-1} = k.$$

The proper time elapsed along γ between the events $p = \gamma(0)$ and $\gamma(t_-)$ is

$$\int_{0}^{t_-} \sqrt{-\mathbf{g}(\gamma(t))(v_{\gamma,\gamma(t)}, v_{\gamma,\gamma(t)})} dt = \int_{0}^{t_-} 1 dt = t_- = k.$$

Solution to Exercise 12.3

As $\mathbf{g}(p)(u,v) = -\dfrac{\sqrt{-\mathbf{g}(p)(v,v)}}{\mathbf{g}(p)(w,v)}\mathbf{g}(p)(w,v) - \dfrac{\mathbf{g}(p)(v,v)}{\sqrt{-\mathbf{g}(p)(v,v)}} = 0$, $u \in v^\perp$.

If: Let $v = cw$ for a $c > 0$. Then $u = -\dfrac{\sqrt{-\mathbf{g}(p)(v,v)}}{\mathbf{g}(p)(cv,v)}cv - \dfrac{v}{\sqrt{-\mathbf{g}(p)(v,v)}} = 0$.

Only if: If $u = 0$, then

$$-\frac{\sqrt{-\mathbf{g}(p)(v,v)}}{\mathbf{g}(p)(w,v)}w - \frac{v}{\sqrt{-\mathbf{g}(p)(v,v)}} = 0. \qquad (\star)$$

Rearranging, $w = cv$, where $c := \dfrac{\mathbf{g}(p)(w,v)}{\mathbf{g}(p)(v,v)}$. Clearly $c \neq 0$.

That $c > 0$: Let the time-orientation be given by $V \in T_0^1M$. Taking scalar product in $(\star)$ with V_p, we get

$$\mathbf{g}(p)(v, V_p) = \frac{\mathbf{g}(p)(v,v)}{\mathbf{g}(p)(w,v)}\mathbf{g}(p)(w, V_p).$$

Note that

$$\begin{aligned}&\mathbf{g}(p)(v,V_p)<0 \quad (v \text{ future-pointing}),\\ &\mathbf{g}(p)(w,V_p)<0 \quad (w \text{ future-pointing}),\\ &\mathbf{g}(p)(v,v)<0 \quad (v \text{ timelike}).\end{aligned}$$

These facts show that $\mathbf{g}(p)(w,v)<0$ too, and so $c=\dfrac{\mathbf{g}(p)(w,v)}{\mathbf{g}(p)(v,v)}\ \left(=\frac{-}{-}\right)>0$.

Solution to Exercise 12.4

We have

$$\begin{aligned} u &= -\frac{\sqrt{-\mathbf{g}(p)(v_{\gamma,p},v_{\gamma,p})}}{\mathbf{g}(p)(v_{\tilde{\gamma},p},v_{\gamma,p})}v_{\tilde{\gamma},p}-\frac{v_{\gamma,p}}{\sqrt{-\mathbf{g}(p)(v_{\gamma,p},v_{\gamma,p})}}\\ &= -\frac{\sqrt{-g(\mathbf{e}_0,\mathbf{e}_0)}}{g(\alpha\,\mathbf{e}_0+\beta\,\mathbf{e}_1,\mathbf{e}_0)}(\alpha\,\mathbf{e}_0+\beta\,\mathbf{e}_1)-\frac{\mathbf{e}_0}{\sqrt{-g(\mathbf{e}_0,\mathbf{e}_0)}}=\frac{\beta}{\alpha}\mathbf{e}_1.\end{aligned}$$

Thus

$$|u|=\sqrt{g\big(\frac{\beta}{\alpha}\mathbf{e}_1,\frac{\beta}{\alpha}\mathbf{e}_1\big)}=\frac{|\beta|}{\alpha}=\frac{|\beta|}{\sqrt{1+\beta^2}}=\frac{1}{\sqrt{\frac{1}{\beta^2}+1}},$$

i.e., $|\beta|=\dfrac{|u|}{\sqrt{1-|u|^2}}$. So $\dfrac{2\widetilde{L}}{2L}=\dfrac{1}{\sqrt{1+\beta^2}}=\sqrt{1-|u|^2}$.

Solution to Exercise 12.5

We have $\tilde{u}^1=\alpha=\dfrac{u^1-\beta}{1-u^1\beta}$, $\tilde{u}^2=0=\dfrac{u^2\sqrt{1-\beta^2}}{1-\beta u^1}$, $\tilde{u}^3=0=\dfrac{u^3\sqrt{1-\beta^2}}{1-\beta u^1}$. Hence $u^2=0$, $u^3=0$. Solving for u^1 from $\alpha=\dfrac{u^1-\beta}{1-u^1\beta}$, yields $u^1=\dfrac{\alpha+\beta}{1+\alpha\beta}$. So

$$u=u^i\mathbf{e}_i=\frac{\alpha+\beta}{1+\alpha\beta}e_1=(\alpha\circledast\beta)\,e_1.$$

By the formula from Exercise 1.2, p.4 (and setting $v=0$ and $u=\alpha$ there),

$$B=\Big(\frac{2\alpha}{\alpha^2+1},\frac{\alpha^2-1}{\alpha^2+1}\Big).$$

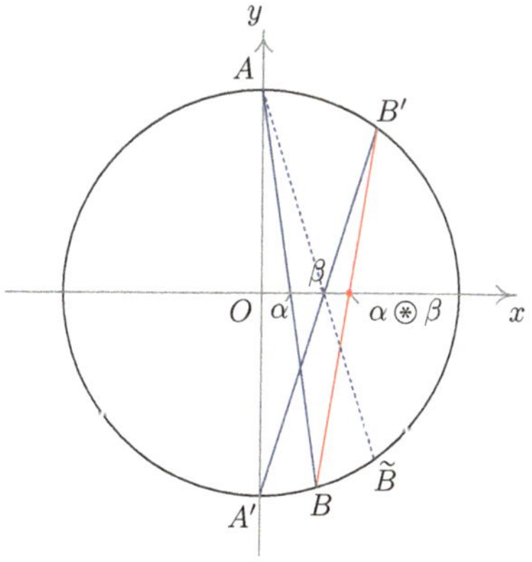

To get the coordinates of B', we note that the reflection $\widetilde{B}$ of B' in the x-axis has the coordinates $\widetilde{B} = \big(\frac{2\beta}{\beta^2+1}, \frac{\beta^2-1}{\beta^2+1}\big)$, and so $B' = \big(\frac{2\beta}{\beta^2+1}, \frac{1-\beta^2}{\beta^2+1}\big)$. So the equation of BB' is

$$\frac{y - \frac{\alpha^2-1}{\alpha^2+1}}{x - \frac{2\alpha}{\alpha^2+1}} = \frac{\frac{1-\beta^2}{\beta^2+1} - \frac{\alpha^2-1}{\alpha^2+1}}{\frac{2\beta}{\beta^2+1} - \frac{2\alpha}{\alpha^2+1}}.$$

To find the x-coordinate of the intersection point of BB' with the x-axis, we set $y = 0$, and solve for x, which yields

$$x = \frac{\alpha+\beta}{1+\alpha\beta} = \alpha \circledast \beta.$$

The commutativity property $\alpha \circledast \beta = \beta \circledast \alpha$ can be seen by reflecting the diagram in the x-axis, as shown in the picture on the left below.

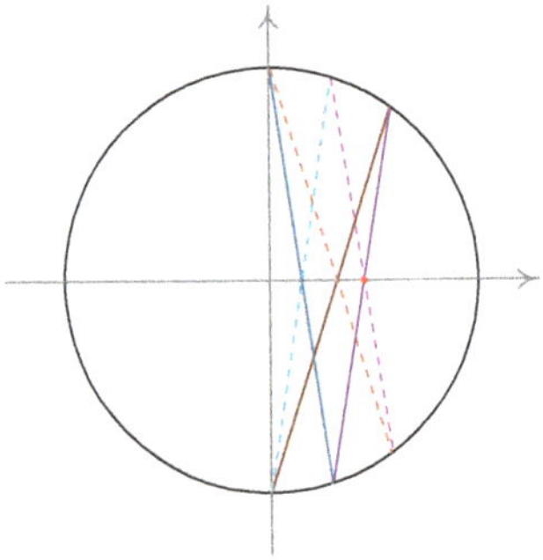

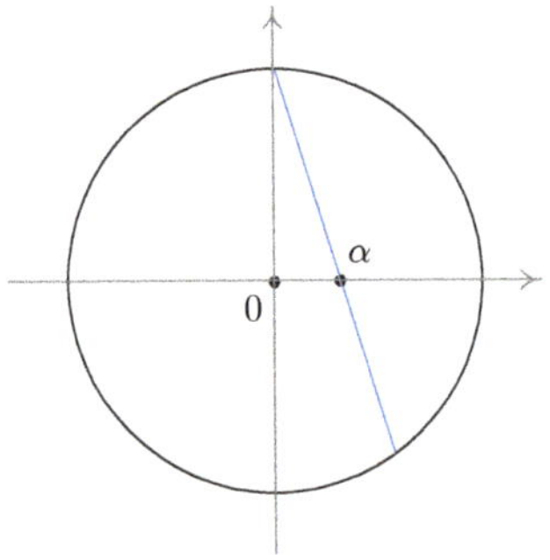

The property $\alpha \circledast 0 = \alpha$ is displayed in the picture on the right above.
The property $\alpha \circledast (-\alpha) = 0$ is shown in the picture on the left below.
Finally, the property $\alpha \circledast 1 = 1$ is shown in the picture on the right below.

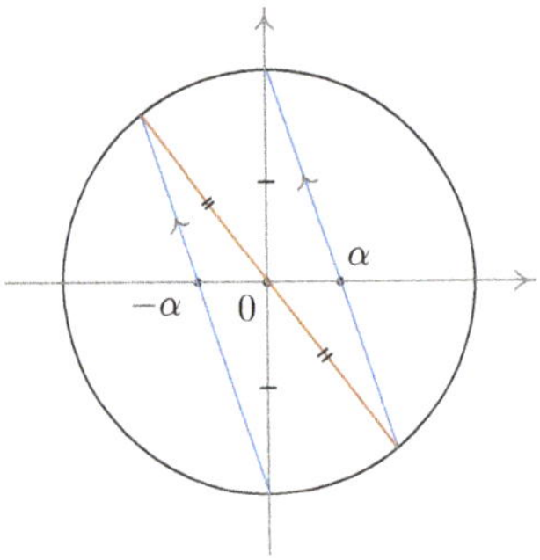

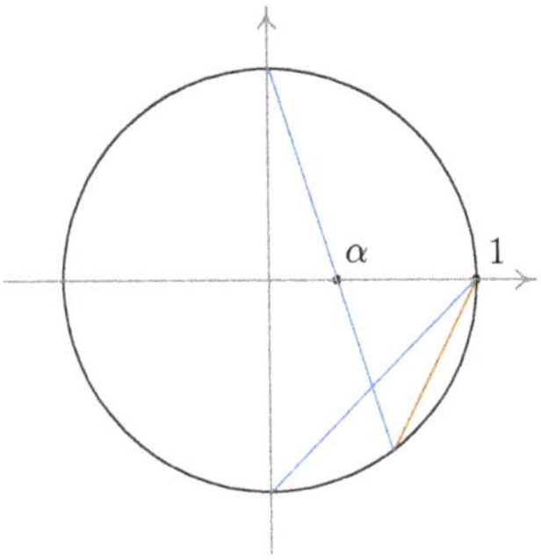

Solution to Exercise 12.6

As $\alpha \in (-1, 1)$, and since $\tanh : \mathbb{R} \to (-1, 1)$ is strictly increasing and onto, there is a unique $a \in \mathbb{R}$ such that $\tanh a = \alpha$. Similarly, let $b \in \mathbb{R}$ be such that $\tanh b = \beta$.

Then

$$\alpha \circledast \beta = \frac{\alpha+\beta}{1+\alpha\beta} = \frac{\tanh a+\tanh b}{1+(\tanh a)(\tanh b)} = \frac{\frac{\sinh a}{\cosh a}+\frac{\sinh b}{\cosh b}}{1+\frac{\sinh a}{\cosh a}\frac{\sinh b}{\cosh b}}$$

$$= \frac{(\sinh a)(\cosh b)+(\sinh b)\cosh a}{(\cosh a)(\cosh b)+(\sinh a)\sinh b} = \frac{\sinh(a+b)}{\cosh(a+b)} = \tanh(a+b).$$

Solution to Exercise 12.7

Let $b \in (0, \infty)$ be such that $\tanh b = \beta$. Let u_k be the relative speed of the kth cart with respect to the table. Then

$$u_2 = \beta \circledast \beta = \tanh(b+b) = \tanh(2b),$$
$$u_3 = \beta \circledast \tanh(2b) = \tanh(b+2b) = \tanh(3b),$$
$$\cdots$$
$$u_k = \beta \circledast \tanh((k-1)b) = \tanh(b+(k-1)b) = \tanh(kb).$$

Thus $u_n = \tanh(nb) = \tanh(n \tanh^{-1}\beta)$. As $b > 0$,

$$\lim_{n\to\infty} u_n = \lim_{n\to\infty} \tanh(n\tanh^{-1}\beta) = \lim_{n\to\infty} \tanh(nb)$$
$$= \lim_{n\to\infty} \frac{\sinh(nb)}{\cosh(nb)} = \lim_{n\to\infty} \frac{(e^{nb}-e^{-nb})/2}{(e^{nb}+e^{-nb})/2}$$
$$= \lim_{n\to\infty} \frac{1-e^{-2nb}}{1+e^{-2nb}} = \frac{1-0}{1+0} = 1.$$

Solution to Exercise 12.8

We have $\widetilde{z}^2 = \mathbf{g}(p)(z, \widetilde{e}_2) = \mathbf{g}(p)(z, e_2) = z^2$. Also, $\widetilde{z}^3 = z^3$, as $\widetilde{e}_3 = e_3$.

As $\widetilde{e}_0 = \widetilde{v} = \dfrac{v+\beta e_1}{\sqrt{1-\beta^2}} = \dfrac{e_0+\beta e_1}{\sqrt{1-\beta^2}}$, we obtain $\widetilde{z}^0 = -\mathbf{g}(p)(z, \widetilde{e}_0) = \dfrac{z^0-\beta z^1}{\sqrt{1-\beta^2}}$.

Similarly, as $\widetilde{e}_1 = \dfrac{\beta e_0 + e_1}{\sqrt{1-\beta^2}}$, we get $\widetilde{z}^1 = \mathbf{g}(p)(z, \widetilde{e}_1) = \dfrac{-\beta z^0 + z^1}{\sqrt{1-\beta^2}}$.

Set $\phi = \tanh^{-1}\beta$. Then $\tanh\phi = \beta$, $\dfrac{\beta}{\sqrt{1-\beta^2}} = \sinh\phi$ and $\dfrac{1}{\sqrt{1-\beta^2}} = \cosh\phi$.

Thus $\begin{bmatrix} \widetilde{z}^0 \\ \widetilde{z}^1 \end{bmatrix} = \begin{bmatrix} \cosh\phi & -\sinh\phi \\ -\sinh\phi & \cosh\phi \end{bmatrix} \begin{bmatrix} z^0 \\ z^1 \end{bmatrix}$, and so $\begin{bmatrix} z^0 \\ z^1 \end{bmatrix} = \begin{bmatrix} \cosh\phi & \sinh\phi \\ \sinh\phi & \cosh\phi \end{bmatrix} \begin{bmatrix} \widetilde{z}^0 \\ \widetilde{z}^1 \end{bmatrix}$.

Solution to Exercise 12.9

We have $\frac{1}{\widetilde{\nu}}\widetilde{v} = v_{p\widetilde{q}} = v_{pq} + \alpha w = \frac{1}{\nu}v + \alpha w$. Thus $\frac{1}{\widetilde{\nu}}\widetilde{v} = \frac{1}{\nu}v + \alpha w$.

Taking the scalar product with w, and noting that w is lightlike (so that $\mathbf{g}(p)(w,w) = 0$), we obtain from the above that

$$\frac{1}{\widetilde{\nu}}\mathbf{g}(p)(w, \widetilde{v}) = \frac{1}{\nu}\mathbf{g}(p)(w, v) + 0.$$

Consequently, $\dfrac{\widetilde{\nu}}{\nu} = \dfrac{\mathbf{g}(p)(w, \widetilde{v})}{\mathbf{g}(p)(w, v)}$.

Solution to Exercise 12.10

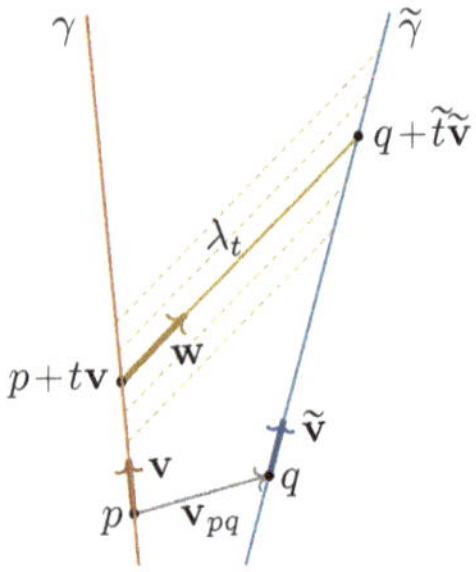

We have $p + t\mathbf{v} + s\mathbf{w} = q + \widetilde{t}\widetilde{\mathbf{v}} = p + \mathbf{v}_{pq} + \widetilde{t}\widetilde{\mathbf{v}}$, and so $t\mathbf{v} + s\mathbf{w} = \mathbf{v}_{pq} + \widetilde{t}\widetilde{\mathbf{v}}$. Taking the scalar product with $\mathbf{w}$, $tg(\mathbf{w}, \mathbf{v}) + 0 = g(\mathbf{w}, \mathbf{v}_{pq}) + \widetilde{t}g(\mathbf{w}, \widetilde{\mathbf{v}})$. Thus $\widetilde{t} = -\dfrac{g(\mathbf{w}, \mathbf{v}_{pq})}{g(\mathbf{w}, \widetilde{\mathbf{v}})} + t\dfrac{g(\mathbf{w}, \mathbf{v})}{g(\mathbf{w}, \widetilde{\mathbf{v}})}$.

Solution to Exercise 12.11

We just consider the case $\theta > 0$ as shown in the following picture, with the light signal $\lambda : \mathbb{R} \to M$ emitted at $\widetilde{q} \in M$ having the form

$$\lambda(t) = \widetilde{q} + kt(\mathbf{e}_0 - \mathbf{e}_1), \quad t \in \mathbb{R},$$

for some fixed nonzero $k \in \mathbb{R}$. (This is the case of the receding emitter.) Suppose the light ray is intercepted by γ at $q \in M$.

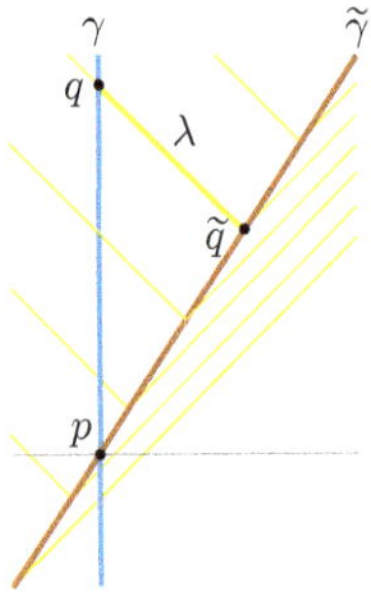

Then $\nu = -g((\cosh\theta)\mathbf{e}_0 + (\sinh\theta)\mathbf{e}_1, k(\mathbf{e}_0 - \mathbf{e}_1)) = k(\cosh\theta + \sinh\theta) = ke^{\theta}$. The frequency of the light signal perceived by the normalised instantaneous observer $v_{\gamma,q}$ at q is $\widetilde{\nu} := -g(\mathbf{e}_0, k(\mathbf{e}_0 - \mathbf{e}_1)) = k$. Hence $\dfrac{\widetilde{\nu}}{\nu} = \dfrac{k}{ke^{\theta}} = e^{-\theta}$.

Let $0 < \theta \ll 1$. Then, in the classical case, we have

$$\frac{\widetilde{\nu}'}{\nu} = \frac{1}{1 + \tanh\theta} \approx 1 - \theta.$$

In Minkowski spacetime, we get $\dfrac{\widetilde{\nu}}{\nu} = e^{-\theta} \approx 1 - \theta$.

The two approximations match, as expected.

Solution to Exercise 12.12

Let the time-orientation be given by $V \in T_0^1 M$. Let the annihilation occur at event $p \in M$. At p, let the velocity of the electron be $v_- \in T_pM$, and that of the positron be $v_+ \in T_pM$. Let the velocity of the photon be $w \in T_pM$. Then $\mathbf{g}(p)(v_-, v_-) = -1 = \mathbf{g}(p)(v_+, v_+)$ and $\mathbf{g}(p)(w, w) = 0$.

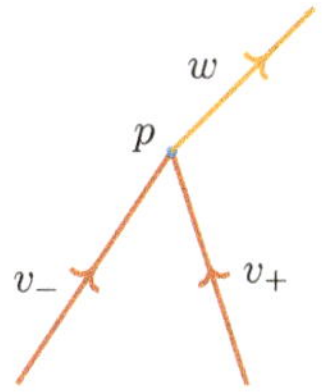

By the law of conservation of energy-momentum, $m_e(v_-)^\flat + m_e(v_+)^\flat = w^\flat$. So $m_e(v_- + v_+) = w$. Hence

$$\begin{aligned} 0 &= \mathbf{g}(p)(w, w) \\ &= \mathbf{g}(p)(m_e(v_- + v_+), m_e(v_- + v_+)) \\ &= m_e^2(-1 - 1 + 2\mathbf{g}(p)(v_-, v_+)), \end{aligned}$$

and so $\mathbf{g}(p)(v_-, v_+) = 1 > 0$. Since v_-, v_+ are timelike and future-pointing, we have $\mathbf{g}(p)(v_-, V_p) < 0$ and $\mathbf{g}(p)(v_+, V_p) < 0$. Exercise 5.5 (p.81) implies that also $\mathbf{g}(p)(v_-, v_+) < 0$, which is a contradiction to $\mathbf{g}(p)(v_-, v_+) = 1 > 0$.

Solution to Exercise 12.13

Let $\{e_1, e_2, e_3\}$ be an orthonormal basis for $v^\perp$ such that (e_1, e_2, e_3, v) is a positively oriented basis for T_pM with respect to the orientation $[\Omega(p)]$ Then (e_1, e_2, e_3) is a positively oriented basis for $v^\perp$ because

$$\omega_v(e_1, e_2, e_3) = \Omega(p)(e_1, e_2, e_3, v) > 0.$$

Let $\nu \in \bigwedge^3 (v^\perp)^*$ denote the volume form on $v^\perp$ induced by the orientation $[\omega_v]$ and $\mathbf{g}(p)|_{v^\perp}$. Then we have that $\omega_v = c\nu$, where $c = \omega_v(e_1, e_2, e_3)$, and $\Omega(p) = C \,\mathrm{vol}_{\mathbf{g}(p)}$, where $C = \Omega(p)(e_1, e_2, e_3, v) = \omega_v(e_1, e_2, e_3) = c$. Moreover, $\mathbf{F}(p)(x, y) = (\star(\mathrm{B}^\flat))(x, y) = \star(\mathrm{B}^\flat \wedge x^\flat \wedge y^\flat) = \star(k\nu) = k$, where k is such that $\mathrm{B}^\flat \wedge x^\flat \wedge y^\flat = k\nu$, that is, $k = (\mathrm{B}^\flat \wedge x^\flat \wedge y^\flat)(e_1, e_2, e_3)$. Thus

$$\begin{aligned} &\mathbf{F}(p)(x, y) = (\mathrm{B}^\flat \wedge x^\flat \wedge y^\flat)(e_1, e_2, e_3) \\ &= \det \begin{bmatrix} \mathbf{g}(p)(\mathrm{B}, e_1) & \mathbf{g}(p)(\mathrm{B}, e_2) & \mathbf{g}(p)(\mathrm{B}, e_3) \\ \mathbf{g}(p)(x, e_1) & \mathbf{g}(p)(x, e_2) & \mathbf{g}(p)(x, e_3) \\ \mathbf{g}(p)(y, e_1) & \mathbf{g}(p)(y, e_2) & \mathbf{g}(p)(y, e_3) \end{bmatrix} = \det \begin{bmatrix} \mathbf{g}(p)(\mathrm{B}, e_1) & \mathbf{g}(p)(x, e_1) & \mathbf{g}(p)(y, e_1) \\ \mathbf{g}(p)(\mathrm{B}, e_2) & \mathbf{g}(p)(x, e_2) & \mathbf{g}(p)(y, e_2) \\ \mathbf{g}(p)(\mathrm{B}, e_3) & \mathbf{g}(p)(x, e_3) & \mathbf{g}(p)(y, e_3) \end{bmatrix} \\ &= (e_1^\flat \wedge e_2^\flat \wedge e_3^\flat)(\mathrm{B}, x, y) = \nu(\mathrm{B}, x, y) = \frac{1}{c}\omega_v(\mathrm{B}, x, y) = \frac{1}{C}\omega_v(\mathrm{B}, x, y) \\ &= \frac{1}{C}\Omega(p)(\mathrm{B}, x, y, v) = \mathrm{vol}_{\mathbf{g}(p)}(\mathrm{B}, x, y, v) = \mathrm{vol}_{\mathbf{g}(p)}(x, y, \mathrm{B}, v). \end{aligned}$$

Claim: If $\widetilde{\mathrm{B}} \in v^{\perp}$ satisfies for all $x, y \in v^{\perp}$, $\mathrm{vol}_{\mathbf{g}(p)}(x, y, \widetilde{\mathrm{B}}, v) = \mathbf{F}(p)(x, y)$, then $\widetilde{\mathrm{B}} = \mathrm{B}$.

Suppose that $w := \widetilde{\mathrm{B}} - \mathrm{B} \neq 0$. Set $f_1 = \alpha\, w$, where $\alpha = 1/\sqrt{\mathbf{g}(p)(w, w)}$. Then we can find vectors $f_2, f_3 \in v^{\perp}$ so that $\{f_1, f_2, f_3\}$ is an orthonormal basis for $v^{\perp}$. As (f_1, f_2, f_3, v) is an orthonormal basis for T_pM, we have that $\mathrm{vol}_{\mathbf{g}(p)}(f_1, f_2, f_3, v) \neq 0$. But

$$\begin{aligned} 0 &= \mathbf{F}(p)(f_2, f_3) - \mathbf{F}(p)(f_2, f_3) \\ &= \mathrm{vol}_{\mathbf{g}(p)}(f_2, f_3, \widetilde{\mathrm{B}}, v) - \mathrm{vol}_{\mathbf{g}(p)}(f_2, f_3, \mathrm{B}, v) \\ &= \mathrm{vol}_{\mathbf{g}(p)}(f_2, f_3, \alpha\, f_1, v) \\ &= \alpha\, \mathrm{vol}_{\mathbf{g}(p)}(f_1, f_2, f_3, v), \end{aligned}$$

and so $\mathrm{vol}_{\mathbf{g}(p)}(f_1, f_2, f_3, v) = 0$, a contradiction.

Solution to Exercise 12.14

We will use the curl-of-curl identity[22] $\vec{\nabla} \times (\vec{\nabla} \times \vec{\mathbf{v}}) = \vec{\nabla}(\vec{\nabla} \cdot \vec{\mathbf{v}}) - \Delta \vec{\mathbf{v}}$ for any smooth map $\vec{\mathbf{v}} : \mathbb{R}^3 \to \mathbb{R}^3$. We have

$$\begin{aligned} \frac{\partial^2 \vec{\mathbf{E}}}{\partial t^2} &= \frac{\partial}{\partial t}\frac{\partial \vec{\mathbf{E}}}{\partial t} = \frac{\partial}{\partial t}(-\vec{\mathbf{j}} + \vec{\nabla} \times \vec{\mathbf{B}}) = -\frac{\partial \vec{\mathbf{j}}}{\partial t} + \frac{\partial}{\partial t}(\vec{\nabla} \times \vec{\mathbf{B}}) = -\frac{\partial \vec{\mathbf{j}}}{\partial t} + \vec{\nabla} \times \frac{\partial \vec{\mathbf{B}}}{\partial t} \\ &= -\frac{\partial \vec{\mathbf{j}}}{\partial t} + \vec{\nabla} \times (-\vec{\nabla} \times \vec{\mathbf{E}}) = -\frac{\partial \vec{\mathbf{j}}}{\partial t} - \vec{\nabla}(\vec{\nabla} \cdot \vec{\mathbf{E}}) + \Delta \vec{\mathbf{E}} \\ &= -\frac{\partial \vec{\mathbf{j}}}{\partial t} - \vec{\nabla}\rho + \Delta \vec{\mathbf{E}}. \end{aligned}$$

Also,

$$\begin{aligned} \frac{\partial^2 \vec{\mathbf{B}}}{\partial t^2} &= \frac{\partial}{\partial t}\frac{\partial \vec{\mathbf{B}}}{\partial t} = \frac{\partial}{\partial t}(-\vec{\nabla} \times \vec{\mathbf{E}}) = -\vec{\nabla} \times \frac{\partial \vec{\mathbf{E}}}{\partial t} = -\vec{\nabla} \times (-\vec{\mathbf{j}} + \vec{\nabla} \times \vec{\mathbf{B}}) \\ &= \vec{\nabla} \times \vec{\mathbf{j}} - \vec{\nabla} \times (\vec{\nabla} \times \vec{\mathbf{B}}) = \vec{\nabla} \times \vec{\mathbf{j}} - \vec{\nabla}(\vec{\nabla} \cdot \vec{\mathbf{B}}) + \Delta \vec{\mathbf{B}} \\ &= \vec{\nabla} \times \vec{\mathbf{j}} - \vec{\nabla} 0 + \Delta \vec{\mathbf{B}} = \vec{\nabla} \times \vec{\mathbf{j}} + \Delta \vec{\mathbf{B}}. \end{aligned}$$

With $\vec{\mathbf{E}}(t, x, y, z) = (\sin(z - t), 0, 0)$ and $\vec{\mathbf{B}}(t, x, y, z) = (0, \sin(z - t), 0)$,

$$\begin{aligned} \vec{\nabla} \cdot \vec{\mathbf{E}} &= \frac{\partial}{\partial x}\sin(z - t) + \frac{\partial}{\partial y}0 + \frac{\partial}{\partial z}0 = 0 \\ \vec{\nabla} \times \vec{\mathbf{E}} &= (0, \cos(z - t) - 0, 0) = -(0, -\cos(z - t), 0) = -\frac{\partial \vec{\mathbf{B}}}{\partial t} \\ \vec{\nabla} \cdot \vec{\mathbf{B}} &= \frac{\partial}{\partial x}0 + \frac{\partial}{\partial y}\sin(z - t) + \frac{\partial}{\partial z}0 = 0 \\ \vec{\nabla} \times \vec{\mathbf{B}} &= (-\cos(z - t), 0, 0) = \frac{\partial \vec{\mathbf{E}}}{\partial t}. \end{aligned}$$

[22] See for instance [Apostol (1969), Example 5, p.445, (12.38)].

Solution to Exercise 12.15

The orthonormality of $(V_i)_p$, $i = 1, 2, 3, 4$, is easily verified. For example,

$$\mathbf{g}(p)((V_4)_p, (V_4)_p) = -f(r)\,(dt_p \otimes dt_p)\Big(\frac{1}{\sqrt{f(r)}}\,\partial_{t,p}, \frac{1}{\sqrt{f(r)}}\,\partial_{t,p}\Big) = -1.$$

Recall the volume form field $\mathrm{vol}_{\mathbf{g}_{S^2}}$ on $(S^2, \mathbf{g}_{S^2})$ as in Example 11.4 (p.249). Then the ordered basis $(\partial_{\theta,\mathbf{p}}, \partial_{\phi,\mathbf{p}})$ is positively oriented with the induced orientation on $T_{\mathbf{p}}S^2$, for all $\mathbf{p} \in U$, and $\mathrm{vol}_{\mathbf{g}_{S^2}}(\mathbf{p})(\partial_{\theta,\mathbf{p}}, \partial_{\phi,\mathbf{p}}) = \sin\theta > 0$. With the orientation $[\Omega]$ on M, where $\Omega := r^2 dr \wedge \mathrm{vol}_{\mathbf{g}_{S^2}} \wedge dt$, for all $p \in \mathcal{U}$,

$$\Omega(p)((V_1)_p, (V_2)_p, (V_3)_p, (V_4)_p) = r^2\sqrt{f(r)}\frac{1}{r}\frac{1}{r(\sin\theta)}(\sin\theta)\frac{1}{\sqrt{f(r)}} = 1 > 0.$$

So for each $p \in \mathcal{U}$, $((V_1)_p, (V_2)_p, (V_3)_p, (V_4)_p)$ is a positively oriented ordered basis for T_pM.

Finally, we show that $\mathbf{F}$ satisfies the Maxwell equations $d\mathbf{F} = 0$ and $d{\star}\mathbf{F} = 0$. (The RHS of the second equation is 0 since $\mathbf{J} = 0$.) That $d\mathbf{F} = 0$ is easily seen, since

$$d\mathbf{F} = d\Big(\frac{e}{r^2}dr \wedge dt\Big) = \Big(\partial_r \frac{e}{r^2}\Big)dr \wedge dr \wedge dt = 0.$$

To show $d{\star}\mathbf{F} = 0$, we first determine ${\star}\mathbf{F}$ using Proposition 11.3 (p.255):

$$\begin{aligned}
{\star}\mathbf{F} &= \frac{e}{r^2} \star \Big(\frac{1}{\sqrt{f(r)}}dr \wedge \sqrt{f(r)}dt\Big) \\
&= \frac{e}{r^2}\,\mathrm{sign}\begin{pmatrix} 1\,2\,3\,4 \\ 1\,4\,2\,3 \end{pmatrix} \mathbf{g}(V_2, V_2)\,\mathbf{g}(V_3, V_3)\,\Omega^2 \wedge \Omega^3 \\
&= \frac{e}{r^2}(1)(1)(1)\,r d\theta \wedge (r\sin\theta)d\phi = e\,(\sin\theta)\,d\theta \wedge d\phi.
\end{aligned}$$

Consequently, $d{\star}\mathbf{F} = e\,(\partial_\theta \sin\theta)\,d\theta \wedge d\theta \wedge d\phi = 0$.

Chapter 13

Solution to Exercise 13.1

It is convenient to use cylindrical coordinates. Let $\rho := m/(\frac{4}{3}\pi R^3)$.

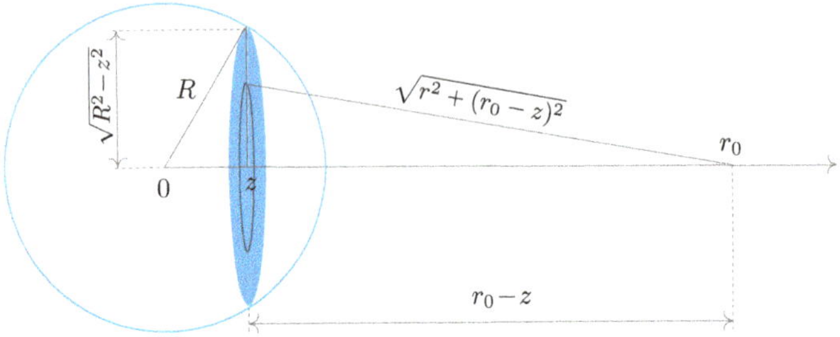

Referring to the picture, we have

$$\begin{aligned}\Phi &= -\int_{\mathbb{R}^3} \frac{\rho}{\|\mathbf{y}-\mathbf{x}\|} dV(\mathbf{y}) = -\int_{-R}^{R}\int_0^{\sqrt{R^2-z^2}} \frac{\rho\, 2\pi r}{\sqrt{r^2+(r_0-z)^2}} dr dz \\ &= -2\pi\rho \int_{-R}^{R} (\sqrt{R^2+r_0^2-2r_0 z} - (r_0-z)) dz \\ &= -2\pi\rho\Big(\sqrt{2r_0}\frac{2}{3}\big(\big(\frac{R^2+r_0^2}{2r_0}+R\big)^{3/2} - \big(\frac{R^2+r_0^2}{2r_0}-R\big)^{3/2}\big) - r_0 \cdot 2R\Big) \\ &= -4\pi\rho\big(\frac{1}{3}\big(\frac{(R+r_0)^3}{2r_0} - \frac{(r_0-R)^3}{2r_0}\big) - r_0 R\big) \\ &= -4\pi\rho \cdot \frac{1}{3} \cdot \frac{R^3}{r_0} = -\frac{m}{r_0}.\end{aligned}$$

Solution to Exercise 13.2

The constant function ν is smooth and everywhere positive. Moreover, we have $\nabla_V V = \nabla_{\partial_t}\partial_t = \Gamma^i_{00}\partial_{x^i} = 0$, $\mathbf{g}(p)(\partial_{t,p}, \partial_{t,p}) = -1$, and $\partial_{t,p}$ is future-pointing (as the time-orientation is given by ∂_t). Finally,

$$\operatorname{div}(\nu V) = \nu_* \, dx^i(\nabla_{\partial_{x^i}} \partial_t) = \nu_* \, dx^i(\Gamma^k_{0i}\partial_{x^k}) = \nu_* \, dx^i(0\,\partial_{x^k}) = \nu_* \, dx^i(0) = 0.$$

Since $\mathbf{T}_{ij} = \mathbf{T}(\partial_{x^i}, \partial_{x^j}) = m\nu_* \mathbf{g}(\partial_t, \partial_{x^i})\mathbf{g}(\partial_t, \partial_{x^j}) = m\nu_*(-\delta_{i0})(-\delta_{j0})$,

$$[\mathbf{T}_{ij}] = \begin{bmatrix} m\nu_* & 0 & 0 & 0 \\ 0 & 0 & 0 & 0 \\ 0 & 0 & 0 & 0 \\ 0 & 0 & 0 & 0 \end{bmatrix}.$$

As the curvature tensor field for the Minkowski spacetime is identically zero, it follows that the Ricci curvature tensor field $\mathbf{Ric} = \mathbf{0}$, as well as the scalar curvature $\mathbf{S} = 0$, and so the field equation becomes $0 - 0 + \Lambda\mathbf{g} = 8\pi\mathbf{T}$.

In the chart used above, this acquires the form

$$\Lambda \begin{bmatrix} -1 & & & \\ & 1 & & \\ & & 1 & \\ & & & 1 \end{bmatrix} = \begin{bmatrix} m\nu_* & & & \\ & 0 & & \\ & & 0 & \\ & & & 0 \end{bmatrix}.$$

This holds if and only if $0=\Lambda=-m\nu_*$. As $m>0$, the field equation holds if and only if $\nu_*=0$.

Solution to Exercise 13.3

(1) Consider a spatial ball B_r with centre $\mathbf{x}$ and radius $r > 0$, and with spherical boundary ∂B_r having an outward normal $\mathbf{n_y}$ at $\mathbf{y} \in \partial B_r$. By the Gauss divergence theorem, we get

$$\int_{B_r} \vec{\nabla}\cdot(\rho\mathbf{v})\,dV = \int_{\partial B_r} \rho\mathbf{v}\cdot\mathbf{n_y}\,dA.$$

The latter must equal the rate of change of mass in B_r, that is,

$$\int_{\partial B_r} \rho\mathbf{v}\cdot\mathbf{n_y}\,dA = -\frac{d}{dt}\int_{B_r} \rho dV = -\int_{B_r} \frac{\partial\rho}{\partial t}\,dV.$$

Consequently,

$$\int_{B_r} \Big(\frac{\partial\rho}{\partial t} + \vec{\nabla}\cdot(\rho\mathbf{v})\Big)\,dV = 0.$$

Passing to the limit as $r \to 0$, we get $\dfrac{\partial\rho}{\partial t} + \vec{\nabla}\cdot(\rho\mathbf{v}) = 0$.

(2) From the continuity equation, $0=\dfrac{\partial\rho}{\partial t}+\vec{\nabla}\cdot(\rho\mathbf{v})=0+\rho\vec{\nabla}\cdot\mathbf{v}$. So $\vec{\nabla}\cdot\mathbf{v}=0$.

For an elemental volume dV, the mass is ρdV, and by Newton's second law of motion, $\rho dV\dfrac{d}{dt}\mathbf{v} = \mathbf{0}$. This yields

$$\frac{\partial v^i}{\partial t} + (\vec{\nabla}v^i)\cdot\mathbf{v} = 0 \quad (i = 1,2,3).$$

For $i \in \{1,2,3\}$, the ith component of the classical momentum in Ω is $\displaystyle\int_\Omega \rho v^i\,dV$, and its rate of change is

$$\frac{d}{dt}\int_\Omega \rho v^i\,dV = \int_\Omega \rho\frac{\partial v^i}{\partial t}\,dV.$$

Using the Euler equation, and noting that $\vec{\nabla}\cdot\mathbf{v} = 0$, we get

$$\frac{d}{dt}\int_\Omega \rho v^i\,dV = -\rho\int_\Omega (\vec{\nabla}v^i)\cdot\mathbf{v}\,dV = -\rho\int_\Omega \vec{\nabla}\cdot(v^i\mathbf{v})\,dV.$$

By the Gauss divergence theorem,

$$\begin{aligned}-\frac{d}{dt}\int_\Omega \rho v^i\,dV &= \int_{\partial\Omega} \rho v^i\,\mathbf{v}\cdot\mathbf{n_y}\,dA(\mathbf{y}) = \int_{\partial\Omega} \rho\mathbf{v}^\flat(\mathbf{e}_i)\mathbf{v}^\flat(\mathbf{n_y})\,dA(\mathbf{y}) \\ &= \int_{\partial\Omega} P(\mathbf{e}_i,\mathbf{n_y})\,dA(\mathbf{y}).\end{aligned}$$

Solution to Exercise 13.4

We use the reversed Cauchy-Schwarz inequality: For all $q \in M$, all causal $X \in T_0^1 M$, and V as in Definition 13.4 (p.298), we have that

$$(\mathbf{g}(q)(V_q, X_q))^2 \geqslant \mathbf{g}(V_q, V_q)\mathbf{g}(q)(X_q, X_q) = -1\,\mathbf{g}(q)(X_q, X_q).$$

Thus

$$\begin{aligned}\mathbf{T}(q)(X_q, X_q) &= (\rho(q) + \mathrm{p}(q))((V^\flat X)(q))^2 + \mathrm{p}(q)\mathbf{g}(q)(X_q, X_q)\\ &= (\rho(q) + \mathrm{p}(q))(\mathbf{g}(q)(V_q, X_q))^2 + \mathrm{p}(q)\mathbf{g}(q)(X_q, X_q)\\ &= \rho(q)\,(\mathbf{g}(q)(V_q, X_q))^2 + \mathrm{p}(q)\big((\mathbf{g}(q)(V_q, X_q))^2 + \mathbf{g}(q)(X_q, X_q)\big)\\ &\geqslant \rho(q)\,0 + \mathrm{p}(q)\,(-\mathbf{g}(q)(X_q, X_q) + \mathbf{g}(q)(X_q, X_q)) = 0.\end{aligned}$$

Consequently, $\mathbf{T}(X, X) \geqslant 0$ (pointwise) for all causal $X \in T_0^1 M$, i.e., $\mathbf{T}$ satisfies the weak energy condition.

Solution to Exercise 13.5

For $\Omega \in T_1^0 M$ and $Y \in T_0^1 M$,

$$\begin{aligned}(V^\flat \otimes V^\flat)^\sharp(\Omega, Y) &= (V^\flat \otimes V^\flat)(\Omega^\sharp, Y) = V^\flat(\Omega^\sharp)V^\flat(Y) = \mathbf{g}(V, \Omega^\sharp)\mathbf{g}(V, Y)\\ &= (\Omega^\sharp)^\flat(V)\mathbf{g}(V, Y) = \Omega(V)\mathbf{g}(V, Y).\end{aligned}$$

Suppose that $(U, \mathbf{x})$ is an admissible chart for M. Then in U, we have that $\mathbf{C}((V^\flat \otimes V^\flat)^\sharp) = dx^i(V)\mathbf{g}(V, \partial_{x^i}) = V^i\mathbf{g}(V, \partial_{x^i}) = \mathbf{g}(V, V^i\partial_{x^i}) = \mathbf{g}(V, V) = -1.$ So $\operatorname{trace}\mathbf{T} = (\rho + \mathrm{p})\operatorname{trace}(V^\flat \otimes V^\flat) + \mathrm{p}\operatorname{trace}\mathbf{g} = -(\rho + \mathrm{p}) + 4\mathrm{p} = 3\mathrm{p} - \rho$.

Solution to Exercise 13.6

Differentiation of both sides of $3\dot{a}^2 - \Lambda a^2 = 8\pi\rho a^2$ (first Friedmann equation) with respect to t gives

$$6\dot{a}\ddot{a} - 2\Lambda a\dot{a} = 8\pi(\dot{\rho}a^2 + 2\rho a\dot{a}). \qquad (\star)$$

On the other hand, multiplying (13.7), namely

$$-2\frac{\ddot{a}}{a} - \frac{\dot{a}^2}{a^2} + \Lambda = 8\pi\,\mathrm{p},$$

throughout by $3\dot{a}a$ yields

$$-6\dot{a}\ddot{a} - 3\frac{\dot{a}^3}{a} + 3\Lambda a\dot{a} = 3(8\pi\,\mathrm{p}a\dot{a}). \qquad (\star\star)$$

Adding $(\star)$ and $(\star\star)$, we obtain

$$-3\dot{a}a\Big(\frac{\dot{a}^2}{a^2} - \frac{\Lambda}{3}\Big) = -3\frac{\dot{a}^3}{a} + \Lambda a\dot{a} = 8\pi a(3\mathrm{p}\dot{a} + 2\rho\dot{a} + \dot{\rho}a). \qquad (*)$$

But, by first Friedmann equation

$$\frac{\dot{a}^2}{a^2} - \frac{\Lambda}{3} = \frac{8\pi\rho}{3}.$$

So, replacing the bracketed expression on the LHS of $(*)$ by $\frac{8\pi\rho}{3}$,

$$-3\dot{a}a\frac{8\pi\rho}{3} = 8\pi a(3\mathrm{p}\dot{a} + 2\rho\dot{a} + \dot{\rho}a).$$

Simplifying and rearranging, it follows that $\dot{\rho} a + 3\rho\dot{a} + 3\mathrm{p}\dot{a} = 0$. Multiplying throughout by a^2,

$$0 = \dot{\rho} a^3 + \rho 3 a^2 \dot{a} + \mathrm{p} 3 a^2 \dot{a} = \frac{d}{dt}(\rho a^3) + \mathrm{p}\frac{d}{dt}(a^3).$$

Solution to Exercise 13.7

Suppose that $V \in T_0^1 M$ is causal. We use the vector fields V_1, V_2, V_3, V_4 from Exercise 12.15 (p.287) to decompose $V = f^i V_i$, where $f^i \in C^\infty(\mathcal{U})$ in the chart $\mathcal{U}$. As V is causal, $\mathbf{g}(V,V) \leqslant 0$ (pointwise). Using the 'orthonormality' of V_1, V_2, V_3, V_4, we obtain $(f^1)^2 + (f^2)^2 + (f^3)^2 - (f^4)^2 \leqslant 0$ pointwise in $\mathcal{U}$. Now

$$\begin{aligned}\mathbf{T}(V,V) &= \frac{e^2}{8\pi r^4}((f^4)^2 - (f^1)^2 + (f^2)^2 + (f^3)^2) \\ &\geqslant \frac{e^2}{8\pi r^4}((f^2)^2 + (f^3)^2 + (f^2)^2 + (f^3)^2) \geqslant 0.\end{aligned}$$

Thus $\mathbf{T}$ satisfies the weak energy condition.

Chapter 14

Solution to Exercise 14.1

We know that $\operatorname{trace}\mathbf{Ric} = \mathbf{S}$ and $\operatorname{trace}\mathbf{g} = 4$. Taking the trace on both sides of the field equation yields $\mathbf{S} - \frac{1}{2}\mathbf{S}\,4 = 8\pi \operatorname{trace}\mathbf{T}$, and so $\mathbf{S} = -8\pi \operatorname{trace}\mathbf{T}$. Substituting this back in the field equation, $\mathbf{Ric} - \frac{1}{2}(-8\pi \operatorname{trace}\mathbf{T})\mathbf{g} = 8\pi\mathbf{T}$. Rearranging, we obtain $\mathbf{Ric} = 8\pi(\mathbf{T} - \frac{1}{2}(\operatorname{trace}\mathbf{T})\mathbf{g})$.

On the other hand, if $\mathbf{Ric} = 8\pi(\mathbf{T} - \frac{1}{2}(\operatorname{trace}\mathbf{T})\mathbf{g})$, then again taking traces, $\mathbf{S} = 8\pi((\operatorname{trace}\mathbf{T}) - \frac{1}{2}(\operatorname{trace}\mathbf{T})4) = -8\pi \operatorname{trace}\mathbf{T}$. Substituting this in the assumed equation, we get $\mathbf{Ric} = 8\pi\mathbf{T} + \frac{1}{2}(-8\pi \operatorname{trace}\mathbf{T})\mathbf{g} = 8\pi\mathbf{T} + \frac{1}{2}\mathbf{S}\mathbf{g}$. Rearranging, we obtain the field equation $\mathbf{Ric} - \frac{1}{2}\mathbf{S}\mathbf{g} = 8\pi\mathbf{T}$.

Solution to Exercise 14.2

Let $V \in T_0^1 M$ be lightlike. Then V is causal, and so by the weak energy condition, $\mathbf{T}(V,V) \geqslant 0$. Also as V is lightlike, $\mathbf{g}(V,V) = 0$. Using the field equation,

$$0 \leqslant 8\pi\mathbf{T}(V,V) = \mathbf{Ric}(V,V) - \tfrac{1}{2}\mathbf{S}\,\mathbf{g}(V,V) + \Lambda\,\mathbf{g}(V,V) = \mathbf{Ric}(V,V).$$

If $\mathbf{T}$ satisfies the strong energy condition, and $V \in T_0^1 M$ is timelike, then by Exercise 14.1, $\mathbf{Ric}(V,V) = 8\pi(\mathbf{T}(V,V) - \frac{1}{2}(\operatorname{trace}\mathbf{T})\mathbf{g}(V,V)) \geqslant 0$.

Solution to Exercise 14.3

In the given global chart $(\mathbb{R}^4, (t,x,y,z) \mapsto (t,x,y,z))$, the component matrix of $\mathbf{g}$ is (with $h := H(x,y,t-z)$):

$$[\mathbf{g}_{ij}] = \begin{bmatrix} -1 & & & \\ & 1 & & \\ & & 1 & \\ & & & 1 \end{bmatrix} + \begin{bmatrix} 2h & 0 & 0 & -2h \\ 0 & 0 & 0 & 0 \\ 0 & 0 & 0 & 0 \\ -2h & 0 & 0 & 2h \end{bmatrix} = \begin{bmatrix} -1+2h & & & -2h \\ & 1 & & \\ & & 1 & \\ -2h & & & 1+2h \end{bmatrix}.$$

To see that Y is lightlike every where, we note that using the basis $(\partial_t, \partial_x, \partial_y, \partial_z)$, the vector field Y is represented by the column vector $(1,0,0,1)$, and so computing

$$\begin{bmatrix} 1 & 1 \end{bmatrix} \begin{bmatrix} -1+2h & -2h \\ -2h & 1+2h \end{bmatrix} \begin{bmatrix} 1 \\ 1 \end{bmatrix} = \begin{bmatrix} -1 & 1 \end{bmatrix} \begin{bmatrix} 1 \\ 1 \end{bmatrix} = -1 + 1 = 0,$$

we conclude that $\mathbf{g}(Y,Y) = 0$. Similarly, we can represent the V_i as column vectors using the basis $(\partial_t, \partial_x, \partial_y, \partial_z)$, and use the component matrix of $\mathbf{g}$ given above to compute

$$[\mathbf{g}(V_i, V_j)] = \begin{bmatrix} -1 & & & \\ & 1 & & \\ & & 1 & \\ & & & 1 \end{bmatrix}.$$

For example, the $(1,1)$ entry follows by noting that

$$\begin{bmatrix} \frac{2h+3}{2\sqrt{2}} & \frac{2h+1}{2\sqrt{2}} \end{bmatrix} \begin{bmatrix} -1+2h & -2h \\ -2h & 1+2h \end{bmatrix} \begin{bmatrix} \frac{2h+3}{2\sqrt{2}} \\ \frac{2h+1}{2\sqrt{2}} \end{bmatrix} = -1.$$

From the form of $[\mathbf{g}(V_i, V_j)]$, it follows that at each point $p \in M$, the set $\{(V_i)_p : i = 1, 2, 3, 4\}$ is an orthonormal basis for T_pM with respect to the scalar product $\mathbf{g}(p)$, and the index of $\mathbf{g}(p)$ is 1. Thus $\mathbf{g}$ is Lorentzian.

The scalar curvature is $\mathbf{S} = \mathbf{g}^{ij}\mathbf{Ric}_{ij}$. To find $[\mathbf{g}^{ij}] = [\mathbf{g}_{ij}]^{-1}$, note

$$\begin{bmatrix} -1+2h & -2h \\ -2h & 1+2h \end{bmatrix}^{-1} = \begin{bmatrix} -1-2h & -2h \\ -2h & 1-2h \end{bmatrix}.$$

Thus

$$[\mathbf{g}^{ij}] = \begin{bmatrix} -1-2h & & & -2h \\ & 1 & & \\ & & 1 & \\ -2h & & & 1-2h \end{bmatrix}.$$

As the only nonzero components of $\mathbf{Ric}$ are the (i,j)-components with $i, j \in \{1, 4\}$, we obtain, with $\Delta h := \partial_x(\partial_x h) + \partial_y(\partial_y h)$, that

$$\begin{aligned} \mathbf{S} &= \mathbf{g}^{11}\mathbf{Ric}_{11} + \mathbf{g}^{14}\mathbf{Ric}_{14} + \mathbf{g}^{41}\mathbf{Ric}_{41} + \mathbf{g}^{44}\mathbf{Ric}_{44} \\ &= (-1-2h)(-\Delta h) + (-2h)(\Delta h) + (-2h)(\Delta h) + (1-2h)(-\Delta h) \\ &= (\Delta h)(1 + 2h - 2h - 2h - 1 + 2h) = (\Delta h)(0) = 0. \end{aligned}$$

For a harmonic (in x, y) h, $\Delta h = 0$. So $\mathbf{Ric} = -(\Delta h)\,(dt - dz) \otimes (dt - dz) = 0$. So the vacuum ($\mathbf{T} = 0$) field equation is satisfied with the cosmological constant $\Lambda = 0$: $\mathbf{Ric} - \frac{\mathbf{S}}{2}\mathbf{g} + \Lambda\mathbf{g} = 0 - \frac{0}{2}\mathbf{g} + 0\,\mathbf{g} = 0 = 8\pi 0 = 8\pi\mathbf{T}$.

Conversely, if the vacuum field equation is satisfied, then we have that $\mathbf{Ric} = \mathbf{Ric} - \frac{0}{2}\mathbf{g} + 0\,\mathbf{g} = \mathbf{Ric} - \frac{\mathbf{S}}{2}\mathbf{g} + \Lambda\mathbf{g} = 8\pi\mathbf{T} = 8\pi 0 = 0$. So $\mathbf{Ric} = 0$. In particular, $\mathbf{Ric}(\partial_t, \partial_t) = 0$, i.e., $-\Delta h = 0$. Thus h is harmonic.

We have $\Delta(x^2 - y^2) = 2 + (-2) = 0$, and $\Delta(xy) = 0 + 0 = 0$. Thus $\Delta h = \frac{1}{2}f(t-z)\Delta(x^2 - y^2) + g(t-z)\Delta(xy) = \frac{1}{2}f(t-z)0 + g(t-z)0 = 0$.

Solution to Exercise 14.4

We can rewrite the metric as

$$\mathbf{g} = -(dt + e^x dy) \otimes (dt + e^x dy) + dx \otimes dx + \frac{e^{2x}}{2} dy \otimes dy + dz \otimes dz.$$

Define $V_1 = \partial_t$, $V_2 = \partial_x$, $V_3 = -\sqrt{2}\,\partial_t + \sqrt{2}\,e^{-x}\,\partial_y$, $V_4 = \partial_z$. (Note that V_3 satisfies $(dt + e^x dy)V_3 = 0$ and $\mathbf{g}(V_3, V_3) = 1$.) Then

$$[\mathbf{g}(V_i, V_j)] = \begin{bmatrix} -1 & & & \\ & 1 & & \\ & & 1 & \\ & & & 1 \end{bmatrix}.$$

Hence $\mathbf{g}$ is Lorentzian. As $\mathbf{g}(\partial_t, \partial_t) = -1$, $V := \partial_t$ is timelike everywhere.

$\mathbf{T}$ is the energy-momentum tensor field of dust: We set $m = \frac{1}{8\pi} > 0$, $\nu \equiv 1$, $V = \partial_t = V_1$. We have $\mathbf{g}(p)(V_p, V_p) = \mathbf{g}(p)(\partial_t, \partial_t) = -1$. It follows also that V_p is future-pointing. Furthermore, we have $\nabla_V V = \nabla_{\partial_t} \partial_t = \Gamma^i_{tt} \partial_{x^i}$. But since $\mathbf{g}_{tj}$ does not depend on t, and since $\mathbf{g}_{tt} = -1$, it follows that

$$\Gamma^i_{tt} = \frac{\mathbf{g}^{ij}}{2}(\partial_t \mathbf{g}_{tj} + \partial_t \mathbf{g}_{jt} - \partial_j \mathbf{g}_{tt}) = \frac{\mathbf{g}^{ij}}{2}(0 + 0 - 0) = 0.$$

So $\nabla_V V = 0$. We have

$$\operatorname{div}(\nu V) = \operatorname{div} V = dx^i(\nabla_{\partial_{x^i}} \partial_t) = dx^i(\Gamma^j_{ti} \partial_{x^j}) = \Gamma^j_{ti} \delta^i_j = \Gamma^i_{ti}.$$

We have already seen $\Gamma^t_{tt} = 0$. One can check that also Γ^x_{tx}, Γ^y_{ty}, Γ^z_{tz} are 0: For example, noting that $\mathbf{g}_{tt} = -1$ and $\partial_t \mathbf{g}_{yy} = \partial_t(-e^{2x}/2) = 0$, we have

$$\begin{aligned}
\Gamma^y_{ty} &= \frac{\mathbf{g}^{yj}}{2}(\partial_t \mathbf{g}_{jy} + \partial_y \mathbf{g}_{tj} - \partial_j \mathbf{g}_{ty}) \\
&= \frac{\mathbf{g}^{yt}}{2}(\partial_t \mathbf{g}_{ty} + \partial_y \mathbf{g}_{tt} - \partial_t \mathbf{g}_{ty}) + \frac{\mathbf{g}^{yy}}{2}(\partial_t \mathbf{g}_{yy} + \partial_y \mathbf{g}_{ty} - \partial_y \mathbf{g}_{ty}) = 0.
\end{aligned}$$

Thus $\mathbf{T}$ is the energy-momentum tensor field of dust $(m, \nu, V) = (\frac{1}{8\pi}, 1, \partial_t)$.

As $\mathbf{Ric} = (\partial_t)^\flat \otimes (\partial_t)^\flat$, its components in the chart $\mathbb{R}^4$ with the chart coordinates $(t, x, y, z) =: (x^1, x^2, x^3, x^4)$, are given by (using $(\partial_t)^\flat = \mathbf{g}_{1k} dx^k$):

$$\mathbf{Ric}_{ij} = (\mathbf{g}_{1k} dx^k \otimes \mathbf{g}_{1\ell} dx^\ell)(\partial_{x^i}, \partial_{x^j}) = \mathbf{g}_{1k} \mathbf{g}_{1\ell} \delta^k_i \delta^\ell_j = \mathbf{g}_{1i} \mathbf{g}_{1j}.$$

The scalar curvature is $\mathbf{S} = \mathbf{g}^{ij} \mathbf{Ric}_{ij} = \mathbf{g}^{ij} \mathbf{g}_{1i} \mathbf{g}_{1j} = \delta^j_1 \mathbf{g}_{1j} = \mathbf{g}_{11} = -1$. So

$$\begin{aligned}
\mathbf{Ric} - \frac{\mathbf{S}}{2}\mathbf{g} + \Lambda \mathbf{g} &= (\partial_t)^\flat \otimes (\partial_t)^\flat - \frac{(-1)}{2}\mathbf{g} + \left(-\frac{1}{2}\right)\mathbf{g} \\
&= (\partial_t)^\flat \otimes (\partial_t)^\flat \\
&= 8\pi\left(\frac{1}{8\pi}\right)(\partial_t)^\flat \otimes (\partial_t)^\flat = 8\pi \mathbf{T},
\end{aligned}$$

that is, the field equation is satisfied with the given cosmological constant and the matter energy-momentum tensor field $\mathbf{T}$.

To see that the given curve is closed, we compute

$$\begin{aligned}
\lim_{s \to \pi} t(s) &= \lim_{s \to \pi} \left(2\sqrt{2} \tan^{-1}\left(e^{-d} \tan \tfrac{s}{2}\right) - \sqrt{2}s\right) = 2\sqrt{2} \tan^{-1}(+\infty) - \sqrt{2}\pi \\
&= 2\sqrt{2}\tfrac{\pi}{2} - \sqrt{2}\pi = 0,
\end{aligned}$$

$$\lim_{s \to -\pi} t(s) = 2\sqrt{2} \tan^{-1}(-\infty) - \sqrt{2}(-\pi) = 2\sqrt{2}\left(-\tfrac{\pi}{2}\right) + \sqrt{2}\pi = 0,$$

and also

$$\begin{aligned}
x(\pm\pi) &= \log((\cosh d) + (-1)\sinh d) = \log e^{-d} = -d \\
y(\pm\pi) &= (\sin(\pm\pi))(\cdots) = 0(\cdots) = 0 \\
z(\pm\pi) &= 0.
\end{aligned}$$

A plot of the curve is shown below.

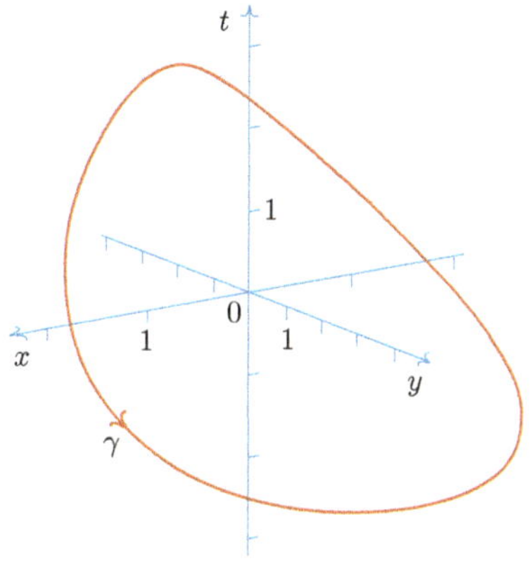

Solution to Exercise 14.5

(1) By Exercise 14.1 (p.306), the vacuum field equation with $\Lambda = 0$ is equivalent to $\mathbf{Ric} = \mathbf{0}$. Looking at the zz and tz components of the Ricci curvature, we obtain (c1) and (c2). We have

$$\begin{aligned} 0 &= \frac{0}{A^2} + \frac{0}{B^2} = \frac{\mathbf{Ric}_{xx}}{A^2} + \frac{\mathbf{Ric}_{yy}}{B^2} \\ &= \frac{A''}{A} - \frac{\ddot{A}}{A} + \frac{1}{AB}(A'B' - \dot{A}\dot{B}) + \frac{B''}{B} - \frac{\ddot{B}}{B} + \frac{1}{AB}(A'B' - \dot{A}\dot{B}) \\ &= \mathbf{Ric}_{zz} - \mathbf{Ric}_{tt} + \frac{2}{AB}(A'B' - \dot{A}\dot{B}) = 0 - 0 + \frac{2}{AB}(A'B' - \dot{A}\dot{B}), \end{aligned}$$

that is, (c3) holds. Knowing (c3), and using $\mathbf{Ric}_{xx} = 0$, we see that (e1) holds. Finally, (c3) and $\mathbf{Ric}_{yy} = 0$ yield (e2).

Now suppose (e1), (e2), (c1), (c2), (c3) hold. We want to show $\mathbf{Ric} = \mathbf{0}$. Equations (c1), (c2) give $\mathbf{Ric}_{zz} = 0$, $\mathbf{Ric}_{tz} = \mathbf{Ric}_{zt} = 0$, respectively. Also, considering A(e1) + $\frac{A}{B}$(c3) shows that $\mathbf{Ric}_{xx} = 0$, and similarly B(e2) + $\frac{B}{A}$(c3) gives $\mathbf{Ric}_{yy} = 0$. Finally, $\frac{1}{A}$(e1) + $\frac{1}{B}$(e2) + (c1) gives $\mathbf{Ric}_{tt} = 0$.

(2) Equations (e1), (e2) hold because

$$\partial_t^2 A - \partial_z^2 A = f''(t-z) - (-1)^2 f''(t-z) = 0,$$
$$\partial_t^2 B - \partial_z^2 B = g''(t-z) - (-1)^2 g''(t-z) = 0.$$

If (c1) is satisfied at t_0, then

$$0 = \frac{\partial_z^2 A}{A} + \frac{\partial_z^2 B}{B} = \frac{(-1)^2 f''(t_0 - z)}{f(t_0 - z)} + \frac{(-1)^2 g''(t_0 - z)}{g(t_0 - z)},$$

for all $z \in \mathbb{R}$, and hence

$$\frac{f''}{f} + \frac{g''}{g} = 0.$$

Then the constraint equations (c1), (c2) hold for all t, since

$$\frac{\partial_z^2 A}{A}+\frac{\partial_z^2 B}{B}=\frac{(-1)^2 f''(t-z)}{f(t-z)}+\frac{(-1)^2 g''(t-z)}{g(t-z)}=0,$$
$$\frac{\partial_z\partial_t A}{A}+\frac{\partial_z\partial_t B}{B}=\frac{(-1)f''(t-z)}{f(t-z)}+\frac{(-1)g''(t-z)}{g(t-z)}=0.$$

Finally,

$$\begin{aligned}&(\partial_t A)(\partial_t B)-(\partial_z A)(\partial_z B)\\&=f'(t-z)g'(t-z)-(-1)f'(t-z)(-1)g'(t-z)=0,\end{aligned}$$

establishing (c3).

(3) Let the curve $\gamma_{\mathbf{p}}: I_{\mathbf{p}} \to U$ be given by $\gamma_{\mathbf{p}}(t)=(t,x_0,y_0,z_0)$ for all $t\in I_{\mathbf{p}}$. Then $\gamma_{\mathbf{p}}^t(t)=t$, $\gamma_{\mathbf{p}}^x(t)=x_0$, $\gamma_{\mathbf{p}}^y(t)=y_0$, and $\gamma_{\mathbf{p}}^z(t)=z_0$. It follows that $\dot\gamma_{\mathbf{p}}^t(t)=1$, and $\dot\gamma_{\mathbf{p}}^x(t)=0=\dot\gamma_{\mathbf{p}}^y(t)=\dot\gamma_{\mathbf{p}}^z(t)$. Thus $v_{\gamma_{\mathbf{p}},\gamma_{\mathbf{p}}(t)}=\partial_{t,\gamma_{\mathbf{p}}(t)}$, and

$$\mathbf{g}(\gamma_{\mathbf{p}}(t))(v_{\gamma_{\mathbf{p}},\gamma_{\mathbf{p}}(t)},v_{\gamma_{\mathbf{p}},\gamma_{\mathbf{p}}(t)})=\mathbf{g}(\gamma_{\mathbf{p}}(t))(\partial_{t,\gamma_{\mathbf{p}}(t)},\partial_{t,\gamma_{\mathbf{p}}(t)})=-1.$$

So the curve γ is parametrised by arclength/proper time. Now we check that the geodesic equation $\nabla_{V_{\gamma_{\mathbf{p}}}}V_{\gamma_{\mathbf{p}}}=0$ is satisfied:

$$\begin{aligned}
\ddot\gamma_{\mathbf{p}}^t(t)+\Gamma_{ij}^t(\gamma_{\mathbf{p}}(t))\dot\gamma_{\mathbf{p}}^i(t)\dot\gamma_{\mathbf{p}}^j(t)&=0+\Gamma_{xx}^t(\gamma_{\mathbf{p}}(t))\dot\gamma_{\mathbf{p}}^x(t)\dot\gamma_{\mathbf{p}}^x(t)\\
&\quad+\Gamma_{yy}^t(\gamma_{\mathbf{p}}(t))\dot\gamma_{\mathbf{p}}^y(t)\dot\gamma_{\mathbf{p}}^y(t)\\
&=0+\Gamma_{xx}^t(\gamma_{\mathbf{p}}(t))0\cdot 0+\Gamma_{yy}^t(\gamma_{\mathbf{p}}(t))0\cdot 0=0,\\
\ddot\gamma_{\mathbf{p}}^x(t)+\Gamma_{ij}^x(\gamma_{\mathbf{p}}(t))\dot\gamma_{\mathbf{p}}^i(t)\dot\gamma_{\mathbf{p}}^j(t)&=0+2\Gamma_{xt}^x(\gamma_{\mathbf{p}}(t))\dot\gamma_{\mathbf{p}}^x(t)\dot\gamma_{\mathbf{p}}^t(t)\\
&\quad+2\Gamma_{xz}^x(\gamma_{\mathbf{p}}(t))\dot\gamma_{\mathbf{p}}^x(t)\dot\gamma_{\mathbf{p}}^z(t)\\
&=2\Gamma_{xt}^x(\gamma_{\mathbf{p}}(t))0\cdot 1+2\Gamma_{xz}^x(\gamma_{\mathbf{p}}(t))0\cdot 0=0,\\
\ddot\gamma_{\mathbf{p}}^y(t)+\Gamma_{ij}^y(\gamma_{\mathbf{p}}(t))\dot\gamma_{\mathbf{p}}^i(t)\dot\gamma_{\mathbf{p}}^j(t)&=0+2\Gamma_{yt}^y(\gamma_{\mathbf{p}}(t))\dot\gamma_{\mathbf{p}}^y(t)\dot\gamma_{\mathbf{p}}^t(t)\\
&\quad+2\Gamma_{yz}^y(\gamma_{\mathbf{p}}(t))\dot\gamma_{\mathbf{p}}^y(t)\dot\gamma_{\mathbf{p}}^z(t)\\
&=2\Gamma_{yt}^y(\gamma_{\mathbf{p}}(t))0\cdot 1+2\Gamma_{yz}^y(\gamma_{\mathbf{p}}(t))0\cdot 0=0,\\
\ddot\gamma_{\mathbf{p}}^z(t)+\Gamma_{ij}^z(\gamma_{\mathbf{p}}(t))\dot\gamma_{\mathbf{p}}^i(t)\dot\gamma_{\mathbf{p}}^j(t)&=0+\Gamma_{xx}^z(\gamma_{\mathbf{p}}(t))\dot\gamma_{\mathbf{p}}^x(t)\dot\gamma_{\mathbf{p}}^x(t)\\
&\quad+\Gamma_{yy}^z(\gamma_{\mathbf{p}}(t))\dot\gamma_{\mathbf{p}}^y(t)\dot\gamma_{\mathbf{p}}^y(t)\\
&=0+\Gamma_{xx}^z(\gamma_{\mathbf{p}}(t))0\cdot 0+\Gamma_{yy}^z(\gamma_{\mathbf{p}}(t))0\cdot 0=0.
\end{aligned}$$

(4) We have

$$\begin{aligned}
\theta^2+2\partial_t\theta&=\Big(\frac{\partial_t A}{A}+\frac{\partial_t B}{B}\Big)^2+2\Big(\frac{\partial_t^2 A}{A}-\frac{(\partial_t A)^2}{A^2}+\frac{\partial_t^2 B}{B}-\frac{(\partial_t B)^2}{B^2}\Big)\\
&=2\Big(\frac{\partial_t^2 A}{A}+\frac{\partial_t^2 B}{B}\Big)-\Big(\frac{\partial_t A}{A}-\frac{\partial_t B}{B}\Big)^2\\
&=0-\Big(\frac{\partial_t A}{A}-\frac{\partial_t B}{B}\Big)^2\leqslant 0.
\end{aligned}$$

Integrating $\partial_t\theta\leqslant-\frac{1}{2}\theta^2$ from t_0 to t, we get $-\dfrac{1}{\theta(t,z)}+\dfrac{1}{\theta(t_0,z)}\leqslant-\frac{1}{2}(t-t_0)$. For $t\in[t_0,t_0-\dfrac{2}{\theta(t_0,z)})=J$, the above gives $\dfrac{1}{\theta(t,z)}\geqslant\dfrac{1}{\theta(t_0,z)}+\dfrac{1}{2}(t-t_0)$.

If θ does not change sign in J, then this inequality shows that $\frac{1}{\theta(t,z)} \nearrow 0$ (i.e., $\theta(t,z) \to -\infty$) as $t \to t_0 - \frac{2}{\theta(t_0,z)} < \infty$, and otherwise θ goes to $-\infty$ even earlier than $t_0 - \frac{2}{\theta(t_0,z)}$. So the spacetime is not geodesically complete (as $\gamma_{\mathbf{p}}$ for $\mathbf{p} \in S$ are not defined for all $t \geqslant t_0$).

Solution to Exercise 14.6

Let $G^{(s)} := [\mathbf{g}_{ij}^{(s)}]$, $Y := [\eta_{ij}]$, $H := [\mathbf{h}_{ij}]$. Then $G^{(s)} \approx Y + sH$. As $Y = Y^{-1}$, we have for small $|s|$,

$$\begin{aligned}(G^{(s)})^{-1} &= (Y + sH)^{-1} = (Y(I + sYH))^{-1} = (I + sYH)^{-1}Y \\ &= (I - sYH + s^2YHYH - + \cdots)Y.\end{aligned}$$

Thus

$$\begin{aligned}\lim_{s\to 0} \frac{(G^{(s)})^{-1} - (G^{(0)})^{-1}}{s} &= \lim_{s\to 0} \frac{(I - sYH + s^2YHYH - + \cdots)Y - Y}{s} \\ &= \lim_{s\to 0} \frac{-sYHY + s^2YHYHY - + \cdots}{s} = -YHY.\end{aligned}$$

Recall that $(\mathbf{g}^{(s)})^{ij}$ is the entry in the ith row and j^{th} column of $G^{(s)}$. Hence

$$\frac{\partial (\mathbf{g}^{(s)})^{ij}}{\partial s}\Big|_{s=0} = [-YHY]_{ij} = -\eta_{ik}\,\mathbf{h}_{k\ell}\,\eta_{\ell j}.$$

Also,

$$\begin{aligned}\frac{\partial \Gamma^{(s)}{}_{ij}^{k}}{\partial s}\Big|_{s=0} &= \frac{\partial}{\partial s}\Big(\frac{(\mathbf{g}^{(s)})^{k\ell}}{2}(\partial_{x^i}(\mathbf{g}^{(s)})_{j\ell} + \partial_{x^j}(\mathbf{g}^{(s)})_{i\ell} - \partial_{x^\ell}(\mathbf{g}^{(s)})_{ij})\Big)\Big|_{s=0} \\ &= \frac{1}{2}\frac{\partial (\mathbf{g}^{(s)})^{k\ell}}{\partial s}\Big|_{s=0}(\partial_{x^i}\eta_{j\ell} + \partial_{x^j}\eta_{i\ell} - \partial_{x^\ell}\eta_{ij}) \\ &\quad + \frac{\eta^{k\ell}}{2}(\partial_{x^i}\mathbf{h}_{j\ell} + \partial_{x^j}\mathbf{h}_{i\ell} - \partial_{x^\ell}\mathbf{h}_{ij}) \\ &= 0 + \frac{\eta^{k\ell}}{2}(\partial_{x^i}\mathbf{h}_{j\ell} + \partial_{x^j}\mathbf{h}_{i\ell} - \partial_{x^\ell}\mathbf{h}_{ij}).\end{aligned}$$

Solution to Exercise 14.7

By Proposition 6.6 (p.130), for all $p \in M$, and all $1 \leqslant i, j \leqslant 4$, we have

$$\mathbf{h}_{ij}^V(p) = \lim_{s\to 0} \frac{(\mathbf{g}^{(s)})_{ij}(p) - \mathbf{g}_{ij}(p)}{s} = \lim_{s\to 0} \frac{(\psi_s^*\mathbf{g})_{ij}(p) - \mathbf{g}_{ij}(p)}{s} = (\mathcal{L}_V\mathbf{g})_{ij}(p).$$

Consequently, $\mathbf{h}^V = \mathcal{L}_V\mathbf{g}$.

Solution to Exercise 14.8

In the global chart $(\mathbb{R}^4, \mathrm{id})$, the matrix of chart induced components of $\mathbf{g}$ is

$$[\mathbf{g}_{ij}] = \begin{bmatrix} -1 & & & \\ & 1+\alpha h\cos(kz-\omega t) & \beta h\cos(kz-\omega t) & \\ & \beta h\cos(kz-\omega t) & 1-\alpha h\cos(kz-\omega t) & \\ & & & 1 \end{bmatrix}.$$

We have $\gamma_{\mathbf{p}}^t(t)=t$, $\gamma_{\mathbf{p}}^x(t)=x_0$, $\gamma_{\mathbf{p}}^y(t)=y_0$ and $\gamma_{\mathbf{p}}^z(t)=0$. Hence $\dot{\gamma}_{\mathbf{p}}^t(t)=1$, $\dot{\gamma}_{\mathbf{p}}^x(t)=0$, $\dot{\gamma}_{\mathbf{p}}^y(t)=0$ and $\dot{\gamma}_{\mathbf{p}}^z(t)=0$, and $\ddot{\gamma}_{\mathbf{p}}^i\equiv 0$, for all $i\in\{t,x,y,z\}$. To show that $\gamma_{\mathbf{p}}$ is a geodesic, we need to show that

$$\begin{aligned}0&=\ddot{\gamma}_{\mathbf{p}}^i(t)+\Gamma_{k\ell}^i(\gamma_{\mathbf{p}}(t))\dot{\gamma}_{\mathbf{p}}^k(t)\dot{\gamma}_{\mathbf{p}}^\ell(t)\\&=0+\Gamma_{tt}^i(\gamma_{\mathbf{p}}(t))\dot{\gamma}_{\mathbf{p}}^t(t)\dot{\gamma}_{\mathbf{p}}^t(t)=\Gamma_{tt}^i(\gamma_{\mathbf{p}}(t))1\cdot 1=\Gamma_{tt}^i(\gamma_{\mathbf{p}}(t)).\end{aligned}$$

We have $2\Gamma_{tt}^i=\mathbf{g}^{ij}(\partial_t\mathbf{g}_{tj}+\partial_t\mathbf{g}_{jt}-\partial_j\mathbf{g}_{tt})=2\mathbf{g}^{ij}\partial_t\mathbf{g}_{tj}$. From the above expressions for $\mathbf{g}_{ij}$, clearly $\partial_t\mathbf{g}_{tj}=0$ for all $j\in\{t,x,y,z\}$. Hence $\Gamma_{tt}^i=0$, and so $\gamma_{\mathbf{p}}$ is a geodesic. Finally, we have

$$\begin{aligned}&\Big(\frac{\widetilde{x}(t)}{R\sqrt{1+h\cos(\omega t)}}\Big)^2+\Big(\frac{\widetilde{y}(t)}{R\sqrt{1-h\cos(\omega t)}}\Big)^2\\&=\frac{d(\phi)^2}{R^2}\Big(\frac{(\cos\phi)^2}{1+h\cos(\omega t)}+\frac{(\sin\phi)^2}{1-h\cos(\omega t)}\Big)\\&=\frac{R^2(1+h\cos(\omega t)\cos(2\phi))}{R^2}\,\frac{(1-h\cos(\omega t)\cos(2\phi))}{(1-h^2(\cos(\omega t))^2)}\\&=\frac{1-h^2(\cos(\omega t))^2(\cos(2\phi))^2}{1-h^2(\cos(\omega t))^2}\approx\frac{1-0}{1-0}=1.\end{aligned}$$

Solution to Exercise 14.9

$$\frac{dE}{dt}=\frac{8G_{\mathrm{N}}m^2\ell^4\omega^6}{5c^5}=\frac{8\times 6.67\times 10^{-11}\times 1^2\times 1^4\times(10^3)^6}{5\times(3\times 10^8)^5}=4.39\times 10^{-35}\,\text{Watts}.$$

Solution to Exercise 14.10

(1) Let $\epsilon^k:=e_k^\flat$. We have

$$\begin{aligned}\int_{\mathbf{y}\in S^2}\mathbf{g}(p)(F_{X_p}v_{\mathbf{y}},v_{\mathbf{y}})\,dA&=\int_{\mathbf{y}\in S^2}\mathbf{g}(p)(F_{X_p}(y^ie_i),y^je_j)\,dA\\&=\epsilon^k(F_{X_p}(e_i))\mathbf{g}(p)(e_k,e_j)\int_{\mathbf{y}\in S^2}y^iy^j\,dA\\&=\epsilon^k(F_{X_p}(e_i))\delta_{kj}\frac{1}{3}\delta^{ij}=\frac{4\pi}{3}\epsilon^i(F_{X_p}(e_i))\\&=\frac{4\pi}{3}\operatorname{trace}F_{X_p}=f_X(p)\int_{\mathbf{y}\in S^2}1\,dA.\end{aligned}$$

(2) Using the results from Exercises 14.1 (p.306) and 13.5 (p.299), we have

$$\begin{aligned}\mathbf{Ric}&=8\pi\big(\mathbf{T}-\tfrac{1}{2}(\operatorname{trace}\mathbf{T})\mathbf{g}\big)=8\pi\big(\mathbf{T}-\tfrac{1}{2}(3\mathrm{p}-\rho)\mathbf{g}\big)\\&=8\pi\big((\rho+\mathrm{p})V^\flat\otimes V^\flat+\tfrac{1}{2}(\rho-\mathrm{p})\mathbf{g}\big).\end{aligned}$$

Using the reversed Cauchy-Schwarz inequality we obtain (pointwise) that $(\mathbf{g}(X,V))^2\geqslant\mathbf{g}(X,X)\mathbf{g}(V,V)=(-1)(-1)=1$. Hence

$$\begin{aligned}f_X&=-\frac{1}{3}\mathbf{Ric}(X,X)=-\frac{8\pi}{3}\big((\mathrm{p}+\rho)(\mathbf{g}(X,V))^2+\tfrac{1}{2}(\rho-\mathrm{p})\mathbf{g}(X,X)\big)\\&\leqslant-\frac{8\pi}{3}\big((\mathrm{p}+\rho)(1)-\tfrac{1}{2}(\rho-\mathrm{p})\big)=-\frac{8\pi}{6}(\rho+3\mathrm{p})<0.\end{aligned}$$

Solution to Exercise 14.11

We have $v_{\gamma,\gamma(\tau)} = t'(\tau)\partial_{t,\gamma(\tau)} + r'(\tau)\partial_{r,\gamma(\tau)}$. Since γ is parametrised by arclength, we have $\mathbf{g}(\gamma(\tau))(v_{\gamma,\gamma(\tau)}, v_{\gamma,\gamma(\tau)}) = -1$ for all $\tau \in I$. Using the expression for the Schwarzschild metric, we obtain

$$-\left(1-\frac{2m}{r}\right)(t')^2 + \left(1-\frac{2m}{r}\right)^{-1}(r')^2 + 0 + 0 = -1. \qquad (\star)$$

The other two equations are obtained by looking at the t- and r-components of the geodesic equation $\ddot{\gamma}^k + \Gamma^k_{ij}\dot{\gamma}^i\dot{\gamma}^j = 0$. Recall the connection coefficient formulae from Example 6.8 (p.121). The t-component of the geodesic equation, $0 = t'' + \Gamma^t_{tt}(t')^2 + 2\Gamma^t_{tr}t'r' + \Gamma^t_{rr}(r')^2$, gives

$$0 = t'' + 0 + \frac{2m}{r^2}\left(1-\frac{2m}{r}\right)^{-1} t'r' + 0.$$

Similarly, the r-component, $0 = r'' + \Gamma^r_{tt}(t')^2 + 2\Gamma^r_{rt}r't' + \Gamma^r_{rr}(r')^2$, gives

$$\begin{aligned}0 &= r'' + \frac{m}{r^2}\left(1-\frac{2m}{r}\right)(t')^2 + 0 - \frac{m}{r^2}\left(1-\frac{2m}{r}\right)^{-1}(r')^2\\ &= r'' - \frac{m}{r^2}\left(-\left(1-\frac{2m}{r}\right)(t')^2 + \left(1-\frac{2m}{r}\right)^{-1}(r')^2\right)\\ &= r'' - \frac{m}{r^2}(-1) \qquad (\text{using } (\star)).\end{aligned}$$

Relative speed of $w = t'\partial_{t,p} + r'\partial_{r,p}$ perceived by v is $u = -\dfrac{w}{\mathbf{g}(p)(w,v)} - v$. So

$$\begin{aligned}u &= -\frac{t'\partial_{t,p} + r'\partial_{r,p}}{\mathbf{g}(p)\left(t'\partial_{t,p} + r'\partial_{r,p}, \sqrt{(1-\frac{2m}{r})^{-1}}\,\partial_{t,p}\right)} - \sqrt{\left(1-\frac{2m}{r}\right)^{-1}}\,\partial_{t,p}\\ &= \frac{r'}{t'}\sqrt{\left(1-\frac{2m}{r}\right)^{-1}}\,\partial_{r,p}.\end{aligned}$$

Thus

$$|u| = \sqrt{\mathbf{g}(p)(u,u)} = \left|\frac{r'}{t'}\right|\sqrt{\left(1-\frac{2m}{r}\right)^{-1}}\sqrt{\mathbf{g}(p)(\partial_{r,p},\partial_{r,p})} = \left|\frac{r'}{t'}\right|\left(1-\frac{2m}{r}\right)^{-1}.$$

For $|u| \ll 1$ and $r \gg 2m$, we obtain $|r'| \ll |t'|$. Equation $(\star)$ now shows that $t' \approx 1$. From the equation for t'' above, we see that if we neglect the small terms, we get $t'' \approx 0$. We note that

$$\frac{dh}{dt} = \frac{1}{t'(h(\cdot))}, \quad \text{and} \quad \frac{d^2h}{dt^2} = -\frac{t''(h(\cdot))\dfrac{dh}{dt}}{(t'(h(\cdot)))^2} = -\frac{t''(h(\cdot))}{(t'(h(\cdot)))^3}.$$

Thus

$$\begin{aligned}\frac{d(r\circ h)}{dt} &= \frac{dr}{dt}\frac{dh}{dt} = r'(h(\cdot))\frac{1}{t'(h(\cdot))}, \quad \text{and}\\ \frac{d^2(r\circ h)}{dt^2} &= r''(h(\cdot))\frac{1}{t'(h(\cdot))}\frac{1}{t'(h(\cdot))} - r'(h(\cdot))\frac{t''(h(\cdot))}{(t'(h(\cdot)))^3}.\end{aligned}$$

Using $t' \approx 1$, $|r'| \ll t'$, and $r \gg 2m$, we get

$$\begin{aligned}
\frac{d^2}{dt^2}(r\circ h) &= r''\frac{1}{(t')^2} - r'\frac{1}{(t')^3}t'' = -\frac{m}{r^2}\frac{1}{(t')^2} - r'\frac{1}{(t')^3}\frac{(-2m)}{r^2}\Big(1-\frac{2m}{r}\Big)^{-1}t'r' \\
&= -\frac{m}{r^2}\Big(\frac{1}{(t')^2} - 2\Big(\frac{r'}{t'}\Big)^2\Big(1-\frac{2m}{r}\Big)^{-1}\Big) \\
&\approx -\frac{m}{r^2}\Big(\frac{1}{1} + 2(0)^2\frac{1}{1-0}\Big) = -\frac{m}{r^2}.
\end{aligned}$$

Solution to Exercise 14.12

With $E := \big(1 - \frac{2m}{r}\big)t'$, we have

$$E' = \frac{2m}{r^2}r't' + \Big(1 - \frac{2m}{r}\Big)t'' = \frac{2m}{r^2}\Big(-\frac{r^2}{2m}\Big(1 - \frac{2m}{r}\Big)t''\Big) + \Big(1 - \frac{2m}{r}\Big)t'' = 0.$$

So E is constant along the geodesic. With $f(r) := 1 - \frac{2m}{r}$, we have

$$-1 = -f(r)(t')^2 + \frac{(r')^2}{f(r)} = -f(r)\frac{E^2}{(f(r))^2} + \frac{(r')^2}{f(r)},$$

and so $(r')^2 = E^2 - f(r) = E^2 - 1 + \frac{2m}{r} =: g(r)$.

Suppose that γ is a radial geodesic parametrised by proper time such that $r'(\tau) > 0$ for all $\tau > 0$. We first show that $r(\tau) \to \infty$ as $\tau \to \infty$. If not, then as $r(\cdot)$ is increasing, it will have a finite limit, say r_*, as $t \to \infty$. But then $r'(\tau) = \sqrt{g(r(\tau))} \to \sqrt{g(r_*)}$ as $\tau \to \infty$. By the mean value theorem $r(n+1) - r(n) = r'(c_n)$ for some $c_n \in (n, n+1)$, and passing to the limit as $n \to \infty$ yields $0 = r_* - r_* = \sqrt{g(r_*)}$, and so $r'(\tau) \to 0$ as $\tau \to \infty$. Using

$$r'' = -\frac{m}{r^2}\Big(f(r)(t')^2 - \frac{(r')^2}{f(r)}\Big) = -\frac{m}{r^2}\Big(f(r)\frac{E^2}{(f(r))^2} - \frac{(r')^2}{f(r)}\Big),$$

we conclude that

$$r''(\tau) \to -\frac{mE^2}{r_*^2 f(r_*)} =: L < 0 \quad \text{as } \tau \to \infty.$$

By the mean value theorem $r'(n+1) - r'(n) = r''(d_n)$ for some $d_n \in (n, n+1)$, and passing to the limit as $n \to \infty$ yields $0 = L$, a contradiction.

Now passing to the limit as $\tau \to \infty$ in $(r')^2 = E^2 - 1 + \frac{2m}{r}$ yields

$$\lim_{\tau\to\infty} (r'(\tau))^2 = E^2 - 1 + 0.$$

But since $(r'(\tau))^2 \geqslant 0$ for all $\tau > 0$, we get $E^2 \geqslant 1$, i.e., $E \geqslant 1$. As E is constant, $f(r(0))t'(0) = E \geqslant 1$. Following the computation in Exercise 14.11, we have

$$\begin{aligned}
|u| &= \Big|\frac{r'(0)}{t'(0)}\Big|\Big(1 - \frac{2m}{r(0)}\Big)^{-1} = \frac{r'(0)}{t'(0)f(r(0))} = \frac{\sqrt{E^2 - f(r(0))}}{E} \\
&= \sqrt{1 - \frac{f(r(0))}{E^2}} \geqslant \sqrt{1 - \frac{f(r(0))}{1^2}} = \sqrt{\frac{2m}{r(0)}}.
\end{aligned}$$

In fact, if $E=1$, then we get $(r')^2=\frac{2m}{r}$, so that

$$r(\tau) = \Big((r(0))^{3/2} + 3\sqrt{\frac{m}{2}}\tau\Big)^{2/3}, \text{ and } t(\tau) = \int_0^\tau \frac{1}{f(r(s))}ds,$$

and so we can explicitly construct an 'escaping' geodesic. Then it can be checked that the equations listed in the first display of Exercise 14.11 (p.321) are satisfied. Hence the escape velocity is equal to

$$|u|_{\text{escape}} = \sqrt{\frac{2m}{r(0)}}. \qquad (*)$$

Classically (see Exercise 15.2, p.338), we can find the escape velocity by equating the kinetic energy to the classical gravitational potential, so that if m_0 is the mass of the particle, then in units where $G_{\mathrm{N}}=1$, we have

$$\frac{1}{2}m_0|u|^2=\frac{m_0 m}{r(0)},$$

giving the same expression given in $(*)$ above.

Solution to Exercise 14.13

Reinstating the constants $c = 3\times 10^8\,\mathrm{m/s}$ ('speed of light'), and the Newton gravitational constant $G_{\mathrm{N}}=6.6\times 10^{-11}\frac{\mathrm{m}^3}{\mathrm{kg}\cdot\mathrm{s}^2}$, the mass m is $m=\frac{c^2}{G_{\mathrm{N}}}\frac{R_{\mathrm{S}}}{2}$, and this will deliver the answer in kilograms if R_{S} is specified in meters, with the above c and G_{N}. So

$$\text{density} = \frac{\text{mass}}{\text{volume}} = \frac{\frac{c^2}{G_{\mathrm{N}}}\frac{R_{\mathrm{S}}}{2}}{\frac{4\pi R_\oplus^3}{3}} \geqslant \frac{\frac{c^2}{G_{\mathrm{N}}}\frac{R_\oplus}{2}}{\frac{4\pi R_\oplus^3}{3}} = \frac{3c^2}{8\pi G_{\mathrm{N}}}\frac{1}{R_\oplus^2} \approx 4\times 10^{12}\,\frac{\mathrm{kg}}{\mathrm{m}^3}.$$

The minimum density of the body is $\rho=4\times 10^{12}\,\frac{\mathrm{kg}}{\mathrm{m}^3}$, and $\rho:\rho_{\text{water}}=4\times 10^9$.

Solution to Exercise 14.14

Recall the expressions on page 320 for the components of **Ric** for a spherically symmetric stationary metric $\mathbf{g}$. Taking $k=\frac{1}{f}$, and $f=1-\frac{2m}{r}+\frac{e^2}{r^2}$,

$$\begin{aligned}
\mathbf{Ric}_{00} &= \frac{ff''}{2}+\frac{f'f}{r} = \frac{e^2 f}{r^4}\\
\mathbf{Ric}_{11} &= -\frac{f''}{2f}-\frac{f'}{rf} = -\frac{e^2}{r^4 f}\\
\mathbf{Ric}_{22} &= -f+1-rf' = \frac{e^2}{r^2}\\
\mathbf{Ric}_{33} &= (\sin\theta)^2\mathbf{Ric}_{22} = \frac{e^2(\sin\theta)^2}{r^2},
\end{aligned}$$

and all other components are identically 0. Thus the scalar curvature is

$$\mathbf{S}=\mathbf{g}^{ij}\mathbf{Ric}_{ij} = -\frac{1}{f}\frac{e^2 f}{r^4}+f\Big(-\frac{e^2}{r^4 f}\Big)+\frac{1}{r^2}\frac{e^2}{r^2}+\frac{1}{r^2(\sin\theta)^2}\frac{e^2(\sin\theta)^2}{r^2} = 0.$$

Hence the field equation with $\Lambda = 0$ reduces to $\mathbf{Ric} = 8\pi\mathbf{T}$. We recall that

$$8\pi\mathbf{T} = \frac{e^2}{r^4}(\Omega^4 \otimes \Omega^4 - \Omega^1 \otimes \Omega^1 + \Omega^2 \otimes \Omega^2 + \Omega^3 \otimes \Omega^3).$$

Also recall the vector fields V_i from Exercise 12.15 (p.287) that are 'dual' to the one form fields Ω^j. We have

$$\begin{aligned}
8\pi\mathbf{T}_{00} &= \frac{e^2}{r^4}(\Omega^4(\partial_t))^2 = \frac{e^2}{r^4}(\Omega^4(\sqrt{f}\,V_4))^2 = \frac{e^2 f}{r^4} = \mathbf{Ric}_{00}\\
8\pi\mathbf{T}_{11} &= -\frac{e^2}{r^4}(\Omega^1(\partial_r))^2 = -\frac{e^2}{r^4}\Big(\Omega^1\Big(\frac{1}{\sqrt{f}}V_1\Big)\Big)^2 = -\frac{e^2}{r^4 f} = \mathbf{Ric}_{11}\\
8\pi\mathbf{T}_{22} &= \frac{e^2}{r^4}(\Omega^2(\partial_\theta))^2 = \frac{e^2}{r^4}(\Omega^2(rV_2))^2 = \frac{e^2}{r^2} = \mathbf{Ric}_{22}\\
8\pi\mathbf{T}_{33} &= \frac{e^2}{r^4}(\Omega^3(\partial_\phi))^2 = \frac{e^2}{r^4}(\Omega^3(r(\sin\theta)V_3))^2 = \frac{e^2(\sin\theta)^2}{r^2} = \mathbf{Ric}_{33}.
\end{aligned}$$

The components of $8\pi\mathbf{T}_{ij}$ for $i \neq j$ are zero, and so they coincide with the components $\mathbf{Ric}_{ij} = 0$. So the field equation is satisfied.

Chapter 15

Solution to Exercise 15.1

We want to evaluate the integral

$$\widetilde{t}(r) = \int_{r_0}^{r} -\frac{1}{1-\frac{a}{r}}\frac{\sqrt{r}}{\sqrt{a}}\,dr = -\int_{r_0}^{r}\frac{r\sqrt{r}}{(r-a)\sqrt{a}}\,dr,$$

where $a := 2m$. We use the substitution $r = u^2$. Then we have

$$\begin{aligned}\widetilde{t}(r) &= -\int_{r_0}^{r}\frac{r\sqrt{r}}{(r-a)\sqrt{a}}\,dr = -\int_{\sqrt{r_0}}^{\sqrt{r}}\frac{u^2 u}{(u^2-a)\sqrt{a}}2u\,du \\ &= -\frac{2}{\sqrt{a}}\int_{\sqrt{r_0}}^{\sqrt{r}}\Big(u^2+a+\frac{a^2}{2\sqrt{a}}\big(\frac{1}{u-\sqrt{a}}-\frac{1}{u+\sqrt{a}}\big)\Big)du \\ &= \frac{2}{\sqrt{a}}\frac{r_0\sqrt{r_0}-r\sqrt{r}}{3}+2\sqrt{a}(\sqrt{r_0}-\sqrt{r})+a\log\Big(\frac{\sqrt{r_0}-\sqrt{a}}{\sqrt{r_0}+\sqrt{a}}\cdot\frac{\sqrt{r}+\sqrt{a}}{\sqrt{r}-\sqrt{a}}\Big).\end{aligned}$$

The plot with the given numerical values is shown below.

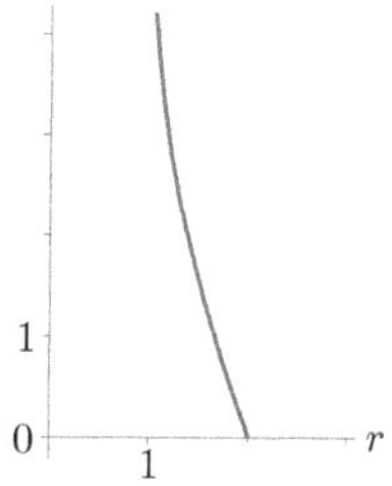

Solution to Exercise 15.2

Setting $\frac{1}{2}\widetilde{m}(v(r))^2 = -V(r) = \frac{G_N\widetilde{m}m}{r}$, we obtain $v(r) = \sqrt{\frac{2G_N m}{r}}$.

If the escape velocity at $r = r_s$ equals the speed of light c, then $r_s = \frac{2G_N m}{c^2}$.

Finally, for the star to not shine, we must have that the escape velocity $v(R)$ at its surface exceeds the speed of light, that is, $v(R) > c$. Since $m = (\text{volume})\cdot(\text{density}) = \frac{4\pi}{3}R^3\rho$, substituting this in the expression for the escape velocity, the inequality $v(R) > c$ gives

$$\sqrt{\frac{2G_N\frac{4\pi}{3}R^3\rho}{R}} > c, \text{ i.e., } R > \sqrt{\frac{3}{8\pi}\frac{c^2}{G_N\rho}}.$$

Solution to Exercise 15.3

When the mass Δm reaches the singularity at $r = 0$, the mass of the black hole becomes $m + \Delta m$, and the Schwarzschild radius increases. Thus the increase in area is

$$\Delta A = 4\pi\Big(\frac{2G_N(m+\Delta m)}{c^2}\Big)^2 - 4\pi\Big(\frac{2G_N m}{c^2}\Big)^2 \approx 4\pi\frac{4(G_N)^2}{c^4}2m\Delta m = 32\frac{\pi(G_N)^2 m}{c^4}\Delta m.$$

So $\Delta S = \alpha k_{\rm B}\dfrac{\Delta A}{\ell_{\rm P}^2} = \dfrac{1}{4}k_{\rm B}\,32\dfrac{\pi(G_{\rm N})^2 m}{c^4}\Delta m \dfrac{1}{\frac{\hbar G_{\rm N}}{c^3}} = \dfrac{8\pi G_{\rm N} k_{\rm B} m}{c\hbar}\Delta m.$

Thus $T_{\rm H} = \dfrac{c^2\Delta m}{\Delta S} = \dfrac{c^2\Delta m}{\frac{8\pi G_{\rm N}k_{\rm B}m}{c\hbar}\Delta m} = \dfrac{c^3\hbar}{8\pi G_{\rm N}k_{\rm B}m}.$

Solution to Exercise 15.4

(1) With $s := \sin\theta$ and $c := \cos\theta$, we have

$$\begin{aligned}
&\mathbf{g}(V,V)\\
&= -(1-\frac{2mr}{r^2+a^2c^2})(r^2+a^2)^2 - \frac{4mra^2s^2}{r^2+a^2c^2}(r^2+a^2) + (r^2+a^2+\frac{2mra^2s^2}{r^2+a^2c^2})s^2a^2\\
&= (r^2+a^2)(s^2a^2-r^2-a^2) + \frac{2mr}{r^2+a^2c^2}((r^2+a^2)^2-2(r^2+a^2)a^2s^2+a^4s^4)\\
&= -(r^2+a^2)(r^2+a^2c^2) + \frac{2mr}{r^2+a^2c^2}(r^2+a^2-a^2s^2)^2\\
&= -(r^2+a^2)(r^2+a^2c^2) + \frac{2mr}{r^2+a^2c^2}(r^2+a^2c^2)^2\\
&= -(r^2+a^2c^2)(r^2+a^2-2mr) = -\rho^2\Delta.
\end{aligned}$$

In M_1 and M_3, we have pointwise that

$$\Delta = r^2 - 2mr + a^2 = (r-r_-)(r-r_+) > 0,$$

and so it follows that $\mathbf{g}(p)(V_p, V_p) < 0$ for all $p \in M_1 \cup M_3$. Thus V is timelike at each point of M_1 and M_3.

(2) If $r \in (r_-, r_+)$, then we have

$$\mathbf{g}(\partial_r, \partial_r) = \frac{\rho^2}{\Delta} = \frac{\rho^2}{(r-r_-)(r-r_+)} < 0.$$

So at each point $p \in M_2$, $\partial_{r,p}$ is timelike.

(3) We have $\mathbf{g}(\partial_t, \partial_t) = -(1-\frac{2mr}{\rho^2}) < 0$ if and only if $\frac{2mr}{\rho^2} < 1$, i.e., $2mr < \rho^2$.

Suppose that for a point p, $r \in (r_+, r_0)$. Let $c := \cos\theta$. We factorise

$$\rho^2 - 2mr = r^2 + a^2c^2 - 2mr = (r - r_0)(r - (m - \sqrt{m^2 - a^2c^2})).$$

Since $r \in (r_+, r_0)$, the first factor on the RHS, $r - r_0$, is negative. Also,

$$r > r_+ = m + \sqrt{m^2 - a^2} > m > m - \sqrt{m^2 - a^2c^2},$$

and so the second factor above is positive. Consequently $\rho^2 < 2mr$, and so it follows that $\partial_{t,p}$ is not timelike. In fact, $\partial_{t,p}$ is spacelike, as

$$\mathbf{g}(p)(\partial_{t,p}, \partial_{t,p}) = -(1 - \frac{2mr}{\rho^2}) = -1 + \frac{2mr}{\rho^2} > -1 + 1 = 0.$$

(4) This is immediate from Exercise 6.33 (p.129), since the components of $\mathbf{g}$ do not depend on t.

Solution to Exercise 15.5

(1) Since the chart-induced components of the metric are independent of t and ϕ, it follows that ∂_t and ∂_ϕ are Killing vector fields for $\mathbf{g}$ in the given chart by the result from Exercise 6.33 (p.129). It follows from Exercise 8.2 (p.153) that for the timelike geodesic γ lying in this chart, $\mathbf{g}(\gamma(\cdot))(\partial_{t,\gamma(\cdot)}, v_{\gamma,\gamma(\cdot)})$ and $\mathbf{g}(\gamma(\cdot))(\partial_{\phi,\gamma(\cdot)}, v_{\gamma,\gamma(\cdot)})$ are constant functions on I. So E and L are constant along γ.

(2) Since E is constant along γ, and $(\sin\theta)^2 = (\sin\frac{\pi}{2})^2 = 1$, we get

$$\big(1-\frac{2m}{r}\big)t' + \frac{2J}{r}\phi' = E. \qquad (\star)$$

As L is constant along γ, we also have

$$-\frac{2J}{r}t' + \big(1+\frac{2m}{r}\big)r^2\phi' = L. \qquad (\star\star)$$

If $L=0$, then the equations $(\star)$ and $(\star\star)$ yield

$$\frac{d\phi}{d\tau} = \frac{E\frac{2J}{r}}{\big(1-(\frac{2m}{r})^2\big)r^2 + \big(\frac{2J}{r}\big)^2}.$$

If $E>0$, then clearly $\dfrac{d\phi}{d\tau}>0$.

(3) As $\phi'(0)=0$, and $\sin\theta = \sin\frac{\pi}{2} = 1$, we get

$$E = \big(1-\frac{2m}{r(0)}\big)t'(0) \quad \text{and} \quad L = -\frac{2J}{r(0)}t'(0).$$

Using the constancy of E and L along γ, and the equations $(\star)$ and $(\star\star)$ again, we obtain

$$\begin{aligned}
\frac{d\phi}{d\tau}(\tau) &= \frac{E\frac{2J}{r(\tau)} + \big(1-\frac{2m}{r(\tau)}\big)L}{\big(1-(\frac{2m}{r(\tau)})^2\big)(r(\tau))^2 + \big(\frac{2J}{r(\tau)}\big)^2} \\
&= \frac{\Big(\big(1-\frac{2m}{r(0)}\big)\frac{2J}{r(\tau)} - \big(1-\frac{2m}{r(\tau)}\big)\frac{2J}{r(0)}\Big)t'(0)}{\big(1-(\frac{2m}{r(\tau)})^2\big)(r(\tau))^2 + \big(\frac{2J}{r(\tau)}\big)^2} \\
&= \frac{2J\big(\frac{1}{r(\tau)} - \frac{1}{r(0)}\big)t'(0)}{\big(1-(\frac{2m}{r(\tau)})^2\big)(r(\tau))^2 + \big(\frac{2J}{r(\tau)}\big)^2}.
\end{aligned}$$

Solution to Exercise 15.6

Consider an infinitesimally small cylindrical part of cross-sectional area A of a spherical shell of inner radius r and thickness dr. Its mass is given by $\rho A dr$, and the gravitational pull on the cylinder is $G_N M(r)\rho A dr/r^2$.

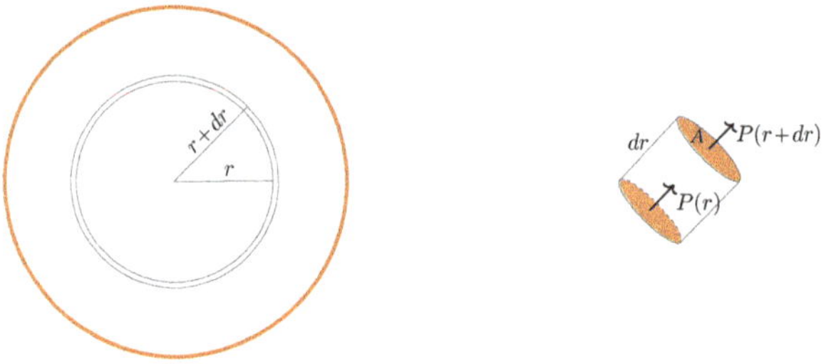

We know that the pressure decreases as the radius increases. So the equilibrium condition yields $(P(r) - P(r+dr))A = G_{\rm N} M(r)\rho A dr/r^2$. Thus,

$$\frac{dP}{dr}(r) = -G_{\rm N}\frac{M(r)\rho}{r^2}. \qquad (\star)$$

We also have

$$\frac{dM}{dr}(r) = 4\pi r^2 \rho. \qquad (\star\star)$$

From $(\star)$ and $(\star\star)$, we obtain

$$\begin{aligned}\frac{1}{r^2}\frac{d}{dr}\Big(\frac{r^2}{\rho}\frac{dP}{dr}\Big) &= \frac{1}{r^2}\frac{d}{dr}\Big(\frac{r^2}{\rho}\Big(-G_{\rm N}\frac{M(r)\rho}{r^2}\Big)\Big) = -G_{\rm N}\frac{1}{r^2}\frac{d}{dr}M(r) = -G_{\rm N}\frac{1}{r^2}4\pi r^2\rho \\ &= -4\pi G_{\rm N}\rho.\end{aligned}$$

Chapter 16

Solution to Exercise 16.1

(1) We have $\mathbf{S} = \boldsymbol{\sigma}^{ij}\mathbf{Ric}_{ij} = \boldsymbol{\sigma}^{ij} f \boldsymbol{\sigma}_{ij} = (\dim S)f$. From the calculation done at the end of the proof of Schur's theorem, we have $\operatorname{div}(f\boldsymbol{\sigma}) = df$. From Prop. 9.6 (p.191), $\operatorname{div}\mathbf{Ric} = \frac{1}{2}d\mathbf{S}$. Hence, using the above, we get

$$df = \operatorname{div}(f\boldsymbol{\sigma}) = \operatorname{div}\mathbf{Ric} = \frac{1}{2}d\mathbf{S} = \frac{1}{2}((\dim S)f) = \frac{\dim S}{2}df.$$

As $\dim S \geqslant 3$, we get $df = 0$, that is, for all $V \in T_0^1 S$, we have $Vf = 0$. By the result from Exercise 3.4 (p.42), we conclude that f is constant.

(2) ('Only if' part.) Suppose that $(S, \boldsymbol{\sigma})$ is a connected Einstein manifold. Let $p \in S$, and let $B = \{e_1, e_2, e_3\}$ be an orthonormal basis for T_pS with respect to $\boldsymbol{\sigma}(p)$. Let $(U, \mathbf{x})$ be a normal chart starting with B. Then for $i \neq j$, $K_p(\{e_i, e_j\}) = \mathbf{R}(p)(e_i^\flat, e_i, e_j, e_j)$, and

$$\begin{aligned}\lambda = \lambda\boldsymbol{\sigma}(p)(e_1, e_1) &= \lambda\boldsymbol{\sigma}(p)(\partial_{x^1,p}, \partial_{x^1,p}) = (\mathbf{Ric}(\partial_{x^1}, \partial_{x^1}))(p)\\ &= \mathbf{R}^1_{111}(p) + \mathbf{R}^2_{211}(p) + \mathbf{R}^3_{311}(p)\\ &= 0 + K_p(\{e_1, e_2\}) + K_p(\{e_1, e_3\}).\end{aligned}$$

Similarly, we also obtain the equations $\lambda = K_p(\{e_1, e_2\}) + K_p(\{e_2, e_3\})$ and $\lambda = K_p(\{e_1, e_3\}) + K_p(\{e_2, e_3\})$. Adding these 3 equations, we get

$$\frac{3}{2}\lambda = K_p(\{e_1, e_2\}) + K_p(\{e_2, e_3\}) + K_p(\{e_3, e_1\}),$$

and subtracting each from this last equation yields

$$K_p(\{e_1, e_2\}) = K_p(\{e_2, e_3\}) = K_p(\{e_3, e_1\}) = \frac{\lambda}{2}$$

As we started with an *arbitrary* orthonormal basis $\{e_1, e_2, e_3\}$, this holds for *any* orthonormal basis. Thus K_p is constant, taking the value $\frac{\lambda}{2}$. The 'if' part follows immediately from Schur's theorem: $\mathbf{Ric} = (2k)\boldsymbol{\sigma}$.

Solution to Exercise 16.2

The arclength of $\gamma|_{(0,r)}$ is $L(r) := \displaystyle\int_0^r \sqrt{\boldsymbol{\sigma}(\gamma(r))(v_{\gamma,\gamma(r)}, v_{\gamma,\gamma(r)})}\, dr$.

As $v_{\gamma,\gamma(r)} = \dfrac{dr}{dr}\partial_{r,\gamma(r)} + \dfrac{d\theta_0}{dr}\partial_{\theta,\gamma(r)} + \dfrac{d\phi_0}{dr}\partial_{\phi,\gamma(r)} = \partial_{r,\gamma(r)}$, we have

$$\boldsymbol{\sigma}(\gamma(r))(v_{\gamma,\gamma(r)}, v_{\gamma,\gamma(r)}) = \frac{1}{1-kr^2}.$$

Thus if $k=0$, then $L(r) = \displaystyle\int_0^r \frac{1}{\sqrt{1-kr^2}}\, dr = \int_0^r \frac{1}{1-0}\, dr = r$.

If $k<0$, then we use the substitution $\sinh u = \sqrt{-k}\,r \in (0, \infty)$ to obtain

$$L(r) = \int_0^r \frac{1}{\sqrt{1-kr^2}}\, dr = \int_0^{\sinh^{-1}(\sqrt{-k}r)} \frac{1}{\cosh u}\frac{\cosh u}{\sqrt{-k}}\, du = \frac{\sinh^{-1}(\sqrt{-k}r)}{\sqrt{-k}}.$$

If $k>0$, then we use the substitution $\sin u = \sqrt{k}r \in (0,1)$ to obtain

$$L(r)=\int_0^r \frac{1}{\sqrt{1-kr^2}}\,dr=\int_0^{\sin^{-1}(\sqrt{k}r)} \frac{1}{\cos u}\frac{\cos u}{\sqrt{k}}\,du=\frac{\sin^{-1}(\sqrt{k}r)}{\sqrt{k}}.$$

By the mean value theorem, for $x>0$, there is a $c_x \in (0,x)$ such that

$$\frac{\sinh^{-1}x}{x}=\frac{\sinh^{-1}x-\sinh^{-1}0}{x-0}=\frac{d\sinh^{-1}}{dx}(c_x)=\frac{1}{\sqrt{1+c_x^2}}<1.$$

Using this, we see that if $k<0$, then $L(r)<\frac{1}{\sqrt{-k}}(\sqrt{-k}r)=r$.

For $x \in (0,1)$, there exists a $c_x \in (0,1)$ such that

$$\frac{\sin^{-1}x}{x}=\frac{\sin^{-1}x-\sin^{-1}0}{x-0}=\frac{d\sin^{-1}}{dx}(c_x)=\frac{1}{\sqrt{1-c_x^2}}>1.$$

Hence, if $k>0$, then $L(r)>\frac{1}{\sqrt{k}}(\sqrt{k}r)=r$.

Solution to Exercise 16.3

We have $v_{\mu,\mu(s)}=r'(s)\,\partial_{r,\mu(s)}$, where $\cdot'=\frac{d}{ds}$. Thus

$$\boldsymbol{\sigma}(\mu(s))(v_{\mu,\mu(s)},v_{\mu,\mu(s)})=(r'(s))^2\frac{(a(t))^2}{1-k(r(s))^2}=1.$$

So the length $d(t)$ of μ in $(S,\boldsymbol{\sigma})$ is

$$d(t)=\int_0^{s_0}\sqrt{\boldsymbol{\sigma}(\mu(s))(v_{\mu,\mu(s)},v_{\mu,\mu(s)})}\,ds=\int_0^{s_0}1\,ds=s_0.$$

Also, using the differential equation for r, we have

$$s_0=\int_0^{s_0}1\,ds=\int_0^{s_0}\frac{a(t)\,r'(s)}{\sqrt{1-k(r(s))^2}}\,ds=\int_{r(0)}^{r(s_0)}\frac{a(t)}{\sqrt{1-kr^2}}\,dr=a(t)\int_0^{r_0}\frac{1}{\sqrt{1-kr^2}}\,dr,$$

where we used the substitution $r=r(s)$, and the given boundary values for r. The θ- and ϕ-components of μ, being constant functions, satisfy $\theta'\equiv 0$ and $\phi'\equiv 0$, and so

$$\theta''+2\Gamma^\theta_{r\theta}\theta' r'+\Gamma^\theta_{\phi\phi}(\phi')^2=0+2\Gamma^\theta_{r\theta}0\,r'+\Gamma^\theta_{\phi\phi}(0)^2=0,$$
$$\phi''+2\Gamma^\phi_{r\phi}\phi' r'+2\Gamma^\phi_{\theta\phi}\phi'\theta'=0+2\Gamma^\phi_{r\phi}0\,r'+2\Gamma^\phi_{\theta\phi}00=0.$$

Finally, the r-component of the geodesic equation is also satisfied, since

$$\begin{aligned}
&r''+\Gamma^r_{rr}(r')^2+\Gamma^r_{\theta\theta}(\theta')^2+\Gamma^r_{\phi\phi}(\phi')^2\\
&=r''+\Gamma^r_{rr}(r')^2+\Gamma^r_{\theta\theta}(0)^2+\Gamma^r_{\phi\phi}(0)^2=r''(s)+\Gamma^r_{rr}(r')^2\\
&=\frac{d}{ds}\frac{\sqrt{1-k(r(s))^2}}{a(t)}+\frac{kr(s)}{1-k(r(s))^2}\Big(\frac{\sqrt{1-k(r(s))^2}}{a(t)}\Big)^2\\
&=\frac{1}{a(t)}\frac{1}{2\sqrt{1-k(r(s))^2}}(-2kr(s)r'(s))+\frac{kr(s)}{(a(t))^2}\\
&=-\frac{1}{a(t)}\frac{1}{\sqrt{1-k(r(s))^2}}kr(s)\frac{\sqrt{1-k(r(s))^2}}{a(t)}+\frac{kr(s)}{(a(t))^2}=0.
\end{aligned}$$

Solution to Exercise 16.4

Differentiating $d(t) = a(t)\int_0^{r_0} \frac{1}{\sqrt{1-kr^2}}\,dr$ with respect to t, we get

$$\dot{d}(t) = \dot{a}(t)\int_0^{r_0} \frac{1}{\sqrt{1-kr^2}}\,dr.$$

Dividing these yields $\frac{\dot{d}(t)}{d(t)} = \frac{\dot{a}(t)}{a(t)} = H(t)$. So $\dot{d}(t) = H(t)\,d(t)$.

Solution to Exercise 16.5

Differentiating $\dot{a}^2 + k = \frac{8\pi}{3}\rho a^2$ gives

$$2\dot{a}\ddot{a} + 0 = \frac{8\pi}{3}(\dot{\rho}a^2 + 2\rho a\dot{a}),$$

and so $\ddot{a} = \frac{8\pi}{3}\frac{(\dot{\rho}a^2 + 2\rho a\dot{a})}{2\dot{a}}$. Substituting in $2\frac{\ddot{a}}{a} + \frac{\dot{a}^2}{a^2} + \frac{k}{a^2} = -8\pi\mathrm{p}$ yields

$$2\frac{8\pi}{3}\frac{(\dot{\rho}a + 2\rho\dot{a})}{2\dot{a}} + \frac{\dot{a}^2}{a^2} + \frac{k}{a^2} = -8\pi\mathrm{p},$$

and using $\frac{\dot{a}^2}{a^2} + \frac{k}{a^2} = \frac{8\pi}{3}\rho$ in the left-hand side of the above now gives

$$2\frac{8\pi}{3}\frac{(\dot{\rho}a + 2\rho\dot{a})}{2\dot{a}} + \frac{8\pi}{3}\rho = -8\pi\mathrm{p}.$$

Rearranging, we obtain $\dot{\rho} + 3(\mathrm{p} + \rho)\frac{\dot{a}}{a} = 0$, as wanted.

Assuming $\mathrm{p} = w\rho$, we have $\dot{\rho} + 3(1+w)\rho\frac{\dot{a}}{a} = 0$, that is,

$$\frac{d}{dt}((\log\rho) + 3(1+w)\log a) = \frac{\dot{\rho}}{\rho} + 3(1+w)\frac{\dot{a}}{a} = 0$$

Thus on I, we have $\log\rho + 3(1+w)\log a \equiv C'$, a constant. Consequently, $\rho(t) = Ca(t)^{-3(1+w)}$, for all $t \in I$, where $C = e^{C'}$.

- For radiation, $w = 1/3$. So $\rho \propto a^{-3(1+\frac{1}{3})} = a^{-4}$.
- For dust, $w = 0$. So $\rho \propto a^{-3(1+0)} = a^{-3}$.
- For the cosmological constant, $w = -1$. So $\rho = Ca^{-3(1-1)} = C$, a constant.

Solution to Exercise 16.6

If $\Lambda > 0$, then $a = \ell\cosh(t/\ell)$, $\dot{a} = \sinh(t/\ell)$ and $\ddot{a} = (1/\ell)\cosh(t/\ell)$. Thus, using $(\cosh(t/\ell))^2 - (\sinh(t/\ell))^2 = 1$, we obtain

$$\frac{\dot{a}^2}{a^2} + \frac{k}{a^2} = \frac{(\sinh(t/\ell))^2}{\ell^2(\cosh(t/\ell))^2} + \frac{1}{\ell^2(\cosh(t/\ell))^2} = \frac{1}{\ell^2} = \frac{\Lambda}{3} = \frac{8\pi}{3}\frac{\Lambda}{8\pi} = \frac{8\pi}{3}\rho,$$

showing that (16.9) holds. Similarly, (16.10) holds since

$$2\frac{\ddot{a}}{a} + \frac{\dot{a}^2}{a^2} + \frac{k}{a^2} = 2\frac{\ell^{-1}\cosh(t/\ell)}{\ell\cosh(t/\ell)} + \frac{(\sinh(t/\ell))^2}{\ell^2(\cosh(t/\ell))^2} + \frac{1}{\ell^2(\cosh(t/\ell))^2} = \frac{3}{\ell^2} = \Lambda = -8\pi\mathrm{p}.$$

If $\Lambda < 0$, then $a = \ell \sin(t/\ell)$, $\dot{a} = \cos(t/\ell)$ and $\ddot{a} = -(1/\ell)\sin(t/\ell)$. Thus, using $(\cos(t/\ell))^2 + (\sin(t/\ell))^2 = 1$, we obtain

$$\frac{\dot{a}^2}{a^2} + \frac{k}{a^2} = \frac{(\cos(t/\ell))^2}{\ell^2(\sin(t/\ell))^2} + \frac{-1}{\ell^2(\sin(t/\ell))^2} = -\frac{1}{\ell^2} = \frac{\Lambda}{3} = \frac{8\pi}{3}\frac{\Lambda}{8\pi} = \frac{8\pi}{3}\rho,$$

showing that (16.9) holds. Similarly, (16.10) holds since

$$2\frac{\ddot{a}}{a} + \frac{\dot{a}^2}{a^2} + \frac{k}{a^2} = 2\frac{(-\ell^{-1}\sin(t/\ell))}{\ell\sin(t/\ell)} + \frac{(\cos(t/\ell))^2}{\ell^2(\sin(t/\ell))^2} + \frac{-1}{\ell^2(\sin(t/\ell))^2} = -\frac{3}{\ell^2} = \Lambda = -8\pi\mathrm{p}.$$

Solution to Exercise 16.7

Using the first Friedmann equation, and the definitions of H, M, L,

$$H^2 = \frac{\dot{a}^2}{a^2} \overset{(\star)}{=} \frac{8\pi}{3}\rho + \frac{\Lambda}{3} = \frac{8\pi}{3}\rho\frac{a^3}{a^3} + \frac{1}{L^2} = \frac{M}{a^3} + \frac{1}{L^2}.$$

Also, using both Friedmann equations, and noting that $\mathrm{p} = 0$, we obtain

$$\dot{H} = \frac{\ddot{a}}{a} - \frac{\dot{a}^2}{a^2} = \frac{\Lambda}{3} - \frac{4\pi}{3}\rho - \frac{8\pi}{3}\rho - \frac{\Lambda}{3} = -\frac{3}{2}\Big(\frac{8\pi}{3}\Big)\rho \overset{(\star)}{=} \frac{3}{2}\Big(\frac{\Lambda}{3} - H^2\Big) = \frac{3}{2}\Big(\frac{1}{L^2} - H^2\Big).$$

With $q := \dfrac{1}{(1/L) - H}$, we have

$$\dot{q} = \frac{\dot{H}}{(\frac{1}{L} - H)^2} = \frac{\frac{3}{2}(\frac{1}{L^2} - H^2)}{(\frac{1}{L} - H)^2} = \frac{\frac{3}{2}(\frac{1}{L} + H)}{\frac{1}{L} - H} = -\frac{3}{2} + \frac{3}{L}\frac{1}{(\frac{1}{L} - H)} = \frac{3}{L}q - \frac{3}{2}.$$

Multiplying by $e^{-\frac{3}{L}t}$ throughout, and rearranging, we obtain

$$\frac{d}{dt}(e^{-\frac{3}{L}t}q) = e^{-\frac{3}{L}t}\dot{q} - \frac{3}{L}e^{-\frac{3}{L}t}q = -\frac{3}{2}e^{-\frac{3}{L}t}.$$

Integrating from $t = 0$ to t, we obtain $e^{-\frac{3}{L}t}q(t) - q(0) = \frac{3}{2}\Big(\dfrac{e^{-\frac{3}{L}t} - 1}{3/L}\Big)$.

Substituting $q(t) = \dfrac{1}{(1/L) - H(t)}$, and solving for $H(t)$, we find

$$H(t) = \frac{1}{L}\frac{(\frac{1}{L} + H(0))e^{\frac{3}{L}t} - (\frac{1}{L} - H(0))}{(\frac{1}{L} + H(0))e^{\frac{3}{L}t} + (\frac{1}{L} - H(0))}. \qquad (*)$$

Using $H^2 = \dfrac{M}{a^3} + \dfrac{1}{L^2}$, we have

$$a(t) = \sqrt[3]{M}\Big((H(t))^2 - \frac{1}{L^2}\Big)^{-\frac{1}{3}}. \qquad (**)$$

Setting $t = 0$ in $H^2 = \dfrac{M}{a^3} + \dfrac{1}{L^2}$, we get $H(0)^2 = \dfrac{M}{A^3} + \dfrac{1}{L^2}$. So

$$\frac{1}{L^2} - (H(0))^2 = -\frac{M}{A^3},$$

$$\frac{1}{L} \pm H(0) = \frac{1}{L} \pm \sqrt{\frac{M}{A^3} + \frac{1}{L^2}} = \frac{1}{L}\Big(1 \pm \sqrt{1 + \frac{ML^2}{A^3}}\Big).$$

Using these, and by substituting for $H(t)$ from $(*)$ into $(**)$, we get

$$a(t) = \frac{A}{2^{2/3}}\Big(\big(1+\sqrt{1+\frac{ML^2}{A^3}}\big)e^{\frac{3}{2L}t} + \big(1-\sqrt{1+\frac{ML^2}{A^3}}\big)e^{-\frac{3}{2L}t}\Big)^{\frac{2}{3}}.$$

Solution to Exercise 16.8

We have $v_{\lambda,\lambda(s)} = t'(s)\partial_{t,\lambda(s)} + r'(s)\partial_{r,\lambda(s)}$, where $\cdot' := \frac{d}{ds}$. As the geodesic λ is null, we have $\mathbf{g}(\lambda(s))(v_{\lambda,\lambda(s)}, v_{\lambda,\lambda(s)}) = 0$ for all $s \in (\alpha,\beta)$, giving

$$-(t'(s))^2 + (a(t(s)))^2(r'(s))^2 = 0.$$

Using this, and the facts that $t' > 0$, $r' < 0$, and $a > 0$, we obtain

$$\begin{aligned} R = R - 0 = r(\alpha) - r(\beta) &= -\int_\alpha^\beta r'(s)\,ds \\ &= \int_\alpha^\beta \frac{t'(s)}{a(t(s))}\,ds \\ &= \int_{t(\alpha)}^{t(\beta)} \frac{1}{a(\tau)}\,d\tau \quad \text{(using the substitution } t(s)=\tau) \\ &= \int_{t_0}^{T} \frac{1}{a(\tau)}\,d\tau. \end{aligned}$$

Since $t_0 > 0$, and $a > 0$, we also get $R = \int_{t_0}^{T} \frac{1}{a(t)}\,dt < \int_0^T \frac{1}{a(t)}\,dt = R_{\mathrm{H}}(T)$.

Solution to Exercise 16.9

(1) We have $v_{\gamma,\gamma(s)} = t'(s)\partial_{t,\gamma(s)} + r'(s)\partial_{r,\gamma(s)}$ for all $s \in J$. As the signal γ is lightlike, we also have $\mathbf{g}(\gamma(s))(v_{\gamma,\gamma(s)}, v_{\gamma,\gamma(s)}) = 0$ for all $s \in J$, and so

$$0 = \mathbf{g}(\gamma(s))(v_{\gamma,\gamma(s)}, v_{\gamma,\gamma(s)}) = -(t'(s))^2 + \frac{(a(t(s)))^2}{1-k(r(s))^2}(r'(s))^2.$$

Since $r'(s), t'(s) > 0$, the above yields $\dfrac{t'(s)}{r'(s)} = \dfrac{a(t(s))}{\sqrt{1-k(r(s))^2}}$.

(2) Let $J = (s_A, s_B)$, and h be the inverse of the map $r : J \to (0, r(s_B))$. Set $\widetilde{t} = t \circ h : (0, r(s_B)) \to (t(s_A), t(s_B))$. Then we have

$$\frac{d\widetilde{t}}{dr}(r) = t'(h(r))h'(r) = \frac{t'(h(r))}{r'(h(r))} = \frac{a(\widetilde{t}(r))}{\sqrt{1-kr^2}}. \qquad (\star)$$

We have $\widetilde{t}(0) = t(s_A)$ and $\widetilde{t}(r(s_B)) = t(s_B)$.

(a) When $k = -\beta^2$ and $a(t) = \beta t$, $(\star)$ becomes $\dfrac{d\widetilde{t}}{dr}(r) = \dfrac{\beta\widetilde{t}(r)}{\sqrt{1+\beta^2r^2}}$.

Integrating, we get $\log\dfrac{t(s_B)}{t(s_A)} = \displaystyle\int_0^{r(s_B)} \frac{\beta}{\sqrt{1+\beta^2r^2}}\,dr = \sinh^{-1}(\beta r(s_B))$.

So $r(s_B) = \beta^{-1}\sinh(\log(1+\mathrm{z})) = \dfrac{1}{2\beta}\big(1+\mathrm{z}-\dfrac{1}{1+\mathrm{z}}\big)$.

Thus $R = a(t(s_B))r(s_B) = t(s_B)\dfrac{\mathrm{z}^2+2\mathrm{z}}{2\mathrm{z}+2}$.

(b) When $k=0$ and $a(t)=e^{Ht}$, $(\star)$ is $\frac{d\widetilde{t}}{dr}(r)=e^{H\widetilde{t}(r)}$. Integrating,

$$r(s_B)=\int_{t(s_A)}^{t(s_B)} e^{-Ht}dt=\frac{e^{-Ht(s_A)}-e^{-Ht(s_B)}}{H}=\frac{1}{H}\Big(\frac{1}{a(t(s_A))}-\frac{1}{a(t(s_B))}\Big).$$

Thus $R=a(t(s_B))r(s_B)=\frac{1}{H}(\mathrm{z}+1-1)=\frac{\mathrm{z}}{H}$.

(c) When $k=0$ and $a(t)=\beta t^p$, $(\star)$ is $\frac{d\widetilde{t}}{dr}(r)=\beta(\widetilde{t}(r))^p$. Integrating,

$$r(s_B)=\int_{t(s_A)}^{t(s_B)}\frac{1}{\beta t^p}dt=\frac{(t(s_B))^{-p+1}-(t(s_A))^{-p+1}}{\beta(1-p)}.$$

Thus

$$\begin{aligned}
R = a(t(s_B))r(s_B) &= \frac{(t(s_B))^p}{1-p}\big((t(s_B))^{-p+1}-(t(s_A))^{-p+1}\big)\\
&= \frac{t(s_B)}{1-p}\Big(1-\big(\frac{t(s_A)}{t(s_B)}\big)^{-p+1}\Big) = \frac{t(s_B)}{1-p}\Big(1-\big(\frac{a(t(s_A))}{a(t(s_B))}\big)^{\frac{1}{p}\cdot(-p+1)}\Big)\\
&= \frac{t(s_B)}{1-p}\Big(1-\big(\frac{1}{1+\mathrm{z}}\big)^{\frac{1}{p}-1}\Big) = \frac{t(s_B)}{1-p}(1-(1+\mathrm{z})^{1-\frac{1}{p}}).
\end{aligned}$$

Notes

Chapter 3: The proof of Prop. 3.2 is based on [Godinho and Natário (2014), see Exercise 12(a), p.33, and its solution, pp.332–333].

Chapter 6: The proof of Thm. 6.3 is based on [O'Neill (1983), Ch. 9, Prop. 21].

Chapter 9: Thm. 9.2 is based on [Moretti (2020), §9.2].

Chapter 12: The diagrammatic procedure in Exercise 12.5 was given in [Kocik (2012)]; see also the exposition [Sasane and Ufnarovski (2016)]. Exercise 12.7 is based on [Lightman, Press, Price and Teukolsky (1975), Prob. 1.4]. The discussion on the momentum of light is based on [Feynman (1963), §34-9]. Lemma 12.2 and Prop. 12.3 are based on [Oloff (2023), §5.1].

Chapter 14: Exercise 14.10 is based on [Sachs and Wu (1977), Prop. 4.3.2].

Chapter 15: The exposition in Section 15.3 is based on [Oloff (2023), §14.3]. The picture on p.341 is based on [Oloff (2023), Fig. 14.7, p.232]. The discussion on the evolution of stars in §15.4 is based on [Venkataraman (1992)].

Chapter 16: Section 16.1 is based on [Oloff (2023), §15.1].

Bibliography

Apostol, T. (1969). *Calculus. Volume II*, 2nd edition (Wiley).

Artin, M. (1991). *Algebra* (Prentice-Hall).

Beem, J., Ehrlich, P., and Easley, K. (1996). *Global Lorentzian Geometry*, 2nd edition (Marcel Dekker).

Bishop, R., and Goldberg, S. (1980). *Tensor Analysis on Manifolds* (Dover).

Choquet-Bruhat, Y. (2009). *General Relativity and the Einstein Equations* (Oxford University Press).

Cooke, R. (2017). *It's About Time. Elementary Mathematical Aspects of Relativity* (American Mathematical Society).

Corless, R., Gonnet, G., Hare, D., Jeffrey, D., and Knuth, D. (1996). On the Lambert W function. *Advances in Computational Mathematics* **5**, 4, pp. 329–359.

Dirac, P. (1974). Cosmological models and the large numbers hypothesis. *Proceedings of the Royal Society of London. A. Mathematical and Physical Sciences* **338**, pp. 439–446.

Dodson, C., and Poston, T. (1997). *Tensor Geometry. The Geometric Viewpoint and its Uses*, 2nd edition (Springer).

Feynman, R. (1963). *Lectures on Physics. Volume 1* (Addison-Wesley).

Feynman, R. (1964). *Lectures on Physics. Volume 2* (Addison-Wesley).

Frobenius, G. (1877). Über das Pfaffsche Problem. *Journal für die Reine und Angewandte Mathematik* **82**, pp. 230–315.

Godinho, L., and Natário, J. (2014). *An Introduction to Riemannian Geometry. With Applications to Mechanics and Relativity* (Springer).

Grave, F., and Mueller, T. (2010). *Catalogue of Spacetimes* (`https://arxiv.org/abs/0904.4184`).

Hakopian, H., and Tonoyan, M. (2004). Partial differential analogs of ordinary differential equations and systems. *New York Journal of Mathematics* **10**, pp. 89–116.

Hall, G. (2004). *Symmetries and Curvature Structure in General Relativity*, Lecture Notes in Physics 46 (World Scientific).

Halmos, P. (1987). *Finite-Dimensional Vector Spaces*, 2nd edition (Springer).

Hartman, R. (2002). *Ordinary Differential Equations*, Classics in Applied Mathematics 38 (SIAM).

d'Inverno, R., and Vickers, J. (2022). *Introducing Einstein's Relativity: A Deeper Understanding*, 2nd edition (Oxford University Press).

Klainerman, S., and Nicolò, F. (2003). *The Evolution Problem in General Relativity*, Progress in Mathematical Physics, 25 (Birkhäuser).

Kocik, J. (2012). Geometric diagram for relativistic addition of velocities. *American Journal of Physics* **80**, 8, p. 737.

Kriele, M. (2001). *Spacetime*, Corrected 2nd printing (Springer).

Landau, L., and Lifshitz, E. (1976). *Mechanics. Course of Theoretical Physics. Volume 1*, 3rd edition (Elsevier).

Lang, S. (1999). *Fundamentals of Differential Geometry* (Springer).

Lee, J. (2013). *Introduction to Smooth Manifolds*, 2nd edition (Springer).

Lee, J. (2018). *Introduction to Riemannian Manifolds*, 2nd edition (Springer).

Lightman, A., Press, W., Price, R., and Teukolsky, S. (1975). *Problem Book in Relativity and Gravitation* (Princeton University Press).

Marsden, J., and T. Ratiu, T. (2007). *Manifolds, Tensor Analysis and Applications*, 3rd edition (Springer).

Mehra, J. (1974). *Einstein, Hilbert, and the Theory of Gravitation. Historical Origins of General Relativity Theory* (D. Reidel).

Misner, C., Thorne, K., and Wheeler, J. (2017). *Gravitation* (Princeton University Press).

Momin, A. (2002). *The Gödel Solution to the Einstein Field Equations* (`http://www.math.toronto.edu/~colliand/426/Papers/A_Monin.pdf`).

Moretti, V. (2020). *Geometric Methods in Mathematical Physics II: Tensor Analysis on Manifolds and General Relativity* (`http://www.science.unitn.it/~moretti/dispense.html`).

Natário, J. (2021). *An Introduction to Mathematical Relativity*, Latin American Mathematical Series (Springer).

Oloff, R. (2023). *The geometry of spacetime – a mathematical introduction to relativity theory*, Graduate Texts in Physics (Springer).

O'Neill, B. (1983). *Semi-Riemannian Geometry. With Applications to Relativity* (Academic Press).

O'Neill, B. (2014). *The Geometry of Kerr Black Holes* (Dover).

Padmanabhan, T. (2016). The atoms of space, gravity and the cosmological constant. *International Journal of Modern Physics. D.*, **25**, 7, 1630020.

Penrose, R. (1969). Gravitational collapse: The role of general relativity. *La Rivista del Nuovo Cimento*, **1**, pp. 252–276.

Rudin, W. (1983). *Principles of Mathematical Analysis*, 3rd edition (McGraw-Hill).

Sasane, A. (2017). *A Friendly Approach to Functional Analysis* (World Scientific).

Sasane, A., and Ufnarovski, V. (2016). Alternative proofs for Kocik's geometric diagram for relativistic velocity addition. *Elemente der Mathematik* **71**, 3, pp. 122–130.

Sachs, R., and Wu, H. (1977). *General Relativity for Mathematicians* (Springer).

Schuller, F. (2015). *Lectures, WE-Heraeus International Winter School on Gravity and Light* (`www.youtube.com/@thewe-heraeusinternational2060`).

Shilov, G. (1975) *Analyse Mathématique. Fonctions de Plusieurs Variables Réeles* (MIR).

Venkataraman, G. (1992) *Chandrasekhar and His Limit* (Universities Press).

Visser, M. (2009) The Kerr Spacetime–A Brief Introduction. In *The Kerr Spacetime*, pp. 3–37 (Cambridge University Press).

Wald, R. (1984). *General Relativity* (University of Chicago Press).

Index

www.ingramcontent.com/pod-product-compliance
Ingram Content Group UK Ltd.
Pitfield, Milton Keynes, MK11 3LW, UK
UKHW021407130225
4587UKWH00019B/156

9 789819 805457